CHEMICAL
PROCESS
PRINCIPLES

CHEMICAL PROCESS PRINCIPLES

PART II

THERMODYNAMICS

OLAF A. HOUGEN

PROFESSOR OF CHEMICAL ENGINEERING
UNIVERSITY OF WISCONSIN

KENNETH M. WATSON

LAKE ZURICH
ILLINOIS

ROLAND A. RAGATZ

PROFESSOR OF CHEMICAL ENGINEERING
UNIVERSITY OF WISCONSIN

Second Edition

CBS Publishers & Distributors Pvt. Ltd.

New Delhi • Bengaluru • Chennai • Kochi • Kolkata • Mumbai
Hyderabad • Nagpur • Patna • Pune • Vijayawada

ISBN: 81-239-1160-2

First Indian Edition: 1995
Reprint: 2001, 2003, 2004

This edition has been published in India by special arrangements with The Authors

Published by:
Satish Kumar Jain for CBS Publishers & Distributors Pvt. Ltd.,
4819/XI Prahlad Street, 24 Ansari Road, Daryaganj, New Delhi - 110002
delhi@cbspd.com, cbspubs@airtelmail.in • www.cbspd.com
Ph.: 23289259, 23266861, 23266867 • Fax: 011-23243014

Corporate Office: 204 FIE, Industrial Area, Patparganj, Delhi - 110 092
Ph: 49344934 • Fax: 011-49344935
E-mail: publishing@cbspd.com • publicity@cbspd.com

Branches:
- ***Bengaluru:*** 2975, 17th Cross, K.R. Road, Bansankari 2nd Stage, Bengaluru - 70 • Ph: +91-80-26771678/79 • Fax: +91-80-26771680 E-mail: cbsbng@gmail.com, bangalore@cbspd.com
- ***Chennai:*** No. 7, Subbaraya Street, Shenoy Nagar, Chennai - 600030 Ph: +91-44-26681266, 26680620 • Fax: +91-44-42032115 E-mail: chennai@cbspd.com
- ***Kochi:*** Ashana House, 39/1904, A.M. Thomas Road, Valanjambalam, Ernakulum, Kochi • Ph: +91-484-4059061-65 Fax: +91-484-4059065 • E-mail: cochin@cbspd.com
- ***Kolkata:*** 6-B, Ground Floor, Rameshwar Shaw Road, Kolkata - 700014 Ph: +91-33-22891126/7/8 • E-mail: kolkata@cbspd.com
- ***Mumbai:*** 83-C, Dr. E. Moses Road, Worli, Mumbai - 400018 Ph: +91-9833017933, 022-24902340/41 • E-mail: mumbai@cbspd.com

Representatives:
- Hyderabad: 0-9885175004
- Nagpur: 0-9021734563
- Patna: 0-9334159340
- Pune: 0-9623451994
- Vijayawada: 0-9000660880

Printed at:
India Binding House, Noida, UP (India)

Preface

"In the following pages certain industrially important principles of chemistry and physics have been selected for detailed study. The significance of each principle is intensively developed and its applicability and limitations scrutinized." Thus reads the preface to the first edition of *Industrial Chemical Calculations*, the precursor of this book. The present book continues to give intensive quantitative training in the practical applications of the principles of physical chemistry to the solution of complicated industrial problems and in methods of predicting missing physicochemical data from generalized principles. In addition, through recent developments in thermodynamics and kinetics, these principles have been integrated into procedures for process design and analysis with the objective of arriving at optimum economic results from a minimum of pilot-plant or test data. The title *Chemical Process Principles* was selected to emphasize the importance of this approach to process design and operation.

The design of a chemical process involves three types of problems, which although closely interrelated depend on quite different technical principles. The first group of problems is encountered in the preparation of the material and energy balances of the process and the establishment of the duties to be performed by the various items of equipment. The second type of problem is the determination of the process specifications of the equipment necessary to perform these duties. Under the third classification are the problems of equipment and materials selection, mechanical design, and the integration of the various units into a coordinated plan.

These three types may be designated as process, unit-operation, and plant-design problems, respectively. In the design of a plant these problems cannot be segregated and each treated individually without consideration of the others. However, in spite of this interdependence in application the three types may advantageously be segregated for study and development because of the different principles involved. Process problems are primarily chemical and physicochemical in nature; unit-operation problems are for the most part physical; the plant-design problems are to a large extent mechanical.

In this book only process problems of a chemical and physicochemical nature are treated, and it has been attempted to avoid overlapping into the fields of unit operations and plant design. The first part deals primarily with the applications of general physical chemistry, thermophysics, thermochemistry, and the first law of thermodynamics. Generalized procedures for estimating vapor pressures, critical constants, and heats of vaporization have been elaborated. New methods are presented for dealing with equilibrium problems in extraction, adsorption, dissolution, and crystallization. The construction and use of enthalpy-concentration charts have been extended to complex systems. The treatment of material balances has been elaborated to include the effects of recycling, by-passing, changes of inventory, and accumulation of inerts.

In the second part the fundamental principles of thermodynamics are presented with particular attention to generalized methods. The applications of these principles to problems in the compression and expansion of fluids, power generation, and refrigeration are discussed. However, it is not attempted to treat the mechanical or equipment problems of such operations.

Considerable attention is devoted to the thermodynamics of solutions with particular emphasis on generalized methods for dealing with deviations from ideal behavior. These principles are applied to the calculation of equilibrium compositions in both physical and chemical processes.

All these principles are combined in the solution of the ultimate problem of the kinetics of industrial reactions. Quantitative treatment of these problems is difficult, and designs generally have been based on extensive pilot-plant operations carried out by a trial-and-error procedure on successively larger scales. However, recent developments of the theory of absolute reaction rates have led to a thermodynamic approach to kinetic problems which is of considerable value in clarifying the subject and reducing it to the point of practical applicability. These principles are developed and their application is discussed for homogeneous, heterogeneous, and catalytic systems. Particular attention is given to the interpretation of pilot-plant data. Economic considerations are emphasized and problems are included in establishing optimum conditions of operation.

In covering so broad a range of subjects, widely varying comprehensibility is encountered. It has been attempted to arrange the material in the order of progressive difficulty. Where the book is used for college instruction in chemical engineering the material of the first part is suitable for second- and third-year undergraduate work. The second part is suitable for third- and fourth-year undergraduate work; the third part for senior and graduate studies.

A few problems were selected from *Chemical Engineering Problems* published by the American Institute of Chemical Engineers (1946), with permission.

The authors wish to acknowledge gratefully the suggestions of Professors Joseph Hirschfelder, R. J. Altpeter, K. A. Kobe, E. N. Lightfoot, R. G. Taecker, and A. L. Lydersen.

OLAF A. HOUGEN
KENNETH M. WATSON
ROLAND A. RAGATZ

Preface to Part II

Thermodynamics

Eleven years have elapsed since the first publication of the "Thermodynamics" section of *Chemical Process Principles.* Changes in this new edition are in recognition of the great advances made in the development of chemical engineering curricula in terms of the basic engineering sciences. With increasing independence of instruction from service courses taught by other branches of engineering, the treatment of thermodynamics in this textbook has been enlarged particularly in the fields of fluid flow and power generation. Principles of fluid flow with consideration of thrust have been extended to the supersonic range as well as to the flow of compressible fluids through ducts and nozzles. Principles in the generation of power include steam turbines, spark-ignition engines, compression-ignition engines, gas turbines, free-piston engines, turbojets, ramjets, and rockets. The thermodynamic problems of these devices are analyzed in terms of basic scientific principles, but problems of mechanical design are not treated.

Generalized methods of thermodynamics have been improved and extended by introduction of a third parameter with presentation of tables covering both gas and liquid phases based on the available *pvT* data of 82 different compounds. The construction of thermodynamic charts for specific compounds by means of generalized tables is fully presented. These developments have been made in the interest of more effective teaching and of solving practical problems without requiring tedious manipulations of complex equations of state.

Tables of vaporization equilibrium constants for multicomponent solutions have been developed by generalized methods with applications to distillation and to composition changes caused by partial condensation or partial evaporation produced by free and isentropic expansion.

Methods are presented for the construction of enthalpy–composition and entropy–composition charts of gaseous and liquid solutions.

For vapor–liquid equilibria in nonideal solutions, the correlation of data by the Redlich–Kister equation is presented with extension to conditions of constant pressure and to conditions of constant temperature. Correlation of vapor–liquid data in binary nonideal systems is presented with application to the separation of azeotropes by changes in temperature, pressure, and by the introduction of a third component. Chapters on chemical equilibria now include electrochemical reactions, complex reactions, and equilibrium composition changes in internal-combustion engines.

The first edition has been nearly entirely rewritten with introduction of 70% new material.

The chapter on thermodynamic properties from molecular structure presented in the previous edition has been omitted for the reason that this aspect of thermodynamics is more adequately treated in treatises on basic sciences.

In this revision, credit is acknowledged for suggestions and aid from the following sources: R. B. Bird, K. C. Chao, R. A. Greenkorn, J. O. Hirschfelder, R. H. Kadlec, A. L. Lydersen, P. S. Myers, R. E. Peck, and O. A. Uyehara.

Olaf A. Hougen
Kenneth M. Watson
Roland A. Ragatz

Contents

PART II THERMODYNAMICS

Table of Symbols

A	area
A	atomic weight
A	component A
A	total work function
A	total work function per mole
a	activity
a	velocity of sound
B	availability function
B	component B
BDC	bottom dead center
Bhp	brake horse power
Bmep	brake mean effective pressure
C	component C
C	concentration per unit volume
°C	degrees Centigrade
C	number of components
C_p	heat capacity at constant pressure
C_v	heat capacity at constant volume
CI	compression ignition
c	specific heat
c	velocity of light
c_p	molal heat capacity at constant pressure
c_v	molal heat capacity at constant volume
D	diameter
d	differential operator
E	electrode potential
E	energy in general

E_K	kinetic energy
e	base of natural logarithms
e	electron
F	degrees of freedom
F	Faraday constant
F	force
$\vec{F}_D$	total drag force
$\vec{F}_E$	external thrust
$\vec{F}_G$	gravitational force
$\vec{F}_T$	total thrust
f	friction factor
f	fugacity
f	weight fraction
G	free energy
G	mass velocity per unit area
G	specific gravity
ΔG	change in free energy
G	free energy per mole
$\bar{\mathrm{G}}$	partial molal free energy
g_c	standard gravitational constant, 32.174 ft/sec^{-2}
(g)	gaseous state
H	enthalpy
H	Henry's constant
H	humidity
H_p	percentage humidity
H_R	relative humidity
ΔH	change in enthalpy
H_c	heat of combustion
H_f	heat of formation
H_r	heat of reaction
H	enthalpy per mole
$\bar{\mathrm{H}}$	partial molal enthalpy
$\Delta\bar{\mathrm{H}}$	change in partial molal enthalpy
I	Integration constant
I	moment of inertia
Ihp	indicated horsepower
Imep	indicated mean effective pressure

13

Thermodynamic Principles

In its original and *restricted* sense thermodynamics deals with the limitations imposed on the conversion of heat or internal energy into useful work. In a *general* sense thermodynamics deals with the limitations imposed on the transformation of energy from one form to another.

The principles of thermodynamics have been extended to the calculation and prediction of the properties of a substance that are related to energy and of the equilibrium conditions that are approached in physical or chemical processes. From these principles properties related to energy can be established from a minimum of experimental data.

A thermodynamic potential constitutes the driving force that causes a spontaneous process to progress toward a state of equilibrium. However, the rate at which the state of equilibrium is approached depends not only on the driving force but also on a resistance factor. Thermodynamic principles alone do not permit evaluation of these resistances or rate constants.

System and Process. In thermodynamic terminology, a *system* refers to a substance or group of substances set apart for study, and a *process* to the changes taking place in the system. Thus, hydrogen, oxygen, and water constitute a system, and the combustion of hydrogen to form water constitutes a process. In this system other processes might be considered, such as the dissolving of the hydrogen and oxygen in the water, or all three processes might be considered.

Open and Closed Systems. A system is *closed* when there can be no exchange of matter with its surroundings and *open* when such exchange occurs. In a multiphase system each separate phase is open, since material is free to enter and leave each phase although the system as a whole may be closed.

Isolation of a System. A closed system may be free to exchange heat and work with its surroundings. Heat can be transferred through the walls of the vessel enclosing the system, and mechanical work can be performed upon or by the system by means of a piston connecting the vessel with some external mechanism. Devices can also be arranged for the removal or addition of electric energy. A closed system is *thermally* *isolated* when the enclosing walls are impervious to the flow of heat, *mechanically isolated* when it is enclosed by rigid walls, and *completely isolated* when neither material nor energy in any form can be added or removed. A closed system in contact with a *heat reservoir* without thermal insulation is free to receive or lose energy by the flow of heat from or to the reservoir. A closed system in contact with a piston is free to receive or deliver work in the movement of the piston. The atmosphere surrounding a closed system may also act as a reservoir for the flow of heat or as a piston for transmitting work from and to the system. The atmosphere may also be made part of the system as in the combustion of a fuel in air. In a consideration of the energy relations of any non-isolated system, the energy of the surroundings must be taken into account as well as that of the system itself.

Availability and Degradation of Energy. It is well known that the various forms of external and internal energies within a system are not all equally available for transformation into mechanical work. Thus, the rotation of a shaft or the translation of a piston represent energies of the highest order, since in the absence of friction they can be completely converted into useful work. Similarly, electric energy can be converted entirely into mechanical work in the absence of electrical resistance and mechanical friction. Actually, because of friction and electrical resistance, part of the mechanical or electric energy of a machine is always converted into heat. It is common experience that, whereas heat can be converted into mechanical or electric energy by means of an engine, only part of the heat that flows into the engine can be recovered as work. Thus, although there is an exact quantitative equivalence among the different forms of energy, there is a marked difference in the availability of these forms for useful work. Heat represents the least available form of energy, and the transformation of other forms into heat represents a degradation of energy.

There are other means of degrading energy and rendering it less available for useful work. Thus, the internal energy of a gas at high pressure is more available for performing useful work than that of the same gas at low pressure and at the same temperature. Even though the internal energy of the gas were the same at both pressures, as in the case for an ideal gas, the availability of the internal energy for further work of expansion is less in the expanded gas. Thus, an isothermal reduction in the pressure of a gas represents a degradation of its internal energy. Similarly, the isothermal mixing of two unlike gases represents a degradation of the internal energy of the system since work is required to separate them, and this work is not recovered upon remixing.

Reversibility. A *reversible* process is defined as one that proceeds under conditions of balanced forces such that the direction of the process

Isfc	indicated specific fuel consumption
i	any component i
J	mechanical equivalent of heat
j	any component except i
K	characterization factor
K	distribution coefficient
K	equilibrium constant
K	vaporization equilibrium constant
°K	degrees Kelvin
K_c	equilibrium constant, concentration units
K_p	equilibrium constant, pressure units
L	mass velocity of liquid per unit area
l	length
ln	natural logarithm
log	logarithm to base 10
(l)	liquid state
M	molecular weight
M_m	mean molecular weight
m	mass
m	slope of equilibrium curve dy^*/dx
N	mole fraction
N	Avogadro number $(6.024)\ 10^{23}$
N_c	cycles per minute
N_M	Mach number
N_R	revolutions per minute
N_{Re}	Reynolds number
n	number of moles
P	pressure (used only in special cases to distinguish pressure of pure components from partial pressure of a component in solution)
p	total pressure
p_c	critical pressure
p_i	partial pressure of component i
p_R	relative pressure
p_r	reduced pressure

q	heat *added to* a system
q_f	heating value of fuels
q_r	rate of heat flow
R	component R
R	gas constant
R	total radius
°R	degrees Rankine
r	radius
r_c	compression ratio
r_e	expansion ratio
r_h	hydraulic radius
r_t	load ratio or cut off ratio
r_v	velocity ratio
S	component S
S	cross-section area
S	entropy
S	humid heat
SC	standard conditions
SI	spark ignition
S_p	percentage saturation
S_r	relative saturation
s	molal entropy
(s)	solid state
$\bar{\mathrm{s}}$	partial molal entropy
T	absolute temperature, °R or °K
T	torque
T_c	critical temperature
T_r	reduced temperature
TDC	top dead center
t	temperature, °F or °C
U	internal energy
U	internal energy per mole
u	velocity
u_j	jet velocity
V	volume
V_r	volume of reactor
v	volume per mole

$\bar{v}$	partial molal volume
v_c	critical molal volume
v_r	reduced volume
W	mass flow rate
W_E	engine air rate
W_F	engine fuel rate
w	weight
w	work *done by* system
w_e	work of expansion *done by* system
w_f	electrical work *done by* system
w_f	energy degraded by friction
w_s	shaft work *done by* system
x	mole fraction in liquid phase
x	quality
y	mole fraction in vapor phase
y^*	mole fraction in vapor phase, equilibrium value
y_p	percentage humidity
y_r	relative humidity in per cent
Z	elevation above datum plane
z	compressibility factor
z	mole fraction in total system
z_c	critical compressibility factor

Greek Symbols

α	Riedel parameter
α	coefficient of compressibility
α	relative volatility
α	angle
α	residual molal volume
(α)	crystal form
α	partial residual molal volume
β	angle
β	coefficient of volumetric expansion
(β)	crystal form
γ	activity coefficient
(γ)	crystal form
Δ	finite change of a property; a positive value indicates an increase

δ	change in moles per mole of reactant
δ	solubility parameter
∂	partial differential operator
ϵ	energy per molecule
η	efficiency
Θ	thermodynamic temperature
θ	angular position
κ	ratio of heat capacities
Λ	heat of vaporization
λ	molal heat of vaporization
λ	wave length
λ_f	heat of fusion per mole
μ	chemical potential
μ	viscosity
ν	frequency, reciprocal seconds
ν	fugacity coefficient
ν	number of molecules
π	total pressure of a mixture, used where necessary to distinguish from p
π	3.1416
ρ	density
ρ_c	critical density
ρ_r	reduced density
Σ	summation
σ	surface tension
τ	time
ϕ	number of phases
ω	wave number
$\dot{\omega}$	Pitzer parameter

Superscripts

*	ideal behavior
*	equilibrium state
°	standard state
′	pseudo state
—	partial molal property
$\widetilde{}$	average property
M	mixing
f	free volume
E	excess property

can be reversed by an infinitesimal change in external conditions. An example of a reversible process is the vaporization of a liquid under its own vapor pressure in a cylinder fitted with a frictionless piston and in contact with an isothermal heat reservoir. At any stage an infinitesimal increase in pressure upon the piston will produce condensation, and an infinitesimal decrease will cause vaporization. It follows that *in a reversible process an infinitesimal change will tend to restore the original conditions, so that no degradation of energy takes place, and the availability of the energy of the combined system and its surroundings remains constant.*

Actual processes are generally to some extent irreversible and result in a decrease in the availability of energy. The flow of heat from one body to another is always an irreversible process. It is impossible to restore heat to the first body except by transfer from a still hotter body or by the degradation of mechanical energy. Similarly, the dropping of an object without restraint to a lower level is an irreversible process.

Another example of an irreversible process is the free expansion of a gas. If a gas is confined in a closed container of volume V_1 and is allowed to expand into a vacuum chamber such that the new total volume is V_2, and if this system is entirely isolated from its surroundings by rigid insulating walls, the gas is said to *expand freely.* No work is performed on the surroundings, no heat flows in or out of the system, and for an ideal gas there is no change in temperature. Hence, the energy content of the expanded gas is the same as that of the original gas. However, in order to return the expanded gas to its original condition the expenditure of external energy as work of compression is required. For an ideal gas all this work manifests itself as heat in compressing the gas, and in isothermal compression this heat is transferred to the surroundings. Thus, in isothermal compression of an ideal gas the energy content remains constant but assumes a higher level of availability.

Other examples of irreversible processes are the mixing of hot and cold fluids, the inelastic deformation of a solid, the magnetic hysteresis of an iron core, the flow of electricity through a resistance, the dissolution of a solid, and spontaneous chemical reactions.

Second Law of Thermodynamics. The concept of reversibility forms the basis for the second law of thermodynamics. One statement of this law is that *all spontaneous processes are to some extent irreversible and are accompanied by a degradation of energy.* A corollary to this statement is that it is impossible for any self-acting machine to transfer energy from a given state to a higher state of availability.

The validity of the second law is confirmed by extensive experimental evidence when applied to gross masses of matter in which large numbers of molecules are present. Under these circumstances the laws of probability are followed in establishing the distribution of molecular and atomic energies.

An important deduction from the second law is that any machine that performs work by transforming molecular energy received at fixed intake conditions and rejecting molecular energy in a more degraded state at fixed discharge conditions will convert a maximum fraction of the energy received into work when it operates reversibly. Any actual self-acting machine must operate with a lower efficiency of transformation than that of a reversible machine. If it were possible to construct a more efficient engine, it could be used to drive the reversible machine in reverse to transfer molecular energy from a given state to the original less degraded state in contradiction of the second law.

The steam engine is an example of a machine that performs work by receiving steam at a high temperature and rejecting it at a low temperature. The maximum efficiency possible with given intake and exhaust conditions is obtained when such an engine operates reversibly. This theoretical maximum efficiency is always less than 100 per cent because of the energy content of the rejected steam.

Entropy

Before undertaking development of the quantitative relationships involved in transformations of energy, it is desirable to define an additional thermodynamic property to serve as a measure of the *unavailability* or *degradation* of the energy of a system. This property is termed *entropy*.

The physical concept of entropy may be expressed as follows: *Entropy is an intrinsic property of matter so defined that an increase in the unavailability of the total energy of a system is quantitatively expressed by a corresponding increase in its entropy.* Since entropy is an intrinsic property, its magnitude is dependent only on the nature of the matter under consideration and the state in which it exists and not on its external position or motion relative to other bodies. Thus, the entropy of an elevated object is no different from that of the same object in a lower position, and the entropy of a rotating flywheel is equal to that of the same flywheel at rest under the same conditions of temperature and pressure. The same is true of other intrinsic properties such as internal energy and enthalpy.

Entropy and Heat. Since entropy is a measure of unavailability of internal energy, it follows that the entropy of a system is increased by the degradation into heat of any higher form of energy that it possesses. This is illustrated by consideration of an isolated system comprising an inelastic object suspended above a rigid plate. If the object is allowed to fall, the total energy content of the system is unchanged, but the potential energy of the object is converted into heat as a result of the inelastic impact on the plate. The temperature of both the object and the plate are correspondingly increased by this increase in internal energy. Dropping the object has not changed the energy content of the system but has degraded the energy to a less available form, and thus the entropy of the system is increased. Similarly, if a rotating flywheel is stopped by a brake, kinetic energy is degraded into heat, and the entropy of the system is increased. However, if the flywheel were stopped in a reversible manner by means of a frictionless transmission which would set another flywheel in motion, there would be no degradation of energy and no change in entropy.

In both of these systems an increase in entropy results from the addition of heat to the system through the degradation of a higher form of energy. However, since entropy is an intrinsic property of matter not influenced by elevation or external movement, it follows that *the entropy of a system is increased by the addition of heat through any mechanism or from any source.* Thus, heat might be added to the system containing the elevated object and the rigid plate, increasing the total energy content of the system until a temperature is reached equal to that attained after the object has been permitted to fall in the system when isolated. Since entropy is independent of the position of the object and the plate, the final entropies of the system are equal, and the increases in entropy accompanying these two processes are also equal, although heat is added to the system by the degradation of its own energy in one case and by increase in its total energy in the other.

Entropy and Temperature. It is evident that the amount of heat added to a system is a partial measure of the magnitude of its increase in entropy. However, the quantity of heat added is not the sole measure of increase in entropy. The unavailability of energy and the entropy of an isolated system are increased by the transfer of heat within the system to a region of lower temperature. Similarly, the addition of heat to a system at a lower temperature results in a greater degradation of energy than at a higher temperature. It follows that *the increase in entropy accompanying the addition of a given amount of heat to a system is increased by lowering the temperature* at which the heat is added.

Entropy and Reversibility. In order to complete the definition of entropy, one more factor must be considered in addition to the amount of heat added and the temperature level at which it is added. For example, a gas may be expanded freely to a lower pressure within a closed system which is completely isolated both thermally and mechanically. This is an irreversible process resulting in degradation of the energy of the system and an increase in its entropy. However, no heat is added, and no work is done, and for an ideal gas the process is isothermal. In this irreversible process the entropy increase is not measured by the addition of heat. The same final state might be reached by expanding the gas through an engine within the system and continuously converting the work done into heat by means of friction. In this case heat is added to the system by the degradation of the mechanical work. The amount of heat added increases as the efficiency of the engine is increased and reaches a maximum when the engine operates reversibly. However, entropy is an intrinsic property of matter, the change of which in any process is dependent only on initial and final states and not on the path. It is evident that in the process under consideration the amount of heat added to the system can be taken as a measure of the increase in entropy only if the nature of the process is specified. This specification is logically taken to correspond to the maximum possible degradation of higher forms of energy into heat. Thus *the increase in entropy of a system is measured by the amount of heat added only when all changes in the intrinsic states of the matter of the system occur reversibly.*

The three requirements for the quantitative definition of entropy are hence satisfied by the following equation:

$$dS = \frac{d'q}{T} \quad \text{(reversible change)} \tag{1}$$

where S = entropy.

The primed differential symbol d' is used for indicating incremental quantities of heat and work. In contrast to the other terms in the energy equation, neither heat nor work are properties of the system, nor can they in general be expressed as a function of the state of the system. An infinitesimal change in a property such as volume can be expressed as dV which is an *exact* differential with respect to the state of the system. However, since neither q nor w are properties, an increment of either is referred to as an *inexact* differential.

An important corollary of equation 1 which follows from the second law and the definition of entropy as a measure of degradation is that the entropy change accompanying an irreversible change is greater than $d'q/T$, or

$$T\,dS > d'q \quad \text{(irreversible change)} \tag{2}$$

Thus, if an isothermal change such as the expansion of a gas occurs reversibly, the gain in entropy of the gas is compensated by a reduction in the entropy of its surroundings resulting from the withdrawal of heat. The combined entropy of the system and its surroundings remains unchanged, provided the loss of heat from the surroundings also proceeds reversibly. If the same transfer of heat from the surroundings is accomplished irreversibly, the combined entropy of the gas and its surroundings increases although the increase in entropy of the gas itself is the same as for the reversible change.

It is important that the significance of the restriction to reversibility in equation 1 be appreciated. This restriction does not apply to the manner in which heat is added to the system or the process employed in the degradation of higher forms of energy. Thus, if a block of steel is heated by a flame, its increase in entropy is measured by the amount of heat added, even though the addition of the heat from the flame to the solid is highly irreversible. Similarly, the restriction does not apply to changes involving only the relative positions or movement of bodies of matter. The process involving the falling object which was discussed previously is irreversible, but the accompanying entropy change is measured by the heat added. Only processes involving changes in the intrinsic state of the matter itself must be reversible for the application of equation 1.

Entropy and Probability. A further physical concept of entropy results from its relationship to probability. The existence of such a relationship is evident from the fact that all spontaneous changes are in the direction of maximum probability and are also accompanied by increases in entropy. It follows from the theory of probability that the most probable state in which a system can exist is that having the least orderly arrangement or greatest randomness. This concept may be illustrated by a large number of black and white balls shaken together in a box. The most probable average arrangement is a uniform distribution of the black and white balls, and the chance of a compartment of the box containing balls of only one color is remote. Thus, a state of randomness, free from orderly arrangement, is the most probable state of any system, and entropy may be looked upon as a measure of randomness which is a minimum for systems in orderly arrangement.

Entropy as an Extensive Property. It is evident that for any given change in state accompanied by an increase in entropy more heat is added to a system of large mass than to a smaller one. Thus, like internal energy or volume, entropy is an *extensive property*, the magnitude of which is dependent on the mass involved. It follows that the entropy of a system is equal to the sum of the entropies of its separate phases.

Entropy and Internal Energy. From a statistical approach it can be shown that entropy is a measure of the distribution of internal energy in matter under equilibrium conditions. For example, for gases, entropy can be accurately calculated by statistical methods from the distribution of internal energy into its various forms of internal and external molecular motion and corresponding quantum states. This distribution is a function of the temperature, pressure, and nature of the system and in multicomponent systems is also a function of concentration. The redistribution of the same amount of internal energy in a given system to a new equilibrium state at a lower temperature, lower pressure, or lower concentration results in an increase in entropy of the system and a reduction in the availability of this internal energy for producing useful work. From spectroscopic measurements and statistical methods of analysis of the magnitude and distribution of internal forms of motion under given state conditions the entropy of gases can be calculated without recourse to direct calorimetric measurements under reversible conditions and without using the concept of the classical Carnot cycle. The statistical approach gives a satisfactory comprehension of entropy as a fundamental extensive property of a system rather than as a nebulous property based on the performance of a reversible engine or reversible transfer of heat.

Calculation of Entropy Changes

In calculating entropy changes certain generalizations must be kept in mind and adhered to. One important general principle is that, when a reversible process occurs in a given system, with accompanying reversible processes in other systems, the summation of the entropy changes of all the participating systems is zero. Conversely, whenever an irreversible process occurs, the summation of the entropy changes for all systems participating in the process is some positive quantity. A second principle is that $dS = d'q/T$ for reversible processes only. A third principle is that entropy is a point function, and therefore the entropy change of a system is determined by the initial and final states of the system, regardless of how the system is brought from the initial to the final state. This principle is particularly useful in analyzing irreversible processes. In such instances, it is not necessary to make a direct analysis of the actual process. Instead, an imaginary reversible process is substituted for the irreversible one with the same terminal states. The entropy change is then calculated by considering the reversible process. These procedures are demonstrated by the following examples.

Energy Absorption by Constant-Temperature Bath; No Chemical Reactions Occurring. In the absence of chemical reaction, the entropy increase accompanying energy absorption by a constant-temperature bath is measured by the amount of thermal energy that is added. If the total heat capacity of the bath is so high in relation to the energy absorbed that the temperature change is infinitesimal,

$$\Delta S_{\text{bath}} = \frac{q}{T} \tag{3}$$

Illustration 1. The stirrer in a large fused salt bath operating at 1000° F is driven by an electric motor. What is the entropy change of the bath, due to the addition of 5 hp-hr of mechanical energy to the bath by the stirrer, assuming no heat losses?

The conversion of the mechanical energy into thermal energy by friction is an irreversible process. From the standpoint of entropy change of the bath, the process is equivalent to transferring a like amount of thermal energy to the bath by conduction from an outside heat source. Therefore,

$$\Delta S_{\text{bath}} = \frac{q}{T} = \frac{5(2544.5)}{1459.7} = 8.72 \text{ Btu per R}^\circ$$

Since, for this process, there is no entropy change except in the bath,

$$\Delta S_{\text{total}} = \Delta S_{\text{bath}} = 8.72 \text{ Btu per R}^\circ$$

Isothermal Phase Changes. Pure substances under fixed pressure conditions undergo phase changes such as allotropic transition, fusion, and vaporization at constant temperatures. If the pressure is maintained constant and if the heat is added slowly, such processes are reversible, and the accompanying entropy changes are calculated by dividing the latent heat of transformation by the equilibrium transformation temperature.

Illustration 2. Ethyl alcohol melts at −114.4° C with a heat of fusion of 25.76 g-cal per gram. Under a pressure of 1 atm, the boiling point is 78.3° C with a latent heat of vaporization of 204.26 g-cal per gram. Calculate the molal entropies of fusion and vaporization.

$$\Delta s_{\text{fusion}} = \frac{\lambda_{\text{fusion}}}{T_{\text{fusion}}} = \frac{(25.76)(46.07)}{273.2 - 114.4} = 7.47 \text{ g-cal/(g-mole)(K}^\circ)$$

$$\Delta s_{\text{vap}} = \frac{\lambda_{\text{vap}}}{T_{\text{vap}}} = \frac{(204.26)(46.07)}{273.2 + 78.3} = 26.77 \text{ g-cal/(g-mole)(K}^\circ)$$

It may be observed that the units of entropy are the same as the units of heat capacity, but the two properties are entirely unlike. It should be noted that 1 g-cal/(g-mole)(K°) = 1 Btu/(lb-mole)(R°).

Heating or Cooling Without Chemical Change. The entropy change in heating or cooling matter in the absence of a chemical reaction is computed by integration of the equation

$$\Delta S = \int_{T_1}^{T_2} \frac{d'q}{T} \tag{4}$$

If the heating or cooling is done under constant-pressure conditions, $d'q = nc_p\, dT$; if it is done under constant-volume conditions, $d'q = nc_v\, dT$.

For constant-pressure conditions, if the heat capacity is independent of temperature,

$$\Delta S = nc_p \int_{T_1}^{T_2} \frac{dT}{T} = nc_p \ln \frac{T_2}{T_1} \tag{5}$$

If the heat capacity varies with temperature, the equation relating heat capacity and temperature is inserted, and the equation is integrated as follows,

$$\Delta S = n\int_{T_1}^{T_2}\left(\frac{c_p}{T}\right)dT = n\int_{T_1}^{T_2}\left(\frac{a + bT + cT^2 + dT^3}{T}\right)dT$$

$$= n\left[a \ln \frac{T_2}{T_1} + b(T_2 - T_1) + \frac{c}{2}(T_2^2 - T_1^2) + \frac{d}{3}(T_2^3 - T_1^3)\right] \quad (6)$$

Illustration 3. Calculate the entropy change of 1 lb of ideal carbon dioxide when heated at constant pressure from 77° to 2200° F.

The equation for the molal heat capacity of carbon dioxide is as follows, where T is in degrees Kelvin:

$$c_p = 5.316 + (1.4285)10^{-2}T - (0.8362)10^{-5}T^2 + (1.784)10^{-9}T^3$$

It is necessary to convert the given Fahrenheit temperatures into the equivalent Kelvin temperatures. Furthermore, it is convenient to base the calculations on 1 g-mole, and then to change to the basis of 1 lb:

$$\Delta s = \int_{298.2°}^{1477.6°}\left[\frac{5.316}{T} + (1.4285)10^{-2} - (0.8362)10^{-5}T + (1.784)10^{-9}T^2\right]dT$$

$$= 18.501 \text{ g-cal/(g-mole)(K°)} = 18.501 \text{ Btu/(lb-mole)(R°)}$$

or

$$18.501/44.01 = 0.4204 \text{ Btu/(lb)(R°)}$$

If the entropy change is to be computed through the use of a mean heat capacity, the following relation holds by definition

$$\Delta s = c'_{pm} \ln (T_2/T_1) \quad (7)$$

Combining equations 6 and 7 gives

$$c'_{pm} = a + \frac{b(T_2 - T_1) + (c/2)(T_2^2 - T_1^2) + (d/3)(T_2^3 - T_1^3)}{\ln (T_2/T_1)} \quad (8)$$

It will be recognized that the entropy–mean heat capacity is slightly different from the enthalpy–mean heat capacity as tabulated in Table 19. In the present instance, the entropy–mean heat capacity is 11.56 g-cal/(g-mole)(K°), whereas the enthalpy–mean heat capacity is 12.24 g-cal/(g-mole)(K°).

The foregoing procedures are valid, regardless of how the heating is conducted in bringing the system from its initial to its final temperature. Even though the heating of the system under consideration may have been conducted in an irreversible manner with high temperature gradients existing during the heating process, the entropy change of the system is the same as if it had been brought from initial to final temperature by a process of reversible heating, whereby only infinitesimal temperature gradients existed at any time during the heating process. Thus,

$$\Delta S = \int_{T_1}^{T_2} \frac{d'q_{\text{rev}}}{T} = n\int_{T_1}^{T_2} \frac{c_p}{T}\,dT \tag{9}$$

Entropy Changes in an Ideal Gas. If one mole of an ideal gas is brought from p_1, v_1, T_1 to p_2, v_2, T_2, the entropy change is determined by following an imaginary reversible path from state 1 to state 2, regardless of how the change may have actually taken place. While there is a wide choice of paths, it will be convenient first to change the temperature at constant pressure and then to change the pressure reversibly and isothermally.

For the first step,

$$\Delta s_1 = \int_{T_1}^{T_2} \frac{c_p}{T}\,dT \tag{10}$$

In the isothermal reversible change of pressure, the first law is applied:

$$d'q = d\text{u} + d'w_e \tag{11}$$

The reversible work of expansion, $d'w_e$, equals $p\,dv$. By definition, an ideal gas has the following characteristics: (1) It follows the simple gas law, $pv = RT$, and (2) its internal energy is a function of temperature alone, and is not affected by pressure. Accordingly,

$$d\text{u} = 0 \tag{12}$$

$$p\,dv + v\,dp = R\,dT = 0 \tag{13}$$

$$p\,dv = -v\,dp = -RT\frac{dp}{p} \tag{14}$$

Substituting in equation 11 yields

$$d'q = -RT\frac{dp}{p} \tag{15}$$

The entropy change for the second step is as follows:

$$\Delta s_2 = \int_{p_1}^{p_2} \frac{d'q}{T} = R\int_{p_1}^{p_2} \frac{dp}{p} = -R\ln\frac{p_2}{p_1} \tag{16}$$

$$\Delta s_{\text{total}} = \Delta s_1 + \Delta s_2 = \int_{T_1}^{T_2} \frac{c_p}{T}\,dT - R\ln\frac{p_2}{p_1} \tag{17}$$

An alternative equation developed by first changing temperature at constant volume and then changing volume reversibly and isothermally is as follows:

$$\Delta s_{\text{total}} = \int_{T_1}^{T_2} \frac{c_v}{T}\,dT + R\ln\frac{v_2}{v_1} \tag{18}$$

Illustration 4. One gram-mole of pure nitrogen at 2 atm and 25° C is allowed to flow into an evacuated container. Its final pressure is 0.5 atm and the temperature is unchanged. Calculate the entropy change.

This process is irreversible, but nevertheless the entropy change may be evaluated by applying equation 17 to the initial and final conditions.

$$\Delta s = -R \ln \frac{p_2}{p_1} = (-1.987)(2.303) \log \frac{0.5}{2.0} = 2.755 \text{ g-cal/(g-mole)(K°)}$$

Chemical Reactions. The entropy changes accompanying chemical reactions are evaluated through the use of absolute entropies, the development of which is discussed on page 520.

$$\Delta S = S_P - S_R \tag{19}$$

Illustration 5. Calculate the entropy change at 25° C for the following reaction

$$CO(g, 1 \text{ atm}) + \tfrac{1}{2}O_2(g, 1 \text{ atm}) \rightarrow CO_2(g, 1 \text{ atm})$$

$$\Delta s = s_{CO_2} - [s_{CO} + \tfrac{1}{2}s_{O_2}]$$

Values of absolute entropies at 25° C are taken from Table 68 and substituted in the foregoing equation.

$$\Delta s = 51.061 - [47.301 + \tfrac{1}{2}(49.003)] = -20.742 \text{ g-cal/(g-mole CO)(K°)}$$

The standard heat of reaction may be calculated from heat of formation data, as follows:

$$\Delta H = (\Delta H_f)_{CO_2} - (\Delta H_f)_{CO} - \tfrac{1}{2}(\Delta H_f)_{O_2} = -94{,}051.8 - (-26{,}415.7) - 0$$

$$= -67{,}636 \text{ g-cal per g-mole CO}$$

This value of heat evolved may not be used to evaluate the entropy change, as the reaction is irreversible.

$$\Delta s \neq \left[\frac{\Delta H}{T} = -\frac{67{,}636}{298.16} = -226.8 \text{ g-cal/(g-mole CO)(K°)}\right]$$

It will be noted that for this process $\Delta H/T$ is not equal to Δs.

If this reaction could be made to proceed reversibly, the heat added would be $T\,\Delta s = (298.16)(-20.742) = -6184$ g-cal per g-mole CO. The energy available for useful work, as, for example, in the generation of electric energy, would then be $-\Delta H + q_{rev} = -(-67{,}636) - 6184 = 61{,}452$ g-cal per g-mole CO.

Adiabatic Mixing of a Hot and a Cold Fluid. When a hot fluid is adiabatically mixed with a cooler portion of the same fluid, the hot portion cools to the final temperature of the mixture, and the other portion heats to the same final temperature. The entropy change of each is evaluated through the use of the equation

$$\Delta S = \int_{T_1}^{T_2} \frac{C_p}{T}\, dT \tag{20}$$

The total entropy change is then obtained by addition of the two individual entropy changes.

Illustration 6. One hundred pounds of hot water at 200° F are mixed with 20 lb of cold water at 60° F. Calculate the entropy change if these two lots of water are mixed under adiabatic conditions. Assume constant unit heat capacity for water.

The final temperature after mixing is evaluated by an energy balance and is found to be 176.7° F, or 636.4° R.

$$\Delta S_1 = 20\int_{519.7^\circ}^{636.4^\circ}\frac{dT}{T} = 20(2.303)\log\frac{636.4}{519.7} = 4.052 \text{ Btu per R}^\circ$$

$$\Delta S_2 = 100\int_{659.7^\circ}^{636.4^\circ}\frac{dT}{T} = 100(2.303)\log\frac{636.4}{659.7} = -3.597 \text{ Btu per R}^\circ$$

$$\Delta S_{\text{total}} = \Delta S_1 + \Delta S_2 = 4.052 - 3.597 = 0.455 \text{ Btu per R}^\circ$$

Illustration 6 can be extended to show that, in any problem of adiabatic mixing of fluids of the same kind which are initially at different temperatures, the entropy of the system will always increase.

Isothermal Mixing of Ideal Gases. Two pure ideal gases, each at pressure p and temperature T, are mixed in such proportions that N_A and N_B represent the respective mole fractions in the final mixture. This is an irreversible process. The entropy change accompanying this process may be evaluated through the following series of reversible operations, applied to N_A moles of pure A and N_B moles of pure B.

Step 1. Expand each pure component reversibly and isothermally from pressure p to a pressure corresponding to its partial pressure in the final mixture:

$$\Delta S_A = -N_A R \ln\frac{N_A p}{p} = -N_A R \ln N_A \tag{21}$$

$$\Delta S_B = -N_B R \ln\frac{N_B p}{p} = -N_B R \ln N_B \tag{22}$$

$$\Delta s_1 = \Delta S_A + \Delta S_B = -N_A R \ln N_A - N_B R \ln N_B \tag{23}$$

Step 2. Introduce each pure component reversibly into a large volume of gas at pressure p and temperature T, in which A and B occur in the ratio of N_A to N_B and in which the respective partial pressures, therefore, are $N_A p$ and $N_B p$. This is accomplished by slowly forcing A through a semipermeable membrane which will permit A to enter the large volume of gas, but will not permit B to diffuse out. Likewise, B is introduced by slowly forcing it through a semipermeable membrane that permits the passage of B but not of A. Since these operations are reversible,

$$\Delta s_2 = 0 \tag{24}$$

$$\Delta s_{\text{total}} = \Delta S_1 + \Delta S_2 = -N_A R \ln N_A - N_B R \ln N_B \tag{25}$$

The foregoing analysis can be applied to systems composed of any number of components.

Illustration 7. Calculate the entropy of 1 lb-mole of air relative to pure oxygen and atmospheric nitrogen at the same temperature and pressure.

$$\Delta s = -(0.2100)(1.987)(2.303) \log 0.21 - (0.7900)(1.987)(2.303) \log 0.79$$

$$= 0.651 + 0.370 = 1.021 \text{ Btu/(lb-mole air)(R}^\circ)$$

Irreversibility of Heat Transfer

The transfer of heat from one body to another is an irreversible process except in the limiting case where the transfer is made under conditions of zero temperature difference between the two bodies. In this hypothetical case, the quantity of heat transferred and stored as internal energy remains at the same temperature, and no reduction in entropy of the system occurs; the rate of heat transfer is zero, and a corresponding infinite size of heat exchange surface is required. The practical problem of heat flow thus is not a concern of thermodynamics but rather a problem in economic design in securing high rates of heat transfer with a corresponding reduction in size of heat exchanger. In the transfer of heat between two fluid streams through the medium of a heat exchanger, reversibility can be approached only by countercurrent flow (Fig. 130*a*), if the flow rates are adjusted so that the temperature difference approaches zero throughout the path of each stream. The net result is the transfer of heat from one fluid to the other with no change in entropy of the system; the entropy gain in one stream is equal to the entropy loss in the other. When a cold stream is heated by a hot stream in parallel flow (Fig. 130*b*) by flow of heat through the walls of a heat exchanger, a finite temperature difference must exist at the entrance of the exchanger, thus resulting in a net increase in entropy even though the two streams leave at the same temperature. In this case the net result is a transfer of heat to a lower temperature level with an increase in entropy of the system; the gain in entropy of the initially cold stream is greater than the loss in entropy of the initially hot stream.

A high rate of heat transfer between two fluids flowing in a heat exchanger is obtained by increasing the temperature drop between the two streams and by increasing the mass velocity of each stream. Increasing the temperature drop increases the irreversibility of the process, whereas increase in mass velocities requires external energy for pumping the fluids. Most of this external energy is dissipated as heat. A compromise is reached by minimizing the total cost charged to the equipment and to the energy required for fluid flow.

In condensing vapor from an inert gas by means of a counterflow heat exchanger (Fig. 130c), the average temperature difference between the two streams may be much greater than at either terminal, thus increasing the irreversibility of the process over countercurrent heat exchange as in Fig. 130a.

Where heat is being transferred from a condensing vapor to a fluid stream (Fig. 130d), a high degree of irreversibility exists owing to the

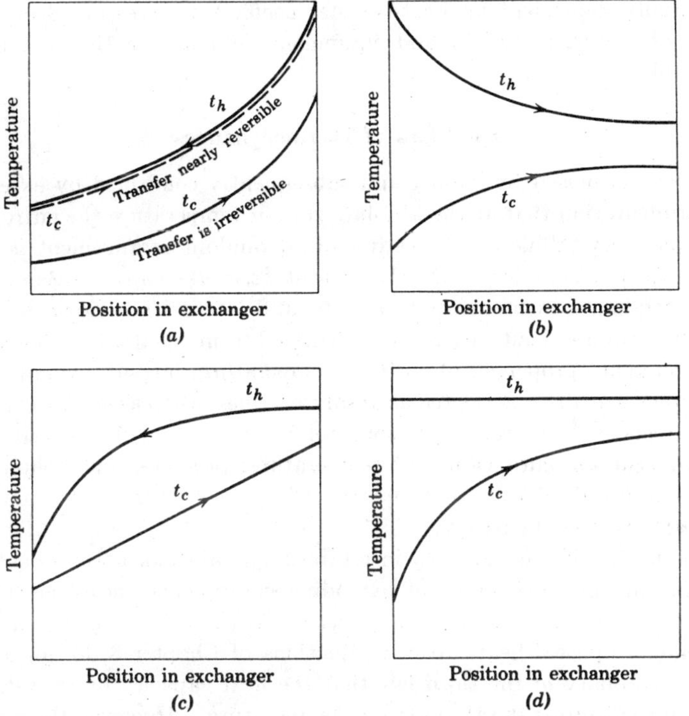

FIG. 130. Irreversibility in heat transfer. (a) Countercurrent heat exchange. (b) Parallel flow heat exchange. (c) Heat exchange from condensing vapors admixed with inert gases. (d) Heat exchange from condensing vapors

large temperature difference between the condensing vapor and the incoming cold fluid. This irreversibility can be diminished if the vapor is expanded reversibly in a series of turbines with intermediate heat exchangers, through which the fluid being heated is passed in counterflow with respect to the vapor operating the turbines. If an infinite number of turbines and heat exchangers were provided, the transfer of heat would be accomplished reversibly. This general procedure of decreasing the irreversibility of heat transfer is employed in regenerative feed-water heating in power plants, as discussed in Chapter 17.

The most economic design of heat exchangers is in minimizing the total hourly costs charged to power, materials, and equipment by securing high rates of heat transfer with corresponding increase in power consumption and decrease in size of equipment. This compromise always results in a net gain of entropy. The problem of economic design is essentially dependent on heat-transfer coefficients, pressure drops, and costs of power, materials, and equipment, and not on thermodynamic principles.

Third Law of Thermodynamics

It was proposed by Nernst and subsequently confirmed by extensive experimentation that at the absolute zero of temperature the entropy of any pure crystalline substance free of all random arrangement is zero. This principle is known as the *third law of thermodynamics*. Accordingly, by extending measurements of specific and latent heats down to 0° K, absolute values of entropy can be calculated from equation 1. Thus, the four reference properties of matter—temperature, pressure, volume, and entropy—can all be presented as absolute values, whereas energy contents can be reported only relative to some reference state. It has already been shown that absolute values of temperature, pressure, and volume are required in formulating the properties of matter. It is shown later that the same is true of entropy.

The utility of the third law in establishing values of absolute entropy brings out the importance of extending calorimetric measurements to temperatures approaching absolute zero. At these low temperatures the common empirical heat-capacity equations of Chapter 8 do not apply. It is a corollary of the third law that the heat capacity of a crystalline substance is zero at the absolute zero temperature. However, the manner in which the heat capacity diminishes with decreasing temperature in the low-temperature range is a unique property, varying widely for different materials.

Thermodynamic Temperature Scale

In the foregoing discussion no consideration has been given to the scale or function employed for the quantitative expression of temperature in the definition of entropy. Although any arbitrary scale may be used to indicate relative hotness or coldness, it is evident that only one type of scale or function can be used to express the temperature as used in equation 1 if entropy as defined by it is to represent a true measure of the unavailability of energy at different temperature levels. This correct function is termed the *thermodynamic temperature scale.*

At this point it is difficult to arrive at the nature of the thermodynamic temperature by direct derivation, and so it will be tentatively assumed that it is the same as the ideal-gas temperature which is defined as $T = pv/R$. Rigorous thermodynamic properties and relations are developed from equation 1 without consideration of the nature of the temperature scale. Any size of temperature unit may be used with no change in the relationships involved except a change in the numerical values of temperature and entropy. A confirmation of the identity of the thermodynamic temperature with the ideal-gas temperature is presented on page 550.

The Energy Properties

Free Energy. A thermodynamic property of great utility is defined by the following relation and termed the *free energy* G;

$$G = H - TS \tag{26}$$

For changes taking place at constant temperature, or for changes where, regardless of path, the initial and final temperatures are the same,

$$\Delta G = \Delta H - T\,\Delta S \tag{27}$$

Where work is available from a process in addition to the mechanical work of expansion, it follows from the first law discussed in Chapter 8 that, for nonflow process,

$$q = \Delta U + w_e + w_f \tag{28}$$

where

w_e = work of expansion done by the system
w_f = *other useful work* done by the system, such as electric energy.

For a reversible process at constant temperature and pressure $q = T\,\Delta S$ and $w_e = p\,\Delta V$. Hence, under these conditions the first law may be written as

$$\Delta U + p\,\Delta V = T\,\Delta S - w_f = \Delta H \tag{29}$$

By combination of 27 and 29, it follows that

$$\Delta G = -w_f \tag{30}$$

Thus, in a process that occurs reversibly at constant temperature and pressure, the useful work is equal to the decrease in the free energy of the system. When the process occurs irreversibly at constant temperature and pressure, $T\,\Delta S > q$, and the useful work is always less than the decrease in the free energy of the system. Therefore, the decrease in free energy accompanying a process at constant temperature and pressure represents the maximum useful work that becomes available only when the process is reversible.

Total Work. Another useful thermodynamic property is defined by the following relation and termed *total work A*:

$$A = U - TS \tag{31}$$

For changes taking place at constant temperature or for changes where, regardless of path, the initial and final temperatures are the same,

$$\Delta A = \Delta U - T\,\Delta S \tag{32}$$

From the first law, for a reversible nonflow process at constant temperature,

$$\Delta U = q - w = T\,\Delta S - w \tag{33}$$

Combining equations 32 and 33 for a reversible process at constant temperature gives

$$\Delta A = -w = -(w_e + w_f) \tag{34}$$

Therefore, the decrease in the total work accompanying a process at constant temperature represents the maximum total work that becomes available only when the process is reversible. For a reaction proceeding at constant volume, $w_e = 0$, and $\Delta A = \Delta G$. The difference between ΔA and ΔG is usually small.

Among different writers there exists confusion concerning the names and symbols used for free energy and total work, and no uniform standard has been adopted.

Differential Energy Properties. If consideration is limited to systems of constant mass and composition and to reversible processes in which the only external force is pressure, differential equations for the four energy properties are developed as follows from their definitions and from equation 1:

$$dU = d'q - d'w_e = T\,dS - p\,dV \tag{35}$$

$$dH = dU + d(pV) = dU + p\,dV + V\,dp \tag{36}$$

$$dA = dU - d(TS) = dU - T\,dS - S\,dT \tag{37}$$

$$dG = dH - d(TS) = dH - T\,dS - S\,dT \tag{38}$$

By combination of the foregoing, four basic differential equations are obtained which express the energy properties in terms of the four reference properties, p, V, T, and S. Thus:

$$dU = T\,dS - p\,dV \tag{39}$$

$$dH = T\,dS + V\,dp \tag{40}$$

$$dA = -S\,dT - p\,dV \tag{41}$$

$$dG = -S\,dT + V\,dp \tag{42}$$

Significance of the Four Energy Properties. Under certain conditions of restraint, changes in internal energy ΔU, enthalpy ΔH, total work ΔA, and free energy ΔG are tangible quantities which represent useful work. For this reason these properties are designated as *energy properties* although in a general sense internal energy alone represents energy under all conditions of restraint. These functions are also referred to as thermodynamic potentials. This latter terminology, however, leads to confusion with the usual concept of a potential as a driving force and hence is not recommended.

Most chemical reactions proceed under no restraint other than that of pressure and dissipate the internal energy of the system as heat with no production of useful work. Exceptional cases are found in the operation of electrolytic cells wherein a part of the energy content of the system is converted into electric energy. As an example of the significance of the four energy properties, a system and process are chosen wherein 1 g-atom of zinc metal is dissolved at constant temperature and pressure in an excess of hydrochloric acid of a given initial concentration and with the release of hydrogen gas. It is assumed possible to connect the zinc as anode through an external circuit with an insoluble cathode placed in the same solution, and to allow the reaction to proceed slowly and reversibly at constant pressure with generation of heat, electric energy, and work against the atmosphere. In the absence of the electric circuit, this reaction proceeds irreversibly with no generation of electric energy.

When the reaction proceeds at constant pressure and irreversibly with no generation of electric energy, the heat given up by the system and flowing into the calorimeter is equal to the decrease in the enthalpy of the system, or $-q_p = -\Delta H$. If the reaction proceeds at constant volume and irreversibly and with no generation of electric energy, the heat given up by the system to the calorimeter is equal to the decrease in internal energy of the system, or $-q_v = -\Delta U$.

If the zinc were connected as anode through an external electric circuit and the reaction allowed to proceed reversibly at constant pressure, the electric energy developed would be equal to the decrease in free energy, or $w_f = -\Delta G$. The work of expansion in the release of hydrogen against the atmosphere is equal to w_e. The total work done is hence $w_f + w_e$. This total work is equal to the decrease in the total work property of the system, or $w_f + w_e = -\Delta A$.

The values of ΔU, ΔH, ΔA, and ΔG depend only on initial and final conditions, and are independent of the path pursued, but the direct calorimetric and electrical measurements of these quantities must be made under definitely prescribed paths.

In summary, for nonflow processes:

At constant volume, $$\Delta U = q - w_f \tag{43}$$

At constant pressure, $$\Delta H = q - w_f \tag{44}$$

In any isothermal reversible process, $$\Delta A = -w_f - w_e \tag{45}$$

In an isothermal reversible process at constant pressure, $$\Delta G = -w_f \tag{46}$$

Illustration 8. A chemical reaction was carried out reversibly in an electrolytic cell at a constant pressure of 2 atm and a constant temperature of 200° C. Electric energy equal to 65,400 g-cal was given off. Five gram-moles of gas were evolved by the reaction.

The same chemical reaction carried out irreversibly at a constant pressure of 2 atm and with starting and ending temperatures of 200° C developed 24,500 g-cal of electric energy, and 45,000 g-cal of heat were absorbed. As before, 5 g-moles of gas were given off.

For both processes, calculate q, w_e, w_f, ΔS, ΔU, ΔH, ΔA, ΔG.

The solution to the problem is presented in Table A. The numbers in brackets indicate the order of calculation. Since S, U, H, A, and G are point properties, the respective increments in these properties are the same for the reversible and the irreversible processes.

Equilibrium in Closed Systems

A system is in equilibrium when its state is such that it can undergo no spontaneous or unaided changes. Such a condition can result only when all forces or potentials that tend to promote change are absent or are exactly balanced against similar opposing forces or potentials. An iron ball resting at the bottom of a spherical bowl is in mechanical equilibrium with the bowl, since the mechanical forces acting on the ball are balanced and the ball is at rest. No alteration in the position of the ball can possibly take place except at the expenditure of mechanical energy received from the outside.

Although the iron ball resting at the bottom of the bowl is in a state of mechanical equilibrium, it is not in a state of chemical equilibrium with the surrounding air. Spontaneous irreversible vaporization and oxidation of the iron can occur to form vapor and rust. These spontaneous processes will proceed, although very slowly, until the iron has been vaporized and oxidized to a condition of complete thermodynamic equilibrium. Thus many conditions of apparent equilibrium actually represent only partial equilibrium with respect to all possible changes. It is necessary to distinguish between conditions of partial and of complete equilibrium.

General Criteria of Equilibrium. Since a principal objective of thermodynamic theory is the development of expressions for equilibrium

TABLE A

	Reversible Reaction	Irreversible Reaction
w_f	[1] $+65{,}400$ g-cal (given)	[2] $+24{,}500$ g-cal (given)
w_e	[4] $w_{e1} = p\,\Delta V = \Delta nRT$ (reversible) $= 5(1.987)(473.2) = +4700$ g-cal	[5] $w_{e2} = w_{e1}$ $= +4700$ g-cal (reversible)
q	[6] $q_1 = \Delta U_1 + w_{e1} + w_{f1}$ (*a*) $q_2 = \Delta U_2 + w_{e2} + w_{f2}$ (*b*) Subtracting *b* from *a*, and, since $\Delta U_1 = \Delta U_2$ and $w_{e1} = w_{e2}$, there results $q_1 = q_2 + w_{f1} - w_{f2}$ $q_1 = 45{,}000 + 65{,}400 - 24{,}500$ $= +85{,}900$ g-cal	[3] $q_2 = +45{,}000$ g-cal (given)
ΔS	[7] $\Delta S_1 = \frac{q_{1,\,rev}}{T} = \frac{85{,}900}{473.2}$ $= +181.5$ g-cal/K°	[8] $\Delta S_2 = \Delta S_1 = +181.5$ g-cal/K°
ΔU	[9] $\Delta U_1 = q_1 - w_{e1} - w_{f1}$ $= 85{,}900 - 4700 - 65{,}400$ $= +15{,}800$ g-cal	[10] $\Delta U_2 = \Delta U_1 = 15{,}800$ g-cal or $\Delta U_2 = q_2 - w_{e2} - w_{f2}$ $= 45{,}000 - 4700 - 24{,}500$ $= +15{,}800$ g-cal
ΔH	[11] $\Delta H_1 = \Delta U_1 + \Delta(pV) = \Delta U_1 + w_{e1}$ $= 15{,}800 + 4700$ $= +20{,}500$ g-cal	[12] $\Delta H_2 = \Delta H_1 = +20{,}500$ g-cal or $\Delta H_2 = \Delta U_2 + w_{e2}$ $= 15{,}800 + 4700$ $= +20{,}500$ g-cal
ΔA	[13] $\Delta A_1 = \Delta U_1 - T\,\Delta S_1 = \Delta U_1 - q_1$ $= 15{,}800 - 85{,}900$ $= -70{,}100$ g-cal or $\Delta A_1 = -w_{e1} - w_{f1}$ $= -4700 - 65{,}400$ $= -70{,}100$ g-cal	[14] $\Delta A_2 = \Delta A_1 = -70{,}100$ g-cal
ΔG	[15] $\Delta G_1 = \Delta H_1 - T\,\Delta S_1 = \Delta H_1 - q_1$ $= 20{,}500 - 85{,}900$ $= -65{,}400$ g-cal or $\Delta G_1 = -w_{f1}$ $= -65{,}400$ g-cal	[16] $\Delta G_2 = \Delta G_1 = -65{,}400$ g-cal

conditions, it is important to derive specifications which must be fulfilled by any system at equilibrium. Such specifications, termed *criteria of equilibrium*, are the foundation for complete relationships among the various properties of a system at equilibrium.

From a thermodynamic standpoint, it is necessary that, in a system at equilibrium, every possible change that might take place to an infinitesimal extent shall be reversible, since any irreversible change would result in a displacement which would destroy the original equilibrium. As was previously pointed out, reversible processes are accompanied by no change in total entropy of the combined system and its surroundings, whereas every spontaneous process is accompanied by an increase in total entropy. Thus, a universal thermodynamic criterion of equilibrium is that, for any change that takes place, the total entropy of the system *and its surroundings* shall be constant. In a completely isolated system the entropy of the system itself is constant at equilibrium. From equation 1, defining entropy, it follows that, where heat is added to a system in which all changes of state are reversible, $d'q = T\,dS$, where S is the entropy of the system itself, not including its surroundings. This expression may also be taken as a criterion of reversibility and equilibrium. Since for all irreversible changes of state $dS > d'q/T$, if any incremental addition of heat to the system is accompanied by an entropy increase equal to and not greater than $d'q/T$, then all thermodynamic processes within the system must be reversible, and it follows that the system is in equilibrium.

Stable and Unstable Equilibrium. A system is in *stable* equilibrium if after a finite displacement it spontaneously returns to its original state when the displacing force is returned to its original value. A round pencil lying in the bottom of a cylindrical trough is a mechanical example of this type of equilibrium. If, however, this pencil is carefully balanced on its sharpened point, it will be in a state of *unstable* equilibrium such that finite displacement does not lead to a spontaneous return to its original conditions. Although, at equilibrium, any infinitesimal change is accompanied by no change in the total entropy of the system and its surroundings, thermodynamically a system is in unstable equilibrium if a finite displacement involves an increase in total entropy. For example, a finite displacement of the pencil balanced on its point results in an irreversible process whereby heat is developed and the entropy of the pencil and of its surroundings increases. In a system in stable equilibrium no finite change can be accompanied by an increase in total entropy. Thus, at stable equilibrium, for any change resulting from a temporarily applied extraneous force, not associated with the system or its normal surroundings,

$$dS_t = 0, \qquad \Delta S_t \leqq 0 \tag{47}$$

where S_t = total entropy of the system and its surroundings. On the other hand, in unstable equilibrium, for any change

$$dS_t = 0, \qquad \Delta S_t > 0 \tag{48}$$

Energy Properties as Criteria of Equilibria. Useful restricted criteria of equilibria for systems under various types of restraint may be derived from the foregoing general criterion and the definitions of the energy properties. For any system at equilibrium, $d'q = TdS$, hence equation 28 may be written in differential form:

$$dU = T\,dS - p\,dV - d'w_f \tag{49}$$

Since $U = H - pV = A + TS = G - pV + TS$, similar equations may be written in terms of the other energy properties. Thus:

$$dH = T\,dS + V\,dp - d'w_f \tag{50}$$

$$dA = -S\,dT - p\,dV - d'w_f \tag{51}$$

$$dG = -S\,dT + V\,dp - d'w_f \tag{52}$$

Each of the equations, 49 through 52, serves as a simple criterion of equilibrium under specified conditions of restraint. Thus, if the system is restrained to conditions of constant entropy and volume, equation 49 becomes $dU = -d'w_f$. The corresponding criteria of equilibrium for various types of restraint are as follows:

Restraint	Criterion of Equilibrium	
Constant entropy and volume	$dU = -d'w_f$	(53)
Constant entropy and pressure	$dH = -d'w_f$	(54)
Constant temperature and volume	$dA = -d'w_f$	(55)
Constant temperature and pressure	$dG = -d'w_f$	(56)

If the system under consideration is subject to no external forces other than pressure, no useful work can be performed, and $d'w_f = 0$. Under this additional restriction equilibrium is achieved when the appropriate differential energy property is equal to zero.

Thermodynamic Formulations

The thermodynamic properties of a system or a process may be classified into four groups as *reference properties*, *energy properties*, *derived properties*, and *path properties*.

The *reference properties*—temperature, pressure, volume, entropy, and composition—are those required to define completely the state of a system. Of these properties temperature, pressure, and composition are intensive; volume and entropy are extensive.

The *energy properties* are extensive properties whose values are known only relative to some arbitrary reference state.

Such properties as specific heat, coefficient of expansion, coefficient of compressibility, and Joule–Thomson coefficient are classified as *derived properties*.

All these properties have intrinsic values that are determined by the existing state of a system and are independent of the path followed in arriving at that state. Properties such as these in which changes are dependent only on the initial and final conditions of the system and are independent of the path followed in producing the changes are termed *point properties.*

As previously pointed out, heat and work including mechanical, radiant, and electrical forms are not properties of a system but are manifestations of changes occurring within the system, and as such may be considered properties of a process rather than of a system. From the previous discussions it is evident that these manifestations are dependent on the particular path followed and may differ widely in two processes, even though the initial and final properties of the systems are identical. For this reason these properties of a process are termed *path properties*.

Thermodynamic Surface. When any two of the reference properties—temperature, pressure, volume, and entropy—of a pure substance or of a substance of fixed composition in any single phase are specified, the relative values of other point properties are definitely established. With reference to one selected pair of properties, such as pressure and temperature, the magnitude of a third property, such as volume, may be represented as a surface in space for each separate phase. If extended to all conditions of pressure and temperature of a pure substance with its several states of aggregation, solid, liquid, and gas, the surface developed assumes a contour of great complexity. Upon such a contour surface, for example, with pressure and temperature as independent variables and volume as the dependent variable, equal values of all other point properties may be designated by lines extending over the contour surfaces. Separate sets of lines may be traced, respectively, for constant values of entropy, internal energy, specific heat, compressibility, and so forth. The relationship of points on one path to those on another is definitely fixed by the contour of the surface.

If a mathematical equation can be established for the surface in question in terms of p, V, and T, then relative values of all related properties may be established by mathematical methods. If the equation is unknown or too complex for mathematical manipulation, the desired relationships may still be established by graphical methods. Ordinarily, only the more important relations are developed in which some desired property is expressed in terms of properties that are easily measured, such as temperature and pressure.

In dealing with solutions and mixtures, the foregoing discussion must be expanded to include the variable of composition. To represent the properties of a solution of fixed composition, a contour surface can be constructed as for a pure compound, but, in covering the entire range of compositions, a graphical presentation by a contour surface is no longer adequate, and mathematical methods must be employed.

In developing the thermodynamic relations that follow, it will ordinarily be futile to attempt to visualize the relations involved.

Exact Differential Equations. Where z is a single-valued continuous function of two independent variables x and y, the total differential of z can be expressed in terms of its partial derivatives with respect to its different independent variables, thus:

$$dz = \left(\frac{\partial z}{\partial x}\right)_y dx + \left(\frac{\partial z}{\partial y}\right)_x dy \tag{57}$$

The properties of a substance fulfill the requirements of z. Equation 57 can be extended mathematically to include any number of variables. An equation of the type of 57 is termed an *exact differential equation.*

The value of the integral of dz between any two conditions A and B of a relationship represented by equation 57 is dependent only on the value of z at these two points and is independent of the path followed between the two states. For a case involving only three variables, this situation is evident from consideration of Fig. 84 (page 332). Thus:

$$\int_A^B dz = z_B - z_A = f(x, y) \tag{58}$$

Since z_A and z_B are wholly defined by the corresponding values of x and y, the value of the integral is unaffected by the path considered in the integration.

If, however, any line is drawn on the surface of Fig. 84 connecting A and B, the length of the line will be dependent on the path followed. A differential equation expressing the length of the line will involve expressions in addition to those defining the surface and is termed an inexact differential equation. In Fig. 84 the length of the wavy line $d'\sigma$ is an inexact differential.

It is a property of an exact differential equation involving two independent variables x and y that

$$\frac{\partial}{\partial y}\left(\frac{\partial z}{\partial x}\right)_y = \frac{\partial}{\partial x}\left(\frac{\partial z}{\partial y}\right)_x \tag{59}$$

or

$$\frac{\partial^2 z}{\partial y\, \partial x} = \frac{\partial^2 z}{\partial x\, \partial y} \tag{60}$$

If more than two independent variables are involved, a relationship similar to equation 59 may be written with respect to any pair of independent variables. Any exact differential equation must satisfy the relations of equation 59.

For example, if dz is expressed by the relation

$$dz = M\, dx + N\, dy \tag{61}$$

where $M = f(x, y)$ and $N = \phi(x, y)$, then equation 61 is an exact differential equation only provided

$$M = \left(\frac{\partial z}{\partial x}\right)_y, \qquad N = \left(\frac{\partial z}{\partial y}\right)_x \tag{62}$$

Under these conditions,

$$\left(\frac{\partial M}{\partial y}\right)_x = \frac{\partial^2 z}{\partial x\, \partial y}, \qquad \left(\frac{\partial N}{\partial x}\right)_y = \frac{\partial^2 z}{\partial y\, \partial x} \quad \text{or} \quad \left(\frac{\partial M}{\partial y}\right)_x = \left(\frac{\partial N}{\partial x}\right)_y \tag{63}$$

If in the exact differential equation,

$$dz = \left(\frac{\partial z}{\partial x}\right)_y dx + \left(\frac{\partial z}{\partial y}\right)_x dy \tag{64}$$

the condition is imposed that z is constant, the equation becomes

$$\left[0 = \left(\frac{\partial z}{\partial x}\right)_y dx + \left(\frac{\partial z}{\partial y}\right)_x dy\right]_z \tag{65}$$

or

$$\left(\frac{\partial z}{\partial x}\right)_y = -\left(\frac{\partial y}{\partial x}\right)_z \left(\frac{\partial z}{\partial y}\right)_x = -\frac{(\partial y/\partial x)_z}{(\partial y/\partial z)_x} \tag{66}$$

If equation 57 is divided by dy, holding w constant, the following is obtained :

$$\left(\frac{\partial z}{\partial y}\right)_w = \left(\frac{\partial z}{\partial x}\right)_y \left(\frac{\partial x}{\partial y}\right)_w + \left(\frac{\partial z}{\partial y}\right)_x \tag{67}$$

Another useful relationship is obtained by dividing both numerator and denominator of a partial derivative by the same differential under the same constant conditions. Thus:

$$\left(\frac{\partial z}{\partial x}\right)_y = \frac{(\partial z/\partial w)_y}{(\partial x/\partial w)_y} = \left(\frac{\partial z}{\partial w}\right)_y\left(\frac{\partial w}{\partial x}\right)_y \tag{68}$$

A useful equation involving two variables q and r in addition to z, x, and y may be derived as follows. Considering z to be a function of q and r and applying the principles of equations 66 and 67, the following is obtained:

$$\left(\frac{\partial z}{\partial q}\right)_y = \left(\frac{\partial z}{\partial q}\right)_r - \left(\frac{\partial z}{\partial r}\right)_q \frac{(\partial y/\partial q)_r}{(\partial y/\partial r)_q} \tag{69}$$

Similarly, if x is considered to be a function of q and r,

$$\left(\frac{\partial x}{\partial q}\right)_y = \left(\frac{\partial x}{\partial q}\right)_r - \left(\frac{\partial x}{\partial r}\right)_q \frac{(\partial y/\partial q)_r}{(\partial y/\partial r)_q} \tag{70}$$

Dividing the first equation by the second gives

$$\left(\frac{\partial z}{\partial x}\right)_y = \frac{(\partial z/\partial q)_r(\partial y/\partial r)_q - (\partial z/\partial r)_q(\partial y/\partial q)_r}{(\partial x/\partial q)_r(\partial y/\partial r)_q - (\partial x/\partial r)_q(\partial y/\partial q)_r} \tag{71}$$

Equations 57 through 71 represent mathematical principles commonly used in the development of thermodynamic relations.

Maxwell Relations. Since equation 39 is an exact equation of the form of equation 57, the relationship of equation 63 permits the derivation of a differential equation relating the coefficients T and p as shown in equation 72. Similar equations result from equating the partial derivatives of the coefficients of equations 40, 41, and 42. Thus:

$$\left(\frac{\partial T}{\partial V}\right)_S = -\left(\frac{\partial p}{\partial S}\right)_V \tag{72}$$

$$\left(\frac{\partial T}{\partial p}\right)_S = \left(\frac{\partial V}{\partial S}\right)_p \tag{73}$$

$$\left(\frac{\partial S}{\partial V}\right)_T = \left(\frac{\partial p}{\partial T}\right)_V \tag{74}$$

$$\left(\frac{\partial S}{\partial p}\right)_T = -\left(\frac{\partial V}{\partial T}\right)_p \tag{75}$$

Equations 72 through 75 are termed the Maxwell relations.[1]

[1] J. Clerk Maxwell, *Theory of Heat*, with corrections and additions (1891), by Lord Rayleigh, Longmans, Green & Co., London (1904).

Equation 74 is the basis of the familiar Clapeyron equation. For the vaporization of a liquid the restriction of constant volume may be omitted because the vapor pressure is independent of volume. Since vaporization is a reversible process at constant temperature, $(\partial S/\partial V)_T = \Lambda/(T\,\Delta V)$, and

$$\frac{dp_s}{dT} = \frac{\Lambda}{T\,\Delta V} = \frac{\Lambda}{T(V_g - V_l)} \tag{76}$$

which is equation 3, page 78.

Illustration 9. From a plot of the following properties per pound of superheated ammonia, show that

$$\left(\frac{\partial T}{\partial V}\right)_S = -\left(\frac{\partial p}{\partial S}\right)_V$$

using the reference conditions, $p = 173$ psia, $t = 185°$ F, $S = 1.300$ Btu/(lb)(R°), and $V = 2.20$ cu ft per lb.

At $S = 1.300$			At $V = 2.20$		
V	t	p	S	t	p
2.461	165.2	150	1.2466	114.8	150
2.339	174.4	160	1.2705	145.1	160
2.229	183.0	170	1.2935	180.3	170
2.200	185.0	173	1.3000	185.0	173
2.132	191.3	180	1.3154	208.0	180
2.043	199.2	190	1.3368	240.3	190
1.960	206.9	200	1.3575	283.1	200

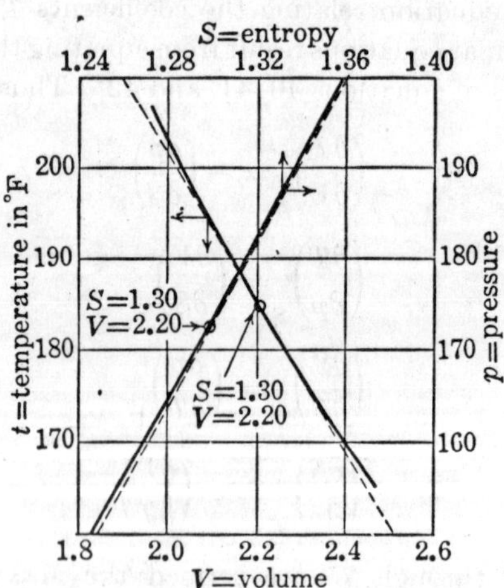

FIG. 131. Evaluation of $(\partial T/\partial V)_S$ and $(\partial p/\partial S)_V$

In Fig. 131, p is plotted against S at $V = 2.20$, and T is plotted against V at $S = 1.300$.

The value of $(\partial T/\partial V)_S$ is the slope of T plotted against V at $S = 1.300$, or

$$\left(\frac{\partial t}{\partial V}\right)_S = \frac{160 - 210}{2.50 - 1.90} = -83.3.$$

The value of $(\partial p/\partial S)_V$ is the slope of p plotted against S at $V = 2.20$. This slope is $(200 - 150)/(1.36 - 1.25) = 454$. To convert this slope into the same units as $(\partial T/\partial V)_S$ requires multiplication by 144/778, where 144 is the conversion factor for pounds per square inch to pounds per square foot and 778 represents the number of foot-pounds equivalent to 1 Btu. The result is 84 in agreement with the Maxwell relation.

Heat Capacity in Terms of Entropy. The two most useful thermal capacities are those at constant pressure and at constant volume, defined as follows:

$$C_p = \left(\frac{\partial' q}{\partial T}\right)_p \tag{77}$$

$$C_v = \left(\frac{\partial' q}{\partial T}\right)_v \tag{78}$$

Since, in reversible heating and cooling, $dS = d'q/T$,

$$C_p = T\left(\frac{\partial S}{\partial T}\right)_p \quad \text{or} \quad \left(\frac{\partial S}{\partial T}\right)_p = \frac{C_p}{T} \tag{79}$$

$$C_v = T\left(\frac{\partial S}{\partial T}\right)_v \quad \text{or} \quad \left(\frac{\partial S}{\partial T}\right)_v = \frac{C_v}{T} \tag{80}$$

Differential Equations for dS. If entropy is considered to be a function of pressure and temperature,

$$dS = \left(\frac{\partial S}{\partial p}\right)_T dp + \left(\frac{\partial S}{\partial T}\right)_p dT \tag{81}$$

Substitutions for $(\partial S/\partial p)_T$ according to equation 75 and for $(\partial S/\partial T)_p$ according to equation 79 give

$$dS = -\left(\frac{\partial V}{\partial T}\right)_p dp + \frac{C_p}{T} dT \tag{82}$$

If entropy is considered to be a function of volume and temperature,

$$dS = \left(\frac{\partial S}{\partial V}\right)_T dV + \left(\frac{\partial S}{\partial T}\right)_v dT \tag{83}$$

Substitutions for $(\partial S/\partial V)_T$ according to equation 74 and for $(\partial S/\partial T)_v$ according to equation 80 give

$$dS = \left(\frac{\partial p}{\partial T}\right)_v dV + \frac{C_v}{T}\, dT$$

$$= -\left[\frac{(\partial V/\partial T)_p}{(\partial V/\partial p)_T}\right] dV + \frac{C_v}{T}\, dT \tag{84}$$

Modified Equations for *dU* and *dH*. The basic differential equations for dU and dH as given by equations 39 and 40 may be modified by substituting for dS according to equation 82 or 84, yielding the following;

$$dU = -T\left(\frac{\partial V}{\partial T}\right)_p dp - p\, dV + C_p\, dT \tag{85}$$

$$dU = -\left[p + T\,\frac{(\partial V/\partial T)_p}{(\partial V/\partial p)_T}\right] dV + C_v\, dT \tag{86}$$

$$dH = \left[V - T\left(\frac{\partial V}{\partial T}\right)_p\right] dp + C_p\, dT \tag{87}$$

$$dH = V\, dp - T\left[\frac{(\partial V/\partial T)_p}{(\partial V/\partial p)_T}\right] dV + C_v\, dT \tag{88}$$

Systematic Derivations of Thermodynamic Relations[2]

In analyzing thermodynamic processes, equations for various partial derivatives relating pressure p, volume V, temperature T, entropy S, internal energy U, enthalpy H, total work A, and free energy G are desired and necessary. For example, in certain liquefaction processes, the cooling produced by the isenthalpic expansion of a gas (the Joule–Thomson type of expansion) is employed to bring about cooling and liquefaction. Accordingly, a general equation for the change of temperature with respect to pressure at constant enthalpy assumes considerable practical importance, as it may be used to determine under what conditions a gas will heat or cool upon free expansion and the temperature change that will occur for a given pressure change. In this particular case, therefore, an equation for the partial derivative $(\partial T/\partial p)_H$ is desired in terms of quantities for which data are readily available.

It may be shown that, ignoring reciprocals, there are 168 partial derivatives involving the eight variables, p, V, T, S, U, H, A, and G.

[2] R. A. Ragatz, *Trans. Chem. Eng. Div. Am. Soc. Eng. Ed.*, 1948, part II, p. 110 revised Jan. 1950, Sept. 1952.

For single-component systems the number of equations relating these 168 partial derivatives is fantastically great. However, if the specification is made that the equations in all instances are to be expressed in terms of C_p (or C_v), $(\partial V/\partial T)_p$, and $(\partial V/\partial p)_T$ (which are easily measurable and for which data may be found in the literature), the problem is no longer of immense scope.

The purpose of this section is to demonstrate systematic methods of procedure that will make it possible to develop an equation for any of the 168 partial derivatives in terms of C_p (or C_v), $(\partial V/\partial T)_p$, and $(\partial V/\partial p)_T$.

Before demonstrating the methods, it is important to point out that this discussion applies to a single-phase system of constant mass and composition. Under these conditions, fixing any two of the eight variables will automatically fix the remaining six. For example, if temperature and volume are fixed, then pressure, entropy, internal energy, enthalpy, total work, and free energy are also fixed. Mathematically, we may write $p = f(T, V)$, $S = f(T, V)$, $U = f(T, V)$, etc.

The working equations required for the derivational procedures to be described are summarized in Table 40.

Table 40. Summary of Equations Used in the Derivation of Thermodynamic Formulas

I. General Mathematical Formulas. If $z = f(x, y)$,

$$dz = \left(\frac{\partial z}{\partial x}\right)_y dx + \left(\frac{\partial z}{\partial y}\right)_x dy = M\,dx + N\,dy \tag{89}$$

$$\left(\frac{\partial z}{\partial y}\right)_w = \left(\frac{\partial z}{\partial x}\right)_y \left(\frac{\partial x}{\partial y}\right)_w + \left(\frac{\partial z}{\partial y}\right)_x \tag{90}$$

$$\left(\frac{\partial z}{\partial x}\right)_y = -\left(\frac{\partial y}{\partial x}\right)_z \left(\frac{\partial z}{\partial y}\right)_x = -\frac{(\partial y/\partial x)_z}{(\partial y/\partial z)_x} \tag{91}$$

$$\left(\frac{\partial z}{\partial x}\right)_y = \left(\frac{\partial z}{\partial w}\right)_y \left(\frac{\partial w}{\partial x}\right)_y = \frac{(\partial z/\partial w)_y}{(\partial x/\partial w)_y} \tag{92}$$

$$\left(\frac{\partial M}{\partial y}\right)_x = \left(\frac{\partial N}{\partial x}\right)_y \tag{93}$$

$$\left(\frac{\partial z}{\partial x}\right)_y = \frac{(\partial z/\partial q)_r(\partial y/\partial r)_q - (\partial z/\partial r)_q(\partial y/\partial q)_r}{(\partial x/\partial q)_r(\partial y/\partial r)_q - (\partial x/\partial r)_q(\partial y/\partial q)_r} \tag{94}$$

If $q = T$, and $r = p$,

$$\left(\frac{\partial z}{\partial x}\right)_y = \frac{(\partial z/\partial T)_p(\partial y/\partial p)_T - (\partial z/\partial p)_T(\partial y/\partial T)_p}{(\partial x/\partial T)_p(\partial y/\partial p)_T - (\partial x/\partial p)_T(\partial y/\partial T)_p} \tag{95}$$

TABLE 40 (*Continued*)

II. The Four Basic Equations and the Modified dU and dH Equations

$$dU = T\,dS - p\,dV \tag{96}$$

$$= -T\left(\frac{\partial V}{\partial T}\right)_p dp - p\,dV + C_p\,dT \tag{97}$$

$$= -\left[p + T\frac{(\partial V/\partial T)_p}{(\partial V/\partial p)_T}\right] dV + C_v\,dT \tag{98}$$

$$dH = T\,dS + V\,dp \tag{99}$$

$$= \left[V - T\left(\frac{\partial V}{\partial T}\right)_p\right] dp + C_p\,dT \tag{100}$$

$$= V\,dp - T\left[\frac{(\partial V/\partial T)_p}{(\partial V/\partial p)_T}\right] dV + C_v\,dT \tag{101}$$

$$dA = -S\,dT - p\,dV \tag{102}$$

$$dG = -S\,dT + V\,dp \tag{103}$$

III. Two dS Equations

$$dS = -\left(\frac{\partial V}{\partial T}\right)_p dp + \frac{C_p}{T}\,dT \tag{104}$$

$$dS = \left(\frac{\partial p}{\partial T}\right)_V dV + \frac{C_v}{T}\,dT \tag{105}$$

$$= -\left[\frac{(\partial V/\partial T)_p}{(\partial V/\partial p)_T}\right] dV + \frac{C_v}{T}\,dT \tag{106}$$

IV. Four Important Entropy Derivatives

$$\left(\frac{\partial S}{\partial V}\right)_T = \left(\frac{\partial p}{\partial T}\right)_V \tag{107}$$

$$\left(\frac{\partial S}{\partial p}\right)_T = -\left(\frac{\partial V}{\partial T}\right)_p \tag{108}$$

$$\left(\frac{\partial S}{\partial T}\right)_p = \frac{C_p}{T} \tag{109}$$

$$\left(\frac{\partial S}{\partial T}\right)_V = \frac{C_v}{T} \tag{110}$$

The 168 partial derivatives involving p, V, T, S, U, H, A, and G may be classified into six groups. A standard method of procedure for each of the six groups is outlined below. In the discussion that follows, the

Greek symbols α, β, and γ represent any one of the functions S, U, H, A, or G, whereas the symbols a, b, and c represent either p or V or T.

Type 1. General type: $(\partial a/\partial b)_c$. Specific example: $(\partial p/\partial T)_V$.

By equation 91,

$$\left(\frac{\partial p}{\partial T}\right)_V = -\left(\frac{\partial V}{\partial T}\right)_p \left(\frac{\partial p}{\partial V}\right)_T = -\frac{(\partial V/\partial T)_p}{(\partial V/\partial p)_T} \tag{111}$$

Exclusive of reciprocals, there are only three partial derivatives of this general type, and all three are related by this single equation.

Type 2. General type: $(\partial \alpha/\partial a)_b$. Specific example: $(\partial G/\partial T)_V$.

Equations for partial derivatives of this type are readily obtainable from equations 97, 100, 102, 103, or 104. In the present example, from equation 103,

$$\left(\frac{\partial G}{\partial T}\right)_V = -S + V\left(\frac{\partial p}{\partial T}\right)_V = -S - V\frac{(\partial V/\partial T)_p}{(\partial V/\partial p)_T} \tag{112}$$

It should be pointed out that, in those instances where $\alpha = U$ or H, the recommended procedure involves the use of the modified dU or dH equations, 97 and 100, rather than the basic equations for dU and dH, 96 and 99.

Partial derivatives of this general type are of importance, as they indicate how p, V, and T affect S, U, H, A, and G. In all, there are 30 partial derivatives of this type, disregarding reciprocals.

Since equations for partial derivatives of this type are easily set up, the general scheme followed in the remaining four cases is to transform the given partial derivative into this type.

Type 3. General type: $(\partial a/\partial b)_\alpha$. Specific example: $(\partial V/\partial T)_S$.

The initial step in the procedure recommended for this type of partial derivative involves expanding according to equation 91. This operation may be termed *expansion without introducing a new variable.* It will be noted that this operation creates two partial derivatives of type 2, which then are handled as previously described.

$$\left(\frac{\partial V}{\partial T}\right)_S = -\left(\frac{\partial S}{\partial T}\right)_V \left(\frac{\partial V}{\partial S}\right)_T = -\frac{(\partial S/\partial T)_V}{(\partial S/\partial V)_T} \tag{113}$$

From equation 104, expressions for $(\partial S/\partial T)_V$ and $(\partial S/\partial V)_T$ are developed and substituted into the above equation, yielding the following:

$$\left(\frac{\partial V}{\partial T}\right)_S = \frac{-(\partial V/\partial T)_p(\partial p/\partial T)_V + C_p/T}{-(\partial V/\partial T)_p(\partial p/\partial V)_T} \tag{114}$$

This may be brought into the final desired form by expanding $(\partial p/\partial T)_V$ according to equation 91, thus obtaining the following:

$$\left(\frac{\partial V}{\partial T}\right)_S = \frac{(\partial V/\partial T)_p{}^2 + \frac{C_p}{T}(\partial V/\partial p)_T}{(\partial V/\partial T)_p} \tag{115}$$

Although the foregoing method represents the standard procedure that may always be relied on to give an equation of the desired type, it should be pointed out that in a few instances a more direct procedure is possible. In those instances (four in all), where $\alpha = S, H, A$, or G, and where a and b correspond to the two *independent* variables on the right side of the differential equation corresponding to α, the following procedure is quicker and more direct than the standard procedure. Suppose, for example, that the equation for $(\partial T/\partial p)_S$ is desired. Using equation 104 and setting dS equal to zero gives

$$\left[0 = -\left(\frac{\partial V}{\partial T}\right)_p dp + \left(\frac{C_p}{T}\right) dT\right]_S \tag{116}$$

Rearranging gives

$$\left(\frac{\partial T}{\partial p}\right)_S = \frac{T}{C_p}\left(\frac{\partial V}{\partial T}\right)_p \tag{117}$$

There are 15 type-3 partial derivatives, exclusive of reciprocals. Some are of considerable importance, particularly those for constant H (Joule–Thomson expansion) and those for constant S (isentropic changes).

Type 4. General type $(\partial\alpha/\partial\beta)_a$. Specific example: $(\partial G/\partial A)_p$.

The initial step in the procedure recommended for this type of partial derivative involves expanding according to equation 92. This operation may be termed *expansion through the introduction of a new variable.* If $a = p$, the new variable may be either V or T; if $a = V$, the new variable may be either p or T; if $a = T$, the new variable may be either p or V. It is apparent that a choice may always be made between two variables. The final result is the same, regardless of which of the two variables is selected. It may be noted that, for partial derivatives of the particular type under discussion, the variable selected has no great influence on the labor involved.

Proceeding now with the specific example, the following operations are carried out.

$$\left(\frac{\partial G}{\partial A}\right)_p = \left(\frac{\partial G}{\partial T}\right)_p\left(\frac{\partial T}{\partial A}\right)_p = \frac{(\partial G/\partial T)_p}{(\partial A/\partial T)_p} \tag{118}$$

It will be noted that this expansion has created two type-2 partial derivatives, which may be evaluated through the use of equations 102 and 103. By substitution, the following is obtained:

$$\left(\frac{\partial G}{\partial A}\right)_p = \frac{S}{S + p(\partial V/\partial T)_p} \tag{119}$$

There are 30 partial derivatives of this type, disregarding reciprocals.

Type 5. General type: $(\partial \alpha/\partial a)_\beta$. Specific example: $(\partial G/\partial p)_A$.

The procedure recommended for handling partial derivatives of this type involves developing an equation analogous to equation 90 from the differential equation for α (G in this case).

$$dG = -S\,dT + V\,dp \tag{103}$$

By dividing by dp and imposing the condition of constancy of A, the following is obtained:

$$\left(\frac{\partial G}{\partial p}\right)_A = -S\left(\frac{\partial T}{\partial p}\right)_A + V \tag{120}$$

It will be noted that, on the right side of the equation, a type-3 partial derivative has resulted, which may be handled as previously described. This involves expanding the partial derivative $(\partial T/\partial p)_A$ according to equation 91, with the following result:

$$\left(\frac{\partial G}{\partial p}\right)_A = S\,\frac{(\partial A/\partial p)_T}{(\partial A/\partial T)_p} + V \tag{121}$$

The two type-2 partial derivatives are then evaluated through the use of equation 102. Substitution in the foregoing equation gives the following:

$$\left(\frac{\partial G}{\partial p}\right)_A = \frac{Sp\,(\partial V/\partial p)_T}{S + p(\partial V/\partial T)_p} + V \tag{122}$$

It may be pointed out that in those instances where $\alpha = U$ or H, and $\beta = S$, a material saving of labor is accomplished by using the unmodified dU or dH equations 96 or 99, instead of the modified ones, 97 or 100. As an extreme case of labor saving, the following may be cited. If it is required to set up an equation for $(\partial U/\partial V)_S$, the use of the unmodified dU equation immediately yields the result $-p$. If the modified dU equation is used, the same final result is obtained but only with the expenditure of a considerable amount of labor.

There are 60 partial derivatives of this type, exclusive of reciprocals. As was the case for type 3, these are of considerable interest, since they are applicable to problems of free expansion, and also to isentropic changes.

Type 6. General type: $(\partial\alpha/\partial\beta)_\gamma$. Specific example: $(\partial G/\partial A)_H$.

The first step of the procedure recommended involves expanding according to the principle of equation 92. In the present example, T is arbitrarily selected as the new variable, though either p or V could have been chosen. The following result is obtained:

$$\left(\frac{\partial G}{\partial A}\right)_H = \frac{(\partial G/\partial T)_H}{(\partial A/\partial T)_H} \tag{123}$$

It will be recognized that this operation has created two type-5 partial derivatives, and each is handled as previously outlined. From equation 103,

$$\left(\frac{\partial G}{\partial T}\right)_H = -S + V\left(\frac{\partial p}{\partial T}\right)_H \tag{124}$$

From equation 102,

$$\left(\frac{\partial A}{\partial T}\right)_H = -S - p\left(\frac{\partial V}{\partial T}\right)_H \tag{125}$$

Substituting yields

$$\left(\frac{\partial G}{\partial A}\right)_H = \frac{-S + V(\partial p/\partial T)_H}{-S - p(\partial V/\partial T)_H} \tag{126}$$

These operations develop two type-3 partial derivatives which are handled as previously described. By omitting the detailed operations, the following equation is obtained:

$$\left(\frac{\partial G}{\partial A}\right)_H = \frac{-S - \dfrac{VC_p}{V - T(\partial V/\partial T)_p}}{-S - p\dfrac{[V - T(\partial V/\partial T)_p](\partial V/\partial T)_p - C_p(\partial V/\partial p)_T}{V - T(\partial V/\partial T)_p}} \tag{127}$$

By further algebraic manipulation, the right side is reduced to the fractional form:

$$\left(\frac{\partial G}{\partial A}\right)_H = \frac{-V(C_p + S) + TS(\partial V/\partial T)_p}{-[S + p(\partial V/\partial T)_p][V - T(\partial V/\partial T)_p] + pC_p(\partial V/\partial p)_T} \tag{128}$$

Some comments again are in order regarding the possibility of saving needless work in certain instances. If $\gamma = S$, and if either α or β equals U or H, it will prove advantageous to use the unmodified dU or dH equations 96 or 99 instead of the modified ones, 97 or 100.

Usually, the choice of a new variable (p, V, or T) involved in the first operation has no great influence on the labor required. However, the general principle may be stated that the least labor is involved by selecting that variable that produces the smallest number of type-3 partial derivatives.

There are 30 partial derivatives of this type, disregarding reciprocals. It may be pointed out that equations for partial derivatives of this type are the most complex encountered. The observation may also be made that partial derivatives of this type are of little practical interest.

According to the methods outlined above, expressions for partial derivatives are obtained in terms of C_p, $(\partial V/\partial p)_T$, and $(\partial V/\partial T)_p$. If it should be considered desirable to express partial derivatives in terms of C_v, $(\partial V/\partial p)_T$, and $(\partial V/\partial T)_p$, equations 98, 101, and 106 should be employed rather than 97, 100, and 104. Otherwise, the procedures are the same.

Heat–Capacity Relations

Heat–Capacity Difference ($C_p - C_v$). From equation 110,

$$C_v = T\left(\frac{\partial S}{\partial T}\right)_v \tag{129}$$

Equation 104 is utilized to set up an expression for $(\partial S/\partial T)_v$.

$$\left(\frac{\partial S}{\partial T}\right)_v = -\left(\frac{\partial V}{\partial T}\right)_p\left(\frac{\partial p}{\partial T}\right)_v + \frac{C_p}{T} \tag{130}$$

Combining, and then substituting for $(\partial p/\partial T)_v$ according to equation 91, gives

$$(C_p - C_v) = T\left(\frac{\partial V}{\partial T}\right)_p\left(\frac{\partial p}{\partial T}\right)_v = -T\left(\frac{\partial V}{\partial T}\right)_p^2\left(\frac{\partial p}{\partial V}\right)_T \tag{131}$$

Heat–Capacity Ratio C_p/C_v. From equations 109 and 110,

$$\frac{C_p}{C_v} = \frac{T(\partial S/\partial T)_p}{T(\partial S/\partial T)_v} \tag{132}$$

The partial derivatives in the numerator and denominator are expanded according to equation 91.

$$\frac{C_p}{C_v} = \frac{-(\partial p/\partial T)_S(\partial S/\partial p)_T}{-(\partial V/\partial T)_S(\partial S/\partial V)_T} = \left(\frac{\partial p}{\partial V}\right)_S\left(\frac{\partial V}{\partial p}\right)_T = -\left(\frac{\partial p}{\partial V}\right)_S\left(\frac{\partial T}{\partial p}\right)_v\left(\frac{\partial V}{\partial T}\right)_p \tag{133}$$

It may be pointed out that it is not possible to secure an expression for (C_p/C_v) in terms of p, V, and T alone; entropy will be involved in at least one of the partial derivatives.

Effect of Pressure and Volume on C_p. By applying the principle of equation 93 to equation 104,

$$\left(\frac{\partial C_p}{\partial p}\right)_T = -T\left(\frac{\partial^2 V}{\partial T^2}\right)_p \tag{134}$$

By applying the principle of equation 92, the following is obtained:

$$\left(\frac{\partial C_p}{\partial V}\right)_T = -T\left(\frac{\partial p}{\partial V}\right)_T\left(\frac{\partial^2 V}{\partial T^2}\right)_p \tag{135}$$

Effect of Pressure and Volume on C_V. By applying the principle of equation 93 to equation 105,

$$\left(\frac{\partial C_v}{\partial V}\right)_T = T\left(\frac{\partial^2 p}{\partial T^2}\right)_v \tag{136}$$

By applying the principle of equation 92 to the foregoing, the following is obtained:

$$\left(\frac{\partial C_v}{\partial p}\right)_T = T\left(\frac{\partial V}{\partial p}\right)_T\left(\frac{\partial^2 p}{\partial T^2}\right)_v \tag{137}$$

Illustration 10. From the van der Waals equation of state, calculate the increase in c_p of CO_2 where the pressure is increased from 1 to 100 atm at 100° C. The van der Waals equation is

$$\left(p + \frac{a}{v^2}\right)(v - b) = RT \tag{a}$$

where v = the molal volume

a, b = the van der Waals constants, characteristic of the gas

For 1 lb-mole, equation 134 is written

$$\left(\frac{\partial c_p}{\partial p}\right)_T = -T\left(\frac{\partial^2 v}{\partial T^2}\right)_p \tag{b}$$

Differentiating equation (a) gives

$$\left(\frac{\partial v}{\partial T}\right)_p = \frac{R}{p - (a/v^2) + (2ab/v^3)} \tag{c}$$

$$\left(\frac{\partial^2 v}{\partial T^2}\right)_p = -R\left[\frac{2av^{-3} - 6abv^{-4}}{(p - av^{-2} + 2abv^{-3})^2}\right]\left(\frac{\partial v}{\partial T}\right)_p \tag{d}$$

Integration of (b) gives

$$\Delta c_p = -\int_1^{100} T\left(\frac{\partial^2 v}{\partial T^2}\right)_p dp \tag{e}$$

Equation (e) is solved by substitution of (d) and graphical integration where $a = 3.60(10^6)$(atm)(cm)^6/(g-mole)^2, b = 42.8 cc/g-mole, R = 82.1 cc-atm/(g-mole)(K°).

Volume v is first obtained from equation (a) at various values of p. Equations (c) and (d) are then solved in turn at various values of p (Table A).

TABLE A

p, atm	v, cc	$\left(\frac{\partial v}{\partial T}\right)_p$	$-T\left(\frac{\partial^2 v}{\partial T^2}\right)_p$
0	∞	∞	0
1	30,590	82.7	0.64
10	2,985	8.55	0.734
25	1,150	3.65	0.931
50	530	2.09	1.52
75	320	1.67	2.8
100	213	1.56	5.07

From graphical integration,

$$\Delta c_p = 194.20 \text{ cc-atm/(g-mole)(C°)} = 4.7 \text{ cal/(g-mole)(C°)}$$

or

$$c_p = 9.7 + 4.7 = 14.4$$

Illustration 11. Prove that $C_p - C_v = (TV\beta^2/\alpha)$ where

$$\beta = \frac{1}{V}\left(\frac{\partial V}{\partial T}\right)_p = \text{coefficient of expansion}$$

$$\alpha = -\frac{1}{V}\left(\frac{\partial V}{\partial p}\right)_T = \text{coefficient of compressibility}$$

From this relation, calculate C_v for mercury, where, at 0° C, $\beta = 0.00018$ per C°, $\alpha = 0.000\,003\,9$ per atm, $C_p = 0.0333$ cal/(gram)(C°), and the density $= 13.596$ grams per cc.

Introducing α and β in equation 131 gives

$$C_p - C_v = \frac{-T\beta^2 V^2}{-\alpha V} = \frac{T\beta^2 V}{\alpha}$$

For mercury at 0° C,

$$C_p - C_v = \frac{273(0.000\,18)^2}{(13.596)(0.000\,003\,9)(41.3)} = 0.0040$$

where 41.3 is the conversion factor from cubic centimeter-atmospheres to calories. Hence,

$$C_v = 0.0333 - 0.0040 = 0.0293 \text{ cal/(gram)(C°)}$$

Other Procedures for Deriving Thermodynamic Relations

Use of a Single General Equation. Equation 95 may be employed to develop the equation for any one of the 168 partial derivatives previously mentioned. It will be noted that the right side of this equation contains partial derivatives, all of which are of type 2. These may all be evaluated through the use of the equations for dS, dU, dH, dA, and dG as given in Table 40. They are then substituted in equation 95.

Bridgman Table. Table 41 developed by Bridgman[3] affords the quickest and simplest procedure for setting up the equation for any partial derivative involving p, V, T, S, U, H, A, and G in terms of C_p, $(\partial V/\partial T)_p$, and $(\partial V/\partial p)_T$. The items entered in this table are based on equation 95 where the special symbol $(\partial z)_y$ is used to designate the numerator, and the special symbol $(\partial x)_y$ the denominator.

$$\left(\frac{\partial z}{\partial x}\right)_y = \frac{(\partial z/\partial T)_p(\partial y/\partial p)_T - (\partial z/\partial p)_T(\partial y/\partial T)_p}{(\partial x/\partial T)_p(\partial y/\partial p)_T - (\partial x/\partial p)_T(\partial y/\partial T)_p} = \frac{(\partial z)_y}{(\partial x)_y} \tag{138}$$

Bridgman worked out expressions for all the items of the type $(\partial z)_y$ that involve the variables p, V, T, S, U, H, A, and G, and then assembled them in Table 41. The procedure followed is shown in Illustration 12.

Illustration 12. Verify the expression for $(\partial S)_V$ given in the Bridgman table. According to the system of notation indicated by equation 138:

$$(\partial S)_V = \left[\left(\frac{\partial S}{\partial T}\right)_p\left(\frac{\partial V}{\partial p}\right)_T - \left(\frac{\partial S}{\partial p}\right)_T\left(\frac{\partial V}{\partial T}\right)_p\right] \tag{a}$$

Substituting equations 109 and 108 in equation (a), gives

$$(\partial S)_V = \frac{C_p}{T}\left(\frac{\partial V}{\partial p}\right)_T - \left(-\frac{\partial V}{\partial T}\right)_p\left(\frac{\partial V}{\partial T}\right)_p$$

$$= \frac{1}{T}\left[C_p\left(\frac{\partial V}{\partial p}\right)_T + T\left(\frac{\partial V}{\partial T}\right)_p^2\right] \tag{b}$$

The use of the Bridgman table follows from equation 138, which indicates the significance of the special notation adopted. Illustration 13 demonstrates the procedure employed.

Illustration 13. Using Table 41, develop an equation for $(\partial A/\partial U)_p$.

$$\left(\frac{\partial A}{\partial U}\right)_p = \frac{(\partial A)_p}{(\partial U)_p} \tag{a}$$

Expressions for $(\partial A)_p$ and $(\partial U)_p$ are selected from Table 41 and substituted in the foregoing equation to yield the following result:

$$\left(\frac{\partial A}{\partial U}\right)_p = \frac{-[S + p(\partial V/\partial T)_p]}{C_p - p(\partial V/\partial T)_p} \tag{b}$$

[3] P. W. Bridgman, *Condensed Collection of Thermodynamic Formulas*, Harvard University Press, Cambridge, Mass. (1926). This was first published in *Phys. Rev.* (2), **3**, 273 (1914). The original article contained two errors involving $(\partial A)_U$ and $(\partial A)_H$. These errors were eliminated in the book published in 1926.

TABLE 41. CONDENSED SUMMARY OF THERMODYNAMIC RELATIONSHIPS

I. Pressure Constant and Pressure Variable

$$(\partial V)_p = -(\partial p)_V = (\partial V/\partial T)_p$$
$$(\partial T)_p = -(\partial p)_T = 1$$
$$(\partial S)_p = -(\partial p)_S = C_p/T$$
$$(\partial U)_p = -(\partial p)_U = C_p - p(\partial V/\partial T)_p$$
$$(\partial H)_p = -(\partial p)_H = C_p$$
$$(\partial A)_p = -(\partial p)_A = -[S + p(\partial V/\partial T)_p]$$
$$(\partial G)_p = -(\partial p)_G = -S$$

II. Temperature Constant and Temperature Variable

$$(\partial V)_T = -(\partial T)_V = -(\partial V/\partial p)_T$$
$$(\partial S)_T = -(\partial T)_S = (\partial V/\partial T)_p$$
$$(\partial U)_T = -(\partial T)_U = T(\partial V/\partial T)_p + p(\partial V/\partial p)_T$$
$$(\partial H)_T = -(\partial T)_H = -V + T(\partial V/\partial T)_p$$
$$(\partial A)_T = -(\partial T)_A = p(\partial V/\partial p)_T$$
$$(\partial G)_T = -(\partial T)_G = -V$$

III. Volume Constant and Volume Variable

$$(\partial S)_V = -(\partial V)_S = (1/T)[C_p(\partial V/\partial p)_T + T(\partial V/\partial T)_p^2]$$
$$(\partial U)_V = -(\partial V)_U = C_p(\partial V/\partial p)_T + T(\partial V/\partial T)_p^2$$
$$(\partial H)_V = -(\partial V)_H = C_p(\partial V/\partial p)_T + T(\partial V/\partial T)_p^2 - V(\partial V/\partial T)_p$$
$$(\partial A)_V = -(\partial V)_A = -S(\partial V/\partial p)_T$$
$$(\partial G)_V = -(\partial V)_G = -[V(\partial V/\partial T)_p + S(\partial V/\partial p)_T]$$

IV. Entropy Constant and Entropy Variable

$$(\partial U)_S = -(\partial S)_U = (p/T)[C_p(\partial V/\partial p)_T + T(\partial V/\partial T)_p^2]$$
$$(\partial H)_S = -(\partial S)_H = -(VC_p/T)$$
$$(\partial A)_S = -(\partial S)_A = (1/T)\{p[C_p(\partial V/\partial p)_T + T(\partial V/\partial T)_p^2] + ST(\partial V/\partial T)_p\}$$
$$(\partial G)_S = -(\partial S)_G = -(1/T)[VC_p - ST(\partial V/\partial T)_p]$$

V. Internal Energy Constant and Internal Energy Variable

$$(\partial H)_U = -(\partial U)_H = -V[C_p - p(\partial V/\partial T)_p] - p[C_p(\partial V/\partial p)_T + T(\partial V/\partial T)_p^2]$$
$$(\partial A)_U = -(\partial U)_A = p[C_p(\partial V/\partial p)_T + T(\partial V/\partial T)_p^2] + S[T(\partial V/\partial T)_p + p(\partial V/\partial p)_T]$$
$$(\partial G)_U = -(\partial U)_G = -V[C_p - p(\partial V/\partial T)_p] + S[T(\partial V/\partial T)_p + p(\partial V/\partial p)_T]$$

VI. Enthalpy Constant and Enthalpy Variable

$$(\partial A)_H = -(\partial H)_A = -[S + p(\partial V/\partial T)_p][V - T(\partial V/\partial T)_p] + pC_p(\partial V/\partial p)_T$$
$$(\partial G)_H = -(\partial H)_G = -V(C_p + S) + TS(\partial V/\partial T)_p$$

VII. Free Energy Constant and Free Energy Variable

$$(\partial A)_G = -(\partial G)_A = -S[V + p(\partial V/\partial p)_T] - pV(\partial V/\partial T)_p$$

Source: P. W. Bridgman, *Condensed Collection of Thermodynamic Formulas*, Harvard University Press, Cambridge, Mass. (1926).

As Bridgman has pointed out, since any given partial derivative may be expressed in terms of C_p, $(\partial V/\partial T)_p$, and $(\partial V/\partial p)_T$, it is theoretically possible to derive an equation relating any four partial derivatives by writing four independent equations, one for each of the partial derivatives, in terms of these three terms. Then, by combining these four equations, these three terms may be eliminated, yielding an equation that contains no partial derivatives other than the four in question. Although this is theoretically possible, in many instances the algebraic operations become extremely involved.

The number of possible equations involving four partial derivatives is extremely large. It has already been pointed out that it is possible to set up 168 partial derivatives (ignoring reciprocals) involving p, V, T, S, U, H, A, and G. These 168 partial derivatives will yield $(3.2)10^7$ equations, each of which contains only four partial derivatives.

Shaw,[4] Lerman,[5] Tobolsky,[6] and Carroll and Lehrman[7] have developed derivational procedures other than those outlined above.

Partial Derivatives of the Energy Properties

The effects of pressure, volume, and temperature on the four thermodynamic energy properties are among the most useful relations. The more important relations are developed herewith.

Application of equation 97 at constant temperature and at constant volume, respectively, results in

$$\left(\frac{\partial U}{\partial V}\right)_T = T\left(\frac{\partial p}{\partial T}\right)_V - p \tag{139}$$

$$\left(\frac{\partial U}{\partial T}\right)_V = C_v \tag{140}$$

Applying equation 100 to constant temperature and constant pressure operations, respectively, results in

$$\left(\frac{\partial H}{\partial p}\right)_T = V - T\left(\frac{\partial V}{\partial T}\right)_p \tag{141}$$

$$\left(\frac{\partial H}{\partial T}\right)_p = C_p \tag{142}$$

[4] A. N. Shaw, *Phil. Trans. Roy. Soc. London, A*, **234**, 299 (1934–35).

[5] F. Lerman, *J. Chem. Phys.*, **5**, 792 (1937).

[6] A. V. Tobolsky, *J. Chem. Phys.*, **10**, 644 (1942).

[7] B. Carroll and A. Lehrman, *J. Chem. Ed.*, **24**, 389 (1947).

In connection with the analysis of chemical equilibria, an equation relating $(\Delta G/T)_p$ with temperature is employed. Such an equation is developed as follows:

From the defining equation for G,

$$\Delta G = \Delta H - \Delta(TS) \tag{143}$$

In chemical equilibria, the reactions are always considered under the same terminal temperatures. Accordingly, $\Delta(TS) = T\,\Delta S$, and

$$\Delta G = \Delta H - T\,\Delta S \tag{144}$$

$$\frac{\Delta G}{T} = \frac{\Delta H}{T} - \Delta S \tag{145}$$

Differentiating with respect to temperature at constant pressure yields

$$\left(\frac{\partial \Delta G/T}{\partial T}\right)_p = \frac{1}{T}\left(\frac{\partial \Delta H}{\partial T}\right)_p - \frac{\Delta H}{T^2} - \left(\frac{\partial \Delta S}{\partial T}\right)_p \tag{146}$$

From equation 99,

$$\left(\frac{\partial H}{\partial T}\right)_p = T\left(\frac{\partial S}{\partial T}\right)_p \tag{147}$$

If this is applied, respectively, to the products and reactants, and if one equation is subtracted from the other,

$$\left[\left(\frac{\partial H}{\partial T}\right)_p\right]_P - \left[\left(\frac{\partial H}{\partial T}\right)_p\right]_R = T\left[\left(\frac{\partial S}{\partial T}\right)_p\right]_P - T\left[\left(\frac{\partial S}{\partial T}\right)_p\right]_R \tag{148}$$

By definition

$$\Delta H = H_P - H_R \tag{149}$$

$$\Delta S = S_P - S_R \tag{150}$$

If the foregoing two equations are differentiated with respect to temperature at constant pressure,

$$\left(\frac{\partial \Delta H}{\partial T}\right)_p = \left[\left(\frac{\partial H}{\partial T}\right)_p\right]_P - \left[\left(\frac{\partial H}{\partial T}\right)_p\right]_R \tag{151}$$

$$\left(\frac{\partial \Delta S}{\partial T}\right)_p = \left[\left(\frac{\partial S}{\partial T}\right)_p\right]_P - \left[\left(\frac{\partial S}{\partial T}\right)_p\right]_R \tag{152}$$

If equations 148, 151, and 152 are combined,

$$\left(\frac{\partial \Delta H}{\partial T}\right)_p - T\left(\frac{\partial \Delta S}{\partial T}\right)_p \tag{153}$$

Accordingly, equation 146 reduces to

$$\left(\frac{\partial \Delta G/T}{\partial T}\right)_p = -\frac{\Delta H}{T^2} \tag{154}$$

Thermodynamic Properties of an Ideal Gas

As previously mentioned, the properties of an ideal gas are defined by the requirements that the pVT relations are exactly expressed by the equation $pv = RT$ and that the internal energy is dependent only on temperature and is independent of pressure and volume.

By differentiation of the ideal-gas equation,

$$p\,dv + v\,dp = R\,dT \tag{155)*$$

or

$$\left(\frac{\partial p}{\partial T}\right)_v = \frac{R}{v} = \frac{p}{T} \quad \text{and} \quad \left(\frac{\partial v}{\partial p}\right)_T = -\frac{v}{p} \quad \text{and} \quad \left(\frac{\partial v}{\partial T}\right)_p = \frac{R}{p} = \frac{v}{T} \tag{156)*$$

General thermodynamic equations that contain partial derivatives involving, p, v, and T may be applied to an ideal gas by substituting for the partial derivatives according to equation 156.

In order to evaluate the entropy of a gas in the ideal state from heat-capacity measurements, it is necessary to start with the crystalline solid at a temperature near the absolute zero, heat it to its fusion point, and then to its boiling point T' under a pressure p', sufficiently low that ideal-gas behavior is obtained. The entropy s' of the saturated vapor at this temperature and pressure is calculated from the low-temperature heat capacities of the crystalline and liquid states and the heats of fusion and vaporization. Equation 104 is then integrated from this limit. Thus:

$$s = \int_{T'}^{T} \frac{c^*_p}{T}\,dT - R \ln \frac{p}{p'} + s' \tag{157)*$$

Table 42 summarizes the more important thermodynamic relations for an ideal gas.

Illustration 14. Assuming that nitrogen behaves ideally over the range indicated, calculate the values of S, U, H, A, and G for 1 lb-mole at 10 atm and 100° F, relative to conditions at 1 atm, and 60° F. Assume $c_v = 5.0$.

$$\Delta U = c_v\,\Delta T = (5.0)40 = 200 \text{ Btu per lb-mole}$$

$$\Delta H = (c_v + R)\,\Delta T = (5.0 + 1.99)40 = 279.6 \text{ Btu per lb-mole}$$

$$\Delta S = c_p \ln \frac{T_2}{T_1} - R \ln \frac{p_2}{p_1} = 6.99 \ln \frac{560}{520} - 1.99 \ln \frac{10}{1} = -4.06 \text{ Btu/(lb-mole)(R°)}$$

Numerical values cannot be secured for ΔA and ΔG unless absolute values of entropy are known.

* Equations whose applicability is restricted to ideal gases are designated by asterisks following the equation numbers.

TABLE 42. THERMODYNAMIC PROPERTIES OF AN IDEAL GAS

The following equations apply only to an ideal gas. It is not correct to utilize them for nonideal gases, liquids, or solids. The symbol v stands for molal volume.

1. First Derivatives

$$\left(\frac{\partial v}{\partial T}\right)_p = \frac{R}{p} = \frac{v}{T} \quad (158)^*$$

$$\left(\frac{\partial p}{\partial T}\right)_V = \frac{R}{v} = \frac{p}{T} \quad (159)^*$$

$$\left(\frac{\partial p}{\partial v}\right)_T = -\frac{p}{v} = -\frac{RT}{v^2} = -\frac{p^2}{RT} \quad (160)^*$$

2. Second Derivatives

$$\left(\frac{\partial^2 v}{\partial T^2}\right)_p = 0 \quad (161)^*$$

$$\left(\frac{\partial^2 p}{\partial T^2}\right)_V = 0 \quad (162)^*$$

3. Effect of Pressure and Volume Change on Internal Energy, Where Temperature is Constant

$$\left(\frac{\partial \text{U}}{\partial p}\right)_T = 0 \quad (163)^*$$

$$\left(\frac{\partial \text{U}}{\partial v}\right)_T = 0 \quad (164)^*$$

Effect of Pressure and Volume Change on Enthalpy, When Temperature is Constant

$$\left(\frac{\partial \text{H}}{\partial p}\right)_T = 0 \quad (165)^*$$

$$\left(\frac{\partial \text{H}}{\partial v}\right)_T = 0 \quad (166)^*$$

5. Effect of Pressure and Volume Change on Heat Capacity, When Temperature is Constant

$$\left(\frac{\partial c_p}{\partial p}\right)_T = 0 \quad (167)^*$$

$$\left(\frac{\partial c_p}{\partial v}\right)_T = 0 \quad (170)^*$$

$$\left(\frac{\partial c_v}{\partial p}\right)_T = 0 \quad (168)^*$$

$$\left(\frac{\partial c_v}{\partial v}\right)_T = 0 \quad (171)^*$$

6. $$c_p - c_v = R \quad (169)^*$$

7. $d\text{U}$, $d\text{H}$, and $d\text{S}$ Equations (Valid for any pvT change)

$$d\text{U} = c_v\,dT \quad (173)^*$$

$$d\text{H} = c_p\,dT \quad (174)^*$$

$$d\text{S} = c_p\frac{dT}{T} - R\frac{dp}{p} \quad (175)^*$$

$$d\text{S} = c_v\frac{dT}{T} + R\frac{dv}{v} \quad (176)^*$$

8. Joule–Thomson, Maxwell-Expansion, and Isentropic Coefficients

$$\left(\frac{\partial T}{\partial p}\right)_H = 0 \quad (177)^*$$

$$\left(\frac{\partial T}{\partial p}\right)_U = 0 \quad (178)^*$$

$$\left(\frac{\partial T}{\partial p}\right)_S = \frac{TR}{pc_p} \quad (179)^*$$

$$\left(\frac{\partial T}{\partial v}\right)_S = -\frac{TR}{vc_v} \quad (180)^*$$

The absolute entropy of nitrogen at 60° F and 1 atm pressure is 45.56 Btu/(lb-mole)(R°). The final entropy after heating and compressing to 100° F and 10 atm is $45.56 + \Delta S = 41.49$.

Then, since $\Delta A = \Delta U - \Delta(TS) = \Delta U - T_2S_2 + T_1S_1$,

$$\Delta A = 200 - 560(41.50) + 520(45.56) = 659 \text{ Btu per lb-mole}$$

Similarly,

$$\Delta G = \Delta H - T_2S_2 + T_1S_1 = 279.6 - 23{,}237 + 23{,}696 = 739 \text{ Btu per lb-mole}$$

Properties of a Perfect Gas. A *perfect gas* may be defined as one that behaves ideally at all conditions and hence is incapable of condensation to either a liquid or a solid phase. It follows that, for the internal energy of such a gas to be independent of volume at all conditions of pressure and temperature, it can possess only translational energy of motion, and its heat capacity must be independent of temperature as well as of pressure, as shown by equation 18, page 251. Thus, for a perfect gas, $c_v = \frac{3}{2}R$ and $c_p = \frac{5}{2}R$, whereas the heat capacity of an ideal gas which is not monatomic varies with temperature. It is evident that no substance fulfills the properties of a perfect gas but that perfect behavior is approached by monatomic gases at low pressures.

For a perfect gas, equation 157 is integrated to

$$S = \tfrac{5}{2}R \ln T - R \ln p + b \tag{181}$$

The integration constant b has been evaluated by methods of statistical mechanics as $(\frac{3}{2}R \ln M - 2.298)$ where M is the molecular weight.

The Thermodynamic Temperature Scale. The equations derived through the introduction of the defining equation for entropy contain a temperature term which is the thermodynamic temperature, and which will be designated as θ in the present discussion. For example,

$$dU = [\theta(\partial p/\partial \theta)_V - p]\,dV + C_v\,d\theta \tag{182}$$

$$(\partial U/\partial V)_\theta = \theta(\partial p/\partial \theta)_v - p \tag{183}$$

This general equation may be applied to an ideal gas. By definition, the internal energy of an ideal gas is a function of temperature alone, and is unaffected by volume. Therefore, $(\partial U/\partial V)_\theta = 0$, and equation 183 reduces to the following

$$p = \theta(\partial p/\partial \theta)_v \tag{184}$$

For constant-volume temperature changes,

$$(dp/p)_V = (d\theta/\theta)_V \tag{185}$$

$$p_2/p_1 = \theta_2/\theta_1 \tag{186}$$

The ideal-gas temperature scale is defined by the relation $pv = RT$ from which

$$p = T(R/v) \tag{187}$$

By differentiation of the ideal-gas law,

$$(R/v) = (\partial p/\partial T)_V \tag{188}$$

By combining equations 187 and 188, the following is obtained:

$$p = T(\partial p/\partial T)_V \tag{189}$$

For constant-volume temperature changes,

$$(dp/p)_V = (dT/T)_V \tag{190}$$

$$p_2/p_1 = T_2/T_1 \tag{191}$$

By combining equations 186 and 191, the following is obtained:

$$\theta_2/\theta_1 = T_2/T_1 \tag{192}$$

This shows that there is a direct proportionality between θ, the thermodynamic temperature, and T, the ideal-gas temperature. It is further specified that the proportionality constant is unity; hence,

$$\theta = T \tag{193}$$

Problems

Values of the properties of steam were taken from Keenan and Keyes, *Thermodynamic Properties of Steam,* John Wiley & Sons (1936).

1. (*a*) Calculate the increase in entropy when 1 lb of ice at 0° F and atmospheric pressure is converted into steam at 220° F and 1 atm pressure.

(*b*) The absolute entropy of water vapor as an ideal gas at 25° C and 1.0 atm is 45.106 g-cal/(g-mole)(K°). Using the data of part (*a*), calculate the absolute entropy of ice at 32° F in Btu/(lb)(R°). Steam at 220° F and 1 atm may be assumed to behave ideally.

2. (*a*) Calculate the entropy change of SO_2 when 1 lb is cooled from 1000° to −150° F, at a pressure of 1 atm. Express the result as Btu/(lb)(R°).

(*b*) Calculate the absolute entropy of solid SO_2 at its melting point under a pressure of 1 atm. Express the result as Btu/(lb)(R°).

Data on SO_2 at 1 atm Pressure

Boiling point = −5.0° C. Latent heat of vaporization = 5960 g-cal/g-mole
Melting point = −75.5° C. Latent heat of fusion = 1769 g-cal/g-mole
Specific heat of liquid SO_2 = 0.310 g-cal/(gram)(K°)
Specific heat of solid SO_2 = 0.229 g-cal/(gram)(K°)

3. (*a*) Calculate the entropy change when 1 lb of ammonia is cooled from 500° to −150° F at a constant pressure of 1 atm. Express the result as Btu/(lb)(R°).
(*b*) Calculate the absolute entropy of solid ammonia at its melting point under 1 atm pressure. Express the result as Btu/(lb)(R°).

Data for Ammonia at 1 atm Pressure

Boiling point = −33.4° C. Latent heat of vaporization = 5581 g-cal/g-mole
Melting point = −77.7° C. Latent heat of fusion = 1352 g-cal/g-mole
Specific heat of liquid = 1.06 g-cal/(gram)(K°)
Specific heat of solid = 0.502 g-cal/(gram)(K°)
Absolute entropy of gas at 25° C = 46.03 g-cal/(g-mole)(K°)

4. One gram-mole of water vapor is heated at constant pressure from 25° to 1600° C. Calculate the following items:

(*a*) Heat absorbed in gram-calories per gram-mole.
(*b*) Enthalpy–mean heat capacity in g-cal/(g-mole)(K°).
(*c*) Increase in entropy in g-cal/(g-mole)(K°).
(*d*) Entropy–mean heat capacity in g-cal/(g-mole)(K°).

5. One gram-mole of carbon dioxide is heated from 25° to 2500° C at a constant pressure of 1 atm. Calculate the following items:

(*a*) Heat absorbed in gram-calories.
(*b*) Enthalpy–mean molal heat capacity in g-cal/(g-mole)(K°).
(*c*) Increase in entropy in g-cal/(g-mole)(K°).
(*d*) Entropy–mean molal heat capacity in g-cal/(g-mole)(K°).

6. Assuming ideal behavior, calculate the entropy of 10 lb of hydrogen gas at 60° F and 10 atm relative to 0° F and 1 atm.

7. When a certain chemical reaction was carried out reversibly in an electrolytic cell at a constant temperature of 20° C and a constant pressure of 1 atm, the electrical work done was equal to 25,700 g-cal. One gram-mole of gas was consumed in the reaction.

When the identical chemical reaction was carried out irreversibly in a calorimeter, the heat absorbed by the system in the calorimeter was 35,900 g-cal. The initial and final temperatures were 20° C, and the pressure was held constant at 1 atm. No electrical work was done.

For both processes, calculate values for q, w_e, w_f, ΔS, ΔU, ΔH, ΔA, ΔG, expressed in gram-calories.

8. Calculate q, w_e, w_f, ΔS, ΔU, ΔH, ΔA, ΔG for each of the following cases. In both instances the chemical reaction is the same, and constant pressure conditions are maintained.

Reversible Case. The chemical reaction was conducted reversibly in an electrolytic cell, at a fixed pressure of 3.5 atm, the temperature being held constant at 150° C. Electric energy equal to 32,500 g-cal was given off. Ten gram-moles of gas were evolved in the reaction.

Irreversible Case. The same chemical reaction was carried out irreversibly, under a constant pressure of 3.5 atm, and with terminal temperatures of 150° C. Electric energy equal to 15,500 g-cal was given off, and 32,500 g-cal of heat were absorbed. As before, 10 g-moles of gas were formed by the chemical reaction.

9. Calculate the values of ΔS, ΔU, ΔA, and ΔG when 1 lb-mole of H_2 gas at 77° F and 1 atm is heated and compressed to 500° F and 100 atm. The initial absolute entropy of hydrogen is 31.211 Btu/(lb-mole)(R°). Assume ideal behavior.

10. Verify the accuracy of the data in Table A by using them in the integrated form of equation 96. In making the test, select as terminal conditions the first and last lines of data.

Table A

p, psi	V, cu ft per lb	t, °F	S, Btu/(lb)(R°)	U, Btu/per lb
280	1.6789	420	1.5238	1121.8
285	1.7047	440	1.5365	1131.6
290	1.7277	460	1.5484	1141.5
295	1.7485	480	1.5596	1150.5
300	1.7675	500	1.5701	1159.5
305	1.7849	520	1.5802	1168.2
310	1.8010	540	1.5898	1176.8
315	1.8159	560	1.5990	1185.4
320	1.8298	580	1.6079	1194.0
325	1.8429	600	1.6166	1202.0
330	1.8552	620	1.6249	1210.4
335	1.8667	640	1.6331	1218.5

Values of U and S are relative to liquid H_2O under its own vapor pressure at 32° F.

11. Verify the accuracy of the data in Table B by applying them to the integrated form of equation 96.

Table B

$V = 1.6789$ cu ft per lb		$S = 1.6331$ Btu/(lb)(R°)	
t, °F	S, Btu/(lb)(R°)	p, psi	V, cu ft/lb
$420 = t_1$	$1.5238 = S_1$	385	1.6759
440	1.5345		$1.6789 = V_1$
460	1.5446	380	1.6932
480	1.5543	375	1.7106
500	1.5635	370	1.7284
520	1.5724	365	1.7468
540	1.5810	360	1.7654
560	1.5893	355	1.7845
580	1.5974	350	1.8046
600	1.6053	345	1.8247
620	1.6128	340	1.8456
640	1.6203	335	$1.8667 = V_2$
660	1.6276		
	$1.6331 = S_2$		
680	1.6347		

12. Test each of the following equations for exactness.

(a) $$dz = 3x^2y^2\,dx + 2x^3y\,dy$$

(b) $$dz = 3x^3y^2\,dx + 4x^3y\,dy$$

(c) $$dz = 5x^2y^5\,dx + 2x^2y^2\,dy$$

(d) $$dz = 2x^2y^3\,dx + 2x^3y^2\,dy$$

(e) $$dz = 3x^3y^5\,dx + 4x^3y^4\,dy$$

13. The volume coefficient of expansion $(\partial V/\partial T)_p/V$ of water at 100° C is 0.00078 per C°. Calculate the change in entropy in cal/(gram)(C°) when the pressure is reduced from 100 to 1 atm at 100° C, neglecting the effect of pressure on volume.

14. Calculate the latent heat of vaporization per pound of water at 160° F from the following data:

$$dp/dT = 0.113 \text{ lb/(sq in.)(R°)}$$

Volume of saturated vapor = 77.29 cu ft per lb and of liquid = 0.017 cu ft per lb.

15. By combination of equation 76 with equation 4–15, page 92, evaluate the heat of vaporization of methylamine at its normal boiling point −6.5° C, assuming that the vapor obeys the ideal-gas law, and neglecting the volume of the liquid.

16. Verify the accuracy of the data in Table C for steam at 300 psia and 500° F, where $V = 1.7675$ cu ft per lb and $S = 1.5701$ Btu/(lb)(R°) by applying them to equation 73.

Table C

$S = 1.5701$ Btu/(lb)(R°) $p = 300$ psia

p, psi	t, °F	V, cu ft per lb	S, Btu/(lb)(R°)
250	460.62	1.5513	1.5126
260	469.01	1.6090	1.5286
270	476.99	1.6638	1.5434
280	484.92	1.7165	1.5572
290	492.63	1.7675	1.5701
300	500.00	1.8170	1.5824
310	507.35	1.8654	1.5941
320	514.37	1.9128	1.6054
330	521.32	1.9594	1.6163
340	528.09		

17. Verify the accuracy of the data in Table D for steam at 300 psia and 500° F, where $V = 1.765$ cu ft per lb and $S = 1.5701$ Btu/(lb)(R°), by applying them to equation 74.

Table D

$t = 500°$ F $V = 1.7675$ cu ft per lb

V, cu ft per lb	S, Btu/(lb)(R°)	t, °F	p, psia
2.151	1.5949	425.82	270
2.063	1.5897	450.03	280
1.9809	1.5846	474.76	290
1.9047	1.5796	500.00	300
1.8338	1.5748	525.80	310
1.7675	1.5701	551.94	320
1.7054	1.5655	578.42	330
1.6472	1.5611	605.21	340
1.5925	1.5567		
1.5410	1.5524		

18. Derive the differential equations for the following relations:

(a) Change of total work function with temperature at constant volume.

(b) Change of enthalpy with entropy at constant pressure.

(c) Change of internal energy with volume at constant entropy.

(d) Change of temperature with volume at constant entropy.

19. Derive general equations for each of the partial derivatives tabulated below. In each case, the right side of the equation must show no partial derivatives other than $(\partial V/\partial T)_p$ and $(\partial V/\partial p)_T$. Furthermore, if a heat-capacity term appears, it should be C_p, not C_v. In each instance:

(a) Derive the equation according to the procedures recommended on pages 535 to 541.

(b) Derive the equation by the method of Bridgman, using Table 41.

(c) Reconcile the two results.

(1) $(\partial p/\partial V)_H$ (6) $(\partial H/\partial V)_T$

(2) $(\partial H/\partial T)_V$ (7) $(\partial G/\partial T)_H$

(3) $(\partial H/\partial A)_V$ (8) $(\partial H/\partial p)_V$

(4) $(\partial H/\partial V)_p$ (9) $(\partial S/\partial A)_H$

(5) $(\partial H/\partial T)_p$ (10) $(\partial H/\partial p)_T$

20. Evaluate the following coefficients for 1 lb-mole of an ideal gas:

(a) $\left(\frac{\partial p}{\partial T}\right)_V$ (b) $\left(\frac{\partial^2 p}{\partial T^2}\right)_V$ (c) $\left(\frac{\partial p}{\partial v}\right)_T$ (d) $\left(\frac{\partial v}{\partial T}\right)_p$

(e) $\left(\frac{\partial^2 v}{\partial T^2}\right)_p$ (f) $\left(\frac{\partial s}{\partial p}\right)_T$ (g) $\left(\frac{\partial s}{\partial v}\right)_T$ (h) $\left(\frac{\partial G}{\partial T}\right)_V$

(i) $\left(\frac{\partial U}{\partial v}\right)_T$ (j) $\left(\frac{\partial U}{\partial p}\right)_T$ (k) $\left(\frac{\partial c_p}{\partial p}\right)_T$ (l) $\left(\frac{\partial c_v}{\partial v}\right)_T$

(m) $\left(\frac{\partial p}{\partial v}\right)_S$ (n) $\left(\frac{\partial T}{\partial p}\right)_H$ (o) $c_p - c_v$

21. For 1 g-mole of a van der Waals gas (Illustration 10, page 542), evaluate the following coefficients:

(a) $\left(\frac{\partial p}{\partial T}\right)_V$ (b) $\left(\frac{\partial^2 p}{\partial T^2}\right)_V$ (c) $\left(\frac{\partial v}{\partial T}\right)_p$ (d) $\left(\frac{\partial^2 v}{\partial T^2}\right)_p$

(e) $\left(\frac{\partial U}{\partial v}\right)_T$ (f) $\left(\frac{\partial U}{\partial p}\right)_T$ (g) $\left(\frac{\partial c_p}{\partial p}\right)_T$ (h) $\left(\frac{\partial p}{\partial v}\right)_S$

22. From the relationships of problem 21 and Table 40, calculate the following changes in the properties of 1 lb-mole of CO_2 gas when the pressure is increased isothermally from 1.0 to 100 atm at 100° C: $a = (3.60)10^6$ (atm)(cc)^2/(g-mole)^2; $b = 42.8$ cc per g-mole.

(a) ΔH (b) ΔU (c) ΔS (d) Δc_p (e) Δc_v

14

Properties of Pure Fluids

At relatively large molal volumes, where the distances between molecules are great, the pressure–volume–temperature relations of gases are given with satisfactory accuracy by the ideal-gas law. Since low pressure and high temperature tend to increase molal volume, decrease of pressure and increase of temperature tend to make a gaseous material approach ideal behavior. If the molal volume is low, the assumption of ideal-gas behavior may cause errors as great as 500%.

Actual Behavior of Gases. The actual pvT relationships of carbon dioxide are shown by the solid lines in Fig. 132. Each curve represents the relationship between the pressure and the molal volume of carbon dioxide at a fixed indicated temperature. The critical point is indicated by C. The double-crosshatched area represents the region of the liquid state. The plain area is the region of a homogeneous fluid which at low pressures is recognized as a gas but at pressures above the critical has continuity with the liquid state along the *critical isotherm*. The single-crosshatched area represents a region in which both liquid and gaseous carbon dioxide are present in equilibrium with each other. The line ACB is designated as the *saturation* or *co-existent line*. Thus, following along the 21.5° C experimental isotherm an increasing pressure is required to cause a reduction in volume until the *saturation curve* CDB is reached. At this point D, the attractive forces between the molecules become sufficiently great to start condensation. Upon removal of heat at constant temperature, the volume of the combined phases is diminished without further increase in pressure until the curve CEA is reached at point E. This represents completion of condensation to the liquid state. A further increase of pressure is accompanied by a relatively small decrease in volume.

Experimentation has demonstrated the following general behavior of pure fluids.

1. As the pressure of a gas is reduced at constant temperature, ideal-gas behavior is approached as the pressure approaches zero, and the product pv approaches a limiting value equal to RT. The slopes of these pv curves at zero pressure depend on the temperature. The *Boyle temperature*

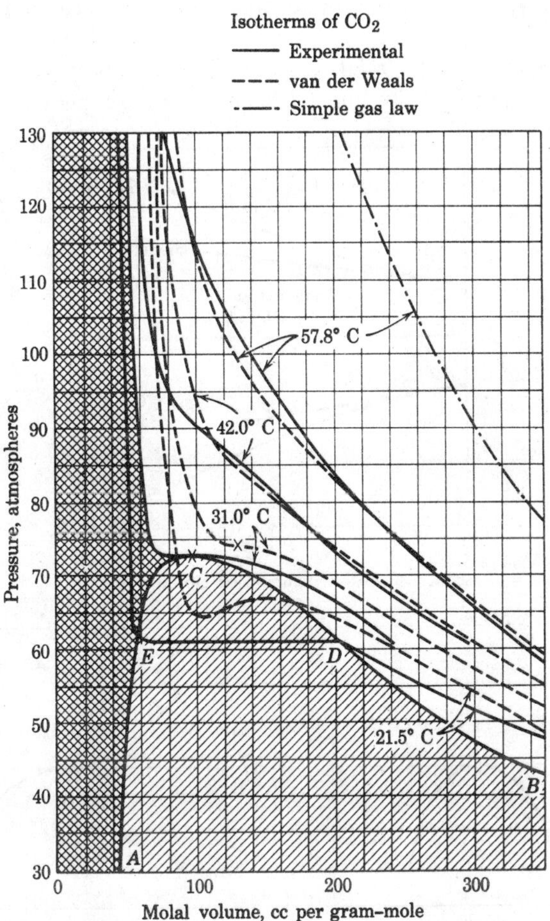

FIG. 132. Isotherms of carbon dioxide

is defined as that temperature where the slope of the pv curve at zero pressure equals zero; that is, $[\partial(pv)/\partial p]_T = 0$ (at $p = 0$). At temperatures above the Boyle point, this slope is greater than zero; at temperatures below the Boyle point, it is less than zero as shown in Fig. 133.

2. As the temperature of a gas is increased at constant pressure, ideal-gas behavior is approached at high temperatures, provided changes in molecular weight due to dissociation are considered.

3. At the critical point on the critical isotherm the first and second

derivatives of pressure with respect to volume are zero, as indicated in Fig. 134.

4. The isometric lines (relating p to T at constant values of molal volume) are approximately linear except at low temperatures and high

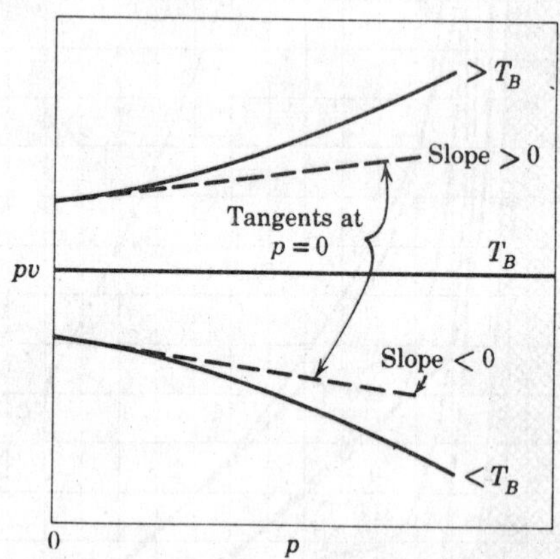

FIG. 133. Identification of Boyle point

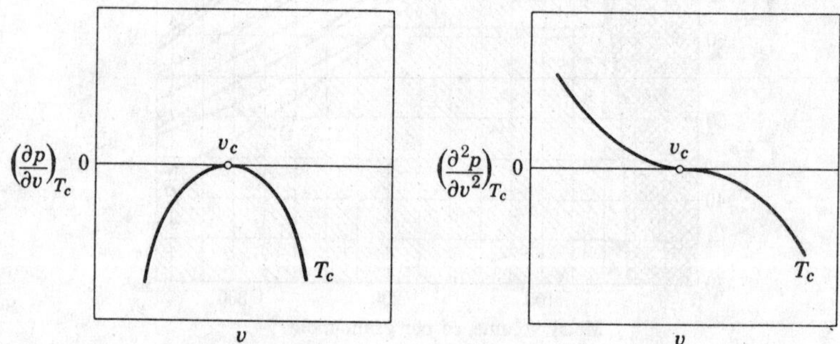

FIG. 134. Behavior of a fluid along the critical isotherm

densities. The limiting slopes as temperature is increased equal R/v. For low densities, the linear relationship holds to lower temperatures and pressures than for low molal volumes.

$$\lim_{T \to \infty} \left(\frac{\partial p}{\partial T}\right)_v = \frac{R}{v}, \qquad \lim_{T \to \infty} \left(\frac{\partial^2 p}{\partial T^2}\right)_v = 0$$

5. The isobars (curves relating volume to temperature at constant pressure) become straight at high temperatures, with slopes equal to R/p.

$$\lim_{T \to \infty} \left(\frac{\partial v}{\partial T}\right)_p = \frac{R}{p}, \qquad \lim_{T \to \infty} \left(\frac{\partial^2 v}{\partial T^2}\right)_p = 0$$

6. The isobars and isotherms are discontinuous in passing from the gas to the liquid state at values of pressure and temperature below the critical values. The isobars above the critical pressure and the isotherms above the critical temperature pass continuously through a region of homogeneous fluid.

Equations of state for gases should meet the general requirements detailed above.

Equations of State for Fluids

It is customary to express the pvT relationships of fluids by empirical equations designed to fit the experimental data. The complexity of such equations and the number of empirical constants required are dependent on the accuracy of experimental measurements, the precision required, and the experimental range of variables.

Such equations represent a convenient condensation of experimental data and are valuable not only for calculating values of pressure, temperature, volume, and density but also for deriving thermodynamic properties therefrom by mathematical procedures. For complex equations of state such calculations become unduly prolonged unless programmed for electronic computers.

For pure gases only two independent variables are required for expressing such relationships. For mathematical simplicity in using equations of state for calculating derived properties it is desirable to express volume as an explicit function of pressure and temperature. Because of greater ease in the correlation of experimental data it is customary to employ pressure as an explicit function of volume and temperature.

Virial Equations of State. Through consideration of intermolecular forces by the method of statistical mechanics, the pvT relationships of gases are presented by so-called virial coefficients. The term *virial* stems from a Latin word *vis* meaning force. Virial coefficients express the deviations from ideality in terms of intermolecular forces. Thus, the term pv/RT for one mole is written as

$$\frac{pv}{RT} = 1 + \frac{\beta}{v} + \frac{\gamma}{v^2} + \cdots \tag{1}$$

where β and γ are the second and third virial coefficients; both are functions of temperature.

The virial coefficients calculated from experimental data are correlated in terms of the intermolecular forces of attraction and repulsion between molecules expressed as a function of their intermolecular distances. For example, for rigid spherical molecules with zero energy of attraction, the virial equation becomes

$$\frac{pv}{RT} = 1 + \frac{b}{v} + 0.625\left(\frac{b}{v}\right)^2 + 0.287\left(\frac{b}{v}\right)^3 + 0.115\left(\frac{b}{v}\right)^4 + \cdots \qquad (2)$$

where b is the excluded volume. The statistical approach, which is beyond the scope of this textbook is valuable in formulating the proper empirical form of an equation of state.[1]

van der Waals' Equation. In developing the ideal-gas law, each molecule is considered to have an available free space, in which it may move about and which is assumed to be equal to the total volume occupied by the gas. This assumption is not correct except under such conditions that the volume of the molecular particles themselves is negligible compared to the total volume. In each mole of gas there is a space of volume $(v - b)$ available for free motion, somewhat less than the total volume. The term b is the *excluded volume* of the particles per mole.

Another term, neglected in the simple kinetic treatment of a gas, is the force of attraction existing between molecules, known as van der Waals' force. This force tends to draw the molecules together and reduces the pressure to less than the value corresponding to ideal behavior. It may be demonstrated from kinetic theory that this reduction is inversely proportional to the square of the molal volume. Thus

$$p = p' - \frac{a}{v^2} \quad \text{or} \quad p' = p + \frac{a}{v^2} \qquad (3)$$

where p' = pressure calculated from the simple kinetic theory. The term p' is the *internal pressure* of the gas. For one mole of gas the van der Waals equation of state then becomes

$$\left(p + \frac{a}{v^2}\right)(v - b) = \tfrac{1}{3}Nmu^2 = RT \qquad (4)$$

The terms a and b, characteristic of each gas, are termed the van der Waals constants. Although this equation is an improvement over the ideal-gas law, it gives poor approximation at high densities. Both terms a and b are actually dependent on temperature and density.

The van der Waals equation gives a simple explanation for the departure of gases from ideal behavior and supports the theory of corresponding states which serves as the basis of the generalized methods presented

[1] J. O. Hirschfelder, C. F. Curtiss, and R. B. Bird, *Molecular Theory of Gases and Liquids*, John Wiley & Sons (1954).

later. However, the van der Waals equation is less accurate than the generalized method, and its application is not reliable.

Relation of the Constants in the van der Waals Equation to the Critical Constants. By applying the van der Waals equation to the critical point and utilizing the principle that at the critical point the first and second derivatives of pressure with respect to volume equal zero, the following three independent equations are obtained:

$$p_c = \frac{RT_c}{v_c - b} - \frac{a}{v_c^2} \tag{5}$$

$$\left(\frac{\partial p}{\partial v}\right)_T = 0 = -\frac{RT_c}{(v_c - b)^2} + \frac{2a}{v_c^3} \tag{6}$$

$$\left(\frac{\partial^2 p}{\partial v^2}\right)_T = 0 = \frac{2RT_c}{(v_c - b)^3} - \frac{6a}{v_c^4} \tag{7}$$

By combining equations 5 through 7, the following equations expressing the critical properties in terms of the van der Waals constants are obtained:

$$p_c = \frac{a}{27b^2} \tag{8}$$

$$v_c = 3b \tag{9}$$

$$T_c = \frac{8a}{27Rb} \tag{10}$$

By combining these equations, the following equations expressing a and b in terms of various critical constants are obtained:

$$a = 3p_c v_c^2 = \frac{27R^2T_c^2}{64p_c} = \tfrac{9}{8} RT_c v_c \tag{11}$$

$$b = \frac{v_c}{3} = \frac{RT_c}{8p_c} \tag{12}$$

Values for the van der Waals constants are generally calculated by means of equations 11 and 12 rather than by the use of pvT data at conditions removed from the critical point. Because critical temperature and pressure data are generally more accurate than critical volume data, the equations showing a and b as functions of p_c and T_c are generally used in preference to those containing v_c. The van der Waals constants given in Table 43 were calculated in this manner.

TABLE 43. VAN DER WAALS AND CRITICAL CONSTANTS OF GASES

a = (atm)(cc/g-mole)^2, b = cubic centimeters per gram-mole
T_c = critical temperature, degrees Kelvin
p_c = critical pressure, atmospheres
ρ_c = critical vapor density, grams per cubic centimeter
v_c = critical molal volume, cubic centimeters per gram-mole

Gases	a	b	T_c	p_c	ρ_c	v_c
Argon	$(1.353)10^6$	32.31	151.2	48.0	0.533	75
Acetylene	$(4.417)10^6$	51.54	309.5	61.6	0.2304	113
Ammonia	$(4.197)10^6$	37.37	405.5	111.3	0.235	72.5
Carbon dioxide	$(3.606)10^6$	42.80	304.2	72.9	0.468	94
Carbon monoxide	$(1.456)10^6$	39.54	133.0	34.5	0.3012	93
Chlorine	$(6.491)10^6$	56.21	417.0	76.1	0.572	124
Ethylene	$(4.508)10^6$	57.50	283.1	50.5	0.226	124
Hydrogen chloride	$(3.672)10^6$	40.85	324.6	81.5	0.419	87
Hydrogen	$(0.2461)10^6$	26.68	33.3	12.8	0.03102	65
Methane	$(2.256)10^6$	42.71	190.7	45.8	0.162	99
Methyl chloride	$(7.470)10^6$	64.80	416.3	65.9	0.353	143
Nitrogen	$(1.351)10^6$	38.64	126.2	33.5	0.311	90
Oxygen	$(1.362)10^6$	31.87	154.4	49.7	0.432	74
Sulfur dioxide	$(6.773)10^6$	56.78	430.7	77.8	0.525	122
Water	$(5.454)10^6$	30.42	647.4	218.3	0.322	56

To convert values of a to (psi)(cu ft/lb-mole)^2, multiply values of a in the table by 0.003771.

To convert values of b to cubic feet per pound-mole, multiply values of b in the table by 0.01602.

TABLE 44. BEATTIE–BRIDGEMAN CONSTANTS

Gas	A_0	a	B_0	b	c	Temperature range, °C	Minimum v, cc/g mole
He	0.0216	0.05984	0.01400	0.0	$(0.004)10^4$	400 to −252	100
H_2	0.1975	−0.00506	0.02096	−0.04359	$(0.0504)10^4$	200 to −244	100
N_2	1.3445	0.02617	0.05046	−0.00691	$(4.20)10^4$	400 to −149	180
O_2	1.4911	0.02562	0.04624	0.004208	$(4.80)10^4$	100 to −117	110
Air	1.3012	0.01931	0.04611	−0.01101	$(4.34)10^4$	200 to −145	125
CO_2	5.0065	0.07132	0.10476	0.07235	$(66.00)10^4$	100 to 0	180
CH_4	2.2769	0.01855	0.05587	−0.01587	$(12.83)10^4$	200 to 0	166

Beattie–Bridgeman Equation of State. One of the most widely used equations of state is that of Beattie and Bridgeman.[2,3]

$$pv^2 = RT\left[v + B_0\left(1 - \frac{b}{v}\right)\right]\left(1 - \frac{c}{vT^3}\right) - A_0\left(1 - \frac{a}{v}\right) \tag{13}$$

where a, b, A_0, B_0, c = empirically determined constants
v = molal volume, liters per gram-mole
p = pressure, atmospheres
T = temperature, degrees Kelvin
$R = 0.08205$ (liter)(atm)/(g-mole)(K°)

This equation contains five constants which are characteristic of each particular gas. The methods of evaluating these constants are discussed by Beattie and Bridgeman and by Deming and Shupe.[4]

In Table 44 are values of the constants of equation 13 for several common gases.

It can be shown that the Beattie–Bridgeman equation qualitatively represents the behavior characteristics of an actual gas. The fulfillment of these various conditions, however, does not insure the accuracy of an equation of state nor validate its use. Even the Beattie–Bridgeman equation with five constants does not fit the experimental data at the critical point. For example, for carbon dioxide the experimental values of the critical constants are $T_c = 31.1°$ C, $p_c = 73.0$ atm, and $v_c = 0.0955$ liter per g-mole. By the Beattie–Bridgeman equation, the calculated value of $v_c = 0.0533$ liter per g-mole, or, solving for critical pressure, $p_c = 108$ atm; both values are seriously in error. The Beattie-Bridgeman equation ordinarily fits experimental values within 0.15% for densities less than 5 moles per liter, and should not be extrapolated to the critical range.

Benedict–Webb–Rubin Equation of State. An equation of state with eight constants has been formulated by Benedict, Webb, and Rubin[5] for the lighter hydrocarbons from experimental data. Thus, where ρ is density in gram-moles per liter,

$$p = RT\rho + \left(B_0RT - A_0 - \frac{C_0}{T^2}\right)\rho^2 + (bRT - a)\rho^3 + a\alpha\rho^6 + c\rho^3\,\frac{1 + \gamma\rho^2}{T^2}\,e^{-\gamma\rho^2} \tag{14}$$

[2] J. A. Beattie and O. C. Bridgeman, *J. Am. Chem. Soc.* **49**, 1665 (1927).

[3] J. A. Beattie and O. C. Bridgeman, *J. Am. Chem. Soc.* **50**, 3133 (1928).

[4] W. E. Deming and L. E. Shupe, *J. Am. Chem. Soc.* **52**, 1382 (1930); **53**, 843, 860 (1931).

[5] M. Benedict, G. W. Webb, and L. C. Rubin, *J. Chem. Phys.*, **8**, 334 (1940).

Values of the constants for four gases are recorded in Table 45. Constants for other gases are reported by others.[6–9]

Table 45. Constants of the Benedict–Webb–Rubin Equation of State

Units: Atmospheres, liters, gram-moles, degrees Kelvin; $R = 0.08205$

	Methane	Ethane	Propane	*n*-Butane
B_0	0.0426000	0.0627724	0.0973130	0.124361
A_0	1.85500	4.15556	6.87225	10.0847
$C_0\ 10^{-6}$	0.0225700	0.179592	0.508256	0.992830
b	0.00338004	0.0111220	0.0225000	0.0399983
a	0.0494000	0.345160	0.947700	1.88231
$c\ 10^{-6}$	0.00254500	0.0327670	0.129000	0.316400
γ	0.0060000	0.0118000	0.0220000	0.0340000
α	0.000124359	0.000243389	0.000607175	0.00110132

Thermodynamic Properties from an Equation of State

The deviations of thermodynamic properties of a fluid from those of an ideal gas can be derived from an equation of state. The procedure will be illustrated by reference to the Beattie–Bridgeman equation. Rearrangement of equation 13 gives

$$\left.\begin{aligned} p &= \frac{\alpha}{v} + \frac{\beta}{v^2} + \frac{\gamma}{v^3} + \frac{\delta}{v^4} \\ \text{where}\quad \alpha &= RT \\ \beta &= -A_0 + B_0RT - \frac{cR}{T^2} \\ \gamma &= aA_0 - bB_0RT - \frac{cB_0R}{T^2} \\ \delta &= \frac{bcB_0R}{T^2} \end{aligned}\right\} \tag{15}$$

The following derivatives needed in these derivations are obtained directly from equation 15:

$$\left(\frac{\partial p}{\partial v}\right)_T = -\frac{\alpha}{v^2} - \frac{2\beta}{v^3} - \frac{3\gamma}{v^4} - \frac{4\delta}{v^5} \tag{16}$$

$$\left(\frac{\partial^2 p}{\partial v^2}\right)_T = \frac{2\alpha}{v^3} + \frac{6\beta}{v^4} + \frac{12\gamma}{v^5} + \frac{20\delta}{v^6} \tag{17}$$

$$\left(\frac{\partial p}{\partial T}\right)_V = \frac{\alpha'}{v} + \frac{\beta'}{v^2} + \frac{\gamma'}{v^3} + \frac{\delta'}{v^4} \tag{18}$$

$$\left(\frac{\partial^2 p}{\partial T^2}\right)_V = \frac{\beta''}{v^2} + \frac{\gamma''}{v^3} + \frac{\delta''}{v^4} \tag{19}$$

[6] E. I. Organick and W. R. Studhalter, *Chem. Eng. Prog.*, **44**, 847 (1948).

[7] M. Benedict, G. W. Webb, and L. C. Rubin, *Chem. Eng. Prog.*, **47**, 419 (1951).

[8] H. H. Stotler and M. Benedict, *Chem. Eng. Prog. Symp. No.* **49** (6), 25 (1953)

[9] L. N. Canjar, R. F. Smith, E. Volianitis, J. F. Galluzzo, and M. Carbacos, *Ind. Eng. Chem.*, **47**, 1028 (1955).

where

$$\left.\begin{aligned}
\alpha' &= R \\
\beta' &= B_0R + \frac{2cR}{T^3} \\
\gamma' &= -bB_0R + \frac{2cB_0R}{T^3} \\
\delta' &= -\frac{2bcB_0R}{T^3} \\
\beta'' &= -\frac{6cR}{T^4} \\
\gamma'' &= -\frac{6cB_0R}{T^4} \\
\delta'' &= \frac{6bcB_0R}{T^4}
\end{aligned}\right\} \tag{20}$$

From equations 13–91 and 18, and 16,

$$\left(\frac{\partial v}{\partial T}\right)_p = -\frac{(\partial p/\partial T)_V}{(\partial p/\partial v)_T} = \frac{\alpha' v^4 + \beta' v^3 + \gamma' v^2 + \delta' v}{\alpha v^3 + 2\beta v^2 + 3\gamma v + 4\delta} \tag{21}$$

In the derivation that follows the asterisk refers to the property of the ideal gas at the same temperature and pressure as the actual fluid.

Enthalpy Departure $(\text{H}^* - \text{H})_T$. From equation 13–100 at constant temperature:

$$d\text{H} = v\,dp - T\left(\frac{\partial v}{\partial T}\right)_p dp \tag{22}$$

The integration of equation 22 using the Beattie–Bridgeman equation requires transformation in terms of dv. Thus:

$$v\,dp = d(pv) - p\,dv \tag{23}$$

and, from equation 13–91,

$$\left[\left(\frac{\partial v}{\partial T}\right)_p dp\right]_T = -\left[\left(\frac{\partial p}{\partial T}\right)_V dv\right]_T \tag{24}$$

Hence

$$d\text{H} = d(pv) + \left[T\left(\frac{\partial p}{\partial T}\right)_V - p\right]_T dv \tag{25}$$

The value of H^* is independent of pressure and corresponds to the limits of integration where $p = 0$, $v = \infty$, and $pv = RT$. Thus, upon integration of equation 25,

$$(\text{H}^*_p - \text{H}_p)_T = RT - pv + \int_{\infty}^{V}\left[p - T\left(\frac{\partial p}{\partial T}\right)_V\right]_T dv \tag{26}$$

Substitution of equation 15 for p and equation 18 for $(\partial p/\partial T)_V$ gives

$$\left.\begin{aligned} (\text{H}^* - \text{H})_T &= RT - pv + \frac{\beta'''}{v} + \frac{\gamma'''}{v^2} + \frac{\delta'''}{v^3} \\ \text{where}\quad \alpha''' &= T\alpha' - \alpha = 0 \\ \beta''' &= T\beta' - \beta = A_0 + \frac{3cR}{T^2} \\ \gamma''' &= \tfrac{1}{2}(T\gamma' - \gamma) = \frac{3cB_0R}{2T^2} - \frac{aA_0}{2} \\ \delta''' &= \tfrac{1}{3}(T\delta' - \delta) = -\frac{bcB_0R}{T^2} \end{aligned}\right\} \tag{27}$$

The numerical values of $(\text{H}^* - \text{H})_T$ calculated from equation 27 using the constants of Table 44 will be in (liter)(atm)/g-mole.

Entropy Departure $(\text{s}^*_p - \text{s}_p)_T$. From equation 13–105 at constant temperature,

$$d\text{s} = \left(\frac{\partial p}{\partial T}\right)_V dv \tag{28}$$

For an ideal gas,

$$d\text{s} = -R\, d \ln v \tag{29}$$

Integration from volume v to infinite volume ($p = 0$) for an actual gas gives

$$\text{s}_p - \text{s}_0 = \int_\infty^V \left(\frac{\partial p}{\partial T}\right)_V dv \tag{30}$$

for an ideal gas

$$\text{s}^*_p - \text{s}^*_0 = +\int_\infty^V R\, d \ln v \tag{31}$$

Since $\text{s}_0 = \text{s}^*_0$,

$$(\text{s}^*_p - \text{s}_p)_T = +\int_\infty^{V^*} R\, d \ln v - \int_\infty^V \left(\frac{\partial p}{\partial T}\right)_V dv \tag{32}$$

Substituting equation 18 into 32 and integrating gives, since $\alpha' = R$,

$$(\text{s}^*_p - \text{s}_p)_T = \frac{\beta'}{v} + \frac{\gamma'}{2v^2} + \frac{\delta'}{3v^3} + R \ln \frac{v^*}{v} \tag{33}$$

The numerical value of $(\text{s}^*_p - \text{s}_p)_T$ is expressed in (liter)(atm)/(g-mole)(K°) in agreement with the units in Table 44.

Heat-Capacity Departure $c_p - c^*_p$. From equations 18, 19, 21, and 13–131 and 136, the following relation for heat-capacity departure may be obtained:

$$\begin{aligned} c_p - c^*_p = {} & \frac{6cR}{T^3}\left(\frac{1}{v} + \frac{B_0}{2v^2} - \frac{bB_0}{3v^3}\right) - R \\ & + T\left(v + B_0 - \frac{B_0 b}{v}\right)\left(\frac{R}{v^2} + \frac{2cR}{v^3T^3}\right)\left(\frac{\alpha'v^4 + \beta'v^3 + \gamma'v^2 + \delta'v}{\alpha v^3 + 2\beta v^2 + 3\gamma v + 4\delta}\right) \end{aligned} \tag{34}$$

Illustration 1. From the Beattie–Bridgeman equation of state, it is desired to calculate the values of $\text{H}^* - \text{H}$ and $c_p - c^*_p$ for carbon dioxide gas at 100° C and 100 atm pressure.

For carbon dioxide, from Table 44:

$$A_0 = 5.0065, \quad b = 0.07235, \quad T = 373.2° \text{ K}$$
$$a = 0.07132, \quad c = 660{,}000$$
$$B_0 = 0.10476, \quad R = 0.08205 \text{ (liter)(atm)/(g-mole)(K°)}$$

From equation 15:

$$\alpha = RT = (0.08205)(373.2) = 30.625$$

$$\beta = RTB_0 - A_0 - \frac{Rc}{T^2} = -2.1871$$

$$\gamma = RTB_0b + aA_0 - \frac{RB_0c}{T^2} = 0.0842$$

$$\delta = \frac{RB_0bc}{T^2} = 0.0029475$$

Hence
$$p = \frac{30.625}{v} - \frac{2.1871}{v^2} + \frac{0.08420}{v^3} + \frac{0.0029475}{v^4} \quad (a)$$

By assuming various values of v, equation (a) may be solved for v graphically or by iteration until a value of p is obtained equal to 100 atm, as shown in Table A. An initial approximate value is obtained by assuming ideal behavior.

Table A

v Assumed	$\frac{30.625}{v}$	$\frac{-2.1871}{v^2}$	$\frac{0.08420}{v^3}$	$\frac{0.0029475}{v^4}$	p calculated
0.2290	133.73	−41.706	7.07	1.07	100.11
0.2295	133.44	−41.523	6.97	1.07	99.96
0.2300	133.15	−41.342	6.92	1.05	99.78

The correct value of v where p is equal to 100 atm is 0.2295 liter.

From equation 20:

$$\alpha' = R = 0.08205$$

$$\beta' = B_0R + \frac{2cR}{T^3} = 0.010681$$

$$\gamma' = -bB_0R + \frac{2cB_0R}{T^3} = -0.00040364$$

$$\delta' = -\frac{2bcB_0R}{T^3} = -0.000015795$$

From equation 27:

$$\alpha''' = 0$$

$$\beta''' = -\beta + \beta'T = 6.1731$$

$$\gamma''' = -\frac{\gamma}{2} + \frac{\gamma'T}{2} = -0.11742$$

$$\delta''' = \frac{(-\delta + \delta'T)}{3} = -0.0029475$$

Substitution of these values into equation 27 gives

$$(H^* - H)_T = 32.10 \text{ (liter)(atm)/g-mole}$$

or

$$(32.10)(24.21) = 777 \text{ cal per g-mole}$$

Evaluation of c_p at 100° C and 100 atm. Substitution of constants into equation 34 gives

$$c_p - c^*_p = 0.2574 \text{ (liter)(atm)/(g-mole)(K°)}$$

or

$$(24.21)(0.2574) = 6.232 \text{ cal/(g-mole)(K°)}$$

Compressibility Factor

The equation of state may be written

$$pV = znRT \tag{35}$$

where z is termed the compressibility factor and is a function of pressure, temperature, and the nature of the gas. The ideal-gas law may be considered as representing a special case in which the compressibility factor is equal to unity.

If values of the compressibility factor of a gas are known, all calculations involving its pvT relationships may be carried out by simple proportionalities derived from equation 35. Thus, applying equation 35 to a given mass of gas at two different conditions gives

$$\frac{p_1V_1}{p_2V_2} = \frac{z_1T_1}{z_2T_2} \tag{36}$$

Illustration 2. One cubic foot of nitrogen at 50° C and 30 atm is compressed to 60 atm and cooled to −50° C. Calculate the final volume, where values of z are given.

Initial conditions:

$p_1 = 30$ atm

$V_1 = 1.0$ cu ft

$T_1 = 50° \text{ C} = 323° \text{ K}$

$z_1 = 1.001$

Final conditions:

$p_2 = 60$ atm

$T_2 = -50° \text{ C} = 223° \text{ K}$

$z_2 = 0.930$

From equation 36:

$$V_2 = \frac{V_1p_1T_2z_2}{p_2T_1z_1}$$

$$V_2 = 1.0\left(\frac{30}{60}\right)\left(\frac{223}{323}\right)\left(\frac{0.930}{1.001}\right) = 0.321 \text{ cu ft}$$

Illustration 3. A steel cylinder having a volume of 5 liters contains 400 grams of nitrogen. Calculate the temperature to which the cylinder may be heated without the pressure exceeding 50 atm where values of z are given.

Moles of nitrogen = 400/28 = 14.28 g-moles

Molal volume = 5000/14.28 = 350 cc

$p = 50$ atm

$z = 0.945$

$R = 82.1$ (cc)(atm)/(g-mole)(K°)

From equation 35:

$$T = \frac{50(350)}{(82.1)(0.945)} = 225° \text{ K} \quad \text{or} \quad -48° \text{ C}$$

Generalized Properties of Fluids

The *reduced conditions* of temperature, pressure, volume, and density are defined on page 87 by the ratios

$$T_r = T/T_c, \quad p_r = p/p_c, \quad v_r = v/v_c, \quad \text{and} \quad \rho_r = \rho/\rho_c \tag{37}$$

Pure substances are said to be in corresponding states when they exist at the same reduced conditions of temperature and pressure.

In 1873 van der Waals first defined the term *reduced condition* and presented the *theorem of corresponding states* that all pure gases manifest the same compressibility factors when measured at the same reduced conditions of pressure and temperature. This concept was extended to liquids by Young in 1899.

According to this principle, the deviations of thermodynamic properties of different pure fluids would manifest the same departure from the properties of these substances in their ideal gaseous state when examined at the same reduced conditions of temperature and pressure. Thus:

$$z = f(p_r, T_r) \tag{38}$$

Based on this principle, generalized charts[10] have been developed for the compressibility factors for gases and properties derived therefrom such as fugacities, and for departures from ideal-gas behavior of enthalpy, entropy, and heat capacity. Watson extended these generalized charts to the liquid state.

Critical Compressibility Factor z_c. According to equation 38, the critical compressibility factor z_c should be the same for all substances; actually these values range from 0.20 to 0.30. The early generalized charts for compressibility factors and derived properties were based on

[10] O. A. Hougen and K. M. Watson, Chemical Process Principles Charts, John Wiley & Sons (1946).

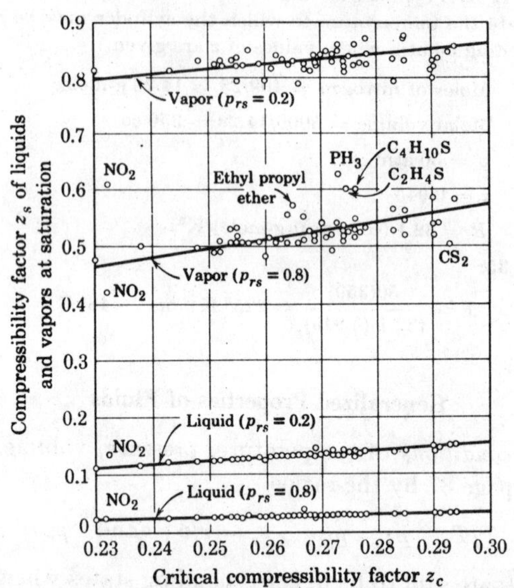

FIG. 135. Correlation of compressibility factors of liquids and vapors with the critical compressibility factor z_c

the behavior of only five to eight compounds. Such charts were found to be deficient in failing to define the conditions of saturation and in showing a wide blurred band for the saturation envelope. This situation called for the introduction of a third parameter. Since the greatest deviation in the saturation range occurs in the value of z at the critical point, the critical compressibility factor z_c was chosen as the third parameter[11,12] and is obtained from the critical constants through the ratio

$$z_c = \frac{p_c v_c}{RT_c} \tag{39}$$

With this modification, the theorem of corresponding states becomes

$$z = f(p_r, T_r, z_c) \tag{40}$$

The merit of z_c as a third parameter is shown in Fig. 135 where the compressibility factors z_s at saturation for both liquids and vapors are plotted against the corresponding values of z_c for 71 different compounds.[11]

[11] A. L. Lydersen, R. A. Greenkorn, and O. A. Hougen, "Generalized Thermodynamic Properties of Pure Fluids," *Univ. Wisconsin Eng. Exp. Sta. Rept.* **4** (Oct. 1955).

[12] H. P. Meissner and R. Seferian, *Chem. Eng. Prog.*, **45**, 579 (1951).

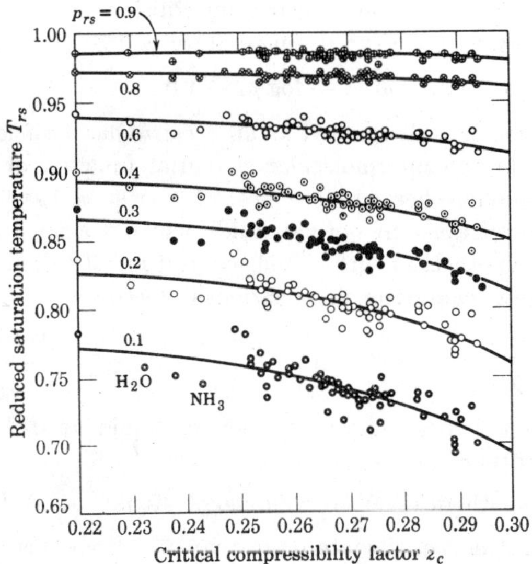

FIG. 136. Effect of z_c on reduced saturation temperatures at various reduced pressures

Values are plotted at only two values of reduced pressure; namely, 0.2 and 0.8. In the liquid state nearly all points for a given saturation pressure fall on the same straight line. In the vapor state greater scattering occurs. In Figure 136 are plotted values of reduced saturation temperature T_{rs} against z_c for various reduced saturation pressures p_{rs} for 71 different compounds.[11] Each point in Figs. 135 and 136 for a given line corresponds to a different compound. The correlation improves as the critical pressure is approached but becomes poor at low reduced pressures. According to the original theorem of corresponding states, all lines in Figs. 135 and 136 would be horizontal.

Several other third parameters were tried,[12] such as the reduced dipole moment, bond length, and normal heat of vaporization, with less favorable results. In using the reduced dipole moment as a third parameter, the deviation of z for the same reduced conditions varies nearly as much for nonpolar as for polar compounds.

Riedel[13] introduced the slope α of the vapor-pressure curve at the critical temperature as the third parameter where

$$\alpha = \frac{d \ln p}{d \ln T} \quad \text{(at the critical point)} \tag{41}$$

[13] L. Riedel, *Chem. Ing. Tech.* **26**, 83, 257, 679 (1954).

A related parameter ω was selected by Pitzer[14] in terms of reduced saturation pressure at a reduced temperature of 0.70 by the relation

$$\omega = -\log p_r - 1.0 \tag{42}$$

The ω parameter is designated as an *acentric factor* since it measures the deviation of the intermolecular potential function of a substance from that of simple spherical molecules. Pitzer et al. have limited their generalized correlations to normal fluids which Riedel has defined in terms of the hypothetical surface tension and molal volume of a liquid. In terms of the ω parameter this relationship becomes

$$\frac{\sigma_0 V_0^{2/3}}{T_c} = 1.86 + 1.18\omega \frac{(\text{dynes/cm})(\text{cc/g-mole})^{2/3}}{\text{K}^\circ} \tag{43}$$

where σ_0 = hypothetical surface tension of liquid at 0° K, dynes per centimeter

V_0 = hypothetical molal volume of liquid at 0° K, cc/g-mole

Compounds that do not deviate more than 5% from this relation were included as normal fluids. Ammonia, water, acetic acid, methyl alcohol, and hydrogen cyanide deviate widely from this relationship.

Curl and Pitzer[15] reported that more reliable correlation could be obtained by the use of the ω parameter than for z_c because of the uncertainty in evaluation of critical density when used in calculating z_c. Much of this uncertainty is eliminated if z_c is determined from experimental liquid density data by the method discussed on page 578, or, where vapor-pressure data are available, improved accuracy of z_c values may be obtained from values of α or ω by the relation

$$z_c = \frac{1}{1.28\omega + 3.41} = \frac{1}{0.26\alpha + 1.90} \tag{44}$$

Table 46 relates the values of the different third parameters.

Table 46. Relation of Different Third Parameters

z_c	α	ω	z_c	α	ω
0.20	11.92	1.242	0.26	7.46	0.341
0.21	11.00	1.056	0.27	6.94	0.230
0.22	10.17	0.887	0.28	6.43	0.126
0.23	9.53	0.756	0.29	5.95	0.030
0.24	8.72	0.591	0.30	5.51	−0.060
0.25	8.08	0.461			

[14] K. S. Pitzer, D. Z. Lippman, R. F. Curl, C. M. Higgins, and D. E. Peterson, *J. Am. Chem. Soc.*, **77**, 3433 (1955).

[15] R. F. Curl and K. S. Pitzer, *Ind. Eng. Chem.*, **50**, 265 (1958).

The ω parameter is limited to normal fluids as defined by equation 42 and would require a fourth parameter to include nonnormal fluids. The z_c parameter has the advantage of representing nearly all pure fluids although with a slight loss of precision for normal fluids.

With the acceptance of z_c as a third parameter Lydersen, Greenkorn, and Hougen[11] assembled the pvT data of 82 different compounds for both gaseous and liquid states and constructed tables for compressibility factors and reduced densities arranged in four groups according to values of z_c based on 8000 values for liquids and gases.

z_c	Representative Compounds
0.232	Water
0.24 to 0.26	Acetone, ammonia, esters, alcohols
0.26 to 0.28	60% of compounds, mostly hydrocarbons
0.28 to 0.30	O_2, N_2, CO, H_2S, CH_4, C_2H_6, A, Ne

Hydrogen and helium do not fit the generalized correlations in their saturation ranges.

The significance of the third parameter z_c is shown in the plot of compressibility factors at saturation z_s in Fig. 137 and for reduced density at saturation ρ_{rs} in Fig. 138 for both liquid and vapor phases. The failure of the original z chart (neglecting the third parameter) to show a definite saturation curve is shown by these graphs. The parameter z_c brings into correlation the saturation envelopes for different substances, including water.

The generalized tables presented in this textbook were calculated from z tables using an electronic calculator.[11] Smoother tabulations could be obtained if machine calculations were made from closed equations of state for z covering both liquid and gaseous states.

Hirschfelder, Buehler, McGee, and Sutton[16] have developed such a generalized reduced equation of state covering both liquid and gaseous states over the entire pressure, temperature, and third-parameter ranges in terms of both parameters z_c and α. Tables of generalized thermodynamic excess functions for pure gases and liquids were also evaluated by these investigators[17] from these equations. Thermodynamic properties derived from closed equations are smoother than those evaluated from tabulated data of compressibility factors because small errors in tabular functions lead to large errors in the first and second derivatives when

[16] J. O. Hirschfelder, R. J. Buehler, H. A. McGee Jr., and J. R. Sutton, *Ind. Eng. Chem.*, **50**, 375 (1958).

[17] J. O. Hirschfelder, R. J. Buehler, H. A. McGee Jr., and J. R. Sutton, *Ind. Eng. Chem.* **50**, 386 (1958).

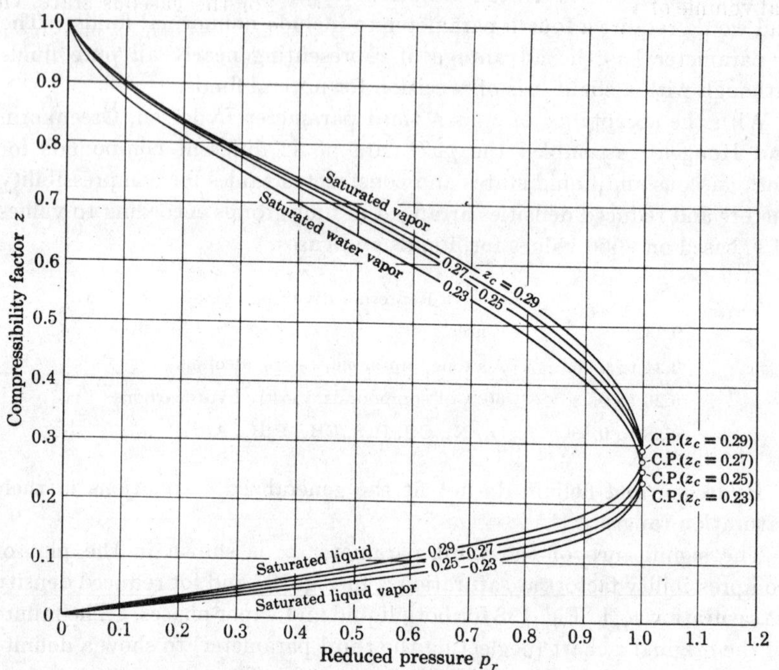

Fig. 137. Compressibility factors of saturated liquids and vapors. C.P. = critical point

determined by numerical differentiation methods. The original tables of Lydersen et al. are retained in this textbook because of the great span of experimental data employed in their construction.

Effect of Parameter z_c. Generalized tables and graphs of thermodynamic properties are herein presented for values of z_c equal to 0.27, since 60% of the pure fluids studied fall in the narrow range of z_c from 0.26 to 0.28. Properties at values of z_c other than at 0.27 may be calculated by the use of deviation or correction terms D, which are also tabulated. The value of a property B' at values of z_c other than at 0.27 including enthalpy departure, entropy departure, internal energy departure, and reduced density, is calculated by the following formula:

$$B' = B + D(z_c - 0.27) \tag{45}$$

where B = property at $z_c = 0.27$

D_a = deviation term for use where z_c is greater than 0.27

D_b = deviation term for use when z_c is less than 0.27

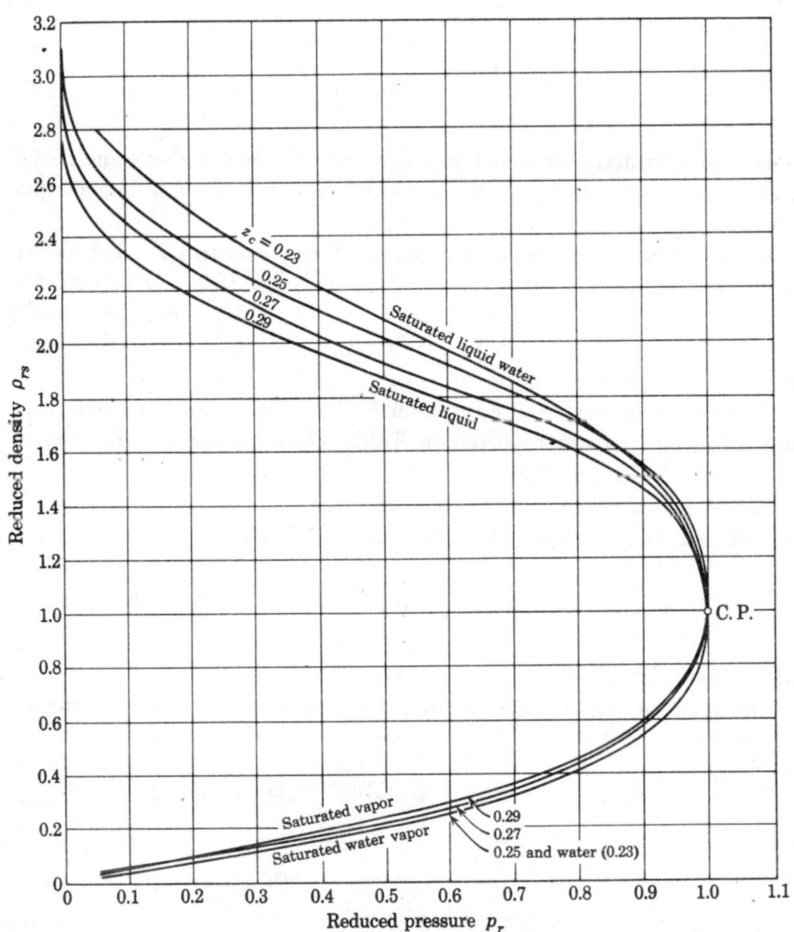

FIG. 138. Reduced densities of saturated liquids and vapors. C.P. = critical point

For the fugacity coefficient a deviation term based on logarithmic values of f/p is required. Thus

$$\left(\frac{f}{p}\right)' = \left(\frac{f}{p}\right) 10^{D(z_c - 0.27)} \tag{46}$$

where again values of D_u and D_b are used for extrapolation above and below $z_c = 0.27$.

For the gaseous state where the ratio $(f/p)'/(f/p)$ is nearly unity, the following formula may be used instead of equation 46:

$$\left(\frac{f}{p}\right)' = \frac{f}{p}[1 + 2.303D(z_c - 0.27)] \tag{47}$$

The use of different interpolation equations for values above and below $z_c = 0.27$ does not imply an actual break in curves at 0.27 but accepts linear interpolations over the two ranges of z_c.

Values of D are tabulated for reduced pressures up to 1.2. For higher pressures the same values of D as recorded at $p_r = 1.2$ may be used for liquids where $T_r < 0.8$, and for vapors where $T_r > 1.2$. For the reduced temperature range from 0.8 to 1.2 extrapolation to obtain D values at $p_r > 1.2$ is unreliable.

Generalized Properties at Saturation. Values of thermodynamic properties at saturation are reported in Table 47 for water and for values of z_c at 0.25, 0.27, and 0.29.

The data on *reduced vapor pressures* were calculated from Riedel's generalized reduced-vapor-pressure equation;[16] thus,

$$\ln p_r = \alpha \ln T_r - 0.0838(\alpha - 3.75)\left(\frac{36}{T_r} - 35 - T_r^6 + 42 \ln T_r\right) \tag{48}$$

where α is related to z_c by equation 44 or Table 46.

The reduced vapor pressure of Gamson and Watson can be further generalized to include the third parameter z_c; thus,

$$\log p_r = -\frac{A(1 - T_r)}{T_r} - 10^{-8.68(T_r - b)^2} \tag{49}$$

where

$$A = 16.25 - 73.85z_c + 90z_c^2 \tag{50}$$

$$b = 1.80 - 6.20z_c \tag{51}$$

Equation 49 is in good agreement with equation 48 and is preferable for extrapolation to temperatures above the critical.

From the third Maxwell relation the *latent heat of vaporization* in terms of the compressibility factor and vapor pressure is given precisely as

$$\frac{\lambda}{T_c} = \frac{RT_{rs}^2(z_{sG} - z_{sL})}{p_{rs}}\left(\frac{\partial p_r}{\partial T_r}\right)_s \tag{52}$$

Values of $(\partial p_r/\partial T_r)_s$ may be obtained from either equation 8–30 or 48. From one known value of λ, values of λ may also be obtained at other temperatures from equation 8–40.

Density of Pure Liquids. Riedel[18] has shown that the molal volume of a saturated liquid extrapolated to 0° K is an easily reproducible value and more reliable than the critical volume as a reference state in defining reduced volumes or reduced densities. For reduced temperatures below 0.7, values of ρ/ρ_0 plotted as the same unique function of reduced temperature for all normal liquids, and, above $T_r = 0.7$, as a function of the acentric factor α. This method is a refinement of the expansion factor correlation introduced by Watson.[18a]

Reduced densities of liquids are tabulated in Table 48 for values of z_c equal to 0.27 along with deviation correction terms D_a and D_b for values of z_c above and below 0.27. These tables were derived from the data of Riedel,[18] using z_c as the third parameter. Saturation values for four values of z_c are given in Table 47 and shown in Fig. 139 for $z_c = 0.27$.

The value of z is related to density by the ratio $z = p/RT\rho$. Since the density of liquids is not greatly influenced by pressure, for liquids z is a measure of the ratio p/T rather than of ρ. For this reason, values of ρ_r rather than of z should be used in calculating liquid densities.

Values of liquid densities can be estimated directly from Table 48 from known values of T_c, p_c, ρ_c, and z_c. Such a procedure is not recommended because of uncertainties in the value of ρ_c. The recommended procedure is to use as a reference state for each liquid a single accurate density measurement usually taken at atmospheric pressure and to employ this value in conjunction with the reduced density table by the following relation,

$$\frac{\rho_2}{\rho_{r2}} = \frac{\rho_1}{\rho_{r1}} = \rho'_c = \frac{p_c}{z'_c R T_c} \tag{53}$$

where ρ_1 = accurate measurement of density at T_{r1} and p_{r1}

ρ_2 = desired density at T_{r2} and p_{r2}

ρ'_c, z'_c = corrected values of ρ_c and z_c.

This method is not sensitive to errors in ρ_c or z_c. The average error in estimating liquid densities by this method is about 1%.

This procedure may be employed to establish improved values of ρ_c and z_c for any pure liquid for which experimental densities are available

Illustration 4. The density of liquid ethyl propionate at 20° C and 1 atm is 0.891 g per cc. Its critical constants are as follows: $z_c = 0.254$, $T_c = 546.1°$ K, $p_c = 33.0$ atm. Estimate the density of the liquid at (*a*) 146° C and 30 atm, (*b*) 257° C, saturated. (*c*) Estimate the critical density and the corresponding value of z_c.

[18] L. Riedel, *Chem. Ing. Tech.* **26**, 259 (1954).
[18a] K. M. Watson, *Ind. Eng. Chem.*, **35**, 398 (1943).

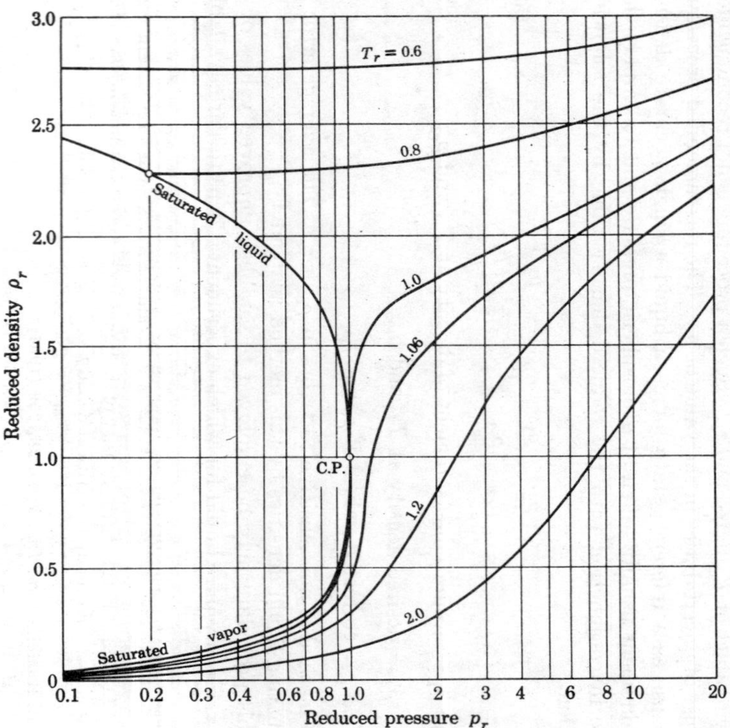

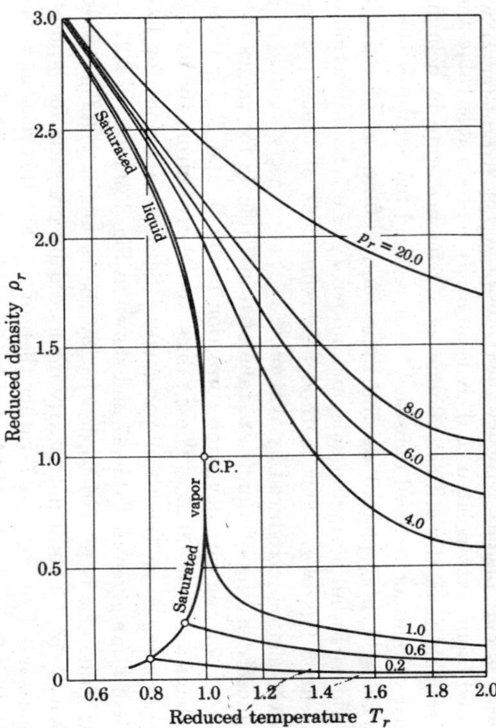

Fig. 139. Reduced densities of gases and liquids; $z_c = 0.27$

At 20° C, 1 atm, $T_{r1} = \dfrac{293.2}{546.1} = 0.537, \qquad p_{r1} = \dfrac{1}{33.0} = 0.0303$

From Table 48, $\rho_{r1} = 3.007$

At 146° C, 30 atm, $T_{r2} = \dfrac{419.2}{546.1} = 0.768, \qquad p_{r2} = \dfrac{30}{33.0} = 0.909$

From Table 48, $\rho_{r2} = 2.518$

At 257° C (saturated), $T_{rs} = \dfrac{530.2}{546.1} = 0.971$

From Table 47, $\rho_{rs} = 1.659$

(*a*) From equation 53,

$$\rho_2 = \left(\frac{\rho_1}{\rho_{r1}}\right)\rho_{r2} = \frac{0.891}{3.007}(2.518) = 0.746 \text{ gram per cc}$$

(*b*)

$$\rho_s = \left(\frac{\rho_1}{\rho_{r1}}\right)\rho_{rs} = \frac{0.891}{3.007}(1.659) = 0.492 \text{ gram per cc}$$

The experimental density is 0.490 gram per cc.

(*c*) The critical density of this liquid in agreement with the first experimental density value is the ratio

$$\frac{\rho_1}{\rho_{r1}} = \frac{0.891}{3.007} = 0.296 \text{ gram per cc}$$

Corresponding value of

$$z'_c = \frac{p_c M}{R T_c \rho_c} = \frac{(33.0)(102.13)}{(82.06)(546.1)(0.296)} = 0.254$$

Generalized Compressibility Factor. From the published pvT data on 82 different compounds, Table 49 on compressibility factor was prepared. In Table 47 values of z_s are given at saturation for both liquid and vapor states, with corresponding values of reduced temperature and pressure at saturation for four values of z_c. For values of p_r greater than 1.2, experimental values were available chiefly for $z_c = 0.27$. Generalized graphs for z covering both gaseous and liquid states are shown in Fig. 140. It will be observed that z values are continuous for all values of T_r above unity. Below the critical isotherm, abrupt discontinuities occur between the two phases.

In the gaseous state, values of z are less than unity for all values of T_r below 2 when reduced pressures are below 8. At values of p_r greater than 8, values of z exceed unity at all temperatures. At p_r equal to about 8, the z isotherms intersect at a value of about unity. The greatest departure of the gas phase from ideality occurs at extremely high pressures and near the critical region where the gas becomes nearly five times as dense as in the ideal gaseous state. For each isotherm, the lowest value of z for the gas phase does not occur at the critical point but at a reduced pressure slightly greater than unity.

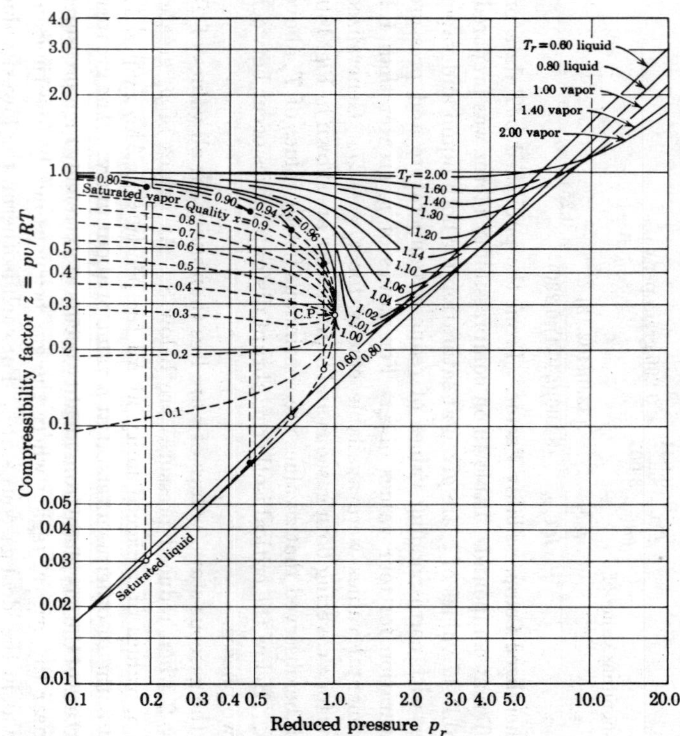

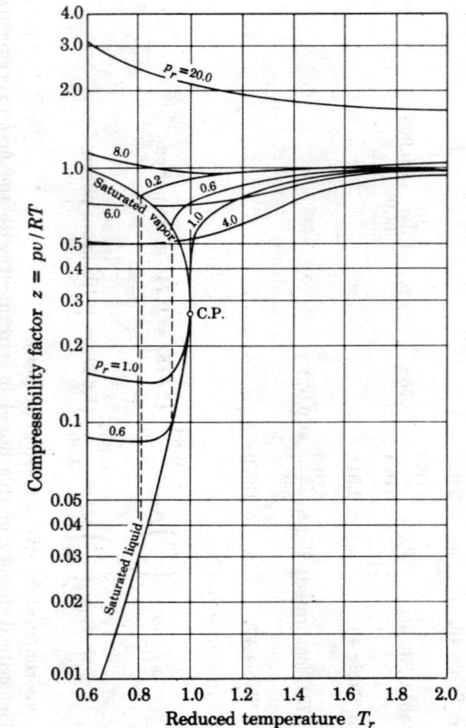

Fig. 140. Compressibility factors of gases and liquids; $z_c = 0.27$

In the liquid state the isotherms for z are nearly linear with the logarithm of pressure. At reduced pressures below 0.1, values of z for the liquid are negligible compared with values of z for the gaseous state. At high values of p_r, values of z for the liquid exceed those of the gas.

The average deviation of experimental values of z from tabulated values varies from zero at low pressures to 2.5% in the critical region and to 2% at high pressures.

Compressibility factors are used directly for calculating pressure, density, or temperature of a gas from any two of these values.

Illustration 5. (*Volume Unknown*). Calculate the volume occupied by 1 lb of methane gas at 500° R and 1015 psia.

$$T_c = 190.7° \text{ K}, \quad p_c = 45.8 \text{ atm}, \quad z_c = 0.290, \quad M = 16.04$$

$$T_r = \frac{500}{(1.8)(190.7)} = 1.457, \qquad p_r = \frac{1015}{(45.8)(14.7)} = 1.508$$

The value of z determined by using values taken from Table 49 and interpolating with respect to z_c, T_r, and p_r, is found to be 0.871.

$$V = \frac{znRT}{p} = \frac{(0.871)(1/16.04)(0.7302)500}{1015/14.70} = 0.287 \text{ cu ft per lb}$$

Illustration 6. (*Pressure Unknown.*) Calculate the pressure in a cylinder having a volume of 360 liters and containing 70 kg of carbon dioxide gas at 62° C.

$$T_c = 304.2° \text{ K}, \quad p_c = 72.9 \text{ atm}, \quad z_c = 0.275, \quad M = 44.01$$

At 62° C, $T_r = 335.2/304.2 = 1.102$

$$z = \frac{pV}{nRT} = \left(\frac{p_cV}{nRT}\right)p_r$$

$$z = \left[\frac{(72.9)360}{(70{,}000/44.01)(0.08205)(335.2)}\right]p_r = 0.600p_r \qquad (a)$$

This equation is used to calculate the z values shown in the second column of Table A. The z values in the third column of the table are obtained using values taken from Table 49, and interpolating with respect to z_c and T_r.

TABLE A

$T_r = 1.102, \qquad z_c = 0.275$

p_r	z from the Equation a	z from Table 49
1.000	0.600	0.704
1.100	0.660	0.659
1.200	0.720	0.625
1.300	0.780	0.582

Two curves are plotted, p_r vs. z, using the two sets of z values from the tabulation above. The two curves intersect at $p_r = 1.099$. Then

$$p = p_cp_r = (72.9)(1.099) = 80.1 \text{ atm}$$

Table 47. Generalized Thermodynamic Properties of Saturated Pure Vapors and Liquids

At Regular Intervals of Reduced Temperature T_{rs} and Reduced Pressure p_{rs}

T_{rs}	p_{rs} (vapor or liquid)				z_s (vapor)				z_s (liquid)			
	Water	$\alpha = 8.08$ $z_c = 0.25$	$\alpha = 6.94$ $z_c = 0.27$	$\alpha = 5.95$ $z_c = 0.29$	$z_c = 0.23$	$z_c = 0.25$	$z_c = 0.27$	$z_c = 0.29$	$z_c = 0.23$	$z_c = 0.25$	$z_c = 0.27$	$z_c = 0.29$
0.50	0.0006	0.00021	0.0010	0.0038	0.996	0.999	0.998	0.996	0.0001	0.001	0.001	0.001
0.55	0.0024	0.00116	0.0038	0.0108	0.989	0.998	0.996	0.989	0.0003	0.002	0.002	0.002
0.60	0.0077	0.0045	0.0113	0.0250	0.978	0.990	0.979	0.982	0.0010	0.002	0.002	0.002
0.65	0.0203	0.0137	0.0279	0.0513	0.959	0.972	0.961	0.949	0.0025	0.006	0.007	0.010
0.70	0.0454	0.0345	0.0590	0.0938	0.929	0.948	0.929	0.912	0.0054	0.009	0.009	0.016
0.75	0.0902	0.0750	0.112	0.158	0.889	0.906	0.891	0.866	0.0105	0.015	0.018	0.027
0.80	0.1637	0.146	0.193	0.249	0.835	0.850	0.840	0.816	0.0200	0.024	0.030	0.042
0.85	0.2784	0.261	0.312	0.370	0.764	0.782	0.776	0.760	0.0321	0.037	0.047	0.062
0.90	0.4430	0.421	0.476	0.543	0.673	0.705	0.701	0.684	0.0522	0.062	0.074	0.091
0.92	0.5258	0.506	0.557	0.607	0.628	0.659	0.661	0.648	0.0639	0.076	0.088	0.105
0.94	0.6202	0.620	0.650	0.690	0.577	0.606	0.608	0.602	0.0780	0.092	0.105	0.122
0.96	0.7328	0.719	0.753	0.783	0.514	0.538	0.545	0.546	0.0994	0.112	0.126	0.144
0.98	0.8561	0.849	0.871	0.887	0.437	0.452	0.464	0.470	0.125	0.143	0.155	0.175
1.00	1.000	1.000	1.000	1.000	0.232	0.250	0.270	0.290	0.232	0.250	0.270	0.290
p_{rs}	T_{rs} (vapor or liquid)				z_s (vapor)				z_s (liquid)			
0.05	0.707	0.719	0.690	0.549	0.921	0.937	0.942	0.944	0.0059	0.007	0.009	0.011
0.10	0.758	0.771	0.740	0.707	0.886	0.895	0.898	0.900	0.0116	0.014	0.015	0.018
0.15	0.792	0.802	0.760	0.746	0.851	0.860	0.864	0.869	0.0172	0.019	0.022	0.025
0.20	0.819	0.826	0.804	0.775	0.820	0.830	0.833	0.839	0.0227	0.028	0.030	0.034
0.25	0.838	0.847	0.825	0.801	0.790	0.805	0.807	0.813	0.0287	0.032	0.037	0.042
0.30	0.858	0.865	0.847	0.823	0.760	0.780	0.783	0.790	0.0347	0.040	0.045	0.051
0.35	0.873	0.881	0.864	0.844	0.730	0.756	0.760	0.766	0.0405	0.046	0.052	0.058
0.40	0.889	0.894	0.879	0.862	0.700	0.732	0.738	0.746	0.0470	0.052	0.060	0.068
0.45	0.902	0.907	0.894	0.878	0.675	0.704	0.713	0.722	0.0535	0.060	0.069	0.076
0.50	0.914	0.919	0.907	0.892	0.650	0.681	0.693	0.698	0.0604	0.069	0.077	0.086
0.55	0.924	0.930	0.918	0.906	0.629	0.652	0.665	0.677	0.0671	0.077	0.088	0.095
0.60	0.936	0.941	0.929	0.919	0.602	0.628	0.641	0.650	0.0749	0.086	0.096	0.103
0.65	0.945	0.948	0.940	0.913	0.571	0.600	0.612	0.624	0.0826	0.094	0.106	0.114
0.70	0.954	0.957	0.950	0.942	0.548	0.570	0.583	0.596	0.0908	0.103	0.114	0.125
0.75	0.963	0.965	0.959	0.952	0.518	0.539	0.553	0.569	0.102	0.113	0.125	0.137
0.80	0.971	0.973	0.967	0.963	0.486	0.505	0.519	0.536	0.110	0.124	0.136	0.150
0.85	0.979	0.980	0.977	0.973	0.450	0.470	0.486	0.503	0.122	0.136	0.148	0.162
0.90	0.986	0.987	0.984	0.983	0.415	0.427	0.443	0.460	0.136	0.152	0.164	0.177
0.95	0.993	0.994	0.993	0.992	0.354	0.378	0.392	0.410	0.159	0.176	0.190	0.202
1.00	1.000	1.000	1.000	1.000	0.232	0.250	0.270	0.290	0.232	0.250	0.270	0.290

Table 47 (*Continued*). Generalized Thermodynamic Properties of Saturated Pure Vapors and Liquids

T_{rs}	ρ_{rs} (vapor)				ρ_{rs} (liquid)				$(f/p)_s$ (vapor or liquid)			
	Water	$z_c = 0.25$	$z_c = 0.27$	$z_c = 0.29$	Water	$z_c = 0.25$	$z_c = 0.27$	$z_c = 0.29$	Water	$z_c = 0.25$	$z_c = 0.27$	$z_c = 0.29$
0.50	0.0005	0.0001	0.0005	0.0029	3.101	3.115	2.937	2.753	0.998	0.999	0.998	0.997
0.55	0.0010	0.0005	0.0025	0.0059	3.047	3.020	2.844	2.666	0.994	0.998	0.997	0.996
0.60	0.0030	0.00295	0.005	0.0110	2.973	2.913	2.746	2.574	0.981	0.996	0.992	0.990
0.65	0.0076	0.00673	0.013	0.0193	2.889	2.800	2.640	2.480	0.960	0.987	0.981	0.977
0.70	0.0162	0.0158	0.027	0.0377	2.775	2.686	2.532	2.374	0.935	0.974	0.965	0.957
0.75	0.0314	0.0331	0.046	0.0665	2.667	2.560	2.411	2.260	0.900	0.950	0.936	0.917
0.80	0.0569	0.0618	0.076	0.109	2.535	2.420	2.284	2.145	0.864	0.902	0.891	0.873
0.85	0.0993	0.107	0.128	0.171	2.370	2.263	2.141	2.010	0.814	0.844	0.837	0.822
0.90	0.1700	0.180	0.208	0.250	2.191	2.076	1.969	1.859	0.765	0.785	0.782	0.778
0.92	0.2113	0.224	0.248	0.294	2.077	1.989	1.890	1.789	0.741	0.763	0.763	0.758
0.94	0.2657	0.280	0.310	0.353	1.965	1.888	1.797	1.707	0.723	0.737	0.740	0.742
0.96	0.3447	0.362	0.394	0.434	1.784	1.765	1.685	1.605	0.696	0.710	0.718	0.720
0.98	0.4644	0.491	0.517	0.558	1.628	1.598	1.535	1.469	0.675	0.683	0.696	0.701
1.00	1.0000	1.000	1.00	1.000	1.000	1.000	1.000	1.000	0.650	0.654	0.665	0.677
p_{rs}	ρ_{rs} (vapor)				ρ_{rs} (liquid)				$(f/p)_s$ (vapor or liquid)			
0.05	0.0178	0.021	0.025	0.0232	2.770	2.66	2.57	2.45	0.934	0.970	0.972	0.974
0.10	0.0340	0.0369	0.040	0.0451	2.651	2.54	2.44	2.34	0.887	0.944	0.946	0.948
0.15	0.0526	0.0551	0.061	0.0667	2.556	2.44	2.36	2.25	0.862	0.913	0.915	0.917
0.20	0.0646	0.0737	0.087	0.0885	2.480	2.37	2.28	2.13	0.840	0.885	0.887	0.890
0.25	0.0858	0.0920	0.100	0.111	2.418	2.30	2.21	2.12	0.821	0.860	0.863	0.866
0.30	0.1082	0.112	0.122	0.133	2.349	2.24	2.15	2.07	0.806	0.838	0.840	0.843
0.35	0.1242	0.133	0.145	0.157	2.297	2.18	2.09	2.02	0.784	0.819	0.821	0.823
0.40	0.1493	0.154	0.167	0.181	2.226	2.13	2.04	1.97	0.772	0.803	0.805	0.807
0.45	0.1716	0.178	0.190	0.206	2.168	2.07	1.99	1.93	0.760	0.787	0.790	0.794
0.50	0.1976	0.202	0.216	0.233	2.105	2.02	1.94	1.88	0.749	0.775	0.777	0.781
0.55	0.2176	0.229	0.241	0.260	2.061	1.98	1.89	1.83	0.737	0.761	0.765	0.771
0.60	0.2525	0.256	0.272	0.291	1.988	1.93	1.85	1.79	0.727	0.749	0.754	0.760
0.65	0.2798	0.288	0.301	0.324	1.935	1.87	1.80	1.74	0.715	0.738	0.742	0.748
0.70	0.3113	0.322	0.340	0.362	1.877	1.82	1.76	1.70	0.705	0.727	0.730	0.738
0.75	0.3567	0.362	0.380	0.401	1.803	1.77	1.71	1.65	0.697	0.716	0.720	0.728
0.80	0.3939	0.409	0.430	0.449	1.734	1.70	1.66	1.59	0.683	0.703	0.710	0.718
0.85	0.4487	0.463	0.483	0.504	1.660	1.64	1.59	1.53	0.673	0.692	0.700	0.708
0.90	0.5107	0.535	0.557	0.578	1.570	1.54	1.50	1.46	0.664	0.680	0.688	0.698
0.95	0.6272	0.633	0.660	0.678	1.445	1.41	1.38	1.36	0.654	0.667	0.677	0.688
1.00	1.0000	1.000	1.000	1.000	1.000	1.00	1.00	1.00	0.650	0.654	0.665	0.677

Table 47 (*Continued*). Generalized Thermodynamic Properties of Saturated Pure Vapors and Liquids

Values are given in Btu/(lb-mole)(R°) or in cal/(g-mole)(K°)

T_{rs}	$\left(\frac{H^*-H}{T_c}\right)_s$ (liquid)				$\left(\frac{H^*-H}{T_c}\right)_s$ (vapor)				$\left(\frac{U^*-U}{T_c}\right)_s$ (vapor)			
	Water	$z_c = 0.25$	$z_c = 0.27$	$z_c = 0.29$	Water	$z_c = 0.25$	$z_c = 0.27$	$z_c = 0.29$	Water	$z_c = 0.25$	$z_c = 0.27$	$z_c = 0.29$
0.50	15.8433	16.75	14.80	11.20	0.0124	0.01	0.01	0.01	0.0082	0.00900	0.008	0.0060
0.55	15.3499	15.30	14.26	10.92	0.0510	0.02	0.02	0.02	0.0393	0.0278	0.024	0.018
0.60	14.8535	14.76	13.74	10.67	0.1237	0.04	0.04	0.04	0.0973	0.0550	0.05	0.0400
0.65	14.3546	14.18	13.13	10.37	0.2458	0.13	0.13	0.13	0.192	0.0938	0.08	0.0741
0.70	13.8141	13.52	12.52	10.10	0.4313	0.25	0.25	0.25	0.333	0.178	0.15	0.128
0.75	13.1920	12.83	11.97	9.78	0.7000	0.50	0.50	0.50	0.468	0.360	0.34	0.300
0.80	12.6778	12.10	11.34	9.43	1.0977	0.72	0.72	0.72	0.780	0.482	0.47	0.428
0.85	12.0226	11.36	10.65	9.00	1.6434	1.25	1.25	1.25	1.245	0.882	0.87	0.845
0.90	11.2719	10.52	9.90	8.48	2.4453	1.93	1.86	1.79	1.861	1.402	1.33	1.22
0.92	10.9441	10.13	9.53	8.24	2.8555	2.29	2.16	2.07	2.176	1.67	1.56	1.43
0.94	10.5530	9.71	9.12	7.97	3.3548	2.74	2.52	2.42	2.564	2.00	1.79	1.68
0.96	10.1170	9.20	8.54	7.65	3.9918	3.36	3.04	2.84	3.065	2.48	2.17	1.97
0.98	9.4723	8.56	7.80	7.16	4.8374	4.23	3.76	3.50	3.741	3.16	2.72	2.47
1.00	7.4780	6.50	5.80	5.40	7.4780	6.50	5.80	5.40	5.953	5.01	4.35	3.99
p_{rs}	$\left(\frac{H^*-H}{T_c}\right)_s$ (liquid)				$\left(\frac{H^*-H}{T_c}\right)_s$ (vapor)				$\left(\frac{U^*-U}{T_c}\right)_s$ (vapor)			
0.05	13.7390	13.37	12.67	10.31	0.4623	0.30	0.22	0.18	0.351	0.211	0.14	0.1062
0.10	13.1933	12.68	12.08	10.04	0.7583	0.55	0.43	0.33	0.610	0.402	0.28	0.188
0.15	12.8432	12.19	11.61	9.80	1.0286	0.73	0.62	0.51	0.820	0.510	0.41	0.315
0.20	12.4800	11.85	11.28	9.60	1.2752	0.92	0.81	0.66	1.02	0.644	0.54	0.410
0.25	12.2239	11.53	10.98	9.45	1.5039	1.10	0.97	0.83	1.183	0.775	0.65	0.531
0.30	11.9119	11.30	10.79	9.29	1.7609	1.30	1.15	1.01	1.336	0.926	0.79	0.665
0.35	11.7419	11.00	10.52	9.13	1.9619	1.50	1.32	1.19	1.500	1.08	0.91	0.797
0.40	11.4744	10.80	10.34	8.93	2.21	1.71	1.52	1.35	1.704	1.24	1.06	0.915
0.45	11.2626	10.55	10.08	8.78	2.4659	1.93	1.72	1.53	1.884	1.40	1.21	1.05
0.50	11.0353	10.31	9.86	8.58	2.7225	2.11	1.91	1.70	2.074	1.53	1.36	1.16
0.55	10.8944	10.12	9.61	8.41	2.9278	2.36	2.10	1.88	2.259	1.72	1.49	1.30
0.60	10.6411	9.85	9.38	8.24	3.2389	2.55	2.28	2.04	2.476	1.86	1.62	1.40
0.65	10.4680	9.70	9.12	8.10	3.4816	2.81	2.50	2.25	2.676	2.06	1.78	1.55
0.70	10.2179	9.46	8.88	7.95	3.7460	3.08	2.74	2.45	2.888	2.27	1.95	1.69
0.75	9.9191	9.21	8.56	7.75	4.0600	3.37	2.96	2.62	3.100	2.49	2.11	1.86
0.80	9.7531	8.97	8.29	7.58	4.4000	3.70	3.24	2.93	3.362	2.75	2.32	2.04
0.85	9.5512	8.72	7.91	7.40	4.7230	4.02	3.55	3.23	3.652	2.99	2.55	2.22
0.90	9.2605	8.46	7.54	7.15	5.1560	4.50	4.02	3.61	4.010	3.38	2.90	2.56
0.95	8.7813	7.92	7.00	6.73	5.8207	5.04	4.52	4.15	4.547	3.81	3.32	2.99
1.00	7.4780	6.50	5.80	5.40	7.4780	6.50	5.80	5.40	5.953	5.01	4.35	3.99

TABLE 47 (*Continued*). GENERALIZED THERMODYNAMIC PROPERTIES OF SATURATED PURE VAPORS AND LIQUIDS

Values are given in Btu/(lb-mole)(R°) or in cal/(g-mole)(K°)

T_r	$(s^* - s)_s$ (vapor)				$(s^* - s)_s$ (liquid)				$\left(\frac{U^* - U}{T_c}\right)_s$ (liquid)			
	Water	I $z_c = 0.25$	II $z_c = 0.27$	III $z_c = 0.29$	Water	I $z_c = 0.25$	II $z_c = 0.27$	III $z_c = 0.29$	Water	I $z_c = 0.25$	II $z_c = 0.27$	III $z_c = 0.29$
0.50	0.0224	0.0180	0.0160	0.0140	31.684	31.50	29.60	22.39	14.850	14.76	13.80	10.21
0.55	0.0800	0.0324	0.0300	0.020	28.020	27.81	25.92	19.85	14.257	14.21	13.17	9.83
0.60	0.1582	0.087	0.081	0.074	24.729	24.59	22.98	17.76	13.663	13.57	12.55	9.48
0.65	0.292	0.176	0.172	0.168	21.993	21.29	20.16	15.91	13.063	12.90	11.85	9.09
0.70	0.481	0.376	0.349	0.320	19.601	19.26	17.81	14.34	12.431	12.14	11.14	8.73
0.75	0.660	0.565	0.536	0.488	17.489	17.01	15.83	12.86	11.717	11.36	10.51	8.33
0.80	1.075	0.699	0.671	0.618	15.565	14.92	13.95	11.51	11.118	10.55	9.79	7.91
0.85	1.540	1.13	1.12	1.06	13.742	13.03	12.[illegible]8	10.17	10.388	9.73	9.04	7.42
0.90	2.184	1.66	1.58	1.49	11.994	11.21	10.51	8.92	9.577	8.84	8.24	6.85
0.92	2.516	1.95	1.81	1.70	11.293	10.47	9.82	8.41	9.233	8.44	7.86	6.60
0.94	2.921	2.31	2.08	1.98	10.585	9.72	9.10	7.89	8.831	8.01	7.45	6.33
0.96	3.440	2.83	2.51	2.31	9.778	8.92	8.24	7.02	8.399	7.51	6.87	6.02
0.98	4.154	3.58	3.14	2.87	8.887	7.95	7.26	6.20	7.768	6.89	6.15	5.55
1.00	6.622	5.66	4.99	3.99	6.622	5.66	4.99	3.99	5.953	5.01	4.35	3.99
p_{r_s}	$(s^* - s)_s$ (vapor)				$(s^* - s)_s$ (liquid)				$\left(\frac{U^* - U}{T_c}\right)_s$ (liquid)			
0.05	0.516	0.363	0.263	0.220	19.300	18.77	18.28	15.50	12.343	11.97	11.31	9.01
0.10	0.751	0.624	0.469	0.357	17.176	16.61	16.15	13.96	11.705	11.19	10.63	8.65
0.15	1.00	0.743	0.618	0.507	15.946	15.21	14.89	12.89	11.297	10.65	10.11	8.35
0.20	1.20	0.884	0.768	0.660	15.080	14.26	13.77	12.01	10.950	10.27	9.73	8.10
0.25	1.359	1.013	0.902	0.745	14.178	13.46	13.02	11.45	10.607	9.91	9.40	7.92
0.30	1.580	1.167	1.01	0.840	13.460	12.85	12.41	10.91	10.266	9.67	9.18	7.73
0.35	1.738	1.333	1.14	1.02	12.967	12.23	11.82	10.42	10.078	9.35	8.90	7.55
0.40	2.009	1.50	1.31	1.14	12.415	11.77	11.37	9.95	9.791	9.13	8.70	7.34
0.45	2.218	1.67	1.46	1.29	11.961	11.27	10.85	9.55	9.566	8.87	8.43	7.17
0.50	2.405	1.81	1.62	1.42	11.487	10.82	10.45	9.13	9.329	8.63	8.21	6.96
0.55	2.600	2.02	1.76	1.56	11.166	10.45	9.95	8.77	9.182	8.43	7.95	6.78
0.60	2.800	2.13	1.89	1.67	10.739	9.99	9.54	8.42	8.921	8.20	7.71	6.60
0.65	3.007	2.33	2.08	1.84	10.405	9.70	9.13	8.13	8.745	8.00	7.45	6.46
0.70	3.212	2.60	2.26	2.00	10.100	9.30	8.73	7.84	8.600	7.76	7.21	6.28
0.75	3.420	2.87	2.44	2.18	9.750	8.98	8.28	7.50	8.260	7.57	6.89	6.12
0.80	3.7[illegible]4	3.12	2.67	2.39	9.366	8.56	7.89	7.22	8.099	7.28	6.63	5.95
0.85	4.090	3.38	2.93	2.63	9.044	8.19	7.40	6.92	7.842	7.04	6.26	5.78
0.90	4.520	3.80	3.25	2.90	8.600	7.82	6.93	6.57	7.568	6.70	5.91	5.54
0.95	5.034	4.27	3.78	3.28	7.978	7.17	6.28	6.05	7.121	6.28	5.40	5.16
1.00	6.622	5.66	4.99	3.99	6.622	5.66	4.99	3.99	5.953	5.01	4.35	3.99

TABLE 48. REDUCED DENSITY OF LIQUIDS ρ_r

T_r	Saturated State			$p_r = 1.0$			$p_r = 2.0$			$p_r = 4.0$			$p_r = 6.0$		
	D_b	ρ_r	D_a	D_b	ρ_r	D_a	D_b	ρ_r	D_a	D_b	ρ_r	D_a	D_b	ρ_r	D_a
0.30	−10.0	3.287	−10.3	−10.0	3.290	−10.3	−10.0	3.294	−10.3	−10.0	3.300	−10.3	−10.0	3.305	−10.3
0.34	−9.8	3.223	−10.1	−9.8	3.227	−10.1	−9.8	3.231	−10.1	−9.8	3.240	−10.1	−9.8	3.245	−10.1
0.38	−9.6	3.156	−9.8	−9.6	3.162	−9.8	−9.6	3.170	−9.9	−9.6	3.180	−9.9	−9.6	3.185	−9.9
0.42	−9.3	3.084	−9.6	−9.3	3.090	−9.6	−9.4	3.099	−9.7	−9.4	3.112	−9.7	−9.4	3.117	−9.7
0.46	−9.1	3.012	−9.4	−9.1	3.020	−9.4	−9.2	3.031	−9.4	−9.2	3.047	−9.5	−9.2	3.054	−9.5
0.50	−8.9	2.937	−9.2	−8.9	2.947	−9.2	−8.9	2.957	−9.2	−9.0	2.975	−9.3	−9.0	2.990	−9.3
0.54	−8.7	2.862	−8.9	−8.7	2.875	−8.9	−8.7	2.888	−9.0	−8.8	2.911	−9.1	−8.9	2.926	−9.1
0.58	−8.4	2.787	−8.7	−8.4	2.800	−8.7	−8.5	2.823	−8.8	−8.6	2.847	−8.9	−8.7	2.870	−8.9
0.62	−8.2	2.704	−8.4	−8.2	2.723	−8.4	−8.3	2.749	−8.5	−8.4	2.776	−8.7	−8.5	2.809	−8.7
0.66	−7.9	2.622	−8.2	−8.0	2.640	−8.2	−8.1	2.674	−8.3	−8.2	2.712	−8.5	−8.3	2.749	−8.6
0.70	−7.7	2.532	−7.9	−7.8	2.562	−7.9	−7.9	2.599	−8.0	−8.0	2.642	−8.2	−8.1	2.686	−8.3
0.74	−7.3	2.438	−7.6	−7.5	2.471	−7.7	−7.6	2.512	−7.9	−7.8	2.573	−8.0	−7.9	2.622	−8.1
0.78	−7.0	2.336	−7.3	−7.2	2.378	−7.5	−7.4	2.423	−7.7	−7.6	2.500	−7.8	−7.7	2.554	−8.0
0.80	−6.8	2.284	−7.1	−7.0	2.329	−7.3	−7.2	2.377	−7.5	−7.4	2.460	−7.7	−7.6	2.520	−7.9
0.82	−6.6	2.231	−6.8	−6.8	2.280	−7.1	−7.0	2.330	−7.3	−7.3	2.420	−7.5	−7.5	2.487	−7.8
0.84	−6.3	2.171	−6.6	−6.4	2.224	−6.9	−6.7	2.281	−7.2	−6.9	2.380	−7.4	−7.3	2.449	−7.7
0.86	−6.0	2.107	−6.3	−6.0	2.161	−6.4	−6.4	2.231	−7.7	−6.6	2.340	−7.0	−7.1	2.416	−7.4
0.88	−5.6	2.043	−6.0	−5.6	2.098	−6.0	−6.0	2.177	−6.3	−6.3	2.299	−6.7	−6.8	2.380	−7.1
0.90	−5.3	1.969	−5.7	−5.3	2.027	−5.7	−5.6	2.122	−5.9	−6.0	2.257	−6.4	−6.6	2.344	−6.9
0.91	−5.1	1.932	−5.4	−5.1	1.990	−5.4	−5.3	2.092	−5.6	−5.8	2.235	−6.2	−6.5	2.325	−6.7
0.92	−4.9	1.890	−5.2	−4.9	1.948	−5.2	−5.1	2.064	−5.3	−5.7	2.214	−6.1	−6.4	2.307	−6.6
0.93	−4.7	1.846	−5.0	−4.7	1.904	−5.0	−4.8	2.033	−5.1	−5.5	2.191	−6.0	−6.3	2.288	−6.5
0.94	−4.5	1.797	−4.7	−4.5	1.855	−4.7	−4.6	2.001	−4.8	−5.4	2.168	−5.7	−6.2	2.268	−6.4
0.95	−4.2	1.745	−4.4	−4.2	1.803	−4.4	−4.3	1.965	−4.5	−5.2	2.145	−5.4	−6.1	2.249	−6.3
0.96	−4.0	1.685	−4.0	−4.0	1.743	−4.0	−4.0	1.931	−4.0	−5.0	2.120	−5.2	−6.0	2.229	−6.2
0.97	−3.4	1.617	−3.6	−3.5	1.667	−3.6	−3.6	1.892	−3.6	−4.8	2.095	−5.1	−5.9	2.208	−6.2
0.98	−3.1	1.535	−3.1	−3.1	1.580	−3.1	−3.2	1.852	−2.8	−4.7	2.072	−4.8	−5.8	2.188	−6.0
0.99	−1.5	1.420	−2.7	−1.5	1.450	−2.6	−1.6	1.810	−2.1	−4.6	2.043	−4.5	−5.6	2.165	−5.9
1.00	−0.0	1.000	0.0	0.0	1.000	0.0	0.0	1.764	−2.0	−4.5	2.016	−4.4	−5.5	2.143	−5.8

Values of ρ_r are recorded for $z_c = 0.27$, At other values of z_c, $\rho'_r = \rho_r + D(z_c - 0.27)$.

TABLE 48 (*Continued*). REDUCED DENSITY OF LIQUIDS ρ_r

T_r	$p_r = 10.0$			$p_r = 15.0$			$p_r = 20.0$			$p_r = 25.0$			$p_r = 30.0$		
	D_b	ρ_r	D_a	D_b	ρ_r	D_a	D_b	ρ_r	D_a	D_b	ρ_r	D_a	D_b	ρ_r	D_a
0.30	−10.0	3.320	−10.4	−10.1	3.325	−10.4	−10.1	3.333	−10.4	−10.1	3.337	−10.5	−10.1	3.343	−10.5
0.34	−9.8	3.255	−10.2	−9.9	3.265	−10.2	−9.9	3.275	−10.2	−9.9	3.282	−10.3	−9.9	3.289	−10.3
0.38	−9.7	3.195	−10.0	−9.7	3.206	−10.0	−9.7	3.215	−10.0	−9.8	3.225	−10.0	−9.8	3.233	−10.0
0.42	−9.5	3.135	−9.8	−9.5	3.147	−9.8	−9.5	3.158	−9.8	−9.6	3.168	−9.8	−9.6	3.177	−9.8
0.46	−9.3	3.075	−9.6	−9.4	3.090	−9.6	−9.4	3.100	−9.6	−9.4	3.113	−9.7	−9.4	3.121	−9.7
0.50	−9.1	3.015	−9.4	−9.2	3.030	−9.5	−9.2	3.041	−9.5	−9.2	3.059	−9.6	−9.2	3.066	−9.6
0.54	−8.9	2.955	−9.2	−9.0	2.973	−9.3	−9.0	2.986	−9.3	−9.1	3.004	−9.4	−9.1	3.015	−9.4
0.58	−8.7	2.896	−9.1	−8.8	2.916	−9.1	−8.8	2.931	−9.2	−8.9	2.951	−9.2	−8.9	2.965	−9.3
0.62	−8.6	2.841	−8.9	−8.7	2.862	−8.9	−8.7	2.881	−9.0	−8.8	2.900	−9.0	−8.8	2.919	−9.1
0.66	−8.4	2.782	−8.7	−8.5	2.811	−8.8	−8.5	2.833	−8.9	−8.6	2.855	−8.9	−8.7	2.875	−9.0
0.70	−8.2	2.730	−8.5	−8.4	2.760	−8.6	−8.4	2.786	−8.7	−8.5	2.810	−8.7	−8.6	2.833	−8.8
0.74	−8.0	2.675	−8.3	−8.2	2.710	−8.4	−8.3	2.740	−8.5	−8.3	2.765	−8.6	−8.4	2.794	−8.7
0.78	−7.9	2.617	−8.2	−8.0	2.659	−8.3	−8.2	2.695	−8.4	−8.2	2.722	−8.5	−8.3	2.753	−8.6
0.80	−7.8	2.587	−8.1	−7.9	2.634	−8.2	−8.1	2.673	−8.3	−8.1	2.702	−8.4	−8.2	2.734	−8.5
0.82	−7.7	2.558	−8.0	−7.9	2.609	−8.1	−8.0	2.650	−8.2	−8.0	2.683	−8.3	−8.2	2.715	−8.4
0.84	−7.5	2.527	−7.9	−7.7	2.583	−8.0	−7.8	2.628	−8.2	−8.0	2.664	−8.3	−8.1	2.698	−8.4
0.86	−7.3	2.496	−7.6	−7.5	2.559	−7.8	−7.7	2.605	−8.0	−8.0	2.645	−8.2	−8.1	2.680	−8.3
0.88	−7.1	2.465	−7.4	−7.4	2.532	−7.7	−7.6	2.585	−7.9	−7.9	2.626	−8.2	−8.0	2.669	−8.2
0.90	−7.0	2.434	−7.2	−7.3	2.506	−7.6	−7.6	2.563	−7.8	−7.9	2.608	−8.1	−8.0	2.647	−8.2
0.91	−6.9	2.418	−7.1	−7.2	2.493	−7.5	−7.5	2.552	−7.7	−7.8	2.598	−8.1	−8.0	2.638	−8.2
0.92	−6.9	2.402	−7.0	−7.2	2.481	−7.4	−7.5	2.541	−7.7	−7.8	2.588	−8.1	−8.0	2.630	−8.2
0.93	−6.8	2.387	−7.0	−7.2	2.470	−7.4	−7.4	2.531	−7.7	−7.8	2.579	−8.1	−7.9	2.622	−8.2
0.94	−6.7	2.370	−6.9	−7.2	2.457	−7.3	−7.4	2.521	−7.6	−7.8	2.570	−8.0	−7.9	2.614	−8.1
0.95	−6.6	2.355	−6.8	−7.1	2.445	−7.3	−7.4	2.511	−7.6	−7.7	2.561	−8.0	−7.8	2.605	−8.1
0.96	−6.6	2.338	−6.6	−7.1	2.433	−7.2	−7.4	2.500	−7.5	−7.7	2.551	−7.9	−7.8	2.596	−8.1
0.97	−6.5	2.322	−6.5	−7.0	2.420	−7.2	−7.3	2.489	−7.5	−7.7	2.542	−7.9	−7.8	2.588	−8.1
0.98	−6.4	2.306	−6.4	−7.0	2.409	−7.1	−7.3	2.480	−7.5	−7.7	2.532	−7.9	−7.8	2.580	−8.0
0.99	−6.3	2.290	−6.3	−6.9	2.396	−7.1	−7.2	2.467	−7.5	−7.6	2.523	−7.9	−7.8	2.571	−8.0
1.00	−6.3	2.274	−6.2	−6.9	2.383	−7.1	−7.2	2.457	−7.4	−7.6	2.514	−7.8	−7.8	2.563	−8.0

Values of ρ_r are recorded for $z_c = 0.27$. At other values of z_c, $\rho'_r = \rho_r + D(z_c - 0.27)$.

TABLE 49. COMPRESSIBILITY FACTORS OF PURE GASES AND LIQUIDS, z

	$p_r = 0.01$			$p_r = 0.05$			$p_r = 0.10$			$p_r = 0.2$			$p_r = 0.3$		
T_{rs}		0.59			0.690			0.740			0.804			0.847	
z sat. gas		0.985			0.942			0.898			0.833			0.783	
z sat. liquid		0.002			0.009			0.015			0.030			0.045	
T_r	D_b	z	D_a	D_b	z	D_a	D_b	z	D_a	D_b	z	D_a	D_b	z	D_a
0.50	0.01	0.002	0.01	0.05	0.009	0.07	0.11	0.0184	0.14	0.22	0.0367	0.27	0.35	0.0551	0.40
0.60	0.20	0.990	0.02	0.05	0.008	0.07	0.10	0.0164	0.12	0.20	0.0328	0.25	0.31	0.0491	0.37
0.70	0.07	0.992	0.02	0.33	0.943	0.35	0.09	0.0152	0.12	0.19	0.0304	0.23	0.29	0.0456	0.34
0.80	0.01	0.993	0.02	0.13	0.960	0.18	0.28	0.920	0.40	0.18	0.0295	0.20	0.28	0.0441	0.31
0.90	0.01	0.994	0.02	0.07	0.973	0.10	0.14	0.947	0.20	0.28	0.890	0.36	0.44	0.825	0.50
0.92	0.01	0.995	0.02	0.07	0.975	0.10	0.13	0.951	0.19	0.26	0.900	0.34	0.40	0.840	0.47
0.94	0.01	0.995	0.02	0.06	0.977	0.10	0.12	0.954	0.18	0.24	0.908	0.33	0.37	0.854	0.44
0.96	0.01	0.995	0.02	0.05	0.978	0.09	0.11	0.958	0.17	0.22	0.915	0.30	0.33	0.868	0.40
0.98	0.01	0.996	0.02	0.05	0.980	0.09	0.10	0.961	0.16	0.21	0.921	0.28	0.31	0.879	0.37
1.00	0.01	0.996	0.02	0.04	0.982	0.09	0.10	0.964	0.15	0.21	0.927	0.24	0.28	0.889	0.34
1.01	0.01	0.996	0.02	0.04	0.983	0.08	0.10	0.966	0.15	0.20	0.930	0.24	0.26	0.894	0.33
1.02	0.01	0.996	0.02	0.04	0.983	0.08	0.10	0.967	0.15	0.19	0.933	0.23	0.25	0.897	0.34
1.03	0.01	0.996	0.02	0.04	0.984	0.08	0.09	0.968	0.14	0.18	0.935	0.22	0.24	0.902	0.32
1.04	0.01	0.996	0.02	0.04	0.985	0.08	0.09	0.970	0.14	0.18	0.938	0.21	0.24	0.905	0.29
1.05	0.00	0.996	0.02	0.04	0.985	0.08	0.08	0.971	0.14	0.17	0.940	0.20	0.23	0.909	0.28
1.06	0.00	0.996	0.02	0.04	0.986	0.08	0.08	0.972	0.14	0.17	0.942	0.20	0.22	0.913	0.26
1.07	0.00	0.996	0.02	0.04	0.986	0.07	0.08	0.973	0.14	0.16	0.944	0.19	0.21	0.916	0.25
1.08	0.00	0.996	0.02	0.04	0.987	0.07	0.08	0.974	0.13	0.16	0.946	0.18	0.20	0.918	0.24
1.09	0.00	0.997	0.01	0.04	0.987	0.07	0.07	0.975	0.12	0.15	0.948	0.17	0.19	0.922	0.24
1.10	0.00	0.997	0.01	0.04	0.988	0.07	0.07	0.976	0.12	0.14	0.950	0.17	0.18	0.924	0.21
1.12	0.00	0.997	0.01	0.04	0.988	0.06	0.06	0.977	0.12	0.13	0.953	0.16	0.17	0.928	0.20
1.14	0.00	0.997	0.01	0.03	0.989	0.06	0.06	0.979	0.11	0.12	0.956	0.14	0.16	0.933	0.19
1.16	0.00	0.997	0.01	0.03	0.990	0.05	0.06	0.980	0.09	0.12	0.960	0.13	0.14	0.937	0.16
1.18	0.00	0.997	0.01	0.03	0.991	0.04	0.06	0.982	0.09	0.12	0.962	0.12	0.12	0.942	0.15
1.20	0.00	0.998	0.01	0.03	0.991	0.03	0.06	0.983	0.07	0.09	0.965	0.10	0.11	0.945	0.13
1.30	0.00	0.998	0.01	0.03	0.993	0.02	0.04	0.987	0.05	0.07	0.974	0.08	0.07	0.960	0.10
1.40	0.00	0.998	0.00	0.02	0.995	0.01	0.03	0.990	0.03	0.05	0.982	0.05	0.06	0.971	0.07
1.50	0.00	0.999	0.00	0.01	0.995	0.01	0.01	0.991	0.02	0.03	0.986	0.03	0.02	0.980	0.04
1.60	0.00	0.999	0.00	0.01	0.996	0.00	0.00	0.992	0.00	0.01	0.988	0.02	0.01	0.986	0.02
1.70	0.00	0.999	0.00	0.00	0.996	0.00	0.00	0.992	0.00	0.00	0.989	0.01	0.00	0.989	0.01
1.80	0.00	0.999	0.00	0.00	0.996	0.00	0.00	0.993	0.00	0.00	0.991	0.01	0.00	0.991	0.01
1.90	0.00	1.000	0.00	0.00	0.996	0.00	0.00	0.993	0.00	0.00	0.992	0.00	0.00	0.993	0.00
2.00		1.000			0.997	0.00	0.00	0.994	0.00	0.00	0.994	0.00	0.00	0.995	

Values of z are recorded for $z_c = 0.27$. At other values of z_c, $z' = z + D(z_c - 0.27)$.

Table 49 (*Continued*). Compressibility Factors of Pure Gases and Liquids, z

T_{rs} z sat. gas z sat. liquid	$p_r = 0.4$			$p_r = 0.5$			$p_r = 0.6$			$p_r = 0.7$			$p_r = 0.8$		
		0.879 0.738 0.060			0.909 0.693 0.077			0.929 0.641 0.096			0.950 0.583 0.114			0.967 0.519 0.136	
T_r	D_b	z	D_a	D_b	z	D_a	D_b	z	D_a	D_b	z	D_a	D_b	z	D_a
0.50	0.46	0.073[illegible]	0.53	0.57	0.0918	0.66	0.70	0.110	0.81	0.81	0.128	0.95	0.93	0.147	1.07
0.60	0.41	0.065[illegible]	0.49	0.52	0.0817	0.60	0.63	0.0980	0.71	0.74	0.113	0.82	0.84	0.130	0.95
0.70	0.39	0.0605	0.45	0.49	0.0758	0.55	0.59	0.0906	0.65	0.69	0.106	0.77	0.79	0.121	0.88
0.80	0.37	0.0583	0.40	0.47	0.0735	0.52	0.57	0.0879	0.62	0.66	0.102	0.73	0.76	0.116	0.85
0.90	0.73	0.763	0.63	0.45	0.0761	0.50	0.55	0.0908	0.60	0.64	0.105	0.71	0.74	0.120	0.82
0.92	0.60	0.783	0.59	0.81	0.710	0.70	0.55	0.0929	0.60	0.65	0.108	0.70	0.74	0.122	0.82
0.94	0.50	0.800	0.55	0.63	0.735	0.64	0.77	0.660	0.73	0.65	0.111	0.70	0.74	0.126	0.82
0.96	0.44	0.817	0.51	0.53	0.760	0.59	0.65	0.700	0.67	0.75	0.613	0.76	0.76	0.133	0.82
0.98	0.39	0.832	0.47	0.46	0.781	0.54	0.54	0.729	0.62	0.62	0.665	0.68	0.70	0.580	0.76
1.00	0.34	0.845	0.42	0.41	0.800	0.48	0.47	0.755	0.54	0.52	0.704	0.60	0.60	0.636	0.65
1.01	0.33	0.852	0.42	0.38	0.809	0.47	0.44	0.765	0.51	0.50	0.718	0.56	0.55	0.659	0.61
1.02	0.30	0.858	0.39	0.36	0.817	0.44	0.41	0.775	0.48	0.45	0.732	0.52	0.50	0.678	0.56
1.03	0.29	0.863	0.37	0.34	0.825	0.42	0.38	0.786	0.46	0.42	0.745	0.50	0.46	0.696	0.54
1.04	0.28	0.869	0.34	0.32	0.832	0.38	0.35	0.794	0.40	0.39	0.755	0.44	0.43	0.710	0.46
1.05	0.27	0.873	0.30	0.30	0.838	0.33	0.33	0.802	0.35	0.36	0.765	0.38	0.39	0.723	0.39
1.06	0.26	0.878	0.29	0.29	0.845	0.32	0.31	0.810	0.33	0.34	0.773	0.35	0.35	0.735	0.36
1.07	0.25	0.883	0.27	0.27	0.850	0.28	0.29	0.817	0.30	0.32	0.781	0.31	0.33	0.745	0.33
1.08	0.24	0.886	0.26	0.26	0.856	0.27	0.28	0.824	0.28	0.30	0.790	0.28	0.31	0.755	0.29
1.09	0.22	0.890	0.25	0.24	0.862	0.25	0.26	0.830	0.25	0.23	0.798	0.25	0.28	0.764	0.25
1.10	0.21	0.89[illegible]	0.22	0.23	0.867	0.22	0.24	0.836	0.22	0.25	0.805	0.22	0.25	0.773	0.23
1.12	0.19	0.900	0.20	0.21	0.876	0.20	0.21	0.848	0.20	0.22	0.818	0.20	0.22	0.789	0.20
1.14	0.18	0.907	0.20	0.18	0.884	0.20	0.19	0.859	0.20	0.19	0.830	0.20	0.19	0.803	0.20
1.16	0.15	0.913	0.19	0.16	0.891	0.20	0.17	0.868	0.20	0.17	0.842	0.20	0.17	0.816	0.18
1.18	0.13	0.918	0.16	0.14	0.898	0.17	0.15	0.877	0.18	0.15	0.852	0.18	0.15	0.830	0.17
1.20	0.12	0.924	0.15	0.13	0.905	0.15	0.13	0.885	0.15	0.14	0.862	0.14	0.14	0.841	0.15
1.30	0.09	0.944	0.11	0.10	0.931	0.11	0.10	0.916	0.11	0.10	0.900	0.11	0.10	0.888	0.12
1.40	0.06	0.959	0.08	0.07	0.949	0.08	0.07	0.937	0.09	0.07	0.928	0.09	0.07	0.920	0.09
1.50	0.04	0.970	0.05	0.04	0.963	0.06	0.05	0.952	0.07	0.05	0.948	0.07	0.05	0.945	0.07
1.60	0.02	0.978	0.03	0.02	0.973	0.04	0.03	0.965	0.05	0.03	0.964	0.06	0.03	0.960	0.06
1.70	0.01	0.983	0.02	0.01	0.980	0.03	0.02	0.974	0.03	0.02	0.974	0.03	0.02	0.970	0.04
1.80	0.00	0.987	0.02	0.00	0.985	0.02	0.01	0.982	0.02	0.01	0.982	0.02	0.01	0.980	0.02
1.90	0.00	0.991	0.01	0.00	0.989	0.01	0.00	0.987	0.01	0.00	0.987	0.01	0.00	0.987	0.02
2.00	0.00	0.994	0.00	0.00	0.993	0.00	0.00	0.992	0.01	0.00	0.992	0.01	0.00	0.989	0.02

Values of z are recorded for $z_c = 0.27$. At other values of z_c, $z' = z + D(z_c - 0.27)$.

Table 49 (*Continued*). Compressibility Factors of Pure Gases and Liquids, z

	$p_r = 0.9$			$p_r = 1.0$			$p_r = 1.05$			$p_r = 1.1$			$p_r = 1.2$		
T_{rs}		0.984			1.000										
z sat. gas		0.443			0.270										
z sat. liquid		0.164			0.270										
T_r	D_b	z	D_a	D_b	z	D_a	D_b	z	D_a	D_b	z	D_a	D_b	z	D_a
0.50	1.05	0.165	1.20	1.17	0.183	1.35	1.22	0.192	1.40	1.28	0.201	1.48	1.40	0.220	1.62
0.60	0.95	0.147	1.05	1.05	0.163	1.17	1.11	0.171	1.23	1.16	0.179	1.28	1.27	0.195	1.39
0.70	0.90	0.136	0.99	1.00	0.151	1.10	1.05	0.158	1.15	1.10	0.165	1.20	1.20	0.180	1.30
0.80	0.86	0.131	0.95	0.95	0.145	1.05	1.00	0.152	1.10	1.05	0.159	1.15	1.15	0.173	1.25
0.90	0.83	0.134	0.92	0.92	0.148	1.02	0.97	0.155	1.07	1.01	0.162	1.11	1.10	0.176	1.20
0.92	0.83	0.137	0.92	0.92	0.151	1.02	0.97	0.158	1.06	1.01	0.165	1.11	1.10	0.179	1.19
0.94	0.83	0.141	0.92	0.93	0.155	1.01	0.98	0.162	1.06	1.01	0.169	1.10	1.10	0.183	1.18
0.96	0.85	0.147	0.92	0.94	0.161	1.01	0.99	0.169	1.05	1.03	0.176	1.09	1.13	0.189	1.17
0.98	0.87	0.161	0.92	0.97	0.174	1.00	1.02	0.182	1.05	1.07	0.189	1.09	1.16	0.202	1.16
1.00	0.70	0.520	0.82	1.00	0.270	1.00	1.06	0.230	1.05	1.14	0.224	1.09	1.20	0.220	1.15
1.01	0.60	0.568	0.68	0.65	0.424	0.75	0.67	0.365	0.79	0.68	0.256	0.83	0.70	0.242	0.88
1.02	0.55	0.600	0.62	0.58	0.509	0.67	0.59	0.447	0.70	0.60	0.374	0.73	0.60	0.295	0.77
1.03	0.49	0.627	0.57	0.51	0.555	0.60	0.52	0.505	0.62	0.52	0.461	0.63	0.52	0.369	0.66
1.04	0.44	0.642	0.49	0.45	0.585	0.51	0.45	0.546	0.53	0.45	0.505	0.54	0.45	0.422	0.56
1.05	0.40	0.670	0.41	0.41	0.611	0.43	0.41	0.577	0.44	0.41	0.541	0.45	0.41	0.478	0.46
1.06	0.37	0.687	0.38	0.38	0.633	0.39	0.38	0.603	0.40	0.38	0.568	0.40	0.38	0.517	0.40
1.07	0.34	0.700	0.34	0.35	0.654	0.35	0.35	0.627	0.35	0.35	0.594	0.35	0.35	0.548	0.36
1.08	0.32	0.715	0.29	0.32	0.671	0.30	0.32	0.647	0.30	0.32	0.616	0.30	0.32	0.573	0.30
1.09	0.28	0.726	0.25	0.28	0.686	0.25	0.28	0.662	0.25	0.28	0.637	0.25	0.28	0.600	0.25
1.10	0.25	0.738	0.23	0.25	0.700	0.23	0.25	0.678	0.23	0.25	0.655	0.23	0.25	0.620	0.23
1.12	0.22	0.756	0.21	0.22	0.723	0.21	0.22	0.704	0.21	0.22	0.686	0.21	0.22	0.654	0.21
1.14	0.19	0.773	0.20	0.19	0.745	0.20	0.19	0.731	0.20	0.19	0.712	0.20	0.19	0.683	0.20
1.16	0.18	0.790	0.18	0.18	0.764	0.18	0.18	0.750	0.18	0.18	0.735	0.18	0.18	0.707	0.20
1.18	0.17	0.805	0.17	0.17	0.780	0.17	0.17	0.771	0.17	0.17	0.756	0.17	0.17	0.730	0.18
1.20	0.15	0.818	0.15	0.15	0.795	0.15	0.15	0.787	0.15	0.15	0.775	0.15	0.15	0.751	0.15
1.30	0.11	0.874	0.12	0.11	0.857	0.12	0.11	0.849	0.12	0.11	0.841	0.13	0.11	0.827	0.13
1.40	0.07	0.912	0.10	0.07	0.899	0.10	0.07	0.890	0.10	0.07	0.888	0.10	0.07	0.875	0.10
1.50	0.05	0.938	0.08	0.05	0.927	0.08	0.05	0.922	0.08	0.05	0.918	0.08	0.05	0.911	0.08
1.60	0.03	0.955	0.06	0.03	0.948	0.07	0.03	0.944	0.07	0.03	0.940	0.08	0.03	0.935	0.08
1.70	0.01	0.968	0.04	0.02	0.964	0.05	0.02	0.958	0.05	0.03	0.956	0.06	0.03	0.951	0.07
1.80	0.00	0.976	0.03	0.00	0.974	0.03	0.00	0.968	0.04	0.00	0.968	0.05	0.02	0.963	0.06
1.90	0.00	0.985	0.02	0.00	0.983	0.02	0.00	0.978	0.02	0.00	0.978	0.03	0.01	0.974	0.05
2.00		0.990	0.02	0.00	0.988	0.02	0.00	0.986	0.02	0.00	0.984	0.03	0.01	0.981	0.03

Values of z are recorded for $z_c = 0.27$. At other values of z_c, $z' = z + D(z_c - 0.27)$.

TABLE 49 (*Continued*). COMPRESSIBILITY FACTORS OF PURE GASES AND LIQUIDS, z

T_r	$p_r = 1.4$ z	$p_r = 1.6$ z	$p_r = 1.8$ z	$p_r = 2.0$ z	$p_r = 4.0$ z	$p_r = 6.0$ z	$p_r = 8.0$ z	$p_r = 10.0$ z	$p_r = 20.0$ z	$p_r = 30.0$ z
0.50	0.256	0.293	0.329	0.365	0.726	1.083	1.439	1.791	3.551	5.28
0.60	0.227	0.259	0.291	0.323	0.640	0.952	1.262	1.568	3.098	4.59
0.70	0.210	0.239	0.268	0.297	0.584	0.862	1.139	1.413	2.769	4.08
0.80	0.201	0.229	0.257	0.284	0.549	0.804	1.056	1.305	2.525	3.70
0.90	0.203	0.230	0.257	0.283	0.532	0.768	1.005	1.233	2.341	3.40
0.92	0.206	0.233	0.259	0.284	0.530	0.763	0.997	1.222	2.310	3.35
0.94	0.210	0.237	0.262	0.287	0.530	0.760	0.991	1.201	2.278	3.30
0.96	0.217	0.242	0.267	0.291	0.531	0.757	0.985	1.202	2.250	3.25
0.98	0.228	0.253	0.276	0.298	0.532	0.755	0.980	1.195	2.224	3.20
1.00	0.234	0.254	0.279	0.306	0.536	0.756	0.975	1.193	2.200	3.15
1.01	0.246	0.262	0.287	0.312	0.538	0.757	0.974	1.188	2.188	3.14
1.02	0.264	0.276	0.296	0.318	0.540	0.758	0.973	1.184	2.175	3.11
1.03	0.288	0.289	0.307	0.326	0.543	0.759	0.972	1.181	2.164	3.08
1.04	0.323	0.305	0.317	0.333	0.546	0.760	0.972	1.177	2.153	3.06
1.05	0.366	0.323	0.332	0.341	0.548	0.761	0.972	1.174	2.142	3.04
1.06	0.403	0.347	0.347	0.351	0.552	0.762	0.971	1.171	2.130	3.02
1.07	0.438	0.370	0.365	0.361	0.554	0.763	0.970	1.168	2.119	3.00
1.08	0.472	0.396	0.380	0.372	0.558	0.764	0.970	1.165	2.109	2.96
1.09	0.507	0.424	0.398	0.386	0.562	0.766	0.970	1.162	2.098	2.95
1.10	0.534	0.455	0.416	0.400	0.565	0.768	0.970	1.160	2.088	2.93
1.12	0.577	0.505	0.454	0.432	0.572	0.772	0.970	1.156	2.068	2.89
1.14	0.315	0.549	0.494	0.466	0.581	0.776	0.970	1.153	2.049	2.85
1.16	0.347	0.588	0.540	0.503	0.589	0.780	0.972	1.151	2.030	2.81
1.18	0.377	0.622	0.583	0.542	0.599	0.786	0.973	1.150	2.013	2.78
1.20	0.705	0.653	0.620	0.573	0.609	0.792	0.975	1.148	1.995	2.74
1.30	0.795	0.768	0.742	0.716	0.687	0.824	0.984	1.144	1.921	2.63
1.40	0.355	0.837	0.819	0.801	0.763	0.863	0.996	1.144	1.862	2.56
1.50	0.394	0.882	0.869	0.852	0.813	0.893	1.012	1.146	1.818	2.49
1.60	0.923	0.914	0.904	0.888	0.852	0.918	1.028	1.150	1.790	2.44
1.70	0.945	0.934	0.929	0.915	0.883	0.940	1.041	1.154	1.767	2.39
1.80	0.960	0.950	0.946	0.935	0.909	0.960	1.052	1.156	1.744	2.33
1.90	0.972	0.965	0.960	0.952	0.932	0.977	1.061	1.158	1.714	2.29
2.00	0.979	0.974	0.971	0.966	0.952	0.993	1.070	1.159	1.691	2.24
3.00	1.000	0.997	0.995	0.986	0.990	1.008	1.068	1.130	1.500	1.84
4.00	1.000	1.000	0.997	0.992	1.000	1.014	1.065	1.120	1.400	1.66
6.00	1.004	1.003	1.000	1.000	1.013	1.024	1.064	1.100	1.300	1.50
8.00	1.008	1.008	1.005	1.005	1.016	1.030	1.063	1.085	1.250	1.40
10.00	1.010	1.010	1.008	1.010	1.020	1.035	1.062	1.080	1.185	1.30
15.00	1.020	1.020	1.020	1.020	1.030	1.045	1.061	1.070	1.140	1.20

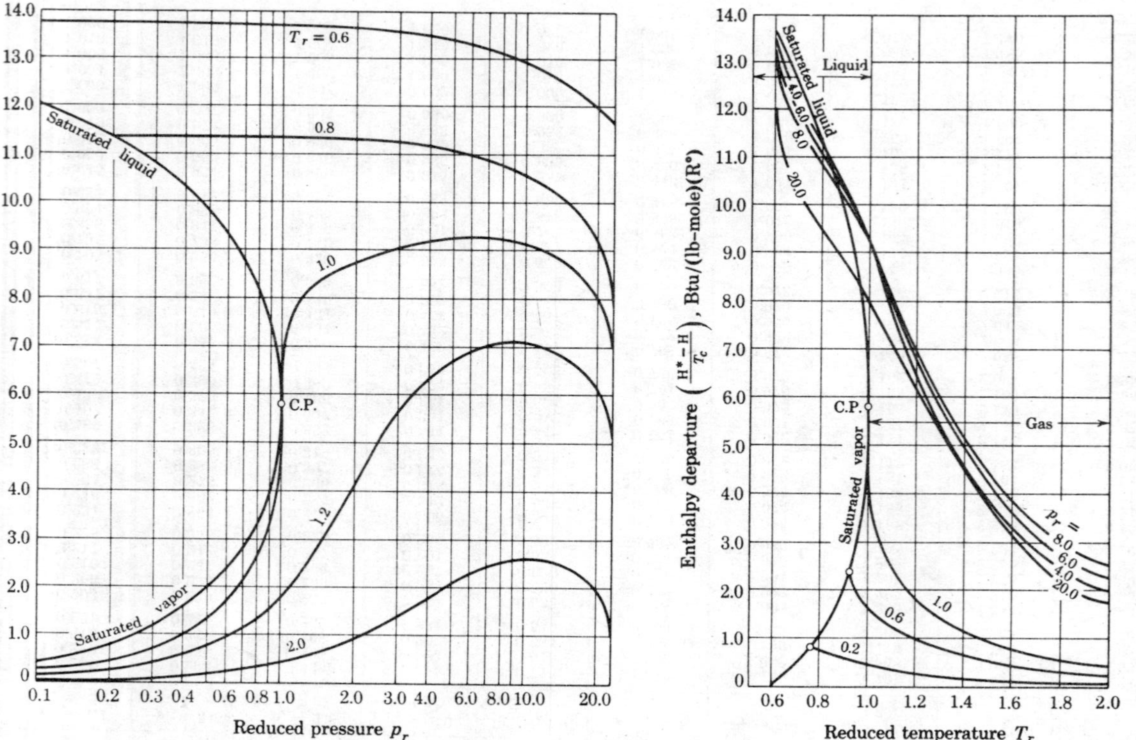

Fig. 141. Enthalpy departure of gases and liquids; $z_c = 0.27$

Enthalpy Departures. The enthalpy of a fluid relative to that of its ideal gas when both are at the same temperature may be obtained by integration of equation 13–141 between the limits of the existing pressure and zero. At zero pressure all fluids are in their ideal gaseous state and the enthalpy H becomes independent of pressure. Thus

$$\text{H}^* - \text{H} = \int_0^p \left[T\left(\frac{\partial v}{\partial T}\right)_p - v \right]_T dp \tag{54}$$

Differentiation of equation 35 gives

$$\left(\frac{\partial v}{\partial T}\right)_p = \frac{Rz}{p} + \frac{RT}{p}\left(\frac{\partial z}{\partial T}\right)_p \tag{55}$$

Combining equations 35, 54, and 55 results in

$$\left(\frac{\partial H}{\partial p}\right)_T = -\frac{RT^2}{p}\left(\frac{\partial z}{\partial T}\right)_p \tag{56}$$

In terms of reduced properties, equation 56 becomes

$$\frac{1}{T_c}\left(\frac{\partial H}{\partial p_r}\right)_{T_r} = -\frac{RT_r^{\,2}}{p_r}\left(\frac{\partial z}{\partial T_r}\right)_{p_r} \tag{57}$$

Integration at constant temperature gives

$$\left(\frac{\text{H}^* - \text{H}}{T_c}\right)_T = \left[RT_r^{\,2} \int_0^{p_r} \left(\frac{\partial z}{\partial T_r}\right)_{p_r} \frac{dp_r}{p_r} \right]_T \tag{58}$$

Values of $(\text{H}^* - \text{H})/T_c$ were calculated over the entire range of reduced temperatures and pressures and are given in Table 50 for both gases and liquids. Values at saturation are given in Table 47 for four values of z_c. Results are shown graphically in Fig. 141 for $z_c = 0.27$.

The discontinuities in the two-phase region for $(\text{H}^* - \text{H})/T_c$ are equal to the latent heat of vaporization divided by T_c. Thus, from the enthalpy of a pure substance at a given temperature in its ideal gaseous state the enthalpy at the same temperature of either the liquid or the gas phase may be obtained.

At values of p_r above unity, the vapor goes from the gaseous to the liquid state without going through any abrupt change in enthalpy. For example, in cooling from a temperature $T_r = 2.0$ to $T_r = 0.60$ at $p_r = 0.2$ the change in $(\text{H}^* - \text{H})/T_c$ is 13.7 units; 11.0 of these units are accounted for by heat of condensation. At a pressure of $p_r = 2$, the change is 13.0 units with no heat of condensation. At values of p_r above unity, latent heats are replaced by increased values of heat capacities.

In Table 47 values of $(H^* - H)/T_c$ for the saturated vapor and saturated liquid are given for four values of z_c.

Illustration 7. Calculate the enthalpy of 1 g-mole of CO_2 gas at 100 atm and 100° C, relative to the ideal gaseous state at 0° K.

From Table 70, at 100° C, $(H^*_{100° C} - H^*_{0° K}) = 2933$ g-cal per g-mole

$$T_c = 304.2° K, \quad p_c = 72.9 \text{ atm}, \quad z_c = 0.275$$

$$T_r = 273.2/304.2 = 1.227$$

$$p_r = 100/72.9 = 1.372$$

From Table 50, $\left(\dfrac{H^* - H}{T_c}\right)_{\substack{100° C \\ 100 \text{ atm}}} = 2.41$

$$(H^* - H)_{\substack{100° C \\ 100 \text{ atm}}} = (2.41)(304.2) = 733 \text{ g-cal per g-mole}$$

$$H_{\substack{100° C \\ 100 \text{ atm}}} - H^*_{0° K} = (H^*_{100° C} - H^*_{0° K}) - (H^*_{\substack{100° C \\ 100 \text{ atm}}} - H_{\substack{100° C \\ 100 \text{ atm}}})$$
$$= 2933 - 733 = 2200 \text{ g-cal per g-mole}$$

It may be noted that, in Illustration 1, the enthalpy correction at 100° C and 100 atm was found to be 777 g-cal per g-mole by the use of the Beattie–Bridgeman equation. Using this value, the enthalpy at 100° C and 100 atm, relative to the ideal state at 0° K, is equal to 2933 − 777, or 2156 g-cal per g-mole.

Illustration 8. Calculate the enthalpy of 1 g-mole of liquid CO_2 at its triple point (75.1 psia and −69.9° F), relative to the ideal gaseous state at 0° K.

From the data of Table 70, $H^*_{(-69.9° F)} - H^*_{0° K} = 1605$ g-cal per g-mole

$$T_c = 304.2° K, \quad p_c = 72.9 \text{ atm}, \quad z_c = 0.275$$

$$T = -69.9° F = 389.8° R = 216.6° K, \qquad T_r = 216.6/304.2 = 0.712$$

$$p = 75.1 \text{ psia} = 5.109 \text{ atm} \qquad p_r = 5.109/72.9 = 0.0700$$

From Table 47, $(H^* - H)/T_c = 11.80$ for saturated liquid at −70° F and 75 psia

$$[H^*_{(-70° F)} - H_{S(-69.9° F)}] = (11.80)(304.2) = 3590 \text{ g-cal per g-mole}$$

$$[H_{S(-69.9° F)} - H^*_{0° K}] = [H^*_{(-69.9° F)} - H^*_{0° K}] - [H^*_{(-69.9° F)} - H_{(-69.9° F)}]$$
$$= 1605 - 3590 = -1985 \text{ g-cal per g-mole}$$

Fugacity Coefficient. A derived thermodynamic property of special importance to be shown later is designated as fugacity and is arbitrarily defined for a pure substance in terms of free energy by the relation

$$(dG)_T = (RT\, d \ln f)_T \tag{59}$$

From equation 13–103,

$$(dG)_T = (v\, dp)_T \tag{60}$$

TABLE 50. ENTHALPY DEPARTURES OF PURE GASES AND LIQUIDS FROM IDEAL-GAS BEHAVIOR, $(H^* - H)/T_c$

Values are given in Btu/(lb-mole)(R°) or in cal/(g-mole)(K°).

Values of $(H^*-H)/T_c$ are recorded for $z_c = 0.27$

	$p_r = 0.01$			$p_r = 0.05$			$p_r = 0.10$			$p_r = 0.20$			$p_r = 0.30$		
T_{rs}		0.59			0.690			0.740			0.804			0.847	
sat. gas		0.04			0.22			0.430			0.810			1.15	
sat. liquid		13.80			12.67			12.08			11.28			10.79	
T_r	D_b	$\frac{H^* - H}{T_c}$	D_a	D_b	$\frac{H^* - H}{T_c}$	D_a	D_b	$\frac{H^* - H}{T_c}$	D_a	D_b	$\frac{H^* - H}{T_c}$	D_a	D_b	$\frac{H^* - H}{T_c}$	D_a
0.50		14.31			14.81	−180.0		14.81	−180.0		14.82	−180.0		14.82	−180.5
0.60	−1.2	0.057	−1.0	−50.0	13.75	−150.0	−51.0	13.75	−150.0	−52.0	13.76	−150.0	−52.0	13.76	−150.0
0.70	−0.9	0.047	−1.0	−4.0	0.23		−44.0	12.52	−121.0	−45.0	12.53	−121.0	−47.0	12.54	−121.0
0.80	−0.4	0.038	−1.0	−2.2	0.19	−1.2	−4.0	0.38	−2.5	−39.0	11.36	−94.0	−41.0	11.37	−93.0
0.90	−0.3	0.030	−1.0	−1.5	0.15	−2.0	−2.0	0.30	−3.0	−4.0	0.64	−5.0	−6.0	1.00	−8.0
0.92	−0.2	0.028	−1.0	−1.0	0.14	−1.6	−2.0	0.28	−3.8	−4.0	0.60	−6.0	−4.0	0.94	−8.0
0.94	−0.2	0.027	−1.0	−0.8	0.135	−1.2	−1.7	0.27	−2.5	−3.7	0.57	−6.5	−5.0	0.89	−8.0
0.96	−0.1	0.025	−1.0	−0.7	0.125	−1.1	−1.5	0.25	−2.0	−3.5	0.54	−5.5	−4.5	0.84	−7.5
0.98	−0.1	0.024	−1.0	−0.7	0.12	−1.1	−1.7	0.24	−2.0	−3.3	0.51	−5.0	−4.0	0.79	−7.5
1.00	−0.1	0.023	−1.0	−0.7	0.115	−1.0	−2.0	0.23	−2.0	−3.0	0.48	−4.5	−3.0	0.74	−7.5
1.01	−0.1	0.022	−1.0	−0.7	0.11	−1.0	−1.7	0.22	−2.0	−2.8	0.47	−4.4	−4.0	0.71	−7.0
1.02	−0.1	0.021	−1.0	−0.7	0.105	−1.0	−1.5	0.21	−2.0	−2.5	0.45	−4.2	−4.0	0.69	−6.5
1.03	−0.1	0.020	−1.0	−0.7	0.10	−1.0	−1.5	0.20	−2.0	−2.5	0.44	−4.0	−3.7	0.66	−6.5
1.04	−0.1	0.020	−1.0	−0.7	0.10	−1.0	−1.5	0.20	−2.0	−2.5	0.43	−3.9	−3.5	0.64	−6.0
1.05	−0.1	0.019	−1.0	−0.7	0.095	−1.0	−1.4	0.19	−2.0	−2.2	0.41	−3.8	−3.3	0.62	−6.0
1.06	−0.1	0.018	−1.0	−0.6	0.09	−1.0	−1.3	0.18	−2.0	−2.0	0.40	−3.7	−3.2	0.60	−6.0
1.07	−0.1	0.018	−1.0	−0.6	0.09	−1.0	−1.3	0.18	−2.0	−1.8	0.39	−3.5	−3.0	0.58	−5.5
1.08	−0.1	0.018	−1.0	−0.6	0.09	−1.0	−1.2	0.18	−2.0	−1.5	0.38	−3.3	−2.8	0.56	−5.5
1.09	−0.1	0.017	−1.0	−0.5	0.085	−1.0	−1.1	0.17	−2.0	−1.2	0.37	−3.2	−2.6	0.55	−5.5
1.10	−0.1	0.016	−1.0	−0.5	0.08	−1.0	−1.0	0.16	−2.0	−1.0	0.36	−3.0	−2.5	0.53	−5.0
1.12	−0.1	0.016	−1.0	−0.5	0.08	−1.0	−0.9	0.16	−2.0	−0.9	0.34	−3.0	−2.4	0.50	−5.0
1.14	−0.1	0.015	−1.0	−0.4	0.075	−1.0	−0.8	0.15	−2.0	−0.8	0.32	−3.0	−2.2	0.48	−5.0
1.16	−0.1	0.014	−1.0	−0.4	0.07	−1.0	−0.8	0.14	−2.0	−0.8	0.30	−3.0	−2.0	0.45	−5.0
1.18	−0.1	0.013	−1.0	0.4	0.067	−1.0	−0.7	0.135	−2.0	−0.7	0.29	−3.0	−1.9	0.42	−5.0
1.20	−0.1	0.013	−1.0	−0.3	0.065	−1.0	−0.7	0.13	−2.0	−0.7	0.27	−3.0	−1.8	0.40	−5.0
1.30	−0.0	0.011	−1.0	−0.3	0.055		−0.6	0.11	−1.5	−0.6	0.21	−2.2	−1.7	0.32	−3.5
1.40	0.0	0.010	−1.0	−0.3	0.05	−1.0	−0.6	0.10	−1.0	−0.6	0.17	−2.0	−1.3	0.26	−3.0
1.50	0.0	0.008	−1.0	−0.2	0.04	−1.0	−0.5	0.08	−1.0	−0.6	0.14	−1.0	−0.9	0.21	−2.0
1.60	0.0	0.007	−1.0	−0.1	0.035	−1.0	−0.3	0.07	−1.0	−0.4	0.12	−1.0	−0.6	0.19	−2.0
1.70	0.0	0.006	−1.0	0.0	0.03	−1.0	−0.1	0.06	−1.0	−0.2	0.10	−1.0	−0.3	0.16	−1.0
1.80	0.0	0.005	−1.0	0.0	0.025	−1.0	0.0	0.05	−1.0	0.0	0.08	−1.0	0.0	0.14	−1.0
1.90	0.0	0.004	−1.0	0.0	0.02	−1.0	0.0	0.04	−1.0	0.0	0.07	−1.0	0.0	0.12	−1.0
2.00	0.0	0.003	−1.0	0.0	0.015	−1.0	0.0	0.03	−1.0	0.0	0.06	−1.0	0.0	0.10	−1.0

Values of $\frac{H^* - H}{T_c}$ are recorded for $z_c = 0.27$. At other values of z_c, $\left(\frac{H^* - H}{T_c}\right)' = \left(\frac{H^* - H}{T_c}\right) + D(z_c - 0.27)$.

Table 50 (*Continued*). Enthalpy Departures of Pure Gases and Liquids from Ideal-Gas Behavior, $(H^* - H)/T_c$

	$p_r = 0.40$			$p_r = 0.50$			$p_r = 0.60$			$p_r = 0.70$			$p_r = 0.80$		
T_{rs}		0.879			0.909			0.929			0.950			0.967	
sat. gas		1.52			1.91			2.28			2.74			3.24	
sat. liquid		10.34			9.86			9.38			8.88			8.29	
	D_b	$\frac{H^* - H}{T_c}$	D_a	D_b	$\frac{H^* - H}{T_c}$	D_a	D_b	$\frac{H^* - H}{T_c}$	D_a	D_b	$\frac{H^* - H}{T_c}$	D_a	D_b	$\frac{H^* - H}{T_c}$	D_a
T_r															
0.50		14.82	−181.5		14.82	−183.0		14.82	−185.0		14.82	−187.0		14.82	−189.0
0.60	−52.0	13.76	−150.0	−53.0	13.76	−150.0	−53.0	13.77	−150.0	−53.0	13.77	−151.0	−53.0	13.77	−152.0
0.70	−48.0	12.55	−121.0	−48.0	12.55	−121.0	−49.0	12.56	−121.0	−49.0	12.56	−120.0	−50.0	12.56	−119.0
0.80	−42.0	11.38	−92.0	−43.0	11.39	−91.0	−45.0	11.39	−89.0	−46.0	11.39	−87.0	−46.0	11.39	−85.0
0.90	−9.0	1.48	−11.0	−33.0	9.90	−69.0	−36.0	9.90	−64.0	−38.0	9.90	−60.0	−38.0	9.91	−56.0
0.92	−7.5	1.36	−11.0	−11.0	1.78	−12.5	−32.0	9.52	−63.0	−35.0	9.52	−56.0	−36.0	9.52	−51.0
0.94	−6.0	1.25	−11.0	−9.0	1.64	−11.0	−14.0	2.12	−12.5	−31.0	9.13	−56.0	−34.0	9.14	−49.0
0.96	−5.0	1.16	−10.0	−7.0	1.52	−11.0	−11.0	1.94	−13.0	−16.0	2.56	−19.0	−30.0	8.56	−49.0
0.98	−5.0	1.08	−10.0	−6.0	1.41	−11.0	−8.0	1.80	−13.0	−12.0	2.30	−17.0	−17.0	2.98	−16.0
1.00	−4.0	1.00	−9.0	−5.0	1.32	−11.0	−7.0	1.66	−13.0	−9.0	2.08	−15.0	−12.0	2.63	−17.0
1.01	−4.0	0.96	−8.5	−5.0	1.28	−11.0	−6.5	1.60	−13.0	−8.5	2.00	−15.0	−11.0	2.49	−17.0
1.02	−4.0	0.93	−8.5	−5.0	1.24	−10.5	−6.0	1.54	−12.5	−8.0	1.92	−14.5	−10.0	2.38	−16.5
1.03	−4.0	0.91	−8.5	−5.0	1.20	−10.5	−5.5	1.49	−12.5	−7.5	1.84	−14.5	−9.0	2.28	−16.5
1.04	−4.0	0.88	−8.0	−5.0	1.16	−10.0	−5.0	1.44	−12.0	−7.0	1.77	−14.5	−8.0	2.19	−16.0
1.05	−4.0	0.85	−8.0	−4.5	1.12	−10.0	−5.0	1.40	−12.0	−6.5	1.71	−14.0	−7.5	2.10	−16.0
1.06	−3.5	0.82	−8.0	−4.0	1.09	−10.0	−4.5	1.36	−12.0	−6.0	1.65	−14.0	−7.0	2.03	−16.0
1.07	−3.5	0.80	−7.5	−4.0	1.06	−9.5	−4.5	1.31	−11.5	−5.5	1.60	−13.5	−6.5	1.95	−15.5
1.08	−3.5	0.78	−7.5	−3.5	1.02	−9.5	−4.5	1.28	−11.5	−5.0	1.54	−13.5	−6.0	1.89	−15.5
1.09	−3.0	0.76	−7.5	−3.0	1.00	−9.0	−4.0	1.24	−11.0	−4.5	1.49	−13.0	−5.5	1.82	−15.0
1.10	−3.0	0.74	−7.0	−3.0	0.97	−9.0	−4.0	1.20	−11.0	−4.0	1.45	−13.0	−5.0	1.76	−15.0
1.12	−2.5	0.70	−6.5	−3.0	0.92	−8.5	−3.5	1.14	−10.5	−4.0	1.37	−12.5	−4.5	1.66	−14.0
1.14	−2.5	0.66	−6.5	−3.0	0.87	−8.5	−3.5	1.08	−10.5	−4.0	1.30	−12.0	−4.5	1.56	−13.0
1.16	−2.0	0.62	−6.5	−2.5	0.82	−8.5	−3.0	1.02	−10.0	−3.5	1.24	−11.5	−4.0	1.48	−12.0
1.18	−2.0	0.59	−6.0	−2.5	0.78	−8.5	−3.0	0.98	−9.5	−3.0	1.18	−11.0	−3.5	1.40	−11.0
1.20	−1.5	0.56	−6.0	−2.0	0.75	−8.0	−2.5	0.93	−9.0	−3.0	1.13	−10.0	−3.5	1.33	−10.0
1.30	−1.5	0.42	−5.0	−2.0	0.60	−6.0	−2.0	0.73		−2.5	0.92		−3.0	1.06	−7.5
1.40	−1.0	0.34	−4.0	−1.5	0.48	−5.0	−1.5	0.59	−5.0	−2.0	0.74	−5.0	−2.5	0.85	−5.0
1.50	−1.0	0.28	−3.0	−1.2	0.38	−4.0	−1.4	0.48	−4.0	−1.6	0.60	−4.0	−1.8	0.70	−4.0
1.60	−0.8	0.24	−2.0	−0.9	0.32	−3.0	−1.1	0.40	−3.0	−1.3	0.50	−3.0	−1.4	0.58	−3.0
1.70	−0.5	0.21	−1.0	−0.6	0.28	−2.0	−0.8	0.35	−2.0	−1.0	0.42	−2.0	−1.0	0.49	−2.0
1.80	−0.1	0.18	−1.0	−0.1	0.24	−1.0	−0.3	0.30	−1.0	−0.3	0.36	−1.0	−0.4	0.41	−1.0
1.90	0.0	0.15	−1.0	0.0	0.21	−1.0	0.0	0.25	−1.0	0.0	0.32	−1.0	−0.1	0.36	−1.0
2.00	0.0	0.13	−1.0	0.0	0.17	−1.0	0.0	0.22	−1.0	0.0	0.28	−1.0	0.0	0.32	−1.0

Values of $\frac{H^* - H}{T_c}$ are recorded for $z_c = 0.27$. At other values of z_c, $\left(\frac{H^* - H}{T_c}\right)' = \left(\frac{H^* - H}{T_c}\right) + D(z_c - 0.27)$.

TABLE 50 (*Continued*). ENTHALPY DEPARTURES OF PURE GASES AND LIQUIDS FROM IDEAL-GAS BEHAVIOR, $(H^* - H)/T_c$

	$p_r = 0.90$			$p_r = 1.00$			$p_r = 1.05$			$p_r = 1.10$			$p_r = 1.20$		
T_{rs}		0.984			1.000										
sat. gas		4.02			5.80										
sat. liquid		7.5[illegible]			5.80										
T_r	D_b	$\frac{H^* - H}{T_c}$	D_a	D_b	$\frac{H^* - H}{T_c}$	D_a	D_b	$\frac{H^* - H}{T_c}$	D_a	D_b	$\frac{H^* - H}{T_c}$	D_a	D_b	$\frac{H^* - H}{T_c}$	D_a
0.50		14.8[illegible]	−191.0		14.81	−192.0		14.80	−192.0		14.80	−192.5		14.80	−193.0
0.60	−52.0	13.77	−154.0	−52.0	13.77	−155.0	−52.0	13.77	−155.0	−51.0	13.77	−155.0	−50.0	13.76	−155.0
0.70	−50.0	12.56	−117.0	−49.0	12.56	−114.0	−49.0	12.56	−111.5	−48.0	12.55	−110.0	−47.0	12.54	−105.0
0.80	−46.0	11.40	−82.0	−45.0	11.40	−78.0	−45.0	11.40	−72.0	−44.0	11.40	−72.0	−43.0	11.40	−66.0
0.90	−38.0	9.92	−52.0	−37.0	9.94	−50.0	−36.0	9.96	−49.0	−36.0	9.98	−48.0	−34.0	10.03	−47.0
0.92	−36.0	9.55	−47.0	−35.0	9.60	−45.0	−34.0	9.61	−44.0	−34.0	9.63	−44.0	−31.0	9.68	−44.0
0.94	−34.0	9.16	−44.0	−33.0	9.19	−41.0	−32.0	9.21	−40.0	−31.0	9.24	−40.0	−28.0	9.29	−40.0
0.96	−31.0	8.60	−42.0	−29.0	8.67	−37.0	−28.0	8.70	−36.0	−27.0	8.74	−36.0	−24.0	8.84	−38.0
0.98	−26.0	7.84	−39.0	−26.0	7.99	−33.0	−25.0	8.02	−32.5	−24.0	8.14	−33.5	−21.0	8.38	−35.0
1.00	−16.0	3.28	−20.0	−20.0	5.80	−25.0	−20.0	6.58	−27.0	−20.0	7.16	−30.0	−18.0	7.68	−32.0
1.01	−14.0	3.08	−20.0	−17.0	4.26	−24.5	−17.5	5.47	−26.5	−17.5	6.68	−29.0	−16.0	7.12	−30.0
1.02	−12.0	2.92	−20.0	−14.0	3.86	−24.0	−15.0	4.73	−26.0	−15.0	5.60	−27.5	−14.0	6.40	−29.0
1.03	−11.0	2.80	−20.5	−12.5	3.56	−23.0	−13.0	4.08	−25.0	−13.0	4.60	−26.0	−12.5	5.48	−28.0
1.04	−10.0	2.68	−20.0	−11.0	3.31	−22.0	−11.0	3.70	−23.5	−11.5	4.10	−24.5	−11.0	4.48	−26.0
1.05	−9.0	2.56	−20.0	−9.0	3.10	−21.0	−9.5	3.41	−22.0	−9.0	3.73	−23.0	−9.5	4.30	−24.0
1.06	−8.0	2.46	−19.5	−8.0	2.93	−20.0	−8.0	3.20	−21.5	−8.0	3.47	−22.5	−8.0	3.92	−23.0
1.07	−7.5	2.36	−19.0	−7.5	2.79	−19.5	−7.5	3.02	−21.0	−7.5	3.26	−22.0	−7.5	3.64	−22.5
1.08	−7.0	2.28	−18.5	−7.0	2.66	−19.0	−7.0	2.87	−20.5	−7.0	3.08	−21.0	−7.0	3.43	−22.0
1.09	−6.5	2.19	−18.0	−6.5	2.52	−18.5	−6.5	2.72	−20.0	−6.5	2.92	−20.0	−6.5	3.26	−21.0
1.10	−6.0	2.11	−17.0	−6.0	2.42	−18.0	−6.0	2.60	−19.0	−6.0	2.79	−19.0	−6.0	3.12	−20.0
1.12	−5.0	1.97	−16.0	−5.0	2.24	−16.5	−5.5	2.40	−17.0	−5.5	2.56	−18.0	−5.5	2.88	−18.0
1.14	−4.5	1.85	−14.0	−4.5	2.09	−14.0	−5.0	2.24	−14.5	−5.0	2.38	−16.0	−5.0	2.68	−16.0
1.16	−4.0	1.74	−12.0	−4.0	1.97	−12.5	−4.5	2.10	−13.0	−4.5	2.24	−14.0	−4.5	2.50	−14.0
1.18	−3.5	1.65	−11.0	−3.5	1.87	−11.5	−4.0	1.99	−12.0	−4.0	2.12	−12.5	−4.0	2.37	−12.5
1.20	−3.5	1.56	−10.5	−3.5	1.78	−11.0	−3.5	1.90	−11.0	−3.5	2.02	−11.0	−3.5	2.25	−11.0
1.30	−3.0	1.2[illegible]	−8.0	−3.0	1.40	−8.0	−3.0	1.49	−8.0	−3.0	1.58	−5.0	−3.0	1.74	−8.0
1.40	−2.5	0.97	−5.0	−2.5	1.13	−5.0	−2.5	1.21	−5.0	−2.5	1.28	−5.0	−2.5	1.43	−5.0
1.50	−1.9	0.78	−4.5	−2.0	0.91	−4.5	−2.0	0.98	−4.5	−2.0	1.04	−4.5	−2.0	1.17	−5.0
1.60	−1.4	0.64	−3.0	−1.5	0.74	−3.0	−1.5	0.79	−3.0	−1.5	0.84	−3.0	−1.5	0.96	−3.0
1.70	−1.1	0.56	−2.0	−1.2	0.63	−2.0	−1.2	0.66	−2.0	−1.2	0.70	−2.0	−1.2	0.80	−2.0
1.80	−0.5	0.48	−1.0	−0.5	0.54	0.0	−0.5	0.57	0.0	−0.5	0.60	0.0	−0.5	0.68	0.0
1.90	−0.1	0.43	−1.0	−0.1	0.48	0.0	−0.1	0.51	0.0	−0.1	0.54	0.0	−0.1	0.58	0.0
2.00	0.0	0.39	−1.0	0.0	0.43	0.0	0.0	0.46	0.0	0.0	0.48	0.0	0.0	0.52	0.0

Values of $\left(\frac{H^* - H}{T_c}\right)$ are recorded for $z_c = 0.27$. At other values of z_c, $\left(\frac{H^* - H}{T_c}\right)' = \left(\frac{H^* - H}{T_c}\right) + D(z_c - 0.27)$.

TABLE 50 (*Continued*). ENTHALPY DEPARTURES OF PURE GASES AND LIQUIDS FROM IDEAL-GAS BEHAVIOR, $(H^* - H)/T$

T_r	$p_r = 1.40$	$p_r = 1.60$	$p_r = 1.80$	$p_r = 2.0$	$p_r = 4.0$	$p_r = 6.0$	$p_r = 8.0$	$p_r = 10.0$	$p_r = 20.0$	$p_r = 30.0$
	$\frac{H^* - H}{T_c}$	$\frac{H^* - H}{T_c}$	$\frac{H^* - H}{T_c}$	$\frac{H^* - H}{T_c}$	$\frac{H^* - H}{T_c}$	$\frac{H^* - H}{T_c}$	$\frac{H^* - H}{T_c}$	$\frac{H^* - H}{T_c}$	$\frac{H^* - H}{T_c}$	$\frac{H^* - H}{T_c}$
0.50	14.80	14.80	14.80	14.80	14.73	14.64	14.52	14.40	13.57	12.57
0.60	13.76	13.75	13.73	13.72	13.48	13.30	13.14	12.96	11.90	10.58
0.70	12.53	12.52	12.51	12.49	12.22	12.03	11.82	11.57	10.36	8.94
0.80	11.40	11.40	11.40	11.38	11.21	11.02	10.84	10.57	9.34	8.00
0.90	10.12	10.20	10.26	10.29	10.26	10.14	10.02	9.83	8.74	7.58
0.92	9.80	9.91	10.00	10.04	10.04	9.97	9.85	9.68	8.62	7.47
0.94	9.48	9.58	9.68	9.76	9.90	9.80	9.69	9.52	8.51	7.37
0.96	9.02	9.20	9.34	9.44	9.60	9.60	9.50	9.37	8.38	7.24
0.98	8.64	8.78	8.96	9.08	9.40	9.40	9.31	9.22	8.24	7.10
1.00	8.16	8.32	8.55	8.69	9.08	9.17	9.12	9.02	8.06	6.94
1.01	7.83	8.00	8.34	8.46	8.94	9.06	9.02	8.92	7.94	6.84
1.02	7.45	7.80	8.11	8.27	8.80	8.95	8.91	8.82	7.82	6.70
1.03	6.89	7.47	7.83	8.01	8.66	8.83	8.80	8.70	7.71	6.60
1.04	6.58	7.10	7.52	7.77	8.51	8.71	8.69	8.60	7.60	6.50
1.05	5.56	6.65	7.16	7.48	8.37	8.60	8.58	8.48	7.51	6.40
1.06	5.12	6.24	6.81	7.16	8.20	8.45	8.45	8.36	7.40	6.30
1.07	4.70	5.68	6.36	6.81	8.03	8.30	8.31	8.22	7.29	6.20
1.08	4.40	5.18	5.91	6.44	7.87	8.16	8.18	8.14	7.20	6.11
1.09	4.11	4.88	5.52	6.09	7.70	8.02	8.04	8.00	7.06	6.01
1.10	3.88	4.63	5.22	5.72	7.53	7.88	7.91	7.90	6.94	5.92
1.12	3.51	4.18	4.72	5.18	7.26	7.66	7.70	7.66	6.74	5.74
1.14	3.23	3.85	4.40	4.80	7.00	7.44	7.50	7.45	6.56	5.55
1.16	3.01	3.56	4.04	4.50	6.74	7.22	7.30	7.23	6.36	5.37
1.18	2.83	3.36	3.82	4.24	6.49	7.00	7.10	7.03	6.18	5.19
1.20	2.68	3.12	3.56	4.00	6.24	6.78	6.89	6.86	6.00	5.00
1.30	2.08	2.42	2.68	3.05	5.18	5.82	5.98	5.96	5.16	4.18
1.40	1.65	1.91	2.16	2.38	4.26	4.94	5.10	5.18	4.44	3.45
1.50	1.35	1.58	1.80	1.93	3.50	4.17	4.36	4.48	3.78	2.80
1.60	1.13	1.32	1.48	1.63	2.91	3.56	3.78	3.90	3.25	2.23
1.70	0.96	1.10	1.26	1.39	2.45	3.08	3.32	3.44	2.80	1.80
1.80	0.79	0.96	1.07	1.18	2.11	2.71	2.97	3.07	2.45	1.43
1.90	0.69	0.82	0.92	1.03	1.91	2.45	2.71	2.79	2.18	1.14
2.00	0.60	0.70	0.79	0.88	1.76	2.28	2.54	2.62	1.98	0.94
3.00	0.84	0.27	0.30	0.32	0.53	0.65	0.67	0.63	−0.25	−1.78
4.00			0.05	0.11	0.28	0.35	0.26	0.14	−0.23	−1.48
6.00			0.01	0.02	0.06	0.14	0.05	0.04	−0.14	−0.48
8.00			0.01	0.01	0.02	0.03	0.02	0.02	−0.07	−0.21
10.00			0.00	0.00	0.00	0.00	0.00	0.00	0.00	−0.03
15.00			0.03	0.03	0.09	0.13	0.16	0.17	+0.14	+0.07

For ideal-gas behavior equation 60 becomes

$$(dG)_T = (RT\, d \ln p)_T \tag{61}$$

For ideal-gas behavior fugacity is made numerically equivalent to pressure, but for substances other than ideal gases fugacity is not equivalent to pressure and should not be so employed.

Equations 59 and 60 combined with the definition of z give

$$\left(\frac{\partial \ln f}{\partial \ln p}\right)_T = z \tag{62}$$

Subtracting 1 from both sides of equation 62 gives

$$\left(d \ln \frac{f}{p}\right)_T = [(z-1)\, d \ln p]_T \tag{63}$$

At $p = 0$, pressure and fugacity become equal. In terms of reduced properties equation 63 becomes

$$\left(\ln \frac{f}{p}\right)_{T_r} = \left[\int_0^{p_r} \frac{(z-1)\, dp_r}{p_r}\right]_{T_r} \tag{64}$$

Equation 64 has been integrated from the generalized values of z over the entire range of reduced temperatures and pressures, and tabulated in Table 51. Values are shown graphically in Fig. 142 for $z_c = 0.27$.

Where a pure liquid and its vapor are in equilibrium at a given temperature, the free energy of vaporization is zero.

$$(\Delta G)_{pT} = \left[RT \ln\left(\frac{f_v}{f_L}\right) = 0\right]_{pT} \tag{65}$$

Hence

$$f_v = f_L \tag{66}$$

Thus, at equilibrium between a pure liquid and its vapor, the fugacity of a given component is the same in each phase. This condition is represented by the saturation line common to both gas and liquid phases as shown in Fig. 142.

Illustration 9. Calculate the fugacity of methane gas at 120° F and 800 psia.

$T_c = 190.7°$ K, $p_c = 45.8$ atm, $z_c = 0.290$

Temperature = 120° F = 48.9° C = 322.1° K

$T_r = 322.1/190.7 = 1.689$

Pressure = 800 psia = 54.42 atm

$p_r = 54.42/45.8 = 1.188$

From Table 51, $f/p = 0.952$

$f = (0.952)800 = 762$ psia

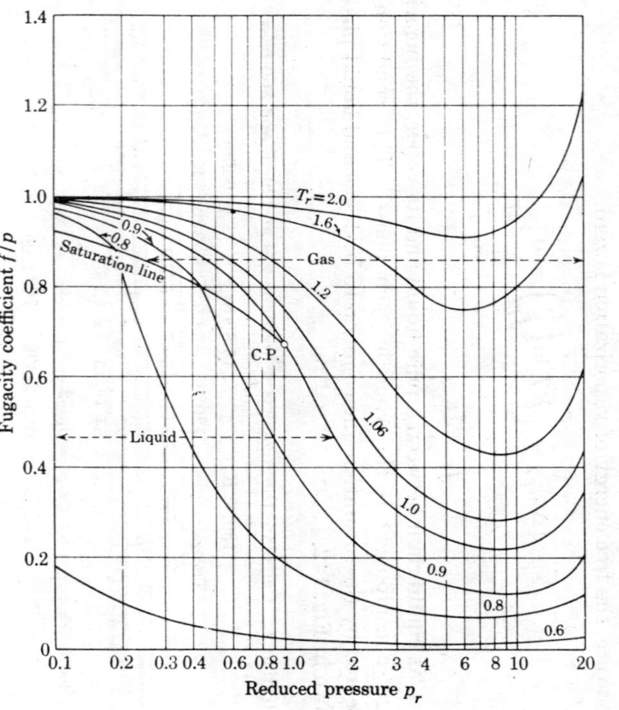

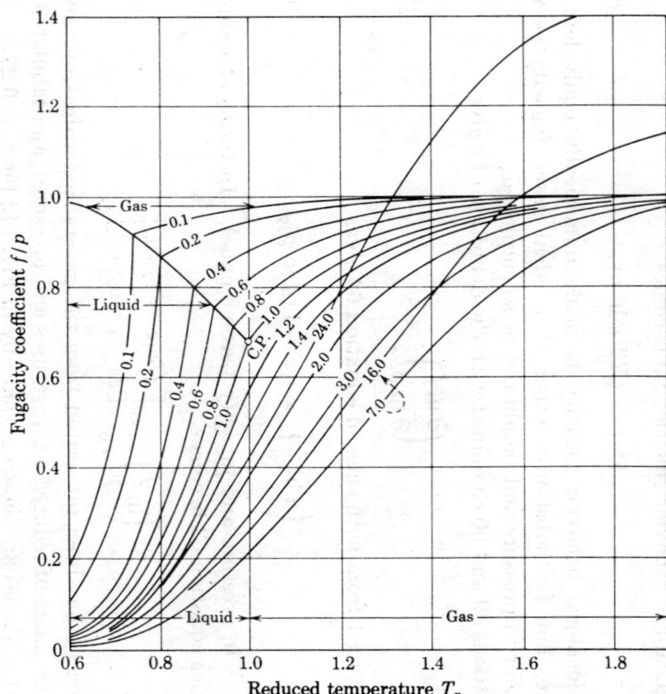

Fig. 142. Fugacity coefficients of gases and liquids; $z_c = 0.27$

TABLE 51. FUGACITY COEFFICIENTS OF PURE GASES AND LIQUIDS, f/p

	$p_r = 0.01$			$p_r = 0.05$			$p_r = 0.10$			$p_r = 0.20$			$p_r = 0.30$		
T_{rs} sat. gas and liquid		0.59 0.993			0.690 0.972			0.740 0.946			0.804 0.887			0.847 0.840	
	D_b	f/p	D_a	D_b	f/p	D_a	D_b	f/p	D_a	D_b	f/p	D_a	D_b	f/p	D_a
T_r 0.50		0.021			0.015			0.0102			0.00518			0.00351	
0.60	0.06	0.994	0.09	7.9	0.17	12.50	8.40	0.111	12.80	11.65	0.0563	13.50	11.0	0.0381	14.00
0.70	0.02	0.995	0.03	0.06	0.975	0.13	3.30	0.640	6.40	4.20	0.325	6.80	4.5	0.220	7.20
0.80	0.00	0.996	0.01	0.02	0.980	0.06	0.04	0.961	0.18	2.85	0.848	2.90	2.53	0.574	3.20
0.90	0.00	0.997	0.01	0.01	0.986	0.03	0.03	0.972	0.11	0.07	0.922	0.23	0.19	0.871	0.32
0.92	0.00	0.997	0.01	0.01	0.987	0.00	0.01	0.975	0.10	0.06	0.929	0.21	0.12	0.882	0.28
0.94	0.00	0.998	0.01	0.01	0.988	0.03	0.01	0.977	0.09	0.05	0.934	0.19	0.10	0.892	0.25
0.96	0.00	0.998	0.01	0.01	0.989	0.03	0.01	0.979	0.08	0.05	0.939	0.17	0.09	0.900	0.22
0.98	0.00	0.998	0.00	0.00	0.990	0.02	0.01	0.981	0.07	0.05	0.943	0.15	0.08	0.907	0.20
1.00	0.00	0.998	0.00	0.00	0.991	0.02	0.02	0.983	0.06	0.05	0.947	0.14	0.08	0.916	0.19
1.01	0.00	0.998	0.00	0.00	0.991	0.02	0.02	0.983	0.06	0.04	0.949	0.13	0.08	0.917	0.17
1.02	0.00	0.998	0.00	0.00	0.991	0.02	0.01	0.983	0.05	0.04	0.951	0.12	0.07	0.919	0.16
1.03	0.00	0.998	0.00	0.00	0.992	0.02	0.01	0.984	0.05	0.04	0.953	0.11	0.07	0.922	0.16
1.04	0.00	0.998	0.00	0.00	0.992	0.02	0.01	0.985	0.04	0.04	0.955	0.10	0.07	0.926	0.15
1.05	0.00	0.998	0.00	0.00	0.992	0.01	0.01	0.985	0.04	0.03	0.956	0.09	0.06	0.928	0.14
1.06	0.00	0.999	0.00	0.00	0.993	0.01	0.01	0.986	0.03	0.03	0.958	0.08	0.06	0.930	0.13
1.07	0.00	0.999	0.00	0.00	0.993	0.01	0.01	0.986	0.03	0.03	0.959	0.08	0.05	0.933	0.13
1.08	0.00	0.999	0.00	0.00	0.993	0.01	0.00	0.986	0.03	0.03	0.962	0.07	0.05	0.935	0.12
1.09	0.00	0.999	0.00	0.00	0.993	0.01	0.01	0.986	0.03	0.02	0.962	0.07	0.04	0.938	0.11
1.10	0.00	0.999	0.00	0.00	0.993	0.01	0.01	0.987	0.02	0.02	0.964	0.06	0.04	0.942	0.10
1.12	0.00	0.999	0.00	0.00	0.994	0.01	0.01	0.988	0.02	0.02	0.967	0.05	0.03	0.943	0.09
1.14	0.00	0.999	0.00	0.00	0.994	0.01	0.01	0.989	0.01	0.02	0.969	0.04	0.03	0.947	0.08
1.16	0.00	0.999	0.00	0.00	0.995	0.00	0.01	0.990	0.01	0.02	0.970	0.04	0.02	0.950	0.07
1.18	0.00	0.999	0.00	0.00	0.995	0.01	0.01	0.991	0.01	0.02	0.973	0.03	0.02	0.954	0.06
1.20	0.00	0.999	0.00	0.00	0.995	0.01	0.01	0.991	0.01	0.01	0.974	0.03	0.02	0.958	0.05
1.30	0.00	0.999	0.00	0.00	0.996	0.01	0.01	0.993	0.01	0.01	0.980	0.02	0.01	0.968	0.03
1.40	0.00	0.999	0.00	0.00	0.997	0.01	0.00	0.995	0.01	0.0	0.985	0.02	0.00	0.976	0.02
1.50	0.00	1.000	0.00	0.00	0.998	0.01	−0.01	0.996	0.01	−0.01	0.987	0.02	−0.02	0.980	0.04
1.60	0.00	1.000	0.00	0.00	0.998	0.01	−0.01	0.997	0.01	−0.02	0.989	0.04	−0.02	0.983	0.05
1.70	0.00	1.000	0.00	0.00	0.998	0.01	−0.01	0.997	0.02	−0.02	0.989	0.05	−0.03	0.984	0.08
1.80	0.00	1.000	0.00	0.00	0.998	0.01	−0.01	0.997	0.03	−0.02	0.990	0.08	−0.04	0.986	0.10
1.90	0.00	1.000	0.00	0.00	0.999	0.01	−0.01	0.998	0.04	−0.03	0.990	0.10	−0.05	0.987	0.13
2.00		1.000	0.00	0.00	0.999	0.01		0.999	0.06		0.992	0.13		0.989	0.15

Values of f/p are recorded for $z_c = 0.27$. At other values of z_c, $(f/p)' = (f/p)(10^{D(z_c-0.27)})$.

TABLE 51 (*Continued*). FUGACITY COEFFICIENTS OF PURE GASES AND LIQUIDS, f/p

	$p_r = 0.40$			$p_r = 0.50$			$p_r = 0.60$			$p_r = 0.70$			$p_r = 0.80$		
T_{rs} sat. gas and liquid		0.879 0.805			0.907 0.777			0.929 0.752			0.950 0.730			0.967 0.710	
	D_b	f/p	D_a	D_b	f/p	D_a	D_b	f/p	D_a	D_b	f/p	D_a	D_b	f/p	D_a
T_r															
0.50		0.0027			0.0022			0.00186			0.00162			0.00144	
0.60	10.90	0.0291	14.40	10.70	0.0236	14.70	10.60	0.0200	15.00	10.50	0.0174	15.20	10.18	0.0155	15.40
0.70	4.70	0.168	7.50	4.75	0.136	7.80	4.80	0.115	8.00	5.00	0.100	8.00	5.03	0.0890	8.00
0.80	2.41	0.437	3.50	2.40	0.354	3.70	2.45	0.300	3.83	2.50	0.261	4.00	2.60	0.231	4.22
0.90	0.21	0.820	0.38	1.32	0.739	1.50	1.20	0.642	1.63	1.11	0.559	1.71	1.15	0.496	1.77
0.92	0.19	0.836	0.34	0.24	0.789	0.40	1.04	0.722	1.12	0.94	0.629	1.40	0.92	0.559	1.43
0.94	0.17	0.849	0.31	0.22	0.806	0.36	0.25	0.762	0.42	0.69	0.699	0.97	0.71	0.622	1.09
0.96	0.15	0.860	0.28	0.21	0.820	0.33	0.28	0.781	0.38	0.35	0.741	0.43	0.60	0.688	0.68
0.98	0.13	0.870	0.25	0.18	0.834	0.31	0.23	0.798	0.35	0.29	0.762	0.39	0.33	0.725	0.43
1.00	0.11	0.879	0.23	0.15	0.846	0.28	0.19	0.812	0.32	0.23	0.779	0.35	0.27	0.745	0.38
1.01	0.11	0.884	0.22	0.14	0.852	0.25	0.18	0.818	0.29	0.22	0.786	0.32	0.25	0.755	0.34
1.02	0.11	0.887	0.21	0.13	0.856	0.24	0.17	0.826	0.28	0.21	0.794	0.31	0.24	0.764	0.33
1.03	0.10	0.891	0.20	0.15	0.860	0.24	0.16	0.831	0.27	0.19	0.802	0.30	0.22	0.772	0.32
1.04	0.10	0.896	0.19	0.12	0.868	0.23	0.15	0.838	0.26	0.18	0.810	0.28	0.20	0.781	0.30
1.05	0.09	0.899	0.18	0.11	0.871	0.22	0.14	0.842	0.25	0.17	0.816	0.27	0.18	0.788	0.29
1.06	0.08	0.903	0.17	0.10	0.876	0.21	0.13	0.848	0.24	0.15	0.822	0.26	0.17	0.795	0.28
1.07	0.08	0.906	0.16	0.09	0.879	0.20	0.13	0.853	0.23	0.14	0.826	0.25	0.16	0.800	0.27
1.08	0.08	0.910	0.15	0.09	0.883	0.18	0.12	0.857	0.21	0.13	0.832	0.23	0.15	0.807	0.25
1.09	0.07	0.913	0.14	0.08	0.887	0.17	0.11	0.863	0.19	0.12	0.838	0.21	0.14	0.813	0.23
1.10	0.06	0.916	0.13	0.08	0.892	0.15	0.10	0.867	0.18	0.12	0.843	0.19	0.13	0.820	0.21
1.12	0.06	0.920	0.12	0.07	0.898	0.14	0.09	0.876	0.17	0.10	0.853	0.18	0.12	0.831	0.19
1.14	0.06	0.926	0.10	0.06	0.905	0.12	0.08	0.884	0.14	0.09	0.863	0.15	0.11	0.843	0.16
1.16	0.05	0.931	0.09	0.06	0.910	0.11	0.07	0.890	0.12	0.08	0.871	0.13	0.10	0.852	0.14
1.18	0.05	0.936	0.08	0.05	0.917	0.09	0.07	0.898	0.10	0.08	0.880	0.11	0.09	0.862	0.12
1.20	0.04	0.939	0.07	0.05	0.920	0.08	0.07	0.906	0.09	0.08	0.889	0.10	0.08	0.872	0.10
1.30	0.02	0.955	0.05	0.02	0.942	0.06	0.03	0.929	0.07	0.04	0.917	0.07	0.05	0.903	0.07
1.40	0.00	0.967	0.03	0.00	0.956	0.04	0.00	0.947	0.04	0.01	0.938	0.05	0.02	0.928	0.05
1.50	−0.02	0.973	0.05	−0.02	0.966	0.06	−0.02	0.958	0.06	−0.01	0.950	0.06	−0.01	0.944	0.06
1.60	−0.03	0.978	0.07	−0.03	0.972	0.07	−0.04	0.962	0.07	−0.03	0.961	0.07	−0.03	0.956	0.07
1.70	−0.04	0.980	0.09	−0.04	0.976	0.09	−0.05	0.972	0.09	−0.05	0.968	0.09	−0.05	0.965	0.09
1.80	−0.06	0.983	0.11	−0.06	0.980	0.11	−0.07	0.977	0.11	−0.07	0.974	0.11	−0.07	0.971	0.11
1.90	−0.07	0.985	0.13	−0.08	0.983	0.13	−0.08	0.981	0.13	−0.08	0.979	0.13	−0.08	0.977	0.13
2.00		0.987	0.15		0.985	0.15		0.983	0.15		0.982	0.15		0.981	0.15

Values of f/p are recorded for $z_c = 0.27$. At other values of z_c, $(f/p)' = (f/p)(10^{D(z_c - 0.27)})$.

TABLE 51 (*Continued*). FUGACITY COEFFICIENTS OF PURE GASES AND LIQUIDS, f/p

T_{rs} sat. gas and liquid	$p_r = 0.90$			$p_r = 1.00$			$p_r = 1.05$			$p_r = 1.10$			$p_r = 1.20$		
		0.984 0.688			1.000 0.665										
	D_b	f/p	D_a	D_b	f/p	D_a	D_b	f/p	D_a	D_b	f/p	D_a	D_b	f/p	D_a
T_r															
0.50		0.00131			0.00120			0.00115			0.00111			0.00104	
0.60	10.24	0.0150	15.60	10.14	0.0134	15.70	10.10	0.0123	15.70	10.00	0.0119	15.80	9.81	0.0110	15.80
0.70	5.20	0.0803	8.00	5.30	0.0734	8.00	5.35	0.0704	8.00	5.40	0.0677	8.00	5.50	0.0619	8.00
0.80	2.60	0.209	4.30	2.70	0.191	4.40	2.70	0.183	4.40	2.75	0.176	4.45	2.80	0.163	4.60
0.90	1.19	0.448	1.77	1.20	0.409	1.88	1.20	0.392	1.91	1.25	0.377	1.98	1.25	0.351	2.12
0.92	0.92	0.504	1.42	0.98	0.461	1.54	1.00	0.442	1.56	1.05	0.425	1.67	1.13	0.396	1.69
0.94	0.71	0.561	1.15	0.78	0.513	1.15	0.81	0.493	1.16	0.84	0.474	1.20	0.90	0.441	1.29
0.96	0.57	0.622	0.86	0.57	0.569	0.82	0.65	0.546	0.93	0.67	0.526	0.97	0.72	0.489	1.04
0.98	0.44	0.678	0.57	0.45	0.621	0.62	0.51	0.606	0.67	0.54	0.574	0.71	0.58	0.535	0.76
1.00	0.32	0.710	0.41	0.36	0.665	0.42	0.39	0.641	0.47	0.41	0.618	0.49	0.46	0.578	0.52
1.01	0.29	0.721	0.37	0.37	0.685	0.40	0.35	0.665	0.43	0.38	0.643	0.45	0.41	0.602	0.48
1.02	0.27	0.733	0.36	0.30	0.699	0.38	0.32	0.681	0.40	0.34	0.664	0.42	0.37	0.620	0.44
1.03	0.25	0.742	0.35	0.28	0.710	0.37	0.29	0.695	0.38	0.31	0.678	0.39	0.33	0.645	0.41
1.04	0.23	0.752	0.33	0.26	0.722	0.34	0.27	0.706	0.35	0.28	0.692	0.36	0.30	0.660	0.38
1.05	0.21	0.760	0.31	0.24	0.731	0.32	0.25	0.717	0.33	0.26	0.703	0.34	0.28	0.674	0.35
1.06	0.19	0.768	0.29	0.22	0.741	0.30	0.23	0.727	0.31	0.24	0.713	0.32	0.26	0.685	0.33
1.07	0.18	0.776	0.28	0.20	0.750	0.29	0.21	0.736	0.29	0.23	0.723	0.30	0.24	0.697	0.31
1.08	0.17	0.783	0.26	0.19	0.759	0.27	0.20	0.746	0.27	0.21	0.733	0.28	0.23	0.706	0.29
1.09	0.16	0.789	0.24	0.18	0.766	0.25	0.19	0.753	0.26	0.20	0.741	0.26	0.21	0.716	0.27
1.10	0.15	0.796	0.22	0.17	0.772	0.22	0.18	0.762	0.23	0.19	0.750	0.23	0.20	0.726	0.24
1.12	0.13	0.809	0.20	0.15	0.787	0.20	0.16	0.776	0.21	0.16	0.766	0.21	0.18	0.745	0.21
1.14	0.12	0.822	0.17	0.13	0.802	0.17	0.14	0.792	0.18	0.14	0.781	0.18	0.16	0.761	0.18
1.16	0.11	0.832	0.14	0.11	0.812	0.14	0.12	0.802	0.15	0.12	0.793	0.15	0.13	0.774	0.16
1.18	0.10	0.844	0.12	0.10	0.825	0.13	0.10	0.816	0.14	0.11	0.807	0.14	0.11	0.790	0.14
1.20	0.09	0.855	0.11	0.09	0.838	0.12	0.09	0.829	0.13	0.10	0.821	0.13	0.10	0.805	0.13
1.30	0.05	0.890	0.07	0.05	0.878	0.07	0.05	0.872	0.08	0.05	0.866	0.08	0.05	0.854	0.08
1.40	0.02	0.920	0.05	0.02	0.910	0.05	0.02	0.905	0.05	0.02	0.903	0.05	0.02	0.892	0.05
1.50	−0.01	0.937	0.06	−0.01	0.931	0.06	−0.01	0.927	0.06	−0.01	0.923	0.06	−0.01	0.916	0.06
1.60	−0.03	0.951	0.07	−0.03	0.946	0.07	−0.03	0.944	0.07	−0.03	0.941	0.07	−0.03	0.935	0.07
1.70	−0.05	0.961	0.09	−0.05	0.957	0.09	−0.05	0.955	0.09	−0.05	0.953	0.09	−0.05	0.949	0.09
1.80	−0.07	0.969	0.11	−0.07	0.966	0.11	−0.07	0.965	0.11	−0.07	0.964	0.11	−0.07	0.961	0.11
1.90	−0.08	0.975	0.13	−0.08	0.973	0.13	−0.08	0.972	0.13	−0.08	0.971	0.13	−0.08	0.970	0.13
2.00		0.980	0.15		0.979	0.15		0.978	0.15		0.977	0.15		0.975	0.15

Values of f/p are recorded for $z_c = 0.27$. At other values of z_c, $(f/p)' = (f/p)(10^{D(z_c - 0.27)})$.

TABLE 51 (*Continued*). FUGACITY COEFFICIENTS OF PURE GASES AND LIQUIDS, f/p

	$p_r = 1.40$	$p_r = 1.60$	$p_r = 1.80$	$p_r = 2.0$	$p_r = 4.0$	$p_r = 6.0$	$p_r = 8.0$	$p_r = 10.0$	$p_r = 20.0$	$p_r = 30.0$
T_r	f/p	f/p	f/p	f/p	f/p	f/p	f/p	f/p	f/p	f/p
0.50	0.00092	0.00084	0.00077	0.00072	0.00052	0.00050	0.00053	0.00061	0.00182	0.00709
0.60	0.00978	0.00884	0.00811	0.00798	0.00520	0.00477	0.00511	0.0058	0.0127	0.0396
0.70	0.0556	0.0502	0.0459	0.0426	0.0286	0.0255	0.0254	0.0270	0.0547	0.144
0.80	0.144	0.130	0.119	0.110	0.0727	0.0636	0.062	0.0648	0.121	0.271
0.90	0.310	0.279	0.255	0.236	0.155	0.135	0.131	0.134	0.222	0.469
0.92	0.349	0.315	0.288	0.267	0.175	0.152	0.146	0.150	0.245	0.508
0.94	0.390	0.351	0.321	0.298	0.196	0.169	0.163	0.167	0.269	0.550
0.96	0.433	0.391	0.358	0.332	0.219	0.189	0.182	0.186	0.296	0.595
0.98	0.474	0.429	0.393	0.365	0.241	0.209	0.201	0.204	0.322	0.638
1.00	0.513	0.464	0.426	0.396	0.263	0.228	0.219	0.223	0.348	0.679
1.01	0.540	0.485	0.446	0.414	0.275	0.238	0.228	0.232	0.358	0.697
1.02	0.559	0.507	0.465	0.433	0.288	0.248	0.238	0.242	0.372	0.721
1.03	0.580	0.529	0.485	0.451	0.303	0.261	0.250	0.255	0.392	0.750
1.04	0.599	0.546	0.505	0.470	0.316	0.272	0.262	0.265	0.403	0.769
1.05	0.615	0.563	0.520	0.485	0.327	0.282	0.272	0.276	0.418	0.786
1.06	0.630	0.580	0.536	0.499	0.339	0.294	0.282	0.286	0.430	0.815
1.07	0.644	0.594	0.552	0.515	0.351	0.304	0.292	0.296	0.444	0.835
1.08	0.658	0.610	0.566	0.531	0.362	0.314	0.302	0.306	0.457	0.855
1.09	0.668	0.622	0.581	0.545	0.374	0.324	0.312	0.315	0.470	0.875
1.10	0.680	0.636	0.594	0.555	0.385	0.334	0.322	0.325	0.484	0.894
1.12	0.701	0.660	0.620	0.585	0.408	0.356	0.342	0.346	0.509	0.934
1.14	0.721	0.682	0.645	0.611	0.431	0.376	0.362	0.366	0.536	0.966
1.16	0.737	0.700	0.664	0.632	0.453	0.396	0.382	0.386	0.561	1.004
1.18	0.756	0.721	0.689	0.658	0.481	0.422	0.408	0.412	0.594	1.051
1.20	0.773	0.741	0.710	0.681	0.504	0.445	0.429	0.434	0.620	1.088
1.30	0.830	0.806	0.783	0.762	0.609	0.548	0.533	0.540	0.756	1.248
1.40	0.874	0.857	0.840	0.824	0.700	0.647	0.634	0.643	0.881	1.420
1.50	0.901	0.888	0.876	0.863	0.768	0.722	0.712	0.724	0.976	1.545
1.60	0.924	0.913	0.903	0.893	0.813	0.766	0.753	0.768	1.030	1.595
1.70	0.942	0.933	0.925	0.918	0.852	0.820	0.817	0.834	1.106	1.688
1.80	0.955	0.949	0.943	0.937	0.886	0.862	0.862	0.882	1.164	1.753
1.90	0.966	0.962	0.958	0.954	0.912	0.893	0.897	0.919	1.201	1.775
2.00	0.971	0.968	0.964	0.960	0.929	0.916	0.923	0.947	1.224	1.784
3.00	1.000	1.000	1.000	0.999	0.989	0.986	0.994	1.016	1.242	1.621
4.00	1.000	1.000	1.000	1.000	0.991	0.989	0.999	1.019	1.204	1.476
6.00	1.000	1.000	1.000	1.000	0.996	0.995	1.005	1.024	1.171	1.372
8.00	1.000	1.000	1.000	1.001	1.001	1.001	1.010	1.027	1.148	1.306
10.00	1.000	1.000	1.003	1.003	1.003	1.004	1.015	1.031	1.146	1.224
15.00	1.000	1.000	1.006	1.008	1.025	1.042	1.057	1.071	1.145	1.220

Illustration 10. Calculate the fugacity of liquid benzene at: (*a*) 428° F, saturated, $p_s = 281$ psia, (*b*) 428° F, 2000 psia.

$$T_c = 562.6° \text{ K}, \qquad p_c = 48.6 \text{ atm}, \qquad z_c = 0.274$$

$$\text{Temperature} = 428° \text{ F} = 220.0° \text{ C} = 493.2° \text{ K}$$

(*a*)

$$T_r = 493.2/562.6 = 0.877$$

$$\text{Pressure} = 281 \text{ psia} = 19.12 \text{ atm}$$

$$p_r = 19.12/48.6 = 0.393$$

From Table 51, using the reduced temperature of 0.877:

$$f/p = 0.806$$

$$f = (0.806)281 = 226 \text{ psia}$$

(*b*)

$$T_r = 0.877$$

$$\text{Pressure} = 2000 \text{ psia} = 136.05 \text{ atm}$$

$$p_r = 136.05/48.6 = 2.800$$

From Table 51, using the logarithmic interpolation formula:

$$f/p = 0.183$$

$$f = (0.183)2000 = 366 \text{ psia}$$

In this instance, even though f/p is far removed from unity, the nonlogarithmic interpolation formula, equation 47, gives the same value of f/p; namely 0.183. This is because z_c does not differ greatly from the standard value of 0.270.

Entropy Departure. By integration of equation 13–104 at constant temperature from pressure p to zero,

$$(\mathrm{s}_p - \mathrm{s}^*_0)_T = -\left[\int_0^p \left(\frac{\partial v}{\partial T}\right)_p dp\right]_T \tag{67}$$

Equation 67 is not directly useful, since it expresses the entropy of a substance relative to its entropy at zero pressure where its value is infinite.

To avoid this difficulty, an equation similar to equation 67 is written for an ideal gas. Thus:

$$(\mathrm{s}^*_p - \mathrm{s}^*_0)_T = -\left[\int_0^p \left(\frac{\partial v}{\partial T}\right)_p dp\right]_T = -\left(R\int_0^p \frac{dp}{p}\right)_T \tag{68}$$

Subtracting equation 67 from equation 68 gives

$$(\mathrm{s}^*_p - \mathrm{s}_p)_T = \int_p^0 \left[\frac{R}{p} - \left(\frac{\partial v}{\partial T}\right)_p\right]_T dp \tag{69}$$

Substitution of equation 55 in equation 69 gives

$$(\mathrm{s}^*_p - \mathrm{s}_p)_T = R\int_p^0 \left[\frac{1-z}{p} - \frac{T}{p}\left(\frac{\partial z}{\partial T}\right)_p\right]_T dp \tag{70}$$

In terms of reduced properties, equation 70 becomes

$$(\mathrm{s}^*_p - \mathrm{s}_p)_T = -R\int_0^{p_r} \frac{(1-z)\,dp_r}{p_r} + RT_r\int_0^{p_r} \left(\frac{\partial z}{\partial T_r}\right)_{p_r} \frac{dp_r}{p_r} \tag{71}$$

Combining equations 71 and 58 gives

$$(\mathrm{S}^*_p - \mathrm{S}_p)_T = -R \int_0^{p_r} (1 - z) \frac{dp_r}{p_r} + \frac{\mathrm{H}^* - \mathrm{H}}{T_r T_c} \tag{72}$$

By combining equations 72 and 64 there results at constant temperature

$$(\mathrm{S}^*_p - \mathrm{S}_p)_T = \left[\left(\frac{\mathrm{H}^* - \mathrm{H}}{T_c}\right)\frac{1}{T_r} + R \ln\left(\frac{f}{p}\right)\right]_T \tag{73}$$

Thus the departure of the entropy of a substance from ideal-gas behavior at constant temperature and pressure is obtained by combining the enthalpy departure with the fugacity coefficient according to equation 73.

Values of entropy departures from ideal-gas behavior are tabulated in Table 52 and shown graphically in Fig. 143 for $z_c = 0.27$.

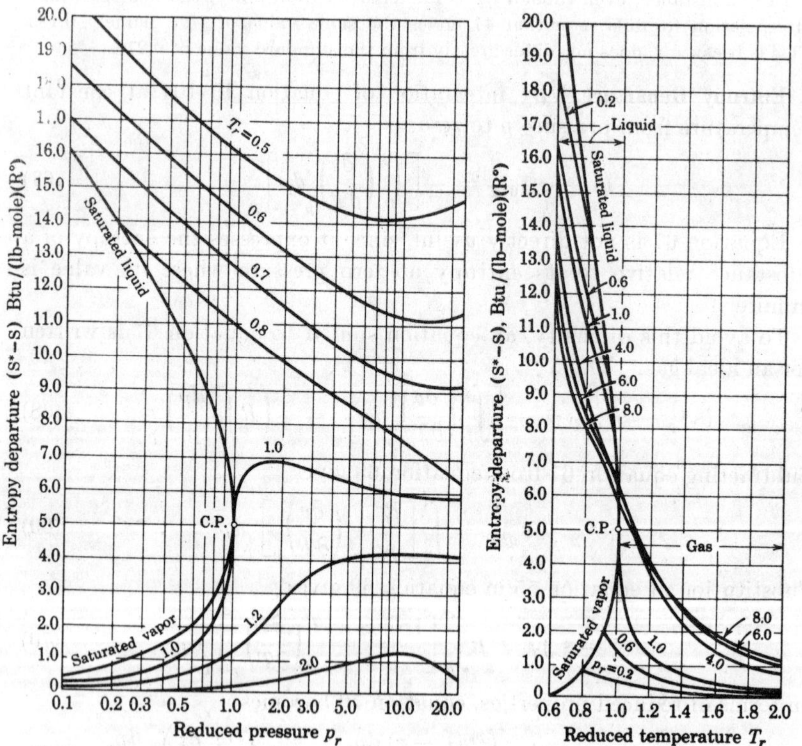

Fig. 143. Entropy departure of gases and liquids, $z_c = 0.27$

TABLE 52. ENTROPY DEPARTURES OF PURE GASES AND LIQUIDS FROM IDEAL-GAS BEHAVIOR, s* − s

Values are given in Btu/(lb-mole)(R°) or in cal/(g-mole)(K°)

	$p_r = 0.01$			$p_r = 0.05$			$p_r = 0.10$			$p_r = 0.20$			$p_r = 0.30$		
T_{rs}		0.59			0.690			0.740			0.804			0.847	
sat. gas								0.469			0.768			1.010	
sat. liquid								16.15			13.77			12.41	
T_r	D_b	s* − s	D_a	D_b	s* − s	D_a	D_b	s* − s	D_a	D_b	s* − s	D_a	D_b	s* − s	D_a
0.50		21.94	−210.0		21.27	−210.0		20.51	−210.0		19.18	−210.0		18.41	−210.0
0.60	−1.4	0.083	−0.3	−40.0	19.44	−183.0	−40.0	18.59	−183.0	−38.0	17.26	−183.0	−37.0	16.49	−183.0
0.70	−0.9	0.057	−0.4	−5.4	0.28	−1.1	−54.0	17.01	−148.0	−54.0	15.69	−148.0	−54.0	14.92	−147.0
0.80	−0.6	0.040	−0.5	−2.6	0.20	−1.2	−5.4	0.40	−4.0	−39.0	13.87	−109.0	−40.0	13.10	−105.0
0.90	−0.4	0.028	−0.3	−1.2	0.14	−1.1	−2.5	0.28	−3.0	−4.6	0.55	−5.2	−7.0	0.83	−8.0
0.92	−0.2	0.026	−0.2	−1.1	0.13	−1.1	−2.1	0.26	−2.8	−3.9	0.50	−5.0	−6.1	0.77	−7.6
0.94	−0.2	0.024	−0.2	−0.9	0.12	−1.1	−1.9	0.24	−2.3	−3.4	0.47	−4.6	−5.3	0.72	−7.0
0.96	−0.2	0.022	−0.2	−0.8	0.11	−1.1	−1.8	0.22	−2.2	−3.0	0.44	−4.4	−4.7	0.66	−6.6
0.98	−0.2	0.021	−0.2	−0.7	0.105	−1.1	−1.6	0.21	−2.1	−2.7	0.40	−4.0	−4.1	0.61	−6.2
1.00	−0.2	0.019	−0.2	−0.7	0.095	−1.1	−1.5	0.19	−2.0	−2.5	0.37	−3.8	−3.4	0.56	−5.6
1.01	−0.1	0.018	−0.2	−0.7	0.09	−1.1	−1.4	0.18	−2.0	−2.4	0.36	−3.8	−3.5	0.53	−5.5
1.02	−0.1	0.017	−0.2	−0.7	0.085	−1.1	−1.3	0.17	−1.9	−2.3	0.35	−3.7	−3.2	0.51	−5.3
1.03	−0.1	0.016	−0.2	−0.7	0.08	−1.1	−1.2	0.16	−1.8	−2.1	0.33	−3.6	−3.1	0.48	−5.1
1.04	−0.1	0.016	−0.2	−0.7	0.08	−1.1	−1.2	0.16	−1.7	−2.0	0.32	−3.5	−3.0	0.46	−4.9
1.05	−0.1	0.015	−0.2	−0.6	0.075	−1.1	−1.1	0.15	−1.6	−1.9	0.30	−3.4	−2.8	0.44	−4.8
1.06	−0.1	0.014	−0.2	−0.6	0.07	−1.0	−1.0	0.14	−1.6	−1.8	0.29	−3.3	−2.6	0.42	−4.6
1.07	−0.1	0.014	−0.2	−0.5	0.07	−1.0	−0.9	0.14	−1.6	−1.7	0.28	−3.2	−2.4	0.40	−4.5
1.08	−0.1	0.014	−0.2	−0.4	0.07	−1.0	−0.8	0.14	−1.5	−1.6	0.27	−3.1	−2.2	0.39	−4.3
1.09	−0.1	0.013	−0.2	−0.3	0.065	−1.0	−0.8	0.13	−1.4	−1.5	0.26	−3.0	−2.1	0.38	−4.2
1.10	−0.1	0.012	−0.2	−0.3	0.06	−1.0	−0.8	0.12	−1.3	−1.4	0.25	−2.9	−2.0	0.36	−4.0
1.12	−0.1	0.012	−0.2	−0.3	0.06	−1.0	−0.7	0.12	−1.2	−1.3	0.24	−2.8	−1.8	0.33	−3.8
1.14	−0.1	0.011	−0.2	−0.2	0.055	−0.9	−0.6	0.11	−1.2	−1.2	0.22	−2.6	−1.7	0.31	−3.5
1.16	−0.1	0.010	−0.2	−0.2	0.05	−0.9	−0.5	0.10	−1.1	−1.1	0.20	−2.4	−1.5	0.29	−3.3
1.18	−0.1	0.010	−0.2	−0.2	0.05	−0.8	−0.5	0.10	−1.1	−1.0	0.19	−2.2	−1.4	0.27	−3.0
1.20	0.0	0.009	−0.2	−0.2	0.045	−0.8	−0.5	0.091	−1.0	−0.9	0.173	−2.0	−1.2	0.25	−2.8
1.30	0.0	0.007	−0.2	−0.2	0.035	−0.6	−0.3	0.071	−0.7	−0.8	0.121	−1.4	−0.7	0.182	−2.0
1.40	0.0	0.006	−0.2	−0.1	0.030	−0.4	−0.2	0.061	−0.4	−0.5	0.091	−0.8	−0.6	0.125	−0.5
1.50	0.0	0.005	−0.1	−0.1	0.023	−0.2	−0.2	0.046	−0.2	−0.3	0.067	−0.6	−0.6	0.100	−0.6
1.60	0.0	0.004	0.0	−0.1	0.019	−0.1	−0.1	0.038	0.0	−0.3	0.053	−0.5	−0.6	0.085	−0.7
1.70	0.0	0.003	0.0	−0.1	0.014	−0.1	−0.1	0.029	0.0	−0.3	0.037	−0.2	−0.5	0.062	−0.3
1.80	0.0	0.002	0.0	−0.1	0.011	0.0	−0.1	0.022	0.0	−0.3	0.024	0.0	−0.5	0.050	0.0
1.90	0.0	0.002	0.0	0.0	0.010	0.0	−0.1	0.020	0.0	−0.3	0.011	+0.1	−0.5	0.037	+0.4
2.00		0.001	0.0		0.007	0.0		0.015	0.0	0.0	0.010	+0.2	0.0	0.028	+0.7

Values of s* − s are recorded for $z_c = 0.27$. At other values of z_c, $(S^* - S)' = (S^* - S) - D(z_c - 0.27)$.

TABLE 52 (*Continued*). ENTROPY DEPARTURES OF PURE GASES AND LIQUIDS FROM IDEAL-GAS BEHAVIOR, s* − s

	$p_r = 0.40$			$p_r = 0.50$			$p_r = 0.60$			$p_r = 0.70$			$p_r = 0.80$		
T_{rs}		0.879			0.907			0.929			0.950			0.967	
sat. gas		1.31			1.62			1.89			2.26			2.67	
sat. liquid		10.85			9.95			9.13			8.28			7.40	
	D_b	s* − s	D_a	D_b	s* − s	D_a	D_b	s* − s	D_a	D_b	s* − s	D_a	D_b	s* − s	D_a
T_r															
0.50		17.89	−210.0		17.48	−210.0		17.15	−210.0		16.87	−210.0		16.64	−210.0
0.60	−37.0	15.95	−183.0	−37.0	15.53	−183.0	−37.0	15.23	−183.0	−38.0	14.95	−183.0	−38.0	14.82	−183.0
0.70	−54.0	14.41	−146.0	−53.0	13.99	−145.0	−51.9	13.66	−143.0	−51.0	13.38	−141.0	−50.0	13.15	−138.0
0.80	−41.0	12.58	−102.0	−43.0	12.18	−98.0	−46.0	11.84	−93.0	−46.0	11.56	−89.0	−45.0	11.32	−85.0
0.90	−10.0	1.20	−11.0	−33.0	10.45	−76.0	−34.0	10.12	−69.0	−35.0	9.84	−62.0	−35.0	9.62	−57.0
0.92	−8.2	1.10	−10.2	−13.0	1.46	−12.9	−28.9	9.70	−65.0	−31.0	9.43	−58.0	−33.0	9.19	−52.0
0.94	−7.0	1.01	−9.2	−10.0	1.32	−11.6	−15.0	1.72	−14.0	−29.7	9.00	−56.0	−31.0	8.78	−50.0
0.96	−6.0	0.91	−8.6	−8.0	1.19	−10.9	−11.0	1.53	−13.4	−16.0	2.07	−15.4	−29.0	8.18	−46.0
0.98	−5.0	0.83	−8.2	−7.0	1.08	−10.2	−9.0	1.39	−12.6	−12.0	1.81	−15.0	−17.0	2.40	−19.0
1.00	−4.6	0.74	−7.8	−6.0	0.99	−9.6	−8.0	1.25	−11.8	−11.0	1.58	−14.0	−13.0	2.05	−17.8
1.01	−4.3	0.71	−7.6	−5.7	0.95	−9.2	−7.5	1.19	−12.4	−10.0	1.50	−13.6	−11.3	1.91	−17.0
1.02	−4.0	0.68	−7.4	−5.4	0.91	−9.0	−7.0	1.13	−11.2	−9.0	1.43	−13.0	−9.6	1.80	−16.2
1.03	−3.8	0.65	−7.2	−5.0	0.87	−8.6	−6.4	1.08	−11.0	−8.0	1.35	−13.0	−8.8	1.70	−15.2
1.04	−3.6	0.63	−7.0	−4.6	0.83	−8.4	−5.9	1.04	−10.6	−7.0	1.28	−12.7	−8.0	1.62	−14.9
1.05	−3.4	0.60	−6.8	−4.3	0.79	−8.0	−5.4	0.99	−10.2	−6.2	1.22	−12.4	−7.3	1.53	−14.6
1.06	−3.2	0.57	−6.5	−4.0	0.77	−7.9	−5.0	0.96	−9.9	−5.4	1.17	−11.9	−6.6	1.46	−14.3
1.07	−3.1	0.55	−6.2	−3.8	0.74	−7.8	−4.6	0.91	−9.6	−5.1	1.12	−11.4	−5.9	1.38	−14.0
1.08	−3.0	0.52	−6.0	−3.6	0.72	−7.6	−4.2	0.88	−9.3	−4.8	1.06	−11.0	−5.3	1.33	−13.5
1.09	−2.9	0.51	−5.8	−3.4	0.68	−7.4	−4.0	0.84	−9.0	−4.5	1.01	−10.6	−5.0	1.26	−13.0
1.10	−2.8	0.50	−5.6	−3.2	0.66	−7.0	−3.8	0.81	−9.0	−4.2	0.98	−10.0	−4.7	1.21	−12.0
1.12	−2.5	0.46	−5.2	−2.9	0.61	−6.6	−3.4	0.75	−8.0	−3.8	0.91	−9.8	−4.1	1.11	−11.4
1.14	−2.2	0.43	−4.9	−2.6	0.57	−6.2	−3.0	0.70	−7.5	−3.4	0.85	−9.0	−3.6	1.03	−10.3
1.16	−2.1	0.40	−4.6	−2.3	0.52	−5.8	−2.9	0.65	−7.0	−3.2	0.80	−8.2	−3.3	0.96	−9.2
1.18	−2.0	0.37	−4.3	−2.1	0.49	−5.4	−2.8	0.62	−6.5	−3.0	0.75	−7.7	−3.1	0.89	−8.3
1.20	−1.5	0.342	−4.0	−1.9	0.46	−5.0	−2.4	0.578	−6.0	−2.5	0.709	−7.2	−3.0	0.84	−7.4
1.30	−1.2	0.240	−2.9	−1.3	0.320	−3.5	−1.4	0.416	−4.1	−1.6	0.534	−4.8	−1.6	0.61	−5.0
1.40	−0.9	0.176	−1.9	−1.0	0.254	−2.0	−1.1	0.311	−2.2	−1.3	0.401	−2.4	−1.3	0.46	−2.6
1.50	−0.6	0.133	−1.4	−0.8	0.184	−1.5	−0.8	0.233	−1.7	−1.0	0.300	−1.6	−1.0	0.35	−1.7
1.60		0.106	−0.9	−0.6	0.144	−1.1	−0.8	0.172	−1.2	−0.9	0.235	−0.8	−0.9	0.268	−0.8
1.70	−0.5	0.083	−0.4	−0.5	0.117	−0.5	−0.8	0.140	−0.6	−0.8	0.183	−0.4	−0.8	0.221	−0.4
1.80	−0.5	0.066	0.0	−0.5	0.093	0.0	−0.8	0.117	0.0	−0.8	0.150	0.0	−0.8	0.171	−0.1
1.90	−0.5	0.049	+0.4	−0.5	0.076	+0.4	−0.8	0.094	+0.4	−0.8	0.126	+0.3	−0.8	0.144	+0.3
2.00		0.039	+0.7		0.055	+0.7		0.076	+0.7		0.104	+0.6		0.122	+0.6

Values of s* − s are recorded for $z_c = 0.27$. At other values of z_c, $(s^* - s)' = (s^* - s) + D(z_c - 0.27)$.

Table 52 (*Continued*). Entropy Departures of Pure Gases and Liquids from Ideal-Gas Behavior, s* − s

	$p_r = 0.90$			$p_r = 1.00$			$p_r = 1.05$			$p_r = 1.10$			$p_r = 1.20$		
T_{rs}		0.984			1.000										
sat. gas		3.25			4.99										
sat. liquid		6.28			4.99										
	D_b	s* − s	D_a	D_b	s* − s	D_a	D_b	s* − s	D_a	D_b	s* − s	D_a	D_b	s* − s	D_a
T_r															
0.50		16.43	−210.0		16.31	−210.0		16.15	−210.0		16.08	−210.0		15.95	−210.0
0.60	−39.0	14.65	−183.0	−40.0	14.43	−183.0	−40.0	14.26	−183.0	−40.0	14.19	−183.0	−42.0	14.02	−183.0
0.70	−48.0	12.95	−135.0	−45.0	12.77	−131.0	−44.0	12.69	−128.0	−43.0	12.60	−125.0	−38.9	12.40	−115.0
0.80	−43.0	11.14	−81.0	−40.0	10.96	−76.0	−40.0	10.89	−73.0	−38.0	10.83	−71.0	−35.0	10.68	−65.0
0.90	−35.0	9.42	−52.0	−35.0	9.26	−47.0	−34.0	9.19	−46.0	−33.0	9.14	−44.0	−29.0	9.05	−43.0
0.92	−33.0	9.02	−47.0	−33.0	8.88	−42.0	−32.0	8.83	−41.0	−31.0	8.78	−40.1	−28.0	8.68	−40.0
0.94	−31.0	8.59	−44.0	−30.0	8.49	−40.0	−29.0	8.40	−39.0	−28.1	8.35	−38.0	−27.0	8.27	−37.0
0.96	−28.0	8.02	−41.0	−28.0	7.95	−36.0	−28.0	7.86	−35.0	−27.0	7.83	−34.0	−25.0	7.79	−34.0
0.98	−25.0	7.23	−38.0	−26.0	7.20	−34.0	−25.3	7.20	−32.0	−25.0	7.20	−32.0	−22.0	7.29	−31.0
1.00	−17.0	2.60	−21.0	−24.0	4.99	−25.0	−24.0	6.10	−28.0	−23.0	6.40	−29.0	−20.0	6.60	−29.5
1.01	−15.0	2.40	−20.0	−17.0	3.46	−23.0	−17.0	4.60	−25.0	−17.0	5.25	−26.0	−15.0	6.05	−28.0
1.02	−13.6	2.24	−19.0	−14.0	3.07	−21.5	−14.0	3.87	−23.2	−13.8	4.40	−24.2	−13.0	5.30	−26.5
1.03	−11.4	2.13	−18.0	−12.0	2.78	−20.6	−11.9	3.24	−21.4	−11.9	3.66	−22.5	−10.5	4.40	−25.0
1.04	−9.2	2.01	−17.4	−10.0	2.54	−19.8	−9.9	2.87	−20.2	−10.0	3.19	−21.7	−8.0	3.78	−23.5
1.05	−8.1	1.89	−16.8	−8.3	2.33	−19.0	−8.3	2.59	−20.0	−8.4	2.84	−21.0	−7.4	3.30	−22.0
1.06	−7.0	1.80	−16.0	−6.7	2.17	−18.0	−6.8	2.39	−18.0	−6.9	2.58	−20.0	−6.8	2.97	−21.0
1.07	−6.5	1.70	−15.2	−6.3	2.04	−17.1	−6.4	2.22	−17.0	−6.5	2.38	−19.0	−6.3	2.71	−20.0
1.08	−6.0	1.63	−14.5	−6.0	1.92	−16.2	−6.0	2.08	−16.0	−6.1	2.23	−18.1	−5.8	2.52	−19.3
1.09	−5.5	1.54	−13.8	−5.5	1.78	−15.4	−5.5	1.93	−15.0	−5.5	2.07	−17.2	−5.4	2.36	−18.6
1.10	−5.0	1.47	−13.0	−5.0	1.69	−15.0	−5.0	1.82	−15.0	−5.0	1.95	−16.0	−5.0	2.21	−17.0
1.12	−4.4	1.34	−12.0	−4.4	1.53	−13.8	−4.4	1.64	−14.4	−4.4	1.78	−15.0	−4.4	2.00	−15.1
1.14	−3.8	1.23	−11.0	−3.8	1.41	−12.4	−3.8	1.50	−12.5	−3.8	1.60	−13.3	−3.8	1.80	−13.5
1.16	−3.5	1.13	−10.0	−3.5	1.29	−11.0	−3.5	1.37	−10.6	−3.5	1.47	−11.6	−3.5	1.65	−12.0
1.18	−3.3	1.06	−9.0	−3.3	1.20	−9.7	−3.3	1.28	−9.6	−3.3	1.37	−10.1	−3.3	1.54	−10.9
1.20	−3.1	0.99	−8.0	−3.1	1.13	−8.4	−3.1	1.21	−8.6	−3.1	1.29	−8.7	−3.1	1.44	−9.9
1.30	−1.8	0.70	−5.3	−2.0	0.82	−6.2	−2.0	0.87	−5.6	−2.0	0.93	−5.6	−2.0	1.03	−6.9
1.40	−1.4	0.53	−2.6	−1.5	0.62	−4.0	−1.5	0.67	−2.6	−1.5	0.71	−2.6	−1.5	0.79	−4.0
1.50	−1.0	0.39	−1.7	−1.0	0.46	−2.9	−1.0	0.50	−1.8	−1.0	0.54	−1.8	−1.0	0.60	−2.9
1.60	−0.9	0.30	−0.8	−0.9	0.35	−1.8	−0.9	0.38	−1.0	−0.9	0.41	−1.0	−0.9	0.46	−1.8
1.70	−0.8	0.25	−0.3	−0.8	0.283	−0.9	−0.8	0.30	−0.6	−0.8	0.32	−0.6	−0.8	0.37	−1.0
1.80	−0.8	0.20	−0.2	−0.8	0.231	−0.1	−0.8	0.25	−0.2	−0.8	0.27	−0.2	−0.8	0.30	−0.2
1.90	−0.8	0.18	−0.4	−0.8	0.200	+0.2	−0.8	0.21	+0.2	−0.8	0.22	+0.2	−0.8	0.25	+0.2
2.00		0.16	−0.6		0.178	+0.5		0.19	+0.5		0.20	+0.5		0.21	+0.5

Values of s* − s are recorded for $z_c = 0.27$. At other values of z_c, $(s - s^*)' = (s^* - s) + D(z_c - 0.27)$.

TABLE 52 (*Continued*). ENTROPY DEPARTURES OF PURE GASES AND LIQUIDS FROM IDEAL-GAS BEHAVIOR, s* − s

	$p_r = 1.40$	$p_r = 1.60$	$p_r = 1.80$	$p_r = 2.00$	$p_r = 4.0$	$p_r = 6.0$	$p_r = 8.0$	$p_r = 10.0$	$p_r = 20.0$	$p_r = 30.0$
T_r	s* − s	s* − s	s* − s	s* − s	s* − s	s* − s	s* − s	s* − s	s* − s	s* − s
0.50	15.71	15.53	15.35	15.22	14.43	14.17	14.05	14.09	14.60	15.30
0.60	13.78	13.56	13.36	13.15	12.06	11.59	11.45	11.35	11.19	11.25
0.70	12.18	11.95	11.77	11.59	10.45	9.91	9.60	9.37	9.04	8.93
0.80	10.37	10.16	9.94	9.81	8.80	8.26	7.99	7.77	7.47	7.41
0.90	8.91	8.78	8.67	8.55	7.69	7.25	7.05	6.92	6.71	6.91
0.92	8.56	8.43	8.40	8.29	7.47	7.07	6.87	6.78	6.60	6.79
0.94	8.21	8.12	8.05	7.98	7.21	6.89	6.70	6.53	6.41	6.62
0.96	7.78	7.71	7.70	7.63	6.97	6.67	6.48	6.40	6.30	6.50
0.98	7.33	7.27	7.27	7.25	6.72	6.45	6.29	6.24	6.15	6.35
1.00	6.82	6.84	6.85	6.84	6.40	6.25	6.13	6.04	5.96	6.17
1.01	6.42	6.49	6.65	6.62	6.27	6.11	5.99	5.93	5.82	6.05
1.02	6.10	6.30	6.43	6.44	6.14	6.01	5.89	5.82	5.69	5.92
1.03	5.53	5.99	6.17	6.20	6.01	5.91	5.80	5.72	5.62	5.83
1.04	4.89	5.63	5.88	5.98	5.88	5.79	5.69	5.62	5.49	5.72
1.05	4.30	5.19	5.52	5.69	5.72	5.68	5.58	5.50	5.40	5.60
1.06	3.89	4.80	5.19	5.38	5.57	5.53	5.45	5.37	5.28	5.51
1.07	3.52	4.28	4.77	5.05	5.41	5.39	5.32	5.22	5.17	5.41
1.08	3.23	3.81	4.35	4.71	5.24	5.25	5.19	5.12	5.14	5.37
1.09	2.97	3.53	3.98	4.38	5.08	5.11	5.06	5.06	5.00	5.26
1.10	2.80	3.31	3.71	4.04	4.92	4.99	4.97	4.96	4.88	5.17
1.12	2.44	2.91	3.27	3.56	4.68	4.78	4.74	4.71	4.66	4.97
1.14	2.20	2.62	2.99	3.23	4.45	4.58	4.56	4.56	4.47	4.81
1.16	2.00	2.36	2.67	2.97	4.27	4.38	4.38	4.33	4.32	4.63
1.18	1.85	2.20	2.50	2.76	4.01	4.21	4.23	4.22	4.21	4.51
1.20	1.72	2.00	2.29	2.57	3.79	4.04	4.06	4.03	4.03	4.32
1.30	1.23	1.43	1.58	1.81	2.95	3.26	3.35	3.37	3.41	3.66
1.40	0.92	1.06	1.20	1.32	2.29	2.66	2.74	2.80	2.90	3.15
1.50	0.69	0.82	0.94	1.00	1.82	2.14	2.23	2.36	2.48	2.74
1.60	0.55	0.64	0.72	0.79	1.39	1.70	1.80	1.90	2.10	2.34
1.70	0.45	0.51	0.59	0.65	1.08	1.42	1.55	1.67	1.85	2.10
1.80	0.36	0.41	0.48	0.53	0.88	1.21	1.36	1.47	1.67	1.92
1.90	0.30	0.35	0.40	0.45	0.81	1.06	1.21	1.31	1.52	1.74
2.00	0.25	0.28	0.32	0.36	0.72	0.97	1.11	1.20	1.39	1.62
3.00	0.08	0.09	0.10	0.11	0.15	0.19	0.22	0.24	0.35	0.32
4.00	0.00		0.01	0.03	0.06	0.01	0.05	0.05	0.31	0.40
6.00				0.00		0.01	0.00	0.05	0.29	0.55
8.00				0.00		0.00	−0.02	0.06	0.27	0.50
10.00				−0.01		−0.01	−0.03	0.06	0.23	0.40
15.00				−0.01		−0.07		0.15	0.28	0.44

A separate table of values of $s^*_p - s_p$ for the saturated vapor and liquid as unique functions of p_{rs} and T_{rs} is given in Table 47 for three values of z_c.

Illustration 11. The absolute entropy of carbon dioxide in the ideal gaseous state at 25° C and 1 atm is 51.061 g-cal/(g-mole)(K°). Calculate the absolute entropy of (*a*) gaseous CO_2 at 100° C and 100 atm, (*b*) liquid CO_2 at its triple point (−69.9° F, 216.6° K, 75.1 psia).

$$T_c = 304.2° \text{ K}, \qquad p_c = 72.9 \text{ atm}, \qquad z_c = 0.275$$

$$c_p = 5.316 + (1.4285)10^{-2}T - (0.8362)10^{-5}T^2 + (1.784)10^{-9}T^3$$

(*a*)

$$T_r = 373.2/304.2 = 1.227$$

$$p_r = 100/72.9 = 1.372$$

From Table 52, $(s^* - s) = 1.51$ g-cal/(g-mole)(K°)

$$s = 51.06 + \int_{298.2}^{373.2} \frac{c_p}{T}\,dT - R \ln \frac{100}{1} - (s^* - s)$$

$$= 51.061 + 2.069 - 9.151 - 1.51 = 42.47 \text{ g-cal/(g-mole)(K°)}$$

(*b*)

$$T_r = 216.6/304.2 = 0.712$$

$$p_r = 75.1/(14.70)(72.9) = 0.0701$$

From Table 47, $(s^* - s)_s$ for saturated liquid is 16.50 g-cal/(g-mole)(K°).

$$s = 51.061 + \int_{298.2}^{216.6} \frac{c_p}{T}\,dT - R \ln \frac{75.1}{14.70} - (s^* - s)$$

$$= 51.061 - 2.699 - 3.241 - 16.50$$

$$= 28.62 \text{ g-cal/(g-mole)(K°)}$$

Internal-Energy Departure. Internal energy may be obtained from enthalpy and compressibility data from the relation

$$H = U + zRT \tag{74}$$

The departure of internal energy from ideal-gas behavior is hence expressed as

$$\frac{U^* - U}{T_c} = \frac{H^* - H}{T_c} - (1 - z)RT_r \tag{75}$$

Values of $(U^* - U)/T_c$ are given in Table 53. Saturation values appear in Table 47. For $z_c = 0.27$, values are shown graphically in Fig. 144.

Heat-Capacity Departures. The molal heat capacity of a substance at constant pressure can be defined in terms of enthalpy for any substance and for an ideal gas as follows:

$$c_p = \left(\frac{\partial H}{\partial T}\right)_p \quad \text{and} \quad c^*_p = \left(\frac{\partial H^*}{\partial T}\right)_p \tag{76}$$

Hence

$$c_p - c^*_p = \frac{\partial\left(\dfrac{H - H^*}{T_c}\right)}{\partial T_r} \tag{77}$$

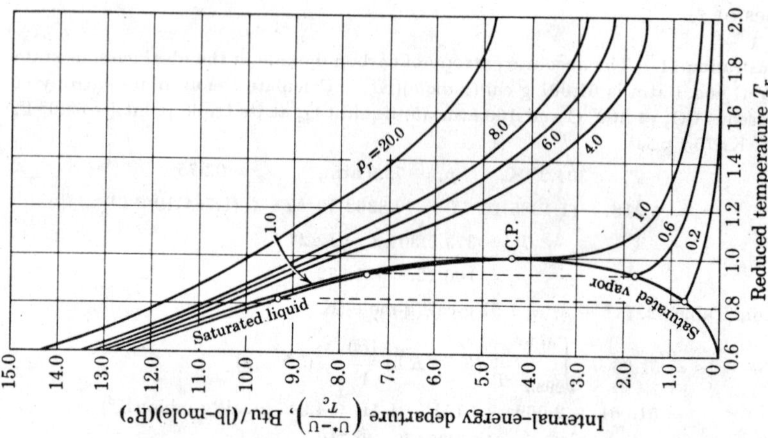

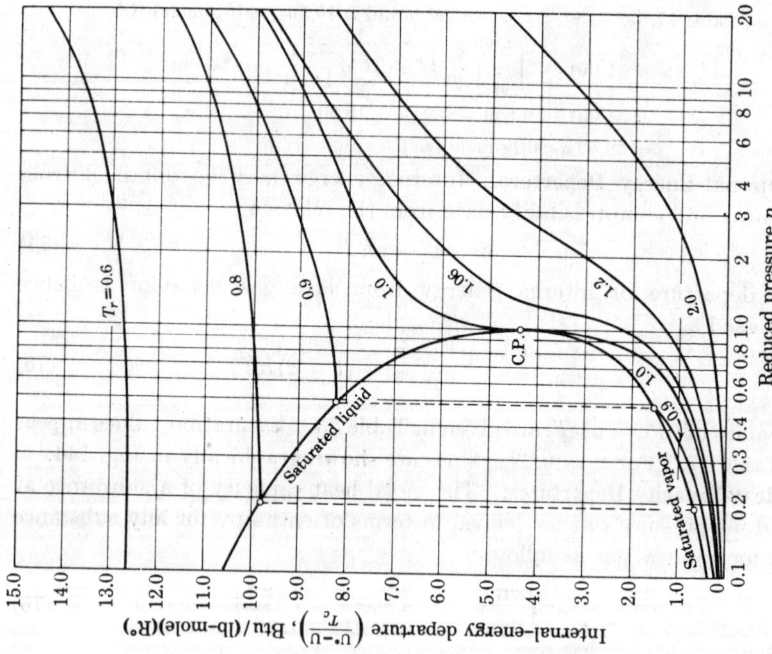

Fig. 144. Internal-energy departure of gases and liquids, $z_c = 0.27$

TABLE 53. INTERNAL-ENERGY DEPARTURES OF PURE GASES AND LIQUIDS, $(U^* - U)/T_c$

Values are in Btu/(lb-mole)(R°) or in cal/(g-mole)(K°)

	$p_r = 0.01$			$p_r = 0.05$			$p_r = 0.1$			$p_r = 0.2$			$p_r = 0.30$		
T_{rs}		0.59			0.690			0.740			0.804			0.847	
sat. gas		0.022			0.14			0.280			0.540			0.790	
sat. liquid		12.71			11.31			10.63			9.73			9.18	
T_r	D_b	$\frac{U^* - U}{T_c}$	D_a	D_b	$\frac{U^* - U}{T_c}$	D_a	D_b	$\frac{U^* - U}{T_c}$	D_a	D_b	$\frac{U^* - U}{T_c}$	D_a	D_b	$\frac{U^* - U}{T_c}$	D_a
0.50		13.80	−180.0		13.81	−180.0		13.82	−180.0		13.84	−180.0		13.86	−182.0
0.60	−0.7	0.045	−0.1	−52.0	12.58	−149.0	−52.0	12.58	−150.0	−52.0	12.61	−150.0	−52.0	12.63	−150.0
0.70	−0.5	0.031	−0.1	−3.9	0.150	−1.1	−53.0	11.15	−122.0	−53.0	11.18	−121.0	−52.0	11.21	−121.0
0.80	−0.3	0.025	−0.1	−2.1	0.125	−1.0	−4.3	0.25	−2.1	−38.0	9.82	−96.0	−39.0	9.85	−95.0
0.90	−0.2	0.020	−0.2	−1.0	0.110	−1.0	−2.0	0.20	−2.1	−3.5	0.44	−6.0	−6.1	0.72	−9.4
0.92	−0.2	0.019	−0.2	−0.8	0.100	−1.0	−1.8	0.19	−2.1	−3.4	0.42	−5.2	−5.3	0.68	−8.2
0.94	−0.2	0.018	−0.2	−0.7	0.090	−1.0	−1.6	0.18	−2.0	−2.9	0.39	−4.5	−4.7	0.64	−7.0
0.96	−0.1	0.017	−0.2	−0.6	0.085	−1.0	−1.4	0.17	−2.0	−2.5	0.37	−4.4	−4.1	0.60	−6.8
0.98	−0.1	0.017	−0.2	−0.5	0.085	−1.0	−1.2	0.17	−2.1	−2.3	0.35	−4.4	−3.7	0.55	−6.6
1.00	−0.1	0.017	−0.2	−0.4	0.085	−1.0	−1.1	0.17	−2.0	−2.1	0.33	−4.1	−3.3	0.52	−6.3
1.01	−0.1	0.016	−0.2	−0.4	0.080	−1.0	−1.0	0.16	−2.0	−2.0	0.32	−4.1	−3.1	0.50	−6.1
1.02	−0.1	0.015	−0.2	−0.4	0.075	−1.0	−1.0	0.15	−2.0	−1.9	0.31	−4.1	−3.0	0.48	−6.1
1.03	−0.1	0.014	−0.2	−0.4	0.072	−1.0	−0.9	0.145	−2.0	−1.8	0.31	−4.0	−2.8	0.46	−6.0
1.04	−0.1	0.014	−0.2	−0.4	0.070	−1.0	−0.9	0.14	−2.0	−1.8	0.30	−4.0	−2.7	0.44	−6.0
1.05	−0.1	0.013	−0.2	−0.4	0.065	−1.0	−0.8	0.13	−2.0	−1.7	0.29	−4.1	−2.6	0.43	−6.0
1.06	−0.1	0.012	−0.2	−0.4	0.062	−1.0	−0.8	0.125	−2.0	−1.6	0.28	−3.9	−2.6	0.41	−5.8
1.07	−0.1	0.012	−0.2	−0.3	0.060	−1.0	−0.7	0.12	−2.0	−1.5	0.27	−3.8	−2.4	0.40	−5.6
1.08	−0.1	0.012	−0.2	−0.3	0.060	−1.0	−0.7	0.12	−1.9	−1.5	0.26	−3.8	−2.3	0.38	−5.5
1.09	−0.1	0.012	−0.2	−0.3	0.060	−1.0	−0.6	0.12	−1.8	−1.4	0.26	−3.9	−2.2	0.38	−5.4
1.10	−0.1	0.011	−0.2	−0.3	0.055	−0.9	−0.6	0.11	−1.8	−1.3	0.25	−3.8	−2.1	0.36	−5.4
1.12	−0.1	0.011	−0.2	−0.3	0.055	−0.8	−0.5	0.11	−1.9	−1.2	0.235	−3.8	−1.9	0.34	−5.4
1.14	−0.1	0.010	−0.2	−0.3	0.050	−0.8	−0.4	0.10	−1.7	−1.2	0.22	−3.5	−1.8	0.33	−5.0
1.16	−0.1	0.009	−0.2	−0.3	0.045	−0.8	−0.4	0.09	−1.6	−1.1	0.21	−3.2	−1.7	0.31	−4.6
1.18	−0.1	0.009	−0.2	−0.3	0.045	−0.7	−0.4	0.09	−1.5	−1.1	0.20	−2.9	−1.6	0.29	−4.3
1.20	−0.1	0.008	−0.2	−0.2	0.040	−0.7	−0.4	0.08	−1.4	−1.0	0.19	−2.6	−1.3	0.27	−4.0
1.30	0.0	0.007	−0.1	−0.1	0.035	−0.4	−0.3	0.07	−1.0	−0.7	0.145	−1.9	−1.0	0.22	−2.8
1.40	0.0	0.005	−0.1	−0.1	0.025	−0.4	−0.2	0.05	−0.7	−0.5	0.12	−1.3	−0.8	0.16	−2.0
1.50	0.0	0.004	0.0	−0.1	0.022	−0.2	−0.2	0.045	−0.4	−0.4	0.10	−0.9	−0.7	0.14	−1.4
1.60	0.0	0.004	0.0	−0.1	0.020	−0.1	−0.1	0.040	−0.2	−0.3	0.08	−0.5	−0.5	0.12	−0.8
1.70	0.0	0.003	0.0	0.0	0.015	0.0	−0.1	0.03	−0.1	−0.3	0.06	−0.3	−0.4	0.10	−0.5
1.80	0.0	0.002	0.0	0.0	0.012	0.0	−0.1	0.025	−0.1	−0.2	0.05	−0.2	−0.3	0.09	−0.2
1.90	0.0	0.002	0.0	0.0	0.010	0.0	−0.1	0.02	−0.1	−0.2	0.04	−0.1	−0.2	0.08	−0.1
2.00		0.001		0.0	0.005			0.01	0.0		0.03		0.0	0.07	0.0
3.00		0.001			0.005			0.01			0.07				
4.00															

Values of $\left(\frac{U^* - U}{T_c}\right)$ are recorded for $z_c = 0.27$. At other values of z_c, $\left(\frac{U^* - U}{T_c}\right)' = \left(\frac{U^* - U}{T_c}\right) + D(z_c - 0.27)$.

TABLE 53 (*Continued*). INTERNAL-ENERGY DEPARTURES OF PURE GASES AND LIQUIDS, $(U^* - U)/T_c$

	$p_r = 0.40$			$p_r = 0.50$			$p_r = 0.60$			$p_r = 0.7$			$p_r = 0.8$		
	D_b	$\frac{U^* - U}{T_c}$	D_a	D_b	$\frac{U^* - U}{T_c}$	D_a	D_b	$\frac{U^* - U}{T_c}$	D_a	D_b	$\frac{U^* - U}{T_c}$	D_a	D_b	$\frac{U^* - U}{T_c}$	D_a
T_{rs}		0.879			0.907			0.929			0.950			0.967	
sat. gas		1.06			1.36			1.62			1.95			2.32	
sat. liquid		8.70			8.21			7.71			7.21			6.63	
T_r															
0.50		13.88	−183.0		13.90	−184.0		13.92	−185.0		13.94	−186.0		13.96	−187.0
0.60	−51.0	12.65	−151.0	−51.0	12.67	−151.0	−51.0	12.70	−151.0	−51.0	12.71	−152.0	−50.0	12.73	−152.0
0.70	−51.0	11.24	−121.0	−51.0	11.27	−121.0	−50.0	11.30	−121.0	−50.0	11.32	−120.0	−49.0	11.34	−119.0
0.80	−41.0	9.88	−93.0	−42.0	9.92	−91.0	−43.0	9.95	−89.0	−43.0	9.97	−87.0	−44.0	9.99	−84.0
0.90	−8.0	0.96	−13.0	−32.0	8.26	−56.0	−32.0	8.28	−55.0	−33.0	8.30	−54.0	−35.0	8.34	−53.0
0.92	−7.0	0.91	−11.2	−10.0	1.24	−13.7	−30.0	7.86	−48.0	−31.0	7.89	−48.0	−32.0	7.92	−47.0
0.94	−6.2	0.84	−9.4	−8.9	1.15	−12.0	−12.0	1.49	−14.2	−28.0	7.47	−39.0	−28.0	7.51	−39.0
0.96	−5.5	0.78	−9.1	−7.8	1.06	−11.5	−10.3	1.37	−13.7	−13.0	1.82	−16.4	−24.0	6.91	−32.0
0.98	−5.1	0.74	−8.8	−6.9	0.99	−11.0	−9.0	1.28	−13.2	−11.0	1.65	−15.4	−14.0	2.16	−18.0
1.00	−4.7	0.69	−8.3	−6.1	0.93	−10.2	−7.8	1.17	−12.4	−9.8	1.49	−14.6	−12.0	1.92	−17.2
1.01	−4.4	0.66	−8.1	−5.8	0.90	−10.2	−7.3	1.13	−12.2	−8.5	1.43	−14.6	−10.9	1.78	−17.0
1.02	−4.2	0.64	−8.0	−5.5	0.87	−10.1	−6.8	1.09	−12.1	−8.0	1.38	−14.3	−9.8	1.73	−16.3
1.03	−3.9	0.63	−8.0	−5.1	0.85	−10.0	−6.3	1.05	−12.0	−7.4	1.32	−14.0	−8.9	1.66	−16.0
1.04	−3.7	0.61	−7.9	−4.7	0.81	−9.7	−5.8	1.01	−11.6	−6.8	1.27	−13.6	−8.0	1.59	−15.7
1.05	−3.5	0.59	−7.8	−4.4	0.78	−9.4	−5.4	0.99	−11.2	−6.2	1.22	−13.2	−7.2	1.52	−15.4
1.06	−3.4	0.56	−7.6	−4.2	0.76	−9.2	−5.0	0.96	−11.0	−5.7	1.18	−12.9	−6.4	1.47	−13.0
1.07	−3.2	0.55	−7.4	−3.9	0.74	−9.0	−4.6	0.92	−10.8	−5.3	1.14	−12.6	−5.9	1.41	−10.6
1.08	−3.0	0.54	−7.2	−3.7	0.73	−8.8	−4.3	0.90	−10.5	−4.9	1.09	−12.3	−5.5	1.36	−12.2
1.09	−2.9	0.52	−7.0	−3.5	0.70	−8.7	−4.3	0.87	−10.3	−4.7	1.05	−12.1	−5.1	1.31	−13.9
1.10	−2.8	0.51	−7.0	−3.3	0.68	−8.5	−4.2	0.84	−10.1	−4.5	1.04	−11.8	−4.8	1.26	−13.4
1.12	−2.5	0.48	−7.0	−3.0	0.64	−8.4	−3.7	0.80	−10.0	−4.1	0.97	−11.5	−4.5	1.19	−13.0
1.14	−2.3	0.45	−6.5	−2.8	0.61	−8.0	−3.3	0.76	−9.3	−3.8	0.92	−10.6	−4.2	1.12	−11.9
1.16	−2.2	0.42	−6.1	−2.7	0.57	−7.7	−3.1	0.72	−8.6	−3.5	0.88	−9.8	−3.8	1.06	−10.9
1.18	−2.1	0.40	−5.7	−2.6	0.54	−7.0	−2.9	0.69	−8.1	−3.2	0.83	−9.2	−3.5	1.00	−10.1
1.20	−2.0	0.38	−5.4	−2.4	0.52	−6.4	−2.7	0.66	−7.6	−3.0	0.80	−8.6	−3.3	0.95	−9.4
1.30	−1.4	0.28	−4.0	−1.8	0.42	−4.6	−1.9	0.51	−5.4	−2.2	0.66	−5.8	−2.5	0.77	−6.1
1.40	−1.1	0.23	−2.5	−1.4	0.34	−3.0	−1.6	0.42	−3.4	−1.8	0.54	−3.6	−2.0	0.63	−3.8
1.50	−0.9	0.19	−1.7	−1.1	0.27	−2.1	−1.3	0.34	−2.4	−1.5	0.45	−2.6	−1.6	0.54	−2.7
1.60	−0.7	0.17	−1.0	−0.9	0.23	−1.2	−1.0	0.29	−1.4	−1.2	0.39	−1.6	−1.3	0.45	−1.7
1.70	−0.5	0.15	−0.7	−0.7	0.21	−0.2	−0.8	0.26	−1.0	−0.9	0.33	−1.1	−1.0	0.40	−1.2
1.80	−0.4	0.13	−0.4	−0.5	0.19	−0.5	−0.6	0.24	−0.6	−0.7	0.30	−0.6	−0.8	0.34	−0.8
1.90	−0.3	0.12	−0.2	−0.4	0.17	−0.3	−0.5	0.20	−0.4	−0.5	0.27	−0.4	−0.6	0.31	−0.5
2.00		0.10	−0.1		0.13	−0.1		0.19	−0.2		0.25	−0.2		0.28	−0.2
3.00		0.12			0.13			0.15			0.16			0.17	

Values of $\left(\frac{U^* - U}{T_c}\right)$ are recorded for $z_c = 0.27$. At other values of z_c, $\left(\frac{U^* - U}{T_c}\right)' = \left(\frac{U^* - U}{T}\right) + D(z_c - 0.27)$.

TABLE 53 (*Continued*). INTERNAL-ENERGY DEPARTURES OF PURE GASES AND LIQUIDS, $(U^* - U)/T_c$

	$p_r = 0.90$			$p_r = 1.0$			$p_r = 1.05$			$p_r = 1.10$			$p_r = 1.20$		
T_r sat. gas sat. liquid		0.984 2.90 5.91			1.000 4.35 4.35										
	D_b	$\frac{U^* - U}{T_c}$	D_a	D_b	$\frac{U^* - U}{T_c}$	D_a	D_b	$\frac{U^* - U}{T_c}$	D_c	D_b	$\frac{U^* - U}{T_c}$	D_a	D_b	$\frac{U^* - U}{T_c}$	D_a
T_r															
0.50		13.98	−188.0		14.00	−189.0		14.02	−190.0		14.02	−190.0		14.04	−191.5
0.60	−50.0	12.75	−152.0	−50.0	12.77	−153.0	−50.0	12.78	−153.0	−50.0	12.79	−153.0	−50.0	12.85	−153.0
0.70	−49.0	11.36	−117.0	−49.0	11.40	−114.0	−49.0	11.39	−112.0	−49.0	11.40	−110.0	−49.0	11.41	−104.0
0.80	−45.0	10.02	−80.0	−46.0	10.04	−76.0	−46.0	10.05	−74.0	−46.0	10.06	−73.0	−46.0	10.09	−69.0
0.90	−35.0	8.37	−52.0	−35.0	8.42	−50.0	−35.0	8.45	−49.0	−36.0	8.49	−48.0	−36.0	8.56	−46.0
0.92	−32.0	7.97	−46.0	−32.0	8.04	−45.0	−33.0	8.07	−44.0	−33.0	8.11	−43.0	−33.0	8.18	−42.0
0.94	−29.0	7.56	−39.0	−29.0	7.61	−39.0	−29.0	7.65	−39.0	−29.0	7.69	−39.0	−30.0	7.77	−39.0
0.96	−25.0	6.97	−33.0	−25.0	7.07	−34.0	−26.0	7.12	−35.0	−26.0	7.17	−35.0	−26.0	7.30	−36.0
0.98	−19.0	6.21	−28.0	−21.0	6.35	−30.0	−21.0	6.43	−31.0	−22.0	6.56	−32.0	−23.0	6.80	−34.0
1.00	−14.2	2.33	−19.8	−16.5	4.35	−23.6	−17.8	5.13	−25.0	−19.0	5.56	−28.0	−21.0	6.14	−33.0
1.01	−12.8	2.21	−19.4	−14.7	3.10	−20.0	−15.9	4.22	−24.4	−15.2	4.81	−26.4	−16.7	5.61	−30.4
1.02	−11.5	2.11	−18.9	−13.0	2.87	−20.7	−14.0	3.64	−23.6	−14.0	4.13	−25.3	−14.0	4.95	−29.2
1.03	−10.2	2.06	−18.4	−11.5	2.65	−21.3	−12.3	3.07	−22.8	−10.6	3.45	−24.3	−11.2	4.16	−28.0
1.04	−9.0	1.96	−18.0	−10.1	2.45	−20.7	−10.7	2.76	−22.0	−10.0	3.07	−23.4	−10.0	3.63	−27.0
1.05	−8.0	1.87	−17.6	−8.8	2.29	−20.0	−9.3	2.53	−21.3	−9.5	2.80	−22.6	−10.0	3.21	−26.0
1.06	−7.0	1.80	−17.0	−7.6	2.16	−19.2	−7.9	2.36	−20.4	−8.1	2.55	−21.6	−8.5	2.92	−24.3
1.07	−6.5	1.73	−16.5	−7.0	2.05	−18.5	−7.3	2.23	−19.5	−7.5	2.37	−20.6	−7.9	2.73	−22.6
1.08	−6.0	1.67	−16.1	−6.5	1.95	−18.1	−6.7	2.09	−19.0	−6.9	2.25	−20.1	−7.3	2.55	−22.1
1.09	−5.6	1.60	−15.8	−6.1	1.84	−17.7	−6.3	1.99	−18.6	−6.5	2.13	−19.6	−6.9	2.41	−21.6
1.10	−5.3	1.54	−15.0	−5.8	1.76	−16.7	−6.0	1.90	−17.5	−6.1	2.03	−18.3	−6.5	2.29	−20.0
1.12	−4.9	1.43	−14.3	−5.3	1.62	−15.8	−5.4	1.74	−16.4	−5.5	1.87	−17.0	−6.9	2.11	−18.4
1.14	−4.5	1.33	−13.1	−4.8	1.52	−14.4	−4.9	1.63	−14.9	−5.0	1.74	−15.6	−7.3	1.96	−16.7
1.16	−4.2	1.26	−12.0	−4.5	1.45	−13.0	−4.5	1.52	−13.4	−5.3	1.63	−14.2	−5.9	1.82	−15.0
1.18	−4.0	1.19	−11.0	−4.2	1.35	−11.8	−4.2	1.45	−12.1	−4.2	1.50	−12.6	−4.5	1.74	−13.0
1.20	−3.7	1.13	−10.1	−3.0	1.29	−10.6	−3.8	1.39	−10.8	−4.0	1.48	−11.0	−4.1	1.66	−11.0
1.30	−2.7	0.89	−6.4	−2.9	1.03	−6.5	−2.9	1.10	−6.5	−3.0	1.18	−6.6	−3.0	1.30	−6.6
1.40	−2.2	0.73	−3.8	−2.4	0.85	−3.8	−2.4	0.91	−3.8	−2.5	0.98	−3.8	−2.5	1.10	−3.8
1.50	−1.8	0.63	−2.8	−1.9	0.70	−2.8	−1.9	0.75	−2.8	−2.0	0.80	−2.9	−2.0	0.89	−2.9
1.60	−1.4	0.50	−1.8	−1.5	0.58	−1.9	−1.5	0.61	−1.9	−1.6	0.66	−2.0	−1.6	0.74	−2.0
1.70	−1.1	0.45	−1.3	−1.2	0.51	−1.3	−1.2	0.52	−1.3	−1.3	0.56	−1.4	−1.2	0.64	−0.6
1.80	−0.8	0.40	−0.8	−0.9	0.45	−0.8	−0.9	0.46	−0.8	−1.0	0.47	−0.9	−0.9	0.55	−0.9
1.90	−0.6	0.37	−0.5	−0.6	0.42	−0.5	−0.6	0.43	−0.5	−0.7	0.45	−0.6	−0.7	0.49	−0.6
2.00		0.35	−0.2		0.38	−0.2		0.40	−0.2		0.42	−0.3		0.45	−0.3
3.00		0.18			0.19			0.20			0.21			0.22	

Values of $\left(\frac{U^* - U}{T_c}\right)$ are recorded for $z_c = 0.27$. At other values of z_c, $\left(\frac{U^* - U}{T_c}\right)' = \left(\frac{U^* - U}{T_c}\right) + D(z_c - 0.27)$.

TABLE 53 (*Continued*). INTERNAL-ENERGY DEPARTURES OF PURE GASES AND LIQUIDS, $(U^* - U)/T_c$

	$p_r = 1.40$	$p_r = 1.6$	$p_r = 1.8$	$p_r = 2.0$	$p_r = 4.0$	$p_r = 6.0$	$p_r = 8.0$	$p_r = 10.0$	$p_r = 20.0$	$p_r = 30.0$
T_r	$\frac{U^* - U}{T_c}$	$\frac{U^* - U}{T_c}$	$\frac{U^* - U}{T_c}$	$\frac{U^* - U}{T_c}$	$\frac{U^* - U}{T_c}$	$\frac{U^* - U}{T_c}$	$\frac{U^* - U}{T_c}$	$\frac{U^* - U}{T_c}$	$\frac{U^* - U}{T_c}$	$\frac{U^* - U}{T_c}$
0.50	14.08	14.12	14.16	14.20	14.47	14.70	14.92	15.15	15.91	16.62
0.60	12.88	12.87	12.89	12.91	13.06	13.24	13.45	13.64	14.40	14.86
0.70	11.44	11.46	11.49	11.51	11.65	11.83	12.01	12.14	12.82	13.23
0.80	10.13	10.17	10.22	10.24	10.50	10.71	10.93	11.06	11.77	12.30
0.90	8.70	8.82	8.93	9.01	9.42	9.73	10.03	10.25	11.14	11.87
0.92	8.35	8.51	8.65	8.73	9.19	9.54	9.85	10.09	11.02	11.76
0.94	7.99	8.16	8.30	8.43	8.93	9.35	9.67	9.92	10.90	11.66
0.96	7.59	7.75	7.94	8.09	8.70	9.14	9.47	9.76	10.76	11.53
0.98	7.15	7.33	7.55	7.71	8.45	8.92	9.29	9.60	10.62	11.39
1.00	6.62	6.84	7.12	7.31	8.13	8.69	9.07	9.40	10.44	11.31
1.01	6.30	6.52	6.91	7.08	7.98	8.57	8.97	9.30	10.32	11.24
1.02	5.92	6.33	6.68	6.88	7.83	8.46	8.86	9.19	10.20	11.12
1.03	5.36	6.02	6.41	6.63	7.70	8.34	8.74	9.07	10.09	11.00
1.04	4.76	5.66	6.11	6.39	7.54	8.21	8.63	8.97	9.98	10.99
1.05	4.23	5.24	5.77	6.11	7.50	8.10	8.52	8.84	9.89	10.78
1.06	3.84	4.87	5.44	5.79	7.22	7.94	8.39	8.72	9.78	10.70
1.07	3.50	4.34	5.01	5.45	7.05	7.80	8.25	8.58	9.67	10.62
1.08	3.26	3.88	4.58	5.09	6.87	7.65	8.12	8.49	9.58	10.51
1.09	2.97	3.63	4.22	4.76	6.71	7.51	7.98	8.35	9.44	10.43
1.10	2.85	3.44	3.94	4.41	6.54	7.37	7.84	8.25	9.32	10.25
1.12	2.58	3.08	3.51	3.92	6.27	7.15	7.63	8.01	9.12	9.94
1.14	2.38	2.83	3.25	3.59	6.02	6.93	7.43	7.80	8.94	9.74
1.16	2.21	2.61	2.98	3.35	5.77	6.71	7.23	7.58	8.73	9.55
1.18	2.09	2.47	2.84	3.17	5.52	6.50	7.04	7.38	8.56	9.36
1.20	1.93	2.29	2.65	2.98	5.25	6.28	6.83	7.21	8.37	9.16
1.30	1.53	1.82	2.01	2.32	4.31	5.37	5.94	6.33	7.54	8.39
1.40	1.25	1.46	1.66	1.83	3.55	4.56	5.10	5.58	6.84	7.78
1.50	1.05	1.23	1.41	1.59	2.99	3.85	4.40	5.42	6.22	7.26
1.60	0.85	0.98	1.18	1.27	2.43	3.30	3.87	4.38	5.76	6.82
1.70	0.62	0.82	1.02	1.10	2.03	2.88	3.46	3.96	5.39	6.48
1.80	0.49	0.70	0.88	0.95	1.78	2.57	3.16	3.63	5.11	6.21
1.90	0.41	0.65	0.77	0.85	1.64	2.33	2.94	3.39	4.89	6.00
2.00	0.52	0.60	0.68	0.75	1.38	2.25	2.82	3.25	4.73	5.86
3.00	0.24	0.25	0.25	0.30	0.48	0.72	1.08	1.35	2.50	3.23
4.00	0.00	0.01	0.03	0.03	0.20	0.35	0.71	1.09	2.87	3.77
6.00	0.01	0.04	−0.01	−0.04	0.01	0.13	0.81	1.23	3.44	5.48
8.00	0.06	0.04	−0.05	−0.05	−0.02	0.03	1.02	1.37	3.90	6.15
10.00	0.16	0.20	−0.05	−0.05	−0.02	0.20	1.23	1.59	4.10	5.93
15.00	0.40	0.62	−0.63	−0.63	0.07	1.47	1.66	2.26	4.19	6.18

Values of $c_p - c^*_p$ can thus be obtained by equation 77 from data of $(H^* - H)/T_c$. Such values have been calculated for $z_c = 0.27$ over a wide range of reduced temperatures and pressures for both liquid and gas shown graphically in Fig. 145. Values of c^*_p where unknown can be estimated by the method of group contributions as given in Chapter 25.

Illustration 12. Calculate the molal heat capacity of CO_2 gas at 100° C and 100 atm pressure. At 100° C, 100 atm, $T_r = 1.23$; $p_r = 1.37$. At 100° C, $c^*_p = 9.6$. From Fig. 145, $c_p - c^*_p = 7.5$

$$c_p = 7.5 + 9.6 = 17.1 \text{ cal/(g-mole)(K°)}$$

Extrapolation of Experimental Data by Generalized Tables. Wherever reliable experimental thermodynamic data are available, these should be used in preference to the tables of generalized properties, and the latter should be used for extrapolation to conditions where experimental data are not available. Several alternative methods are available for such extrapolations. The choice among them is dependent on the function under consideration, the range of conditions in which experimental values are available, and the accuracy with which the critical properties are known.

One such procedure is described on page 577 for calculating the density of a liquid at any condition of temperature or pressure from one accurate density measurement made at a known temperature and pressure. In general, where one experimental value of any property is known at a given temperature and pressure, its value at other temperatures and pressures may be estimated by use of a correction ratio in conjunction with values from the generalized tables. Thus;

$$\beta_{2,\text{ext}} = \beta_2\left(\frac{\beta_{1,\text{exp}}}{\beta_1}\right) \tag{78}$$

where $\beta_{2,\text{ext}}$ = desired extrapolated value at T_2 and p_2

$\beta_{1,\text{exp}}$ = experimental value at T_1 and p_1

β_1 = value from table, T_{r1} and p_{r1}, and the given z_c

β_2 = value from table, T_{r2} and p_{r2}, and the given z_c

The correction ratio $\beta_{1(\text{exp})}/\beta_1$ may be assumed to be constant at an average value based on the various values of $\beta_{1(\text{exp})}$ that are available. This method may be used for improving estimated values of any tabulated property, z, $(H^* - H)/T_c$, $S^* - S$, $(U^* - U)/T_c$, as well as ρ. For the fugacity coefficient f/p, the extrapolation is better done with the ratio of log β values rather than with β.

It is evident that application of equation 78 to the critical point may lead to values of β_2 different from those on which z_c is based. For example,

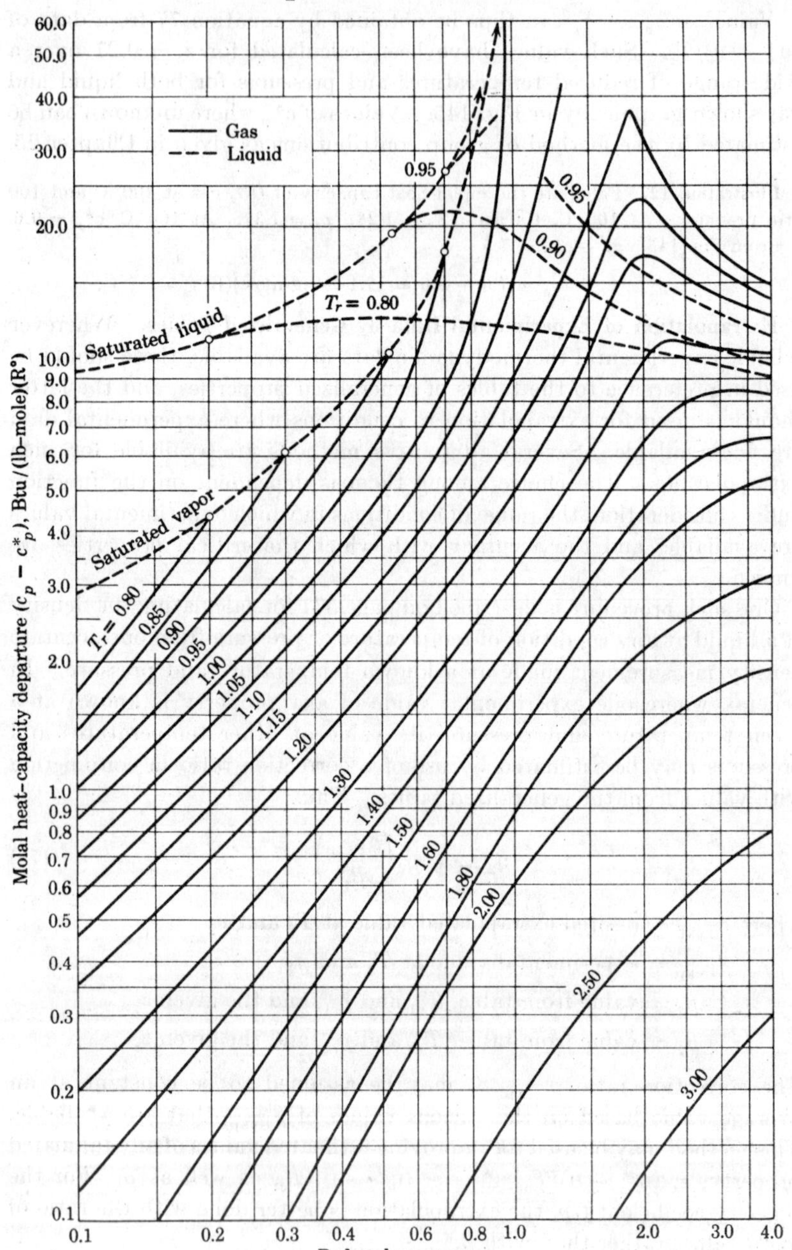

Fig. 145. Departure of molal heat capacities of gases and liquids from ideal-gas behavior, $z_c = 0.27$

if the correction ratio is not 1.0, equation 78 leads to a value of ρ_r which is not 1.0 where T_r and p_r equal 1.0.

An alternative method for using the generalized tables in extrapolation of experimental data is to utilize the experimental values to evaluate the apparent or effective value of z_c which will result in agreement between the experimental and generalized values. For extrapolation of a function β the apparent value of z_c may be termed $z_{c\beta}$. The generalized tables are then used for evaluation of values of β_2 corresponding to T_{r2}, p_{r2}, and $z_{c\beta}$. This method is preferable for liquid densities if the critical volume is known with high accuracy and it is desired to maintain agreement both with experimental density data at low temperatures and with the critical volume. In this case values of ρ_r are read from the tables corresponding to $z_{c\beta}$, but z_c or ρ_c are used to calculate densities.

Thermodynamic Charts by Generalized Methods

Tables and charts of thermodynamic relationships, such as enthalpy–temperature, temperature–entropy, enthalpy–entropy, and pressure–enthalpy of a specific pure fluid may be constructed from generalized data in conjunction with the heat capacity of the gas at zero pressure as a function of temperature and the critical properties of the specific compound.

Enthalpy–Temperature Chart. The construction of the enthalpy–temperature chart proceeds as follows with reference to Fig. 146.

The construction of this chart will be illustrated by establishing ten values of enthalpies at three temperature levels: namely, at T_1 and T_2 below the critical temperature for ideal-gas behavior, and at temperature T_2 for the saturated gas and saturated liquid, and for the liquid at a pressure exceeding saturation. The enthalpy of the gas for ideal behavior at the critical point and for the actual gas at the critical point are also established. By similar procedures any other points on the diagram are calculated.

The reference state at point 1 is arbitrarily taken as the saturated liquid at a pressure p_1 and temperature T_1 with corresponding reduced pressure and temperature of p_{r1} and T_{r1}. A value of zero or any other number may be arbitrarily assigned to the value of enthalpy at this point. Point 2, the enthalpy of the saturated vapor at temperature T_1, is obtained by adding to H_1 the heat of vaporization λ_1 obtained from Table 47 at the reduced saturation pressure p_{rs1}; thus $H_2 = H_1 + \lambda_1$.

Where the reduced pressure at point 1 is below 0.1, it is better to obtain λ_1 by the procedure outlined in Illustration 8–2, page 275, or equation 52. Where an experimental value of the latent heat of vaporization such as at the normal boiling point is known, the latent heat at temperature T_{r1} can also be obtained by equation 8–40, page 281.

Point 3* corresponds to the molal enthalpy of the ideal gas at temperature T_1. H^*_3 is obtained by adding to H_2 the enthalpy correction obtained from Table 47 corresponding to p_{r1} and T_{r1}. Thus, $H^*_3 = H_2 + (H^*_3 - H_2)$.

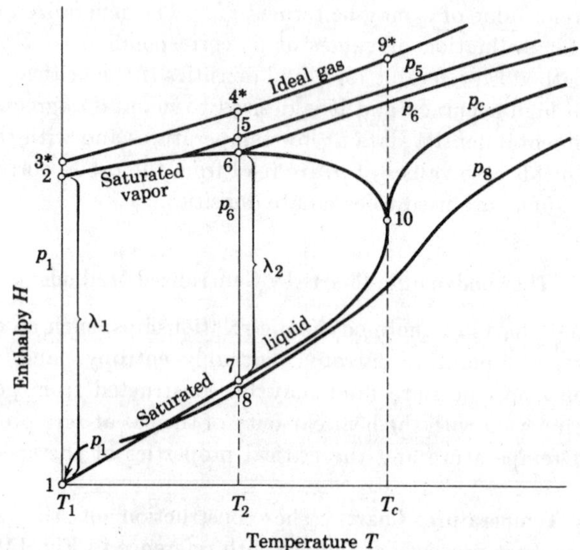

Fig. 146. Construction of enthalpy–temperature chart

Point 4* represents the enthalpy of the ideal gas at temperature T_2, and is obtained by adding to H_3 the value $\int_{T_1}^{T_2} c^*_p \, dT$ where c^*_p is the molal heat capacity for ideal-gas behavior as a function of temperature, $H^*_4 = H^*_3 + \int_{T_1}^{T_2} c^*_p \, dT$.

The enthalpy H_5 of the gas at temperature T_2 below the critical and any desired pressure p_5 below saturation can then be obtained directly by subtracting from H^*_4 the value of $H^*_4 - H_5$ obtained from Table 50 for the gas corresponding to T_{r2} and p_{r5}; $H_5 = H^*_4 - (H^*_4 - H^*_5)$. At point 6, H_6 corresponds to the enthalpy of the saturated vapor and is obtained from H^*_4 by subtracting the value of $H^*_4 - H_6$ obtained from Table 50 corresponding to a reduced saturation pressure of p_{r6}; thus $H_6 = H^*_4 - (H^*_4 - H_6)$. The vapor pressure p_{r6} is obtained from Table 47 at a reduced saturation temperature T_{r2} and the value of z_c corresponding to the specific compound. Point 7 corresponds to the enthalpy of the saturated liquid at a reduced saturation pressure of p_{r6} and is obtained by subtracting from H^*_4 the values of $H^*_4 - H_7$ obtained from Table 47 corresponding to T_{r2} and p_{r6}, and the value of z_c corresponding to the specific compound; $H_7 = H^*_4 - (H^*_4 - H_7)$.

Point 8 corresponds to the enthalpy of a liquid at a pressure p_8 above its saturation pressure. H_8 is obtained by subtracting from H^*_4 the value of $H^*_4 - H_8$ obtained from Table 50 for the liquid at T_{r2} and p_{r8}. Values of H at any other temperature and pressure are similarly obtained. The value of H_{10} at the critical point is obtained from a value of H^*_9 at point 9 corresponding to ideal-gas behavior at its critical temperature. H^*_9 is obtained from H^*_3 by adding the value of the integral $\int_{T_1}^{T_c} c^*_p\, dT$. At the critical point, H_{10} is then obtained from point 9 by subtracting the value of $H^*_9 - H_{10}$ obtained from Table 50 corresponding to values of p_{rc} and T_{rc}, each equal to unity; thus $H_{10} = H^*_9 - (H^*_9 - H_{10})$. Tables 47 and 50 should be interpolated for the value of z_c of the compound under investigation.

The enthalpy–temperature chart is of particular value in obtaining data for establishing energy balances in flow processes or in processes proceeding at constant pressure. The chart is also useful in establishing the temperature drop in throttling of fluids.

Temperature–Entropy Chart. The temperature–entropy chart (Fig. 147) is constructed in a manner similar to the enthalpy–temperature

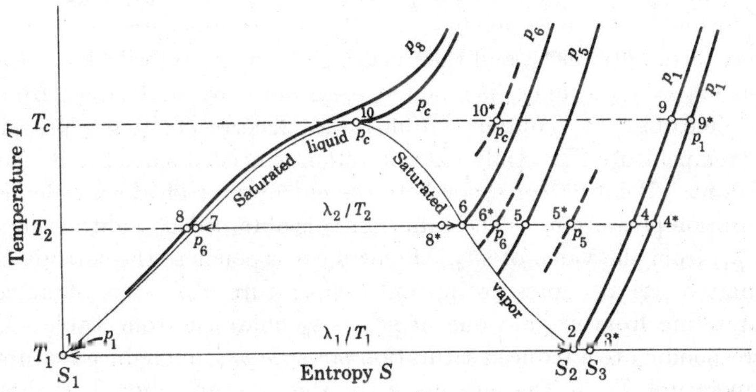

Fig. 147. Construction of temperature–entropy chart

chart, with additional steps for the effect of pressure on the entropy of an ideal gas. For illustration, values of entropy at the same ten conditions are established. Point 1 corresponding to the entropy of the saturated liquid may also be used as the reference state. Absolute values of entropies may be obtained from the published values of entropy at 298° K and 1 atm for ideal-gas behavior. Point 2 represents the molal entropy of the saturated gas at temperature T_1 and pressure p_1. This value is obtained by adding to s_1 the entropy of evaporation λ_1/T_1 obtained from Table 47 corresponding to the reduced saturation pressure p_{rs1}.

Where the reduced pressure at point 1 is below 0.1, it is better to obtain λ_1 from the experimental values of λ at the normal boiling point by use of equation 8–40.

The entropy s^*_3 of the ideal gas at temperature T_1 and pressure p_1 is then obtained by adding to s_2 the entropy correction $s^*_3 - s_2$ from Table 47 corresponding to reduced pressure p_{rs1} and reduced temperature T_{rs1}. In obtaining the pressure correction for entropy, two corrections must be made; one for change in pressure under ideal behavior, and a second for departure from ideal behavior at the given temperature and pressure. For ideal-gas behavior one isothermal line for enthalpies suffices for all pressures, whereas for entropies a different line is required for each pressure. Point 4* represents the entropy s^*_4 at T_2 and p_1 for ideal behavior. s^*_4 is obtained by adding to s_3 the usual integration of heat-capacity data for ideal-gas behavior. Thus:

$$s^*_4 = s^*_3 + \int_{T_1}^{T_2} \frac{c^*_p\, dT}{T}$$

The entropy of the gas at temperature T_2 and any pressure p_5 below saturation is obtained by first making a pressure correction for ideal behavior to obtain s^*_5, and then making a correction for lack of ideality. Thus $s^*_5 = s^*_4 - \ln (p_5/p_1)$, and s_5 is obtained by subtracting from s^*_5 the value of $s^*_5 - s_5$ obtained from Table 52, corresponding to pressure p_1 and temperature T_2. Only one correction is needed since the pressure is constant. Point 6* corresponds to the entropy for ideal-gas behavior at saturation pressure p_6. The value of s^*_6 is obtained by subtracting $R \ln (p_6/p_1)$ from the value of s^*_4. Point 6 corresponds to the entropy of the saturated gas at a pressure p_6 and temperature T_2. s_6 is obtained by subtracting from s^*_6 a value of $s^*_6 - s_6$ obtained from Table 47 and corresponding to a reduced saturation pressure p_{rs6} and reduced saturation temperature T_{rs2}. The entropy s_7 of the saturated liquid is obtained by subtracting from s_6 the entropy of condensation obtained from Table 47 corresponding to the reduced saturation pressure p_{rs6}. The entropy s_8 of the liquid at a pressure above its saturation pressure is obtained by subtracting from s^*_8 the value of $s^*_8 - s_8$ obtained from Table 52 at a reduced pressure of p_{r8} and reduced temperature T_{r2}. Point s^*_8 is obtained by subtracting from s^*_4 the value $R \ln (p_8/p_1)$;

$s^*_8 = s^*_4 - R \ln (p_8/p_1)$. Similarly, values of s are obtained at other temperatures and pressures for both liquid and gas. Point 9* corresponding to the ideal gas at the critical temperature is obtained by adding $\int_{T_1}^{T_c} \frac{c^*_p \, dT}{T}$ to s^*_3. The entropy s_{10} at the critical point is then obtained by subtracting from s^*_9 the term $R \ln (p_c/p_1) + (s^*_{10} - s_{10})$ where the last term is obtained from Table 52 where both p_r and T_r are unity.

The temperature–entropy chart is of particular value in following the temperature changes in isentropic processes.

Mollier Diagram. The enthalpy–entropy chart (Mollier diagram) is constructed directly from values of H and S as described under the two preceding sections. The same corresponding ten points are plotted in Fig. 148.

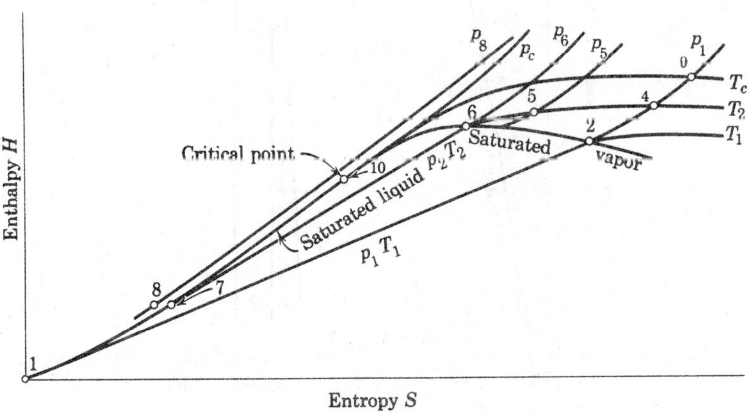

FIG. 148. Construction of enthalpy–entropy (Mollier) chart

The enthalpy–entropy chart is of particular value in establishing energy requirements in flow processes and temperature changes in isentropic and isenthalpic processes.

Pressure–Enthalpy Chart. The pressure–enthalpy chart (Fig. 149) is constructed directly from values of enthalpy as described under the temperature–enthalpy chart. It is convenient to plot pressure on a logarithmic scale. This chart with constant entropy lines is particularly useful in analyzing the heat loads and temperature changes in refrigeration cycles.

All the data presented in the foregoing charts can be presented on any single chart by superimposing isometric lines for all parameters not indicated by the coordinate scales. Such charts contain such a confusing web of lines as to be unreadable. The separate charts are more convenient for particular uses.

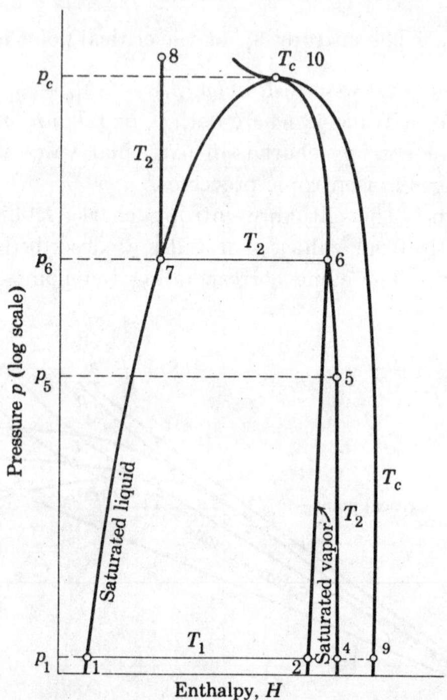

Fig. 149. Construction of log p–enthalpy chart

Illustration 13. Calculate the molal enthalpy and the molal entropy of carbon tetrachloride at the following ten prescribed conditions. The numbers correspond with those in Figs. 146 and 147.

(1) Saturated liquid at 60° F (519.7° R; 288.7° K).
(2) Saturated vapor at 60° F.
(3) Ideal gas at 60° F and the corresponding saturation pressure.
(4) Ideal gas at 440° F and the saturation pressure corresponding to 60° F (0.09650 atm = 1.4182 psia).
(5) Gas at 440° F, 200 psia.
(6) Saturated vapor at 440° F (899.7° R; 499.8° K).
(7) Saturated liquid at 440° F.
(8) Liquid at 440° F and 1000 psia.
(9) Ideal gas at the critical temperature and the saturation pressure corresponding to 60° F (0.09650 atm = 1.4182 psia).
(10) Gas or liquid at the critical temperature and pressure.

The following data are available:

Critical temperature = 556.4° K = 1001.5° R = 541.8° F.

Critical pressure = 45.0 atm = 661.3 psia.
Critical compressibility factor = 0.272.
Molal heat capacity of ideal gas,

$$c^*_p = 12.24 + (3.400)10^{-2}T - (2.995)10^{-5}T^2 + (8.828)10^{-9}T^3$$

where T is in degrees Kelvin.

Vapor pressure is given by the relation

$$\log p_r = \frac{-A(1 - T_r)}{T_r} - e^{-20(T_r - b)^2}$$

where $A = 2.7989$ and $b = 0.158$.

In calculating enthalpies and entropies for the conditions specified above, saturated liquid at 60° F is selected as the reference state, and zero values of enthalpy and entropy are arbitrarily assigned to that state.

At 60° F, saturated:

$$p_s = 0.09650 \text{ atm} = 1.4182 \text{ psia}$$

$$p_r = 0.002145, \qquad T_r = 0.519$$

At 440° F, saturated:

$$p_s = 21.702 \text{ atm} = 318.9 \text{ psia}$$

$$p_r = 0.482 \qquad T_r - 0.898$$

The molal heat of vaporization is given by the relation.

$$\lambda = 2.303(z_G - z_L)RT_c[A + 40T_r^2(T_r - b)e^{-20(T_r-b)^2}$$

At 60° F, from Table 8–26A, $\quad z_G - z_L = 0.996$

$$\lambda = (2.303)(0.996)(1.987)(1001.5)$$

$$\times [2.7989 + 40(0.519)^2(0.519 - 0.158)e^{-20(0.519-0.158)^2}]$$

$$= 14{,}091 \text{ Btu per lb-mole}$$

The latent heat of vaporization may also be estimated through the use of enthalpy deviation values for saturated liquid and saturated gas taken from Table 47.

$$\lambda = \left[\left(\frac{\text{H}^* - \text{H}}{T_c}\right)_{sL} - \left(\frac{\text{H}^* - \text{H}}{T_c}\right)_{sG}\right]T_c = (14.25 - 0.014)(1001.5)$$

$$= 14{,}257 \text{ Btu per lb-mole}$$

This value is not considered as reliable as the preceding value, as p_{rs} is considerably below 0.1.

(1) Saturated liquid at 60° F (reference state):

$$\text{H}_1 = 0, \qquad \text{S}_1 = 0$$

(2) Saturated vapor at 60° F:

$$\text{H}_2 = \lambda_2 = 14{,}091 \text{ Btu per lb-mole}$$

$$\text{S}_2 = \lambda_2/T_2 = 14{,}091/519.7 = 27.114 \text{ Btu/(lb-mole)(R°)}$$

(3) Ideal gas at 60° F and the saturation pressure:
From Table 47,

$$(\text{H}^*_3 - \text{H}_2)/T_c = 0.014, \qquad (\text{S}^*_3 - \text{S}_2) = 0.021$$

$$\text{H}^*_3 = 14{,}091 + (0.014)(1001.5)$$

$$= 14{,}091 + 14 = 14{,}105 \text{ Btu per lb-mole}$$

$$\text{S}^*_3 = \text{S}_2 + 0.021 = 27.114 + 0.021 = 27.135 \text{ Btu/(lb-mole)(R°)}$$

(4) Ideal gas at 440° F and the saturation pressure corresponding to 60° F (0.09650 atm = 1.4182 psia):

$$\text{H}^*_4 = \text{H}^*_3 + 1.8\int_{228.7° \text{K}}^{499.8° \text{K}} c^*_p \, dT$$

$$= 14{,}105 + (1.8)(4529.8) = 14{,}105 + 8154 = 22{,}259 \text{ Btu per lb-mole}$$

$$\text{S}^*_4 = \text{S}^*_3 + \int_{288.7° \text{K}}^{499.8° \text{K}} (c^*_p/T)\, dT$$

$$= 27.135 + 11.699 = 38.834 \text{ Btu/(lb-mole)(R°)}$$

(5) Gas at 440° F, 200 psia:

$$T_r = 0.898, \qquad p_r = 0.302$$

From Tables 50 and 52;

$$(\text{H}^*_4 - \text{H}_5)/T_c = 1.00, \qquad (\text{S}^*_5 - \text{S}_5) = 0.83$$

$$\text{H}_5 = \text{H}^*_4 - (1.00)T_c = 22{,}259 - (1.00)(1001.5)$$

$$= 22{,}259 - 1002 = 21{,}257 \text{ Btu per lb-mole}$$

$$\text{S}_5 = \text{S}^*_5 - 0.83$$

$$\text{S}^*_5 - \text{S}^*_4 = -R \ln (p_5/p_1)$$

$$\text{S}_5 = \text{S}^*_4 - R \ln (p_5/p_1) - 0.83$$

$$= 38.834 - 4.5757 \log (200/1.4182) - 0.83$$

$$= 38.834 - 9.834 - 0.83 = 28.170 \text{ Btu/(lb-mole)(R°)}$$

(6) Saturated vapor at 440° F: From Table 47, considering saturated vapor,

$$(\text{H}^*_4 - \text{H}_6)/T_c = 1.83, \qquad (\text{S}^*_6 - \text{S}_6) = 1.55$$

$$\text{H}_6 = \text{H}^*_4 - (1.83)T_c = 22{,}259 - (1.83)(1001.5)$$

$$= 22{,}259 - 1833 = 20{,}426 \text{ Btu per lb-mole}$$

$$\text{S}_6 = \text{S}^*_6 - 1.55$$

$$\text{S}^*_6 - \text{S}^*_4 = -R \ln (p_6/p_1)$$

$$\text{S}_6 = \text{S}^*_4 - R \ln (p_6/p_1) - 1.55$$

$$= 38.834 - 4.5757 \log (318.9/1.4182) - 1.55$$

$$= 38.834 - 10.762 - 1.55 = 26.522 \text{ Btu/(lb-mole)(R°)}$$

(7) Saturated liquid at 440° F: From Table 47, considering the saturated liquid,

$$(\text{H}^*_4 - \text{H}_7)/T_c = 9.79, \qquad (\text{S}^*_6 - \text{S}_7) = 10.42$$

$$\text{H}_7 = \text{H}^*_4 - (9.79)T_c = 22{,}259 - (9.79)(1001.5)$$

$$= 22{,}259 - 9805 = 12{,}454 \text{ Btu per lb-mole}$$

$$\text{S}_7 = \text{S}^*_6 - 10.42$$

$$\text{S}^*_6 - \text{S}^*_4 = -R \ln (p_6/p_1)$$

$$\text{S}_7 = \text{S}^*_4 - R \ln (p_6/p_1) - 10.42$$

$$= 38.834 - 4.5757 \log (318.9/1.4182) - 10.42 = 17.652 \text{ Btu/(lb-mole)(R°)}$$

At this point somewhat better accuracy could be secured through the use of a latent heat of vaporization value calculated in the same way as at 60° F. Such a procedure gives a latent heat of vaporization of 8007 Btu per lb-mole. The tables at this point give the following value:

$$\lambda_7 = \left[\left(\frac{\text{H}^* - \text{H}}{T_c}\right)_{sL} - \left(\frac{\text{H}^* - \text{H}}{T_c}\right)_{sG}\right] T_c$$

$$= (9.79 - 1.83)(1001.5) = (7.96)(1001.5) = 7972 \text{ Btu per lb-mole}$$

(8) Liquid at 440° F and 1000 psia:

$$T_r = 0.898, \qquad p_r = 1000/661.3 = 1.512$$

From Tables 50 and 52:

$$(\text{H}^*_4 - \text{H}_8)/T_c = 10.10, \qquad (\text{s}^*_8 - \text{s}_8) = 8.78$$

$$\text{H}_8 = \text{H}^*_4 - (10.10)T_c = 22{,}259 - (10.10)(1001.5)$$

$$= 22{,}259 - 10{,}115 = 12{,}144 \text{ Btu per lb-mole}$$

$$\text{s}_8 = \text{s}^*_8 - 8.78$$

$$\text{s}^*_8 - \text{s}^*_4 = -R \ln (p_8/p_1)$$

$$\text{s}_8 = \text{s}^*_4 - R \ln (p_8/p_1) - 8.78$$

$$= 38.834 - 4.5757 \log (1000/1.4182) - 8.78$$

$$= 38.834 - 13.033 - 8.78 = 17.021 \text{ Btu/(lb-mole)(R°)}$$

(9) Ideal gas at the critical temperature and the saturation pressure corresponding to 60° F (0.09650 atm = 1.4182 psia):

$$\text{H}^*_9 = \text{H}^*_3 + 1.8\int_{288.7°\text{ K}}^{556.4°\text{ K}} c^*_p \, dT$$

$$= 14{,}105 + (1.8)(5839.4) = 14{,}105 + 10{,}511 = 24{,}616 \text{ Btu per lb-mole}$$

$$\text{s}^*_9 = \text{s}^*_3 + \int_{288.7°\text{ K}}^{556.4°\text{ K}} (c^*_p/T)\, dT = 27.135 + 14.181 = 41.316 \text{ Btu/(lb-mole)(R°)}$$

(10) Gas or liquid at the critical temperature and pressure:

$$T_r = 1, \qquad p_r = 1$$

From Tables 50 and 52;

$$(\text{H}^*_9 - \text{H}_{10})/T_c = 5.76, \qquad (\text{H}^*_{10} - \text{H}_{10}) = 4.89$$

$$\text{H}_{10} = \text{H}^*_9 - 5.76T_c = 24.616 - (5.76)(1001.5)$$

$$= 24{,}616 - 5769 = 18{,}847 \text{ Btu per lb-mole}$$

$$\text{s}_{10} = \text{s}^*_{10} - 4.89$$

$$\text{s}^*_{10} - \text{s}^*_9 = -R \ln (p_c/p_1)$$

$$\text{s}_{10} = \text{s}^*_9 - R \ln (p_c/p_1) - 4.89$$

$$= 41.316 - 4.5757 \log (661.3/1.4182) - 4.89$$

$$= 24.215 \text{ Btu/(lb-mole)(R°)}$$

Summary

	Temperature, °F	Pressure, psia	Enthalpy, Btu per lb-mole	Entropy, Btu/ (lb-mole)(R°)
1. Saturated liquid	60	1.418	0	0
2. Saturated vapor	60	1.418	14,091	27.114
3. Ideal gas	60	1.418	14,105	27.135
4. Ideal gas	440	1.418	22,259	38.834
5. Superheated gas	440	200	21,257	28.170
6. Saturated vapor	440	318.9	20,426	26.522
7. Saturated liquid	440	318.9	12,454	17.652
8. Compressed liquid	440	1000	12,144	17.021
9. Ideal gas at critical temperature	541.8	1.418	24,616	41.316
10. Gas or liquid at critical point	541.8	661.3	18,847	24.215

Specific thermodynamic charts for ammonia and benzene constructed by generalized methods are presented herewith:

Fig. 150. Pressure–volume chart for ammonia
Fig. 151. Enthalpy–temperature chart for ammonia
Fig. 152. Temperature–entropy chart for ammonia
Fig. 153. Enthalpy–entropy chart for ammonia
Fig. 154. Enthalpy–temperature chart for benzene
Fig. 155. Temperature–entropy chart for benzene
Fig. 156. Enthalpy–entropy chart for benzene

Internal-Energy, Free-Energy, and Total-Work Properties. As was pointed out in Chapter 13, absolute values of the three properties U, G, and A are never known. However, relative values referred to any selected reference state may be calculated by the methods of the preceding sections.

The change in internal energy accompanying any change in conditions is derived from the corresponding enthalpy change. Thus, from the definition of enthalpy,

$$\Delta U = \Delta H - \Delta(pV) \tag{79}$$

Applying equation 79 to determining internal energy relative to a reference state of zero enthalpy, designated by a zero subscript, gives

$$U = H - (pV - p_0V_0) \tag{80}$$

Similarly, by definition,

$$\Delta G = \Delta H - \Delta(TS) \tag{81}$$

or, expressing free energy relative to the reference state gives

$$G = H - (TS - T_0S_0) \tag{82}$$

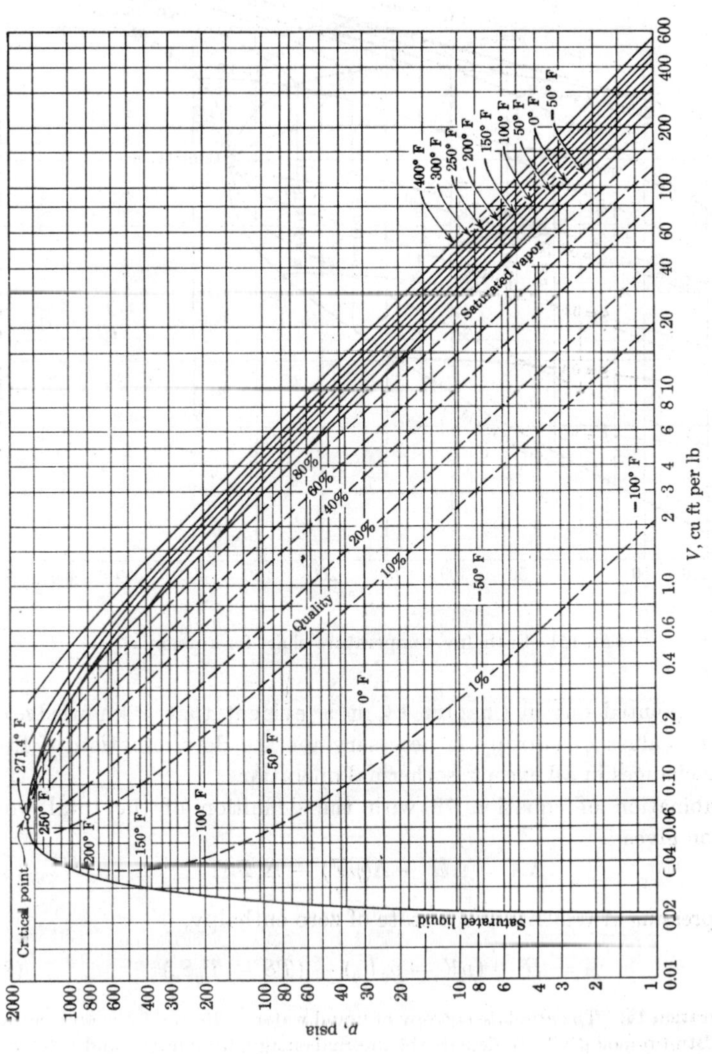

FIG. 150. Pressure–volume chart for ammonia.

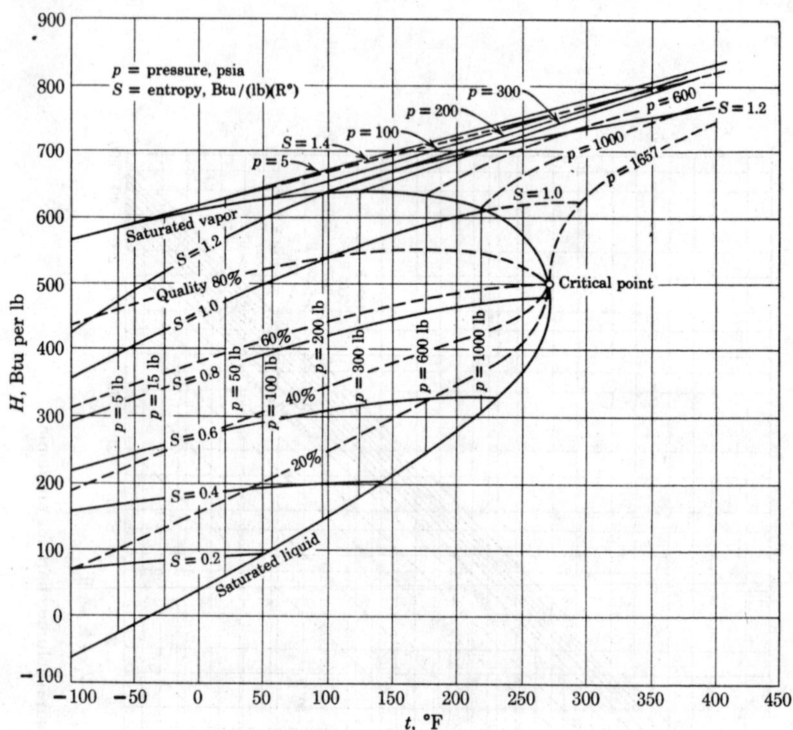

Fig. 151. Enthalpy–temperature chart for ammonia

As was pointed out in Chapter 13, it is evident that a knowledge of absolute values of entropy is necessary for the determination of free-energy changes in all except isothermal operations.

Combination of equation 79 with the definition of the total work function gives

$$\Delta A = \Delta H - \Delta(pV) - \Delta(TS) \tag{83}$$

or, expressing A relative to the state of zero enthalpy,

$$A = H - (pV - p_0V_0) - (TS - T_0S_0) \tag{84}$$

Illustration 14. The absolute entropy of liquid water at 25° C (77° F) and 1 atm is 16.716 Btu/(lb-mole)(R°). Calculate the internal energy, free energy, and total work function, in Btu per pound of saturated water vapor at an absolute pressure of 125 psia, relative to the saturated liquid at 32° F. Use the enthalpy and entropy data of the steam tables, assume an average heat capacity of liquid water between 32 and 77° F as 1.0 Btu per lb per F°, and neglect the effect of pressure on the entropy of the saturated liquid in this range.

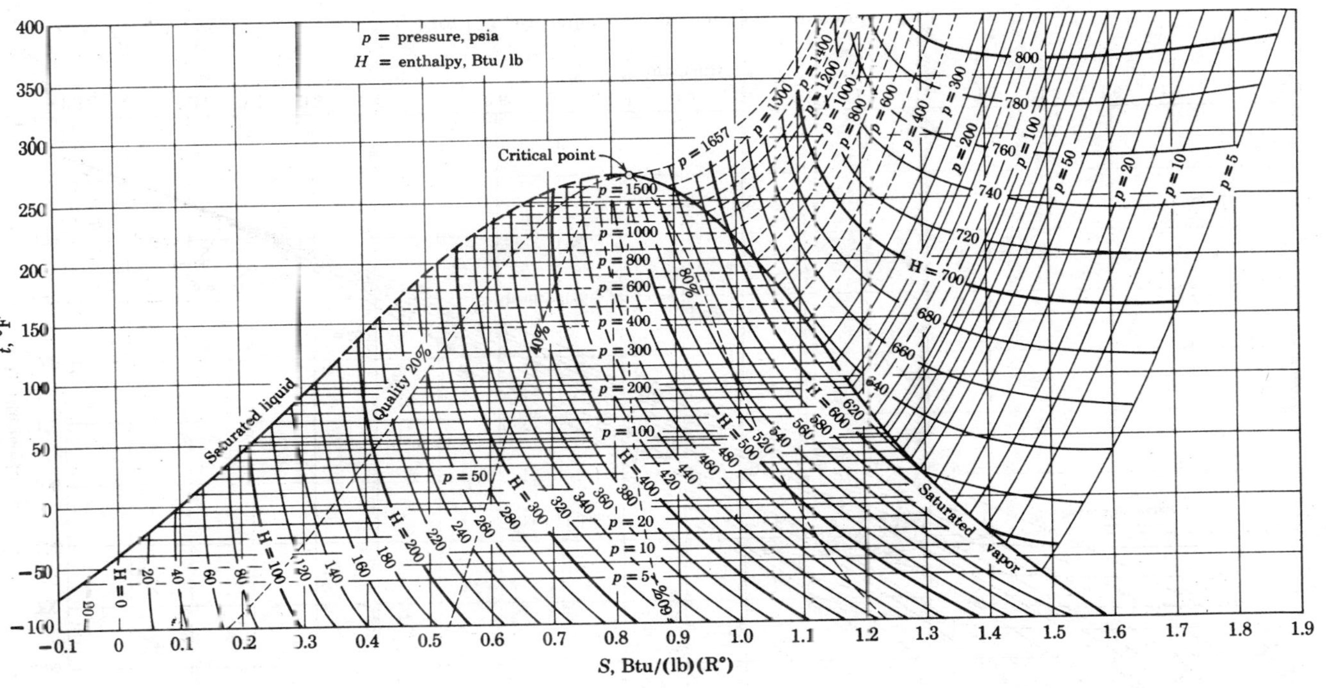

Fig. 152. Temperature–entropy chart for ammonia

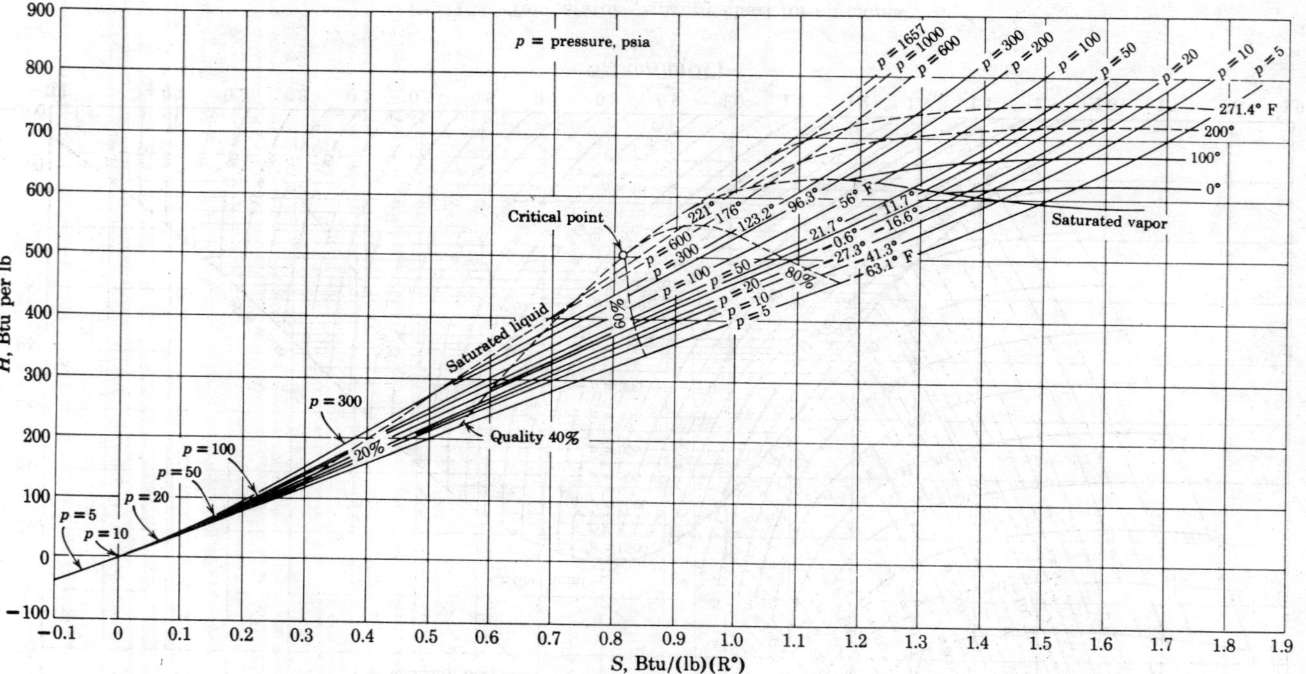

Fig. 153. Enthalpy–entropy chart for ammonia

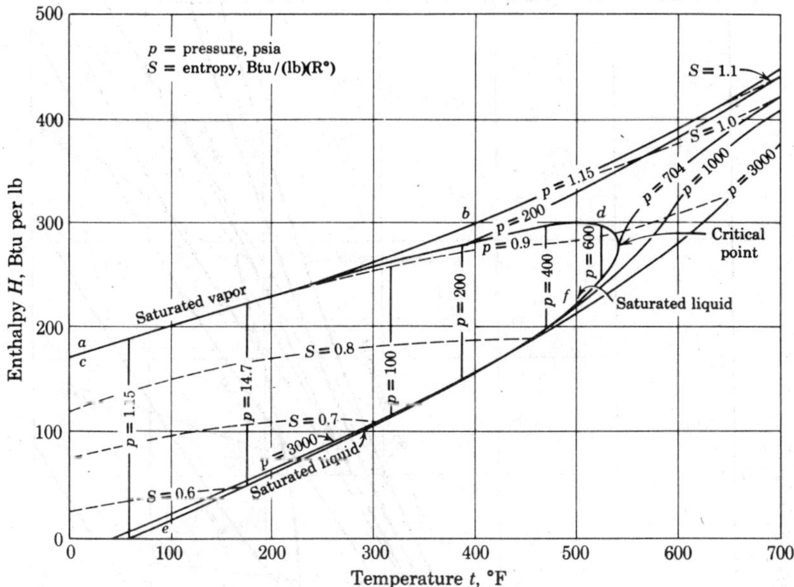

FIG. 154. Enthalpy–temperature chart for benzene

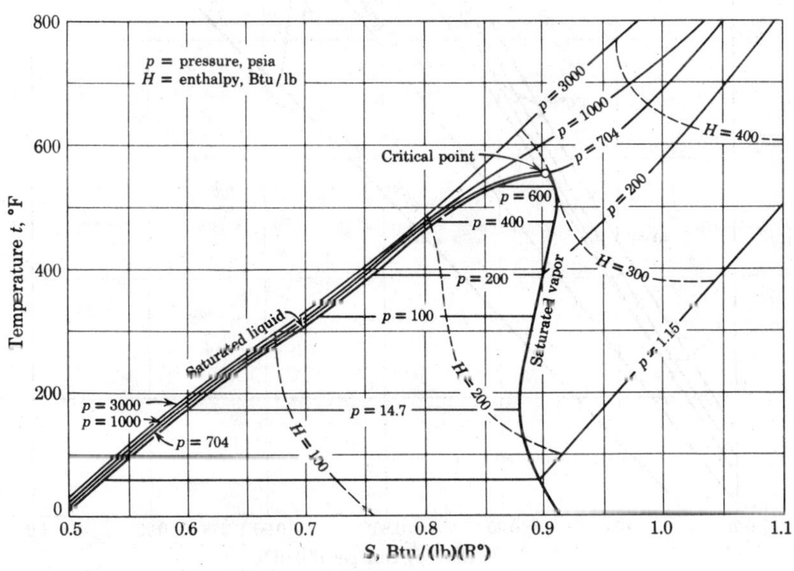

FIG. 155. Temperature–entropy chart for benzene

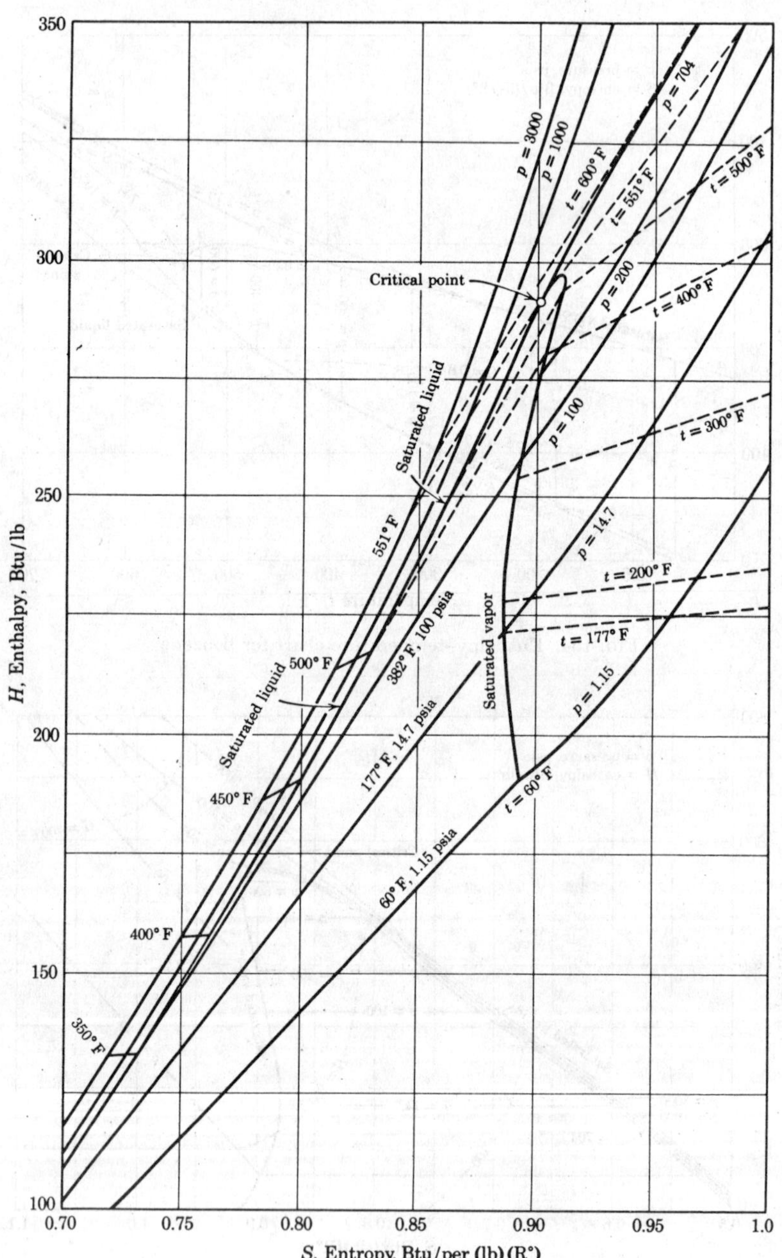

Fig. 156. Enthalpy–entropy chart for benzene

Solution: The absolute entropy of the reference state is calculated from equation 13–5.

$$S_0 = \frac{16.716}{18.016} - 1.0 \ln \left(\frac{536.7}{491.7}\right) = 0.9278 - 0.0875 = 0.8403 \text{ Btu/(lb)(R°)}$$

From the steam tables, at the reference state:

$$p_0 = 0.08854 \text{ psia}$$

$$V_0 = 0.01602 \text{ cu ft per lb}$$

For the saturated vapor at 125 psia and 344.33° F.

$$V = 3.587 \text{ cu ft per lb}$$

$$H = 1191.1 \text{ Btu per lb}$$

$$S = 0.8403 + 1.5844 = 2.4247 \text{ Btu/(lb)(R°)}$$

Equation 80, converting pV from foot-pounds to Btu gives

$$U = 1191.1 - [125(3.587) - (0.08854)(0.01602)] \frac{144}{778}$$

$$= 1191.1 - 83.0 = 1108.1 \text{ Btu per lb}$$

From equation 82,

$$G = 1191.1 - (804.03)(2.4247) + (491.7)(0.8403)$$

$$= 1191.1 - 1949.5 + 413.2 = -345.2 \text{ Btu per lb}$$

From equation 84 and the preceding calculations of $\Delta(pV)$ and $\Delta(TS)$,

$$A = 1191.1 - 83.0 - 1536.3 = -428.2 \text{ Btu per lb}$$

As is evident from Illustration 14, the relative free energy and maximum work may decrease with increase in temperature. Complete tables or charts of these functions could be established either from the generalized correlations of properties or from experimental data. However, such data are of limited value, and the principal application of free energy is in the prediction of equilibria in chemical reactions or solutions.

Problems

1. For the production of liquid oxygen it is desired that the gas be compressed to a pressure of 100 atm at a temperature of −90° C. Oxygen at a pressure of 14.5 psia and a temperature of 22° C is compressed to these conditions. Calculate the volume of compressed gas resulting from 100 cu ft of the original gas (*a*) using the van der Waals equation, (*b*) using data from the Table 49.

2. Using the Beattie–Bridgeman equation, calculate the pressure exerted by 30 liters of air, measured at 20° C under a pressure of 740 mm Hg, when compressed into a volume of 200 cc at a temperature of 0° C. It may be assumed that at the lower pressure the ideal-gas law is applicable.

3. For ethylene gas the value of pV at 20° C and 100 atm is 0.3600 referred to a value of unity at 0° C and 1 atm. The density of ethylene gas at standard conditions is 1.2604 grams per liter. Calculate the volume of 1 g-mole at 20° C and 100 atm.

4. Calculate the volume occupied by 1 g-mole of carbon dioxide gas at its critical state when the product $pV = 0.3131$ referred to 1.00 at 0° C and 1 atm.

5. From the data of the *International Critical Tables*, calculate the compressibility factors of ethylene at a temperature of 20° C and at pressures ranging from 0 to 500 atm. Plot these compressibility factors against pressure in atmospheres and also against molal volumes in cubic centimeters per gram-mole.

6. From the data of problem 5, calculate the volume occupied by 50 grams of ethylene at a temperature of 20° C and a pressure of 5000 psia.

7. Calculate the temperature of CO_2 gas when 70 kg are confined in a volume of 90 liters at 200 atm.

8. In a multistage compressor, carbon dioxide is compressed from a volume of 2 cu ft at a pressure of 100 psia and a temperature of 22° C, to a volume of 0.4 cu ft at a temperature of 30° C. Calculate the pressure necessary, using Table 49..

9. Calculate the volume occupied by 10 lb of chlorine when compressed to a pressure of 125 psia at a temperature of 30° C, using Table 49.

10. Methyl chloride for use in domestic refrigerators is sold in small cylinders having volumes of 0.15 cu ft. Calculate the weight of methyl chloride gas contained in a cylinder if the absolute pressure is 68 psia and the temperature 20° C, using Table 49. ($t_c = 143.1°$ C, $p_c = 65.9$ atm.)

11. For high-pressure distribution in long pipe lines it is proposed to compress natural gas (methane) to an absolute pressure of 500 psia. Calculate its density in pounds per cubic foot at this pressure and a temperature of 80° F, using Table 49.

12. Sulfur dioxide is compressed from a pressure of 40 psia and a temperature of 150° F to a pressure of 190 psia at a temperature of 160° F. Calculate the ratio of the initial to the final volume, using Table 49.

13. The absolute entropy of carbon dioxide in the ideal state at 25° C and 1 atm is 51.06 entropy units per mole. For a temperature of 40° C and a pressure of 50 atm, calculate the absolute entropy, the enthalpy in calories per gram-mole referred to the ideal gas at 0° C, and the molal heat capacity, using Tables 50 and 52, and Fig. 145.

14. Liquid sulfur dioxide has a density of 92.42 lb per cu ft at a temperature of 0° F and an absolute pressure of 10.35 psia. Calculate the density of the liquid at 150° F and 1000 psia from Table 48.

15. Saturated liquid sulfur dioxide exerts a vapor pressure of 136.5 psia at a temperature of 130° F and has an enthalpy of 181.24 Btu per lb and an entropy of 0.32472 unit per lb referred to the saturated liquid at −40° F. Calculate the enthalpy and entropy of the compressed liquid at 130° F and 3000 psia, using the data of problem 14.

16. Sulfur dioxide exists as a saturated liquid at 130° F and 136.5 psia. At these conditions calculate the molal heat capacity of the saturated liquid and the heat capacity of the liquid at constant pressure, using the heat capacity of the ideal-gas (Table C, Appendix) and the generalized correlations. Also calculate the heat capacity of the liquid at 130° F and 3000 psia.

17. Construct temperature–enthalpy, temperature–entropy, and enthalpy–entropy charts for methylamine in Btu per pound referred to the saturated liquid at 0° F. Evaluate the properties of the saturated vapor and liquids at reduced temperatures of 0.7, 0.8, 0.85, 0.9, 0.95, 0.98, and 1.0. Establish constant-pressure lines at absolute pressures of 14.7, 100, 500, 1000, and 2000 psia by calculations at the afore-mentioned temperatures and also at $T_r = 1.1$ and 1.2.

The properties of methylamine are as follows:

Normal boiling point, $-6.5°$ C Saturated density at $-11°$ C, 0.699 g per cc
Heat of vaporization at $-6.5°$ C, 388 Btu per lb

The vapor pressure may be obtained from equation 4–16, which is most conveniently used by plotting a curve relating vapor pressure on a logarithmic scale to temperature on a uniform or reciprocal scale.

18. For methane gas at 122° F and 1000 psia, calculate the values of U, H, S, C_p, G, and A, per pound-mole relative to the actual gas at 32° F and 1 atm pressure by the generalized method and also by the Benedict–Webb–Rubin equation of state. The absolute molal entropy of methane gas in the ideal state at 25° C and 1 atm is 44.50 g-cal/(g-mole)(K°). The other necessary data are contained in Tables 18, page 255, and 45.

19. Benning and McHarness[19] determined the following constants for dichlorofluoromethane $CHCl_2F$ (molecular weight 102.9): In equation 13,

$$A_0 = 20.54 \qquad a = -0.179$$
$$B_0 = 0.286 \qquad b = 0.497$$
$$c = 0 \qquad R = 0.08205$$
$$p = \text{atmospheres} \qquad T = \text{degrees Kelvin}$$
$$v = \text{liters per gram-mole}$$

For vapor pressure,

$$\log p = 38.2974 - \frac{2367.41}{T} - 13.0295 \log T + 0.0071731T$$

$$p = \text{atmospheres} \qquad T = \text{degrees Kelvin}$$

Heat capacity of liquid at 1 atm,

$$C_p = 0.2471 + 0.000189t$$
$$C_p = \text{calories/(gram)(C)} \qquad t = \text{degrees centigrade}$$

Heat capacity of vapor at 1 atm,

$$c_p = 13.65 + 0.0249t \text{ cal/(g-mole)(C°)}$$

Density of saturated liquid ($-40°$ C to $+70°$ C),

$$\rho = 1.4256 - 2.316(10^{-3})t - (2.6)(10^{-6})t^2 \text{ grams per cc}$$

where t is in degrees centigrade. Calculate by rigorous methods:

(*a*) The heat of vaporization at 10° F.

(*b*) The enthalpy and entropy, referred to the saturated liquid at $-40°$ F, of the saturated vapor and the saturated liquid at $+120°$ F.

20. Calculate the density (in pounds per cubic foot) and the entropy (in Btu/(lb-mole)(R°)), of chlorine under the following conditions:

(*a*) Saturated gas at 710° R and 740 psia.

(*b*) Saturated liquid at 710° R and 740 psia.

(*c*) Supercooled liquid at 710° R and 2000 psia.

[19] A. F. Benning and R. C. McHarness, *Ind. Eng. Chem.*, **31**, 912 (1939).

Heat of vaporization at normal boiling point = 8880 Btu per lb-mole. Absolute entropy of saturated liquid at 420° R and 11.5 psia = 30.5 Btu/(lb-mole)(R°).

21. The enthalpy of saturated liquid propane at 190° F and 522 psia is 102.3 Btu per lb. Calculate the enthalpy of liquid propane at

(*a*) 190° F and its critical pressure.

(*b*) 190° F and 1500 psia.

The density of liquid propane at 70° F, 200 psia, is 31.5 lb per cu ft.

22. Estimate the molal heat of vaporization of methylamine at its normal boiling point, −6.5° C. The vapor pressure is given as follows:

$$\log_{10} p = -\frac{2.9589}{T_r} + 7.7066 - e^{-20(T_r - 0.239)}$$

p is in millimeters of mercury.

23. From the data of Table 7B, page 93 and Table 51, calculate the fugacity in pounds per square inch of pure ethane gas at a temperature of 200° F and an absolute pressure of 1500 psia.

24. Calculate the fugacities of the following gases:

(*a*) Oxygen at 60° F and 100 atm.

(*b*) Ammonia at 80° C and 40 atm.

(*c*) Carbon dioxide at 150° F and a gage pressure of 2000 psia.

25. Calculate the fugacity of liquid ethylene in contact with its saturated vapor at 0° C and 40.6 atm.

26. Calculate the fugacity of liquid chlorine in contact with a mixture of hydrogen and its own vapor at a temperature of 122° F and an absolute pressure of 1000 psia. The vapor pressure of liquid chlorine at this temperature is 14.1 atm, and its density is 1.557 grams per cc at −33.6° C.

15

Expansion and Compression of Fluids

The most useful form of energy into which it is generally desirable to convert other forms is the mechanical work of a rotating shaft or a moving piston. *Mechanical work* is defined as the energy that is transferred by the effect of a force acting through a distance and is equal to the product of that force times the distance of action. Thus, when a fluid that is confined under pressure undergoes a change in volume, work is done as the result of the force of pressure moving through the distance corresponding to the volume change. Similarly, a flowing fluid may perform work through changes of volume, of elevation, or of external kinetic energy.

As pointed out in Chapter 8, work is a form of energy that is incapable of storage as such but is in transition from one form of stored energy to another. Mechanical energy in this transitory form is termed *shaft work*, as distinguished from the electrical and other forms of work which may be included in the work terms of a complete energy balance. Shaft work is capable of transmission as such and may pass from one part of a system to another through either solid media such as shafts, pistons, gears, and belts, or through fluid media such as hydraulic or pneumatic couplings and drives. However, no transmission device is perfect, and some work is always lost by degradation to a lower form of energy, made manifest as heat resulting from friction.

Differential Energy Balance. In Chapter 8 a general energy balance was established for a complex chemical process involving several interconnected systems. By applying this energy equation to an infinitesimal change in the energy of a single system of constant mass m and constant composition, the following differential form results,

$$d(pV) + d\left(\frac{mu^2}{2g_c}\right) + d\left(mZ\frac{g_L}{g_c}\right) + d(mE_\sigma) + dU = d'q - d'w \qquad (1)$$

where the energy terms represent, respectively, flow work, external kinetic energy, potential energy, surface energy, internal energy, heat added from the surroundings, and work done upon the surroundings. The work term $d'w$, in general, includes all possible forms of work except flow work, such as shaft, electrical, magnetic, and radiant. Equation 1 may also be written as

$$dE = d'q - d'w \tag{2}$$

where E is the total energy of the system including all the terms represented by the left side of equation 1.

In the absence of electrical, magnetic, and radiant forms of work, $d'w$ represents mechanical or shaft work only. If changes in potential and external kinetic energies are also negligible, the shaft work results only from changes in flow work and from work of expansion. The evaluation of shaft work in such systems requires separate consideration of the special cases of nonflow and flow processes, in both isothermal and isentropic systems for reversible and irreversible processes.

Shaft Work in a Nonflow Process. In a nonflow process, as defined on page 244, flow is absent in the initial and final states of the process, and changes in surface, kinetic, and potential energies are generally negligible. If only work of expansion is performed, equation 1 then becomes

$$d'q = d'w_s + dU \tag{3}$$

where $d'w_s$ represents only mechanical work accomplished as the result of a volume change. In a reversible process $d'q = T\,dS$ and $dU = T\,dS - p\,dV$. Hence, for such a process equation 3 becomes

$$d'w_s = p\,dV$$

or

$$w_{s,\,\text{nonflow}} = \int_1^2 p\,dV \quad \text{(reversible)} \tag{4}$$

Equation 4 is restricted to mechanically reversible nonflow processes. Where the expansion is not mechanically reversible, $\int p\,dV$ is equal to the sum of the useful shaft work plus that lost in friction. In applying equation 4 to actual expansion or compression operations, the pressure used in the equation is that actually exerted on the face of the piston.

Shaft Work in a Flow Process. In applying equation 1 to a reversible flow process, $d(pV) = p\,dV + V\,dp$, $dU = T\,dS - p\,dV$, and $d'q = T\,dS$. Where forms of work other than shaft work are negligible, equation 1 becomes

$$d'w_s = -V\,dp - d\left(\frac{mu^2}{2g_c}\right) - d\left(mZ\,\frac{g_L}{g_c}\right) - d(mE_\sigma) \tag{5}$$

Where changes in kinetic, potential, and surface energies are negligible, equation 5 reduces to $d'w_s = -V\,dp$ or

$$w_{s,\,\text{flow}} = -\int_1^2 V\,dp = \int_2^1 V\,dp \tag{6}$$

Equation 6 represents the shaft work performed by any reversible flow process with negligible kinetic-, potential-, and surface-energy changes and with no electrical, radiant, or magnetic work.

Equations 4 and 6 may be used for calculating the work performed

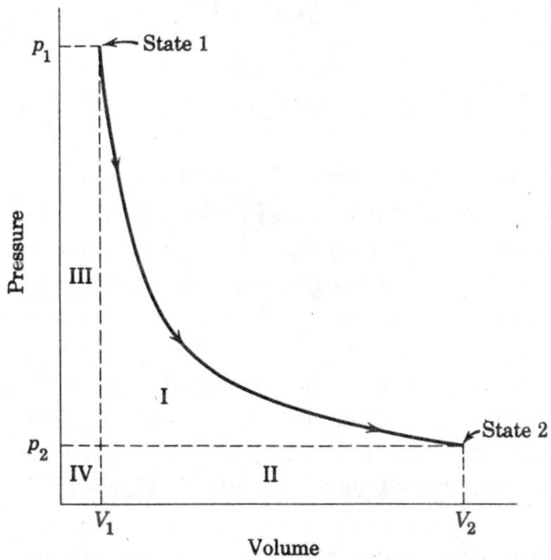

FIG. 157. Shaft work under flow and nonflow conditions

by the volume or pressure changes of a fluid from its state properties, only provided these changes take place reversibly.

The relation between reversible shaft work under flow and nonflow conditions in the absence of electrical, magnetic, or radiant forms of work and with negligible changes in potential, surface, and external kinetic energies is shown in Fig. 157, where the substance expands reversibly from p_1, V_1 to p_2, V_2.

$$w_{S,\,\text{nonflow}} = \int_1^2 p\,dV = \text{area I} + \text{area II} \tag{7}$$

$$w_{S,\,\text{flow}} = -\int_1^2 V\,dp = \int_2^1 V\,dp = \text{area I} + \text{area III} \tag{8}$$

Shaft Work from the Energy Properties. For reversible processes under the conditions defined for equations 4 and 6, the shaft work performed may be expressed as changes of the four energy functions. Thus, from equations 13–39 through 42, combined with equations 4 and 6,

$$\left(\frac{\partial A}{\partial V}\right)_T = -p \quad \text{or} \quad -(\Delta A)_T = \int_1^2 (p\,dV)_T = (w_s)_{T,\,\text{nonflow}} \tag{9}$$

$$\left(\frac{\partial G}{\partial p}\right)_T = V \quad \text{or} \quad -(\Delta G)_T = -\int_1^2 (V\,dp)_T = (w_s)_{T,\,\text{flow}} \tag{10}$$

$$\left(\frac{\partial U}{\partial V}\right)_S = -p \quad \text{or} \quad -(\Delta U)_S = \int_1^2 (p\,dV)_S = (w_s)_{S,\,\text{nonflow}} \tag{11}$$

$$\left(\frac{\partial H}{\partial p}\right)_S = V \quad \text{or} \quad -(\Delta H)_S = -\int_1^2 (V\,dp)_S = (w_s)_{S,\,\text{flow}} \tag{12}$$

From equations 9–12, the work of reversible expansion of a fluid under isothermal or isentropic conditions may be determined either by a pressure-volume integration or as the change of the appropriate energy function. The pV integration must follow the actual reversible path of the change, whereas the thermodynamic energy functions are state properties which are independent of path and are determined directly from the properties of the initial and final states. Both methods of calculating work may be employed with thermodynamic data based on either experimental measurements or generalized correlations. The choice of the method is determined by the type of data available. However, in general, it is more convenient to evaluate the thermodynamic energy functions at the terminal states of the system than to carry out pV integrations from equations of state.

Isothermal Nonflow Expansion. For the reversible isothermal expansion of any substance under nonflow conditions, from equation 9 and the definitions of the thermodynamic energy functions,

$$(w_s)_{T,\,\text{nonflow}} = \int_1^2 p\,dV = -\Delta A = -\Delta U + T\,\Delta S \tag{13}$$

Equation 13 is restricted to reversible isothermal nonflow operations involving only shaft work, and it neglects changes in kinetic, potential, and surface energies.

For ideal gases, $p\,dV$ is evaluated by differentiating the ideal-gas equation at constant temperature. Thus, for one mole,

$$p\,dv = -v\,dp = -RT\frac{dp}{p} \tag{14}^*$$

Substitution of equation 14 in equation 13 and integration gives

$$(w_s)_{T,\,\text{nonflow}} = -RT \ln \frac{p_2}{p_1} \tag{15}^*$$

For actual gases, $pv = zRT$. Differentiation at constant temperature gives

$$p\,dv + v\,dp = RT\,dz \tag{16}$$

Hence

$$(dw_s)_{T,\,\text{nonflow}} = p\,dv = -v\,dp + RT\,dz = -RTz\frac{dp}{p} + RT\,dz \tag{17}$$

or

$$(w_s)_{T,\,\text{nonflow}} = -RT\int_{p_1}^{p_2} z\,d\ln p + RT(z_2 - z_1) \tag{18}$$

The integral of equation 18 can be evaluated graphically by plotting values of z as a function of $\log p$ and determining the area under the curve between the given pressure limits. Values of z may be obtained from Table 49 or from experimental measurements. Where an equation of state such as the Beattie–Bridgeman is available, the integration may be carried out analytically.

The foregoing graphical integration can be avoided by evaluating the work from the internal-energy and entropy terms of equation 13. If the generalized correlations are employed, it is convenient to rearrange equation 13 as follows, for 1 mole of fluid:

$$w_{S,\,\text{nonflow}} = -[(\text{U}^*_1 - \text{U}_1) + (\text{U}^*_2 - \text{U}^*_1) - (\text{U}^*_2 - \text{U}_2)] + T[(\text{S}^*_1 - \text{S}_1) + (\text{S}^*_2 - \text{S}^*_1) - (\text{S}^*_2 - \text{S}_2)]$$

$$w_{S,\,\text{nonflow}} = -T_c\left(\frac{\text{U}^*_1 - \text{U}_1}{T_c} - \frac{\text{U}^*_2 - \text{U}_2}{T_c}\right) + T\left[(\text{S}^*_1 - \text{S}_1) - R\ln\frac{p_2}{p_1} - (\text{S}^*_2 - \text{S}_2)\right] \tag{19}$$

The internal energy and entropy deviation terms in equation 19 are obtained from Tables 53 and 52.

The complication of liquefaction is not encountered in isothermal expansion. If the system is completely gaseous at initial conditions, expansion under isothermal conditions will take the gas further from saturation; its superheat is increased.

Isothermal Flow Expansion. For the reversible isothermal expansion of any fluid under flow conditions, from equation 10,

$$(w_s)_{T,\,\text{flow}} = -\int_1^2 V\,dp = -\Delta G = -\Delta H + T\,\Delta S \tag{20}$$

Equation 20 is restricted to reversible isothermal flow conditions involving only shaft work and neglects changes in kinetic, potential, and surface energies.

For an ideal gas, per mole,

$$(w_s)_{T,\,\text{flow}} = -RT \ln \frac{p_2}{p_1} \tag{21)*$$

Thus, for an ideal gas the work of reversible isothermal expansion is the same for flow as for nonflow conditions. This results from the fact that at constant temperature the change in flow work is zero for an ideal gas, or $d(pV) = 0$.

For an actual gas, per mole, $-v\,dp = -z(RT\,dp)/p$ or

$$(w_s)_{T,\,\text{flow}} = -RT \int_1^2 z\,d \ln p \tag{22}$$

Equation 22 is evaluated graphically by the method described for equation 18 by using generalized compressibility factors or an equation of state. The graphical or analytical integration can be avoided by evaluating the work from the entropy and enthalpy terms of equation 20. If the generalized correlations of Tables 50 and 52 are used, equation 20 is rearranged as follows:

$$(w_s)_{T,\,\text{flow}} = -\left(\frac{\text{H}^*_1 - \text{H}_1}{T_c} - \frac{\text{H}^*_2 - \text{H}_2}{T_c}\right) T_c + T\left[(\text{S}^*_1 - \text{S}_1) - R \ln \left(\frac{p_2}{p_1}\right) - (\text{S}^*_2 - \text{S}_2)\right] \tag{23}$$

Illustration 1. Calculate the work of reversible isothermal expansion when 1 lb-mole of ethylene gas initially at 564° R and 50 atm expands to 1 atm under nonflow and flow conditions. From Table 7B, $z_c = 0.270$, $T_c = 283.1°$ K $=$ 509.6° R, $p_c = 50.5$ atm.

(*a*) Assuming *ideal behavior*, the shaft work is identical for nonflow and for flow conditions, as indicated by equations 15 and 21.

$$(w^*_s)_T = -RT \ln (p_2/p_1) = +(1.987)564(2.303) \log 50$$

$$= 4384 \text{ Btu per lb-mole}$$

(*b*) For *actual behavior under nonflow conditions*, shaft work may be determined by employing equation 18. This requires that $\int_1^2 z\,d \log p$ be evaluated. This is accomplished by plotting values of z against values of $\log p$ and determining the area under the curve between the limits of log 50 and log 1. The necessary data for making such a plot are obtained from Table 49 and are given here in Table A.

TABLE A

p_r	p, atm	log p	z
0.0198	1.0	0.000	0.995
0.0400	2.02	0.305	0.990
0.0600	3.03	0.481	0.986
0.1000	5.05	0.703	0.976
0.2000	10.10	1.004	0.951
0.3000	15.15	1.180	0.925
0.4000	20.20	1.305	0.896
0.5000	25.25	1.402	0.870
0.6000	30.30	1.481	0.840
0.7000	35.35	1.548	0.810
0.8000	40.40	1.606	0.779
0.9000	45.45	1.658	0.744
0.9901	50.00	1.699	0.712

The area under the curve is −1.5895, with the sign negative because the integration is from log 50 to log 1. Substitution in equation 18 then gives

$$w_s = -(1.987)564(2.303)(-1.5895) + (1.987)564(0.995 - 0.712)$$
$$= (1.987)564[(2.303)(1.5895) + (0.995 - 0.712)]$$
$$= 4419 \text{ Btu per lb-mole}$$

The shaft work under nonflow conditions may also be evaluated through the use of equation 19.

$$T_{r1} = 1.107, \quad p_{r1} = 0.990, \quad T_{r2} = 1.107, \quad p_{r2} = 0.0198$$

From Tables 53 and 52,

$$\frac{\text{U}^*_1 - \text{U}_1}{T_c} = 1.69, \qquad \frac{\text{U}^*_2 - \text{U}_2}{T_c} = 0.022$$

$$(\text{S}^*_1 - \text{S}_1) = 1.61, \qquad (\text{S}^*_2 - \text{S}_2) = 0.024$$

Substitution in equation 19 gives

$$w_s = -(1.69 - 0.022)(509.6) + 564[1.61 + (1.987)(2.303) \log 50 - 0.024]$$
$$= -850 + 5279 = 4429 \text{ Btu per lb-mole}$$

This result is in good agreement with that obtained by graphical integration. The small difference may be ascribed to errors inherent to graphical integration and also to small inconsistencies in the deviation tables for compressibility factor, internal energy, and entropy.

(c) For *actual behavior under flow conditions*, shaft work may be evaluated by equation 22, which involves the same integral as in part (*b*); hence

$$w_s = -(1.987)564(2.303)(-1.5895) = 4102 \text{ Btu per lb-mole}$$

Shaft work may also be determined by using equation 23. From Table 50,

$$\frac{H^*_2 - H_2}{T_c} = 0.032$$

$$\frac{H^*_1 - H_1}{T_c} = 2.33$$

$$w_s = (2.33 - 0.032)(509.6) - 564[1.61 + (1.987)(2.303) \log 50 - 0.024]$$
$$= 1171 - 5279 = 4108 \text{ Btu per lb-mole}$$

Where thermodynamic charts are available for any particular system, the reversible work can be obtained directly. The applications of these charts for evaluating the reversible work term under isothermal, isentropic, flow, and nonflow conditions are shown in Illustrations 2 and 4.

Illustration 2. From the data of Figs. 150 through 153, calculate the work performed in the reversible isothermal expansion of 1 lb of ammonia initially at 200 psia and 150° F to 15 psia under nonflow and flow conditions.

From Fig. 150,

$$V_1 = 1.740, \qquad V_2 = 25.5 \text{ cu ft per lb}$$
$$p_1V_1 = (200)144(1.740) = 50{,}100 \text{ ft-lb per lb}$$
$$p_2V_2 = (15)144(25.5) = 55{,}000 \text{ ft-lb per lb}$$
$$\Delta(pV) = 4900 \text{ ft-lb per lb}$$

From Fig. 151,

$$H_1 = 671 \text{ Btu per lb}$$
$$H_2 = 697 \text{ Btu per lb}$$
$$\Delta H = 26 \text{ Btu per lb or } 20{,}200 \text{ ft-lb per lb}$$

From Fig. 152,

$$S_1 = 1.244 \text{ Btu/(lb)(R°)}$$
$$S_2 = 1.575 \text{ Btu/(lb)(R°)}$$
$$\Delta S = 0.331 \text{ Btu/(lb)(R°)} = 257.5 \text{ ft-lb/(lb)(R°)}$$
$$T\,\Delta S = 610(257.5) = 157{,}000 \text{ ft-lb per lb}$$

Under nonflow conditions, by equation 13,

$$w_s = -\Delta H + \Delta(pV) + T\,\Delta S = -20{,}200 + 4900 + 157{,}000 = 141{,}700 \text{ ft-lb per lb}$$

Under flow conditions, by equation 20,

$$w_s = -\Delta H + T\,\Delta S = -20{,}200 + 157{,}000 = 136{,}800 \text{ ft-lb per lb}$$

Isentropic Nonflow Expansion. For reversible isentropic expansion under nonflow conditions, from equation 11,

$$(w_s)_{S,\text{ nonflow}} = \int_1^2 p\,dV = -\Delta U \tag{24}$$

Equation 24 applies to isentropic nonflow conditions where changes in kinetic, potential, and surface energies are negligible and where only mechanical work is involved. Calculation of the work is complicated by the fact that the final temperature or pressure after expansion is unknown and must be evaluated.

For an ideal gas, an equation relating p and T in an isentropic, nonflow, reversible expansion is developed as follows. By applying equation 24 in differential form to one mole of material, the following is obtained;

$$d(w_s)_{S,\,\text{nonflow}} = p\,dV = -d\text{H} + d(pv) = -d\text{H} + p\,dv + v\,dp \qquad (25)$$

For one mole of an ideal gas,

$$d\text{H}^* = v\,dp = \frac{RT}{p}\,dp \qquad (26)^*$$

From equation 13–174, page 549:

$$d\text{H}^* = c^*_p\,dT \qquad (27)^*$$

Combining equations 26 and 27 gives

$$c^*_p \frac{dT}{T} = R\frac{dp}{p} \qquad (28)^*$$

Where moderate temperature changes are involved, it is generally satisfactory to assume a constant mean value of c^*_{pm}. Then equation 28 is integrated to give

$$c^*_{pm} \ln\left(\frac{T_2}{T_1}\right) = R\ln\left(\frac{p_2}{p_1}\right) \qquad (29)^*$$

or

$$T_2 = T_1\left(\frac{p_2}{p_1}\right)^{R/c^*_{pm}} \qquad (30)^*$$

From equation 13–173, page 549, and equations 24 and 30:

$$(w^*_s)_{S,\,\text{nonflow}} = c^{*\prime}_{vm}(T_1 - T_2) = [c^{*\prime}_{pm} - R]T_1\left[1 - \left(\frac{p_2}{p_1}\right)^{R/c^*_{pm}}\right] \qquad (31)^*$$

It should be noted that c^*_{pm} and $c^{*\prime}_{pm}$ are different values which result from differential equations involving $d \ln T$ and dT, respectively.

The corresponding volume changes are obtained by combining the ideal-gas law with equation 30. Thus;

$$V_2 = V_1\left(\frac{p_2}{p_1}\right)^{(R/c^*_{pm})-1} \quad \text{or} \quad \left(\frac{V_2}{V_1}\right)^{c^*_{pm}/(c^*_{pm}-R)} = \frac{p_1}{p_2} \qquad (32)^*$$

Equations 30, 31, and 32 are frequently written in terms of κ, the ratio of the heat capacities at constant pressure and constant volume. For an ideal gas,

$$\kappa = \frac{c^*_p}{c^*_p - R} = \frac{c^*_v + R}{c^*_v} \qquad (33)^*$$

Combination of equation 33 with equations 30, 31, and 32, respectively, gives

$$\frac{T_2}{T_1} = \left(\frac{p_2}{p_1}\right)^{(\kappa-1)/\kappa} = \left(\frac{V_1}{V_2}\right)^{\kappa-1} \qquad (34)^*$$

$$(w^*_s)_{S,\,\text{nonflow}} = \frac{RT_1}{\kappa - 1}\left[1 - \left(\frac{p_2}{p_1}\right)^{(\kappa-1)/\kappa}\right] \qquad (35)^*$$

$$\left(\frac{V_2}{V_1}\right)^{\kappa} = \left(\frac{p_1}{p_2}\right) \qquad (36)^*$$

Equations 29 through 36 are restricted to expansions of ideal gases involving small changes in heat capacity. Where the heat-capacity terms are not satisfactorily represented by constant mean values, empirical equations relating heat capacity to temperature should be used in the integration of equation 28.

For *actual gases*, work is evaluated more readily from the terminal state properties using equation 24 than from the pV integration. Since the expansion is isentropic, $\Delta \text{s} = 0$, or

$$\Delta\text{s} = \text{s}_2 - \text{s}_1 = (\text{s}^*_1 - \text{s}_1) + (\text{s}^*_2 - \text{s}^*_1) - (\text{s}^*_2 - \text{s}_2) = 0 \qquad (37)$$

From equation 13–17, for one mole,

$$\text{s}^*_2 - \text{s}^*_1 = \int_{T_1}^{T_2} \frac{c^*_p}{T}\, dT - R \ln\left(\frac{p_2}{p_1}\right) \qquad (38)$$

where c^*_p, the molal heat capacity for ideal behavior, is a function of temperature only.

Combination of equations 37 and 38 gives

$$(\text{s}^*_1 - \text{s}_1) + \int_{T_1}^{T_2} \frac{c^*_p}{T}\, dT - R \ln\left(\frac{p_2}{p_1}\right) - (\text{s}^*_2 - \text{s}_2) = 0 \qquad (39)$$

The integral of equation 39 is expressed in terms of the empirical heat-capacity equation for the ideal gas. The equation is then solved for the final temperature corresponding to a specified final pressure and initial state. This is best done by assuming a series of final temperatures and plotting the corresponding values of the left side of the equation. That value of T_2 which reduces equation 39 to zero represents the correct final temperature. For ideal gases the two $(\text{s}^* - \text{s})$ terms become zero.

Once the initial and final states are defined, the reversible work is calculated directly from equation 24, rearranged for convenience in using the generalized table of internal-energy deviations.

$$(w_s)_{S,\,\text{nonflow}} = -\left[\left(\frac{\text{u}^*_1 - \text{u}_1}{T_c}\right)T_c + \int_{T_1}^{T_2} c^*_v\, dT - \left(\frac{\text{u}^*_2 - \text{u}_2}{T_c}\right)T_c\right]$$

$$= -\left[\left(\frac{\text{u}^*_1 - \text{u}_1}{T_c}\right)T_c + \int_{T_1}^{T_2} c^*_p\, dT - R(T_2 - T_1) - \left(\frac{\text{u}^*_2 - \text{u}_2}{T_c}\right)T_c\right] \qquad (40)$$

Where liquefaction occurs, temperature T_2 is the saturation temperature corresponding to the known final pressure p_2. Under these conditions, the $(s^*_2 - s_2)$ term of equation 39 is replaced by

$$(s^*_2 - s_2) = x(s^*_2 - s_2)_{sG} + (1 - x)(s^*_2 - s_2)_{sL} \tag{41}$$

where x is the *quality* of the mixture, or the fraction remaining in the gaseous state. If the numerical value of x obtained from equation 39, modified by equation 41, is less than 1, liquefaction does occur, and the final temperature T_2 is in fact the saturation temperature corresponding to p_2. If the numerical value of x is greater than 1, no liquefaction occurs.

If liquefaction occurs, shaft work is evaluated through equation 40, with the $u^*_2 - u_2$ term replaced by

$$u^*_2 - u_2 = x(u^*_2 - u_2)_{sG} + (1 - x)(u^*_2 - u_2)_{sL} \tag{42}$$

Isentropic Flow Expansion. *For an ideal gas,* since heat capacity is independent of pressure, equation 12 may be written

$$(w^*_s)_{S,\,\text{flow}} = -\int_1^2 V\,dp = -\Delta H = -\int_1^2 C^*_p\,dT \qquad (43)^*$$

Equation 43 applies to the work in reversible isentropic expansion under flow conditions and neglects changes in external kinetic, potential, and surface energies. Equation 43 combined with the ideal-gas law for one mole gives

$$-\int_1^2 R\,d\ln p = -\int_1^2 c^*_p\,d\ln T \qquad (44)^*$$

If a constant mean value of heat capacity is assumed, equation 44 integrates to give equation 30, showing that for ideal gases the temperature changes in isentropic expansion are the same for flow and for nonflow processes.

Combination of equations 30 and 44 for 1 mole gives

$$(w^*_o)_{S,\,\text{flow}} = c^{*\prime}_{pm}(T_1 - T_2) = c^{*\prime}_{pm}T_1\left[1 - \left(\frac{p_2}{p_1}\right)^{R/c^*_{pm}}\right] \qquad (45)^*$$

It may be noted that equation 45 differs from 31 in the appearance of $c^{*\prime}_{pm}$ instead of $c^{*\prime}_{vm}$ as the multiplying coefficient.

The temperature and volume changes of isentropic flow expansion are expressed in terms of the heat-capacity ratio by equations 34 and 36. By a similar development,

$$(w^*_s)_{S,\text{ flow}} = \frac{RT_1\kappa}{\kappa - 1}\left[1 - \left(\frac{p_2}{p_1}\right)^{(\kappa-1)/\kappa}\right] \tag{46*}$$

For an actual gas under *flow conditions* the procedure is similar to that followed in the isentropic nonflow case. Equations 37, 39, and 41 are all directly applicable to the flow case since they express the condition of constant entropy. Thus, it may be concluded that, for any fluid, the temperature changes in isentropic expansion are the same for flow and for nonflow processes.

The problem of determining the final temperature and quality is handled exactly as for the nonflow case. Once the final temperature and quality have been determined, shaft work is evaluated through the application of equation 12. For a gas superheated at final conditions,

$$(w_s)_{S,\text{ flow}} = -\Delta H = -\left[\left(\frac{H^*_1 - H_1}{T_c}\right)(T_c) + \int_{T_1}^{T_2} c^*_p\, dT - \left(\frac{H^*_2 - H_2}{T_c}\right)T_c\right] \tag{47}$$

Where liquefaction occurs, the term $(H^*_2 - H_2)$ in equation 47 is replaced by

$$H^*_2 - H_2 = x(H^*_2 - H_2)_{sG} + (1 - x)(H^*_2 - H_2)_{sL} \tag{48}$$

When equations 24 and 43 are compared, it is evident that the work in isentropic flow expansion is greater than that of a nonflow operation by the amount of $-\Delta(pV)$.

Illustration 3. Calculate the work of isentropic expansion of 1 lb-mole of ethylene gas when expanded from an initial pressure of 50 atm and 564° R to a final pressure of 1 atm under nonflow and flow conditions.

(*a*) **Ideal behavior, no liquefaction assumed.** The heat-capacity equation for ideal ethylene in the temperature range below 200° F is given in Table 18, page 255.

$$c^*_p = 7.95 + (8.13)10^{-11}T^{3.85}$$

Since the expansion is isentropic,

$$\Delta s = 0 = \int_{T_1}^{T_2}\left[\frac{7.95}{T} + (8.13)10^{-11}T^{2.85}\right]dT - R\ln\frac{1}{50}$$

$$= 7.95\ln(T_2/564) + (2.11)10^{-11}[T_2^{3.85} - (564)^{3.85}] + (1.987)(2.303)\log 50$$

This equation is solved for T_2, using a trial-and-error procedure, giving $T_2 = 235°$ R.

The reversible work of expansion is then calculated as follows for nonflow and for flow conditions, assuming no condensation:

$$(w^*_s)_{S,\,\text{nonflow}} = -\Delta \text{U} = -\int_{T_1}^{T_2} c^*_v\, dT = \int_{T_2}^{T_1} c^*_p\, dT + R(T_2 - T_1)$$

$$= \int_{235}^{564} [7.95 + (8.13)10^{-11}T^{3.85}]\, dT - (1.987)(564 - 235)$$

$$= 2980 - 654 = 2326 \text{ Btu per lb-mole}$$

$$(w^*_s)_{S,\,\text{flow}} = -\Delta \text{H} = -\int_{T_1}^{T_2} c^*_p\, dT = \int_{T_2}^{T_1} c^*_p\, dT$$

$$= \int_{235}^{564} [7.95 + (8.13)10^{-11}T^{3.85}]\, dT = 2980 \text{ Btu per lb-mole}$$

(*b*) **Actual behavior.** The possibility of liquefaction must be considered. If liquefaction does occur, the final temperature will be the saturation temperature corresponding to 1 atm, namely, 305° R. The properties at the initial and final states then would be as shown in Table A.

TABLE A

	Gas at State 1	Gas at State 2 (if Saturated)
Temperature, °R	564	305
Reduced temperature T_r	1.107	0.599
Pressure p, atm	50	1
Reduced pressure p_r	0.990	0.0198
z	0.712	0.979
$(\text{H}^* - \text{H})/T_c$, Btu/(lb-mole)(R°)	2.33	0.04
$(\text{s}^* - \text{s})$, Btu/(lb-mole)(R°)	1.61	0.08
$(\text{U}^* - \text{U})/T_c$, Btu/(lb-mole)(R°)	1.69	0.05

The molal latent heat of vaporization λ at 1 atm and 305° R is 5822 Btu per lb-mole. The use of an experimental latent heat of vaporization requires slight modification of the equations for Δs, ΔH, and ΔU as indicated below.

On the assumption that liquefaction has resulted from the expansion, the following equation is written and solved for x_2, the final quality of the mixture:

$$\Delta \text{s} = 0 = (\text{s}^*_1 - \text{s}_1) + \int_{T_1}^{T_{2s}} \frac{c^*_p}{T}\, dT - R \ln \frac{p_2}{p_1} - (\text{s}^*_2 - \text{s}_2)_{sG} - (1 - x_2)\frac{\lambda_2}{T_2}$$

$$= 1.61 - \int_{305}^{564} \left[\frac{7.95}{T} + (8.13)10^{-11}T^{2.85}\right] dT + (1.987)(2.303) \log 50$$

$$- 0.08 - (1 - x_2)(5822/305)$$

$$= 1.61 - 5.636 + 7.774 - 0.08 - (1 - x_2)(5822/305)$$

$$x_2 = 0.8078$$

Since the value of x_2 is between 0 and 1, the assumption of liquefaction is valid. The value of 0.8078 for x_2 holds for both the nonflow and the flow cases.

Shaft work is evaluated from the relations given below, ignoring the negligibly small volume of the liquid phase at final conditions:

$$(w_s)_{S,\,\text{nonflow}} = -\Delta U = -\left[\left(\frac{U^*_1 - U_1}{T_c}\right) T_c + \int_{T_1}^{T_{2s}} c^*_p \, dT - R(T_2 - T_1) - \left(\frac{U^*_2 - U_2}{T_c}\right)_{sG} T_c - (1 - x_2)(\lambda - z_{2sG} R T_{2s})\right]$$

$$= -\left\{(1.69)(509.6) - \int_{305}^{564} [7.95 + (8.13)10^{-11} T^{3.85}] \, dT + (1.987)(564 - 305) - 0.1922[5822 - (0.979)(1.987)305]\right\}$$

$$= -(861.2 - 2410.2 + 514.7 - 25.2 - 1005.4) = 2065 \text{ Btu per lb-mole}$$

$$(w_s)_{S,\,\text{flow}} = -\Delta H = -\left[\left(\frac{H^*_1 - H_1}{T_c}\right) T_c + \int_{T_1}^{T_{2s}} c^*_p \, dT - \left(\frac{H^*_2 - H_2}{T_c}\right)_{sG} T_c - (1 - x_2)\lambda_2\right]$$

$$= -\left\{(2.33)(509.6) - \int_{305}^{564} [7.95 + (8.13)10^{-11} T^{3.85}] \, dT - (0.04)(509.6) - (0.1922)5822\right\}$$

$$= -(1187.4 - 2410.2 - 20.4 - 1119.0) = 2362 \text{ Btu per lb-mole}$$

(*c*) **Ideal behavior, liquefaction considered.** The calculations parallel those of part (*b*), except that all deviation terms are omitted. The following results are obtained:

$$x_2 = 0.8880$$

$$(w_s)_{S,\,\text{nonflow}} = 2480 \text{ Btu per lb-mole}$$

$$(w_s)_{S,\,\text{flow}} = 3062 \text{ Btu per lb-mole}$$

Illustration 4. Calculate from Figs. 150 through 153 the work of isentropic expansion of 1 lb of ammonia from an initial absolute pressure of 200 psia and 150° F to a final absolute pressure of 15 psia under nonflow and flow conditions.

From Fig. 152, following a vertical line from the initial conditions to the final pressure gives

$$t_2 = -27° \text{ F}, \qquad x_2 = \frac{1.243 - 0.032}{1.394 - 0.032} = 89.0\%$$

Similarly, from Fig. 153 it is found that

$H_1 = 671$ Btu per lb

H_2 (saturated gas) $= 602.4$ Btu per lb

H_2 (saturated liquid) $= 13.6$ Btu per lb

H_2 (mixture) $= x(602.4) + (1 - x)(13.6) = 537.6$ Btu per lb

From Fig. 150, at the final temperature, pressure, and quality:

V_2 (saturated gas) $= 17.6$ cu ft per lb

$$xp_2V_2 = (0.89)(15)144(17.6) = 33{,}800 \text{ ft-lb per lb}$$

The volume of the liquid phase may be neglected.

For the initial conditions, $V_1 = 1.74$ (Fig. 150):

$$p_1V_1 = (200)144(1.74) = 50{,}100 \text{ ft-lb per lb}$$

Using the data of Illustration 2 for the initial state, we get

$$\Delta H = 538 - 671 = -133 \text{ Btu per lb or } -103{,}300 \text{ ft-lb per lb}$$

$$\Delta(pV) = 33{,}800 - 50{,}100 = -16{,}300 \text{ ft-lb per lb}$$

For nonflow conditions, from equation 24:

$$w_s = -\Delta H + \Delta(pV) = 103{,}300 - 16{,}300 = 87{,}000 \text{ ft-lb per lb}$$

For flow conditions, from equation 12:

$$w_s = -\Delta H = 103{,}300 \text{ ft-lb per lb}$$

Free Expansion

The unrestrained expansion of a gas is known as free expansion. Under conditions of no restraint no work is done, and under adiabatic conditions no heat is added. Free expansion under flow conditions is commonly known as throttling or as the Joule–Thomson effect.

This effect is measured experimentally by expanding the gas slowly under steady flow through a well-insulated porous plug; in this way potential work is lost, and no heat is allowed to enter or leave the system through the walls. The fluid flows reversibly into and out of the process, but the expansion step is completely irreversible. Since no heat is added or lost, the process takes place under conditions of constant enthalpy. However, a change in internal energy results if any change in flow energy pV occurs.

For an ideal gas, since internal energy and enthalpy are independent of pressure and dependent upon temperature only, no change in temperature occurs in free expansion under either nonflow or flow conditions, and values of ΔH, $\Delta(pV)$, and ΔU are zero. However, an increase in entropy results which is equal to $-R \ln p_2/p_1$ by equation 13–17, page 515, and the values of ΔA and ΔG are each equal to $-T\,\Delta S$.

For nonideal gases, the temperature change in Joule–Thomson expansion is of great experimental and practical value. Such data are used for establishing the deviations from ideal behavior of real gases, particularly the effects of pressure on enthalpy and entropy. Free expansion under flow conditions is made use of industrially in the cooling and liquefaction of gases.

Free expansion under nonflow conditions, known as the Maxwell effect, is of little value. Because of the small heat capacity of a gas compared with that of the vessel, adiabatic conditions cannot be approached, and experimental measurements are of uncertain meaning.

In the foregoing discussion it was assumed that kinetic-energy changes were negligible. Where this is not true, the general energy-balance equation including kinetic-energy changes must be employed. Changes in both internal energy and temperature may result from the free expansion of an ideal gas at high velocities in either a flow or a nonflow system.

Joule–Thomson Expansion. The temperature change that results from the free expansion of a real gas under flow conditions is determined through the use of the principle that ΔH for such a process is zero, when changes in kinetic energy, potential energy, and surface energy are negligible. The following equation results:

$$\Delta H = 0 = \left(\frac{H^*_1 - H_1}{T_c}\right) T_c + \int_{T_1}^{T_2} c^*_p \, dT - \left(\frac{H^*_2 - H_2}{T_c}\right) T_c \qquad (49)$$

Under certain conditions partial liquefaction of the gas may occur, in which case, the $H^*_2 - H_2$ term of equation 49 is replaced by equation 48.

The usual procedure followed in order to determine whether or not liquefaction occurs is to assume that it does take place, in which case the final temperature T_2 would be the saturation temperature corresponding to p_2. Equation 49, modified by equation 48, is then used to solve for x_2. If the value of x_2 is below 1, condensation does occur; if the value of x_2 is over 1, the assumption of condensation is proved to be incorrect. If no condensation occurs, the correct value of T_2 must be determined by a trial-and-error procedure, using equation 49.

After the final temperature and quality have been determined, the changes in the other state properties may be established through the use of the following equations. As written, the equations for ΔS and ΔU are for the case in which liquefaction occurs. They may readily be adapted to the case where no liquefaction occurs by setting x equal to 1, and dropping the subscript s that designates saturation at final conditions.

$$\Delta S = (S^*_1 - S_1) + \int_{T_1}^{T_2} \frac{c^*_p}{T} dT - R \ln \frac{p_2}{p_1} - x(S^*_2 - S_2)_{sG} - (1 - x)(S^*_2 - S_2)_{sL} \qquad (50)$$

$$\Delta U = \left(\frac{U^*_1 - U_1}{T_c}\right) T_c + \int_{T_1}^{T_2} c^*_p \, dT - R(T_2 - T_1) - x\left(\frac{U^*_2 - U_2}{T_c}\right)_{sG} T_c - (1 - x)\left(\frac{U^*_2 - U_2}{T_c}\right)_{sL} T_c \qquad (51)$$

$$\Delta A = \Delta U - \Delta(TS) = \Delta U - (T_2 S_2 - T_1 S_1) = \Delta U - [T_2(S_1 + \Delta S) - T_1 S_1] \qquad (52)$$

$$\Delta G = \Delta H - \Delta(TS) = 0 - (T_2 S_2 - T_1 S_1) = -[T_2(S_1 + \Delta S) - T_1 S_1] \qquad (53)$$

It will be noted that evaluation of ΔA and ΔG requires that the absolute entropy at state 1 or state 2 be known.

An alternative method of evaluating final temperature when no condensation occurs may be deduced from a consideration of Fig. 158, which represents a portion of the enthalpy–departure chart. In this figure, points 1 and 2 represent, respectively, the initial and final state points for a Joule–Thomson expansion. Since enthalpy is a point function, the

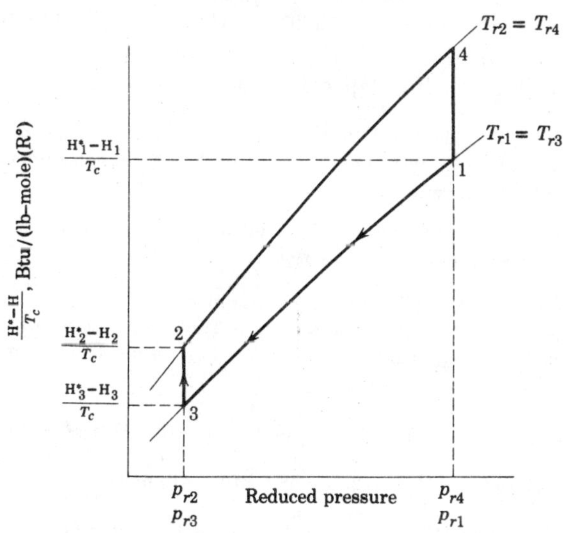

FIG. 158. Analysis of Joule–Thomson expansion

net enthalpy change between points 1 and 2 is zero, regardless of path. By following path 132, the following equation is obtained

$$T_2 = T_1 - \frac{[(H^*_1 - H_1)/T_c - (H^*_3 - H_3)/T_c]T_c}{c_{pm}} \tag{54}$$

where c_{pm} is the real mean molal heat capacity between T_2 and T_1 at the final pressure p_2. Under most conditions, p_2 is so low that c_{pm} is quite close to c^*_{pm}, the ideal molal heat capacity.

Consideration of equation 54 indicates the validity of the following rule, which is convenient for quickly determining whether heating or cooling occurs in a Joule–Thomson expansion with a pressure drop of finite magnitude: Locate state point 1 on the enthalpy–departure chart. Then travel along the T_{r1} isotherm until the ordinate at p_{r2} is intersected, thus locating state point 3. If state point 3 is at a lower level on the chart than state point 1 (this condition is shown in Fig. 158), cooling occurs; if state point 3 is at a higher level than state point 1, heating occurs; if state points 3 and 1 are at the same level, no temperature change occurs.

As a corollary, it may be stated that, if state point 1 lies on an isotherm of the enthalpy–departure chart at a point where the slope is positive, an infinitisimal pressure drop produces cooling. If the state point lies on an isotherm at a point where the slope is negative, heating occurs when the pressure drops infinitesimally. The so-called Joule–Thomson inversion points are those state points that fall at the maxima of the various isotherms on the enthalpy–departure plot.

Illustration 5. Calculate the final temperature when methane undergoes a Joule–Thomson expansion from 1400 psia and 40° F to a final pressure of 14 psia.

(*a*) **Solution based on equation 49:**

$$p_c = 45.8 \text{ atm}, \qquad T_c = 343.3° \text{ R}, \qquad z_c = 0.290$$

$$c^*_p = 7.95 + (6.4)10^{-12}T^4$$

with c^*_p on a molal basis and T in degrees Rankine.

Using values of $(\text{H}^* - \text{H})/T_c$ selected from Table 50, a series of ΔH values is calculated for various assumed values of T_2. For assumed final temperatures of −40° F and −50° F, the following are obtained.

$$(\Delta\text{H})_{-40°\text{F}} = (2.096)(343.3) + \int_{499.7°\text{ R}}^{419.7°\text{ R}} [7.95 + (6.4)10^{-12}T^4]\,dT - (0.006)(343.3)$$

$$= 719.6 - 659.2 - 2.1 = 58.3 \text{ Btu per lb-mole}$$

$$(\Delta\text{H})_{-50°\text{F}} = (2.096)(343.3) + \int_{499.7°\text{ R}}^{409.7°\text{ R}} [7.95 + (6.4)10^{-12}T^4]\,dT - (0.013)(343.3)$$

$$= 719.6 - 740.6 - 4.5 = -25.5 \text{ Btu per lb-mole}$$

By interpolation to $\Delta\text{H} = 0$, the final temperature is found to be −47.0° F; hence, $\Delta T = -87.0$ F°.

(*b*) **Solution based on equation 54:**

$$(\text{H}^*_1 - \text{H}^*_1)/T_c = 2.096, \qquad (\text{H}^*_3 - \text{H}_3)/T_c = -0.002$$

A preliminary value of final temperature is calculated, using the ideal heat capacity at 40° F, which is 8.349 Btu/(lb-mole)(R°)

$$\Delta T = -\frac{[(\text{H}^*_1 - \text{H}_1)/T_c - (\text{H}^*_3 - \text{H}_3)/T_c]T_c}{c_{pm}}$$

$$= -\frac{(2.096 + 0.002)(343.3)}{8.349} = -86.3 \text{ R°}$$

$$T_2 = 413.4° \text{ R} = -46.3° \text{ F}$$

Inspection of the heat-capacity departure chart, Fig. 145, shows that, at both initial and final conditions, the heat-capacity correction is so small that it is not shown on the chart.

A somewhat better value of T_2 is secured by taking the arithmetic average of the ideal heat capacities at 40° F and −46.3° F and using this to recalculate ΔT

$$c^*_{pm} = \frac{8.349 + 8.137}{2} = 8.243$$

$$\Delta T = -\frac{(2.096 + 0.002)(343.3)}{8.243} = -87.4 \text{ F}^\circ$$

$$T_2 = 412.3^\circ \text{ R} = -47.4^\circ \text{ F}$$

The foregoing discussion covers procedures based on the use of the generalized charts. For many pure substances, as, for example, water, ammonia, carbon dioxide, and the lower normal paraffins, accurate thermodynamic charts are available which make it possible to solve problems pertaining to Joule–Thomson expansions with greater speed and accuracy. The procedure simply requires following lines of constant enthalpy; the Mollier, or *HS* diagram, is ideally adapted to such problems.

Illustration 6. The following data were obtained in determining the quality of steam in a throttling calorimeter:

Original pressure of wet steam = 100 psia
Final pressure of steam = 14.3 psia
Final temperature of steam = 248° F

Determine the original quality x_1 of the wet steam by the use of the H–S chart for steam.

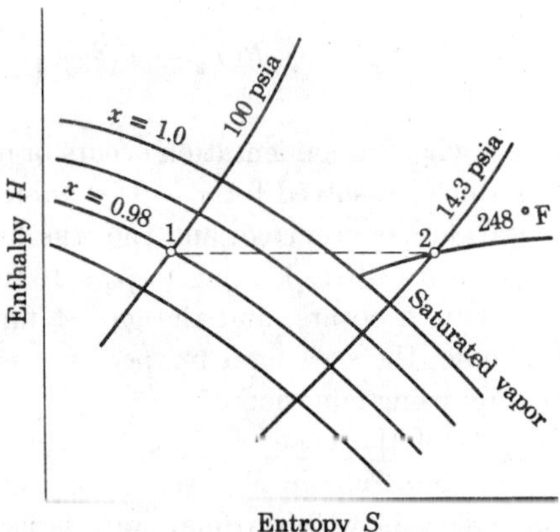

FIG. 159. Temperature change in throttling using Mollier diagram

The method is indicated in Fig. 159. State point 2 is located on the chart. Then a horizontal line is drawn until the isobar corresponding to p_1 is intersected, thus locating state point 1. The location of this point in the network of constant-quality lines indicates $x_1 = 0.98$

If continuous free expansion is carried out at high velocities, the changes in kinetic energy are not negligible. In this case, from the general energy-balance equation 1, page 246, if potential and surface energy changes are negligible, a decrease in enthalpy occurs which is equal to the increase in kinetic energy. Thus, from an energy balance where no condensation occurs:

$$\Delta \text{H} = \frac{-m(u_2^2 - u_1^2)}{2\, g_c} = \left(\frac{\text{H}^*_1 - \text{H}_1}{T_c}\right) T_c + \int_{T_1}^{T_2} c^*_p\, dT - \left(\frac{\text{H}^*_2 - \text{H}_2}{T_c}\right) T_c \tag{55}$$

The final temperature T_2 may be evaluated from equation 55 and the other state properties may be calculated from equations 50 through 53.

Equation 55 applies to the case where no condensation occurs; suitable changes may be made as shown previously to adapt the equation to the case where condensation does occur.

Maxwell Expansion. As previously pointed out, adiabatic free expansion under nonflow conditions is characterized by constant internal energy.

$$\Delta \text{U} = 0 \tag{56}$$

Equation 56 may be expressed in terms of the generalized internal-energy departures to give, in the absence of condensation,

$$\Delta \text{U} = \left(\frac{\text{U}^*_1 - \text{U}_1}{T_c}\right) T_c + \int_{T_1}^{T_2} c^*_p\, dT - R(T_2 - T_1) - \left(\frac{\text{U}^*_2 - \text{U}_2}{T_c}\right) T_c \tag{57}$$

In order to determine whether condensation occurs or not, equation 57, modified as indicated above, is solved for x_2. If a value between 0 and 1 is obtained, condensation has occurred, and the true final temperature is the saturation temperature corresponding to p_2. If the value of x_2 is more than 1, no liquefaction occurs, and the gas at final conditions is superheated. In this case, the true final temperature is obtained by a trial-and-error procedure, using equation 57.

Thermodynamic charts of the type of Figs. 150 through 153 do not permit direct solution of problems in Maxwell expansion. A trial-and-error procedure is necessary in which a final state is assumed, and the assumption is tested by means of equation 57.

From a knowledge of T_2, the changes in the other state properties may be evaluated from equations 50 through 53.

Illustration 7. Calculate the changes in state properties in the free expansion of ethylene gas from 564° R and 50 atm to 1 atm under flow and nonflow conditions.

The absolute entropy of ethylene at 77° F and 1 atm is 52.45 Btu/(lb-mole)(R°). The initial absolute entropy of the ethylene gas is 43.60 Btu/(lb-mole)(R°).

$$T_1 = 564° \text{ R}, \qquad p_1 = 50 \text{ atm}$$
$$T_{r1} = 1.107, \qquad p_{r1} = 0.990$$
$$z_1 = 0.712$$
$$(\text{s}^*_1 - \text{s}_1) = 1.61 \text{ Btu/(lb-mole)(R°)}$$
$$(\text{H}^*_1 - \text{H}_1)/T_c = 2.33 \text{ Btu/(lb-mole)(R°)}$$
$$(\text{U}^*_1 - \text{U}_1)/T_c = 1.69 \text{ Btu/(lb-mole)(R°)}$$

(*a*) **Flow conditions (Joule–Thomson expansion).** A series of values for T_2 is assumed. Values of ΔH are calculated in going from the initial state to 1 atm and the various assumed values of T_2, using the equation

$$\Delta\text{H} = \left(\frac{\text{H}^*_1 - \text{H}_1}{T_c}\right)T_c + \int_{T_1}^{T_2} c^*_p\, dT - \left(\frac{\text{H}^*_2 - \text{H}_2}{T_c}\right)T_c$$

Assumed T_2	ΔH
420° R	−271.3 Btu per lb-mole
440	−90.4
460	+98.3
480	+291.9

By interpolation $\Delta\text{H} = 0$ when $T_2 = 450°$ R. This represents the correct final temperature. It is above the saturation temperature corresponding to 1 atm; hence no condensation occurs.

Using the known value of T_2, the following are obtained from Tables 49, 50, 52, and 53;

$$z_2 = 0.988, \qquad (\text{H}^*_2 - \text{H}_2)/T_c = 0.062$$
$$(\text{s}^*_2 - \text{s}_2) = 0.060, \qquad (\text{U}^*_2 - \text{U}_2)/T_c = 0.042$$
$$v_2 = z_2RT_2/p_2 = (0.988)(0.7302)450/1 = 324.7 \text{ cu ft per lb-mole}$$

$$\Delta\text{s} = (\text{s}^*_1 - \text{s}_1) + \int_{T_1}^{T_2} \frac{c^*_p}{T}\, dT - R\ln\frac{p_2}{p_1} - (\text{s}^*_2 - \text{s}_2)$$
$$= 1.61 - \int_{450}^{564}\left[\frac{7.95}{T} + (8.13)10^{-11}T^{2.85}\right]dT + (1.987)(2.303)\log 50 - 0.060$$
$$= 1.61 - 2.275 + 7.774 - 0.060 = 7.049 \text{ Btu/(lb-mole)(R°)}$$

$$\Delta\text{U} = \left(\frac{\text{U}^*_1 - \text{U}_1}{T_c}\right)T_c + \int_{T_1}^{T_2} c^*_p\, dT - R(T_2 - T_1) - \left(\frac{\text{U}^*_2 - \text{U}_2}{T_c}\right)T_c$$
$$= (1.69)(509.6) - \int_{450}^{564} c^*_p\, dT - 1.987(450 - 564) - (0.042)(509.6)$$
$$= 861.2 - 1152.5 + 226.5 - 21.4 = -86 \text{ Btu per lb-mole}$$

$$\Delta\text{H} = 0$$
$$\Delta(T\text{s}) = T_2\,\Delta\text{s} + \text{s}_1\,\Delta T$$
$$= 450(7.049) + (43.60)(450 - 564) = -1798 \text{ Btu per lb-mole}$$
$$\Delta\text{A} = \Delta\text{U} - \Delta(T\text{s}) = -86 - (-1798) = 1712 \text{ Btu per lb-mole}$$
$$\Delta\text{G} = \Delta\text{H} - \Delta(T\text{s}) = 0 - (-1798) = 1798 \text{ Btu per lb-mole}$$

(*b*) **Nonflow conditions (Maxwell expansion).** The procedure followed in determining T_2 is similar to that followed for the Joule–Thomson expansion, except that the criterion is $\Delta U = 0$ instead of $\Delta H = 0$.

Assumed T_2	ΔU
420° R	−302.7 Btu per lb-mole
440	−159.0
460	−11.0
480	+141.2

By interpolation, $T_2 = 461°$ R. This is the correct final temperature. It is greater than the saturation temperature corresponding to 1 atm; hence there is no condensation.

Using $T_2 = 461°$ R, the following values are obtained from Tables 49, 50, 52, and 53:

$$z_2 = 0.989, \qquad (H^*_2 - H_2)/T_c = 0.058$$

$$(S^*_2 - S_2) = 0.054, \qquad (U^*_2 - U_2)/T_c = 0.039$$

Using these values and the same general equations for the Joule–Thomson case the following values result:

$$v_2 = 332.9 \text{ cu ft per lb-mole}$$

$$\Delta S = 7.281 \text{ Btu/(lb-mole)(R°)}$$

$$\Delta U = 0$$

$$\Delta H = 108 \text{ Btu per lb-mole}$$

$$\Delta A = 1134 \text{ Btu per lb-mole}$$

$$\Delta G = 1242 \text{ Btu per lb-mole}$$

In both parts (*a*) and (*b*), real behavior was considered. The ideal cases are comparatively simple, since an ideal gas undergoes no temperature change in either a Joule–Thomson expansion or a Maxwell expansion. This may be proved readily by developing equations for $(\partial T/\partial p)_H$ and $(\partial T/\partial p)_U$. When such equations are applied to an ideal gas through the use of the equation $pv = RT$, both expressions reduce to zero. If the temperature remains unchanged, there will be no condensation, and it becomes a simple matter to calculate v_2, ΔS, ΔH, ΔU, ΔA, and ΔG.

In Table 54 are summarized the changes in point and path properties when ethylene gas is expanded from 50 atm and 564° R to 1 atm under various conditions of restraint when both ideal and actual behaviors are assumed. It may be observed that for both ideal and actual behaviors the changes in point properties under reversible conditions are identical for both flow and nonflow conditions, but work and heat terms are dependent on the path and conditions of flow. For all reversible conditions $q = \int T\,dS$, and each of the work terms is identified with one of the energy functions. In free expansion, the heat and work terms are zero, and the changes in point properties for actual behavior are different for flow and for nonflow conditions. It is interesting to note that the Joule–Thomson expansion cools ethylene gas to 450° R without condensation, whereas by isentropic expansion the gas is cooled to 305° R with 19% condensed.

Table 54. Thermodynamic Changes in the Expansion of Ethylene Gas

Initial Conditions: 564° R and 50 atm, $s_1 = 43.60$ Btu/(lb-mole)(R°), $s^*_1 = 45.21$ Btu/(lb-mole)(R°)
Properties of Ethylene: $T_c = 509.6°$ R, $p_c = 50.5$ atm, $z_c = 0.270$, $c^*_p = 7.95 + (8.13)10^{-11}T^{3.85}$, where c^*_p is on a molal basis, and T is in degrees Rankine.

For ideal behavior, $v_1 = 8.24$ cu ft per lb-mole. For real behavior, $v_1 = 5.86$ cu ft per lb-mole.

Units T in degrees Rankine; v in cubic feet; Δs in Btu/(lb-mole)(R°); q, w_s, ΔH, ΔG, Δ(Ts) in Btu per pound-mole

Conditions of Expansion	x_2	T_2	v_2	Δs	q	w_s	ΔU	ΔH	ΔA	ΔG	Δ(Ts)
Ideal behavior											
1. Isothermal, nonflow, reversible	1.000	564	411.8	7.77	4385	4385	0	0	−4385	−4385	4385
2. Isothermal, flow, reversible	1.000	564	411.8	7.77	4385	4385	0	0	−4385	−4385	4385
3. Isentropic, nonflow, reversible											
(*a*) Assume no condensation	1.000	235	171.6	0	0	2326	−2326	−2980	12,548	11,894	−14,874
(*b*) Take account of condensation	0.888	305	197.9	0	0	2480	−2480	−3062	9229	8647	−11,709
4. Isentropic, flow, reversible											
(*a*) Assume no condensation	1.000	235	171.6	0	0	2980	−2326	−2980	12,548	11,894	−14,874
(*b*) Take account of condensation	0.888	305	197.9	0	0	3062	−2480	−3062	9229	8647	−11,709
5. Free expansion, nonflow (Maxwell)	1.000	564	411.8	7.77	0	0	0	0	−4385	−4385	4385
6. Free expansion, flow (Joule–Thomson)	1.000	564	411.8	7.77	0	0	0	0	−4385	−4385	4385
Actual behavior											
1. Isothermal, nonflow, reversible	1.000	564	409.8	9.36	5279	4429	850	1171	−4429	−4108	5279
2. Isothermal, flow, reversible	1.000	564	409.8	9.36	5279	4108	850	1171	−4429	−4108	5279
3. Isentropic, nonflow, reversible	0.8078	305	176.3	0	0	2065	−2065	−2362	9227	8930	−11,292
4. Isentropic, flow, reversible	0.8078	305	176.3	0	0	2362	−2065	−2362	9227	8930	−11,292
5. Free expansion, nonflow (Maxwell)	1.000	461	332.9	7.28	0	0	0	108	1134	1242	−1134
6. Free expansion, flow (Joule–Thomson)	1.000	450	324.7	7.05	0	0	−86	0	1712	1798	−1798

Adiabatic Filling and Discharging of Pressure Vessels

Two cases are of particular interest. In the first case, a pressure vessel is discharging gas through a throttle valve. In the second case, a pressure vessel is being filled with gas which is supplied at a fixed temperature and pressure through a throttle valve.

In analyzing the two examples, it will be necessary to assume no heat interchange between the gas and the walls of the pressure vessels. Actually, such interchange of heat cannot be avoided, hence the results of the calculations are to be regarded only as limiting values which are approached to a greater or lesser degree, according to the circumstances involved.

Each example will first be analyzed on the assumption of ideal behavior and constancy of heat capacity. Though it is somewhat questionable whether or not the refinements of considering departure from ideality and variation in heat capacity are justified by the added work required, the procedure will nevertheless be outlined for each case.

Case 1. The original temperature T_{A1} and pressure p_{A1} of the gas in the vessel are known, as well as the volume of the pressure vessel. The throttle valve is opened, and gas is released until the pressure in the vessel drops to a specified value p_{A2}. The problem is to determine the temperature of the gas left in the vessel, and its amount.

During discharge, the gas within the pressure vessel is undergoing isentropic expansion. If ideal behavior is assumed, the final temperature may be determined as follows:

$$T_{A2} = T_{A1}\left(\frac{p_{A2}}{p_{A1}}\right)^{(\kappa-1)/\kappa} \tag{58}$$

The residual gas n_{A2} is determined by the ideal-gas law.

If departure from ideality is to be considered, the following equations are employed:

$$\Delta s_A = 0 = (s^* - s)_{A1} + \int_{T_{A1}}^{T_{A2}} \frac{c^*_p}{T}\,dT - R\ln\left(\frac{p_{A2}}{p_{A1}}\right) - (s^* - s)_{A2} \tag{59}$$

$$n_{A2} = \frac{p_{A2}V_A}{z_{A2}RT_{A2}} \tag{60}$$

By trial-and-error procedures, the value of T_{A2} satisfying equation 59 is determined. This value is then used in equation 60 to determine the amount of gas left in the pressure vessel.

Case 2. Gas at a constant known pressure p_0 and a constant known temperature T_0 is forced through the throttle valve into pressure vessel A until the pressure in the vessel rises to a specified value p_{A2}. The original pressure and temperature of the gas in the vessel are known, as well as the volume of the pressure vessel.

Regardless of whether or not ideal behavior is specified, the following energy balance equation is required:

$$n_{A1}\text{U}_{A1} + (n_{A2} - n_{A1})(\text{U}_0 + p_0 v_0) = n_{A2}\text{U}_{A2} \qquad (61)$$

Introducing enthalpies,

$$n_{A1}\text{H}_{A1} - p_{A1}V_A + (n_{A2} - n_{A1})\text{H}_0 = n_{A2}\text{H}_{A2} - p_{A2}V_A \qquad (62)$$

If ideal behavior and constant heat capacity are assumed, equation 62 may be rewritten in the following manner, if all enthalpies are expressed relative to T_{A1}:

$$0 - p_{A1}V_A + (n_{A2} - n_{A1})c^*_p(T_0 - T_{A1}) = n_{A2}c^*_p(T_{A2} - T_{A1}) - p_{A2}V_A \qquad (63)$$

From the ideal-gas law,

$$n_{A2} = \frac{p_{A2}V_A}{RT_{A2}} \qquad (64)$$

By combining equations 63 and 64, an expression for T_{A2} is obtained which is as follows:

$$T_{A2} = T_0 \frac{\kappa p_{A2}}{p_{A2} + p_{A1}(\kappa - 1) + p_{A1}\kappa(T_0 - T_{A1})/T_{A1}} = \frac{T_0 \kappa}{1 + (p_{A1}/p_{A2})[(\kappa T_0/T_{A1}) - 1]} \qquad (65)$$

This equation simplifies for special conditions. For example, if $T_0 = T_{A1}$, the equation reduces to the following:

$$T_{A2} = T_{A1} \frac{\kappa p_{A2}}{p_{A2} + p_{A1}(\kappa - 1)} \qquad (66)$$

If it is further specified that the pressure vessel is completely evacuated at the start,

$$T_{A2} = \kappa T_0 \qquad (67)$$

To account for departure from ideality, equation 62 is employed. All enthalpies are expressed relative to the initial state of the gas in the pressure vessel. Substitution for n_{A2} is made from the relation $n_{A2} = p_{A2}V_A/z_{A2}RT_{A2}$. The equation then has the following form:

$$0 - p_{A1}V_A + \left(\frac{p_{A2}V_A}{z_{A2}RT_{A2}} - n_{A1}\right)\text{H}_0 = \frac{p_{A2}V_A}{z_{A2}RT_{A2}}\text{H}_{A2} - p_{A2}V_A \qquad (68)$$

In this equation, H_0 is calculated as follows:

$$\text{H}_0 = \left(\frac{\text{H}^* - \text{H}}{T_c}\right)_{A1} T_c + \int_{T_{A1}}^{T_0} c^*_p \, dT - \left(\frac{\text{H}^* - \text{H}}{T_c}\right)_0 T_c \qquad (69)$$

A similar expression is used to evaluate H_{A2}.

By using trial-and-error procedures, that value of T_{A2} satisfying equation 68 is determined. Finally, n_{A2} is computed from the relation $n_{A2} = p_{A2}V_A/z_{A2}RT_{A2}$.

Compression of Gases

Gases are compressed for purposes of transportation, storage, and lique-

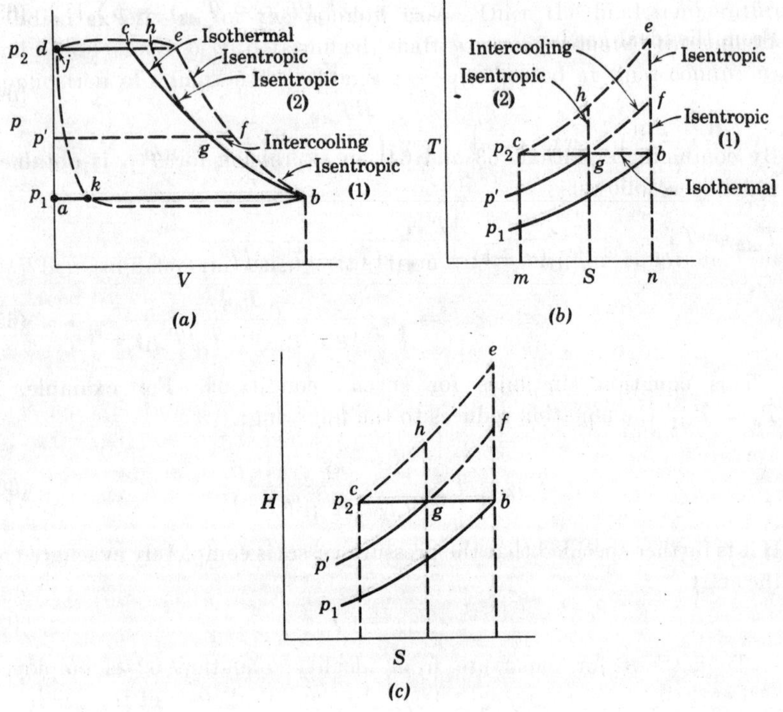

Fig. 160. Compression of gases

faction, or to maintain high or low pressures in chemical and metallurgical processes. Adiabatic compression is always accompanied by an increase in temperature; this temperature rise is not the objective of the operation, except in special applications such as the ignition of fuel in the Diesel engine or in vapor-compression evaporation. Where a high temperature of a compressed gas is desired, the temperature rise resulting from compression is incidental to the operation, since a temperature increase is produced more economically through direct heating than by the dissipation of valuable shaft work.

Mechanical gas compressors are of three general types:

1. Turbocompressors and fans.
2. Positive-displacement rotary blowers.
3. Reciprocating compressors.

Types 2 and 3 operate with a sequence of intermittent operations including a nonflow step. Thus, with reference to Fig. 160, a charge of gas is taken in at the suction pressure p_1 along line *ab* on the pV diagram and compressed to the final pressure p_2 along *be* if the operation is isentropic. The compression step is nonflow in character. It is immediately followed by discharge of the compressed gas along line *ed*. Since the gas flows into and out of the apparatus under the restraint of pressure, the net effect of the over-all operation is a flow process, even though an intermediate nonflow step is involved. The over-all work requirements and changes in thermodynamic properties accordingly follow the equations developed for expansion or compression under flow conditions.

Turbocompressors and Fans. Turbocompressors and fans operate in a continuous manner. Although within the machine high velocities and kinetic energies may be developed, the design is generally such that these intermediate kinetic energies are largely transformed to flow energy at discharge. Kinetic energies are generally negligible compared with the work of compression if the ratio of discharge to suction pressure is greater than 1.1. Under such conditions the theoretical work requirements may be calculated from equation 47 if conditions of reversibility are approached.

Where small pressure changes are involved, changes in the kinetic energies of the suction and discharge streams may constitute an appreciable portion of the work requirement. These effects are taken into account in isentropic compression by combination of equations 5, and 43, with changes in potential and surface energies being neglected.

$$\begin{aligned} -(w_s)_S &= \int_1^2 V\,dp + \frac{m(u_2^2 - u_1^2)}{2g_c} \\ &= (H_2 - H_1)_S + \frac{m(u_2^2 - u_1^2)}{2g_c} \end{aligned} \tag{70}$$

Similarly, for isothermal compression, by combining equations 5 and 20,

$$-(w_s)_T = (H_2 - H_1)_T - T(S_2 - S_1) + \frac{m(u_2^2 - u_1^2)}{2g_c} \tag{71}$$

Thus, the work input is increased by the difference between the kinetic energies of the discharge and suction streams.

The work of isothermal compression for an ideal gas is less than the work of isentropic compression by the amount of heat dissipated during isothermal compression.

Compressor Efficiencies. In Fig. 160 are diagrams showing the thermodynamic changes accompanying various types of compression between given suction and discharge pressures. The ideal work of compression is represented by the area to the left of the compression curve on the pV diagram. Thus, for isentropic compression the ideal work corresponds to area *abed*. For isothermal compression the corresponding work requirement is represented by area *abcd*. It is evident that isothermal operation requires less power. In general, efforts are made to design compressors to approach isothermal operation at the lowest feasible temperature. In this way power requirements are reduced but at the expense of additional cooling equipment. The optimum design is that corresponding to minimum over-all costs of compression and cooling.

Large-size compressors and fans ordinarily approximate adiabatic operation and would behave in accordance with equation 70 if the operation were reversible. However, irreversibility results from fluid friction and turbulence as well as from mechanical friction in the equipment, both of which increase the work requirement. Fluid friction also increases the final temperature of the fluid above the value calculated from equation 70. The departure of the operation from ideal reversibility is expressed by *an isentropic compression efficiency factor* η_S which is defined as the percentage ratio of the ideal work requirement, calculated from equation 70 for the pressure increase obtained, to the total shaft work input to the compressor. Thus, if kinetic-energy changes are negligible,

$$\eta_S = \frac{(H_2 - H_1)_S}{-w_{S,\text{total}}}(100) \tag{72}$$

where H_2 is the enthalpy of the discharged gas at the calculated temperature resulting from isentropic compression.

Inasmuch as work of compression is a minimum in a reversible isothermal operation, another index of performance, termed *the isothermal compression efficiency* η_T is also used.

$$\eta_T = \frac{\text{work of reversible isothermal compression}}{\text{total shaft-work input to the compressor}}(100)$$

If kinetic energy changes are negligible,

$$\eta_T = \frac{(G_2 - G_1)_T}{-w_{S,\text{total}}}(100) \tag{73}$$

The over-all efficiency includes compression efficiency and the effectiveness of the cooling system.

The over-all efficiency η_S may be separated into the mechanical efficiency η_m and the indicated efficiency η_{IS}. Thus

$$\eta_S = \eta_m \eta_{IS} \tag{74}$$

Where kinetic energy is negligible,

$$\eta_{IS} = (H_2 - H_1)_S / -w_i \tag{75}$$

$$-w_i = \text{indicated work of compression}$$

If heat losses are negligible,

$$-w_i = H'_2 - H_1$$

where H'_2 is the actual enthalpy of the compressed gas.

$$\eta_m = w_i / w_{S,\text{total}} \tag{76}$$

Mechanical efficiencies are generally in the range of 0.8 to 0.85 for reciprocating compressors and from 0.95 to 0.98 for turbocompressors. Indicated efficiencies vary from 0.7 to 0.9.

Ideal-Gas Tables.

For problems involving isentropic pressure changes of nearly ideal gases, improved accuracy and much simplification result from use of tables relating temperatures, pressure, and enthalpy in such changes. From equation 29, for isentropic changes in an ideal gas:

$$\left(\ln \frac{p}{p_o}\right)_S = \frac{1}{R}\left(\int_{T_o}^{T} c_p \, d \ln T\right)_S \tag{77}$$

It follows that, if T_o and p_o are chosen as reference states, values of *relative pressures* $p_R = p/p_o$ may be calculated as unique single-valued functions of T. Extensive tables have been developed for air[1] using 0° R as the reference temperature at which enthalpy is taken arbitrarily as zero. The value of p_o is taken as 1.0 at a temperature of 0° C (491.7° R). Table 55 presents an abridged form of these values.

At any selected temperature in Table 55, the corresponding value of p_R represents the pressure from which isentropic expansion to unit pressure would result in a final temperature of 0° C (491.7° R). From equation 77

[1] J. H. Keenan and Joseph Kaye, *Gas Tables*, John Wiley & Sons (1948).

it is evident that p_R is dimensionless and may be taken as corresponding to any selected units. Thus, for example, at $T_R = 650°$ R, $H = 155.50$ and $p_R = 2.655$. These values signify that, if air at a temperature of 650° R and a pressure of 2.655 psia is expanded to 1.0 psia, the final temperature will be 0° C (491.7° R), and the accompanying enthalpy change will be 155.50 − 117.49 = 38.01 Btu per lb. The same results apply if the initial pressure is 2.655 atm and the final pressure is 1.0 atm or if the initial pressure is any value p_1 and the final pressure is $p_1/2.655$.

Similarly, if air at 650° R is expanded from any selected pressure p_1, for example 100 psia, to any final pressure p_2, for example 20 psia, the value of p_{R1} is 2.655 and the final value of $p_{R2} = p_{R1}(p_2/p_1) = (2.655)(0.20) = 0.5310$. Interpolating in Table 55, the corresponding $T_2 = 410.3°$ R and $\Delta H = 155.50 - 95.60 = 59.90$ Btu per lb.

Illustration 8. Air is compressed from 80° F and a pressure of 1 atm to a gage pressure of 300 psi. Assuming ideal-gas behavior and isentropic compression, calculate the final temperature and the work of compression by using (*a*) Table 55, (*b*) an average heat-capacity ratio, $\kappa = 1.395$, and the formulas applicable to the isentropic compression of an ideal gas.

(*a*) $$p_{r2} = p_{r1}(p_2/p_1) = 1.3834(314.7/14.7) = 29.62$$

From Table 55, $T_2 = 1270°$ R

$$w_s = -\Delta H = H_1 - H_2 = 128.99 - 309.07 = -180.08 \text{ Btu per lb}$$

(*b*) $$T_2 = T_1\left(\frac{p_2}{p_1}\right)^{(\kappa-1)/\kappa} = 539.7\left(\frac{314.7}{14.7}\right)^{0.395/1.395} = (539.7)(2.3812) = 1285° \text{ R}$$

$$w_s = -\Delta H = -c_p(T_2 - T_1) = -R\left(\frac{\kappa}{\kappa - 1}\right)T_1\left[\left(\frac{p_2}{p_1}\right)^{(\kappa-1)/\kappa} - 1\right]$$

$$= -(1.987)\frac{1.395}{0.395}(539.7)(2.3812 - 1) = -5230.9 \text{ Btu per lb-mole}$$

or

$$-5230.9/28.970 = -180.56 \text{ Btu per lb}$$

For standard air, the value of $\kappa = 1.395$ is recommended for calculations involving moderate temperatures and pressures. Correction curves have been developed expressing deviations from the standard κ as a function of the pressure ratio. However, for most purposes use of the air tables is more convenient and accurate.

Reciprocating Compressors. As previously shown, an ideal reciprocating compressor would operate on the cycle *abed* of the pV diagram of Fig. 160 if the compression were isentropic. In such a case, the work requirement would be expressed by equation 70. Actual compressors of industrial sizes closely approximate adiabatic operation but deviate from reversibility as a result of fluid and mechanical friction. A further complication is

Table 55. Air Tables*

H = enthalpy, Btu/lb air, relative to 0° R

$$\frac{p_{R1}}{p_{R2}} = \left(\frac{p_1}{p_2}\right)_s$$

T°, R	H	p_R	T°, R	H	p_R
			450	107.5	0.7329
			460	109.90	0.7913
			470	112.30	0.8531
			480	114.69	0.9182
			490	117.08	0.9868
			491.7	117.49	1.0000
100	23.74	0.003841	500	119.48	1.0590
110	26.13	0.005358	510	121.87	1.1349
120	28.53	0.007260	520	124.27	1.2147
130	30.92	0.009601	530	126.66	1.2983
140	33.31	0.012436	540	129.06	1.3860
150	35.71	0.015824	550	131.46	1.4779
160	38.10	0.019822	560	133.86	1.5742
170	40.49	0.02449	570	136.26	1.6748
180	42.89	0.02990	580	138.66	1.7800
190	45.28	0.03612	590	141.06	1.8899
200	47.67	0.04320	600	143.47	2.005
210	50.07	0.05122	610	145.88	2.124
220	52.46	0.06026	620	148.28	2.249
230	54.85	0.07037	630	150.68	2.379
240	57.25	0.08165	640	153.09	2.514
250	59.64	0.09415	650	155.50	2.655
260	62.03	0.10797	660	157.92	2.801
270	64.43	0.12318	670	160.33	2.953
280	66.82	0.13986	680	162.73	3.111
290	69.21	0.15808	690	165.15	3.276
300	71.61	0.17795	700	167.56	3.446
310	74.00	0.19952	710	169.98	3.623
320	76.40	0.22290	720	172.39	3.806
330	78.78	0.24819	730	174.82	3.996
340	81.18	0.27545	740	177.23	4.193
350	83.57	0.3048	750	179.66	4.396
360	85.97	0.3363	760	182.08	4.607
370	88.35	0.3700	770	184.51	4.826
380	90.75	0.4061	780	186.94	5.051
390	93.13	0.4447	790	189.38	5.285
400	95.53	0.4858	800	191.81	5.526
410	97.93	0.5295	810	194.25	5.775
420	100.32	0.5760	820	196.69	6.033
430	102.71	0.6253	830	199.12	6.299
440	105.11	0.6776	840	201.56	6.573

* Abridged from Table 1 of *Gas Tables*, by Joseph H. Keenan and Joseph Kaye, copyright, 1948, by Joseph H. Keenan and Joseph Kaye, published by John Wiley & Sons (1948).

TABLE 55. AIR TABLES (*Continued*)

$T°$ R	H	p_R	$T°$ R	H	p_R
850	204.01	6.856	1250	304.08	27.96
860	206.46	7.149	1260	306.65	28.80
870	208.90	7.450	1270	309.22	29.67
880	211.35	7.761	1280	311.79	30.55
890	213.80	8.081	1290	314.36	31.46
900	216.26	8.411	1300	316.94	32.39
910	218.72	8.752	1310	319.53	33.34
920	221.18	9.102	1320	322.11	34.31
930	223.64	9.463	1330	324.69	35.30
940	226.11	9.834	1340	327.29	36.31
950	228.58	10.216	1350	329.88	37.35
960	231.06	10.610	1360	332.48	38.41
970	233.53	11.014	1370	335.09	39.49
980	236.02	11.430	1380	337.68	40.59
990	238.50	11.858	1390	340.29	41.73
1000	240.98	12.298	1400	342.90	42.88
1010	243.48	12.751	1410	345.52	44.06
1020	245.97	13.215	1420	348.14	45.26
1030	248.45	13.692	1430	350.75	46.49
1040	250.95	14.182	1440	353.37	47.75
1050	253.45	14.686	1450	356.00	49.03
1060	255.96	15.203	1460	358.63	50.34
1070	258.47	15.734	1470	361.27	51.68
1080	260.97	16.278	1480	363.89	53.04
1090	263.48	16.838	1490	366.53	54.43
1100	265.99	17.413	1500	369.17	55.86
1110	268.52	18.000	1510	371.82	57.30
1120	271.03	18.604	1520	374.47	58.78
1130	273.56	19.223	1530	377.11	60.29
1140	276.08	19.858	1540	379.77	61.83
1150	278.61	20.51	1550	382.42	63.40
1160	281.14	21.18	1560	385.08	65.00
1170	283.68	21.86	1570	387.74	66.63
1180	286.21	22.56	1580	390.40	68.30
1190	288.76	23.28	1590	393.07	70.00
1200	291.30	24.01	1600	395.74	71.73
1210	293.86	25.76	1610	398.42	73.49
1220	296.41	24.53	1620	401.09	75.29
1230	298.96	26.32	1630	403.77	77.12
1240	301.52	27.13	1640	406.45	78.99

TABLE 55. AIR TABLES (*Continued*)

$T°$ R	H	p_R	$T°$ R	H	p_R
1650	409.13	80.89	2050	518.61	192.31
1660	411.82	82.83	2060	521.39	196.16
1670	414.51	84.80	2070	524.18	200.06
1680	417.20	86.82	2080	526.97	204.02
1690	419.89	88.87	2090	529.75	208.06
1700	422.59	90.95	2100	532.55	212.1
1710	425.29	93.08	2110	535.35	216.3
1720	428.00	95.24	2120	538.15	220.5
1730	430.69	97.45	2130	540.94	224.8
1740	433.41	99.69	2140	543.74	229.1
1750	436.12	101.98	2150	546.54	233.5
1760	438.83	104.30	2160	549.35	238.0
1770	441.55	106.67	2170	552.16	242.6
1780	444.26	109.08	2180	554.97	247.2
1790	446.99	111.54	2190	557.78	251.9
1800	449.71	114.03	2200	560.59	256.6
1810	452.44	116.57	2210	563.41	261.4
1820	455.17	119.16	2220	566.23	266.3
1830	457.90	121.79	2230	569.04	271.3
1840	460.63	124.47	2240	571.86	276.3
1850	463.37	127.18	2250	574.69	281.4
1860	466.12	129.95	2260	577.51	286.6
1870	468.86	132.77	2270	580.34	291.9
1880	471.60	135.64	2280	583.16	297.2
1890	474.35	138.55	2290	585.99	302.7
1900	477.09	141.51	2300	588.82	308.1
1910	479.85	144.53	2310	591.66	313.7
1920	482.60	147.59	2320	594.49	319.4
1930	485.36	150.70	2330	597.32	325.1
1940	488.12	153.87	2340	600.16	330.9
1950	490.88	157.10	2350	603.00	336.8
1960	493.64	160.37	2360	605.84	342.8
1970	496.40	163.60	2370	608.68	318.0
1980	499.17	167.07	2380	611.53	355.0
1990	501.94	170.50	2390	614.37	361.3
2000	504.71	174.00	2400	617.22	367.6
2010	507.49	177.55			
2020	510.26	181.16			
2030	513.04	184.81			
2040	515.82	188.54			

introduced by the necessity of providing a certain *clearance volume* between the cylinder head and the piston at its extreme position. This clearance volume is represented by *dj* in Fig. 160. At the end of the stroke, the clearance volume is filled with compressed gas which expands to *k*, reducing the effective suction volume to *kb*.

The *volumetric efficiency* η_v is defined as the percentage ratio of the volume of gas, measured at p_1, actually taken into the cylinder during the suction stroke to the volumetric displacement of the piston. The volumetric efficiency determines the capacity of a machine for a specified duty while the compression efficiency determines the total work input required per unit quantity of gas compressed. The total power required for a machine of given displacement operating between specified pressures is increased by increased volumetric efficiency and reduced by increased compression efficiency.

Compression efficiencies are reduced by mechanical friction and by throttling through the suction and discharge valves. As a result of these pressure drops, the actual suction pressure is somewhat below p_1 and the discharge pressure above p_2, as indicated by the dotted lines between *ab* and *je* on Fig. 160. The increased work resulting from the throttling represents a major loss, and proper design of valve openings is of great importance. Compression efficiencies are increased by cooling the gas during compression. This will cause the operation to approach isothermal conditions, as indicated in the pV diagram of Fig. 160. Cylinders of small compressors are frequently water-cooled, but water jackets are not highly effective on large machines because of the small ratio of surface to volume. In some designs cooling water is injected into the gas during compression.

Volumetric efficiency is reduced by pressure drop through the suction valves and by increased clearance volume. The problems of compressor design have been discussed in detail by York[2] who concludes that the gas retained in the clearance volume of large machines approximates isentropic expansion as the piston starts on the suction stroke. In this case the ideal work of compression per unit quantity of gas discharged is unaffected by the clearance volume which may be looked on as containing an isolated quantity of gas alternately compressed and expanded in a reversible manner without any net consumption of power. Reduction in volumetric efficiency due to clearance can be calculated if isentropic expansion of the clearance volume on the intake stroke is assumed, but further reduction due to pressure drops can be determined only by direct measurement.

Variation of the clearance volume offers an effective means of controlling

[2] R. York Jr., *Ind. Eng. Chem.*, **34**, 355 (1942).

the capacity of a compressor driven at a constant speed. Machines are constructed with clearance pockets of variable volume which permit operating the machine at reduced capacities without waste of power. An increase of the clearance volume greatly reduces the volumetric efficiency with relatively little effect on the compression efficiency. Volumetric efficiencies of industrial compressors range from 50 to 90%.

Illustration 9. A single-stage compressor is to compress 800 cu ft per min of ammonia gas at 0° F and pressure of 15.0 psia to 75 psia. Calculate the actual horsepower required to drive the compressor and the piston displacement in cubic feet per minute, assuming an isentropic compression efficiency of 76% and a volumetric efficiency of 85%.

From Fig. 153 and equation 12, for isentropic compression at the special conditions;

$$-(w_s)_S = (H_2 - H_1)_S = 715 - 617 = 98 \text{ Btu per lb}$$

The ideal discharge temperature is 200° F.

At 0° F, 15 psia, the specific volume of ammonia gas from Fig. 150 is 18.92 cu ft per lb.

$$\text{Compression rate} = 800/18.92 = 42.3 \text{ lb per min}$$

$$\text{Theoretical horsepower} = \frac{(42.3)(98)(778)}{(33{,}000)} = 97.8 \text{ hp}$$

$$\text{Actual horsepower} = \frac{97.8}{0.76} = 129 \text{ hp}$$

$$\text{Piston displacement} = \frac{800}{0.85} = 940 \text{ cu ft per min}$$

Multistage Compression. Because of mechanical difficulties associated with high ratios of discharge to suction pressure, a single-stage reciprocating compressor is ordinarily not designed for a compression ratio greater than 5 to 1. Turbocompressors are limited to much lower ratios per stage. In order to develop high over-all compression ratios, it is necessary to use multistage compression in which one stage discharges to the suction of the next.

An important advantage of multistage operation is that it permits *intercooling* of the gas between stages and thus approaches the same suction temperature for each stage. This results in a reduction of the work of compression and approaches isothermal conditions where several stages of small compression ratios are employed. A two-stage operation with intercooling is indicated in Fig. 160. Isentropic compression in the first stage along bf is followed by cooling at constant pressure p' along fg. The final pressure p_2 is reached by isentropic compression in the second stage along gh with a final temperature intermediate between that of e resulting from single-stage isentropic compression and the isothermal value of c.

It is evident that the work required for the two-stage operation is less than that of the single stage by the area *fghe*. If more stages with intercooling were employed, the work would be still further reduced, approaching the ideal isothermal case. For multistage compression, the *equivalent compression efficiency* is defined as the ratio of the work for multistage isentropic compression with complete intercooling between stages divided by the actual shaft-work input.

The total work in a multistage compression is the sum of the work in the individual stages. Thus, from equation 46, if a constant value of κ and a constant suction temperature T_1 are assumed and kinetic-energy change is neglected, the total work per mole for two-stage compression of an ideal gas becomes

$$-w^*_s = \left(\frac{RT_1\kappa}{\kappa - 1}\right)\left\{\left[\left(\frac{p'}{p_1}\right)^{(\kappa-1)/\kappa} - 1\right] + \left[\left(\frac{p_2}{p'}\right)^{(\kappa-1)/\kappa} - 1\right]\right\} \quad (78)^*$$

If the intermediate pressure is p', the work required will be a minimum value when dw^*_s/dp' is zero. The optimum intermediate pressure is determined by equating dw^*_s/dp' derived from equation 78 to zero. Thus,

$$\frac{dw^*_s}{dp'} = \left(\frac{RT\kappa}{\kappa - 1}\right)\left[\left(\frac{1}{p_1}\right)^{(\kappa-1)/\kappa}\left(\frac{\kappa - 1}{\kappa}\right)(p')^{-1/\kappa} + (p_2)^{(\kappa-1)/\kappa}\left(\frac{1 - \kappa}{\kappa}\right)(p')^{(1-2\kappa)/\kappa}\right] = 0 \quad (79)^*$$

$$p'^{[2(\kappa-1)]/\kappa} = (p_1p_2)^{(\kappa-1)/\kappa}$$

$$p' = \sqrt{p_1p_2}$$

Rearrangement of equation 79 gives

$$\frac{p'}{p_1} = \frac{p_2}{p'} = \left(\frac{p_2}{p_1}\right)^{1/2} \quad (80)^*$$

Thus, for best operation, each stage should have the same compression ratio, equal to the square root of the over-all compression ratio. Where s stages are employed, it may be demonstrated similarly that optimum results are obtained when the compression ratio in each stage is equal to

$$\frac{p''}{p'} = \left(\frac{p_2}{p_1}\right)^{1/s} \quad (81)^*$$

If equation 81 is combined with an equation of the form of 78, it follows that the work per stage is constant, and the over-all work per mole is given by

$$-w^*_s = \left(\frac{sRT_1\kappa}{\kappa - 1}\right)\left[\left(\frac{p_2}{p_1}\right)^{(\kappa-1)/s\kappa} - 1\right] \quad (82)^*$$

By means of equation 82, the number of stages for optimum economy of operation may be estimated by consideration of the increased equipment and maintenance costs which accompany an increased number of stages with a resultant reduction in power requirement.

Illustration 10. Air at 75° F and 1 atm is to be compressed to 100 psia at a rate of 60 lb per min. A two-stage compressor is to be used, with an intermediate pressure of 38.34 psia and intercooling to 75° F. For each of the two stages, $\eta_I = 0.83$ and $\eta_M = 0.98$. Calculate the discharge temperature of the air and the total power requirement, using Table 55.

$$H_1 = 127.79 \text{ Btu per lb}$$

$$p_{R2} = 1.3395(38.34/14.70) = 3.494$$

$$H'_2 = [(168.23 - 127.79)/0.83] + 127.79 = 176.49 \text{ Btu per lb}$$

$$T'_2 = 736.7^\circ \text{ R} = 277^\circ \text{ F}$$

Since the compression ratio is the same in the two stages, the power inputs are equal, and the total power is twice that in the first stage.

$$\text{Power} = (2)(60)(60)(176.49 - 127.79)/2545(0.98) = 141 \text{ hp}$$

A single-stage compressor performing the same duty and having the same efficiencies as the two-stage compressor would require 163 hp. The use of the two-stage compressor with intercooling results in a decrease of 13.5% in the power requirement.

For compression of nonideal gases the problem of calculating optimum intermediate pressures involves trial-and-error methods. A simplified method was proposed by York[2] which may be used where economy is of great importance. However, for many cases the over-all work requirement is not greatly affected by small variations of the intermediate pressures from the optimum values, and equation 81 may be used for establishing the intermediate pressures. Once these pressures are established, the work and volumetric capacity for each stage are calculated from equation 47 and the pVT relations. Equation 81 may lead to serious departure from optimum pressure conditions when vapor mixtures are dealt with if large amounts of condensate are separated in the intercoolers. Where this condition is encountered, the intermediate pressures may be adjusted to equalize the work in different stages. If only two stages are involved, a series of intermediate pressures may be assumed and the corresponding over-all work requirements calculated; the optimum intermediate pressure corresponds to the minimum of a curve relating work to intermediate pressure.

Ordinarily in isentropic compression of gases there is no tendency for condensation in the compressor. This is evident from the TS diagram for ammonia, (Fig. 152). Adiabatic compression of the saturated vapor results in superheating. However, gases of high molal heat capacity behave quite differently, as may be seen from the TS diagram for benzene in Fig. 155. If saturated benzene vapor at 100° F is isentropically compressed, it is first superheated, and then at a pressure of 655 psia it becomes saturated, and condensation begins. Formation of condensate in compressors is frequently hazardous, and great care must be exercised in designing compression systems for vapors of this type. Formation of condensate in the intercoolers of multistage compressors is common when vapor mixtures are being compressed, and facilities for its separation from the uncondensed gas passing to the next stage are provided.

Problems

1. Calculate from the changes in the energy functions, by each of the following methods, the work done, the heat added, and the changes in T, v, s, u, h, a, and g, in the reversible expansion to 1 atm of 1 lb-mole of chlorine originally at a pressure of 500 psia and 250° F. The initial absolute entropy of ideal chlorine at 25° C and 1 atm is 53.286 Btu/(lb-mole)(R°). Carry out these calculations (1) assuming ideal behavior, (2) taking into account deviations from ideal behavior by the generalized correlation. Express work and the energy functions in Btu, and give the fraction of gas liquefied where condensation occurs.

(*a*) Nonflow isothermal.
(*b*) Flow isothermal.
(*c*) Nonflow isentropic.
(*d*) Flow isentropic.
(*e*) Maxwell expansion.
(*f*) Joule–Thomson expansion.

The physical properties of chlorine are as follows:

$$\text{Normal boiling point} = 430^\circ \text{ R}$$
$$\lambda = 8780 \text{ Btu per lb-mole at } 430^\circ \text{ R}$$
$$\rho_L = 1.56 \text{ grams per cc at } 430^\circ \text{ R}$$

2. Repeat the calculations of parts 2*a* and 2*b* of problem 1 by graphical integration of the pV changes based on the generalized compressibility-factor correlation.

3. Propane at a pressure of 200 psia and 120° F is expanded through a valve to a pressure of 39 psia. Using the generalized thermodynamic correlations and the vapor-pressure equation from Table 8, page 95, calculate the temperature on the downstream side of the valve and the quality or degrees of superheat of the vapor above saturation before and after expansion.

4. The expansion referred to in problem 3 might also be carried out isentropically in an engine. Calculate the temperature and degrees of superheat or quality after expansion in this manner and the ideal work performed per pound-mole.

5. Hydrogen gas at 20 atm and 80° F is passed through a throttle valve into a cylinder until the pressure in the cylinder rises to 10 atm. The hydrogen initially in the cylinder was at 1 atm and 80° F. The volume of the cylinder is 1 cu ft. Calculate the final temperature and weight of the hydrogen in the cylinder, assuming ideal behavior, and assuming no heat transfer to the cylinder.

6. A tank of 5000 cu ft capacity is to be used for the storage of ethylene gas. The tank is initially filled with ethylene gas at 1 atm pressure and 80° F. It is connected with a supply of ethylene gas at a constant pressure of 100 atm and 80° F, and the gas is allowed to flow into the tank until the pressure is equalized at 100 atm. Calculate the final temperature of the gas in the tank, assuming no heat transfer to the walls of the tank.

(a) Assuming ideal gas behavior.

(b) Assuming actual gas behavior.

7. Air from the atmosphere at 1 atm and 70° F is allowed to flow into an evacuated tank of 387 cu ft capacity. Calculate the temperature of the air in the cylinder when it is filled with air at 1 atm pressure, assuming no transfer of heat to the tank, and assuming ideal behavior.

8. Pressure vessel A, having a volume of 10 cu ft, originally contains nitrogen at 50 atm and 530° R. Pressure vessel B, which has a volume of 15 cu ft, originally contains nitrogen at 1 atm and 540° R. The two pressure vessels are connected by a pipe in which there is a throttle valve. This valve is opened, and gas is allowed to flow from vessel A into vessel B until the pressure in vessel A drops to 25 atm. Assuming no heat interchange between the gas and the vessel walls, calculate T_{A2}, n_{A2}, p_{B2}, T_{B2}, and n_{B2}:

(a) Assuming ideal behavior and constant heat capacity.

(b) Taking account of departure from ideality and variation in heat capacity.

9. A single-stage compressor is required to compress 450 cu ft per min of CO_2 measured at 60° F and 14.5 psia from a pressure of 5 psia and 80° F to a pressure of 20 psia. Assuming isentropic compression, calculate the horsepower required to drive the compressor, the required piston displacement in cubic feet per minute, and the discharge temperature, assuming a volumetric efficiency of 77% and a compression efficiency of 83%.

10. The compression duty of problem 9 might also be handled by a two-stage machine with intercooling to 80° F. Calculate the horsepower required to drive the compressor in this operation and the discharge temperature, assuming the efficiencies given in problem 9.

11. Anhydrous HCl ($t_b = -85°$ C) at a pressure of 30 psia and 80° F is to be compressed to 450 psia at a rate of 100 cu ft per min measured at 60° F and 14.5 psia. A two-stage machine is to be used with intercooling to 80° F. The compression efficiency based on isentropic compression in each stage is 80%, and the volumetric efficiency 65%. The intermediate suction pressure may be calculated from equation 80.

Assuming isentropic compression in each stage, calculate:

(a) The horsepower required.

(b) The required piston displacement, cubic feet per minute in each stage.

(c) The discharge temperature.

12. A compressor having a 75% mechanical efficiency is to compress 1.5 lb-moles of ammonia per minute from a pressure of 15 psia and 100° F to 200 psia. The operation is to be in two stages, each of which may be assumed to be isentropic with intercooling to 115° F. By use of Figs. 150 through 153, calculate the power required to drive the compressor when it is designed for the optimum intermediate pressure. Calculate the discharge temperature from each stage and the heat-transfer duty of the intercooler, and compare the optimum intermediate pressure with that calculated from equation 80.

16

Thermodynamics of Fluid Flow

Only highly restricted applications of thermodynamic principles to problems of fluid flow are possible without involvement in the complexities of the science of *fluid mechanics*. A general approach to fluid mechanics requires consideration of an elementary volume of fluid having both *scalar properties* which possess magnitude but not direction and *vector properties* which must be defined by direction as well as by magnitude. Examples of scalar properties are mass, internal energy, and viscosity while vector quantities are flow rate, force, and momentum. In an elementary volume of a fluid the vector properties have components in the three rectangular coordinates. Both vector and scalar quantities may vary with time.

For an elementary volume the following three fundamental differential equations are developed:

1. *The equation of continuity* is essentially a formulation of the principle of conservation of matter or a mass balance which equates the rate of change of mass in an elementary volume to the net differential rates of three-dimensional flow across its boundaries.

2. *The equation of motion* expresses the principle of conservation of momentum and comprises a momentum and force balance in which the rate of change of momentum of an elementary volume is equated to the vectorial forces on the element and the net flow of momentum across its boundaries.

3. *The equation of energy* is an energy balance which equates the rate of change in the total energy of the elementary volume to the vectorial changes that result from the flow of energy across the boundaries of the element.

The general development and use of these relationships are beyond the scope of this treatment but are available in texts on fluid mechanics and transport phenomena.[1,2]

[1] S. Goldstine, *Modern Developments in Fluid Dynamics*, Oxford University Press (1938).

[2] R. B. Bird, W. E. Stewart, and E. N. Lightfoot, *Notes on Transport Phenomena*, John Wiley & Sons (1958).

Unidirectional Steady-State Flow. When a fluid flows between or around irregular boundary surfaces, a complex pattern of stream lines is established with movement in different directions at different velocities. Even in the simple case of flow through a cylindrical pipe, *velocity profiles* occur across the stream with the maximum velocity in the direction of flow at the center and large velocity gradients at the walls.

At low rates of flow, the velocity profile is established by friction between adjacent layers of fluid, and the individual particles follow the flow lines in what is termed *laminar* or *stream-line flow*. As the total flow rate is increased, a *critical velocity* is reached at which the flow becomes *turbulent* and individual fluid particles pursue erratic transverse paths as part of their movement in the direction of flow.

Under conditions of laminar flow in a circular tube, the velocity falls off parabolically from the center line with an average velocity based on the total volumetric flow rate equal to 0.5 of the maximum along the center line. In turbulent flow the turbulent core of the fluid stream moves at a substantially uniform velocity, and large velocity gradients are restricted to a thin laminar layer near the surface of the pipe. In turbulent flow the ratio of the average to the maximum velocity in cylindrical conduits varies from about 0.75 at velocities near the critical to 0.83 at high rates of flow.

It has been shown by Bird[3] that the commonly used engineering equations for *unidirectional steady-state flow* may be derived directly from the three general differential equations of fluid mechanics. Such simplified equations should include proper averages to account correctly for the velocity profiles.

The following simplified derivation was suggested by Peck.[4]

Equation of Continuity. Figure 161 shows a representative section of conduit of circular cross section of radius R and area s through which a fluid is in steady-state flow from left to right with a mass rate of flow **W**. The rate of flow through an elementary cross-sectional area ds located at radial position r and angular position θ in a cross-sectional plane L is given by

$$d\mathbf{W} = u\rho \, ds \tag{1}$$

where u = velocity of flow. If the density ρ is constant over the cross section, the total mass rate of flow past plane L is then

$$\mathbf{W} = \rho \int_0^{2\pi} \int_0^{R} ur \, dr \, d\theta = \rho \bar{u} s \tag{2}$$

[3] R. B. Bird, *Chem. Eng. Sci.*, **6**, 123 (1957).

[4] R. E. Peck, private communication (1958).

where

$$\bar{u} = \frac{\int_0^{2\pi} \int_0^R ur\, dr\, d\theta}{\mathbf{s}}$$

If a second cross-sectional plane is considered at $L + dL$, the law of conservation of mass requires that $\mathbf{W}_L = \mathbf{W}_{L+dL}$ and $d\mathbf{W}/dL = 0$. Then

$$d(\rho \bar{u} \mathbf{s}) = 0 \quad \text{or} \quad \Delta(\rho \bar{u} \mathbf{s}) = 0 \tag{3}$$

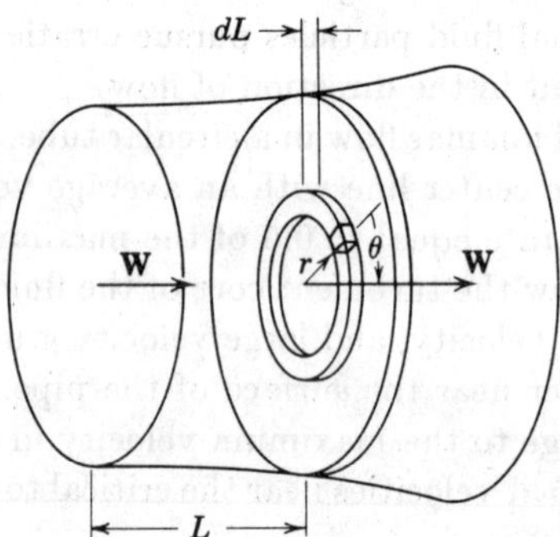

Fig. 161. Element of a diverging conduit

Equation 1 is the differential form of the equation of continuity expressing the principle that in steady-state flow the mass rate of flow is the same at all cross sections.

Equation of Motion. A flowing stream carries a momentum equal to the product of its mass rate of flow times its velocity. At the elementary cross section ds of Fig. 161, the mass rate of flow is $\rho u\, ds$, and the momentum flowing to the section becomes $\rho u^2\, ds/g_c$. Similarly, at a distance dL to the right of plane L the rate of flow of momentum is $\rho u^2\, ds/g_c + [\partial(\rho u^2\, ds)/g_c\, \partial L]\, dL$, and the change of rate of flow of momentum in element $ds\, dL$ is equal to $[\partial(\rho u^2\, ds)/g_c\, \partial L]\, dL$. The fluid contained within element $ds\, dL$ has mass $\rho\, ds\, dL$ and momentum $u\rho\, ds\, dL/g_c$. Its rate of change of momentum is then $ds\, dL[\partial(\rho u)/\partial \tau]/g_c$.

The principle of conservation of momentum which follows from Newton's second law of motion requires that the force accelerating a body of matter is equal to its time rate of change of momentum. In considering the frictionless flow of a fluid in a horizontal plane through element $ds\, dL$, the rate of change of momentum is equal to the sum of the change in rate of flow of momentum in the element plus the rate of change of momentum of the fluid contained within the element. The resultant force on the fluid in the direction of flow is the difference between the pressures on the upstream and downstream faces of the element. Thus:

$$\frac{d(\rho u^2\, ds)}{g_c} + \frac{dL\, ds}{g_c}\frac{\partial(\rho u)}{\partial \tau} = -d(p\, ds) \tag{4}$$

In the case of steady-state flow $\partial(\rho u)/\partial\tau = 0$, and equation 4 becomes

$$\frac{\partial(\rho u^2\, ds)}{g_c \partial L} + \frac{\partial(p\, ds)}{\partial L} = 0 \tag{5}$$

Equation 5 may be integrated over the entire cross section of the conduit.

However, when the walls are included in the momentum force balance, the force $-dF_D$ exerted in the direction of flow *on* the fluid *by* the walls must be included in the equation where F_D is the drag force exerted by the fluid on the conduit in the direction of flow. Thus, if ρ and p are assumed constant across the section,

$$\frac{\partial}{\partial L}\left(\frac{\rho}{g_c}\int_0^{2\pi}\int_0^R u^2 r\, dr\, d\theta\right) + \frac{\partial}{\partial L}\left(p\int_0^{2\pi}\int_0^R r\, dr\, d\theta\right) = -\frac{\partial F_D}{\partial L}$$

or

$$\frac{\partial(\rho \overline{u^2} s)}{g_c\, \partial L} + \frac{\partial(ps)}{\partial L} + \frac{\partial F_D}{\partial L} = 0 \tag{6}$$

where

$$\overline{u^2} = \frac{\displaystyle\int_0^{2\pi}\int_0^R u^2 r\, dr\, d\theta}{s}$$

Combining equations 6 and 2 gives

$$\frac{d(\mathbf{W}\overline{u^2}/\bar{u})}{g_c} + d(ps) + d\mathbf{F}_D = 0 \tag{7}$$

If flow is not in a horizontal plane, the total force on the element of fluid is obtained by adding to the force producing change of momentum and drag the forces exerted on the fluid by gravity in the direction of flow. Thus, for the general case of unidirectional flow, equation 7 may be written

$$\frac{d(\mathbf{W}\overline{u^2}/\bar{u})}{g_c} + d(ps) + d\mathbf{F}_D - d\mathbf{F}_g = 0 \tag{8}$$

or

$$\Delta\frac{\mathbf{W}\overline{u^2}/\bar{u}}{g_c} + \Delta(ps) + \mathbf{F}_D - \mathbf{F}_g = 0$$

where $\mathbf{W}$ = mass rate of flow

$\overline{u^2}$ = average of u^2 across the stream

p = pressure

$\mathbf{s}$ = a vector quantity in the direction of flow and having the magnitude of cross-sectional area

$\mathbf{F}_D$ = total drag force exerted by the fluid on the conduit in the direction of flow

$\mathbf{F}_g$ = external force such as gravity exerted on the fluid in the direction of flow

The drag force exerted by the walls on the fluid is opposite in sign but equal in magnitude to that exerted by the fluid on the walls. In dealing with viscous fluids the drag force $\mathbf{F}_D$ includes the force exerted on the conduit as a result of fluid friction.

The boldface symbols indicate vector quantities all in unidirectional flow. Velocity $\bar{u}$ is also a vectorial quantity corresponding to the vector indicated by $\mathbf{W}$. Where the fluid flows horizontally, the gravitational force producing flow is zero; where the fluid flows vertically downward, the gravitational force is in the direction of flow.

With reference to Fig. 161, the term $d(\mathbf{W}\overline{u^2}/\bar{u}g_c)$ represents the change of momentum of the fluid between the terminal sections of the element. This incremental change results from an incremental force of equal magnitude exerted *on the fluid* in the direction of flow. The term $d(p\mathbf{s})$ is the change in pressure force on the fluid between the terminal sections, and $d\mathbf{F}_g$ is the sum of the incremental external forces such as gravity *on* the fluid in the element. For a fluid flowing downward, the gravitational force in the direction of flow is $d\mathbf{F}_g = d(\rho\, d\mathbf{s}\, dL)$.

The *total drag force* $\mathbf{F}_D$ is the summation of all forces exerted by the fluid on the conduit in the direction of flow. It is the sum of the components in the direction of flow of the pressure force on the walls, termed the *form drag*, plus the force exerted on the walls as a result of fluid friction, or the *friction drag*. With reference to Fig. 161, the pressure force on the element of conduit in the direction of flow is $-p\, d\mathbf{s}$. Thus, if $\mathbf{F}_F$ is the frictional drag force, $d\mathbf{F}_D = -p\, d\mathbf{s} + d\mathbf{F}_F$, and equation 8 may be written for a length dL as

$$d\left(\frac{\mathbf{W}}{g_c}\frac{\overline{u^2}}{\bar{u}}\right) + \mathbf{s}\, dp + d\mathbf{F}_F - d\mathbf{F}_g = 0 \tag{9}$$

Equation of Energy. An expression for the change in kinetic energy

E_K per unit time in the element $ds\,dL$ of Fig. 161 may be written as follows for the case of steady-state flow:

$$dE_K = \frac{d(\rho u^3\,ds)}{2g_c} \tag{10}$$

Integrating over the cross-sectional area of section L,

$$dE_K = \frac{d\rho \int_0^{2\pi}\int_0^R u^3 r\,dr\,d\theta}{2g_c} = \frac{d(\rho \overline{u^3}\mathbf{s})}{2g_c} \tag{11}$$

where

$$\overline{u^3} = \frac{\int_0^{2\pi}\int_0^R u^3 r\,dr\,d\theta}{\mathbf{s}}$$

Combining equations 11 and 3, gives

$$dE_K = \frac{d(\mathbf{W}\overline{u^3}/\bar{u})}{2g_c} \tag{12}$$

Equation 12 may be used to write a complete energy balance per unit mass of fluid of the form of equation 8–1, page 246, for the elementary section $\mathbf{s}\,dL$. Thus, if surface-energy changes are negligible,

$$\frac{d(\overline{u^3}/\bar{u})}{2g_c} + dU + d(pV) + dZ - d'q + d'w = 0 \tag{13}$$

where U = internal energy per unit mass
V = specific volume
Z = potential energy per unit mass or elevation
$d'q$ = heat added
$d'w$ = external work done by fluid

Equation 13 is the general equation of energy for steady-state unidirectional fluid flow.

Equation of Mechanical Energy. The friction induced by fluid flow results in a degradation of mechanical energy dw_F by its conversion to heat which appears as an increase in internal energy and entropy under adiabatic conditions or which is lost to the surroundings in isothermal flow. In general, the sum of this frictional degradation of mechanical energy plus the heat added to the system from an external source is equal to $T\,dS$ which, from equation 13–35, is equal to $dU + p\,dV$. Hence, $dU = d'q + dw_F - p\,dV$, which may be substituted in equation 13 to give an expression containing only mechanical energy terms. Thus, for an element of length dL,

$$V\,dp + d\left(\frac{\overline{u^3}}{2g_c\bar{u}}\right) + dZ + dw_F + d'w = 0 \tag{14}$$

where dw_F = energy degraded by friction.

It has been common engineering practice to neglect the differences in the velocity averages and to assume that $\overline{u^2}/\bar{u} = \bar{u}$ and that $\overline{u^3}/\bar{u} = \bar{u}^2$. These assumptions lead to significant errors where widely varying velocity profiles are encountered. In the kinetic-energy term of equations 10 and 14 use of $\bar{u}^2$ instead of $\overline{u^3}/\bar{u}$ gives a result that is 0.5 of the correct value for laminar flow in a circular pipe. Fortunately kinetic energy is of relatively low magnitude in this low-velocity range. In turbulent flow the corresponding errors are generally less than 10%.

Equation 14 can be arrived at by integration of the general equation of motion without recourse to the energy equation.[3] If restricted to a frictionless fluid performing no external work, equation 14 is the differential form of the classical Bernoulli equation. The original restricted form of the equation was derived from Newton's laws of motion before the first law of thermodynamics was recognized.

Equations 3, 8, 13, and 14 are the four fundamental equations of unidirectional steady-state flow. They may be used individually or in combination to establish the forces, work, and changes in state properties accompanying fluid flow. More general expressions for three-dimensional unsteady-state conditions are developed in advanced treatises on fluid mechanics and transport phenomena.[2]

Flow with Friction in Conduits. The friction factor f between a fluid and a solid boundary surface is defined by the following expression:

$$d\mathbf{F}_F = f\left(\frac{u^2}{2g_c V}\right) dA \tag{15}$$

where $\mathbf{F}_F$ = frictional drag force on the solid surface in the direction of flow

f = friction factor or drag coefficient, (force)/(area)(kinetic energy per unit volume)

A = area of contact

$u^2/(2g_c V)$ = kinetic energy per unit volume of fluid

In steady-state flow in a conduit of uniform cross-sectional area s, the degradation of mechanical energy by friction dw_F is equal to the work resulting from action of the frictional drag force through the distance corresponding to the flow of a unit mass of fluid. Thus:

$$dw_F = \frac{V}{s}\,dF_F = \frac{f}{s}\left(\frac{u^2}{2g_c}\right) dA \tag{16}$$

The physical significance of equation 16 may be visualized by considering that the fluid is stationary and the elementary section of conduit is moved along it at a velocity u by a force $-\mathbf{F}_F$ through distance V/s performing work dw_F which is converted into heat in the fluid. Since only relative motion is involved, this situation is equivalent to the actual case in which the conduit is fixed by force $-\mathbf{F}_F$ and the fluid moved through it.

If the perimeter of contact of the conduit is y, the area of contact dA is equal to $y\,dL$, and equation 16 may be written as

$$dw_F = f\frac{y}{s}\left(\frac{u^2}{2g_c}\right) dL = \left(\frac{u^2}{2g_c}\right)\frac{f\,dL}{r_h} \tag{17}$$

where r_h = hydraulic radius = s/y. For a circular section of diameter D, $r_h = D/4$. Equation 17 is termed the *Fanning equation*, and the dimensionless group $f\,dL/r_h$ is termed the *friction parameter* descriptive of the system.

Combining equations 14 and 17, neglecting velocity profiles, gives

$$V\,dp + d\left(\frac{u^2}{2g_c}\right) + dZ + \left(\frac{u^2}{2g_c}\right)\frac{f\,dL}{r_h} + d'w = 0 \tag{18}$$

The use of the hydraulic radius r_h is an approximation for most shapes. More complex relationships are required for rigorous expression of degradation by friction in conduits of varying noncircular cross sections.

Since $u = WV/s$ equation 18 may be written as

$$V\,dp + \frac{W^2}{g_c}\left(\frac{V\,dV}{s^2} - \frac{V^2\,ds}{s^3}\right) + dZ + \left(\frac{W^2V^2}{s^2 2g_c}\right)\frac{f\,dL}{r_h} + d'w = 0 \tag{19}$$

Friction Factor. It may be demonstrated[1,2] by the principles of dimensional analysis and fluid mechanics that the friction factor f is a function of a dimensionless parameter N_R termed the *Reynolds number*,

$$N_R = \frac{4r_h u\rho}{\mu} = \frac{4r_h G}{\mu} = \frac{DG}{\mu} \tag{20}$$

where μ = viscosity, (force)(time)/(length)^2 or (mass)/(length)(time)
G = mass velocity of flow
D = diameter of circular cross section

The unit of viscosity in the metric system is the *poise*, 1.0 (dyne)(sec)/cm^2 or 1.0 gram/(cm)(sec). Data are commonly tabulated in terms of the centipoise = 0.01 poise, which is equal to 2.42 lb$_m$/(ft)(hr).

The theoretical relationship between f and N_R has been developed through the concept of viscosity for laminar flow through a smooth circular tube. This is the classical Poiseuille equation. In the turbulent-flow region the nature of the tube surface also affects the friction factor. Figure 162 shows the friction factor f as a function of Reynolds number in commercial pipes. It will be noted that the transition from laminar to turbulent flow occurs in the range $N_R = 2000 - 3000$, for adiabatic conditions. For turbulent flow in smooth straight circular pipes, the friction factor may be represented by the relation[5]

$$f = 0.00140 + 0.125(N_R)^{-0.32} \tag{21}$$

Flow of Liquids. For liquids, or for gases flowing under conditions of nearly constant specific volume, equation 19 may be integrated by assuming that V and f are constant at average values. Thus, for conduits of uniform cross-sectional area, since ds and dV are zero, if no external work is performed,

$$\Delta Z + V_{\text{avg}}\,\Delta p + \left(\frac{u^2}{2g_c}\right)_{\text{avg}} \frac{f_{\text{avg}} L}{r_h} = 0 \tag{22}$$

The pressure changes accompanying the flow of liquids and gases in pipes are readily calculated from equation 22 and Fig. 162.

Illustration 1. A solution having a viscosity of 14 centipoises and a density of 1.12 grams per cc is to be pumped at a rate of 15 gal per min through a pipe line 200 ft long and delivered at a pressure of 40 psig at an elevation of 35 ft above the pump. Calculate the pressure at the pump if the line is sized to give a velocity of 10 ft per sec.

$\mu = 14(2.42) \quad = 33.88\ \text{lb}_m/(\text{ft})(\text{hr})$

$\rho = (62.4)(1.12) \quad = 69.9$ lb per cu ft

Rate of flow $= 15(60)/(7.48) \quad = 120.3$ cu ft per hr

$s = (120.3)/10(3600) \quad = (3.34)10^{3}$ sq ft

$D = [4(3.34)10^{3}\pi]^{1/2} \quad = (6.52)10^{-2}$ ft

$N_R = (6.52)10^{-2}(36{,}000)70/(33.88) = 4850$

$f = 0.00140 + 0.125(4850)^{-0.32} \quad = 0.00962$

Substituting in equation 22 gives

$$-V\,\Delta p = 35 + 10^2(0.00962)200(4)/2(32.2)(6.52)10^{-2} = 218 \text{ ft}$$

$$-\Delta p = 218(70)/144 = 106 \text{ psi}$$

$$p_1 = 106 + 40 = 146 \text{ psig}$$

Thrust and Impulse

The basic equations 3, 8, 13, and 14 are rigorously applicable only to unidirectional steady-state flow. In the somewhat more general case of

[5] T. B. Drew, E. C. Koo, and W. H. McAdams, *Trans. Am. Inst. Chem. Engrs.*, **28**, 56 (1932).

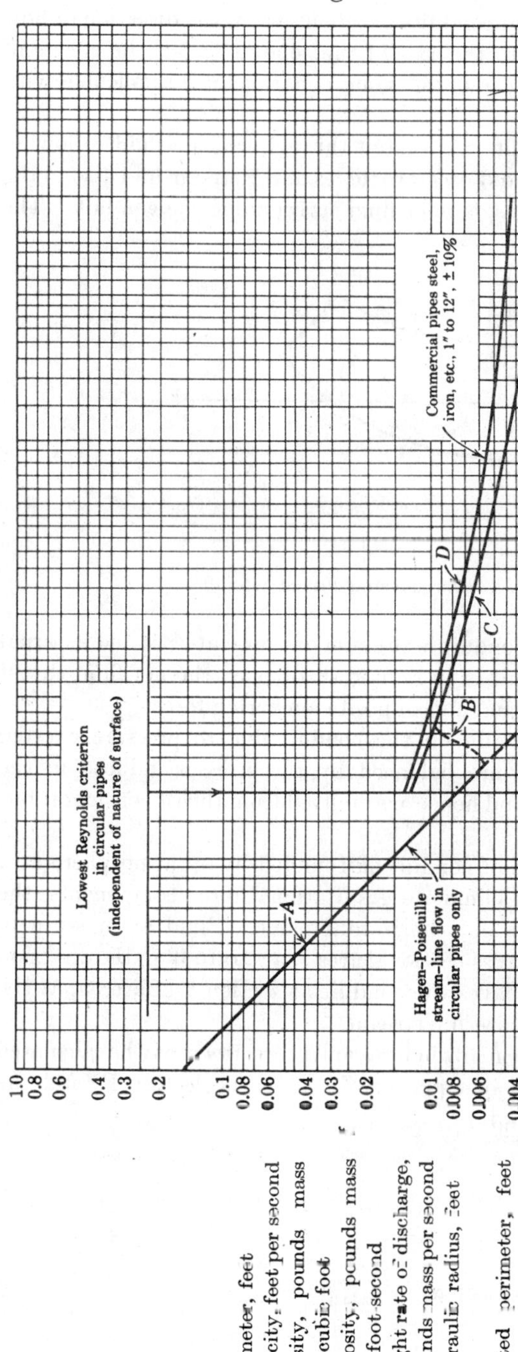

D = diameter, feet
V = velocity, feet per second
ρ = density, pounds mass per cubic foot
μ = viscosity, pounds mass per foot-second
w = weight rate of discharge, pounds mass per second
R_H = hydraulic radius, feet
G = $V\rho$
L_p = wetted perimeter, feet

Fig. 162. Fanning friction factors f. Straight ducts only

Perry's *Chemical Engineers' Handbook*, Fig. 23, p. 382 (1950) McGraw-Hill Book Co. (with permission).

co-axial steady-state flow in conduits or jets in which the direction of flow is approximately parallel to the axis of the conduit or jet, the effects of transverse flow are minor, and the same equations are applicable with little error.

Total Thrust on a Conduit. A conduit through which a fluid is flowing is subject to internal drag and pressure forces of the stream and also to the external forces of the atmosphere or fluid surrounding the conduit. The

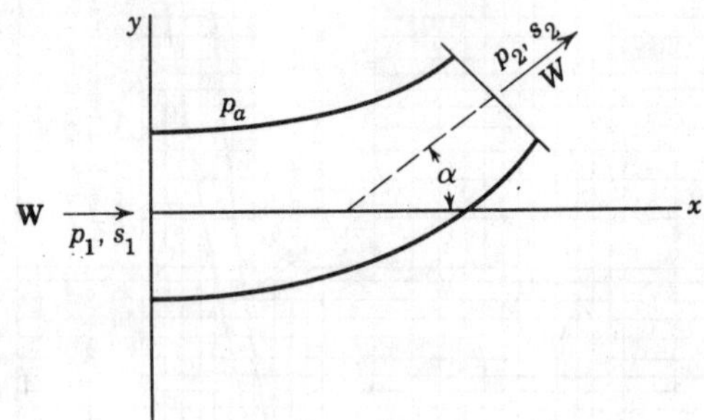

FIG. 163. Convergent and curved conduit

net *thrust* or force on the conduit is the sum of these internal and external fluid forces. If the conduit is to remain stationary, the net fluid thrust must be opposed by an equal and opposite anchoring force.

The forces exerted radially on the conduit due to pressure exerted internally and externally are balanced by the stresses induced in the material of construction, and hence are omitted from thrust considerations in this discussion.

In Fig. 163 is a section of conduit surrounded by an atmosphere at a uniform pressure p_a. To eliminate gravitational force components, the conduit is assumed to lie in a horizontal plane. Fluid at pressure p_1 enters section 1 having a cross-sectional area s_1 at a rate W in the direction of the x coordinate and leaves the conduit at section 2 of area s_2 at the angle α from that of the entering stream.

The internal force components in the x and y directions may be calculated from the integrated form of equation 8. Thus, neglecting $\mathbf{F}_g$ and the difference between $\overline{u^2}/\bar{u}$ and $\bar{u}$,

$$F_{Dx} = -\frac{W}{g_c}(u_2 \cos \alpha - u_1) - (p_2 s_2 \cos \alpha - p_1 s_1) \tag{23}$$

$$F_{Dy} = -\frac{W}{g_c}(u_2 \sin \alpha) - p_2 s_2 \sin \alpha \tag{24}$$

where F_{Dx} and F_{Dy} = components in the x and y directions of the total internal drag force.

The net external force acting on the section of conduit results from the absence of external pressure on sections s_1 and s_2. The resulting unbalanced external forces are given by

$$F_{Ex} = p_a(s_2 \cos \alpha - s_1) \tag{25}$$

$$F_{Ey} = p_a(s_2 \sin \alpha) \tag{26}$$

where F_{Ex} and F_{Ey} = components in the x and y directions of the net external forces on the conduit

p_a = external pressure

The total thrust on the conduit $\mathbf{F}_T$ is the sum of the internal forces

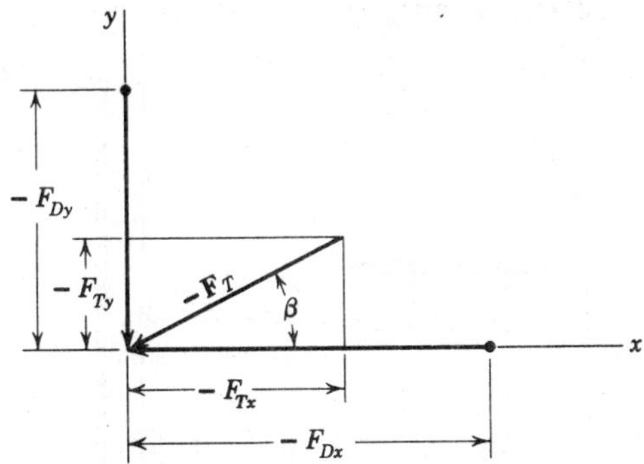

FIG. 164. Thrust components on a conduit

$\mathbf{F}_D$ plus the external $\mathbf{F}_E$. Thus, combining equation 23 with equation 25, and equation 24 with equation 26,

$$F_{Tx} = -\frac{W}{g_c}(u_2 \cos \alpha - u_1) - [(p_2 - p_a)s_2 \cos \alpha - (p_1 - p_a)s_1] \tag{27}$$

$$F_{Ty} = -\frac{W}{g_c}(u_2 \sin \alpha) - (p_2 - p_a)s_2 \sin \alpha \tag{28}$$

The resultant thrust components are indicated diagrammatically in Fig. 164. The resultant total thrust $\mathbf{F}_T$ has a magnitude equal to $[(F_{Tx})^2 + (F_{Ty})^2]^{1/2}$ and acts on the conduit at an angle β to the x axis such that $\sin \beta = F_{Ty}/\mathbf{F}_T$ or $\tan \beta = F_{Ty}/F_{Tx}$.

The total thrust associated with a fluid stream is the driving force of the turbine, water wheel, jet aircraft, and rockets as well as other devices such as the rotating lawn sprinkler. Thrust components for such systems are calculated by the method of equations 27 and 28, after using equation 19 to establish the pressure–velocity relationships throughout the system.

In analyzing devices in which thrust constitutes the driving force, it is convenient to refer to *specific impulse* which is the thrust per unit flow rate and to *total impulse* which is the product of thrust times the duration of its action. Thus

$$\text{Specific impulse} = I_s = \frac{\mathbf{F}_T}{\mathbf{W}} \tag{29}$$

$$\text{Total impulse} = I_T = \int_{\tau_1}^{\tau_2} \mathbf{F}_T \, d\tau \tag{30}$$

where τ = time. Specific and total impulses are of particular value in analyzing the performance of rocket engines and propellants.

Thrust of Open Jets. In an open jet which travels only a short distance

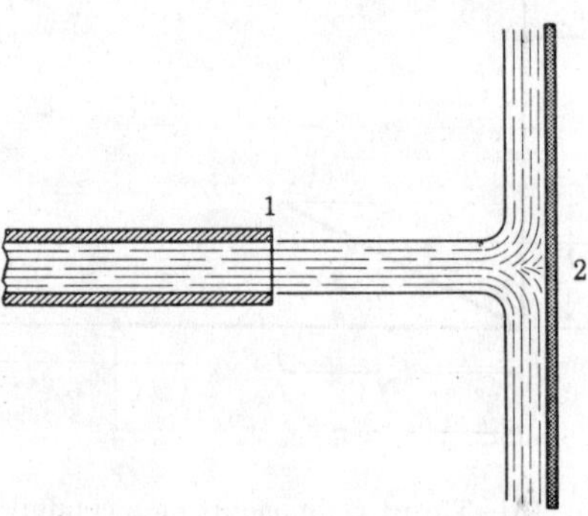

Fig. 165. Thrust of an open jet at right angles to a plate

from a nozzle, it may be assumed that pressure together with velocity and cross section in the direction of flow remain constant. If gravitational force is negligible, equation 8 simplifies and integrates to

$$\mathbf{F}_D = -\frac{\mathbf{W}}{g_c}(u_2 - u_1) \tag{31}$$

If an open jet strikes a flat plate at right angles, as indicated in Fig. 165, the thrust on the plate follows directly from equation 31, since u^2 in the direction of flow is zero. Thus,

$$\mathbf{F}_D = \frac{\mathbf{W}}{g_c} u_1 \tag{32}$$

If an open jet is deflected through an angle α by a tangentially placed blade as in Fig. 166, forces on the blade are produced in both the x and y directions. Thus:

$$F_{Dx} = -\frac{W}{g_c}(u_1 \cos \alpha - u_1) = \frac{W}{g_c} u_1(1 - \cos \alpha) \tag{33}$$

$$F_{Dy} = -\frac{W}{g_c} u_1 \sin \alpha \tag{34}$$

If the jet is reversed in direction by the blade, $\alpha = 180°$ and

$$F_{Dx} = \frac{2W}{g_c} u_1 \tag{35}$$

The impulse turbine and water wheel are operated by the forces produced

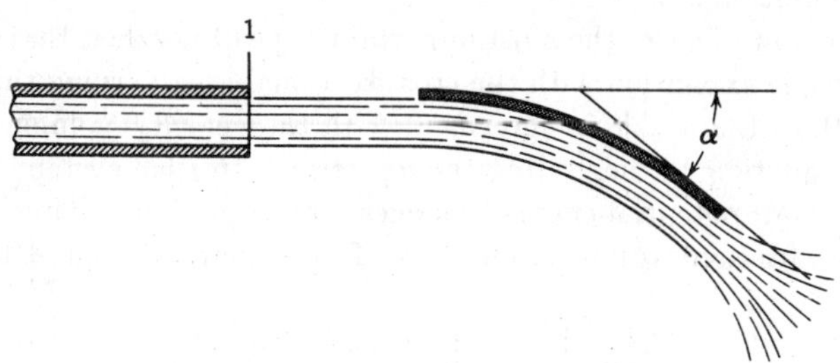

FIG. 166. Thrust on a curved blade

by change of direction of jets directed from nozzles onto closely spaced blades carried on the periphery of a wheel, as shown in Fig. 177. In this case the blades move at a velocity u_B, and equations 33 and 34 are applicable if the relative velocity $(u_1 - u_B)$ is substituted for u_1.

Flow through Nozzles

Nozzles and orifices are of importance in metering, power generation, propulsion, and pumping applications. For simplicity, flow equations are developed for the reversible flow of an incompressible fluid and of an ideal gas. Intermediate situations may be treated by deviation corrections applied to these two cases.

The case of the *incompressible fluid* may be analyzed by considering a stream flowing at pressure p_1 out of a conduit of cross sectional area s_1 through an orifice or nozzle of cross section s_2, which opens into a region of lower pressure p_2. Assuming that no difference in elevation is involved and that no work is done, and neglecting friction and the difference between $\overline{u^3}/\bar{u}$ and u^2, equation 14 may be written

$$u_2^2 - u_1^2 = 2g_c \int_{p_2}^{p_1} V\,dp = 2g_c V(p_1 - p_2) \tag{36}$$

If the square of the entrance velocity u_1 is negligible, the velocity through the nozzle is expressed by

$$u_2 = [2g_c V(p_1 - p_2)]^{1/2} \tag{37}$$

It will be noted that the velocity of flow is increased indefinitely as p_1 is increased, approaching infinity as the pressure difference becomes infinite.

Liquids of low viscosities follow equation 37 with accuracy satisfactory for many purposes at moderate flow rates corresponding to low-pressure differences. At higher rates the effect of fluid friction must be taken into consideration.

For the flow of gases the most important type of nozzle is that designed for isentropic expansion with the cross-sectional area varying throughout the length of the nozzle in agreement with the velocity–volume relationships of equation 14, neglecting the w_F term. In such systems no work is done, and potential-energy changes are negligible. Thus, for the isentropic, frictionless flow of any fluid, from equations 36 and 15–12:

$$\Delta\left(\frac{u^2}{2g_c}\right) = -\int_{p_1}^{p_2} V\,dp = -(\Delta H)_S \tag{38}$$

or

$$u_2 = [2g_c(H_1 - H_2)_S + u_1^2]^{1/2} \tag{39}$$

Equation 39 may be used to relate the conditions at any point 2 downstream in an isentropic nozzle to those at any point 1 upstream. For the isentropic expansion of an ideal gas (equations 15–34):

$$\frac{T_2}{T_1} = \left(\frac{p_2}{p_1}\right)^{(\kappa-1)/\kappa} = \left(\frac{V_1}{V_2}\right)^{\kappa-1} \tag{40}$$

Combining equations 38 and 40, since

$$C_p = \frac{\kappa R}{(\kappa - 1)M}$$

$$\frac{u_2^2 - u_1^2}{2g_c} = \frac{\kappa R T_1}{M(\kappa - 1)}\left[1 - \left(\frac{p_2}{p_1}\right)^{(\kappa-1)/\kappa}\right] = \frac{\kappa R T_1}{M(\kappa - 1)}\left[1 - \left(\frac{V_1}{V_2}\right)^{\kappa-1}\right] \tag{41}$$

$$u_2 = \left\{\frac{2g_c \kappa R T_1}{M(\kappa - 1)}\left[1 - \left(\frac{p_2}{p_1}\right)^{(\kappa-1)/\kappa}\right] + u_1^2\right\}^{1/2} \tag{42}$$

It will be noted that a finite velocity is attained when a gas is expanded into a vacuum with an infinite pressure ratio, $p_1/p_2 = \infty$. Thus:

$$u_{2,\max} = \left[\frac{2g_c\kappa RT_1}{M(\kappa - 1)} + u_1^2\right]^{1/2} \tag{43}$$

The maximum velocity that an ideal gas can attain by isentropic expansion is a function only of its heat capacity, molecular weight, and initial temperature. With no initial velocity, air at a temperature of 70° F would accelerate ideally to a velocity of 2530 ft per sec if expanded into a vacuum.

The general form of the isentropic nozzle becomes evident from consideration of such a nozzle designed to discharge gas from a large reservoir at a uniform high pressure p_1 into a region of low pressure p_2. At the entrance to the nozzle the velocity of flow is negligible. As expansion proceeds and the velocity of flow increases, it is evident that the area of the nozzle must be reduced in a converging form. However, as expansion proceeds still further, relatively large increases in volume accompany the reduction in pressure, and the cross-sectional area must be enlarged in a diverging section in order to accommodate the increased specific volume. This type of nozzle was developed by de Laval and characteristically comprises a *converging section* and a *diverging section* separated by a *throat.*

Critical Velocity and Pressure Ratio. The relationships among cross-sectional area, velocity, pressure, volume, and temperature in a de Laval nozzle may be derived from equation 41. At any point in the nozzle, from the continuity equation,

$$V = \frac{su}{W} \tag{44}$$

Substituting for V in equation 41 gives

$$\frac{u_2^2 - u_1^2}{2g_c} = \frac{\kappa RT_1}{M(\kappa - 1)}\left[1 - \left(\frac{s_1u_1}{s_2u_2}\right)^{\kappa-1}\right] \tag{45}$$

Solving for s_2 and rearranging,

$$s_2 = \frac{s_1u_1}{u_2}[1 - C'(u_2^2 - u_1^2)]^{1/(1-\kappa)} \tag{46}$$

where $C' = M(\kappa - 1)/(2g_c\kappa RT_1)$. The limiting velocity at the throat of the nozzle occurs where $ds_2/du_2 = 0$. Differentiating equation 46 yields

$$\frac{ds_2}{du_2} = -2C's_1u_1\left(\frac{1}{1-\kappa}\right)[1 - C'(u_2^2 - u_1^2)]^{\kappa/(1-\kappa)}$$

$$-\frac{s_1u_1}{u_2^2}[1 - C'(u_2^2 - u_1^2)]^{1/(1-\kappa)} = 0 \tag{47}$$

or.

$$\frac{2C'}{1-\kappa} = -\frac{1}{u_2^2}[1 - C'(u_2^2 - u_1^2)] \tag{48}$$

If the upstream velocity at the inlet to the nozzle is zero, equation 48 gives the following relationships:

$$-u_t^2 = \frac{1-\kappa}{2C'}(1 - C'u_t^2) \tag{49}$$

where u_t = velocity at the throat, or

$$u_t = \left[\frac{\kappa - 1}{(\kappa+1)C'}\right]^{1/2} = \left[\frac{2g_c\kappa RT_i}{M(\kappa+1)}\right]^{1/2} \tag{50}$$

Equation 50 leads to the important conclusion that, *when a nozzle operates with an ideal gas under a pressure differential sufficiently great to result in expansion into the diverging region of flow, the velocity at the throat of the nozzle is a constant, dependent only on the heat capacity, molecular weight, and temperature of the gas, and is independent of the pressures or pressure difference involved.*

By combining equations 50 and 41, relationships among the pressures, temperatures, and volumes at the inlet and throat result for critical flow at the throat and negligible inlet velocity. Thus, since $(u_t^2M)/(2g_c\kappa RT_i) = 1/(\kappa+1)$,

$$\frac{1}{\kappa+1} = \frac{1}{\kappa-1}\left[1 - \left(\frac{p_t}{p_i}\right)^{(\kappa-1)/\kappa}\right] = \frac{1}{\kappa-1}\left[1 - \left(\frac{V_i}{V_t}\right)^{\kappa-1}\right]$$
$$= \frac{1}{\kappa-1}\left(1 - \frac{T_t}{T_i}\right) \tag{51}$$

Then,

$$\frac{p_t}{p_i} = \left(\frac{2}{\kappa+1}\right)^{\kappa/(\kappa-1)} \tag{52}$$

$$\frac{V_i}{V_t} = \left(\frac{2}{\kappa+1}\right)^{1/(\kappa-1)} \tag{53}$$

$$\frac{T_t}{T_i} = \frac{2}{\kappa+1} \tag{54}$$

Equation 52 is of particular interest in that it defines the minimum pressure ratio necessary to attain divergent flow conditions. This *critical pressure ratio* p_t/p_i is independent of temperature and pressure levels. At all pressure ratios below the critical, constant throat velocities given by equation 50 are attained in ideal nozzles. At higher-pressure ratios the conditions of divergent flow are not reached, and the maximum velocities are less than the critical throat velocities.

Combining equations 50 and 54 gives an expression for the critical throat velocity in terms of conditions at the throat. Thus:

$$u_t = [(g_c \kappa R T_t)/M]^{1/2} \tag{55}$$

Velocity of Sound. From classical physics it may be shown that the velocity of sound in any fluid is determined by the isentropic compressibility of the fluid. Thus:

$$a = \left[g_c \left(\frac{\partial p}{\partial \rho} \right)_S \right]^{1/2} \tag{56}$$

where a = velocity of sound. The isentropic pressure–density relationships of an ideal gas are expressed by pV^κ = constant (page 648). Differentiation gives,

$$\left(\frac{\partial p}{\partial \rho} \right)_S = \frac{\kappa p}{\rho} \tag{57}$$

Since for an ideal gas $\rho = pM/RT$, substituting 57 in 56 gives

$$a = [(g_c \kappa R T)/M]^{1/2} \tag{58}$$

Thus, for an ideal gas the velocity of sound depends only on the heat capacity, molecular weight, and temperature of the gas, and is independent of pressure.

Illustration 2. Calculate the velocity of sound in air at 50° F, assuming ideal gas behavior.

$$M = 29; \quad \kappa = \frac{7.0}{7.0 - 2} = 1.4; \quad R = 1542 \text{ ft-lb/(lb-mole)(R°)}$$

$$a = \left[\frac{(32.2)(1.4)1542(510)}{29} \right]^{1/2} = [(1.22)10^6]^{1/2} = 1105 \text{ ft per sec}$$

Equations 55 and 58 are identical. It follows that *the velocity of an ideal gas in the throat of an isentropic nozzle operating at less than the critical pressure ratio is equal to the velocity of sound at the temperature of the throat.* Sonic velocities are closely approached in the throats of several types of nozzles.

It will be noted from equation 56 that the velocity of sound in an incompressible fluid is infinite. As was pointed out on page 692, the velocity of flow of an incompressible fluid through a nozzle also can be infinite, corresponding to a critical pressure ratio of zero. Liquids having compressibility characteristics intermediate between those of the perfect gas and the incompressible fluid will approach sonic throat velocities which are greater than those in gases.

Mach Number. In problems involving the flow of gases, it is frequently convenient to use a dimensionless velocity parameter termed the *Mach number*, which is defined as the ratio of the velocity of flow to the sonic velocity at the local conditions. Thus, for an ideal gas,

$$N_M = \frac{u}{a} = \frac{u}{[(g_c \kappa RT)/M]^{1/2}} \tag{59}$$

where N_M = Mach number.

From the discussion of the foregoing section it follows that $N_M = 1.0$ at the throat of a nozzle operating at less than the critical pressure ratio.

Since the acoustic velocity is the rate of propagation of a small disturbance under ideal reversible conditions, it follows that it is the maximum velocity at which any small molecular disturbance can be propagated in a fluid. However, large disturbances of explosive violence in which large changes in pressure and density are involved may travel at rates several times the velocity of sound. Such disturbances, termed *shock waves*, are discussed further in connection with the flow of fluids in ducts.

When a solid object moves through a stationary fluid, all forces and energy transitions are determined by relative velocities and are independent of the reference coordinates on which absolute values are expressed. It follows that the same effects are produced if a solid moves through a fluid as if the solid is stationary and the fluid is moved past it. This is the basis of the extensive use of *wind tunnels* for studying the forces on objects in flight by fixing the object in a stationary mount and blowing air past it at flight velocities.

When a fluid moves past an object at subsonic speeds, $N_M < 1.0$, the flow disturbances created by the object are propagated upstream and influence the flow pattern of the fluid as it approaches the object. This results in a corresponding steady-state flow pattern at upstream points.

If the velocity of flow is increased to supersonic values, $N_M > 1.0$, small flow disturbances created by the object cannot be propagated upstream, and the flow pattern of the fluid arriving at the object is unaffected by the presence of the object. As a result, the pattern of flow around the object and the forces on it are quite different, owing to absence of the "clearing-away" effects of upstream propagation in subsonic flow.

As flow velocities are increased from subsonic to sonic values, unstable conditions are encountered in the *transonic* range where N_M = 0.85 to 1.2. As the sonic velocity is approached, the upstream flow patterns established at subsonic speeds are not maintained, and may collapse with a sudden change of forces on the object constituting a shock. A large disturbance of this type can be projected forward even at supersonic speeds, establishing new flow patterns which may again collapse, producing another shock. In this manner a series of shock waves can be generated which result in dissipation of kinetic energy as heat and severe stresses on the object.

The foregoing phenomena are responsible for the so-called *sonic barrier* encountered in flight and the violent disturbances created as aircraft pass through the transonic speed range.

Expansion Nozzles. Nozzles used for the expansion of compressible fluids may be classified as subsonic, sonic, or supersonic, depending on the velocity of the discharge.

A subsonic nozzle operates with a pressure ratio greater than the critical and maintains a discharge pressure equal to that of the gas into which it is discharging. The velocity of flow is dependent on both the upstream and downstream pressures in accordance with equation 42 for ideal gases.

If the inlet pressure on a subsonic orifice is increased, or the discharge pressure reduced until the critical pressure ratio is reached, the discharge velocity will become sonic, and the discharge pressure will equal the pressure of the gas into which discharge is occurring.

If the pressure into which a sonic orifice is discharging is reduced, giving a pressure ratio less than critical, the rate of flow through the orifice will be unchanged at sonic velocity. The discharge pressure will also remain at the throat pressure corresponding to the critical-pressure ratio and will be higher than the pressure of the gas into which discharge is occurring. Expansion will be completed in unstable turbulent flow in the discharge atmosphere.

If a sonic nozzle which has a discharge pressure higher than that of the discharge space is provided with a diverging section, expansion of the stream can be continued without major turbulence, and supersonic velocities will result. For *optimum expansion* and maximum velocity, the discharge area of the diverging section should be such that the discharge pressure equals that of the atmosphere into which discharge occurs. If the discharge opening is smaller than the optimum, the nozzle is termed *underexpanding*, and, if larger, it is *overexpanding*. In the latter case, pressures less than that of the discharge space are produced, which may result in unstable discharge conditions and formation of shock waves.

The relationships among the pressures, velocities, temperatures, and areas in ideal nozzles may be calculated from equations 38, 52, and 55. However, the performance of real nozzles differs from the ideal because of friction, turbulence, radial flow, and heat transfer. The corrections to be applied to ideal performance values in order to account for these factors are expressed as empirical coefficients and efficiencies. These corrections depend on the design and shape of the nozzle, the characteristics of the fluid, and the operating conditions.

The *energy conversion efficiency* of a nozzle is defined as the ratio of the kinetic energy in the discharge of a real nozzle to that of an ideal nozzle supplied with the same fluid at the same initial state and expanded to the same exit pressure. Thus

$$\eta_E = \frac{(u'_2)^2}{(u_2)^2} = \frac{H_1 - H'_2}{H_1 - H_2} \tag{60}$$

where the primed quantities refer to the actual nozzle and the unprimed to the ideal.

The velocity correction factor is defined as the square root of the energy efficiency. Thus

$$\eta_u = (\eta_E)^{1/2} \tag{61}$$

The discharge efficiency or coefficient is the ratio of the mass flow rate in an actual nozzle to the flow rate through an ideal nozzle of the same throat area, expanding the same fluid from the same initial conditions to the same discharge pressure.

$$\eta_d = \frac{W'}{W} = \frac{W'}{s_t u_t / V_t} \tag{62}$$

For well-designed nozzles values of η_E generally are in the range of 0.70 to 0.96, while the corresponding values of η_u are 0.85. Discharge coefficients of sonic and supersonic nozzles range from 0.96 to greater than 1.0 if loss of heat occurs in the nozzle. The discharge coefficient of a subsonic nozzle may be considerably greater than 1.0 as a result of the pressure in the throat being less than that of the discharge space.

Illustration 3. A converging-type nozzle having a throat area of 1.0 sq in. discharges air from a large vessel at a temperature of 1500° F into a space which may be controlled at a pressure of either 1.0 atm or 0.1 atm. The pressure in the vessel may be varied from 0.1 atm to 100 psia.

Calculate the following properties as a function of inlet pressure for both receiving pressures, assuming $\eta_d = 0.98$ and $\kappa = 1.395$

(*a*) Discharge velocity in feet per second.
(*b*) Discharge temperature in degrees Fahrenheit.
(*c*) Discharge pressure in pounds per square inch absolute.
(*d*) Mass rate of flow in pounds per second.

Receiver pressure p_2.	0.1 atm	1.0 atm
Critical pressure ratio (equation 52),		
$p_t/p_i = (2/2.395)^{(1.395/0.395)}$	0.53	0.53

Critical flow conditions.

Critical inlet pressure, $p_2/0.53$	2.78	27.8 psia
Discharge pressure, $p_t = p_2$	1.47	14.7 psia
From Table 55 ($T = 1960°$ R):		
p_{R1}	160.37	160.37
H_1	493.64	493.64 Btu per lb
$p_{R2} = (p_{R1})(0.53)$	85.0	85.0
H_2	414.5	414.5 Btu per lb
T_2	1670	1670° R
$u_2 = [2g_c(H_1 - H_2)]^{1/2}$	1990	1990 ft per sec
V_2	420	42 cu ft per lb
$W = (s_t u_2 \eta_d)/V_2$	0.032	0.322 lb per sec

Noncritical flow conditions.

	Below Critical Inlet Pressure	Above Critical Inlet Pressure	
Receiver Pressure, p_0	1.0 atm	0.1 atm	1.0 atm
p_1	20	60	60 psia
$p_2 = p_t = p_1/0.53 =$	14.7	$0.53p_1 = 31.8$	31.8 psia
p_{R1}	160.37	160.37	160.37
H_1	493.64	493.64	493.64 Btu per lb
$p_{R2} = (p_{R1})(p_2)/(p_1)$	118	85	85
H_2	453.96	414.5	414.5 Btu per lb
T_2	1816	1670	1670° R
$u_2 = [2g_c(H_1 - H_2)]^{1/2}$	1410	1990	1990 ft per sec
V_2	45.5	19.5	19.5 cu ft per lb
W_2	0.21	0.692	0.692 lb per sec

In a similar manner calculations are made at other inlet pressures. The results are shown graphically by the solid lines of Fig. 167. It will be noted that, as the inlet pressure is increased with either receiver pressure, the discharge velocity rises to a constant sonic value of 1990 ft per sec at the critical pressures. Similarly, the discharge temperature drops to a constant value of 1210° F. Further increase in inlet pressure does not change the discharge temperature or velocity. As the inlet pressure is increased beyond the critical, the discharge pressure and the mass rate of flow increase linearly.

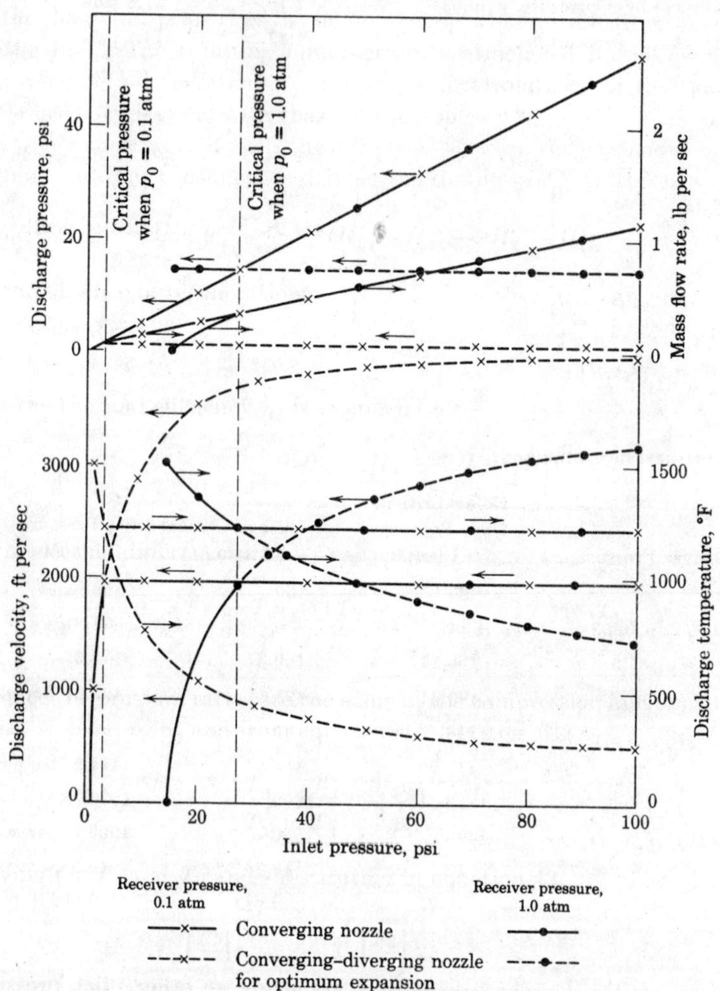

Fig. 167. Discharge characteristics of a 1.0-sq in. nozzle operating on air at 1500° F

Illustration 4. The nozzle of Illustration 3 is equipped with diverging discharge sections designed for optimum expansion to the receiving pressures. The forms of the nozzles are such that the energy conversion efficiency as given by,

$$\eta_E = 1.0 - 0.0012(p_t/p_2) \tag{a}$$

For inlet pressures above the critical, repeat the calculations of Illustration 3 for such optimum expansion and also determine the corresponding optimum area ratios s_2/s_t.

Receiver pressure	0.1	1.0 atm
Inlet pressure	60	60 psia
p_2	1.47	14.7 psia
p_2/p_1	0.0245	0.245
p_{R1}	160.37	160.37
H_1	493.64	493.64 Btu per lb
$p_{R2} = p_{R1}(p_2/p_1)$	3.93	39.3
T_2 (ideal)	726	1368° R
H_2 (ideal)	173.98	334.63 Btu per lb
ΔH (ideal)	−320.66	159.01 Btu per lb
$p_t/p_2 = (60)(0.53)/p_2$	21.6	2.16
$\eta_E = 1 - 0.0012 p_t/p_2$	0.974	0.9974
$\Delta H'$ (actual) $= (0.974)(-320.66)$	−313	−158.7 Btu per lb
H'_2 (actual) $= 493.64 - 313$	180.64	334.94 Btu per lb
T'_2 (actual)	755	1370° R

The values of p_R and H_R corresponding to T'_2 in the air table are of no significance because the actual expansion is not isentropic.

V'_2	190	34.5 cu ft per lb
$u'_2 = 2(32.2)(305.3)/778$	3960	2820 ft per sec
$\dfrac{s_2}{s_t} = \dfrac{u_t}{u'_2}\dfrac{V'_2}{V_t} = \dfrac{1990(190)}{3960(19.5)}$	4.9	1.25

The conditions corresponding to other inlet pressures are calculated in a similar manner. The results are shown by the broken lines of Fig. 167.

Comparison of the broken with the solid lines in Fig. 167 in the region above the critical pressures shows the effects of adding diverging sections to a nozzle. The discharge velocities are markedly increased with the greatest improvement occurring at the lower receiver pressure. With the low-pressure receiver, velocities are obtained approaching the ideal maximum of 4250 ft per sec calculated from equation 43. Although below the critical pressure ratio, receiver pressure does not affect the mass rate of flow, lower pressures are advantageous in all energy conversions involving nozzles if suitable provision is made for expansion to supersonic velocities.

Equations 38 through 59 also permit calculation of the variation in conditions along the length of a nozzle. Such calculations are carried out by selecting a series of pressures ranging from the inlet to the discharge pressure and calculating the corresponding values of u, V, and s. The results of such a series of calculations[6] are shown graphically in Fig. 168 for the nozzle of a rocket engine with inlet conditions of 300 psia and 4000° R and a discharge pressure of 14.7 psia. The propellant consumption is 2.2 lb per sec, producing gases for which $\kappa = 1.30$, $C_p = 0.359$ Btu/(lb)(R°), and $M = 24.0$.

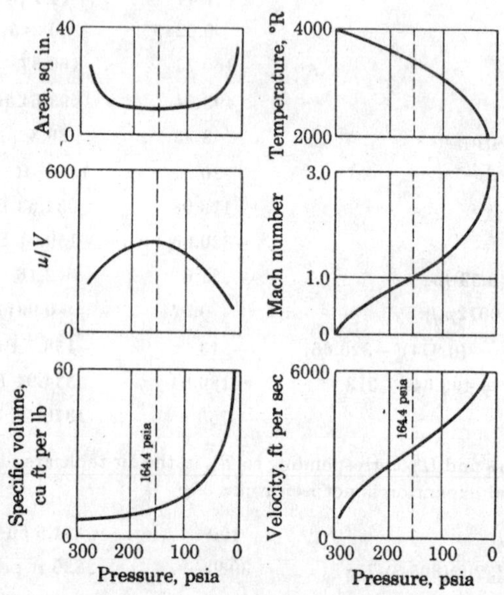

Fig. 168. Typical variation of velocity, area, temperature, specific volume, and Mach number with pressure in a rocket-engine nozzle. (From G. P. Sutton, *Rocket Propulsion Elements*, Fig. 3–6, p. 57, 1956, 2d ed., John Wiley & Sons, with permission.) Throat conditions: area = 0.98 sq in., $V = 9.48$ cu ft per lb, $u = 3050$ ft per sec, pressure = 164.4 psia, temperature = 3480° R

The relationships of Fig. 167 are ideally independent of the shape of the nozzle, which is generally determined by practical considerations of manufacturing difficulty and energy losses. In many cases both the converging and diverging sections may be conical with a rounded inlet and connected by a rounded section at the throat. It has been found empirically that the optimum divergent cone half-angle is generally between 12° and 18°. Larger angles are desirable in the converging section. If the nozzle is designed for uniform pressure drop per unit length, the curves of Fig. 167 represent variations with distance from the inlet.

Compression Nozzles. By reversing the direction of flow through a converging nozzle, it becomes a *diffuser* or compression nozzle which can

[6] G. P. Sutton, *Rocket Propulsion Elements*, 2d ed., John Wiley & Sons (1956).

effectively convert kinetic energy into pressure energy by approximately isentropic compression. Equations 38 through 42 are directly applicable to diffuser problems in which values of p_2/p_1 are greater than 1.0.

The limiting case in which a fluid stream is ideally diffused to zero velocity and pressure p_T is termed *stagnation*. Since $u_2 = 0$, equation 41 becomes

$$u_1^2 = -\frac{2g_c\kappa RT_1}{M(\kappa - 1)}\left[1 - \left(\frac{p_T}{p_1}\right)^{(\kappa-1)/\kappa}\right] \tag{63}$$

Solving for p_T and combining with equation 59 gives

$$p_T = p_1\left[\frac{u_1^2M(\kappa - 1)}{2g_c\kappa RT_1} + 1\right]^{\kappa/(\kappa-1)} = p_1\left[\frac{N_{M1}^2(\kappa - 1)}{2} + 1\right]^{\kappa/(\kappa-1)} \tag{64}$$

From equation 40,

$$T_T = \frac{u_1^2M(\kappa - 1)}{2g_c\kappa R} + T_1 = \frac{u_1^2}{2g_cC_p} + T_1 = T_1\left(1 + \frac{N_M^2(\kappa - 1)}{2}\right) \tag{65}$$

The maximum pressure p_T and the maximum temperature T_T attained in isentropic diffusion are termed the *stagnation pressure and temperature* or the *total pressure and temperature*. These quantities remain constant throughout an isentropic flow process. The maximum pressure ratio p_T/p_1 is termed the ideal pressure rise.

Illustration 5. Air at 100° F and a pressure of 14.7 psia is flowing in a duct at a velocity of 800 ft per sec. Calculate the stagnation temperature and pressure.

Taking $\kappa = 1.395$ in equation 64:

$$p_T = 14.7\left[\frac{(800)^2(2.9)(0.395)}{2(32.2)(1.395)1542(560)} + 1\right]^{1.395/0.395} = 20.2 \text{ psia}$$

From equation 65:

$$t_T = \frac{(800)^2 29(0.395)}{(64.4)(1.395)1542} + 100 = 153^\circ \text{ F}$$

Since stagnation corresponds to the total conversion of kinetic energy into enthalpy, the stagnation temperature and pressure are readily obtained from Table 55.

$$\text{Kinetic energy} = \frac{(800)^2}{(2)(32.2)(778)} = 12.78 \text{ Btu per lb}$$

$T_1 = 560°$ R, $\quad T_T = 613.2°$ R

$H_1 = 133.86$, $\quad H_T = 133.86 + 12.78 = 146.64$

$p_{R1} = 1.5742$, $\quad p_{RT} = 2.164$

$$t_T = 613.2 - 460 = 153.2^\circ \text{ F}$$

$$p_T = 14.7(2.164/1.5742) = 20.2 \text{ psia}$$

In real diffuser nozzles a portion of the kinetic energy lost is transformed into pressure energy while the remainder is converted into heat. For an ideal gas, since enthalpy is independent of pressure, the discharge temperature of a diffuser is equal to the ideal value, which corresponds to the change in velocity produced and is independent of inefficiencies due to turbulence and friction.

The efficiency of a diffuser is defined as the ratio of the isentropic pressure energy increase produced by a specified reduction in velocity to the accompanying loss in kinetic energy. Thus, η_E, the conversion efficiency, is expressed by

$$\eta_E = \frac{\int_{p_1}^{p'_2} V\,dp_s}{(u_1^2 - u_2^2)/2g_c} = \frac{(\Delta H_p)_S}{(\Delta H_u)_S} \tag{66}$$

where

$(\Delta H_p)_S =$ enthalpy change corresponding to ideal isentropic change from pressure p_1 to p'_2, the pressure produced in the actual diffuser

$(\Delta H_u)_S =$ ideal enthalpy change = the kinetic-energy change resulting from the change of velocity from u_1 to u_2

The *ram efficiency* η_R of a diffuser is defined as the ratio of the pressure rise actually produced in the diffuser to the rise that would have resulted from isentropic stagnation. Thus;

$$\eta_R = \frac{p_2 - p_1}{p_T - p_1} \tag{67}$$

where $p_T =$ stagnation pressure of diffuser inlet stream.

In general, the conversion efficiencies of diffusers are somewhat lower than those of expanding nozzles. In the subsonic range, efficiencies of the order of 0.85 to 0.95 are obtained. In the supersonic range, efficiency is highly dependent on specific-design and operating-condition relationships, and at Mach numbers of 3 to 5 will be of the order of 0.50 to 0.75 with good design.

Illustration 6. A missile is in flight at a Mach number of 3 and an altitude of 70,000 ft where the standard atmosphere has the following characteristics:

Temperature	−68° F
Pressure	0.0447 atm
Velocity of sound	974 ft per sec

A diffuser nozzle in the nose of the missile decelerates the air to a velocity of 300 ft per sec with an energy conversion efficiency of 70%. Calculate the discharge temperature and pressure and the pressure ratio produced by the diffuser.

$u_1 = 3(974)$	2922 ft per sec
p_1	0.658 psia
T_1	392° R

The enthalpy change accompanying the deceleration is,

$$(\Delta H)_S = -\frac{(u_2^2 - u_1^2)}{2g_c} = -\frac{(300)^2 - (2922)^2}{2(32.2)778} = 169 \text{ Btu per lb}$$

From Table 55:

p_{R1}	0.4527
H_1	93.61 Btu per lb
$H_2 = 93.61 + 169$	262.61 Btu per lb
T_2	1086° R
p_{R2} (ideal)	16.6
p_2/p_1 (ideal) $= (16.6)/(0.4527)$	36.6
$(\Delta H_p)_S = \eta_E(\Delta H_u)_S = 169(0.7)$	118
H'_2 (actual) $= 93.61 + 118$	211.61
p'_{R2} (actual)	7.77
$(p_2/p_1)' = (7.77)/(0.4527)$	17.2
$p_2 = (17.2)(0.658)$	11.3 psia

The temperature corresponding to H'_2 is of no significance because the actual change is not isentropic. The actual discharge temperature is 1086° R, or 626° F.

The results of Illustration 6 demonstrate the importance of diffuser design in high-speed flight. A 30% reduction in efficiency of the diffuser reduced the pressure ratio to 17.2 from an ideal value of 36.6.

Flow of Compressible Fluids in Ducts

When a compressible fluid flows with friction through a duct, the changes occurring must conform to each of the four fundamental equations 3, 8, 13, and 14 of steady-state flow. In addition, changes in state of the fluid must conform to its equation of state and to the restrictions such as adiabatic or isothermal conditions imposed on the flow. The resulting relationship which expresses the state properties of the fluid under the specified flow conditions is termed the *equation of condition.*

The conditions of flow may be adiabatic, characterized by the specification that $d'q = 0$, isothermal, characterized by the specification $dT = 0$, or the general case of nonadiabatic, nonisothermal flow, in which heat is transferred to the fluid and temperature varies.

Fanno Lines. Of particular interest is the case of adiabatic flow which is approximated in many situations of practical importance. For adiabatic flow in a horizontal duct of uniform cross-sectional area, the energy equation 13 may be written as follows, if differences in the average velocities are neglected:

$$dH + d\left(\frac{u^2}{2g_c}\right) = dH + \frac{G^2}{g_c} V\,dV = 0 \tag{68}$$

If it is assumed that the fluid is an ideal gas, the changes in enthalpy and entropy are as follows:

$$dH = C_p\,dT = \frac{R\kappa}{M(\kappa - 1)}\,dT = \frac{\kappa}{\kappa - 1}\,d(pV) \tag{69}$$

$$dS = \frac{R\kappa\,dT}{M(\kappa - 1)T} - \frac{R}{M}\frac{dp}{p} \tag{70}$$

Equations 69 and 70 may be integrated between an initial state 1 and any subsequent state 2 and used simultaneously with equation 68 and the ideal-gas law to establish an *HS* relationship of flow of the form shown by the solid line in Fig. 169. This curve is termed a *Fanno line* and is characteristic of the specified mass velocity G and initial conditions of state p_1, V_1, and T_1. It is evaluated by selecting arbitrary values of V_2 and calculating the corresponding values of H_2 from the integrated form of equation 68. Corresponding values of p_2 and T_2 are obtained from equation 69, and values of S_2 from equation 70.

As a fluid, at initial flow conditions 1, progresses down a duct, its state follows the Fanno line, and its pressure, temperature, and density diminish while velocity and entropy increase in accordance with equations 68 through 70. At point e, the Fanno line reaches a maximum value of entropy at which, for a small change, $dS = 0$ and $dH = V\,dp$. Substituting this value in equation 68 gives

$$\left(\frac{\partial p_l}{\partial V_l}\right)_S = -\frac{G^2}{g_c} = -\frac{u_l^2}{g_c V_l^2} \tag{71}$$

For an isentropic change in an ideal gas, $pV^\kappa = C$ and $dp/dV = -\kappa C/V^{\kappa+1}$ or $V^2(dp/dV) = -\kappa CV/V^\kappa = -\kappa pV$. Substituting in equation 71 gives

$$u_l^2 = \kappa g_c p_l V_l = \frac{\kappa g_c R T_l}{M} \tag{72}$$

Since equations 72 and 55 are identical, it follows that sonic velocity or a Mach number of 1.0 corresponds to the maximum entropy on a Fanno diagram.

Thus, when a compressible fluid flows adiabatically at subsonic velocities ($N_M < 1.0$) in a uniform duct, the velocity increases with diminishing pressure until a limiting velocity u_l is reached where $N_M = 1.0$. Further increase in velocity or reduction in pressure cannot occur spontaneously because it would be accompanied by a decrease in entropy.

Flow at supersonic velocities can be produced in a de Laval nozzle and introduced into a uniform duct. In this case the initial conditions of

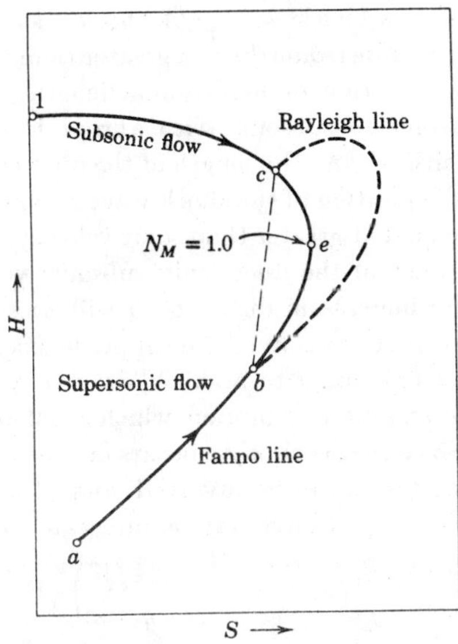

FIG. 169. *H–S* relationships of adiabatic flow in uniform ducts

flow correspond to point *a* on Fig. 169, and the spontaneous changes in the duct cause the state of the fluid to progress along the Fanno line *abe* with increase in pressure, density, and temperature, and decrease in velocity. A limiting condition of minimum velocity is reached at *e* where $N_M = 1.0$. Further reduction in velocity would require a decrease in entropy.

Adiabatic flow in a uniform duct at subsonic velocities is expansion flow with reduction in pressure while at supersonic velocities compression flow occurs with increase in pressure. In both cases sonic velocity is reached as a limit.

When a fluid enters a duct at a specified subsonic velocity, there is a maximum duct length which corresponds to reduction of the pressure to a value producing a sonic discharge velocity. Under these conditions further reduction of the pressure of the space into which discharge occurs has no influence on the flow in the duct. If the maximum duct length is exceeded, *choking* occurs, and the specified inlet velocity is not maintained.

Shock Waves. When a fluid enters a duct at supersonic velocities, stable, continuous flow conditions can be maintained only in ducts of length equal to or less than the limiting length which will result in a sonic discharge velocity. If the limiting duct length is exceeded, a discontinuity of flow conditions occurs with an abrupt increase in pressure, which reduces the downstream velocity to the subsonic range. This phenomenon is termed a shock wave. A shock wave may maintain a stationary position in a supersonic duct of greater than the limiting length and corresponds to a section of infintesimal length in which velocity changes from supersonic to subsonic with abrupt increase in pressure, temperature, and density. As the length of the duct is increased beyond the limiting value, the position of the shock wave moves toward the inlet, thus traveling in the fluid at greater than sonic velocity. When the shock wave stands at the inlet of the duct, only subsonic velocities will exist in the duct. Further increase in duct length will cause the shock wave to move into the diverging section of the supply nozzle.

The conditions corresponding to a shock wave may be developed by consideration of the equation of motion which must be satisfied by all flow phenomena. Since the shock wave occurs in a section of infinitesimal length, no finite drag forces can be involved, and, if transverse velocity gradients are neglected, equation 8 may be integrated and written as

$$p_2 - p_c = -\frac{W}{g_c s}(u_2 - u_c) = -\frac{1}{g_c}\left(\frac{W}{s}\right)^2 (V_2 - V_c) \tag{73}$$

Equation 73 may be used simultaneously with equations 69 and 70 to establish an *HS* relationship corresponding to specified initial conditions of flow in the same manner that the Fanno line was plotted in Fig. 169. The resulting curve is termed the Rayleigh line and is plotted in Fig. 169 for the initial conditions corresponding to point c. This line is defined by the specification of a pressure change $p_2 - p_c$ under such conditions that $F_D = 0$. Under most conditions such a change is not adiabatic, but at point b, where the Rayleigh line intersects the Fanno line in the supersonic region, both the conditions of adiabaticity and absence of drag force are satisfied. It follows that direct changes between b and c can occur without addition or removal of heat. However, it will be noted that the entropy at c is greater than at b. It follows that only a *compression shock wave* can occur spontaneously. A shock wave established at conditions b increases the pressure and decreases the velocity to the conditions at c, but the reverse phenomenon does not occur.

Consideration of shock waves is of major importance in all problems involving velocities in the supersonic range. As previously pointed out, particularly difficult situations arise in the transonic range of flight velocities and in the design of compression nozzles.

Adiabatic Flow in Uniform Ducts. Since adiabatic flow with friction is not isentropic, the pvT relationships of reversible adiabatic expansion are not applicable. The fundamental specification is that $q = 0$. If no facilities for work are present, changes in potential and surface energies are negligible, and cross-sectional area is constant, where $\overline{u^3}/\bar{u}$ is equivalent to u^2, equation 13 becomes

$$dU + d(pV) = -\frac{du^2}{2g_c} = -\frac{G^2V\,dV}{g_c} = dH \tag{74}$$

Also

$$dH = C_p\,dT = \frac{R\kappa}{(\kappa - 1)}\frac{dT}{M} = \frac{\kappa}{\kappa - 1}\,d(pV) \tag{75}$$

Combining equations 74 and 75 yields

$$d(pV) = -\left(\frac{\kappa - 1}{\kappa}\right)\left(\frac{G^2}{g_c}\right)V\,dV \tag{76}$$

Integrating gives

$$pV + \left(\frac{\kappa - 1}{\kappa}\right)\frac{G^2V^2}{2g_c} = \text{constant} = p_1V_1 + \left(\frac{\kappa - 1}{\kappa}\right)\frac{G^2V_1^{\,2}}{2g_c} \tag{77}$$

Equations 76 and 77 are forms of the *equation of condition* which must be followed in adiabatic flow in horizontal uniform ducts. Equation 77 may be rearranged to express the relationships among the pressure, temperature, and volume ratios in such flow.

$$\frac{p_2V_2}{p_1V_1} = \frac{T_2}{T_1} = 1 + \left[\frac{(\kappa - 1)G^2V_1^{\,2}}{\kappa 2g_cp_1V_1}\right]\left[1 - \left(\frac{V_2}{V_1}\right)^2\right] \tag{78}$$

Equation 78 also may be written in terms of Mach number, since

$$N_{M1}^2 = G^2V_1^{\,2}/\kappa g_cp_1V_1$$

$$\frac{p_2V_2}{p_1V_1} = \frac{T_2}{T_1} = 1 + \left(\frac{\kappa - 1}{2}N^2_{M_1}\right)\left[1 - \left(\frac{V_2}{V_1}\right)^2\right] \tag{79}$$

Equations 78 and 79 are alternative forms of the condition equation.

From equation 19, where $dZ = 0$, $d'w = 0$, and $ds = 0$, since mass velocity $G = W/s = u/V$ and $V\,dp = d(pV) - p\,dV$, there results

$$\frac{G^2V\,dV}{g_c} + d(pV) - pV\frac{dV}{V} + \frac{G^2V^2f}{2g_cr_h}\,dL = 0 \tag{80}$$

Substitution of equation 76 for $d(pV)$ and equation 77 for pV into equation 80 and multiplying by $2\kappa g_c/G^2V^2$ gives

$$2\frac{dV}{V} - \left[\frac{2\kappa g_c p_1 V_1}{G^2} + (\kappa - 1) V_1^2\right]\frac{dV}{V^3} + (\kappa - 1)\frac{dV}{V} + \frac{\kappa f}{r_h} dL = 0 \quad (81)$$

Integrating from V_1 to V_2 and $L = 0$ to L gives

$$(\kappa + 1)\ln\frac{V_2}{V_1} + \frac{1}{2}\left[\frac{2\kappa g_c p_1 V_1}{G^2} + (\kappa - 1)V_1^2\right]\left(\frac{1}{V_2^2} - \frac{1}{V_1^2}\right) + \frac{\kappa f L}{r_h} = 0 \quad (82)$$

or

$$\frac{fL}{r_h} = \frac{1}{2\kappa}\left[\frac{2\kappa g_c p_1 V_1}{G^2 V_1^2} + (\kappa - 1)\right]\left[1 - \left(\frac{V_1}{V_2}\right)^2\right] + \frac{\kappa + 1}{2\kappa}\ln\left(\frac{V_1}{V_2}\right)^2 \quad (83)$$

Equation 83 completely expresses specific-volume ratios in terms of initial conditions and the friction parameter. It also may be written in terms of initial Mach number:

$$\frac{fL}{r_h} = \frac{1}{2\kappa}\left[\frac{2 + (\kappa - 1) N^2_{M_1}}{N^2_{M1}}\right]\left[1 - \left(\frac{V_1}{V_2}\right)^2\right] + \frac{\kappa + 1}{2\kappa}\ln\left(\frac{V_1}{V_2}\right)^2 \quad (84)$$

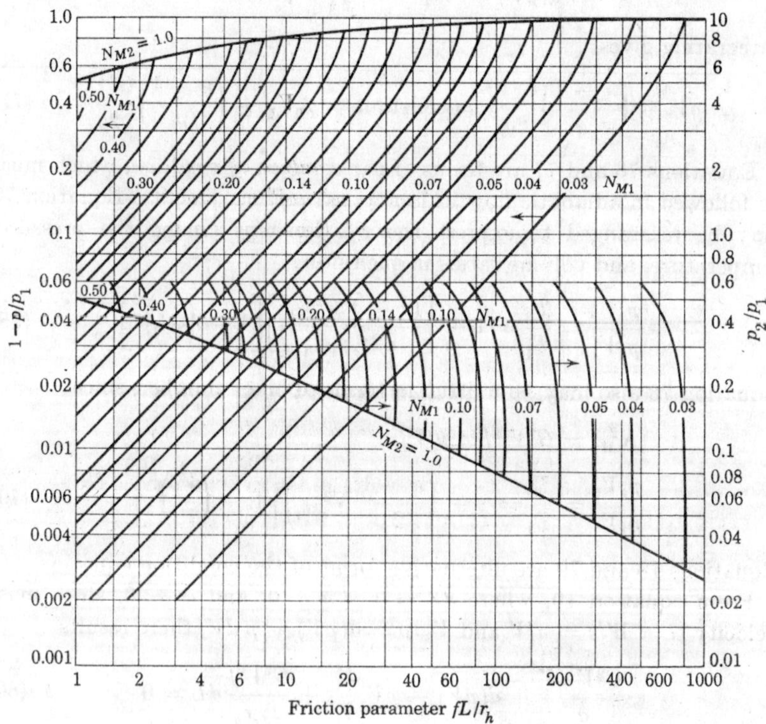

Fig. 170. Pressure-drop ratios for adiabatic flow of air in ducts ($\kappa = 1.395$)

Simultaneous solution of equations 78 and 83 or 79 and 84 to eliminate V_1/V_2 yields pressure ratios p_2/p_1 or temperature ratios T_2/T_1 as functions of fL/r_h. Because of the complex forms of the equations, they are best solved by graphical methods or used to construct charts such as Figs. 170, and 171. From these figures may be read the temperature and pressure ratios corresponding to any initial Mach number and friction parameter. The corresponding Mach number then follows from

$$N_{M2} = N_{M1}\left(\frac{T_2}{T_1}\right)^{1/2}\frac{p_1}{p_2} \tag{85}$$

The velocity may be calculated from

$$u_2 = u_1\left(\frac{T_2}{T_1}\right)\left(\frac{p_1}{p_2}\right) = N_{M1}\left(\frac{\kappa g_c R T_1}{M}\right)^{1/2}\left(\frac{T_2}{T_1}\right)\left(\frac{p_1}{p_2}\right) \tag{86}$$

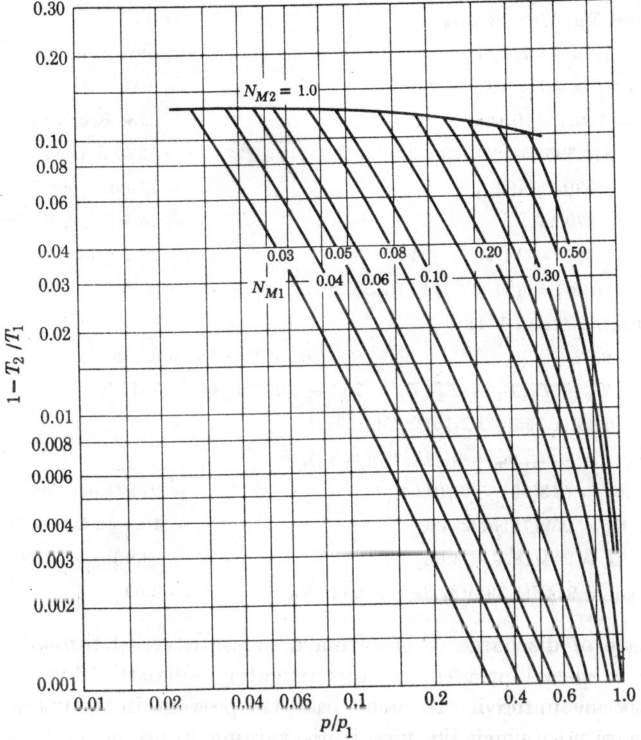

FIG. 171. Temperature-drop ratios for adiabatic flow of air in ducts (κ = 1.395)

Limiting pressure and temperature drops corresponding to sonic discharge velocity are plotted in Figs. 170 and 171. Limiting friction parameters may be read directly from these curves and the corresponding limiting duct length calculated directly therefrom.

Illustration 7. Air is forced through a cylindrical duct having a diameter of 2 in. and a length of 20 ft. The discharge pressure is 1 atm, and the inlet temperature is 300° F. Assuming adiabatic flow and a constant friction factor $f = 0.005$, calculate the following:

(*a*) Limiting pressure ratio.
(*b*) Maximum useful inlet pressure.
(*c*) Maximum mass velocity.
(*d*) Linear velocity at the outlet with 50% of the maximum pressure drop.
(*e*) Mass velocity in *d*.
(*f*) Mach number at the outlet of *d*.

Friction parameter $fL/r_h = (0.005)20(4)12/2 \qquad = 2.4$

(*a*) From Fig. 170: $(p/p_1)_l \qquad = 0.37$

(*b*) $(p_1)_{max} = 14.7/0.37 \qquad = 39.7$ psia

(*c*) From Fig. 171: $(T_1 - T_l)/T_1 \qquad = 0.108$

$T_l = 760(1 - 0.108) \qquad = 678°$ R or 218° F

$u_l = [(1.395)(32.2)1542(678)/29]^{1/2} \qquad = 1270$ ft per sec

$V_l = 359(678/492)/29 \qquad = 17$ cu ft per lb

$G = 1270/17 \qquad = 74.8$ lb/(sq ft)(sec)

(*d*) $p_1 = [(39.7 - 14.7)/2] + 14.7 \qquad = 27.2$ psia

$(p_1 - p_2)/p_1 = (27.2 - 14.7)/27.2 \qquad = 0.46$

From Figs. 170 and 171:

$N_{M1} = 0.38 \qquad (T_1 - T_2)/T_1 = 0.038$

$T_2 = T_1 - T_1(T_1 - T_2)/T_1 = 760 - (0.038)760 = 731°$ R

$u_1 = 0.38[(1.395)(32.2)1542(760)/29]^{1/2} \qquad = 513$ ft per sec

Since $\rho_2 u_2 = \rho_1 u_1$, $u_2 = u_1(T_2/T_1)(p_1/p_2)$, or

$u_2 = 513(731/760)/(0.54) \qquad = 914$ ft per sec

(*e*) $V_2 = 359(731/492)/29 \qquad = 18.4$ cu ft per lb

$G = 914/18.4 = 49.6 \qquad = 49.6$ lb/(sq ft)(sec)

(*f*) $N_{M2} = 914/[(1.395)(32.2)(1.542)73/29]^{1/2} \qquad = 0.69$

The rate of discharge of gas from a pressure vessel through a pipe presents a special problem of approximately adiabatic flow which is frequently encountered. In such a case, the pressure in the vessel p_o and the pressure into which the pipe is discharging p_a are generally known, but the pressure at the inlet to the pipe p_1 is not known. It may be assumed that the gas enters the pipe as a result of isentropic expansion in accordance with equation 41, which may be written in terms of the Mach number. Thus, from equations 40, 41, and 59 between states o and 1, if $u_o = 0$

$$N^2_{M1} = \frac{2}{\kappa - 1}\left[\left(\frac{p_o}{p_1}\right)^{(\kappa-1)/\kappa} - 1\right] \tag{87}$$

If the length of the pipe is greater than the limiting value, the discharge pressure $p_2 = p_a$ and values of p_1 and N_{M1} may be determined by graphical solution of equation 87 in conjunction with Figs. 170 and 171. Assumed values of p_1 and corresponding values of N_{M1} are selected and are read from Fig. 170 and calculated from equation 87. The correct value of p_1 corresponds to zero difference between the N_{M1} values. Temperature T_1 is then evaluated from equation 40 applied to states o and 1.

If the length of the pipe is less than the limiting value, $p_2 > p_a$, and the value of N_{M1} can be read from the limiting pressure drop line of Fig. 170 together with the ratio p_2/p_1. The pressure at the inlet may then be calculated from equation 87, and the mass rate of flow obtained from the density corresponding to p_1 and T_1.

Convenient charts for the direct solution of problems of the foregoing type have been developed by Lapple.[7]

Isothermal Flow in Uniform Ducts. For isothermal flow, the equation of condition follows from the ideal-gas law. Thus

$$d(pV)_T = 0 \tag{88}$$

For horizontal ducts with no facilities for work, dZ and $d'w$ are zero. For this case, substituting equation 88 in equation 80 and dividing by V^2 results in

$$\frac{G^2}{g_c}\frac{dV}{V} - \frac{RT}{M}\frac{dV}{V^3} + \frac{fG^2}{2g_c r_h}\,dL = 0 \tag{89}$$

Integration of equation 89 over a length L gives

$$\frac{G^2}{g_c}\ln\frac{V_2}{V_1} + \frac{RT}{2M}\left(\frac{1}{V_2^2} - \frac{1}{V_1^2}\right) + \frac{fG^2L}{2g_c r_h} = 0 \tag{90}$$

Rearranging and substituting $V = u/G$, gives,

$$\frac{fL}{r_h} = \frac{RTg_c}{M}\frac{1}{u_1^2}\left[1 - \left(\frac{u_1}{u_2}\right)^2\right] - \ln\left(\frac{u_2}{u_1}\right)^2 \tag{91}$$

or, in terms of pressures,

$$\frac{fL}{r_h} = \frac{Mg_c p_1^2}{G^2RT}\left[1 - \left(\frac{p_2}{p_1}\right)^2\right] - \ln\left(\frac{p_1}{p_2}\right)^2 \tag{92}$$

[7] C. E. Lapple, *Trans. Am. Inst. Chem. Engrs.*, **39**, 385 (1943).

Equation 91 also may be written in terms of initial Mach number,

$$N_M = u_1/(\kappa g_c RT/M)^{1/2}$$

$$\frac{fL}{r_h} = \frac{1}{\kappa N_{M1}^2}\left[1 - \left(\frac{p_2}{p_1}\right)^2\right] - \ln\left(\frac{p_1}{p_2}\right)^2 \quad (93)$$

The pressure–velocity relationships in ducts of specified dimensions may be evaluated by solution of equations 91 through 93.

Equation 92 indicates that, if the pressure ratio p_2/p_1 were zero and $\ln(p_1/p_2)$ infinite, the mass velocity G would be zero. It follows that at some intermediate value of p_2/p_1 the mass velocity must reach a maximum where $dG/dp_2 = 0$. In differentiating equation 92, let $(p_2/p_1)^2 = A$, $fL/r_h = B$, and $RT/Mg_cp_1{}^2 = C$. Then

$$CG^2 = \frac{1 - A}{B - \ln A} \quad (94)$$

$$\frac{C\,dG^2}{dA} = \frac{-(B - \ln A) + (1 - A)(1/A)}{(B - \ln A)^2} = 0 \quad (95)$$

$$\frac{1 - A}{B - \ln A} = A = CG^2 \quad (96)$$

or, if p_l is the value of p_2 at the pressure ratio of maximum flow,

$$\left(\frac{p_l}{p_1}\right)^2 = \frac{RTG^2}{Mg_cp_1{}^2} = \frac{V_1G^2}{g_cp_1} \quad (97)$$

The relationship between limiting pressure ratio and the friction parameter is obtained by combining equations 92 and 97:

$$\left(\frac{fL}{r_h}\right)_l = \left(\frac{p_1}{p_l}\right)^2\left[1 - \left(\frac{p_l}{p_1}\right)^2\right] - \ln\left(\frac{p_1}{p_l}\right)^2 \quad (98)$$

Since $p_lV_l = RT/M$ and $G = u_l/V_l$, equation 97 may be written

$$u_l = \left(\frac{g_cRT}{M}\right)^{1/2} = \frac{a}{\sqrt{\kappa}} \quad (99)$$

where a is the isentropic velocity of sound. Thus, if isothermal flow is maintained in a gas having a value of κ greater than 1.0, the maximum velocity will be less than sonic at the outlet of a duct operating at the limiting pressure ratio. It was pointed out by Lapple[7] that isothermal flow may be treated as a limiting case of adiabatic flow for which $\kappa = 1.0$. This situation is approached by gases of high heat capacity. In this case sonic velocity is approached at limiting flow.

Flow with Friction and Heat Transfer. In the general case, fluid flow in pipe lines is accompanied by both friction and flow of heat with neither isothermal nor adiabatic conditions.

It has been demonstrated both theoretically and experimentally that the temperature at the inside surface of the boundary film approaches the stagnation temperature of the main stream. As a result, any heat transfer between the fluid and the walls is governed by the difference between the wall temperature and the stagnation temperature of the gas, not its stream temperature. Thus, for an incremental section,

$$\frac{d'q}{\tau} = \frac{hy}{r_h}(T_T - T_w)\,dL \tag{100}$$

where h = heat-transfer coefficient per unit of wall area

y = perimeter

T_w = wall temperature

By combining equations 100 and 65 with the general dynamic-flow equation and the energy-balance equation, Thompson[8] has developed general differential equations for flow with heat transfer. Integrations were performed for two limiting types of flow: one isentropic and the other at constant Mach number. Actual flow problems are solved by treating them as comprising a sequence of small steps which are alternately at constant entropy and at constant Mach number. Charts were developed to facilitate such calculations.

It is a corollary of equation 100 that, if the wall temperature is held constant at the initial temperature of a flowing gas, no heat is transferred to the gas, even though its stream temperature may fall far below the wall temperature. Instead, heat is lost from the gas in the downstream sections where friction increases the total temperature.

Ejectors

An *ejector* or *jet pump* is a device in which the momentum and kinetic energy of a high-velocity fluid stream are employed to entrain and compress a second fluid stream. Steam ejectors are widely used to maintain subatmospheric pressures in distillation, evaporation, and refrigeration and to induce flow of gases in stacks and ducts. In heating and air conditioning, high-pressure streams of air are used to produce circulation of secondary air, and, in aircraft, engine exhaust gases are discharged through ejectors to pump cooling air through the engines. Liquid jet

[8] A. S. Thompson, *J. Appl. Mech.*, **17**, 19 (1950).

pumps are used in deep-well applications and for the handling of corrosive and high-temperature liquids.

An ejector comprises the following four components which may be analyzed individually by application of equations 3, 8, 13 and 14.

(*a*) A high-pressure nozzle provides for acceleration of the primary or propelling fluid.

(*b*) A secondary-fluid inlet section provides for acceleration of the secondary fluid prior to entrainment by the primary fluid.

(*c*) A mixing section produces intermixing of the secondary and primary streams with further acceleration of the former and deceleration of the latter.

(*d*) A diffuser provides for deceleration of the mixed streams with increase in pressure.

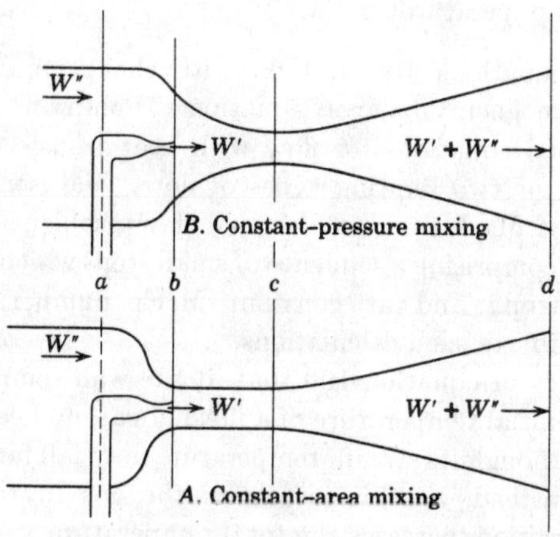

Fig. 172. Ejector types

Ejectors may be classified on the basis of the conditions maintained in the mixing section. In *constant-pressure mixing*, the mixing section is of converging form to maintain constant pressure by increasing the velocity of the mixed streams. In *constant-area mixing*, the pressure of the mixed streams is allowed to rise in a section of uniform cross section. In Fig. 172 these two types are shown diagrammatically in the *Venturi form*, in which the area of the mixing section or throat is less than that of the discharge section. The simplest form of ejector is a constant-area mixing type known as a *draft tube* in which the high-pressure nozzle is placed at the axis of a straight pipe. Thus the secondary-inlet, mixing, and discharge sections are all of uniform cross section.

For most applications the constant-area-mixing Venturi-type ejector is preferred although the constant-pressure and combination types are also used. Much empiricism is involved in the design of efficient ejectors, because of the complex nature of the relationships involved, and the dependence of the friction losses on the details of mechanical design. These problems are reviewed by Lapple and associates[9] who summarize typical design and performance factors.

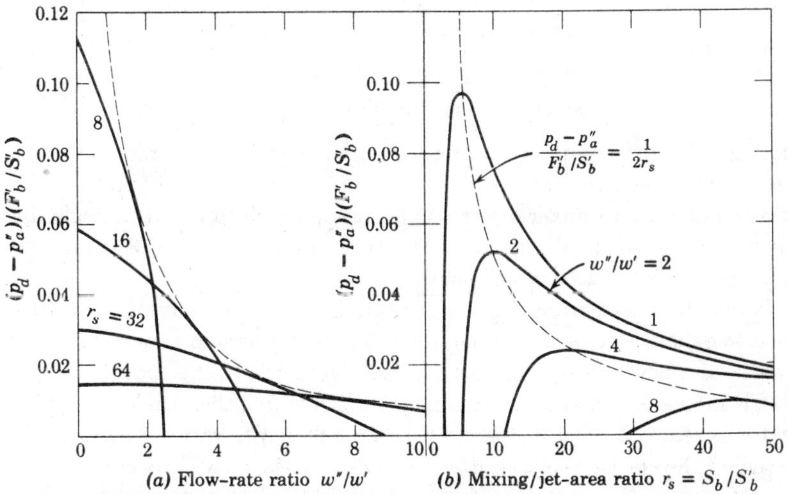

FIG. 173. Performance characteristics of low-compression ejectors. Discharge energy ($u_d^2/2g_c$) negligible; density ratio, $\rho''_b/\rho'_b = 1.0$

Ejector Performance. The typical performance characteristics of an ejector are shown by the curves of Fig. 173 which were plotted from a simplified equation developed by McClintock and Hood.[10] This equation is restricted to constant-area mixing ejectors, in which changes in specific volume of the secondary and mixed fluids are negligible, but the general form of the relationships is typical of all ejectors.

The dimensionless equation of McClintock and Hood[10] expresses the compression produced by the ejector as the ratio of the pressure rise of the secondary fluid $p_d - p''_a$ to the momentum per unit area of the primary jet F'_b/s'_a. In this notation the subscript letters refer to the positions indicated in Fig. 172, primed quantities refer to the primary fluid,

[9] C. E. Lapple et. al., *Fluid and Particle Mechanics*, University of Delaware, Newark (1951).

[10] F. A. McClintock and J. H. Hood, *J. Aero. Sci.*, **13**, 559 (1946).

double primed to the secondary, and unprimed to the mixed fluids. The compression factor is related to the following ratios:

$$\frac{\text{Area of diffuser exit}}{\text{Area of mixing section}} = \frac{s_d}{s_b} \tag{101}$$

$$\frac{\text{Area of mixing section}}{\text{Area of primary jet}} = \frac{s_b}{s'_b} = r_s \tag{102}$$

$$\frac{\text{Secondary flow rate}}{\text{Primary flow rate}} = \frac{w''}{w'} \tag{103}$$

$$\frac{\text{Density of secondary fluid}}{\text{Density of primary fluid}} = \frac{\rho''}{\rho'} \tag{104}$$

In Fig. 173, the compression factor is related to the flow-rate ratio and the mixer-to-jet area ratio for the case in which the density ratio is 1.0 and the diffuser-to-mixer area ratio is large enough to result in negligible kinetic energy in the discharge. The two sections of Fig. 173 are merely cross-plots of the same relationship.

It will be noted from Fig. 173*a* that, if an ejector having $r_s = 16$ is used to exhaust a closed vessel, a maximum compression factor of 0.058 is produced when the secondary flow rate is zero. If the secondary flow rate is increased, thereby increasing the ratio w''/w', then the compression factor is reduced until it reaches zero at a flow-rate ratio of 5.2. If the design of the ejector is changed to make $r_s = 8$, the maximum compression factor at zero flow ratio is increased to 0.113, but the maximum flow ratio at zero compression factor is reduced to 2.5. Curves of different values of r_s define the envelope indicated by the broken line, which establishes the maximum compression factor obtainable for a specified flow ratio. Such maximum compression is obtainable only by employing an optimum area ratio r_s.

The effect of the area ratio r_s is clearly shown by the cross-plot (Fig. 173*b*). Thus, if it is desired to handle a flow-rate ratio of 2.0, the maximum compression factor of 0.52 is obtained with the optimum area ratio $r_s = 0.96$. McClintock and Hood[10] showed that the maximum compression factor is approximately equal to $1/2r_s$ as indicated by the broken curve of Fig. 173*b*.

If the diffuser-to-mixer area ratio is reduced, the compression factor is reduced, whereas, if the density ratio is increased, the compression factor is increased. However, the approximate relationship between compression factor and optimum mixer-to-jet area ratio is little affected by these changes.

Thermodynamic analyses of ejector operation may be developed by successive application of the equations of continuity, momentum, and energy to the individual operations of the nozzle, the mixing section, and the diffuser. However, much empiricism is involved in determining the coefficients or efficiencies involved in these steps.

The over-all efficiency of an ejector is the ratio of the work of isentropic compression of the secondary fluid from the inlet to the discharge pressure, divided by the work of isentropic expansion of the primary fluid from its initial to the discharge pressure. Thus:

$$\eta = \frac{w''(H''_d - H''_a)_S}{w'(H'_a - H'_d)_S} \tag{105}$$

In general, ejector efficiencies are low, of the order of 15 to 30%.

A comprehensive analysis of the problems of ejector design was developed by Flugel[11] who considered both constant-pressure and constant-area mixing of liquids, gases, and condensible vapors in both super- and subsonic velocity ranges. In addition to refined general analyses, simplified relationships are presented for arriving at design factors.

McClintock and Hood[10] developed a general equation applicable to low-compression ejectors in which specific-volume changes are small. A single empirical coefficient expressed as a function of the diffuser–mixer area ratio serves to give good agreement between theoretical and observed performance.

Kroll[12] presents convenient charts for the design of low-compression steam–air and air–air jet pumps discharging at a pressure of 1.0 atm.

Keenan, Neuman, and Lustwerk[13] developed equations for both constant-pressure and constant-area mixing ejectors in which the primary and secondary fluids have the same molecular-weight and heat-capacity ratio. Good correlation of experimental data was obtained.

Takashima[14] developed relationships and design charts for the optimum performance of ejectors handling various gases. Good agreement was found with experimental data.

Problems of the selection and design of high-compression multistage steam ejectors were reviewed by Frumerman[15] and by Fondrk.[16] In addition manufacturers of such equipment publish bulletins and catalogs

[11] G. Flugel, "The Design of Jet Pumps", *NACA Tech. Mem.*, **982**, Washington (1941).

[12] A. E. Kroll, *Chem. Eng. Prog.*, **1**, 21 (1947).

[13] J. H. Keenan, E. P. Neuman, and F. Lustwerk, *J. Appl. Mech.*, **17**, 299 (1950).

[14] Y. Takashima, *Chem. Eng.* (*Japan*), **19**, 440 (1955).

[15] R. Frumerman, *Chem. Eng.*, **63**, 196 (1956).

[16] V. V. Fondrk, *Chem. Eng. Prog.*, **49**, 3 (1953).

containing performance data on their equipment and methods for its proper selection and use.

Single Fluid Ejectors. A simple thermodynamic analysis may be made of the over-all performance of an ejector designed for constant-pressure mixing if only a single fluid is involved in the primary and secondary streams, and if the pressure drop between the source of the secondary fluid and the mixing section is neglected. This method depends on empirical knowledge of mixing efficiency. An example is a steam ejector used to exhaust water vapor from an evaporator or refrigerator.

In such an ejector of the type shown in Fig. 172*b*, W' lb of high-pressure steam per second at pressure p' expands through the nozzle into a mixing chamber at a pressure p_{bc}. The efficiency of transforming enthalpy loss into kinetic energy is the nozzle efficiency η'_{ab}. The water vapor to be exhausted at pressure p_{bc} is admitted to section b at a rate of W'' lb per sec. The impact of one stream into the other may be considered empirically as resulting in a further degradation of kinetic energy of the expanding steam. The efficiency of transforming energy in mixing is designated as η'_{bc}. The resultant degradation of mechanical energy due to the inefficiency of the nozzle in expansion and due to mixing are both expressed in terms of the isentropic change in enthalpy of the high-pressure stream. Thus,

$$\text{Mechanical energy degraded in expansion} = (1 - \eta'_{ab})(H'_a - H'_b)_S \qquad (106)$$

$$\text{Mechanical energy degraded in mixing} = (1 - \eta'_{bc})(H'_a - H'_b)_S \qquad (107)$$

$$\text{Total mechanical energy degraded} = (2 - \eta'_{ab} - \eta'_{bc})(H'_a - H'_b)_S \qquad (108)$$

The actual enthalpy H'_c of the expanded primary steam in the mixture at c is given by

$$H'_c = (2 - \eta_n - \eta_m)(H'_a - H'_b)_S + (H'_b)_S \qquad (109)$$

These operations are shown graphically on the *HS* diagram of Fig. 174 for Illustration 8.

The mixed vapors flow through the diffuser at a rate of $W = W' + W''$ lb per sec, leaving at pressure p_d. In the diffuser, kinetic energy is transformed into pressure head with an efficiency of compression designated as η_{cd} based on the isentropic increase in enthalpy of the combined streams. Thus:

$$\eta_{cd} = \frac{(H_d - H_c)_s}{(H^*_d - H_c)} = \frac{(u_c^2 - u_d^2)}{2g_c(H^*_d - H_c)} \qquad (110)$$

where H^*_d is the actual enthalpy of the discharge.

If the kinetic energy of the discharge stream is negligible, the actual increase in enthalpy of the mixed stream in the diffuser is equal to the loss of enthalpy of the primary stream in the nozzle and mixer. Then, the number of pounds of low-pressure steam compressed per pound of high-pressure steam is given by

$$\frac{W' + W''}{W'} = \frac{H'_a - H'_c}{H^*_d - H_c} \tag{111}$$

If both inlet and discharge kinetic energies are negligible, an over-all energy balance for the ejector may be written in terms of enthalpies. Thus,

$$W'H'_a + W''H''_a = (W' + W'')H^*_d \tag{112}$$

The mass-flow-rate ratio of an ejector operating between specified pressures p'_a, p''_{abc}, and p_d may be calculated by graphical solution of equations 110 through 112. A series of values of H_c is selected, and the corresponding values of H^*_d are calculated from equation 110, and values of W/W' from equation 111. These values of W/W' are then used in equation 112 to calculate alternate values of H^*_d. Curves relating the selected values of H_c to the alternate values of H^*_d intersect at the correct values of H_c and H^*_d from which the correct flow-rate ratio is obtained.

The variation in area in the mixing section to maintain constant pressure could be calculated from a momentum-force balance. However, the importance of friction losses requires empirical approaches to the design of such equipment for high-compression duty.

Illustration 8. Saturated steam at 120 psia is expanded in a steam ejector to compress a secondary stream of steam, saturated at 4 psia, to a pressure of 14.7 psia. Calculate the number of pounds of the high-pressure steam required per pound intake of low-pressure steam.

Nozzle expansion efficiency η_n	= 0.95
Mixing efficiency η_m	= 0.80
Compression efficiency η_c	= 0.90

The desired portion of the HS diagram for steam is shown in Fig. 174:

H'_a = enthalpy of saturated steam at p_1 = 120 psia	= 1190 Btu per lb
$(H'_b)_S$ = enthalpy of steam after isentropic expansion to 4 psia	= 959 Btu per lb
	231 Btu per lb

Actual enthalpy of expanded and mixed primary steam:

$H'_c = (2 - 0.95 - 0.80)231 + 959$	= 1017 Btu per lb

Mechanical energy available for compression:

$H'_a - H'_c = 1190 - 1017$ $= 173$ Btu per lb

$H_c =$ enthalpy of the mixture at 4 psia (assumed) $= 1060$ Btu per lb

$H_d =$ enthalpy after isentropic compression to 14.7 psia $= 1149$ Btu per lb

$$(H_d - H_c)_S = 89$$

$$H^*_d - H_c = \frac{89}{0.90} = 99 \text{ Btu per lb}$$

$$H^*_d = 1060 + 99 = 1159 \text{ Btu per lb}$$

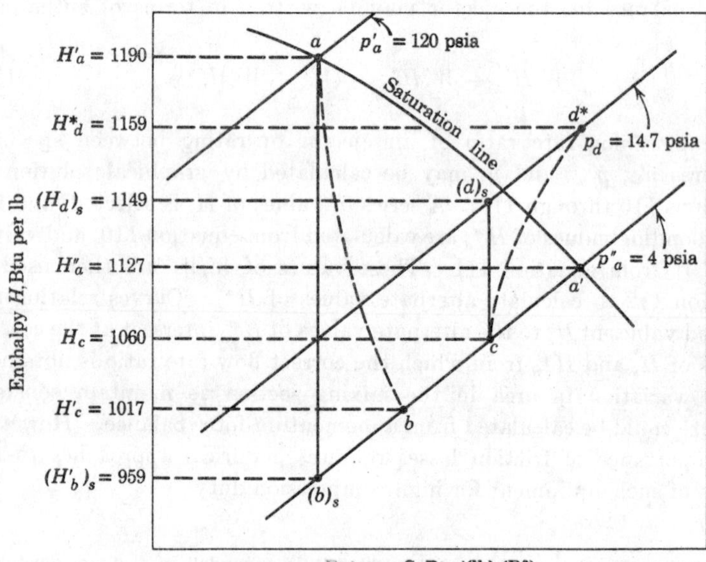

Fig. 174. *HS* diagram of a steam ejector

The number of pounds of mixture compressed per pound of primary steam (equation 112):

$$\frac{W' + W''}{W'} = \frac{173}{99} = 1.75$$

or

$$W''/W' = 0.75$$

The over-all energy balance verifies the assumed value of H_c, provided both calculated values of H^*_d agree.

From equation 112:

$$1(1190) + (0.75)1127 = 1.75H_d$$

or

$$H^*_d = 1160$$

Problems

1. Air at 120° F and 740 mm Hg is passed at a velocity of 30 ft per sec through a duct having a rectangular cross section 3 in. by 12 in. Calculate the pressure drop in inches of water per foot of duct length.

2. Crude oil having a viscosity of $(2.7)10^{-3}$ lb/(ft)(sec) and a density of 0.85 gram per cc is to be pumped through a pipe line at a rate of 1500 bbl (1 bbl = 42 gal) per day. If pumping stations are to be 50 miles apart, calculate the line size required if the pressure drop between stations is 300 psi.

3. A cylindrical smoke stack has an inside diameter of 6.0 ft and a height of 70 ft. Flue gases having a molecular weight of 31.0 are passed up the stack at a velocity of 20 ft per sec and a temperature of 845° F. The barometric pressure at the base of the stack is 746 mm Hg, and the atmospheric temperature is 93° F. Calculate the draft at the base of the stack in inches of water difference between the pressure of the atmosphere and that of the flue gases. The viscosity of the stack gases may be taken as 310 micro poises.

4. A fire hose has an internal diameter of 2 in. and is attached to a nozzle with an opening of 0.75 sq in. The pressure at the inlet of the nozzle is 120 psig. Neglecting fluid friction, calculate the total thrust on the nozzle when the stream is directed horizontally.

5. Calculate the thrust on a flat wall if the stream from the nozzle of problem 4 is directed perpendicularly against it.

6. Steam leaves the nozzles of a turbine at an absolute velocity of 1100 ft per sec and is directed against blades moving away at a velocity of 500 ft per sec. The steam enters the blades at an angle of 12° from the tangential line of motion of the blades and is deflected through an angle of 156° without loss of velocity. Calculate the tangential specific thrusts on the blades in the direction of motion and also the axial specific thrust perpendicular to the plane of motion.

7. A sonic orifice having a throat diameter of 0.1 in. is to be used for regulating the flow of air at 80° F into a process operating at variable pressures below 10 psig. If the discharge efficiency of the nozzle is 0.97, calculate the calibration curve relating mass rate of flow to inlet pressure over the range from the useful minimum to 300 psig.

8. A stationary nozzle for a gas turbine is designed for optimum expansion of gases from 1400° F and 60 psig to a pressure of 19 psig. Assuming that the inlet velocity is negligible, that the gases follow the air tables, that the energy conversion efficiency of the nozzle is 0.92, and that all inefficiencies occur in the diverging section, calculate

(*a*) The temperature and velocity of the discharge jet.

(*b*) The pressure, temperature, and velocity at the throat.

(*c*) The ratio of discharge area to throat area.

(*d*) The throat area required for a discharge rate of 0.010 lb per sec if the discharge coefficient is 0.98.

(*e*) The thrust in pounds developed on a blade against which the jet of part (*d*) is directed, if the blade is moving away from the nozzle at a velocity equal to 0.5 times the velocity of the jet, and if the blade causes the jet to be deflected through an angle of 180° without loss of velocity.

(*f*) The horse power developed by the blade of part (*e*).

9. A rocket engine comprises a vessel containing a propellant substance which develops gases at high temperature and pressure. The gases are discharged through a nozzle designed for optimum expansion to the pressure of the atmosphere. The chamber of such an engine operates at 4100° F and 275 psia, producing gases for which $\kappa = 1.28$, $M = 24$, and $C_p = 0.35$ Btu/(lb)(R°). If the energy conversion efficiency of the nozzle is 0.87, calculate, for discharge pressures of 1.0 and 0.1 atm:

(*a*) The discharge velocities and temperatures.

(*b*) The specific thrust in pounds per pound of flow rate per second.

(*c*) The temperatures, pressures, and velocities at the throat, assuming that all inefficiencies occur in the diverging section.

(*d*) The ratios of discharge areas to throat areas.

10. Calculate the stagnation temperature and pressure of a stream of carbon dioxide at 105° F and 20 psia flowing at a velocity of 800 ft per sec.

11. Air at a temperature of 0° F and a pressure of 6.0 psia enters a diffuser at a velocity of 1050 ft per sec where it is ideally decelerated to a velocity of 270 ft per sec. Calculate

(*a*) The temperature and pressure at the outlet of the diffuser.

(*b*) The ram efficiency of the diffuser.

12. Carbon dioxide enters an adiabatic, uniform duct at a Mach number of 0.05, a temperature of 150° F, and a pressure of 50 psia. Calculate the outlet temperature, pressure, and linear velocity if the friction parameter of the duct is such that the specific volume of the gas at the outlet is 8.0 times the inlet value. Also calculate the value of the friction parameter.

13. Air enters a 4-in.-diameter adiabatic transmission line at 110° F, 115 psia, and a Mach number of 0.03. Using Figs. 170 and 171, calculate the length of line in which the pressure will drop to 14.7 psia if the friction factor is assumed constant at 0.004. Also calculate the outlet temperature and velocity.

14. Air at 60° F and 58 psia is traveling in an adiabatic duct 8 in. in diameter at a velocity of 100 ft per sec. Calculate the limiting duct length and the corresponding pressure, temperature, and velocity of the outlet stream.

15. Solve problem 13 for the case of isothermal flow.

16. Solve problem 14 for the case of isothermal flow.

17. A large pressure vessel is filled with air at 35 psia and 80° F. Calculate the mass rate of flow of air from the vessel through a straight pipe 50 ft long and 1 in. in inside diameter discharging into the atmosphere (p_a = 14.7 psia). Also calculate the temperature at the outlet of the pipe. (Assume f_{avg} = 0.004.)

18. Solve problem 17 for a pipe length of 2.0 ft.

19. A water–water jet pump is to pump from a well at a depth of 65 ft and deliver it to the surface at a pressure of 40 psig. It is desired to maintain a flow rate ratio of 4.0. Assuming Fig. 173 to be applicable, calculate:

(*a*) The optimum mixer–jet area ratio.

(*b*) The required momentum per unit area of the primary jet.

(*c*) The pressure at the mixing section inlet.

(*d*) The static pressure of the primary water stream if its kinetic energy is negligible and the energy efficiency of the primary nozzle is 0.85.

20. Air at a pressure of 20 psig and a temperature of 215° F is to be used in an ejector to exhaust carbon monoxide from a vessel at a pressure of 1.0 atm and discharge it at a pressure of 1.1 atm. The energy efficiency of the primary nozzle is 0.90.

Neglecting changes in the densities of the low-pressure streams, and assuming Fig. 173 to be applicable, calculate the optimum mixer–jet area ratio and the maximum flow-rate ratio.

21. Saturated steam at a pressure of 200 psia is used in a constant-pressure mixing ejector to compress saturated water vapor from 4 psia to 14.7 psia. If the efficiencies of the ejector are the same as those of Illustration 8, calculate the number of pounds of steam required per pound of vapor compressed.

17

Vapor Power Plants

In order to produce useful work from the chemical energy stored in fuels, it is necessary, in our present state of knowledge, first to convert the chemical energy into heat by combustion. The heat released is transformed into mechanical work in an *engine*. This mechanical work may be used directly or may be converted into electric energy. The losses of useful energy accompanying these successive transformations provide great incentive for the development of processes for the efficient transformation of chemical energy directly into electric energy.

For transformation of chemical into mechanical energy two general methods are in widespread use. In *vapor-expansion engines* the working fluid is a vapor generated at high pressure in a fuel-fired boiler furnace. The expanded vapor is condensed and returned to the boiler to complete the cycle. The steam power plant is the familiar example of this method. In *internal-combustion engines* the combustion of the fuel is performed within the engine itself, and the products of combustion comprise the working fluid which motivates the engine.

A vapor-generation power plant comprises the following sequence of essential components: (1) furnace, (2) boiler, (3) engine, (4) condenser, (5) feedwater pump. The over-all efficiency and performance of the plant are determined by the individual components. The problems of combustion in furnaces are discussed in Chapter 11.

Cyclic Processes

A series of operations so conducted that all changes are periodically repeated in the same order is a *cyclic process*. This may comprise a sequence of either nonflow or flow operations or a combination of the two types, either repeated on the same substance or involving a new mass of substance for each cycle. For example, a reciprocating air compressor produces a repeated series of nonflow operations in which the conditions at a selected point of any cycle duplicate those at the same point of any other cycle. However, a new mass of gas is involved in each cycle, and a flow result is produced by the sequence of nonflow steps.

On the other hand, in a steam-power plant the medium may be continually recirculated through a series of operations such as feed-water heating, evaporation, superheating, expansion, and condensation. Such a cycle is essentially composed of flow operations but may involve nonflow steps such as the expansion of steam in a reciprocating engine.

Nonflow cyclic operations repeated on the same mass of substance have no present practical applications but are of great value in establishing the theoretical behavior of cyclic processes in general.

Reversible Cycles. A cyclic process that is wholly reversible must of necessity be composed of individual steps, each of which is in itself reversible. This requires that at no point in the cycle may there be degradation of higher forms of energy to heat through mechanical or fluid friction, nor may there be any other irreversible steps such as free expansion of a fluid or the transfer of heat under a finite temperature difference.

The Carnot Cycle. It has been previously stated that the maximum shaft work accomplished as a result of any change in state of a fluid is obtained when the change takes place reversibly without friction or turbulence. The Carnot engine has been conceived as a hypothetical device which gives this maximum achievement in a cyclic process and hence is used as a standard in evaluating the efficiency and performance of all actual mechanical cycles for transforming heat or internal energy into mechanical work.

In the Carnot engine, a gas is contained in a cylinder equipped with a frictionless piston connected to a frictionless receiver of mechanical energy. The cylinder walls and piston are impervious to heat, but the cylinder head is alternately placed in contact with a thermal insulator plate and with a highly conducting plate. In the first stage of the cycle, that of isothermal expansion, an amount of heat q_1 is supplied to the gas through the conducting cylinder head from a source of heat maintained at a constant temperature T_1. This heat flows into the fluid as a result of an infinitesimal temperature drop. The gas is allowed to expand isothermally and reversibly. The reversible work of expansion from state a to state b is $\int_a^b p\,dV$.

The first stage is followed by a reversible adiabatic or isentropic expansion which is accomplished theoretically by attaching the nonconducting head to the cylinder and permitting expansion to continue. Work of adiabatic expansion is performed at the expense of the internal energy when the fluid expands from state b to state c and is equal to $\int_b^c p\,dV$. In this stage the temperature decreases to T_2.

A receiver of heat is now provided by replacing the impervious cylinder head with a conducting plate in contact with the receiver at a constant temperature T_2. In the third stage, an isothermal compression is performed wherein work of compression $\int_d^c p\,dV$ is performed on the gas (work done by gas $= \int_c^d p\,dV$), and an amount of heat $-q_2$ flows into the receiver at T_2. In the fourth and last stage, the cylinder head is made nonconducting, and compression is continued under isentropic conditions from state d to the original state a with a reversible work of compression done on the gas equal to $\int_a^d p\,dV$ (work done by gas $= \int_d^a p\,dV$).

Since by completing the cycle the fluid is returned to its original state, its net internal energy change is zero. The same is true of all other point properties. Since no heat was received or lost during either adiabatic change, it follows from the law of conservation of energy that the net work done w_{net} must equal q_1+q_2. This is equal to the difference between the amount of heat received from the reservoir at temperature T_1 and the amount rejected to the receiver at temperature T_2. Thus:

$$w_{\text{net}} = \int_a^b p\,dV + \int_b^c p\,dV + \int_c^d p\,dV + \int_d^a p\,dV = \sum \int p\,dV = q_1 + q_2 \quad (1)$$

The change in entropy of the fluid during isothermal expansion is $\Delta S_1 = S_b - S_a = q_1/T_1$, and in isothermal compression is $\Delta S_2 = S_d - S_c = q_2/T_2$. Since the change in entropy over the entire cycle is zero, and the change during each reversible adiabatic stage is zero,

$$\Delta S_{\text{total}} = \Delta S_1 + \Delta S_2 = 0 \quad (2)$$

$$\frac{q_1}{T_1} + \frac{q_2}{T_2} = 0 \quad (3)$$

$$\frac{q_2}{q_1} = -\frac{T_2}{T_1} \quad (4)$$

$$\frac{q_2 + q_1}{q_1} = \frac{T_1 - T_2}{T_1} \quad (5)$$

Since $q_1 + q_2 = w_{\text{net}}$, it follows that

$$\frac{w_{\text{net}}}{q_1} = \frac{q_1 + q_2}{q_1} = \frac{T_1 - T_2}{T_1} \quad (6)$$

The ratio w_{net}/q_1 expressed by equation 6 is termed the *thermodynamic efficiency* of the cycle. It will be recognized that an efficiency of 100%

can be realized only when the temperature of the receiver is at absolute zero or heat is added at infinite temperature.

The operation and efficiency of the Carnot cycle are shown graphically by straight lines on the temperature–entropy diagram (Fig. 175). During the first stage, the reversible isothermal expansion at temperature T_1 is represented by the horizontal line *ab*; the heat removed from the source is represented by the area $S_a abS_b$ or $T_1 \Delta S_1$. During the second stage, the

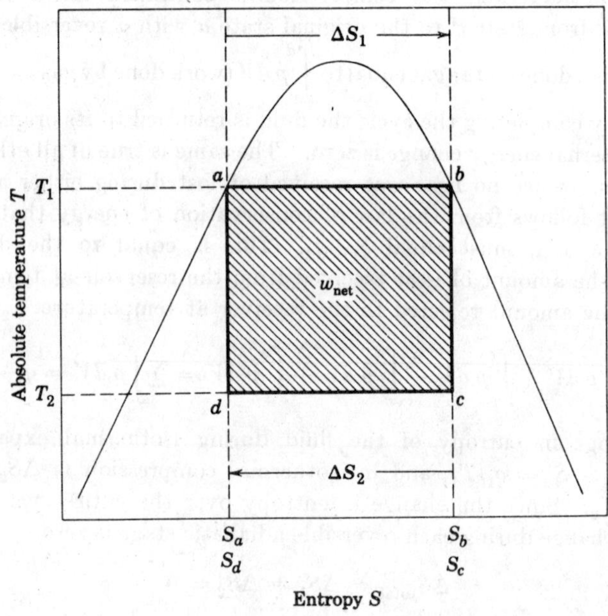

Fig. 175. Carnot cycle

reversible adiabatic expansion is represented by the vertical line *bc* with no change in entropy and with no heat added. In the third stage, heat is rejected at T_2 along horizontal line *cd* and is represented by area $S_a\ dcS_b$ or $-T_2 \Delta S_2$. In the last stage, the adiabatic compression moves along a vertical line *da* of constant entropy and with no heat rejected. The net work done w_{net} is represented by area *abcd* equal to q_1+q_2. The thermodynamic efficiency of the cycle is represented by the ratio of area *abcd* to area $abcS_bS_ad$.

The Carnot cycle may also be operated by starting with a saturated liquid and vaporizing it during the first stage. This condition is indicated on Fig. 175 by the curved line which represents the temperature–entropy relationship of the saturated vapor and liquid of the working fluid. On such a vapor cycle two phases exist in the cylinder at all points except a, where the system is completely liquefied, and b, where it is completely vaporized. Other methods of operation involve varying degrees of liquefaction at the various points of the cycle, but in all cases the energy transformations are expressed by equations 2 through 6.

The concept of the Carnot engine permits visualization of the significance of the *Carnot principle* and the second law of thermodynamics. The Carnot principle involves two basic conclusions which follow directly from equations 1 through 6 and the second law of thermodynamics.

1. No self-acting engine which operates by the absorption of heat at one constant temperature and by the rejection of heat at another lower constant temperature can be more efficient than an engine operating on a reversible cycle with absorption and rejection of heat at only these same two temperature levels. If such an engine of higher efficiency existed, it could be used to drive a reversible Carnot engine operating as a heat pump. The net result of the two engines would then be the transfer of heat from a lower to a higher temperature in contradiction of the second law.

2. The thermodynamic efficiencies of all reversible cycles which absorb heat at the same constant temperature and reject heat at the same constant lower temperature must be equal, regardless of the nature of the system or the fluid. This conclusion follows from equation 6, which expresses efficiency independent of system or fluid characteristics. Although this equation was derived by consideration of the Carnot engine, it is applicable to any other reversible engine operating between two constant temperatures.

Availability of Thermal Energy. The Carnot principle furnishes a basis for quantitatively expressing the availability of thermal energy in performing mechanical work. Thus, equation 6 expresses the maximum fraction of the heat energy q_1 that can be converted into work by an ideal machine when receiving heat from temperature T_1 and rejecting heat at temperature T_2. The quantity q_1+q_2 may thus be termed the available portion of q_1 with respect to temperature T_2. Similarly, q_2 is the unavailable portion with respect to temperature T_2. An expression for the fractional unavailability of heat energy is obtained by rearranging equation 6

$$\text{Fraction unavailable} = \frac{-q_2}{q_1} = \frac{T_2}{T_1} \tag{7}$$

The temperature of rejection is ordinarily fixed by the temperature of the atmosphere or of the available cooling water. As the temperature level of the energy source is reduced, the availability of its heat energy is decreased, and, when the temperature of the source becomes equal to the rejection temperature, heat energy becomes wholly unavailable even in an ideal engine.

Any system undergoing a state change may be considered as a source of heat for conversion into work by means of a Carnot engine where heat is rejected to the surrounding atmosphere or to the available cooling water at temperature T_o. If $-q$ is the heat given up by the system and transferred to the engine, then $-q = q_1$ and equation 6 becomes

$$w = -q\left(\frac{T_1 \quad T_o}{T_1}\right) \tag{8}$$

Where the system changes in temperature from T_1 to T_o,

$$w = -\int_{T_1}^{T_o}\left(\frac{T_1 - T_o}{T_1}\right) dq \tag{9}$$

or

$$w = -\left(q - T_o\int_{T_1}^{T_o}\frac{dq}{T_1}\right) = -\left[q - T_o\,\Delta S\right]_{T_1}^{T_o} \tag{10}$$

At constant pressure the heat lost by the system is equal to its decrease in enthalpy, $-q = -\Delta H$

$$w = -\left[\Delta H - T_o\,\Delta S\right]_{T_1}^{T_o} \tag{11}$$

Since both H and S are point properties and T_o is constant, the term $-[\Delta H - T_o\,\Delta S]$ is also a point property, representing the available energy B of the process, or the *availability function*,

$$B = -\left[\Delta H - T_o\,\Delta S\right]_{T_1}^{T_o} \tag{12}$$

For an actual change of a flow system from temperature T_1 to T_2 where changes in kinetic and potential energies are negligible, the change in the availability function is

$$\Delta B = \left[\Delta H - T_o\,\Delta S\right]_{T_1}^{T_2} \tag{13}$$

Where changes of kinetic and potential energy occur, these forms of energy are entirely available for work; under these circumstances the change in the availability function in a flow system is

$$\Delta B = \left[\Delta H - T_o\,\Delta S + \frac{\Delta u^2}{2g_c} + \frac{g}{g_c}\,\Delta Z\right]_{T_1}^{T_2} \tag{14}$$

Illustration 1. Calculate the available energy lost in the free expansion of ethylene gas through a throttling valve from 50 atm and 564° R to 1 atm under adiabatic-flow conditions relative to a temperature T_0 of 540° R on the basis of 1 lb-mole of ethylene gas.

From Table 54, for free expansion:

$$\Delta\text{H} = 0 \quad \text{and} \quad \Delta\text{s} = 7.05 \text{ Btu/(lb-mole)(R°)}$$

Hence $$T_0\,\Delta\text{s} = 540(7.05) = 3807 \text{ Btu per lb-mole}$$

or $$\Delta\text{B} = (\Delta\text{H} - T_0\,\Delta\text{s}) = -3807 \text{ Btu per lb-mole}$$

The decrease in the availability function amounting to 3807 Btu per lb-mole represents the energy potentially available for useful work which is lost by the system in free expansion.

Ideal Power-Plant Efficiencies

The great amount of chemical energy stored in fuels is released by the highly irreversible process of combustion. As yet no practical method has been devised for the reversible release of this energy as work. The heat generated by combustion at high temperature is transferred to a working fluid in an engine at a much lower temperature and converted more or less irreversibly into mechanical work. The resultant losses caused by these irreversible operations are shown in Illustration 2.

Illustration 2. Calculate the theoretical work obtained from the oxidation of 1 lb of carbon (graphite) to carbon dioxide at a constant pressure of 1 atm: (1) with reversible oxidation, (2) with high-temperature combustion supplying heat to an infinite number of Carnot engines, (3) with generation of superheated steam at 1050° F for operation of a single Carnot engine rejecting heat at 77° F.

Assume that, in each instance, the oxygen used is obtained from air at 25° C and 1 atm, and that only the theoretical amount required for oxidation of the carbon to CO_2 is supplied.

Case 1. The oxidation is carried out reversibly, at a constant temperature of 25° C and a constant pressure of 1 atm. Under these conditions,

$$w_f = -\Delta G° = -\Delta H° + T\,\Delta S°$$

$$\Delta H° = (\Delta H°_f)_{CO_2} = -94{,}051.8 \text{ g-cal per g-atom carbon}$$

From data presented in Chapter 25, it may be shown that the entropy change accompanying the combustion is

$$\Delta S° = 51.061 - (1.3609 + 49.003) = 0.6971 \text{ g-cal/(g-atom carbon)(K°)}$$

Hence,

$$w_f = -(-94{,}051.8) + (298.16)(0.6971) = 94{,}259.8 \text{ g-cal per g-atom carbon or}$$
$$4.140 \text{ kw-hr per lb carbon}$$

This result represents the maximum amount of work obtainable where no irreversible steps are involved.

Case 2. The carbon is oxidized adiabatically, thus producing combustion products at an elevated temperature. The sensible heat of the combustion products is used to supply heat to an infinite number of Carnot engines operating at successively lower temperatures, but with each engine rejecting heat at 25° C. The combustion products are cooled to a final temperature of 25° C.

$$\text{Work done by each Carnot engine} = d'w = d'q\left(\frac{T - T_o}{T}\right) = d'q\left(1 - \frac{T_o}{T}\right)$$

$$d'q = -c_p\,dT$$

$$d'w = -c_p\,dT + T_o\frac{c_p}{T}\,dT$$

$$w_{\text{total}} = -\int_{T_f}^{T_o} c_p\,dT + T_o\int_{T_f}^{T_o}\frac{c_p}{T}\,dT = \int_{T_o}^{T_f} c_p\,dT - T_o\int_{T_o}^{T_f}\frac{c_p}{T}\,dT$$

where T_f is the flame temperature of the combustion products, T_o equals 298.16° K, and c_p is the heat capacity of the gaseous products per gram-atom of carbon burned.

Under adiabatic conditions,

$$\int_{T_o}^{T_f} c_p\,dT = -(\Delta H_f)_{CO_2} = -(-94{,}051.8) = 94{,}051.8 \text{ g-cal per g-atom carbon}$$

By procedures demonstrated in Chapter 9, page 354, the flame temperature was found to be 2458° K.

Integration between 298.16 and 2458° K, using the high-range heat-capacity equations for CO_2 and N_2 from Table C, p. xxiii, Vol. I, gives

$$\int_{T_o}^{T_f}\frac{c_p}{T}\,dT = 87.413 \text{ g-cal/(g-atom carbon)(K°)}$$

$$w_{\text{total}} = 94{,}051.8 - (298.16)87.413 = 68{,}003 \text{ g-cal per g-atom carbon}$$

$$= 2.987 \text{ kw-hr per lb carbon}$$

$$\text{Thermal efficiency} = \frac{\text{work output}}{\text{heat absorbed}} = \frac{68{,}003}{94{,}051.8} = 0.723$$

In comparing this result with the preceding, it will be noted that the work output is much less. This is because the process of combustion is irreversible in this case.

Case 3. The carbon is burned with air as in case 2, but the heat, before being employed to develop work, is transferred to a boiler generating superheated steam at 1050° F(838.7° K). The steam is used to supply heat to a Carnot engine that rejects heat at 77° F (25° C).

Sensible heat in combustion products above 25° C	94,052 g-cal
Heat lost in stack gases at 1050° F	20,551
Heat transferred to boiler, and supplied to Carnot engine	73,501 g-cal

$$\text{Carnot efficiency} = \frac{1509.7 - 536.7}{1509.7} = 0.6445$$

$$\text{Work output} = (0.6445)(73{,}501) = 47{,}371 \text{ g-cal per g-atom carbon}$$

$$= 2.081 \text{ kw-hr per lb carbon}$$

As compared with case 2, an additional irreversible step is involved, namely, the transfer of heat to the boiler, with consequent reduction of work output. Furthermore, no use is made of the stack gases discarded at 1050° F.

SUMMARY

	Irreversible Steps	Kw-hr per lb Carbon	% of Maximum	Heat Rate,* Btu per kw-hr
Case 1	None	4.140	100.0	—
Case 2	Combustion	2.987	72.2	4716
Case 3	Combustion and heat transfer to boiler	2.081	50.3	6770

$$* \text{ Heat rate} = \frac{\text{Btu of heat developed in combustion}}{\text{kw-hr of mechanical work developed by engine}}$$

Steam Turbines

The axial-flow steam turbine comprises a group of circularly distributed stationary nozzles which direct steam jets on *blades* or *buckets* mounted radially on the periphery of a rotating wheel as shown in Fig. 176. Ordinarily the blades are short in proportion to the radius of the wheel, and the nozzles are approximately rectangular in cross section, to conform to approximately rectangular passages between the blades. Radial views of turbine nozzles and blades are shown diagrammatically in Fig. 177. Multiple expansions are obtained by use of several stages in series, with the exhaust from the blades of one stage flowing directly into the nozzles of the next. The wheels of all stages are mounted on a single shaft, and the nozzles of all stages are supported from a continuous housing. A very compact machine results, which can be built economically with ten or more stages for optimum use of high-pressure steam and vacuum exhaust.

Since condensate droplets would seriously erode the high-velocity nozzles and blades, turbines must be operated under such conditions that the exhaust steam does not contain more than 5 to 10% of liquid.

Three types of turbines are in general use, differing in the design and arrangement of the blades. In the simple *impulse* or *de Laval* turbine (Fig. 177*a*) the blades are ideally designed to produce only a reversal in direction of steam flow without drop in pressure. In this type the entire pressure and enthalpy drops and the velocity increase produced by the stage occur in the nozzles. The wheel is propelled by the thrust resulting from the change in momentum accompanying the reversal of direction of the high-velocity steam. Multistage applications of this type are termed *Rateau* turbines.

The *two-row impulse* or *Curtis* turbine stage, Fig. 177*b*, is designed to permit efficient operation at lower blade and wheel speeds than are required

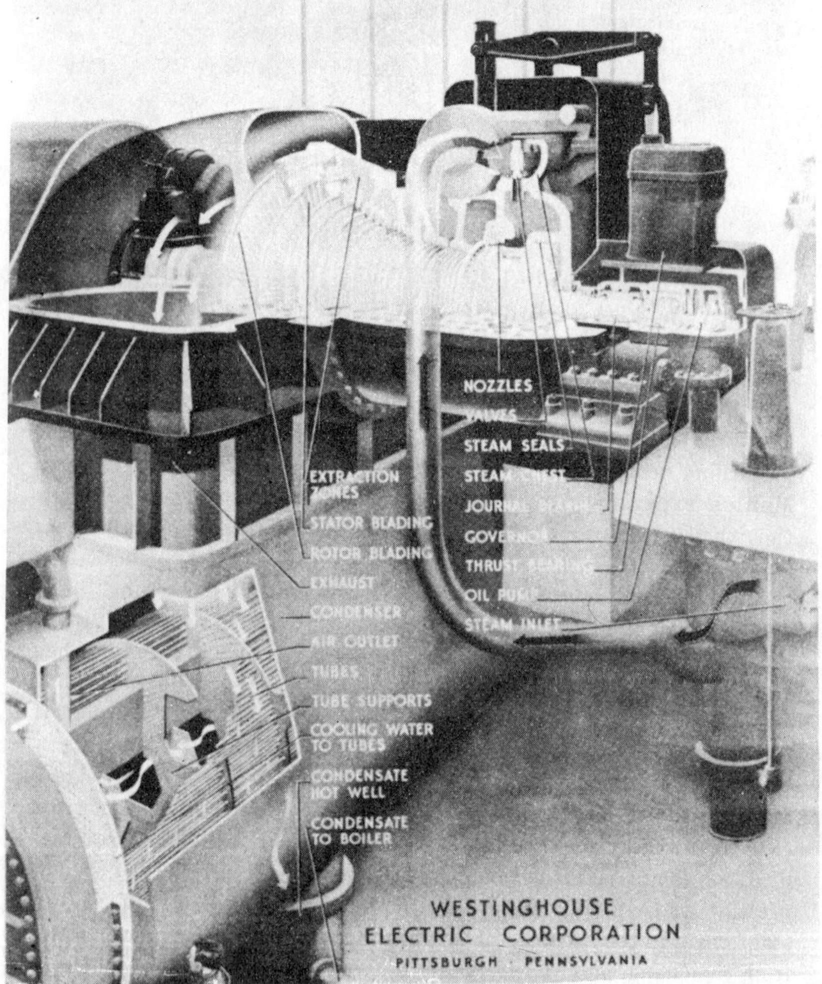

Fig. 176. Steam turbine and condenser

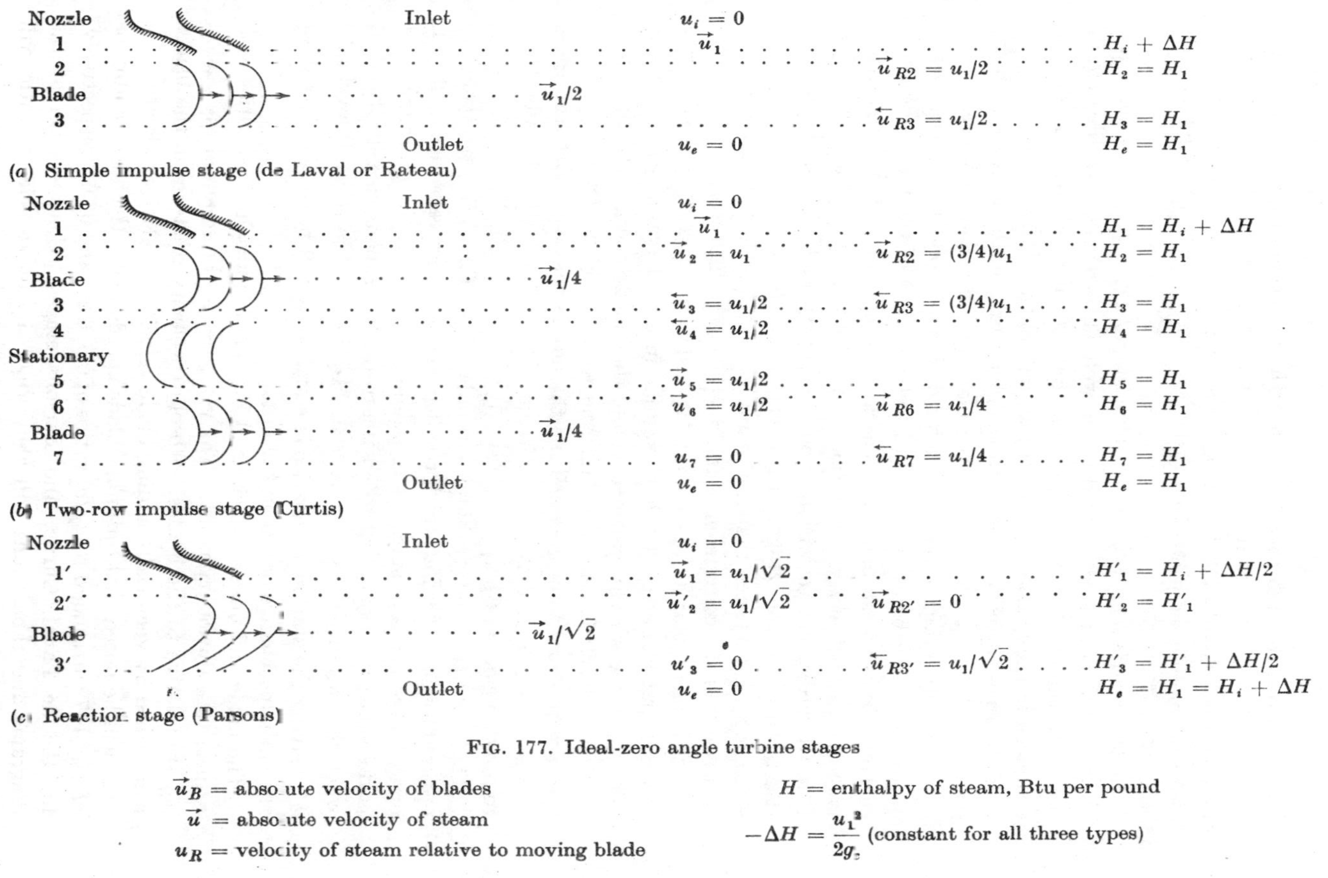

Fig. 177. Ideal-zero angle turbine stages

$\vec{u}_B$ = absolute velocity of blades
$\vec{u}$ = absolute velocity of steam
u_R = velocity of steam relative to moving blade
H = enthalpy of steam, Btu per pound
$-\Delta H = \frac{u_1^2}{2g_c}$ (constant for all three types)

in the Rateau type. All expansion ideally occurs in the nozzles with no pressure drop through the blades, but the direction of flow of the high-velocity steam is reversed twice by employing three rows of blades, two moving and one stationary. Thus, the total thrust is divided between two wheels, and blade speeds may be lower, as discussed in the following section.

In the *reaction* or *Parsons* turbine stage, Fig. 177*c*, the blades are designed to produce pressure drop as well as to reverse the direction of flow. In the ideal reaction stage, one half of the total enthalpy drop is produced in the nozzles, and the other half in the blades.

Ideal Zero-Angle Turbines. In the ideal turbine, the nozzle directs its jet in the direction of motion of the blade and the blade produces a 180° reversal of direction. Actually such a design is impossible, but the *nozzle angles* of real turbines are so small that analysis of the ideal zero-angle turbine serves to establish the general characteristics of turbine performance. Ideal velocity and enthalpy distributions are tabulated in Fig. 177, to correspond to various points of entry and exit in the flow of steam through the stage. In every case conditions for maximum efficiency must be such that the steam leaves the stage with zero velocity as a result of the conversion of its kinetic energy into work.

It is evident that an optimum relationship must exist between nozzle velocity and blade speed. For example in a de Laval stage no work is done when the wheel is stationary, and the kinetic energy is dissipated in friction and exit losses. Similarly, no work would be done if the blade speed were equal to or greater than the nozzle velocity. Between these two extremes there exists an optimum value of the ratio of blade speed to nozzle velocity, u_B/u_1.

The velocities and enthalpies tabulated in Fig. 177 correspond to optimum values of u_B/u_1 which result in zero exhaust velocities. In each of the three types of stages it is assumed that the enthalpy change ΔH is the same, corresponding to an isentropic full-expansion velocity of u_1 when the entrance velocity u_i is zero. In the reaction stage, the velocity u'_1 leaving the stationary nozzles is equal to $u_1/\sqrt{2}$ because only half of the enthalpy change occurs in the nozzles.

The blade and nozzle velocities in Fig. 177 are absolute velocities with respect to a stationary point. The velocities of the steam in the blades, u_R are relative to the blades. Absolute velocity of steam in the blades is obtained by vectorially adding blade velocity to relative velocity.

It will be noted that optimum blade speeds under the ideal conditions of Fig. 177 correspond to u_B/u_1 ratios of 0.5, 0.25, and 0.707, respectively, for the de Laval, Curtis, and Parsons stages. Consideration of the diagrams and tabulations will show that any other velocity ratio will result in exit velocities greater than zero with corresponding loss in efficiency. The relationships between efficiency and velocity ratio are shown Fig. 178.[1] Although these curves were derived for zero-angle ideal turbines, they may be taken as representing the *propulsion efficiencies* η_p of real turbines for analyzing performance characteristics at varying speeds.

In Fig. 178 are also plotted the relative torques of the three types of

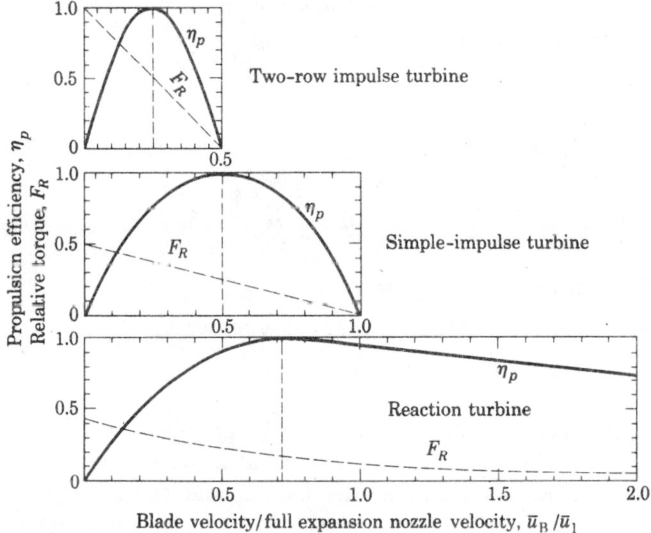

FIG. 178. Efficiency and torque of ideal turbines

turbine stages, assuming the wheel diameters to be equal. The force F on a turbine wheel is equal to $W\ \Delta u/g_c$ where W is the mass rate of steam flow and Δu is the vectorial change in velocity. Since all three types produce the same power at optimum velocity ratios, the corresponding torques are inversely proportional to the optimum blade velocities. Of particular interest is the variation of torque with speed. For both impulse types the torque varies linearly with speed, increasing from zero at twice the optimum speed to 2.0 times optimum torque when the turbine is stalled at zero speed. In the reaction stage, the torque at stall is 2.42 times the optimum torque, and, at speeds above the optimum, torque drops off less rapidly. This torque multiplication characteristic of the turbine at low speeds is particularly advantageous in variable-speed

[1] J. H. Keenan, *Thermodynamics*, p. 133, John Wiley & Sons, (1941), with permission.

applications such as the propulsion of vehicles where high-starting torques are required.

It is evident from Fig. 178 that the reaction turbine has inherent advantages for variable-speed operation, in that it gives maximum torque multiplication and maintains efficiency and torque over wider speed ranges.

Turbine Performance. The indicated efficiency η_I of the steam turbine is the ratio of the work done on the wheel to the isentropic enthalpy drop accompanying expansion of the steam from the inlet state to the outlet pressure. Thus,

$$\eta_I = \frac{w_S}{H_i - H_e} \tag{15}$$

or, neglecting heat losses,

$$\eta_I = \frac{H_i - H'_e}{H_i - H_e}$$

where H_i = enthalpy of inlet steam

H_e = isentropic enthalpy at exhaust pressure

H'_e = actual enthalpy of exhaust

The inefficiencies in the turbine may be grouped into three classes. The nozzle and blade efficiency η_B is dependent on the nozzle angle and efficiency, the friction losses in the blades, and the leakage past the blades and nozzles. The mechanical efficiency η_M is determined by mechanical friction and is generally high. The propulsion efficiency is primarily a function of velocity ratio, as indicated in Fig. 178. Thus, the over-all efficiency is given by

$$\eta_T = \eta_I \eta_M = \eta_B \eta_P \eta_M \tag{16}$$

An alternative method for taking into account mechanical and heat losses is to express them as Btu lost per pound of steam passing through the turbine. Such losses typically range from 10 to 20 Btu per lb.

In general, fluid friction losses are most serious in the Curtis turbine, whereas the leakage problem is greatest in the Parsons type. When cost, flexibility, and efficiency are all considered, each type has advantages for certain applications. Actual turbines are generally combinations of the impulse and reaction types. Some expansion is frequently provided for in the blades of an impulse turbine, and a multistage turbine may combine high-pressure impulse stages with low-pressure reaction stages.

The power output of simple turbines is generally regulated by throttling the intake steam with resultant loss in efficiency. The forms of the *HS condition* curves of throttled and unthrottled multistage turbines are indicated in Fig. 179. It will be noted that the curve representing the condition of the steam in the turbine is a series of steps corresponding to movement through the nozzles and blades. In an impulse stage the enthalpy actually increases in the blades at substantially constant pressure

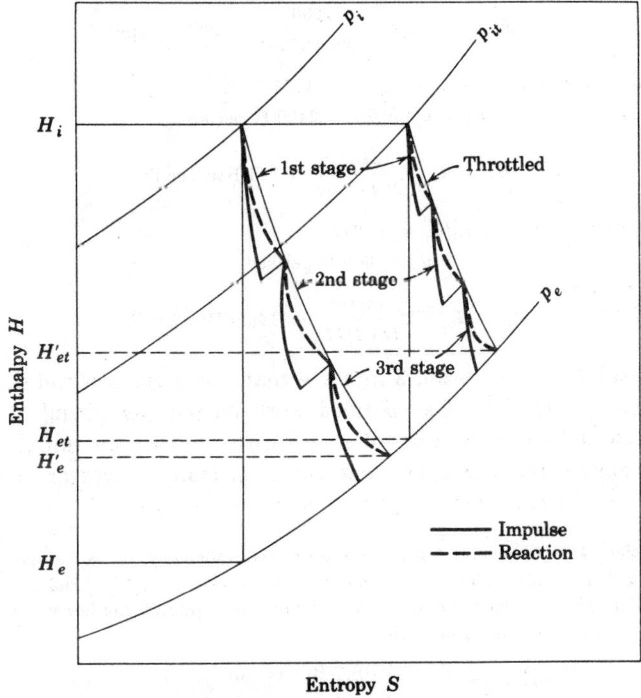

FIG. 179. Condition curves for three-stage turbine

owing to friction. The reduction in work and efficiency resulting from throttling to an intake pressure p_{it} are indicated by the difference between $(H_i-H'_e)$ and $(H_i-H'_{et})$ for the throttled operation.

An alternative method of regulation used in multistage turbines is to shut off the steam flow to part of the nozzles in the first stage. The nozzles are connected in groups, each of which has an independent shut-off valve. In this manner, throttling losses are eliminated, but efficiencies will still drop at part load conditions as a result of increased proportions of friction and leakage losses.

Illustration 3. The peripheral velocity of a turbine wheel of specified diameter is limited by structural stress considerations to 600 ft per sec. Assuming ideal zero-angle performance, calculate the enthalpy drop per stage at the optimum u_B/u_1 ratio for each of the following types:

(*a*) Simple impulse.
(*b*) Two-row impulse.
(*c*) Reaction.

(*a*) Simple impulse, optimum $u_B/u_1 = 0.5$:

$$u_1 = 600/0.5 = 1200 \text{ ft per sec}$$

$$H_i - H_e = u_1^2/2g_c = \frac{(1200)^2}{(64.4)(778)} = 28.7 \text{ Btu per lb}$$

(*b*) Two-row impulse, optimum $u_B/u_1 = 0.25$:

$$u_1 = 600/0.25 = 2400 \text{ ft per sec}$$

$$H_i - H_e = \frac{(2400)^2}{(64.4)778} = 115 \text{ Btu per lb}$$

(*c*) Reaction, optimum $u_B/u_1 = 0.707$:

$$u'_i = 600 \text{ ft per sec}$$

$$H_i - H_e = 2\,\frac{(600)^2}{(64.4)778} = 14.39 \text{ Btu per lb}$$

The results of Illustration 3 indicate that, for a given wheel diameter and speed, production of a specified work output per pound of steam requires the minimum number of stages with the two-row impulse type. For the simple impulse type, four times as many stages are required ideally, and, for the reaction type, eight times as many.

Illustration 4. A turbine having an indicated efficiency of 70% operates on steam at 600 psia and 1000° F, and exhausts at a pressure of 1.0 psia. Calculate the superheat of the exhaust steam and the steam rate in pounds per horsepower-hour. From the steam tables or Mollier chart:

$$H_i = 1516.7 \text{ Btu per lb}$$
$$S_i = 1.7147 \text{ Btu/(lb)(R°)}$$

Ideal isentropic expansion would result in saturated exhaust having a quality of 85.7% and the following properties:

$$H_e = 958 \text{ Btu per lb}$$
$$S_e = 1.7147 \text{ Btu/(lb)(R°)}$$
$$H_i - H_e = 1516.7 - 958 = 558.7 \text{ Btu per lb}$$
$$H'_e = H_i - \eta_I(H_i - H_e) = 1516.7 - (0.70)(558.7) = 1125.6 \text{ Btu per lb}$$

This enthalpy corresponds to a temperature of 148.7° F, or a superheat of 47° F above the saturation temperature of 101.74° F.

$$\text{Steam rate} = \frac{2545}{H_i - H'_e} = \frac{2545}{391} = 6.51 \text{ lb per hp-hr}$$

To obtain the low steam rate of Illustration 4 on the basis of the specification of Illustration 3 would require from 4 to 27 stages, depending on the type used.

Power-Plant Cycles

In a steam power plant, part of the heating value of the fuel is transferred to water in a boiler; the steam generated expands against the piston of an engine or against the blades of a turbine. The expanded vapors either

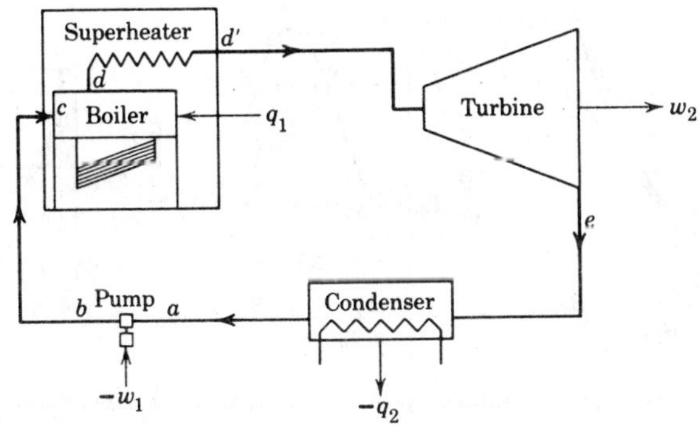

FIG. 180. Rankine-cycle power plant

are exhausted to the atmosphere and discarded, or they may be condensed in a low-temperature heat exchanger, and the condensate pumped back to the boiler for re-use. A flow diagram of these operations is shown in Fig. 180. When both the expansion of the steam and pumping of the condensate occur reversibly, the *Rankine* cycle is followed.

Rankine Cycle. This cycle differs from the Carnot vapor cycle, in that each stage is carried out in a different apparatus, and particularly in that condensation is completed at the low temperature and pressure of the condenser instead of by isentropic compression of the mixed phases. The condensate is pumped back to the boiler to mix irreversibly with the hot liquid already present.

The changes taking place in a Rankine cycle operating without superheating are shown in the *HS*, *TS*, and *pV* diagrams of Fig. 181. In the first step, the feed water is pumped from pressure p_2 to p_1 with an increase in enthalpy from H_a to H_b, as shown on the *HS* diagram, and with a negligible rise in temperature, shown as a to b, on the *TS* diagram. The water enters the boiler subcooled: that is, at a temperature below that corresponding to its saturation pressure. The entropy remains constant during pumping, since the process is assumed to be isentropic, as indicated by the vertical line ab on the HS diagram.

The work done on the feed water in pumping is equal to $\int_{p_a}^{p_b} V_f\,dp$. Since water is nearly incompressible, the work input becomes

$$-w_1 = (p_b - p_a)V_f \tag{17}$$

where V_f is the average specific volume of the feedwater. The work of pumping is represented on the pV diagram as the area $mban$. If the

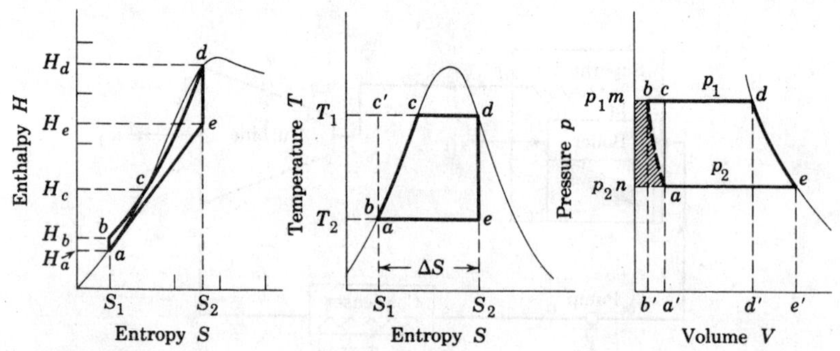

FIG. 181. Rankine-cycle diagrams, without superheating

pumping operation is reversible, and changes in kinetic and potential energies are negligible, the work done on the fluid is equal to its increase in enthalpy, or

$$-w_1 = H_b - H_a \tag{18}$$

In the second step, the water is heated along line bcd, becomes saturated at point c and temperature T_1, and then vaporizes at constant temperature T_1 and pressure p_1 to form saturated vapor at d. The TS diagram shows the wide departure from isentropic conditions in heating the liquid from T_2 to T_1, as represented by the slope of the line bc and the area bcc'. The gain in enthalpy is represented by $H_d - H_b$ on the HS diagram. If kinetic-energy terms are neglected, the heat supplied is equal to $H_d - H_b$ or $\int_1^2 T\,dS$, which is represented on the TS diagram by the area $S_1abcdeS_2$. Thus,

$$q_1 = H_d - H_b = \int_{S_1}^{S_2} T\,dS \tag{19}$$

In the third stage, the vapor expands isentropically from d to e. If the expansion is a flow process with negligible changes in kinetic energy, the work done is equal to the decrease in enthalpy as represented by $H_d - H_e$

on the HS diagram. On the pV diagram the work done during this stage of isentropic expansion is represented by the area *mden*.

In the condensation stage, the enthalpy loss is represented by $H_e - H_a$ on the HS diagram and as area S_1aeS_2 on the TS diagram, both of which equal the heat $-q_2$ rejected by the fluid.

The net work of the cycle represents the engine work minus the small work input to the pump. Since the process is cyclic, the net work done is equal to the heat absorbed less the heat rejected if kinetic-energy changes are negligible. Thus,

$$w_{\text{net}} = -\int_{p_1}^{p_2} V\,dp - (p_1 - p_2)V_f = (H_d - H_e) - (H_b - H_a) = q_1 + q_2 \tag{20}$$

The net work done during the cycle is represented by $(H_d - H_e) - (H_b - H_a)$ on the HS diagram, by the area *abde* on the pV diagram, and by the area *abcde* on the TS diagram.

The thermodynamic efficiency of the Rankine cycle is represented by

$$\frac{w_{\text{net}}}{q_1} = \frac{q_1 + q_2}{q_1} = \frac{(H_d - H_e) - (H_b - H_a)}{H_d - H_b} \tag{21}$$

or, if pump work is neglected,

$$\frac{w_{\text{net}}}{q_1} = \frac{H_d - H_e}{H_d - H_a} \tag{22}$$

The significance of the efficiency of the Rankine cycle is visualized from a study of the TS diagram where it is represented by the ratio of the area *abcde* to the area S_1bcdeS_2. Any change in conditions that will enlarge the former area with respect to the latter will improve the efficiency. It is evident that the efficiency of the Rankine cycle is lower than that of the Carnot cycle operating between the same temperature

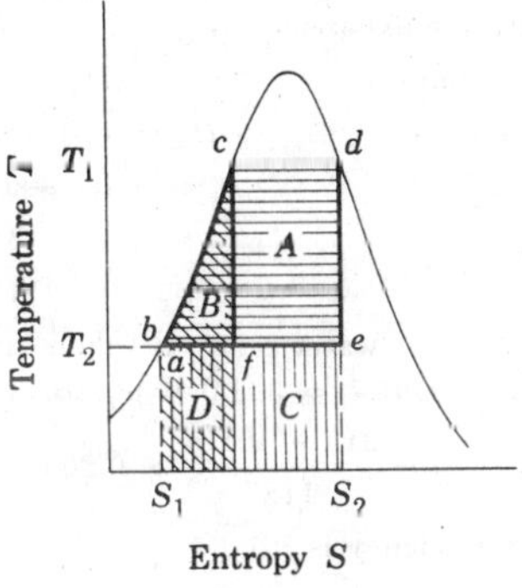

FIG. 182. Comparison of Carnot and Rankine cycles

levels. This results from the fact that, in the Rankine cycle, a portion of the heat absorbed is delivered to the process at intermediate temperatures along the line bc, representing the heating of the liquid to saturation at temperature T_1. This difference is shown in Fig. 182 where the efficiency of the Carnot cycle is represented as $A/(A+C)$ and of the Rankine cycle as $(A+B)/(A+B+D+C)$.

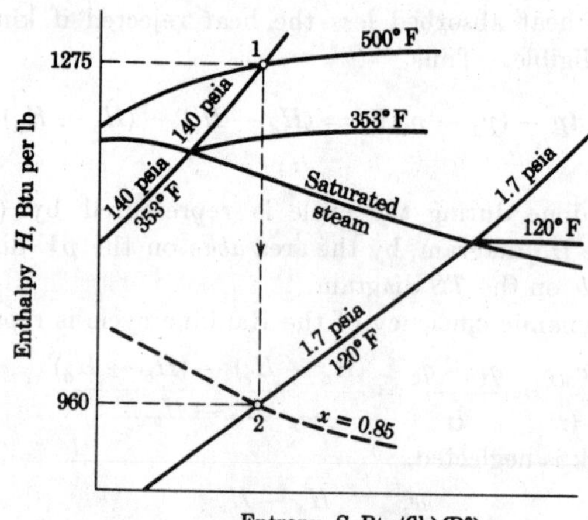

Fig. 183. Mollier diagram for water vapor

Illustration 5. Calculate the thermodynamic efficiency of an ideal Rankine cycle, operated with steam at a pressure of 140 psia and a temperature of 500° F. The pressure in the condenser is 1.7 psia, and the temperature of the condensate is 120° F. Enthalpies are obtained from a Mollier diagram as shown in Fig. 183.

Solution:

Enthalpy before expansion	$H_1 =$	1275 Btu per lb
Enthalpy after isentropic expansion	$H_2 =$	960 Btu per lb
Ideal work done by expansion	$H_1 - H_2 =$	315 Btu per lb
Enthalpy of condensate,	$H_w = 120 - 32 =$	88 Btu per lb
Heat added in boiler,	$H_1 - H_w =$	1187 Btu per lb

Thermodynamic efficiency, $\eta = \dfrac{315}{1187} = 0.265$

If the work added by the feed-water pump is taken into account, $(p_1 - p_0)/\rho_w = (140 - 1.7)(144/61.7) = 323$ ft-lb per lb, or $323/778 = 0.415$ Btu per lb

$$\eta = \frac{315 - 0.415}{1187} = 0.265$$

The corresponding Carnot efficiency is

$$\frac{960 - 580}{960} = 0.396$$

Improvement of Efficiency of Idealized Cycles. It has already been pointed out that, because of the irreversible character of combustion and of heat transfer between the combustion gases and the fluid in the boiler, it is impossible to convert all the heat developed by combustion into mechanical work. The following methods improve the efficiency of the ideal Rankine cycle, and bring about an increase in the over-all efficiency of any actual steam generating plant.

(*a*) *Decreasing the Exhaust Pressure from* p_o *to* p'_o (Fig. 184). This increases the thermodynamic efficiency from the ratio $A/(A+B+D)$ to

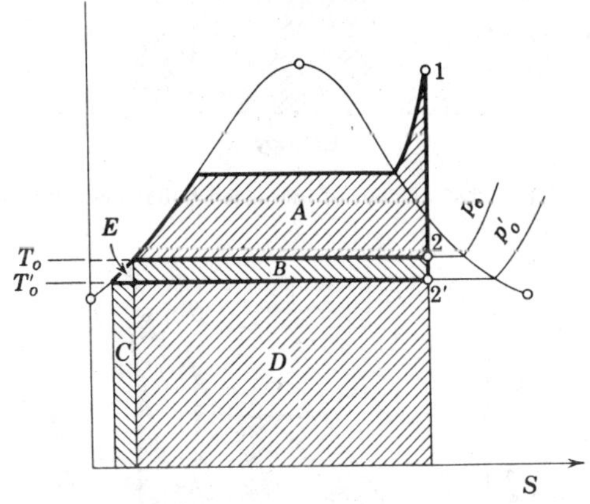

FIG. 184. Effect of decreased condenser pressure

the ratio $(A+B+E)/(A+B+E+C+D)$ as represented by the areas in Fig. 184. The lower temperature T'_o corresponding to pressure p'_o is limited by the temperature of the cooling water available for the condenser and by the size of the condenser.

(*b*) *Increasing the Inlet Pressure* from p_1 to p'_1 (Fig. 185). This increases the thermodynamic efficiency from $(A+B)/(A+B+D+E)$ to $(C+A)/(C+A+D)$. The highest pressure of practical interest for steam is approximately 5000 psia, which is above the critical pressure, 3206 psia.

(*c*) *Reheating.* In order to reduce the moisture content of the expanded steam to a safe limit, *reheating* after partial expansion may be employed as shown by line 23 of Fig. 186. While a reduction of moisture content in expansion has beneficial effects on engine performance, the efficiency of an idealized cycle with reheating generally is lower than without reheating.

(*d*) *Increasing the Superheat Temperature* from T_1 to T'_1 (Fig. 187), increases the efficiency from $A/(A+C)$ to $(A+B)/(A+B+C+D)$. The

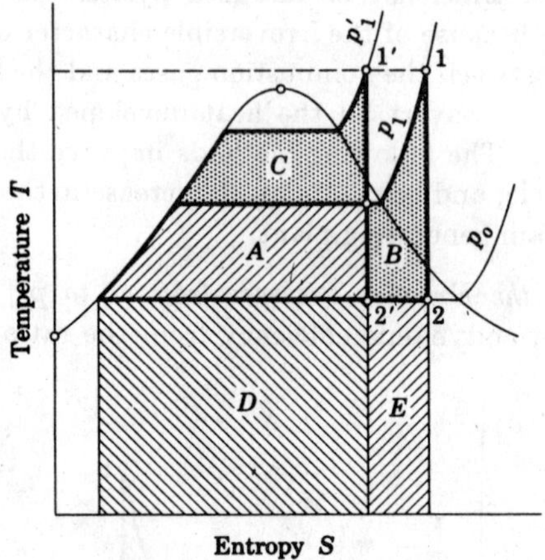

FIG. 185. Effect of increased boiler pressure

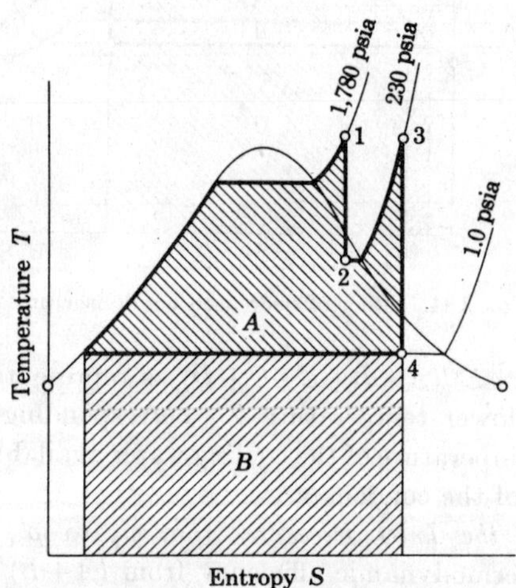

FIG. 186. Cycle with reheating of the steam after partial expansion

highest temperature is limited by the strength of the available materials of construction; about 1050° F is at present the upper practical limit.

(e) *Regenerative Heating of Feed Water.* If the condensate at condenser temperature is pumped directly into the boiler, it is heated to boiler temperature in an irreversible manner. If a steam turbine were constructed with a hollow casing through which the condensate passed in counterflow with respect to the steam within the turbine, and if

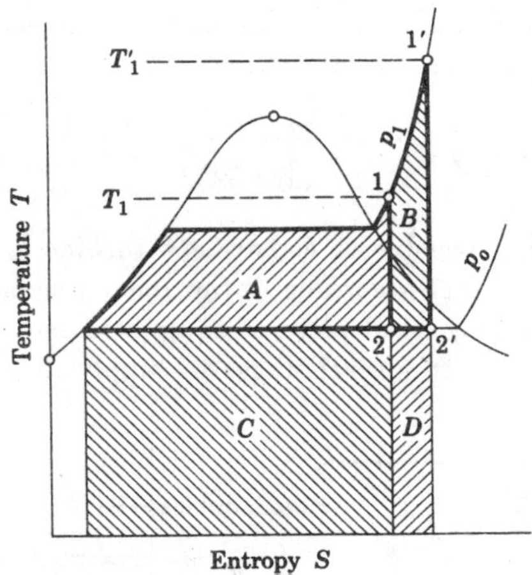

FIG. 187. Effect of increased superheating

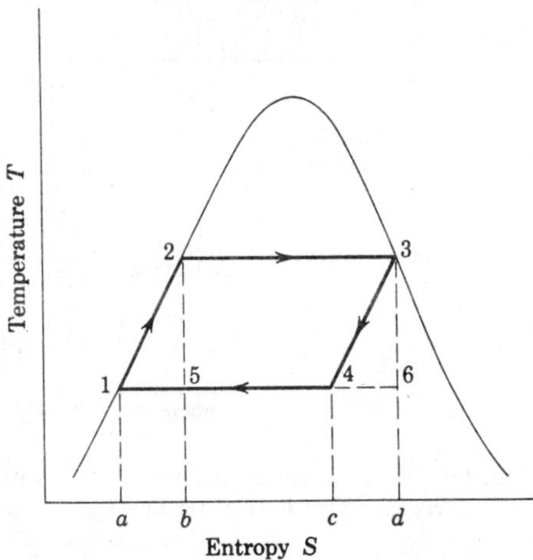

FIG. 188. Regenerative feed-water heating

infinitesimal temperature gradients existed between the water and the steam across the separating wall, the feed water would be reheated to boiler temperature reversibly along path 12 (Fig. 188). The steam, while expanding reversibly in the turbine, would decrease in entropy because of the transfer of heat to the condensate and its path would be along line 34. Since the heat absorbed by the feed water will equal the heat given up by the steam, line 34 will be parallel to line 12, and area

$c43d$ will equal area $a12b$. Thus, the only heat received from outside sources is shown by area $b23d$ and the only heat rejected to the outside is shown by the area $a14c$ (equal to area $b56d$):

$$\text{Ideal cycle efficiency} = \frac{\text{area } b23d - \text{area } b56d}{\text{area } b23d} = \text{Carnot efficiency}$$

Actually, it is not feasible to construct a turbine for transfer of heat from steam to feed water according to the above scheme. However, the action of such a hypothetical piece of equipment may be approximated by bleeding off steam at several stages to supply a series of feed-water

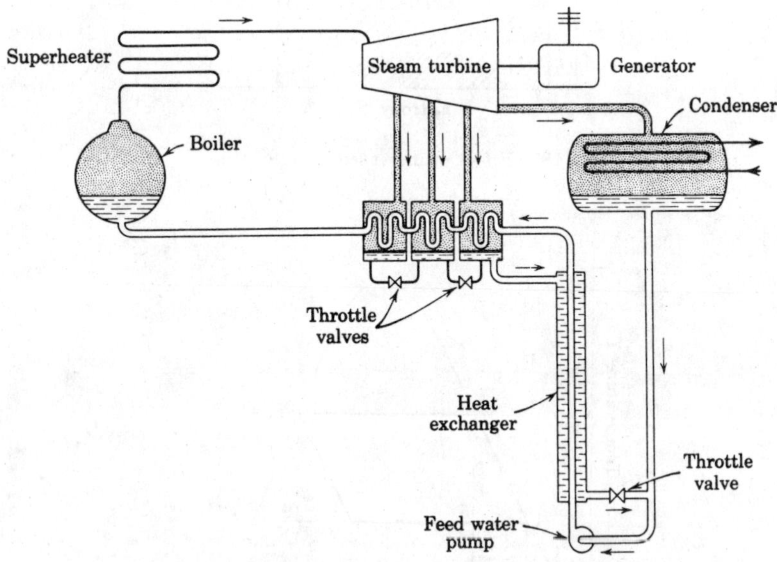

Fig. 189. Diagram for steam power plant with regenerative feed-water heating

heaters, each operating at a different temperature, as shown in Fig. 189. The operation of such a plant cannot conveniently be shown on a temperature–entropy diagram, since such a diagram is based on 1 lb of steam flowing through each part of the cycle with no material being withdrawn. In the equipment shown in Fig. 189, different masses of material flow through various parts of the equipment. The details of analysis of a regenerative feed-water heating system of the type shown in Fig. 189 are to be found in textbooks on mechanical engineering thermodynamics.[2]

[2] V. M. Faires, *Applied Thermodynamics*, p. 278, Macmillan Co. (1938).

(*f*) *Multiple-Fluid Cycles.* Suggestions have been made for improving efficiency by using vapors other than steam, and using different media at different temperature levels. The only cycle of this type operating commercially uses a dual cycle with mercury and water. Fig. 190 shows the *TS* diagram of this cycle. Mercury is evaporated at 932° F, expanded in a turbine, and condensed at 482° F. The heat of condensation of 9.73 lb mercury is sufficient to evaporate 1 lb water at 455° F, and also

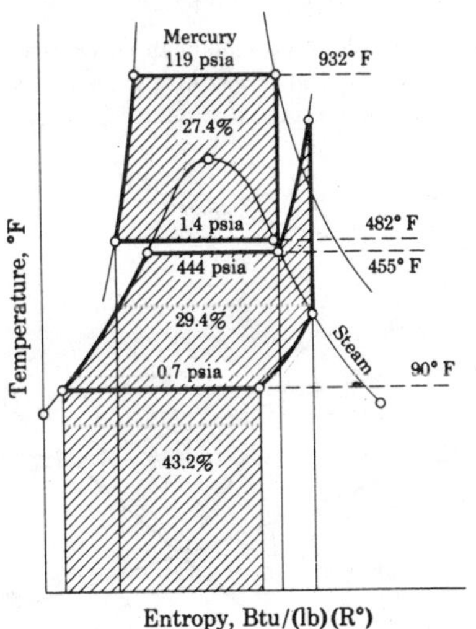

Fig. 190. Mercury–steam cycle. Diagram for 1 lb steam, 9.73 lb mercury

to complete reheating the feed water to steam boiler temperature. The steam is generated at a pressure of 444 psia, all the heat of vaporization being supplied by the condensing mercury. The steam is superheated in the same furnace that supplies heat to the mercury boiler. The steam, after expansion through a turbine, is condensed at 90° F and 0.7 psia. A regenerative system of feed-water heating is employed to heat the condensate partway to steam boiler temperature; the rest of the heating of the feed water is accomplished by the condensing mercury. Therefore, no heat for feed-water heating is drawn from the products of combustion. The power developed by the mercury turbine is 27.4% of the total heat absorbed by the steam and mercury from the products of combustion, while the power developed by the steam turbine is 29.4% of the total heat absorbed. The total cycle efficiency is therefore 56.8%, and 43.2% of the total heat absorbed is rejected in the steam condenser. The Carnot efficiency between 932 and 90° F is $(1392 - 550)/1392 = 0.606$.

Process Steam. Great savings in power production costs can be obtained where process steam is required for heating purposes. The steam is generated at high pressure and expanded in a turbine or engine operating at an exhaust pressure suitable for the heating requirements. If the total amount of exhaust steam is needed for processing, the additional fuel required for power generation can be reduced to less than 5000 Btu per kw-hr measured at the generator.

Illustration 6. A chemical plant uses 5000 lb saturated steam per hour at 20 psia for process heating. A proposal is made to generate power by increasing the boiler pressure from 20 psia to 400 psia and expanding the steam from 400 to 20 psia

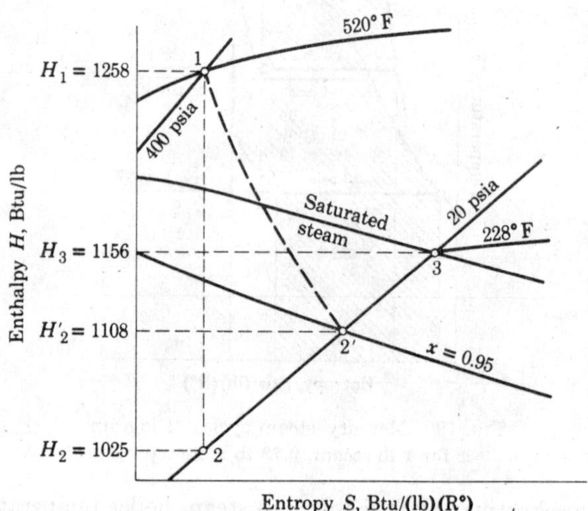

FIG. 191. Mollier diagram for Illustration 7

through a low-cost steam turbine before using it for process heating. The condensate from process heating is supplied to the boiler feed-water pump at 20 psia, saturated. Estimate the following:

(*a*) The temperature required in superheating to secure a quality of 0.95 in the exhaust steam to be used for process heating.

(*b*) The shaft power generated by the turbine, if its indicated efficiency is 64% and its mechanical efficiency is 92%.

(*c*) The additional heat necessary if the power generating equipment is installed, expressed as Btu per kilowatt-hour generated.

(*a*) The state of the exhaust steam is shown by point 2′ in Fig. 191, where the pressure line for 20 psia intersects the quality line, $x = 0.95$. $H'_2 = 1108$ Btu per lb.

From equation 15,

$$\eta_I = \frac{H_1 - H'_2}{H_1 - H_2} \qquad (a)$$

Then, $(H_1 - 1108)/(H_1 - H_2) = 0.64$, giving

$$0.36H_1 + 0.64H_2 = 1108 \qquad (b)$$

Numerical values of H_1 and H_2 are read from the Mollier diagram at constant entropy between the 400 and 20 psia lines, until by trial and error the values fit equation (*b*). Thus, $H_1 = 1258$, $H_2 = 1025$, and the corresponding temperature of the superheated steam is 520° F.

(*b*) The heat supplied to the steam boiler where power is not generated is as follows:

$$q_1 = 5000(H_3 - H_w) = 5000(1156 - 196) = 4{,}800{,}000 \text{ Btu per hr}$$

Where power is generated, W pounds of steam per hr are required. Hence,

$$W(H'_2 - H_w) = W(1108 - 196) = q_1 = 4{,}800{,}000 \text{ Btu per hr}$$

or

$$W = 5263 \text{ lb steam per hr}$$

Indicated work done by the steam turbine,

$$w_i = W(H_1 - H'_2) = 5263(1258 - 1108) = 789{,}000 \text{ Btu per hr}$$

$$\text{Shaft work, } w_s = (789{,}000)(0.92) = 726{,}000 \text{ Btu per hr}$$

or

$$720{,}000/3412 = 213 \text{ kw}$$

(*c*) Heat added to steam boiler if power is generated,

$$q'_1 = W(H_1 - H_w) = 5263(1258 - 196) = 5{,}590{,}000 \text{ Btu per hr}$$

Heat added to steam boiler when power is not generated

$$= 4{,}800{,}000 \text{ Btu per hr}$$

Increase in heat when power is generated

$$q' - q_1 = 790{,}000 \text{ Btu per hr}$$

or

$$790{,}000/213 = 3710 \text{ Btu per kw-hr}$$

If the boiler efficiency is estimated at 80%, the extra fuel consumption corresponds

to 3710/0.80 = 4640 Btu per kw-hr

If the boiler is fired with a fuel oil having a heating value of 18,400 Btu per lb, the added fuel demand for the power is

$$4640/18{,}400 = 0.252 \text{ lb oil per kw-hr.}$$

The low heat rate of 4640 Btu per kw-hr shown in Illustration 6 results from the fact that all the exhaust steam was used for process heating. The situation is less favorable if only part of the steam from the engine is used for process heating.

The over-all performance of various types of power generating plants is indicated by Table 56.[3]

Table 56. Typical Over-all Thermal Performance of Fuel-Burning Power Plants

Type of Plant	Heat Rate, Btu per kw-hr	Thermal Efficiency
All stationary steam plants, average	25,000	0.14
Central station steam plants, average	16,000	0.21
Best record, large central-station steam plant	10,100	0.34
Small noncondensing industrial steam plant	35,000	0.10
Small condensing industrial steam plant	20,000	0.17
By-product power steam plant	4500–5000	0.75
Diesel plant	11,500	0.30
Natural gas engine plant	14,000	0.24
Gasoline engine plant	16,000	0.21
Producer gas engine plant	18,000	0.19

Problems

1. Coke having a heating value of 12,600 Btu per lb and a carbon content of 88.5% is burned in a boiler furnace to generate superheated steam at a temperature of 900° F. Assuming that the furnace is operated with the theoretical quantity of dry air at 77° F to produce stack gases at 900° F with no losses in the ashes and no radiation losses, calculate the work in kilowatt-hours per pound of coke that would be developed by a Carnot engine rejecting heat at 77° F.

2. Hot gases leave the reactor of a catalytic process during the regeneration period at a pressure of 65 psia and a temperature of 900° F. The gases are expanded isentropically through a turbine exhausting at 14.7 psia. Assuming the gases to have the thermodynamic properties of air and ideal zero-angle behavior for the turbine, calculate

(*a*) The exhaust temperature,

(*b*) The work done in Btu per pound of gas,

(*c*) The optimum blade velocities for single-stage turbines of the de Laval, Curtis, and Parsons types.

3. A turbine is to operate on steam at 900 psia and 950° F. Assuming a 70% indicated efficiency, calculate the exhaust pressure that will result in a dry saturated exhaust and the corresponding steam rate in pounds per horsepower-hour.

4. A Rankine-cycle power plant generates dry saturated steam at a pressure of 400 psia which is isentropically expanded through engines to an exhaust pressure of 25 psia for use in process heating. It may be assumed that the condensate is returned to the boilers at its saturation temperature under the exhaust pressure by an isentropic pump.

(*a*) From the steam tables, calculate the pounds of dry saturated low-pressure steam available for process heating per 100 lb of high-pressure steam generated.

[3] T. Baumeister in J. H. Perry's *Chemical Engineers Handbook*, 3d ed. p. 1630, McGraw-Hill Book Co. (1950).

(*b*) Calculate the ideal thermodynamic efficiency of the cycle. Compare this result with the Carnot efficiency.

(*c*) Calculate the ideal water rate of the engine.

5. Determine the effect on the results of problem 4 of superheating the steam 100° F

6. A single-stage turbine has an output of 120 hp, steam is supplied at 120 psia, 30° F superheat, and is exhausted to a condenser operating at 27 in. Hg vacuum. The indicated efficiency of the turbine is 0.80, and the mechanical efficiency is 0.90. Estimate the steam consumption of the turbine in pounds per hour.

7. Steam is expanded in four stages in a turbine from an initial state of 350 psia and 150° F superheat to successive pressures of 250, 100, and 30, and 2 psia. Calculate the jet velocities for each expansion, assuming nozzle efficiencies of 90%.

18

Internal-Combustion Engines

The internal-combustion engine, in which the products of combustion of the fuel comprise the working fluid, has become the most important source of mobile power. Currently produced automobiles, locomotives, and aircraft are exclusively powered by internal-combustion engines. In the fields of stationary power plants and ship propulsion internal-combustion engines compete with the steam turbine, deriving energy from carbonaceous and nuclear fuels. The development of internal-combustion engines has profoundly influenced the course of civilization, and great effort is currently directed toward further improvements.

Internal-combustion engines may be classified according to the method of transforming the energy of the fluid into mechanical work. On this basis, there are three general types: piston, turbine, and jet. In the first group are the familiar spark-ignition and compression-ignition piston engines. The combustion gas turbine is the second type while the third type comprises ram-jet and rocket engines. The turbojet engine widely used in aircraft propulsion is a combination of the turbine and jet types.

Spark-Ignition Engines

Spark-ignition engines are of two general types: the four-stroke cycle or *four-cycle* engines and the two-stroke cycle or *two-cycle* engines. These two types are shown diagrammatically in Fig. 192.

In the four-cycle engine, four strokes of the piston are required to complete a cycle. On the *intake stroke*, the piston moves downward with the intake valve open and the exhaust valve closed. A mixture of air with atomized and vaporized fuel in approximately constant proportions is taken into the cylinder from the *carburetor*. During the *compression stroke*, the air–fuel mixture is compressed with both valves closed. At or before the *top dead center* (TDC) position, the air–fuel charge is ignited by a timed spark, and the power or *expansion stroke* begins with both valves closed. As *bottom dead center* (BDC) is approached, the exhaust valve is opened and the cylinder pressure drops to the exhaust level. During the *exhaust stroke*, the piston moves upward with the exhaust valve open, completing the cycle. The cylinder walls are cooled throughout the entire cycle either by liquid circulating through cylinder jackets or by cooling fins over which cold air is blown.

The output and speed of a spark-ignition engine are regulated by a throttle between the carburetor and the *intake manifold* leading to the intake valve. The load on the engine at any speed is indicated by the *manifold pressure*. At full load with wide open throttle (WOT), the

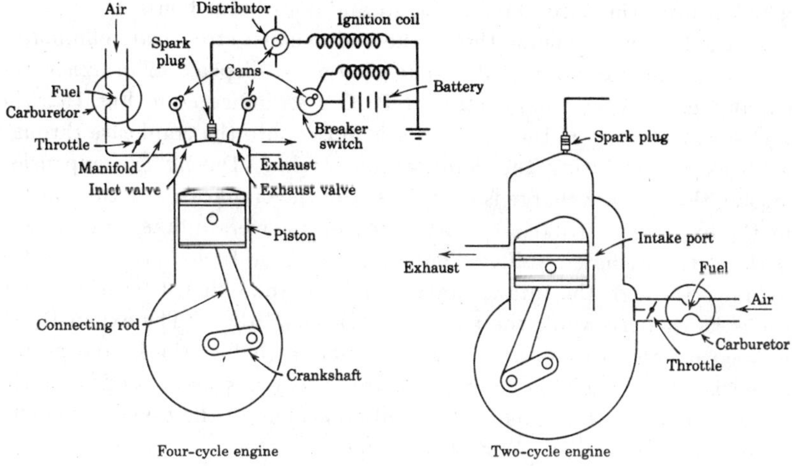

FIG. 192. Spark-ignition engines

manifold pressure is approximately atmospheric. At light loads the manifold pressure is reduced, dropping to approximately 4 to 5 psia at *idling conditions* of no load.

The performance of the four-cycle engine is dependent on precise timing of the valve sequences and the ignition spark. The valves are actuated by cams on a camshaft which is geared to the crankshaft. For this reason the valve timing is ordinarily fixed, independent of load or speed, except in some very large engines.

The ignition spark occurs in a *spark plug* in which a gap is bridged by high voltage current from the *ignition coil*. The timing of the spark is determined by a *breaker switch*, which periodically interrupts the current flowing from the battery to the primary of the coil and produces a high-voltage surge in the secondary. In a multicylinder engine the coil voltage surge is directed to the spark plug of the proper cylinder by a rotary switch termed the *distributor*.

The breaker switch and the distributor are actuated by cams which are geared to the crankshaft. However, an adjustment is provided to permit variation of the timing of the spark from an *advanced* position prior to TDC on the compression stroke to a *retarded* position near TDC. In modern engines the spark timing is automatically varied with engine speed and load to maintain optimum power and efficiency. As speed increases, the spark timing is advanced by a *centrifugal advance* device, which varies the time of opening of the breaker points. In order to keep the engine from knocking as load is increased, the spark is retarded by a *vacuum-advance* device actuated by the pressure in the intake manifold which rotates the entire breaker-point and distributor housing.

In the two-cycle engine, the intake and exhaust strokes are eliminated by using the precompressed intake charge to displace or *scavenge* the exhaust gases when the piston is at BDC. As indicated in Fig. 192, the fuel–air mixture from the carburetor is drawn into the crankcase through a check valve during the compression stroke. During the expansion stroke, the air–fuel charge is compressed in the crankcase. Near the end of the expansion stroke, the piston travel uncovers intake and exhaust *ports* in the cylinder walls. The compressed air–fuel charge enters the cylinder through the intake ports, expelling the exhaust gases through the exhaust ports while the piston is at or near BDC. The exhaust port is opened first so that there is a flow outward before the intake port is opened. As the piston moves upward, the ports are closed, and the compression stroke begins. The ignition system of the two-cycle engine is similar to that of the four-cycle engine.

It is evident that the two-stroke cycle engine has the advantage of a high power-to-weight ratio since it has a power stroke each revolution. Also, it is simple in construction since it has no mechanically timed valves. However, these advantages are offset by the loss of a portion of the intake charge with the exhaust gases, resulting in somewhat lower efficiencies. For this reason the two-cycle spark-ignition engine has only limited application, such as for small outboard boat engines, lawn-mower engines, and the like, where low cost and weight are more important than efficiency. All further discussion of the spark-ignition engine deals with the four-cycle type.

Engine Performance. The performance characteristics of internal-combustion engines are shown on a quantitative basis through evaluation of the following terms.

Indicated mean effective pressure (Imep) is defined as the work per cycle divided by the volumetric displacement of the piston, when the work per cycle is determined from an experimentally determined "indicator card" which shows the pressure–volume changes of the working fluid in going through a complete cycle. Thus, by definition,

$$\text{Imep} = p_m = w_s/LA \tag{1}$$

where w_s = work done on the piston per cycle as shown by the indicator card, in (ft)(lb_f)

L = length of piston stroke (distance between TDC and BDC positions), in feet

A = cross-sectional area of piston, in square inches

With the units specified for w_s, L, and A, Imep will be in lb_f per sq in.

Indicated horsepower (Ihp) is given by the following equation

$$\text{Ihp} = \frac{w_s N_c}{33{,}000} = \frac{p_m LAN_c}{33{,}000} \tag{2}$$

where N_c represents the number of cycles per minute.

Indicated thermal efficiency η_I is defined by the following relation:

$$\eta_I = \left(\frac{w_s}{\text{heating value of fuel supplied per cycle}}\right) \tag{3}$$

Both the numerator and the denominator must be given in the same units. The gross (higher) heating value is generally used in preference to the net (lower) heating value.

The indicated horsepower represents the power delivered to the piston. The useful power delivered by the engine, termed the *brake horsepower* (Bhp), is less because of the mechanical inefficiency of the engine.

Mechanical efficiency η_M is defined as

$$\eta_M = \frac{\text{Bhp}}{\text{Ihp}} \tag{4}$$

Another term used in analyzing reciprocating engine performance is the *brake mean effective pressure* (Bmep), which is defined through the relation

$$\text{Bhp} = \frac{(\text{Bmep})LAN_c}{33{,}000} \tag{5}$$

or

$$\text{Bmep} = \frac{(\text{Bhp})33{,}000}{LAN_c} \tag{6}$$

From equations 4, 5, and 6, it follows that

$$\eta_M = \frac{\text{Bmep}}{\text{Imep}} \tag{7}$$

Brake efficiency, also termed *over-all efficiency*, is the product of the indicated thermal efficiency and the mechanical efficiency:

$$\eta_B = \eta_I \eta_M \tag{8}$$

Torque developed by an engine under various operating conditions is one of its important characteristics, and is determined as follows. The engine operates a drum which is provided with an adjustable brake. In order to prevent the brake from rotating with the drum, a tangential force F must be applied in the direction opposite to the rotation of the drum. When the engine is operating at constant speed, this restraining force F is equal and opposite to the force of the engine. For one revolution, this force acts through a distance $2\pi r$, where r is the distance from the center of the rotating drum to the point where the force measurement is made, and the work done by the engine is equal to $2\pi rF$. The product rF, termed *brake torque*, represents the turning moment of the engine which is a measure of the force it can exert on any type of transmission device. The following equation results from the foregoing considerations.

$$\text{Bhp} = \frac{2\pi rFN_R}{33{,}000} = \frac{2\pi(\text{brake torque})N_R}{33{,}000} \tag{9}$$

where N_R represents the revolutions per minute of the brake drum. Rearranging equation 9, and combining with equation 5, the following results:

$$\text{Brake torque} = \frac{33{,}000\ (\text{Bhp})}{2\pi N_R} = \frac{(\text{Bmep})LAN_c}{2\pi N_R} \tag{10}$$

Indicated torque is computed from an equation similar to equation 10, except that Ihp and Imep values are used in place of Bhp and Bmep.

From equation 10 it is apparent that the torque of an engine of fixed dimensions is directly proportional to the brake mean effective pressure. Torque is of importance as an indication of the ability of an engine to produce acceleration of any mass to which it is connected. For an engine of specified horsepower, the lower the engine speed the higher the torque.

An important measure of the over-all efficiency or economy of an internal-combustion engine is the *specific fuel consumption*, expressed as pounds of fuel per horsepower-hour. This may be either on an indicated horsepower basis Isfc or on a brake horsepower basis Bsfc. A figure for Bsfc is obtained by dividing the experimentally determined fuel consumption rate in pounds per hour by the experimentally determined value for brake horsepower. From the definition of indicated thermal efficiency and the conversion factor, 1 hp-hr = 2544.48 Btu, the following equation for Isfc is obtained.

$$\text{Isfc} = \frac{2544.48}{(\text{HV fuel})\eta_I} \tag{11}$$

As stated previously in connection with the discussion of indicated thermal efficiency, it is customary to use the gross (higher) heating value in preference to the net (lower) heating value.

As indicated on page 762, a gross heating value of 20,100 Btu per lb may be taken as typical of commercial gasolines. Using this value, equation 11 reduces to the following

$$\text{Isfc} = 0.1266/\eta_I \tag{12}$$

Air-Standard Otto Cycle. The operation of a spark-ignition engine approximates that of a heat engine following the hypothetical Otto cycle which comprises two isentropic and two isochoric steps. The pV and TS diagrams of the Otto cycle are shown by the solid lines in Fig. 193*a*. In order to avoid confusion of the lines in the low-pressure region, the diagrams in Fig. 193*a* are not drawn to accurate linear scale. In the ideal cycle with wide-open throttle both the intake and exhaust strokes are represented by line 16.

The ideal Otto cycle is generally analyzed as an *air-standard cycle*; in which it is assumed that air is the working fluid, and that the air behaves ideally and has a constant specific heat. It is assumed that along line 23 heat is transferred to the air, and that along line 41 heat is transferred from the air. Isentropic operation along lines 12 and 34 is also assumed. The simplified calculations based on these assumptions give a good background against which to compare the performance of actual engine cycles.

In analyzing the performance characteristics of internal-combustion engines, the *compression ratio* r_c plays a prominent role. By definition,

$$r_c = V_B/V_T \tag{13}$$

An expression for the ideal indicated efficiency η_I in terms of compression ratio is obtained as follows:

$$\eta_I = \frac{(\text{work done by air in step 34}) - (\text{work done on air in step 12})}{\text{heat added in step 23}}$$

$$= \frac{n_1(\mathrm{u}_3 - \mathrm{u}_4) - n_1(\mathrm{u}_2 - \mathrm{u}_1)}{n_1(\mathrm{u}_3 - \mathrm{u}_2)} = \frac{n_1c_v(T_3 - T_4) - n_1c_v(T_2 - T_1)}{n_1c_v(T_3 - T_2)}$$

$$= \frac{(T_3 - T_2) - (T_4 - T_1)}{T_3 - T_2} = 1 - \frac{T_4 - T_1}{T_3 - T_2} \tag{14}$$

Temperature may be eliminated from equation 14 by expressing T_4, T_3, and T_2 in terms of T_1 through the following relations:

$$T_2 = T_1(V_B/V_T)^{\kappa-1} = T_1r_c^{\kappa-1} \tag{15}$$

$$T_3 = T_2(p_3/p_2) = T_1r_c^{\kappa-1}(p_3/p_2) \tag{16}$$

$$T_4 = T_3(V_T/V_B)^{\kappa-1} = T_3(1/r_c^{\kappa-1}) = T_1(p_3/p_2) \tag{17}$$

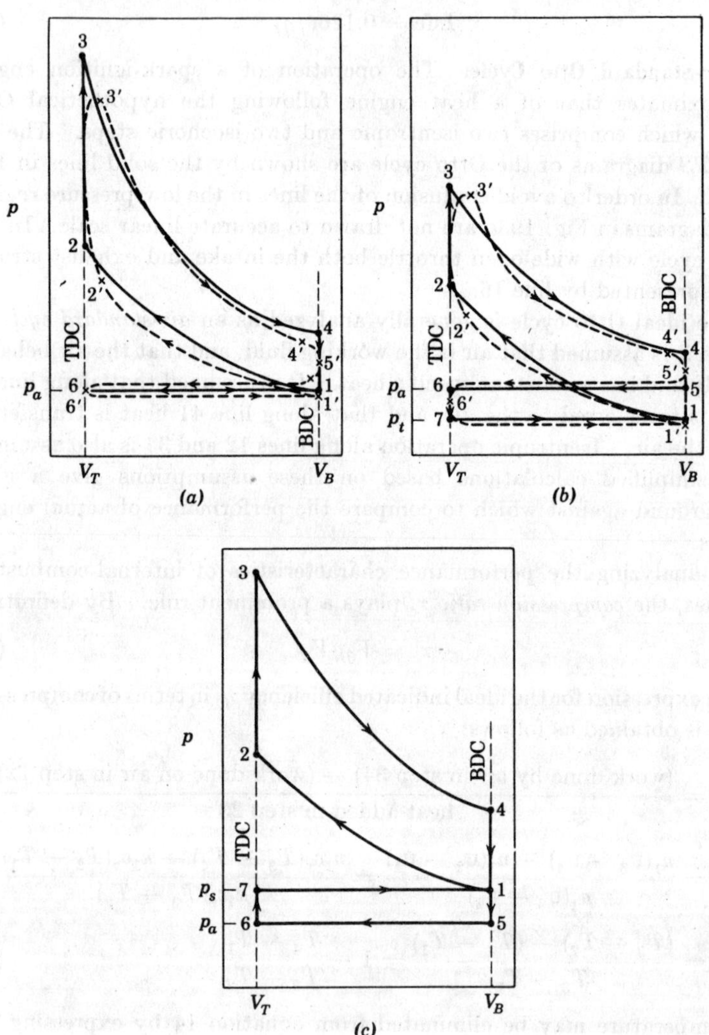

Fig. 193. Indicator diagrams of Otto cycle and four-cycle SI engine. (a) Wide open throttle (WOT). (b) Part throttle. (c) Supercharged (WOT)

When substitutions are made in equation 14, the following results:

$$\eta_I = 1 - \frac{1}{r_c^{\kappa-1}} \tag{18}$$

It is interesting to note that the indicated efficiency of the ideal Otto cycle depends only on the compression ratio r_c. The relation between η_I and r_c is shown in Fig. 194.

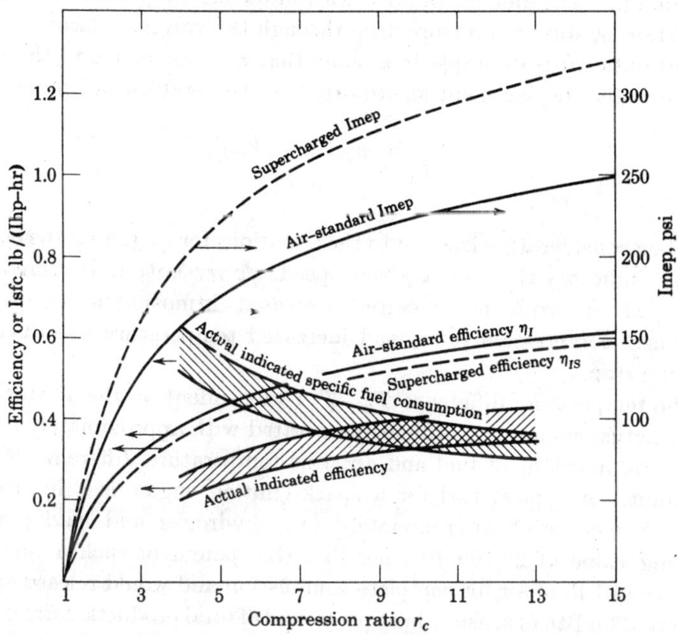

FIG. 194. Air-standard and actual indicated efficiencies, fuel consumption, and Imep for spark-ignition engines. Air-standard values are based on $\kappa = 1.35$, $T_3 - T_2 = 5000$ F°, $p_a = 14.7$ psia, $T_a = 518°$ R. Supercharger ratio $r_s = 1.5$. Actual efficiencies based on total heating value

The net shaft work done per cycle w_s is given from the following equation, which follows from the definition of ideal indicated efficiency.

$$w_s = \eta_I n_1 c_v(T_3 - T_2) = \eta_I n_1 \frac{R}{\kappa - 1}(T_3 - T_2) \tag{19}$$

Indicated mean effective pressure and indicated horsepower may be calculated using equations 1 and 2. Indicated torque is computed from equation 10, modified by using Ihp and Imep.

It will be noted that, in order to evaluate w_s and the other items dependent on w_s such as Imep, Ihp, and indicated torque, it is necessary to determine n_1. This problem is complicated by the presence of warm gases in the clearance volume at the end of the exhaust stroke. These gases mix with the intake gases to comprise n_1. The temperature of the clearance gases is difficult to evaluate because of the rapid cooling at the end of the exhaust stroke where the cooling-surface/volume ratio is at a maximum. The amount of air drawn in during the intake stroke is also uncertain because of pressure drop through the valves. Because of these uncertainties it is desirable to assume that n_1 corresponds to the volume of the piston displacement at atmospheric temperature and pressure.

$$n_1 = \frac{p_a(V_B - V_T)}{RT_a} \tag{20}$$

When consideration is given to the equations for η_I and n_1, the equation for w_s indicates that, at a given speed, power output is increased by increased atmospheric pressure, decreased atmospheric temperature, increased compression ratio, and increased temperature rise during the heating step.

The temperature difference $T_3 - T_2$ is dependent on the air–fuel ratio. Since actual engines are ordinarily operated with approximately stoichiometric proportions of fuel and air, this temperature difference is nearly constant. A typical fuel for a spark-ignition engine is a hydrocarbon mixture containing approximately 14% hydrogen and having a total heating value of 20,100 Btu per lb. One pound of such a fuel would require 14.7 lb of air for complete combustion and would release approximately 1200 Btu of sensible heat per pound of total products, corresponding to an ideal temperature rise of approximately 5000° F in the products for which $M = 29.0$, $C_V = 0.24$ Btu/(lb)(R°), and $\kappa = 1.29$ at the average temperature.

Actual Spark-Ignition Engine Cycles. The broken lines of Fig. 193 indicate the actual pV course of a spark-ignition engine at wide-open throttle. The temperature at 1′ is higher and the pressure lower than ideal because hot exhaust gases in the clearance volume mix with the fresh charge and pressure drop occurs in the intake valve. On the compression stroke 1′2′, heat is lost to the cylinder walls, and ignition occurs in advance of TDC. As engine speed is increased, the ignition spark is advanced to allow time for completion of the combustion reactions. Thus, the heat of the fuel is released over a period of crank rotation from several degrees before TDC to several degrees after, when the maximum pressure is reached at 3′. The maximum pressure reached is lower than ideal because chemical equilibria and limited rates of reaction do not permit complete combustion at this point. Combustion continues during the expansion stroke as the temperature drops and equilibria become more favorable, but heat is steadily lost to the cylinder walls. The exhaust valve is opened at 4′, ahead of BDC and the pressure during the exhaust stroke is higher than ideal because of pressure drop in the outlet valve. Combustion is not complete and the exhaust gases contain carbon monoxide and unburned hydrocarbons.

Thermodynamic charts for the products of combustion of SI engines have been developed[1,2] to take into account variations in composition, heat capacities, and other properties in mixtures at chemical equilibrium. From these charts refined calculations may be made of ideal cycle performance.[3,4]

Pressure drop through the valves reduces both power and efficiency. By reducing the inlet pressure below p_a, the mass m of air–fuel mixture that is taken in is reduced. The *volumetric efficiency* η_V is defined as the ratio of the mass of charge actually taken in to the ideal value of equation 20, assuming atmospheric temperature and pressure. In addition to reducing m, pressure drop through both the inlet and exhaust systems also develops an area of *negative work* 5′6′1′ in the indicator diagram. This *pumping work* must be algebraically added to the ideal work per cycle shown by equation 19.

As a result of these inefficiencies and losses, the efficiencies of actual spark-ignition engines, as shown in Fig. 194, range from 60 to 70% of air-standard efficiencies when operated at wide-open throttle and speeds of optimum economy. At part throttle or nonoptimum speeds, much lower efficiencies are obtained. Actual Imep typically ranges from 70 to 80% of the air-standard Imep at the speed of maximum torque. The performance characteristics of a representative modern automobile engine at wide-open throttle are shown as a function of speed in Fig. 195. The friction horsepower is the difference between the indicated and brake horsepower, and the over-all efficiency is the product of the indicated and mechanical efficiencies. Mechanical efficiencies vary from more than 90% for short-stroke engines at low speeds to less than 60% for long-stroke engines at high speeds.

[1] R. L. Hershey, J. E. Eberhardt, and H. C. Hottel, *Soc. Auto. Engrs. J.*, **39**, 409 (1936).

[2] H. C. Hottel, G. C. Williams, and C. N. Satterfield, *Thermodynamic Charts for Combustion Processes*, John Wiley & Sons (1949).

[3] J. H. Keenan, *Thermodynamics*, John Wiley & Sons (1941).

[4] E. F. Obert, *Internal Combustion Engines*, 2d ed., p. 488, International Textbook Co., Scranton, Pa. (1952).

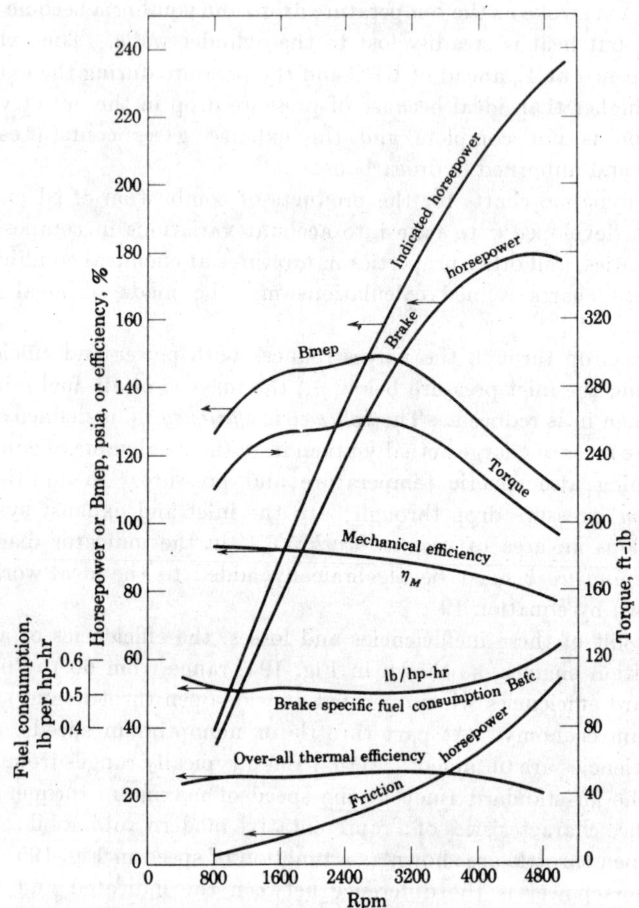

Fig. 195. Performance of a 1955 Chevrolet V-8 engine

Bore	3.75 in.	Compression ratio	8 : 1
Stroke	3.00 in.	Engine weight	531 lb
Displacement	265 cu in.		

(R. F. Sanders, *Soc. Auto. Engrs. J.*, *63*, 400, 1955, with permission.)

It will be noted in Fig. 195 that the minimum specific fuel consumption occurs at a speed much lower than that of maximum power. This leads to the general conclusion that a large engine at wide-open throttle operating at a low speed through suitable gears has lower fuel consumption than a smaller engine operating at wide-open throttle at higher speed of maximum power.

Throttling. The speed and power output of the SI engine are regulated by throttling the air–fuel mixture that leaves the carburetor. In Fig. 193*b*, the solid lines show the ideal indicator diagram of a throttled engine. Throttling introduces negative work of pumping equal to $(p_a - p_t)(V_B - V_T)$ into the ideal cycle.

Taking account of this in evaluating the net work output, the following equation for η_I may be deduced in a manner similar to the WOT case previously discussed:

$$\eta_I = \left(1 - \frac{1}{r_c^{\kappa-1}}\right) - \frac{(p_a - p_t)(\kappa - 1)(V_B - V_T)}{n_1 R(T_3 - T_2)} \tag{21}$$

As stated before, the approximate value of n_1 is given by an equation similar to 20, thus,

$$n_1 = \frac{p_t(V_B - V_T)}{RT_a} \tag{22}$$

The work per cycle, indicated mean effective pressure, indicated horsepower, and indicated torque are computed using the same equations as for the WOT case. Throttling reduces work per cycle and power output by reducing both volumetric and thermal efficiencies. At a high value of the throttling ratio p_a/p_1, the indicated efficiency η_I falls to zero, and the ideal engine becomes inoperative.

It is evident that better fuel economy is achieved in operating a small engine at wide-open throttle than in a large engine throttled back to the same power output.

Supercharging. One method of increasing Imep and power output per unit piston displacement is by raising the compression ratio as shown in Fig. 194. An alternative method termed *supercharging* is to use a rotary compressor or blower between the throttle and the intake manifold to precompress the air–fuel mixture. If the supercharger is mechanically driven by the engine, the resultant indicator diagram is shown in Fig. 193*c*. With a supercharger, the intake manifold pressure may be greater or less than atmospheric, depending on the throttle position, and the volumetric efficiency may be greater than unity.

Supercharging increases the net work per cycle by $(p_s - p_a)(V_B - V_T)$ corresponding to area 1567 on the indicator diagram. The ratio p_s/p_a may be termed the supercharger ratio r_s. Then the added net work per cycle done on the piston becomes $p_a(r_s - 1)(V_B - V_T)$. However, from this gain it is necessary to subtract the work of isentropic compression under flow conditions, of the air supplied per cycle. Taking due account of the various work effects, the following equation for indicated efficiency is obtained:

$$\eta_I = \left(1 - \frac{1}{r_c^{\kappa-1}}\right) + \frac{(p_s - p_a)(V_B - V_T)(\kappa - 1)}{n_1(T_3 - T_2)R} - \frac{(n_1 - n_6)\kappa T_a}{n_1(T_3 - T_2)}[(r_s)^{(\kappa-1)/\kappa} - 1] \quad (23)$$

To evaluate n_1 and n_6 accurately, T_1 and T_6 must be known. Although the necessary equations can be developed, they are cumbersome, and numerical solutions are tedious. This difficulty may be avoided by making two simplifying approximations in equation 23. The first involves taking $(n_1 - n_6)/n_1$ as equal to 1 in the third term on the right side of the equation. The other involves evaluating n_1 through the following approximation equation:

$$n_1 = \frac{p_s(V_B - V_T)}{RT_s} \quad (24)$$

where T_s represents the temperature of the air delivered by the supercharger to the engine.

If these two simplifying assumptions are made, equation 23 reduces to the following:

$$\eta_I = \left(1 - \frac{1}{r_c^{\kappa-1}}\right) + \frac{T_a}{T_3 - T_2}[\kappa(1 - r_s^{-1/\kappa}) - r_s^{(\kappa-1)/\kappa} + r_s^{-1/\kappa}] \quad (25)$$

The right side of equation 25 has two major terms, the second of which is relatively small and has a negative value. This indicates that, in supercharging, there is a small constant reduction in indicated efficiency for constant values of T_a, T_3, T_2, and r_s.

Once a figure for indicated thermal efficiency has been established, values of work per cycle, indicated mean effective pressure, indicated horsepower, and indicated torque may be calculated, using the same equations as for the previous two cases.

Supercharger ratios are generally in the range of 1–2. The effects of a supercharger ratio of 1.5 on the efficiency and Imep of an ideal air-standard cycle are shown by broken lines in Fig. 194. It will be noted that, with this degree of supercharging, corresponding to an absolute manifold pressure of 45 in. Hg at sea level, the Imep corresponding to a 10 compression ratio with atmospheric intake is achieved at a compression ratio of only 5.15. However, the indicated efficiency is only 0.407 compared to 0.553.

Supercharging is most advantageously used on aircraft engines to obtain high power outputs at the low pressures of high altitudes. Aircraft engines are generally limited to compression ratios of approximately 8.5 and supercharged for increased output. Loss of efficiency is avoided by use of an exhaust-gas turbine to drive the supercharger.

Detonation. As the compression pressure of an engine is raised, the temperature at the end of the compression stroke approaches the *auto-ignition* level which will produce spontaneous ignition of the air–fuel mixture. In a normally operating engine, ignition is initiated at the spark plug, and a flame front moves progressively across the combustion chamber with a smooth release of energy and increase in pressure. At high-compression ratios, the pressure and temperature increases accompanying initial combustion may be sufficient to cause the entire remaining charge to ignite simultaneously in advance of the normal flame propagation. This sudden release of energy, termed *detonation*, will produce an audible *knock* which may damage the engine.

It has been found that the detonation characteristics of an engine are largely dependent on the nature of the fuel, and the development of the high-compression engine has been paced by the development of fuels of improved *antiknock* properties.

Another problem limiting the use of high-compression ratios is the formation of deposits in the combustion chamber. Such deposits increase the compression ratio, interfere with the functioning of the spark plug, and reduce heat transfer to the cooling jacket or fins. Flakes of hot deposits produce *surface ignition*, which deviates from the normal ignition timing and produces knock and overheating.

The formation of combustion-chamber deposits is determined by the characteristics of both the fuel and the lubricant used in the engine. The lubricating oils used in high-compression engines contain inorganic, ash-forming additives which serve as detergents and dispersants to remove varnish-like gums and sludge from the engine parts and keep them suspended in the oil. These materials, together with carbonaceous and lead compound residues from combustion of the fuel and lubricant, tend to form hard coatings in the combustion chamber and on the top of the piston.

Compression ratios are currently limited to 10–11 by antiknock and deposit characteristics of available fuels and lubricants. Higher ratios will be possible with fuels of better antiknock properties and lower deposit contributions. Synthetic lubricants are known which greatly reduce deposits from this source.

Fuel Requirements. Spark-ignition engines are generally operated on gasoline, a complex mixture of hydrocarbons ranging from butane to compounds boiling up to about 425° F. Higher-boiling materials would not vaporize adequately in the intake manifold and would contribute to excessive combustion-chamber deposits. A well-formulated gasoline has a Reid vapor pressure of 10 to 12 psia at 100° F, high enough to give easy starting but low enough to avoid *vapor lock* due to vaporization in the fuel system. It is free from nonvolatile, varnish-like *gum* materials and from unstable compounds which form gum in storage. It is low in sulfur to minimize engine corrosion and interference with antiknock additives.

Antiknock is the most important property of a gasoline. It has been found that highly branched paraffin hydrocarbons have excellent antiknock properties while normal paraffins are very poor in this respect. As a result the antiknock properties of gasolines are expressed in terms of empirically defined *octane numbers*. The octane number of a gasoline is the percentage by volume of iso-octane (2,2,4-trimethylpentane) in admixture with normal heptane which produces the same antiknock characteristics as the gasoline in question, when both are run in a special test engine. Thus, normal heptane has zero octane number while iso-octane is 100.

The octane number of a gasoline may be increased by the addition of small quantities of tetraethyl lead which serves as a knock suppressor. Practically all gasoline contains tetraethyl lead. However, the effectiveness of lead varies greatly with the hydrocarbon composition of the fuel.

The gasoline problem is complicated by the fact that antiknock performances of the hydrocarbons vary with engine speed, inlet air temperature, pressure, and engine design. The characteristics of the principal hydrocarbon series are summarized in Table 57. A good gasoline is a blend of hydrocarbon types which will satisfy the *octane requirement* of the engine at all speeds and not cause excessive *octane requirement increase* (ORI) by deposit formation.

TABLE 57. HYDROCARBON ANTIKNOCK CHARACTERISTICS

	Octane Numbers		Tetraethyl Lead
	Low Speed	High Speed	Response
Isoparaffins	Good	Excellent	Good
Normal paraffins	Poor	Poor	Excellent
Naphthenes	Good	Good	Good
Aromatics	Excellent	Good	Poor
Olefins	Good	Poor	Poor

The octane requirement of an engine depends on its compression ratio and intake manifold pressure as well as on the design of the combustion chamber and valves. A typical relationship is shown in Fig 196, in which Imep is plotted against octane requirement with lines of manifold pressures and compression ratios. It will be noted that, for a given octane requirement, the Imep and power output are increased by supercharging to a high manifold pressure with corresponding reduction of the compression ratio.

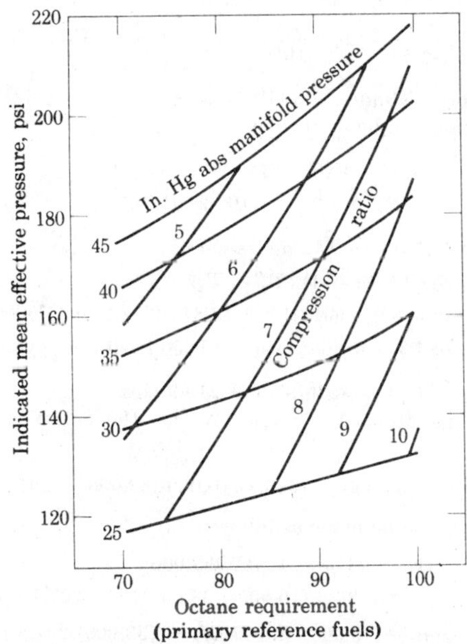

FIG. 196. Relationships among Imep, compression ratio, manifold pressure, and octane number for a one-cylinder engine

(E. F. Obert, *Internal Combustion Engines*, International Textbook Co., Scranton Pa., 1952, with permission. Original data by M. Roensch, Ethyl Corporation)

Illustration 1. A spark-ignition engine has eight cylinders, having a total piston displacement of 300 cu in. The compression ratio is 10. Assume that the air-standard four-stroke Otto cycle is followed with an average κ of 1.35 and that air is supplied at 14.70 psia and 70° F. The temperature rise $T_3 - T_2$ is 5000 F°.

(*a*) If the engine is operating at wide-open throttle, calculate the ideal performance characteristics listed below. The approximation equation for n_1 may be employed in making the solution.

1. η_I, the indicated thermal efficiency.
2. w_s, the indicated shaft work per cycle, in (ft)(lb$_f$).
3. Imep, the indicated mean effective pressure, in pounds per square inch.
4. Ihp, the indicated horsepower of the engine, when operating at 2500 rpm.
5. Indicated torque of the engine, in foot-pounds, when operating at 2500 rpm.
6. Isfc, the indicated specific fuel consumption, in pounds of gasoline per Ihp-hr, assuming that the gasoline has a gross heating value of 20,100 Btu per lb.

7. T_2, the compression temperature, in degrees Rankine.
8. p_2, the compression pressure, in pounds per square inch.
9. p_3, the maximum pressure, in pounds per square inch.

(b) Repeat the calculations of part (a) for operation under throttled conditions, with the manifold pressure at 10 in. Hg. Use the approximation equation for n_1.

(c) Repeat the calculations of part (a), for operation under supercharged conditions, with the manifold pressure at 38 in. Hg. Use the approximation equation for n_1.

Solution to Part (a). Refer to Fig. 193 for points on the cycle.

1. $$\eta_I = 1 - \frac{1}{r_c^{\kappa-1}} = 1 - \frac{1}{10^{1.35-1.00}} = 1 - 0.4467 = 0.5533$$

2. Displacement per cylinder = 300/8 = 37.5 cu in. = 0.021701 cu ft. The approximation equation will be used to evaluate n_1.

$$n_1 = \frac{p_a(V_B - V_T)}{RT_a} = \frac{(14.696)(0.021701)}{(10.731)(529.7)}$$

$$= (0.5611)10^{-4} \text{ lb-moles per cycle}$$

$$w_s = \eta_I n_1[R/(\kappa - 1)](T_3 - T_2)$$

$$= (0.5533)(0.5611)10^{-4}[(1.9872)(778.2)/0.35]5000$$

$$= 685.3 \text{ ft lb}_f \text{ per cycle} = 8230 \text{ in. lb}_f \text{ per cycle}$$

3. Imep $= w_s/(V_B - V_T) = 8230/37.5 = 219.3$ psia

4. At 2500 rpm, the number of cycles N_c for the entire engine is equal to (2500/2)8 = 10,000.

$$\text{Ihp} = w_s N_c/33{,}000 = (685.8)(10{,}000)/33{,}000 = 207.7 \text{ hp}$$

The evaluation may also be made as follows:

$$\text{Ihp} = (\text{Imep})(V_B - V_T)N_c/33{,}000$$

$$= (219.5)(37.5)(10{,}000)/(33{,}000)12 = 207.7 \text{ hp}$$

5. $$\text{Indicated torque} = (33{,}000)(\text{Ihp})/2\pi N_r = (33{,}000)(207.8)/2\pi(2500)$$

$$= 436.3 \text{ ft-lb}_f$$

The calculation may also be made as follows:

$$\text{Indicated torque} = (\text{Imep})(V_B - V_T)N_c/2\pi N_R$$

$$= (219.5)(37.5)(10{,}000)/2\pi(2500)$$

$$= 5239 \text{ in. lb}_f = 436.3 \text{ ft-lb}_f$$

6. For a gasoline having a gross heating value of 20,100 Btu per lb:

$$\text{Isfc} = 0.1266/\eta_I$$

$$\text{Isfc} = 0.1266/0.5533 = 0.2288 \text{ lb gasoline per hp-hr}$$

7. Accurate evaluation of T_2 requires that T_1 be known. As an approximation, it is assumed that $T_1 = T_a = 529.7°$ R.

$$T_2 = T_1(V_B/V_T)^{\kappa-1} = T_1 r_c^{\kappa-1} = (529.7)(2.2387) = 1186° \text{ R}$$

8. $$p_2 = p_1 r_c^{\kappa} = (14.696)(10)^{1.35} = (14.696)(22.387) = 329 \text{ psia}$$

9. $$p_3 = p_2(T_3/T_2) = p_2[(T_2 + 5000)/T_2]$$

$$= 329[(1186 + 5000)/1186] = 1716 \text{ psia}$$

This is an approximation, as it is based on an assumed value of 529.7° R for T_1.

Solution to Parts (*b*) *and* (*c*). The calculations are similar to those for part (*a*). Indicated efficiencies are calculated by equations 21 and 25. Summarized results for the three parts are given below.

SUMMARY

	WOT	Throttled	Supercharged
1. η_I	0.5533	0.4795	0.5525
2. w_s, ft-lb$_f$ per cycle	685.3	198.5	817.0
3. Imep, psia	219.3	63.5	261.4
4. Ihp	207.7	60.2	247.6
5. Indicated torque, ft-lb$_f$	436.3	126.4	520.1
6. Isfc, lb gasoline per Ihp-hr	0.2288	0.2640	0.2291
7. T_2, degrees Rankine	1186	1186	1261
8. p_2, psia	329	110	418
9. p_3, psia	1716	574	2074

Compression–Ignition Engines

With a sufficiently high compression ratio and a suitable fuel, auto-ignition conditions are produced in a reciprocating internal-combustion engine, and an electric ignition system is unnecessary. In a *compression–ignition* (CI) engine, only air is taken into the engine on the intake stroke and compressed to ignition conditions. The fuel is then atomized directly into the combustion chamber at a controlled rate by an *injector* pump. Thus a four-cycle CI engine is similar in general arrangement to the SI engine except that the carburetor is replaced by an injector and the ignition system is eliminated.

The CI engine has no throttle and always takes in a full charge of air. Its speed and power output are regulated by varying the amount of fuel injected, usually by changing the effective stroke of the injection pump. As a result, the fuel–air ratio varies from a minimum at idle to a maximum at full load.

The two-stroke cycle is attractive in CI engines because no fuel is contained in the intake air. Two-cycle engines of the general design shown in Fig. 192 have been extensively used. A modification which is effective in producing improved scavenging is shown in Fig. 197. Air from a supercharger is introduced through ports in the lower cylinder walls. The compressed air forces the exhaust gases out through a mechanically timed valve in the head of the cylinder. The combination of the two-stroke cycle with supercharging produces effective scavenging at high speeds and results in high power-to-weight ratios.

The original compression–ignition engine was designed to operate on

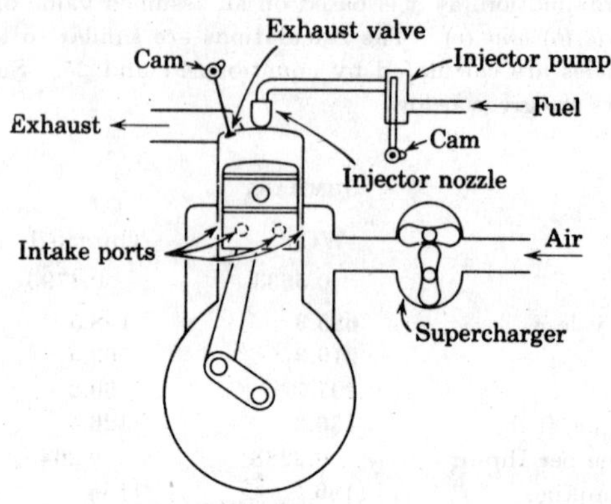

FIG. 197. Two-cycle supercharged compression–ignition engine

the *Diesel cycle* shown in *pV* coordinates in Fig. 198*a*. This cycle comprises isentropic compression and expansion with fuel injection and combustion at constant pressure. Large, low-speed engines approximate this cycle in actual operation, and all compression–ignition engines are popularly termed *Diesels*. However, modern CI engines more closely

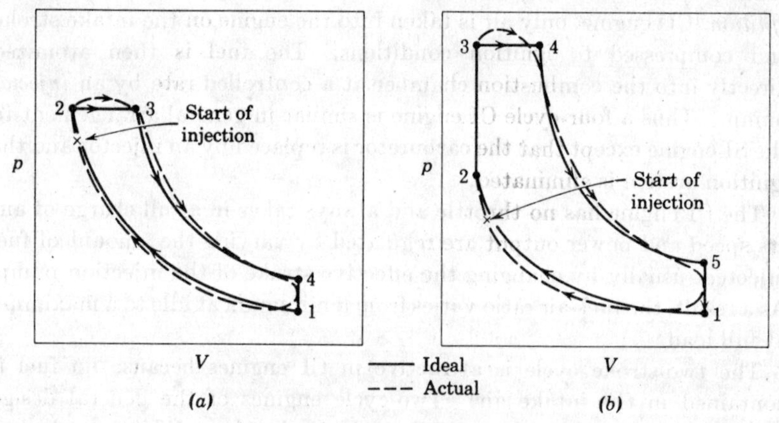

FIG. 198. Compression–ignition engine cycles
(*a*) Diesel cycle (*b*) Dual cycle

approximate the *dual cycle* shown in Fig. 198*b*. By advancing the start of fuel injection and selecting a suitable fuel, a portion of the combustion is caused to occur at essentially constant volume, followed by completion at constant pressure. In this manner higher Imep and power outputs are achieved but with the disadvantage of higher peak cylinder pressures.

Air-Standard Diesel Cycle. The Diesel cycle may be analyzed on an air-standard basis in a manner similar to the analysis for the Otto cycle. In addition to the compression ratio V_B/V_T, the *cut-off ratio* r_t, also called the *load ratio*, enters the analysis, and is defined by the relation

$$r_t = V_3/V_2 \tag{26}$$

It may also be pointed out that in the Diesel cycle, unlike in the Otto cycle, the expansion ratio r_e differs from the compression ratio r_c, and is defined as follows:

$$r_e = V_4/V_3 \tag{27}$$

These three volumetric ratios are related through the following equation:

$$r_c = (r_t)(r_e) \tag{28}$$

An expression for the indicated thermal efficiency η_I is developed from the defining equation

$$\eta_I = \left(\frac{w_s}{\text{heat absorbed in step 23}}\right) \tag{29}$$

where w_s is the net work done by the air in steps 12, 23, and 34.

$$\eta_I = \frac{n_1(\text{U}_3 - \text{U}_4) + p_2(V_3 - V_2) - n_1(\text{U}_2 - \text{U}_1)}{n_1(\text{H}_3 - \text{H}_2)}$$

$$= \frac{n_1c_v(T_3 - T_4) + n_1R(T_3 - T_2) - n_1c_v(T_2 - T_1)}{n_1c_p(T_3 - T_2)}$$

$$= \frac{n_1c_v(T_3 - T_2) + n_1R(T_3 - T_2) - n_1c_v(T_4 - T_1)}{n_1c_p(T_3 - T_2)}$$

$$= \frac{n_1(c_v + R)(T_3 - T_2) - n_1c_v(T_4 - T_1)}{n_1c_p(T_3 - T_2)} = 1 - \frac{1}{\kappa}\left(\frac{T_4 - T_1}{T_3 - T_2}\right) \tag{30}$$

Temperature is eliminated by expressing T_4, T_3, and T_2 in terms of T_1 through the following equations:

$$T_2 = T_1(V_T/V_B)^{\kappa-1} = T_1r_c^{\kappa-1} \tag{31}$$

$$T_3 = T_2(V_3/V_2) = T_2r_t = T_1r_c^{\kappa-1}r_t \tag{32}$$

$$T_4 = T_3(V_3/V_4)^{\kappa-1} = T_3(1/r_e)^{\kappa-1} = T_3(r_t/r_c)^{\kappa-1} = T_1r_t^{\kappa} \tag{33}$$

Substitution for T_2, T_3, and T_4 in equation 30 gives the following:

$$\eta_I = 1 - \frac{r_t^{\kappa} - 1}{\kappa r_c^{\kappa-1}(r_t - 1)} = 1 - \frac{1}{r_c^{\kappa-1}}\left[\frac{r_t^{\kappa} - 1}{\kappa(r_t - 1)}\right] \tag{34}$$

Net shaft work per cycle is obtained by rearranging equation 29 and inserting an expression for the heat absorbed in step 23.

$$w_s = \eta_I n_1 c_p(T_3 - T_2) = \eta_I n_1 R\left(\frac{\kappa}{\kappa - 1}\right)(T_3 - T_2) \tag{35}$$

on the basis of the approximation used in equation 20:

$$n_1 = \frac{p_a(V_B - V_T)}{RT_A} \tag{36}$$

After n_1 has been determined, shaft work per cycle, indicated mean effective pressure, indicated horsepower, and indicated torque may be calculated as previously shown for the Otto cycle.

The foregoing procedures are adapted to those cases in which $T_3 - T_2$ instead of r_t is known. If r_t is known instead of $T_3 - T_2$, it is unnecessary to use an approximation equation for n_1, as w_s may be evaluated from r_t without the need for knowing n_1. This is accomplished as follows:

From equations 31 and 32,

$$T_3 - T_2 = T_1 r_c^{\kappa-1}(r_t - 1) \tag{37}$$

From the ideal-gas law,

$$n_1 = \frac{p_a V_B}{RT_1} \tag{38}$$

By substituting in the equation for w_s, the following is obtained:

$$w_s = \eta_I n_1 c_p(T_3 - T_2) = \eta_I p_a V_B\left(\frac{\kappa}{\kappa - 1}\right) r_c^{\kappa-1}(r_t - 1) \tag{39}$$

It will be noted that equation 34 differs from equation 18 only by the term in brackets, which is a function of the load ratio and is always greater than unity under finite loads. It follows that, *at a given compression ratio, the Otto cycle has a higher efficiency, Imep, and work output than the Diesel cycle.* At low-load conditions where r_t approaches 1.0, the efficiency of the Diesel approaches that of the Otto cycle or actually exceeds it, because of the absence of throttling. The dual cycle is intermediate between the Otto and the Diesel cycles.

The high efficiencies generally associated with compression-ignition engines result from the use of high compression ratios, generally in the range of 13 to 17.

A representative Diesel fuel, when burned with theoretical proportions of air, will produce an ideal temperature rise of approximately 3700° F at constant pressure. This may be taken as the maximum value of $T_3 - T_2$ at full load in the ideal Diesel cycle. On this basis, full- and half-load efficiencies are calculated from equation 34 and plotted in Fig. 199. Corresponding ideal Otto-cycle efficiencies are also plotted and are approached by the Diesel cycle at no load. Also plotted in Fig. 199 are the maximum pressures of the two cycles.

It will be noted that an ideal Otto cycle at 7 compression ratio has the same efficiency as an ideal full-load Diesel cycle at 16 compression ratio.

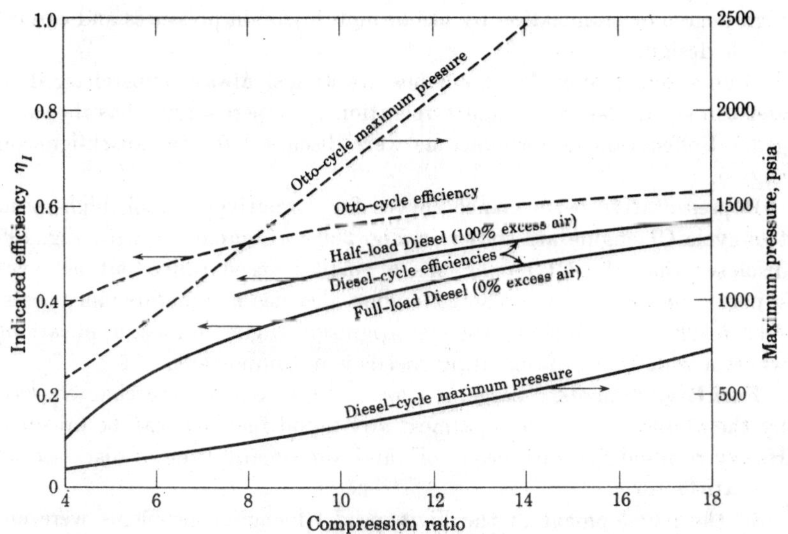

FIG. 199. Ideal air-standard efficiencies and maximum pressures for Otto and Diesel cycles. ($p_a = 14.7$ psia, $T_a = 518°$ R, $\kappa = 1.35$. $T_3 - T_2$: Otto = 5000 F°; Diesel, full-load = 3700 F°, half-load = 1850 F°)

However, if the Diesel is operated with 100% excess air, its efficiency becomes equal to that of an Otto cycle at 9.5 compression ratio. The comparison becomes more favorable to the Diesel cycle if the Otto cycle is throttled. Whereas the efficiency of the Otto cycle is reduced by throttling to part load, that of the Diesel is increased by reducing fuel injection.

Actual Compression–Ignition Cycles. As previously mentioned, modern CI engines depart widely from the classical Diesel cycle. Through early and rapid injection of fast-burning fuels, they approach the dual or Otto cycles in their thermodynamic relationships. Timed, high-pressure fuel injection minimizes the detonation problem encountered in the SI engine and permits use of high compression ratios with fuels available at relatively low cost.

Compression–ignition engines have been built in sizes up to 20,000 Bhp and with speeds ranging from less than 100 to over 6000 rpm. Many of the high-speed engines employ a *precombustion chamber*, into which the fuel is injected. The precombustion chamber is connected to the main combustion chamber by a restricted passageway and serves to promote dispersion and vaporization of the fuel. In *open-chamber* engines, the fuel is injected into the main combustion chamber with care to secure a high degree of atomization by use of high injection pressures and careful nozzle design.

High-speed, two-cycle CI engines are almost always supercharged to assist in the intake and exhaust operations. Supercharging has the same general effects on performance as were discussed for the spark-ignition engine.

Representative performance curves for a relatively small, high-speed two-cycle CI engine are shown in Fig. 200.[4] Comparison with Fig. 195 discloses that the CI engine has a smaller speed range but achieves superior fuel economy. The CI engine also has low maintenance costs. As a result, it is displacing the spark-ignition engine in most applications where a wide range of operating speeds is not important.

Fuel Requirements. Large, low-speed Diesel engines are characterized by the ability to operate on almost any liquid fuel that can be injected. Heavy residual fuel oils, crude oil, and almost any type of distillate oil are satisfactory.

In the development of the high-speed CI engine, problems were encountered as a result of the *ignition-delay* characteristics of certain fuels. Ignition delay is the time interval between the start of injection and ignition. It is a function of compression pressure and temperature, becoming less at higher pressure. Ignition-delay characteristics of fuel are expressed in terms of the *cetane number* which is defined as the percentage of cetane ($nC_{16}H_{34}$) in admixture with α-methylnaphthalene that produces the same ignition-delay characteristics as the fuel in question when both are used in a special test engine. Thus cetane has a cetane number of 100 and a short ignition delay while α-methylnaphthalene is zero with a long ignition delay.

The hydrocarbon types having high cetane numbers are in general those having low octane numbers. Thus, for compression-ignition engines the normal paraffins are the best fuels and the aromatics the poorest.

In addition to ignition delay, the smoking and cold-starting characteristics of CI fuels are important. These characteristics roughly follow the cetane number but are also dependent on viscosity and volatility. Low sulfur content is also important in minimizing lubrication and maintenance problems.

High-speed CI engines of the types used in trucks and tractors are currently designed to operate on a fuel of 40 to 50 cetane number and a boiling range of approximately 400 to 700° F. Lower-boiling materials

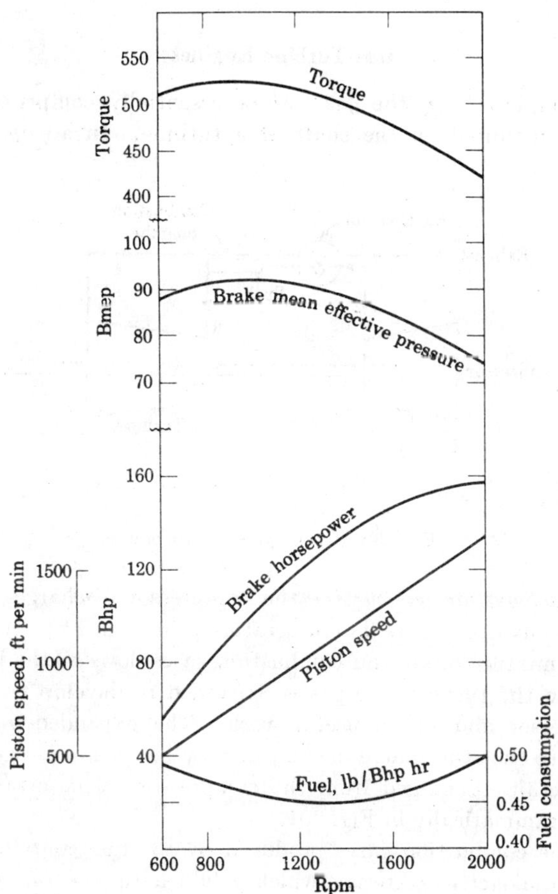

FIG. 200. Performance of General Motors two-stroke-cycle Diesel engine (E. F. Obert, *Internal Combustion Engines*, International Text Book Co., p. 488, 1950)

are not objectionable, but are generally sold as gasoline at higher prices. Railway, marine, and stationary engines tend to use heavier fuels of lower cetane numbers but must balance increased maintenance against reduced fuel costs.

The quality of the fuel as well as the design and condition of the engine determine the extent to which fuel injection can be increased and excess air diminished without excessive smoke in the exhaust. Minimum excess air requirements vary from the order of 50% for large low-speed engines down to 10 to 15% for small high-speed units.

Gas-Turbine Engines

In its simplest form the *gas turbine* essentially comprises a rotary compressor mounted on the shaft of a turbine. In an *open-cycle gas*

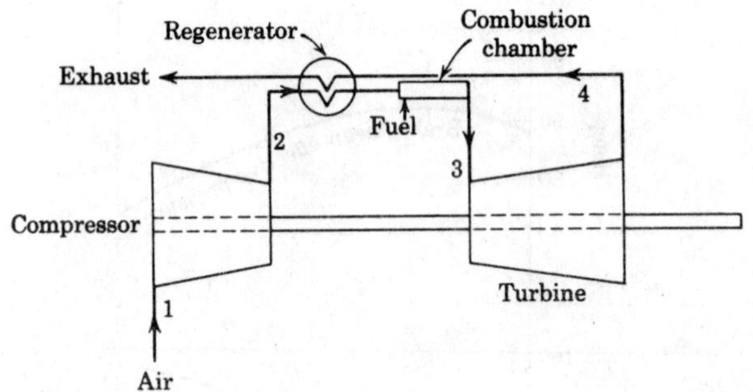

FIG. 201. Open-cycle gas turbine

turbine or *combustion gas turbine* the compressor discharges air into a combustion chamber where its temperature is increased by combustion of fuel. The mixture of air and combustion gases flows to the inlet of the turbine. In the turbine, the gas is expanded to develop power to run the compressor and to do useful work. The expanded gas may be discharged to the atmosphere directly or may be used in a regenerator to preheat the air discharged from the compressor. This arrangement is shown diagrammatically in Fig. 201.

Where the combustion gases produced by the fuel contain corrosive, erosive, or radioactive materials which would harm the turbine, it may be desirable to employ a *closed-cycle* operation. The compressor and turbine operate on a working gas which is recycled to the compressor inlet from an exhaust-gas cooler. Heat is supplied to the cycle by heat transfer from combustion gases. Thus, two additional heat exchangers are required for closed-cycle operation.

For automotive applications where maximum flexibility of power output is desired, the so-called *free turbine* is employed. In this modification, the turbine mounted on the compressor shaft has capacity only to drive the compressor. A second turbine, mounted on an independent shaft, is employed to perform all useful work. The gases from the combustion chamber pass first to the compressor turbine and thence to the power turbine. This arrangement permits the air compressor to be operated at full speed, if desired, when the power turbine is stopped. As a result, the speed of the power turbine can be accelerated more rapidly than that of the single-shaft turbine, in which the relatively massive compressor and compressor–turbine components must be simultaneously accelerated and pressure must be built up.

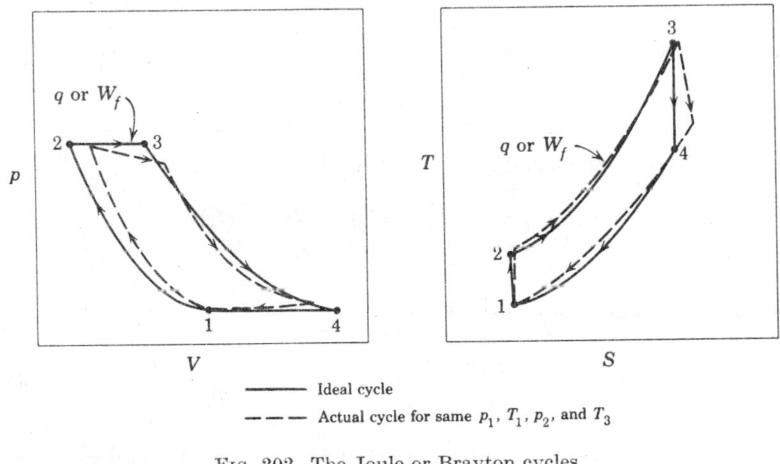

FIG. 202. The Joule or Brayton cycles
q = heat added in Joule cycle
W_f = fuel added in Brayton cycle

The offsetting disadvantage of a constant compressor speed is higher fuel consumption under idling conditions. As a result compromise designs are attractive in which the idling compressor speed is of the order of 25 to 30% of the full-load speed.

Joule and Brayton Cycles. The ideal operating cycle of the gas turbine comprises an isentropic compression, constant-pressure heating, isentropic expansion, and constant-pressure cooling. This is termed the Joule cycle in closed-cycle application or the Brayton cycle in open-cycle operation. It is shown diagrammatically in Fig. 202 in both pV and TS coordinates. The solid lines indicate the ideal, and the broken lines the actual cycle.

The thermal efficiency of a gas turbine is the ratio of the net shaft work performed to the heat of combustion of the fuel burned. In the ideal case, the heat of combustion of the fuel equals the heat added in step 23. Since the gas turbine operates at high air fuel ratios of the order of 100, it is customary in considering open-cycle problems to neglect differences in the mass and properties of the air stream which result from injection of the fuel. The resulting minor errors are included in the efficiency factors of actual operations. On this basis, calculations for the Brayton cycle become identical with those of the Joule cycle.

The net shaft work done is the difference between the work done by the turbine and that consumed in driving the compressor. In the ideal case,

$$w_s = w_t + w_c = (H_3 - H_4) - (H_2 - H_1) \tag{40}$$

In the absence of regeneration,

$$q = H_3 - H_2 \tag{41}$$

and the thermal efficiency η_I is expressed by

$$\eta_{I,\,\text{ideal}} = \frac{(H_3 - H_4) - (H_2 - H_1)}{H_3 - H_2} \tag{42}$$

or, neglecting differences in heat capacities,

$$\eta_I = \frac{(T_3 - T_4) - (T_2 - T_1)}{T_3 - T_2} = 1 - \frac{T_4 - T_1}{T_3 - T_2} \tag{43}$$

Since the pressure ratios are the same in the compression and expansion steps, $p_2/p_1 = p_3/p_4$ and from equaticn 15-34, page 648:

$$\frac{T_2}{T_1} = \frac{T_3}{T_4} = \left(\frac{p_2}{p_1}\right)^{(\kappa-1)/\kappa} \tag{44}$$

Solving for T_1 and T_4 and substituting in equation 43:

$$\eta_I = 1 - \frac{T_3\left(\frac{p_2}{p_1}\right)^{(1-\kappa)/\kappa} - T_2\left(\frac{p_2}{p_1}\right)^{(1-\kappa)/\kappa}}{T_3 - T_2} = 1 - \left(\frac{p_2}{p_1}\right)^{(1-\kappa)/\kappa}$$

$$= 1 - \left(\frac{p_1}{p_2}\right)^{(\kappa-1)/\kappa} \tag{45}$$

Equation 45 shows that the efficiency of an ideal nonregenerative gas turbine depends only on the pressure ratio and is unaffected by temperature.

If ideal regeneration is added to the turbine, the exhaust gases are cooled from T_4 to T_2 with a corresponding increase in the temperature of the air to the combustion chamber. Then the heat supplied by the fuel becomes

$$q_f = q - (H_4 - H_2) = (H_3 - H_2) - (H_4 - H_2) = H_3 - H_4 \quad (46)$$

and

$$\eta_{IR,\text{ideal}} = \frac{(H_3 - H_4) - (H_2 - H_1)}{H_3 - H_4} = 1 - \frac{H_2 - H_1}{H_3 - H_4} \simeq 1 - \frac{T_2 - T_1}{T_3 - T_4} \quad (47)$$

Substituting for T_2 and T_4 from equation 44,

$$\eta_{IR,\text{ideal}} = 1 - \frac{T_1}{T_3}\left(\frac{p_2}{p_1}\right)^{(\kappa-1)/\kappa} \quad (48)$$

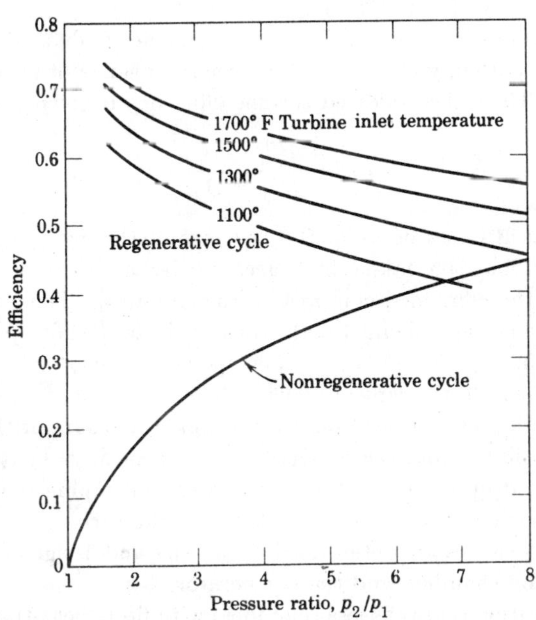

FIG. 203. Ideal gas-turbine thermal efficiencies, $T_1 = 70°$ F

Thus, when ideal regeneration is added to the gas turbine, the efficiency becomes temperature-dependent, and the effect of pressure ratio is reversed compared to the simple nonregenerative turbine. Efficiencies from equations 45 and 48 are shown graphically in Fig. 203 for an air temperature of 70° F.

Figure 203 shows that, without regeneration, the gas turbine is limited to efficiencies of less than 50%, even with high-pressure ratios and no losses. It is evident that regeneration is most effective at low-pressure ratios. At high-pressure ratios, where T_4 becomes less than T_2, regeneration reduces efficiency.

Performance Characteristics. In real gas-turbine cycles numerous losses reduce the efficiencies below ideal values. Of greatest importance are:

1. Compressor efficiency η_C is defined as the ratio of the enthalpy change in ideal isentropic compression to the actual enthalpy change. Thus, this is an indicated efficiency given by the following:

$$\eta_C = \frac{(H_2 - H'_1)_S}{H'_2 - H'_1} \tag{49}$$

Values of η_C range from 75 to 85%.

2. If heat losses are neglected, turbine efficiency is defined as the ratio of the actual enthalpy change to the enthalpy change of ideal isentropic expansion. Thus, the indicated turbine efficiency is given as,

$$\eta_T = \frac{H'_3 - H'_4}{(H'_3 - H_4)_S} \tag{50}$$

3. Mechanical losses occur in the bearings of the compressor and the turbine that may be taken into account by an over-all mechanical efficiency of the combined compressor–turbine unit. Thus, the over-all mechanical efficiency is defined as the ratio of the brake horsepower output of the turbine to its indicated horsepower output. Generally this efficiency is high, in the range of 0.96 to 0.98.

4. Combustion efficiencies, due to incomplete burning of the fuel and heat losses from the combustion chamber, are typically about 95%.

5. Pressure drop occurs in the regenerator and piping, both between the compressor and the turbine and between the turbine outlet and the exhaust: these losses are determined by the size and design of the piping, the combustion chamber, and the regenerator.

6. Regeneration is always less than ideal with finite heat-transfer areas. The optimum size of a regenerator must be arrived at by balancing fuel savings against equipment and maintenance costs. The size and cost of a regenerator per unit of heat transferred are largely determined by the temperature difference producing the heat flow. Neglecting heat losses and changes in heat capacity,

$$\Delta t = T'_4 - T'_R \tag{51}$$

where T'_4 = temperature of air from turbine, entering the regenerator

T'_R = temperature of air from compressor, leaving the regenerator.

The value of Δt that can be economically justified is a function of the cost of fuel and the cost per unit area of regenerator surface.

Regeneration efficiency is expressed as the ratio of the enthalpy increase of the air from the turbine in the regenerator to the total enthalpy difference between T'_4 and T'_2. Thus

$$\eta_R = \frac{H'_R - H'_2}{H'_4 - H'_2} \tag{52}$$

where H'_R is the enthalpy of the air from the compressor leaving the regenerator. The efficiencies of regenerator installations may vary widely, depending on the specific requirements involved.

The effects of the foregoing losses are shown diagrammatically by the broken lines in Fig. 202. Compressor inefficiency causes V_2, T_2, and S_2 to be greater than ideal. Pressure drop between the compressor and turbine causes p_3 to be lower and S_3 greater than ideal. Turbine inefficiencies and pressure drop in the exhaust system cause p_4, T_4, and S_4 to be greater than ideal.

The performance characteristics of a gas-turbine cycle are readily calculated by consideration of the individual steps of the cycle, taking into account the losses accompanying each step. The calculations are most conveniently carried out by means of Table 55.

Illustration 2. A gas turbine operates in air at 70° F and 1.0 atm with a turbine inlet temperature of 1500° R and a pressure ratio p_2/p_1 of 3.0. The component efficiencies are as follows:

Compressor efficiency, 0.80
Turbine efficiency, 0.85
Combustion efficiency, 0.95
Temperature difference in regenerator, 200° F

The pressure drop between the compressor and the turbine is 2 psi, and that between the turbine and the exhaust is 1.0 psi.

Calculate:

(*a*) The *indicated* thermal efficiency η_I, with and without the regenerator.

(*b*) The *indicated* horsepower output for an air-flow rate of 1 lb per sec.

(*c*) The regenerator efficiency, neglecting heat losses.

(*d*) The regenerator area per horsepower, assuming a heat-transfer rate of 1000 Btu/(sq ft)(hr).

Solution: Refer to Figs. 201 and 202. Table 55 in the original unabridged form was used in the solution.

The primed H terms in the solution below represent actual enthalpies, whereas the unprimed H terms represent enthalpies that would result under isentropic conditions.

Compressor Inlet Conditions (Point 1)

$$T_1 = 70^\circ \text{ F} = 529.7^\circ \text{ R}, \qquad H'_1 = 126.6 \text{ Btu per lb}$$
$$p_1 = 14.696 \text{ psia}, \qquad p_{R1} = 1.2958$$

Compressor Outlet Conditions (Point 2)

$$p_2 = (14.696)(3) = 44.088 \text{ psia}$$

$$p_{R2} = p_{R1}\frac{p_2}{p_1} = (1.2958)\frac{44.088}{14.696} = 3.8874$$

Reference to Table 55 gives $H_2 = 173.44$ Btu per lb.
From equation 49,

$$H'_2 = \frac{(H_2 - H'_1)_S}{\eta_C} + H'_1 = \frac{173.44 - 126.6}{0.80} + 126.6 = 185.2 \text{ Btu per lb}$$

Reference to Table 55 gives $T_2 = 772.6^\circ$ R.

Turbine Inlet Conditions (Point 3)

$$T_3 = 1500^\circ \text{ F} = 1959.7^\circ \text{ R}, \qquad H'_3 = 493.5 \text{ Btu per lb} \qquad p_{R3} = 160.271$$
$$p_3 = 44.09 - 2.00 = 42.09 \text{ psia}$$

Turbine Outlet Conditions (Point 4)

$$p_4 = 14.696 + 1.000 = 15.696 \text{ psia}$$

$$p_{R4} = p_{R3}\frac{p_4}{p_3} = 160.271\,\frac{15.696}{42.09} = 59.770$$

Reference to Table 55 gives $H_4 = 376.2$ Btu per lb.
From equation 50,

$$H'_4 = H'_3 - \eta_T(H'_3 - H_4) = 493.5 - 0.85(493.5 - 376.2) = 393.8 \text{ Btu per lb}$$

Reference to Table 55 gives $T_4 = 1592.7^\circ$ R.

(*a*) **Thermal efficiency without regenerator.**

Work done by turbine $= H'_3 - H'_4 = 493.5 - 393.8 = 99.7$ Btu per lb
Work done by compressor $= H'_2 - H'_1 = 185.2 - 126.6 = 58.6$ Btu per lb
Net work done $= 41.1$ Btu per lb

$$\text{Heating value of fuel} = \frac{H'_3 - H'_2}{\eta_{\text{combustion}}} = \frac{493.5 - 185.2}{0.95}$$

$$= \frac{308.3}{0.95} = 324.6 \text{ Btu per lb air throughput}$$

$$\eta_I = \frac{41.1}{324.6} = 0.127$$

(*b*) **Thermal efficiency with regenerator.**

Regenerator exit temperature $T_R = T_4 - T = 1592.7 - 200 = 1392.7°$ R

H'_R $= 341.0$ Btu per lb

Total enthalpy increase between compressor exit and turbine inlet

$$= H'_3 - H'_2 = 493.5 - 185.2 = 308.3 \text{ Btu per lb}$$

Enthalpy increase in regenerator

$$= H'_R - H'_2 = 341.0 - 185.2 = 155.8 \text{ Btu per lb}$$

Enthalpy increase in combustion unit $= 152.5$ Btu per lb

Heating value of fuel $= \dfrac{152.5}{0.95} = 160.6$ Btu per lb

$$\eta_I = \frac{41.1}{160.6} = 0.256$$

(*c*) **Regenerator efficiency.**

$$\eta_R = \frac{H'_R - H'_2}{H'_4 - H'_2} = \frac{341.0 - 185.2}{393.8 - 185.2} = \frac{155.8}{208.6} = 0.746$$

(*d*) **Horsepower output for an air-flow rate of 1 lb per sec.**

$$\text{Ihp} = \frac{(41.1)778}{550} = 58.2$$

(*e*) **Regenerator area per horsepower.**

Heat transferred in regenerator $= H'_R - H'_2 = 155.8$ Btu per lb

$$\text{Regenerator area} = \frac{155.8}{(1000/3600)(58.2)} = 9.64 \text{ sq ft}$$

It will be noted that, under the conditions of Illustration 2, the thermal efficiency of the nonregenerative gas turbine is only 12.7% compared to the ideal value of 26.6%. Addition of regeneration increases the efficiency to 25.6% compared to the ideal value of 63%.

The specifications of Illustration 2 are unfair to the nonregenerative case because the pressure drops specified are largely associated with the regenerator. A nonregenerative turbine would be designed to minimize these losses. For example, if pressure drops were negligible, the thermal efficiency of the nonregenerative case of Illustration 2 would be increased from 12.7 to 15.7%, and the horsepower output would be increased from 58.2 to 71.7.

In order to indicate the effects of pressure ratio and turbine inlet temperature on gas-turbine performance, the calculations of Illustration 2 were carried out at pressure ratios varying from 1.5 to 8 at turbine inlet temperatures of 1500 and 1700° F. The results are shown in Fig. 204

For the nonregenerative curves, it was assumed that pressure drops were negligible.

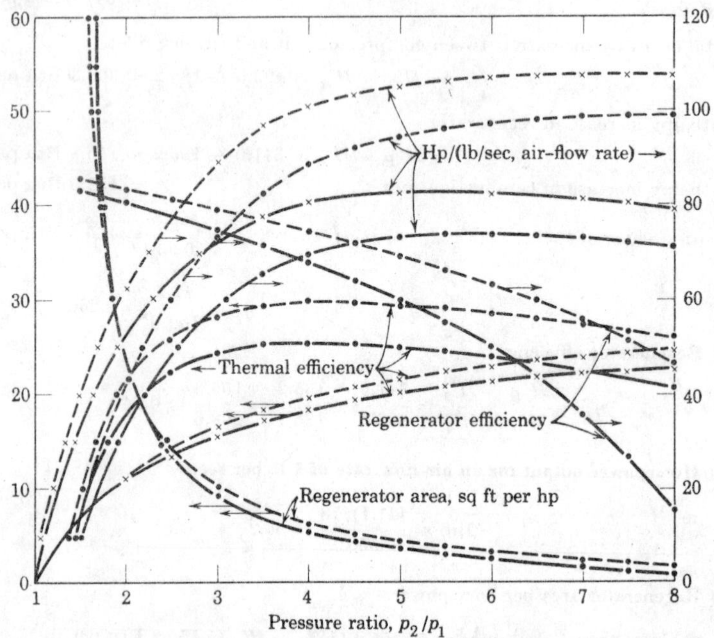

FIG. 204. Performance characteristics of a gas-turbine engine

Compressor efficiency	0.80
Turbine efficiency	0.85
Combustion efficiency	0.95

For regenerative cases:

$$p_2 - p_3 = 2.0 \text{ psi}$$
$$p_4 - p_1 = 1.0 \text{ psi}$$

For nonregenerative cases: Negligible pressure drop
Temperature difference in regenerator: $\Delta t = 200$ F°
Regenerator heat-transfer rate: 1000 Btu/(sq ft)(hr)

Turbine inlet temperature, t_3:	Regenerative	Nonregenerative
1500° F	—·—	—×—
1700° F	–·–·–	–×––

Study of Fig. 204 discloses many of the problems in the application of the gas turbine. In order to obtain reasonable efficiencies in a nonregenerative cycle, high-pressure ratios of the order of 8 are required. The low efficiencies as compared to the ideal cycle result from the fact that the net work is the difference between two large values: the total work of the turbine and the work required for compression. In general, the work of compression is considerably larger than the net work, and inefficiencies in either the compressor or the turbine seriously reduce the difference left for useful work.

Addition of an economical regenerator increases the efficiency to a maximum in the pressure ratio range of 3 to 5. With a gas temperature of 1500° F, the maximum efficiency is 25.5%, while at 1700° F it is increased to 30%. Turbine inlet temperatures are currently limited to the 1500-to-1700° F range by properties of the available materials of construction.

A major problem in some applications of the gas turbine is the poor efficiency at part load or low speed. If speed is maintained, the temperature at the turbine inlet is reduced at part load. As indicated in Fig. 204, the efficiency drops markedly since the compressor work is unchanged, and, at low net power outputs, most of the fuel is required for the inefficiencies of the compressor and turbine. If speed is reduced as well as load, the pressure ratio is reduced, and the turbine inlet temperature is less affected. The resultant reduction in efficiency depends to a large extent on the low-speed characteristics of the compressor and turbine.

The gas turbine has great potential for both stationary and mobile power units. It is capable of high power outputs from a small, low-cost unit, without the complications of steam generation and condensation, with its attendant cooling water requirements. Much development work is in progress in the aircraft, locomotive, automotive, and power-generation fields. Improvements are possible through better materials of construction for higher temperatures, better compressor and turbine designs and cycle modifications such as multistage compression with intercooling.

The most important commercialization of the gas turbine is in the *turboprop* aircraft. A nonregenerative gas turbine is used to drive a propeller through reduction gears. For aircraft applications, refinements are justified which have increased compressor and turbine efficiencies considerably above those assumed in Illustration 2, with resultant improvement in over-all efficiency.

The aircraft application of the gas turbine is attractive because of good power-to-weight ratios and low frontal areas which reduce drag on the aircraft. It is also possible by suitable design to realize improved efficiencies from lower inlet temperatures encountered in high-altitude flight. For example at 30,000 ft, the standard atmospheric pressure is 4.38 psia, and the temperature is 411° R. This reduction in inlet temperature would increase the efficiency of the nonregenerative engine of Illustration 2 from 15.7 to 18.3% where pressure drops are negligible. Further advantages result from effective use of the diffuser or ram pressure

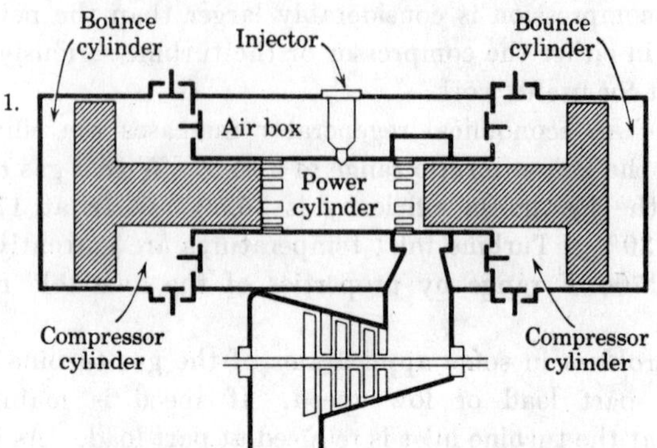

Pistons in starting position. Intake and exhaust ports open. All valves closed.

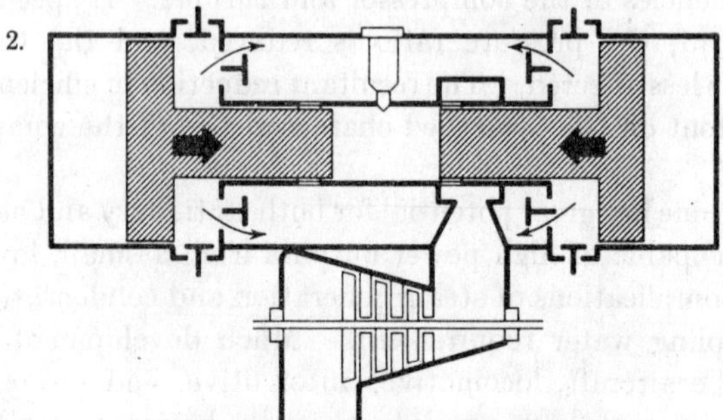

Starting air pressure is admitted to bounce cylinders. Pistons move inward, closing ports, compressing air in power cylinder, and forcing scavenging air from compressor cylinder into air box.

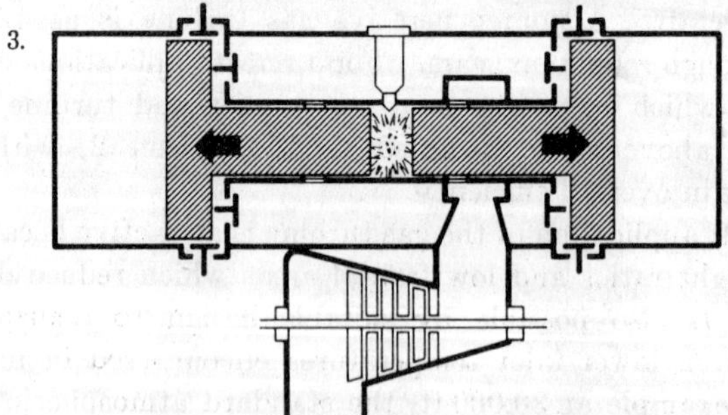

Pistons complete inward travel. Fuel is injected into power cylinder. Combustion starts and power stroke begins.

Fig. 205. Free-piston gas-turbine operation
(A. F. Underwood, *Soc. Auto. Engrs. J.*, **64**, 60, 1956, with permission.)

4.

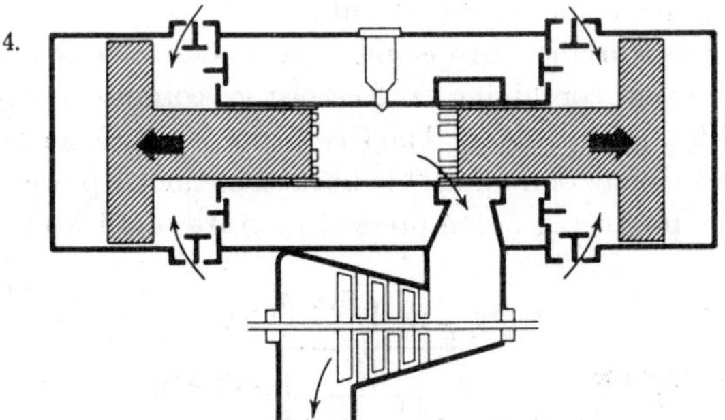

Pistons continue outward travel. Air in bounce cylinders is compressed to store energy for return stroke. Compressor intake valves open. Exhaust ports open and gas is admitted to turbine.

5.

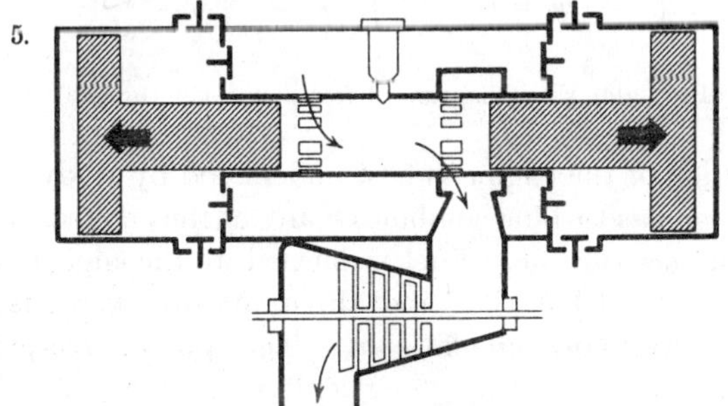

Further outward movement of pistons opens intake ports, completing power stroke. Air from air box scavenges power cylinder and escapes to turbine.

6.

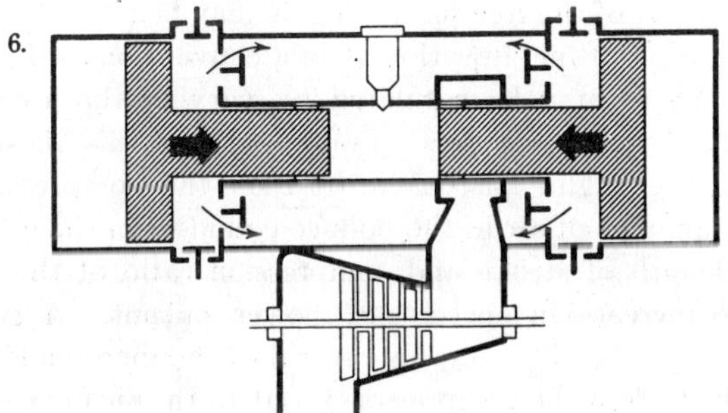

Pressure in bounce cylinder moves pistons inward, starting next cycle.

Fig. 205 (*continued*)

accompanying high-speed flight and of the thrust of the exhaust gases, which is wasted in stationary applications.

Free-Piston Gas Turbine. An engine that is desirable for many application results from combining a free-piston compression-ignition air compressor with a gas turbine. The free-piston compressor serves as the *gasifier* for the turbine, and all fuel is burned in the CI power cylinder of the compressor, producing a compressed mixture of air and combustion

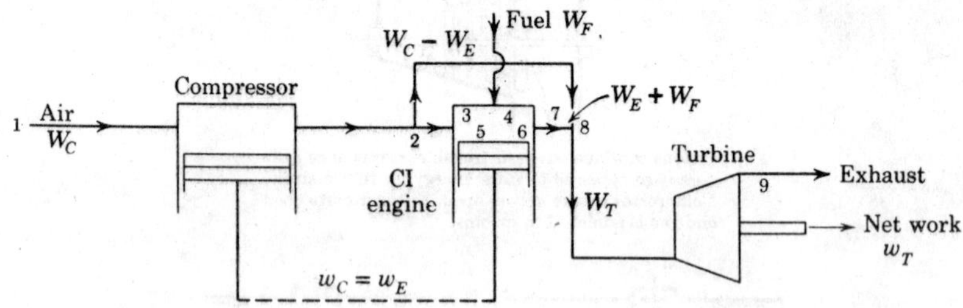

Fig. 206. Flow diagram of free-piston gas turbine

gases. Operation of the engine is best understood by study of Fig. 205.[5]

The free-piston gas turbine combines many of the advantages of the CI engine and the gas turbine. Fuel is burned at the high temperatures and pressures of the CI engine, and all compressor power is developed under these conditions of high efficiency. The opposed free-piston design shown in Fig. 205 is relatively vibrationless, mechanically simple, and low in maintenance costs. All net work is performed by the turbine, and good over-all efficiencies are obtained without the complications of regeneration.

The power output of the free-piston engine may be regulated by control of rate and duration of fuel injection as in a conventional CI engine. In addition, the engine may be regulated by varying the amount of air contained in the bounce cylinders. Compression of this air supplies the force to drive the engine pistons inward on the compression stroke. Increasing the amount of air in the bounce cylinders in the control range increases the length of stroke and compression ratio of the CI engine, with resultant increase in speed and power output. A governor or *stabilizer* may be provided to suitably control the bounce cylinder pressure by bleeding air in from the compressor or out to the turbine.

Another regulating device which is effective for part-load operation is a *recirculation* valve which returns compressed air to the compressor inlet. Such recirculation increases the temperature and reduces the

[5] A. F. Underwood, *S.A.E. Journal*, **64**, 60 (1956).

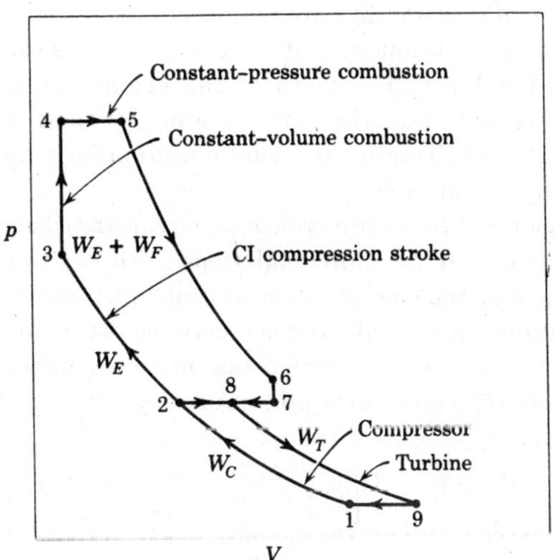

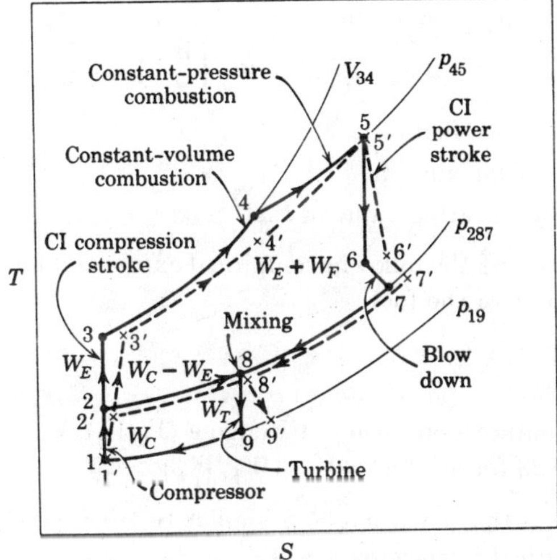

Fig. 207. pV and TS diagrams of free-piston gas-turbine engine (based on unit weight of fluid in each step)

W_C = compressor air-flow rate

W_E = engine air-flow rate

W_F = fuel flow rate

W_T = turbine gas flow rate $W_C + W_F$

density of the inlet air. The amount of air compressed and the load of the compressor are thereby reduced, while the temperatures and speeds necessary for engine operability are maintained.

Thermodynamic cycle analyses of the free-piston gas turbine have been developed by London.[6] The flows of the engine are shown diagrammatically in Fig. 206, and the corresponding pV and TS diagrams in Fig. 207. In the TS diagram, the solid lines represent the ideal and the broken lines the actual cycle.

The performance of the entire cycle may be calculated as the summation of the performances of the individual steps. All net work is performed by the turbine, and the engine power output must equal the compressor power absorption. Such calculations may be based on average heat capacity ratios κ or may be carried out more accurately by means of Table 55. Thus, if η_C is the over-all compressor efficiency, the compressor work w_C is given by

$$w_C = -\frac{W_C R T_a \kappa}{\eta_C M(\kappa - 1)}\left[\left(\frac{p_2}{p_1}\right)^{(\kappa-1)/\kappa} - 1\right] = -\frac{W_C(H_2 - H'_1)}{\eta_C} \tag{53}$$

If η_I is the indicated and η_M the mechanical efficiency of the CI engine, the engine work w_E is given by

$$w_E = W_F q_F \eta_I \eta_M = W_E\left(\frac{W_F}{W_E}\right) q_F \eta_I \eta_M \tag{54}$$

where W_F = engine fuel rate
W_E = engine air rate
q_F = net heating value of fuel

London[6] suggests the following empirical expression for η_I based on the *net* heating value of the fuel

$$\eta_I = 1 - \frac{1}{r_C{}^b} \tag{55}$$

where r_C = compression ratio of CI engine (V_2/V_3)
$b = 0.23$ for air–fuel ratios $(W_E/W_F) > 25$

It will be noted that equation 55 is similar to the Otto-cycle equation 18 with the empirical b replacing $\kappa - 1$.

The ratio of compressor-to-engine air rates is obtained by combining equations 53 and 54, since $w_E + w_C = 0$.

$$\frac{W_C}{W_E} = \frac{W_F q_F \eta_I \eta_M \eta_C M(\kappa - 1)}{W_E R T_a \kappa[(p_2/p_1)^{(\kappa-1)/\kappa} - 1]} = \frac{W_F q_F \eta_I \eta_M \eta_C}{W_E(H_2 - H'_1)} \tag{56}$$

[6] A. L. London, *Trans. Am. Soc. Mech. Engrs.*, **77**, 197 (1955); **78**, 1757 (1956), with permission.

Heat is rejected to the cooling fluid of the CI engine at a rate $-q$ which may be expressed as q^*, a fraction of the heating value of the fuel. Thus,

$$q^* = \frac{-q}{q_F W_F} \tag{57}$$

If the enthalpy of the fuel is neglected, an over-all energy balance of the compressor-engine section is given by

$$W_C H_1 + W_F q_F = (W_C + W_F) H'_8 - q \tag{58}$$

Combining equations 57 and 58 and rearranging gives

$$\frac{W_C}{W_E} = \frac{(W_F/W_E)[q_F(1 - q^*) - H'_8]}{(H'_8 - H'_1)} \tag{59}$$

An expression for the enthalpy of the gases entering the turbine is obtained by equating 56 and 59:

$$\frac{q_F \eta_I \eta_M \eta_C}{H_2 - H'_1} = \frac{q_F(1 - q^*) - H'_8}{H'_8 - H'_1} \tag{60}$$

Rearranging and solving for H'_8,

$$H'_8 = \frac{(1 - q^*) + H'_1 \eta_I \eta_M \eta_C/(H_2 - H'_1)}{1/q_F + \eta_I \eta_M \eta_C/(H_2 - H'_1)} \tag{61}$$

The first term in the denominator of equation 61 is small in comparison to the second. It follows that the enthalpy and temperature of the gas to the turbine are little affected by the heating value of the fuel or the engine air–fuel ratio. The turbine inlet temperature is increased by increasing the compressor pressure ratio, resulting in a greater $H_2 - H'_1$, and is reduced by increasing the fractional heat rejection q^*. Reducing the efficiency terms in equation 61 also increases the turbine temperature.

The net work of the engine is the work of the turbine. Thus, if the indicated efficiency of the turbine is η_T, the indicated work is given as;

$$w_T = \frac{W_T \eta_T R T_{8'\kappa}}{M(\kappa - 1)} \left[1 - \frac{1}{(p'_B/p_1)^{(\kappa-1)/\kappa}}\right] = W_T \eta_T (H'_8 - H_9) \tag{62}$$

It follows from equations 61 and 62 that both the temperature and the power output of the turbine are primarily dependent on the pressure ratio of the compressor which determines both $H_2 - H'_1$ and $H'_8 - H_9$.

The performance of a specified engine is readily calculated from equations 53 to 62. Ordinarily specifications include η_C, η_M, η_T, W_E/W_F, r_C, q_F, q^*, p_1, T_1, p_2, and the pressure difference $p_2 - p'_8$. The engine efficiency is then calculated from equation 55, the turbine inlet temperature from equation 61, the air-rate ratio from equation 59, and the fuel rate from equation 58. The over-all thermal efficiency is the turbine work divided by the heating value of the fuel.

Illustration 3. A free-piston gas turbine is to be built to operate at a compressor pressure ratio of 9 with the following specifications:

$$q_F = 18{,}300 \text{ Btu per lb (net)}, \qquad p_1 = 14.7 \text{ psia}$$
$$T_1 = 530° \text{ R}, \qquad \eta_C = 0.80$$
$$\eta_M = 0.90, \qquad r_C = 10$$
$$W_E/W_F = 30, \qquad q^* = 0.2$$
$$(p_2 - p_8)/p_2 = 0.1, \qquad \eta_T = 0.80$$
$$W_c = 3000 \text{ lb/hr}$$

Calculate:

(*a*) Engine efficiency, η_I.
(*b*) Engine air rate, W_E, lb/hr.
(*c*) Fuel rate, W_F, lb/hr.
(*d*) Turbine gas rate, W_T, lb/hr.
(*e*) The compressor and engine power.
(*f*) Turbine inlet temperature and pressure.
(*g*) Turbine power output, hp.
(*h*) Indicated specific fuel consumption, lb/(Ihp)(hr).
(*i*) Over-all thermal efficiency, based on net heating value.

Solution: Refer to Figs. 206 and 207.

Table 55 in the original unabridged form was used in the solution.

The primed H terms in the solution below represent actual enthalpies, whereas the unprimed H terms represent enthalpies that would result under isentropic conditions.

Compressor Inlet Conditions (Point 1):

$$T_1 = 70° \text{ F} = 529.7° \text{ R}, \qquad H'_1 = 126.6 \text{ Btu per lb}$$
$$p_1 = 14.696 \text{ psia}, \qquad p_{R1} = 1.2958$$

Compressor Outlet Conditions (Point 2):

$$p_2 = 9p_1 = (9)(14.696) = 132.26 \text{ psia}$$

$$p_{R2} = p_{R1}\frac{p_2}{p_1} = (1.2958)\frac{132.26}{14.696} = 11.662$$

Reference to the air table gives $H_2 = 237.4$ Btu per lb.

(*a*) $$\eta_I = 1 - \frac{1}{r^b} = 1 - \frac{1}{10^{0.23}} = 1 - 0.5889 = 0.4111$$

(*b*) **Engine air rate.** Substitutions are made in equation 56, which is then solved for W_E.

$$\frac{3000}{W_E} = \frac{(18{,}300)(0.4111)(0.90)(0.80)}{30(237.4 - 126.6)}$$

$$W_E = 1841 \text{ lb per hr}$$

(*c*) **Engine fuel rate.**

$$W_F = W_E(1/30) = 1841/30 = 61.36 \text{ lb per hr}$$

(*d*) **Turbine gas rate.**

$$W_T = W_C + W_F = 3000 + 61.36 = 3061 \text{ lb per hr}$$

(*e*) **Compressor and engine power.** Substitution in equation 53 gives the following:

$$w_C = -\frac{W_C(H_2 - H'_1)}{\eta_C} = -\frac{3000(237.4 - 126.6)}{0.80}$$

$$= -415{,}520 \text{ Btu per hr}$$

$$\text{Power input to compressor} = \frac{(415{,}520)778}{(33{,}000)60} = 163.3 \text{ hp}$$

Equation 54 may be used to evaluate the work output of the engine, which is identical with the work input to the compressor. The use of equation 54 gives the following:

$$w_e = W_f q_f \eta_I \eta_M = (61.36)(18{,}300)(0.4111)(0.90) = 415{,}520 \text{ Btu per hr}$$

The power output of the engine then equals 163.3 hp.

(*f*) **Inlet conditions of gas turbine (Point 8).** Substitution in equation 61 gives the following:

$$H'_8 = \frac{(1 - q^*) + H'_1\eta_I\eta_M\eta_C/(H_2 - H'_1)}{1/q_F + \eta_I\eta_M\eta_C/(H_2 - H'_1)}$$

$$= \frac{(1 - 0.2) + (126.6)(0.4111)(0.90)(0.80)/(237.4 - 126.6)}{1/18{,}300 + (0.4111)(0.90)(0.80)/(237.4 - 126.6)}$$

$$= 417.4 \text{ Btu per lb}$$

Reference to Table 55 indicates $T_8 = 1681°$ R and $p_{R8} = 86.9637$.

$$(p_2 - p_8)/p_2 = r_p$$

$$p_8 = p_2(1 - r_p) = (132.26)(1 - 0.1) = 119.04 \text{ psia}$$

(*g*) **Turbine power output.**

$$p_{R9} = p_{R8}\frac{p_9}{p_8} = 86.9637\frac{14.696}{119.04} = 10.7303$$

Reference to Table 55 indicates $H_9 = 232.2$ Btu per lb.

$$w_T = \eta_T W_T(H'_8 - H_9) = (0.80)3061(417.4 - 232.2)$$

$$= 453{,}570 \text{ Btu per hr}$$

$$\text{Power output} = \frac{(453{,}570)778}{(33{,}000)60} = 178.2 \text{ Ihp}$$

(*h*) **Indicated specific fuel consumption.**

$$\text{Isfc} = \frac{W_F}{\text{Ihp}} = \frac{61.36}{178.2} = 0.344 \text{ lb fuel per Ihp-hr}$$

(*i*) **Over-all indicated thermal efficiency.**

$$\eta_{\text{over-all}} = (453{,}570)/(61.36)18{,}300 = 0.404$$

Based on the specifications used in Illustration 3, London[6] has calculated performance as a function of pressure ratio. The results are shown graphically in Fig. 208.

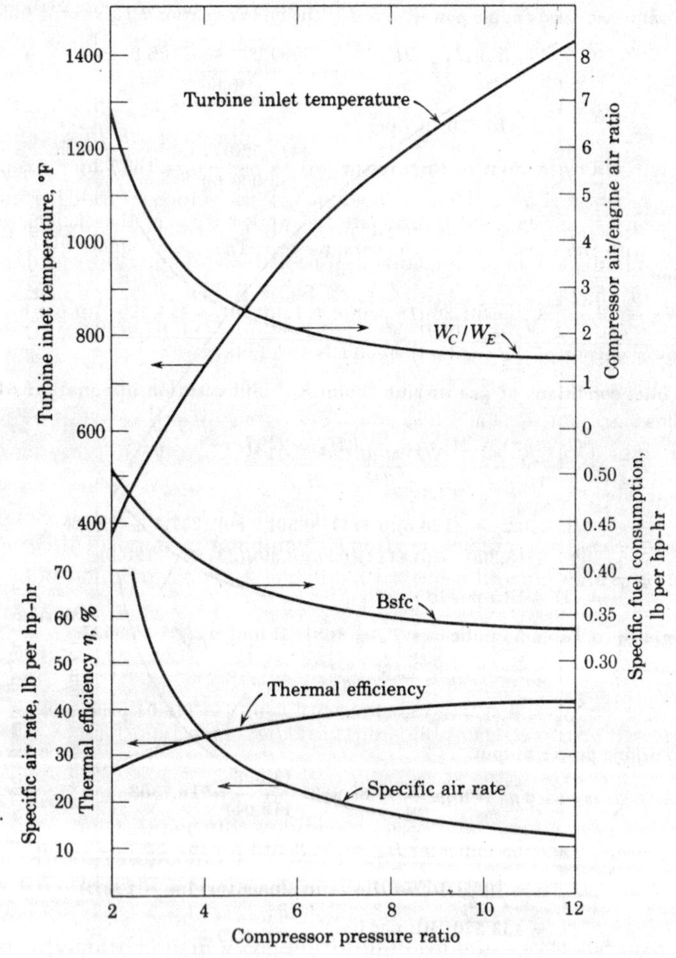

FIG. 208. Free-piston gas-turbine engine performance (specifications of Illustration 3)

Free-piston gas-turbine engines of over 1000 hp have been used in ship propulsion and for stationary power plants with excellent performance and maintenance records. Experimental models are being studied for application to tractors, trucks, locomotives, and passenger cars.

Fuel Requirements. Combustion gas turbines can be operated on a wide variety of fuels including heavy residual fuel oil. However, fuel limitations are encountered when it is attempted to reach maximum efficiency, power-to-weight ratio, or flexibility. These difficulties are most serious in the rotary compressor engines where turbine temperatures must be high and are minimized by the low turbine temperatures of the free-piston type.

For turboprop aircraft engines a clean, distillate oil boiling in the 100-to-700° F range is generally used. An additional specification may be a maximum permissible content of aromatics, which tend to cause trouble by slow burning rates and deposit formation. Gasoline is a satisfactory gas-turbine fuel if it does not contain tetraethyl lead. The combustion products of lead are corrosive and erosive to high-temperature turbine parts.

Turbojet Engines

The turbojet engine is a modification of the gas turbine engine in which the compressor–turbine combination performs no net mechanical work. The useful output of the engine is entirely in the form of the pressure

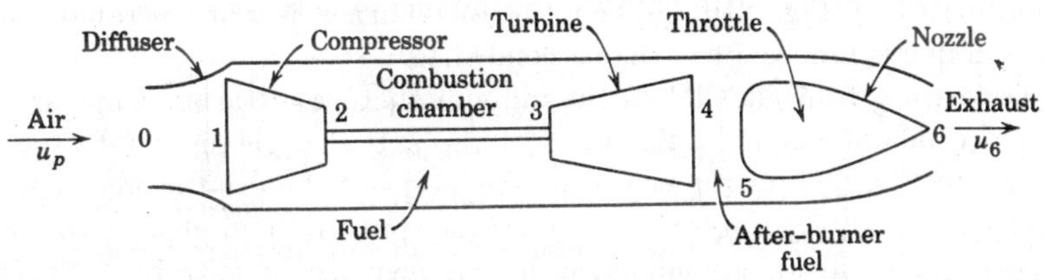

FIG. 209. Turbojet engine

and kinetic energies of the exhaust, which may be converted into propulsive thrust by expansion through a nozzle. The thrust may be increased without increase in turbine temperatures by burning additional fuel in an *afterburner* at the turbine outlet. The components of the engine are shown diagrammatically in Fig. 209.

Air enters the diffuser at flight velocity u_p and is compressed from the pressure and temperature of the atmosphere p_oT_o to the compressor inlet conditions p_1T_1. The compressor raises the pressure and temperature to

p_2, T_2, and combustion of fuel increases the temperature to the turbine inlet conditions p_3, T_3. The turbine expands the gases to outlet conditions p_4, T_4. For maximum thrust, fuel may be added in the afterburner, raising the temperature to T_5 at the nozzle inlet. In the nozzle, the gases are expanded to exhaust conditions p_6, T_6 with a velocity $\bar{u}_j$ in the supersonic range. In jet-engine terminology it is customary to express all velocities relative to the engine.

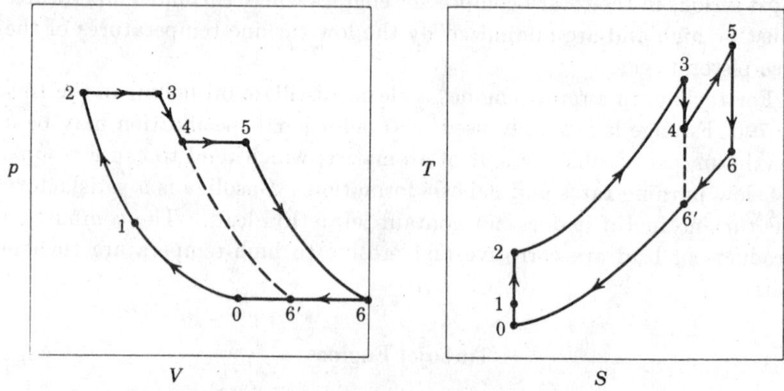

Fig. 210. Ideal turbojet-engine cycle

Ideal Cycle. The ideal cycle of the turbojet engine is composed of isentropic and isobaric steps, as shown graphically in pV and TS coordinates in Fig. 210. When the afterburner is not operated, the broken line is followed to exhaust conditions 6′.

Departures from ideality in the individual steps of turbojet operation produce modifications in the steps of the actual cycle similar to those shown in Fig. 202 for the corresponding steps of the gas-turbine engine. Inefficiencies in the diffuser cause a smaller pressure and greater volume increase than ideal, accompanied by an increase in entropy. Nozzle inefficiencies lead to an exhaust temperature greater than ideal, accompanied by an increase in entropy.

Jet-Propulsion Efficiencies. The only useful work done by a jet results from its thrust acting through a distance. If the jet engine is fixed at rest by a force equal and opposite to its thrust, no work is done, no power is developed, and the efficiency in the usual sense is zero. It is evident that the power and efficiency developed by a jet engine are dependent entirely on the restraining forces, which limit the velocity of its self-propulsion. Maximum efficiency corresponds to a propulsion velocity equal to the velocity of the jet, which could be achieved only with zero restraining forces.

Because of the dependence of power output on the type of service, the usual methods of evaluating output and efficiency have little significance in analyzing the performance characteristics of a jet engine.

The *internal efficiency* or *thermal efficiency* of a jet engine is defined as the ratio of the increase in kinetic energy ΔE_κ to the heating value of the fuel consumed. Thus, since the air enters with a kinetic energy $\bar{u}_p^2/2g_c$, if differences in mass are neglected,

$$\eta_I = \frac{\Delta E_\kappa}{q_f} = \frac{u_j^2 - u_p^2}{2g_c q_f} \tag{63}$$

where η_I = internal efficiency

q_f = heating value of fuel per pound of exhaust gas

The useful propulsive work w_p performed per pound of gas in a nozzle is the product of the specific thrust and the propulsion velocity Fu_p/W. Thus, with an optimum expanding nozzle, for the divergence angle of zero, from equation 16–31,

$$w_p = \left(\frac{F}{W}\right)u_p = \frac{(u_j - u_p)u_p}{g_c} = \frac{u_j^2}{2g_c}[2(1 - r_u)r_u] \tag{64}$$

where w_p = useful propulsive work per pound of gas, (ft)(lb$_f$)/lb$_m$

u_p = propulsion velocity of turbojet engine relative to air, feet per second

F = thrust, lb$_f$

W = exhaust gas-flow rate, lb$_m$ per sec

r_u = velocity ratio, u_p/u_j

u_j = jet velocity, feet per second

The terms in brackets in equation 64 represent the fraction of the total kinetic energy of the jet that is converted into useful propulsive work. It is a function only of the velocity ratio and is shown graphically by curve A in Fig. 211. It will be noted that, for an engine producing a specified jet velocity, the propulsion work done per pound of exhaust gas reaches a maximum where the velocity ratio is 0.5. The maximum work done is 50% of the relative kinetic energy of the exhaust. If the velocity of propulsion is increased beyond a ratio of 0.5, the useful work done declines, reaching zero when the propulsion velocity equals the jet velocity. The factor $2(1 - r_u)r_u$ is a measure of the power output of a jet engine operating with a fixed jet velocity.

The *propulsion efficiency* η_P is defined as the ratio of the propulsion work to the internal increase in kinetic energy ΔE_κ.

$$\eta_P = \frac{w_j}{\Delta E_\kappa} = \frac{2(u_j - u_p)u_p}{u_j^2 - u_p^2} = \frac{2u_p}{u_j + u_p} \tag{65}$$

or, in terms of the velocity ratio r_u,

$$\eta_P = \frac{2r_u}{1 + r_u} \tag{66}$$

The *over-all efficiency* η of a jet engine is defined as the ratio of the useful propulsion work to the heat of combustion of the fuel. Thus,

$$\eta_{\text{over-all}} = \frac{w_p}{q_f} = \frac{(u_j - u_p)u_p}{g_c q_f} \tag{67}$$

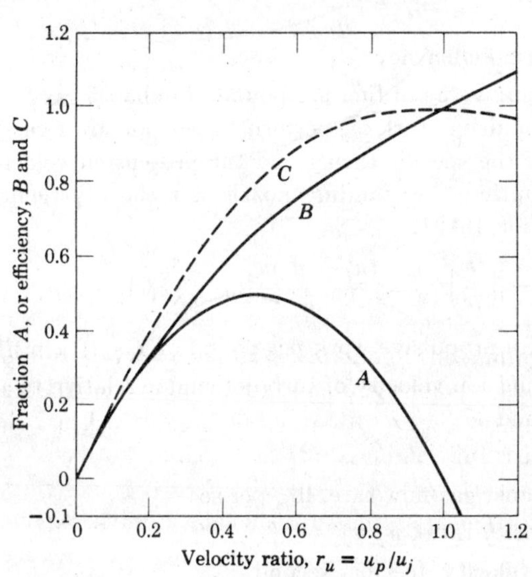

Fig. 211.

A = fraction of total internal jet energy converted to useful work in jet propulsion $= 2(1 - r_u)r_u$
B = propulsion efficiency of jet engine $= 2r_u/(1 + r_u)$
C = propulsion efficiency of rocket engine $= 2r_u/(1 + r_u^2)$

By combination of equations 63, 65, and 67:

$$\eta = \eta_I \eta_P \tag{68}$$

The propulsion efficiency of a jet engine is indicated by curve B of Fig. 211. The efficiency increases over the entire range of velocity ratios, reaching a value of 1.0 at a velocity ratio of 1.0. If the velocity ratio becomes greater than unity, as in a rapid descent in which the force of gravity augments the thrust, the propulsion efficiency becomes greater than 1.0. It follows that, as the propulsion velocity of a jet engine operating at a fixed jet velocity is increased, the power output and efficiency both increase up to a velocity ratio of 0.5, beyond which the efficiency continues to increase but the power output decreases to zero at $r_c = 1.0$. At still higher velocities, the power output becomes negative, and the engine serves as a brake.

Actually the jet velocity of a turbojet engine, operating at fixed turbine inlet temperature and speed, is increased with increased propulsion velocity because of the increased compressor inlet pressure produced by the diffuser. As a result, the actual velocity ratio of maximum power output is above 0.5, depending on the efficiency characteristics of the diffuser.

Turbojet Performance. The performance of a turbojet engine is readily calculated if the efficiencies of the component steps of the cycle are known. The methods are similar to those employed for the gas-turbine engine.

Illustration 4. A turbojet engine operates with a compressor pressure ratio of 6.0, a turbine inlet temperature of 1600° F and no afterburner. Calculate the performance of this engine in flight at 600 miles per hour at sea level (58° F, 1 atm) and at an altitude of 40,000 ft (−68° F, 0.185 atm) with the following specifications:

Ram efficiency of diffuser	0.88
Compressor efficiency	0.83
Combustion efficiency	0.98
Turbine efficiency	0.86
Nozzle energy efficiency	0.85
p_3-p_2, p_5-p_4, p_6-p_0	Negligible

Determine:

(*a*) Internal efficiency.
(*b*) Propulsion efficiency.
(*c*) Over-all efficiency.
(*d*) Specific thrust.
(*e*) Horsepower output per pound of gas flow per second.
(*f*) Nozzle discharge area per pound gas flow per second.
(*g*) Horsepower per square foot of nozzle discharge area.

Solution: Refer to Figs. 209 and 210.

Table 55 in the original unabridged form was used in the solution.

The primed H terms in the solution represent actual enthalpies, whereas the unprimed H terms represent enthalpies that would result under isentropic conditions.

Detailed calculations are shown only for the first case.

Conditions of Surrounding Air (Point o).

$$T_o = 58° \text{ F} = 517.7° \text{ R}, \qquad H'_o = 123.7 \text{ Btu per lb}$$

$$p_o = 14.696 \text{ psia}, \qquad p_{Ro} = 1.1960$$

Flight Velocity:

$$u_p = 600(5280)/3600 = 880 \text{ ft per sec}$$

Conditions at Compressor Inlet (Point 1). As a result of the action of the diffuser, the velocity of the air relative to the compressor inlet is reduced to a relatively low value; hence the kinetic energy at point 1 may be neglected. If it is further assumed that the diffuser operates adiabatically, the following equation may be used to calculate the enthalpy at point 1:

$$H'_1 = H'_o + u_p^2/2g_c$$
$$= 123.7 + (880)^2/2(32.2)778 = 123.7 + 15.5 = 139.2 \text{ Btu per lb}$$

Reference to Table 55 shows that $T_1 = 582.1°$ R and $p_{R1} = 1.8032$. If the compression occurring in the diffuser were isentropic, the pressure developed at point 1 would be calculated as follows:

$$p_1 = p_o \frac{p_{R1}}{p_{Ro}} = 14.696 \frac{1.8032}{1.1960} = 22.16 \text{ psia}$$

However, the diffuser does not operate isentropically; hence the actual pressure at point 1, p'_1, is obtained by using the equation for ram efficiency:

$$\eta_R = \frac{p'_1 - p_o}{p_1 - p_o}$$

$$p'_1 = \eta_R(p_1 - p_o) + p_o = 0.88(22.16 - 14.70) + 14.70 = 21.26 \text{ psia}$$

Conditions at Compressor Outlet (Point 2).

$$p_2 = (21.26)(6) = 127.6 \text{ psia}$$

$$p_{R2} = p_{R1}\frac{p_2}{p_1} = (1.8032)6 = 10.8192$$

From Table 55, $H_2 = 232.3$ Btu per lb.

The actual enthalpy at the compressor outlet is obtained from the equation for compressor efficiency, rearranged as follows:

$$H'_2 = \frac{H_2 - H'_1}{\eta_C} + H'_1 = \frac{232.3 - 139.2}{0.83} + 139.2 = 251.1 \text{ Btu per lb}$$

From Table 55, $T_2 = 1041°$ R.

Conditions at Turbine Inlet (Point 3).

$$T_3 = 1600° \text{ F} = 2059.7° \text{ R}, \qquad H'_3 = 521.3 \text{ Btu per lb}$$
$$p_3 = p_2 = 127.6 \text{ psia}, \qquad p_{R3} = 196.046$$

Conditions at Turbine Outlet (Point 4). Since the shaft-work output of the turbine is equal to the shaft-work input to the compressor, the following equation holds:

$$H'_3 - H'_4 = H'_2 - H'_1$$
$$H'_4 = H'_3 - H'_2 + H'_1 = 521.3 - 251.1 + 139.2 = 409.3 \text{ Btu per lb}$$

From Table 55, $T_4 = 1650.8°$ R.

The value of p_R corresponding to $T_4 = 1650.8°$ R cannot be used to calculate the pressure at the turbine outlet since expansion in the turbine is not isentropic.

However, the turbine outlet pressure may be established through the use of the equation for turbine efficiency, rearranged as follows:

$$H_4 = H'_3 - \frac{H'_3 - H'_4}{\eta_T} = 521.3 - \frac{521.3 - 409.3}{0.86} = 391.1 \text{ Btu per lb}$$

From Table 55, $p_{R4} = 68.7512$.

$$p_4 = p_3 \frac{p_{R4}}{p_{R3}} = 127.6 \frac{68.7512}{196.046} = 44.74 \text{ psia}$$

Conditions at Inlet of Throttle Section (Point 5).

$$T_5 = T_4 = 1650.8^\circ \text{ R}, \qquad H'_5 = H'_4 = 409.3 \text{ Btu per lb}$$

$$p_5 = p_4 = 44.74 \text{ psia}, \qquad p_{R5} = 81.042$$

Note that the value of p_{R5} corresponds to $T = 1650.8$, and is not equal to the p_{R4} value employed above to evaluate the turbine outlet pressure.

Conditions at Outlet of Throttle Section (Point 6).

$$p_{R6} = p_{R5} \frac{p_6}{p_5} = 81.042 \frac{14.696}{44.74} = 26.62 \text{ psia}$$

From Table 55, $H_6 = 299.9$ Btu per lb.

The actual enthalpy at point 6 is established from the nozzle efficiency equation, rearranged as follows:

$$H'_6 = H'_5 - \eta_N(H'_5 - H_6)$$
$$= 409.3 - 0.85(409.3 - 299.9) = 316.3 \text{ Btu per lb}$$

From Table 55, $T_6 = 1297.6^\circ$ R.

$$u_j = [(H'_5 - H'_6)2g_c(778)]^{1/2}$$
$$= [(409.3 - 316.3)2(32.2)778]^{1/2} = 2159 \text{ ft per sec}$$

(*a*) **Internal efficiency.**

$$\eta_I = \frac{(u_j^2/2g_c - u_p^2/2g_c)(1/778)}{(H'_3 - H'_2)/\eta_{\text{combustion}}}$$

$$= \frac{[(2159)^2/2(32.2) - (880)^2/2(32.2)](1/778)}{(521.3 - 251.1)/0.98} = 0.281$$

(*b*) **Propulsion efficiency.**

$$\eta_p = \frac{2r_u}{1 + r_u} = \frac{2(880/2159)}{1 + 880/2159} = 0.579$$

(*c*) **Over-all efficiency.**

$$\eta_{\text{over-all}} = (\eta_I)(\eta_P) = (0.281)(0.579) = 0.1629$$

(*d*) **Specific thrust or impulse.**

$$\frac{F}{W} = \frac{u_j - u_p}{g_c} = \frac{2159 - 880}{32.2} = 39.7 \text{ (lb}_f\text{)(sec)/lb}_m$$

(*e*) **Horsepower per lb gas flow per second.**

$$\text{Hp}/W = (F/W)u_p/550 = (39.7)880/550 = 63.5 \text{ (hp)(sec)/lb}_m$$

(*f*) **Nozzle discharge area per lb gas flow per second.**

$$V_6 = \left(\frac{1}{29}\right)359\,\frac{1297.6}{491.7} = 32.67 \text{ cu ft per lb}$$

$$A/W = V_6/u_j = 32.67/2159 = 0.01513 \text{ (sq ft)(sec)/lb}_m$$

(*g*) **Horsepower per square foot of nozzle discharge area.**

$$\text{Hp}/A = 63.5/0.01513 = 4200 \text{ hp per sq ft}$$

Similar calculations are carried out for an altitude of 40,000 ft, assuming the temperature to be −68° F and the pressure to be 0.185 atm. The results of both sets of calculations are tabulated below.

Summary

	Case I	Case II
Flight speed, miles per hr	600	600
Altitude, above sea level, ft	0	40,000 ft
Air temperature, °F	58	−68
Air pressure, atm	1	0.185
Jet velocity, ft per sec	2159	2494
(*a*) Internal efficiency	0.281	0.329
(*b*) Propulsion efficiency	0.579	0.522
(*c*) Over-all efficiency	0.163	0.171
(*d*) Specific thrust, $(\text{lb}_f)(\text{sec})/\text{lb}_m$	39.7	50.1
(*e*) Hp per lb gas flow per sec	63.5	80.2
(*f*) Nozzle discharge area, sq ft per lb gas flow per sec	0.01513	0.0693
(*g*) Hp per sq ft of nozzle discharge area	4200	1158

The results of Illustration 4 point out the advantages of the turbojet engine for high-speed, high-altitude flight. Even though the specifications on which the case is based are conservative, a creditable over-all efficiency of 17.1% is indicated at an altitude of 40,000 ft. The improved efficiency at high altitude results largely from the low inlet temperature with correspondingly reduced work of compression and a more favorable expansion ratio for the nozzle.

The air-flow rate in a given engine operating at a fixed propulsion velocity and compressor speed is approximately proportional to the density of the air at inlet conditions. The ratio of the density at 40,000 ft to that at sea level is 0.245. On this basis, if the compressor speeds were the same, the air-flow rate in the engine operating at sea level would be 4.1 times greater than at 40,000 ft. Introducing this factor gives approximate comparison of the performance and design requirements of an engine for the two conditions.

Thus, if the results of case II of Illustration 4 are multiplied by 0.245, values of relative thrust, power, and nozzle discharge area are obtained which may be compared to those of case I as follows:

	I	II
	Sea Level	40,000 ft
Relative thrust	39.7	(50.1)(0.245) = 12.3
Relative horsepower	63.5	(80.2)(0.245) = 19.6
Relative nozzle area	0.0151	(0.0693)(0.245) = 0.0170

It follows that the power and thrust at 40,000 ft are approximately 31% of the sea-level values at the same speed. If the high-altitude flight were at higher speeds, this difference would be reduced, owing to increased ram effect and improved propulsion efficiency. It will be noted that the nozzle discharge areas are approximately independent of altitude.

Afterburner Effects. The afterburner may be used for increasing the thrust of a turbojet engine operating at a fixed turbine inlet temperature and speed. One application of the afterburner is to provide temporary maximum thrust for emergencies in operation of military aircraft. Optimum use of the afterburner in this manner requires a nozzle of variable discharge area.

Jet versus Propeller Aircraft. The foregoing discussion permits qualitative evaluation of the general characteristics of the jet-propelled aircraft as compared to the propeller type, driven either by a piston or by a gas-turbine engine. The thrust which propels the propeller-driven aircraft results from the momentum of the propeller slipstream which is limited to subsonic velocities. Thus both types of aircraft are actually jet-propelled: one by a large mass of air at low velocity and the other by a smaller mass at high velocity.

The propulsion efficiency of the propeller aircraft is represented by curve *B* of Fig. 211 in terms of the ratio of propeller stream to flight velocity. However, the efficiency of the propeller itself must be considered, and this drops off sharply as sonic velocities are approached. For this reason the propeller craft is inherently limited to maximum flight velocities in the range of 400 to 500 miles per hour. In the upper portion of this range, propulsion efficiencies are such that the over-all efficiency of the jet and propeller aircraft are of comparable magnitude. At higher speeds the jet becomes markedly superior.

Much development of the turbojet aircraft is in progress, and component efficiencies substantially above those specified in Illustration 4 have been realized.

Ramjet Engines

From the foregoing discussion of the turbojet engine it is evident that, once a high propulsion velocity is attained, thrust and power can be produced without compressor and turbine. The resulting engine comprising only a diffuser, combustion chamber, and nozzle is termed the *ramjet* and is shown diagrammatically in Fig. 212.

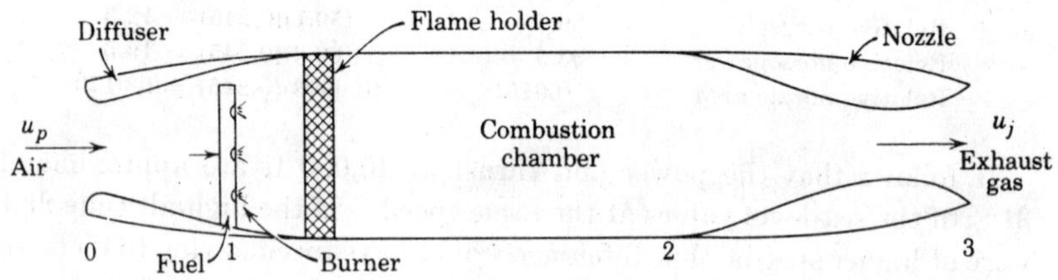

FIG. 212. Ramjet engine

From the standpoint of mechanical simplicity, the ramjet is ideal. However, it must be accelerated to operational velocity by some auxiliary source of power such as a rocket engine. Many difficulties are encountered in achieving efficient and stable operation. It is difficult to maintain combustion in the relatively high velocity and unstable effluent of the diffuser. Elaborate devices to produce turbulence are introduced as *flame holders* to minimize the possibility of extinction of the flame.

Much development work is being done on the ramjet, and it has attractive possibilities for missiles and high-speed aircraft.

Ramjet Performance. The ramjet cycle is identical in appearance with that of the gas-turbine cycle shown graphically in Fig. 202. The diffuser replaces the compressor, and the nozzle replaces the turbine. The same general methods are employed to calculate the performance of the engine, using the efficiency expressions defined for the turbojet.

Illustration 5. Assume that the turbojet engine of Illustration 4, part II, is replaced by a ramjet having the same diffuser, combustion, and nozzle efficiencies. The temperature of the gases leaving the combustion chamber is 1600° F.

Calculate the performance characteristics at an altitude of 40,000 ft and a flight speed of 600 miles per hour, with the air at −68° F and 0.185 atm. Repeat these calculations for a flight speed of 2000 miles per hour at the same altitude and air conditions, assuming that the ram efficiency of the diffuser drops to 80% at the higher speed.

The following subscripts are used to indicate conditions in the ramjet cycle:

0 = diffuser inlet
1 = diffuser outlet
2 = combustion-chamber outlet
3 = nozzle outlet

Solution: Refer to Fig. 212.

Table 55 in the original unabridged form was used in the solution.

The primed H terms in the solution represent enthalpies that would result under actual conditions.

Detailed calculations are shown only for the first case.

Conditions of Surrounding Air (Point o):

$$T_o = -68^\circ \text{ F} = 391.7^\circ \text{ R}, \qquad H'_o = 93.54 \text{ Btu per lb}$$

$$p_o = 0.185 \text{ atm} = 2.719 \text{ psia}, \qquad p_{Ro} = 0.4515$$

Flight Velocity:

$$u_p = 600(5280)/3600 = 880 \text{ ft per sec}$$

Conditions at End of Diffuser (Point 1):

The method of analysis is the same as for the diffuser in Illustration 4, and hence will not be explained in detail.

$$H'_1 = H'_o + u_p^2/2g_c = 93.54 + (880)^2/2(32.2)778$$
$$= 93.54 + 15.46 = 109.0 \text{ Btu per lb}$$

$$T_1 = 456.2^\circ \text{ R}, \qquad p_{R1} = 0.7690$$

$$p_1 = p_o \frac{p_{R1}}{p_{Ro}} = 2.719 \frac{0.7690}{0.4515} = 4.631 \text{ psia}$$

$$p'_1 = \eta_R(p_1 - p_o) + p_o = 0.88(4.631 - 2.719) + 2.719 = 4.402 \text{ psia}$$

Conditions at Nozzle Inlet (Point 2):

$$T_2 = 1600^\circ \text{ F} = 2059.7^\circ \text{ R}, \qquad H'_2 = 521.3 \text{ Btu per lb}$$

$$p_2 = p'_1 = 4.402 \text{ psia}, \qquad p_{R2} = 196.046$$

Conditions at Nozzle Outlet (Point 3):

$$p_{R3} = p_{R2} \frac{p_3}{p_2} = 196.046 \frac{2.719}{4.402} = 121.106$$

From Table 55, $H_3 = 457.2$ Btu per lb.

The actual enthalpy is calculated using the equation for nozzle efficiency, rearranged as follows:

$$H'_3 = H'_2 - \eta_N(H'_2 - H_3) = 521.3 - 0.85(521.3 - 457.2) = 466.8 \text{ Btu per lb}$$

From Table 55, $T_3 = 1863^\circ$ R.

$$u_j = [(H'_2 - H'_3)2g_c(778)]^{1/2} = [(521.3 - 466.8)(64.4)778]^{1/2} = 1652 \text{ ft per sec}$$

(*a*) **Internal efficiency.**

$$\eta_I = \frac{\Delta E_k}{q_F} = \frac{(u_j^2 - u_p^2)/2g_c(778)}{(H'_2 - H'_1)/\eta_{\text{combustion}}}$$

$$= \frac{[(1652)^2 - (880)^2]/2(32.2)778}{(521.3 - 109.0)/0.98} = 0.0928$$

(*b*) **Propulsion efficiency.**

$$\eta_p = \frac{2r_u}{1 + r_u} = \frac{2(880/1652)}{1 + 880/1652} = 0.695$$

(*c*) **Over-all efficiency.**

$$\eta_{\text{overall}} = \eta_I \eta_P = (0.0928)(0.695) = 0.0645$$

(*d*) **Specific thrust.**

$$\frac{F}{W} = \frac{u_j - u_p}{g_c} = \frac{1652 - 880}{32.2} = 24.0(\text{lb}_f)(\text{sec})/\text{lb}_m \text{ air}$$

(*e*) **Horsepower per lb gas flow per second.**

$$\text{Hp}/W = (F/W)u_p/550 = (24.0)(880)/550 = 38.4 \text{ (hp)(sec)/lb}_m$$

(*f*) **Nozzle discharge area per lb of gas flow per second.**

$$V_3 = \left(\frac{1}{29}\right) 359 \left(\frac{14.696}{2.719}\right) \frac{1863}{492} = 253.5 \text{ cu ft per lb}$$

$$A/W = V_3/u_j = 253.5/1652 = 0.1534 \text{ (sq ft)(sec)/lb}_m$$

(*g*) **Horsepower per square foot of nozzle discharge area.**

$$\text{Hp}/A = 38.4/0.1534 = 250 \text{ hp per sq ft}$$

Similar calculations are carried out at a flight speed of 2000 miles per hour. The results of both sets of calculations are tabulated below.

SUMMARY

	Case I	Case II
Flight speed, miles per hr	600	2000
Altitude above sea level, ft	40,000	40,000
Air temperature, °F	−68	−68
Air pressure, atm	0.185	0.185
Jet velocity, ft per sec	1652	3728
(*a*) Internal efficiency	0.0928	0.411
(*b*) Propulsion efficiency	0.695	0.881
(*c*) Over-all efficiency	0.0645	0.362
(*d*) Specific thrust $(\text{lb}_f)(\text{sec})/\text{lb}_m$ air	24.0	24.7
(*e*) Hp per lb gas flow per sec	38.4	131.7
(*f*) Nozzle discharge area, sq ft per lb of gas flow per sec	0.1534	0.0367
(*g*) Hp per sq ft of nozzle discharge area	250	3589

Comparing the turbojet and the ramjet at the same altitude and a speed of 600 miles per hour in Illustrations 4 and 5 shows that, if both are limited to the same maximum temperature, the performance of the ramjet is inferior in every respect. The over-all efficiency is only 6.45% as compared to 17.1%, and the horsepower per square foot of jet area is 250 as compared to 1158. However, if the speed of the ramjet is increased to 2000 miles per hour, its efficiency increases to 36.2%, and the horsepower per square foot of jet area becomes 3589.

It is evident that the ramjet has great potential for very high-speed propulsion at high altitudes. It has the further advantage of being able to operate at higher gas temperatures than the turbojet because no delicate mechanical parts are involved. At present the development of the ramjet is in the experimental stage for missiles, but it may find aircraft applications in conjunction with rocket engines for take-off.

The V-1 missile employed by the Germans in World War II used a modified ramjet engine termed a *pulse jet*. In this modification, the operation of the combustion chamber is stabilized by self-actuating check valves at the inlet to the diffuser. In flight, these valves repetitively open to admit a charge of air and close, owing to the back pressure developed as it is heated in the combustion chamber. This occurs at a frequency determined by the design and results in the characteristic noise of the "buzz bomb."

Rocket Engines

The foregoing discussion of jet propulsion has involved what is termed *duct propulsion*, wherein the surrounding fluid is passed through the device, accelerated, and ejected to provide propulsion thrust. *Rocket propulsion* differs in that the matter ejected to provide thrust is stored within the device and no fluid is ducted through it. Thus rocket propulsion can be applied either in a fluid or in a vacuum.

The *propellant* of a rocket must be a complete fuel which can liberate energy without extraneous matter. Solid, liquid, and gaseous propellants have been used singly and in various combinations. The most familiar is gun powder which is extensively used in military rocket applications. The rate of release of the energy of rocket powder is controlled by the size and shape of the grains. Relatively slow, uniform release is obtained with grains which may be several feet in length with intricate cross-sectional designs.

Liquid propellants are generally used for the sustained operations required in long-range rockets. A monopropellant contains an oxidizing agent and a fuel in a single material. It may be a liquid mixture of several compounds or a single compound such as nitromethane. In either case it must be stable at ambient conditions but decompose with liberation of heat at high temperatures.

A *bipropellant* rocket engine employs two propellant components which are stored separately and mixed in the combustion chamber. The majority of the successful long-range rockets have used bipropellant systems comprising an oxidizer and a fuel.

A rocket engine is simple in principle, comprising a feed system, a combustion chamber, and a nozzle. The feed system may employ compressed gas to force the propellants to the combustion chamber. In large applications, propellant pumps may be used, driven by a small turbine operating on gases produced in its own gas generator.

One of the serious problems in rocket design is the cooling of the combustion chamber to permit the use of temperatures of the order of 5000° F in contact with materials of construction generally limited to much lower temperatures. Cooling is accomplished by various schemes, sometimes using the liquid-propellant components as coolants on their way to the combustion chamber.

Rocket engines are widely used in missiles and for auxiliary aircraft power. Recently the one-man helicopter has been developed with rocket engines at the tips of the rotors. This is an example of the application of the rocket engine to the production of rotative shaft work.

Rocket-Engine Efficiencies. Since no fluid is ducted into the rocket engine, its operation is independent of propulsive velocity. An expression for thrust is obtained by setting $u_1 = 0$ and $\alpha = 0$ in equation 16–27, page 689. Thus,

$$F = \frac{Wu_j}{g_c} + A_j(p_j - p_o) = WI_s + A_j(p_j - p_a) \tag{69}$$

where F = thrust
u_j = jet velocity
W = rate of propellant flow
I_s = specific impulse $= u_j/g_c = \dfrac{F}{W}$ where $p_j = p_o$
A_j = nozzle discharge area
p_j = nozzle discharge pressure
p_o = atmospheric pressure

In a well-designed nozzle $p_j = p_o$, and the pressure-thrust term disappears.

The jet velocity of a rocket is obtained from a simplification of equation 16–42 page 692. Thus,

$$u_j = \left\{\frac{2g_c\kappa RT_1}{M(\kappa - 1)}\left[1 - \left(\frac{p_o}{p_1}\right)^{(\kappa-1)/\kappa}\right]\right\}^{1/2} = \left\{\frac{2g_c c_p T_1}{M}\left[1 - \left(\frac{p_o}{p_1}\right)^{R/c_p}\right]\right\}^{1/2} \tag{70}$$

where p_o = atmospheric pressure
p_1 = combustion chamber pressure
c_p = molal heat capacity of exhaust gases
T_1 = combustion-chamber temperature

Efficiencies of rocket engines are expressed in terms similar to those developed on pages 797–805 for the turbojet engine. The internal efficiency η_I is defined as follows;

$$\eta_I = \frac{u_j^2}{2g_c(-\Delta H)} \tag{71}$$

where ΔH = standard heat of reaction at 25° C, corresponding to the products produced in the chamber, per unit mass of propellant.

The propulsion efficiency is defined as the ratio of the propulsion work to the sum of the propulsion work and the unused kinetic energy of the exhaust jet. In ideal jet propulsion, with the propulsion speed equal to the jet velocity, the vehicle would merely move through stationary air, and the absolute velocity of the jet would be zero. Thus the residual energy of the jet which is wasted in actual propulsion is its absolute kinetic energy relative to a fixed point. Then, considering unit mass flow rate of propellant and assuming no pressure thrust,

$$\eta_P = \frac{Fu_p}{Fu_p + (u_j - u_p)^2/2g_c} = \frac{u_j u_p}{u_j u_p + (u_j - u_p)^2/2} = \frac{2u_j u_p}{u_j^2 + u_p^2} \tag{72}$$

where u_p = propulsion velocity, or, in terms of the velocity ratio $r_u = u_p/u_j$,

$$\eta_P = \frac{2r_u}{1 + r_u^2} \tag{73}$$

The variation of η_P with r_u is shown graphically by curve C of Fig. 211. It will be noted that in the rocket the propulsive efficiency reaches a maximum of 1.0 and then declines at higher values of r_u.

The over-all efficiency is defined as the ratio of the propulsion work to the heat of reaction and kinetic energy of the fuel. Thus, for unit mass flow rate and no pressure thrust,

$$\eta_T = \frac{Fu_p}{-\Delta H + u_p^2/2g_c} = \frac{u_j u_p}{(-\Delta H)g_c + u_p^2/2} \tag{74}$$

It will be noted that, with these definitions, the over-all efficiency is *not* the product of the internal efficiency times the propulsion efficiency.

Rocket-Engine Performance. Since the rocket engine must carry its oxidizer as well as its fuel, the problems of maximum performance per unit mass and per unit volume of propellant become of paramount importance. High performance per unit mass results from high specific impulse corresponding to a high jet velocity. The factors determining jet velocity and specific impulse are apparent from equation 70. The following are desirable:

1. Low molecular weight of exhaust gases.
2. Low heat-capacity ratio or maximum molal heat capacity of the exhaust gases.
3. Maximum combustion temperature.
4. Maximum combustion-chamber pressure.

Thus, the ratio of the molal heat capacity to the molecular weight of the exhaust gases is an important consideration in evaluating propellants. A high combustion-chamber temperature requires a propellant producing a high flame temperature as a result of a high heat of reaction relative to the heat capacity of its products. In modern cooled combustion chambers, temperatures in excess of 4500° F can be utilized. In this temperature range, chemical equilibria are of major importance in determining actual heats of reaction and flame temperatures as discussed in Chapter 26.

High performance per unit volume of propellant is obtained from a fuel having both a high specific impulse I_s and a high bulk density d. Because of the importance of the volume of a rocket system, the product dI_s is frequently used as a basis for comparing the desirability of various propellant combinations.

The optimum pressure of the rocket combustion chamber is determined by consideration of the increased weight of the chamber in relation to the increased thrust obtained. In general, pressures of the order of 300 to 500 psia are used. In this range, an increase of 100% in pressure will increase the thrust by less than 10%.

In Table 58 are calculated theoretical performance characteristics of several liquid-propellant combinations of current interest.[7] These values correspond to a chamber pressure of 500 psia and a nozzle exhaust pressure of 14.7 psia with optimum expansion. It is assumed that combustion is adiabatic, and that the combustion products expand isentropically as ideal gases, with an infinite ratio of chamber area to throat area. It is also assumed that the composition of the products is frozen at the equilibrium corresponding to the conditions of the combustion chamber. Taking into account the shift in equilibrium compositions, accompanying expansion and cooling would increase the theoretical specific impulse by 3 to 10%.

[7] Selected from a pamphlet published by Rocketdyne, a division of North American Aviation, Canoga Park, Calif.

TABLE 58 THEORETICAL PERFORMANCE OF SEVERAL ROCKET PROPELLANT COMBINATIONS

r = pounds of oxidizer per pound of fuel
κ = specific heat ratio of products
d = bulk density of combination at 80° F
$d = (r + 1)/(r/d_o + 1/d_f)$
T_c = theoretical combustion temperature, degrees Fahrenheit
M_c = average molecular weight of combustion products
I_s = theoretical specific impulse, seconds

Oxidizer	Fuel	r	T_c	κ	M_c	d	I_s	dI_s
Brominepentafluoride	Ammonia	6.0	6660	1.34	29	1.80	236	425
Fluorine	Ammonia	2.6	7270	1.33	19	1.15	306	352
	Diborane	5.0	7880	1.30	21	1.07	310	332
	Hydrazine	2.0	7740	1.33	19	1.30	316	411
	Hydrogen (max dI_s)	19.0	8530	1.34	18	0.75	338	254
	Hydrogen (max I_s)	4.5	5000	1.33	8.9	0.32	374	120
	JP-4 (gasoline)	2.6	7100	1.33	24	1.19	282	336
Hydrogen peroxide (99.6% H_2O_2)	Ethyl alcohol, 92.5%	4.0	4600	1.20	23	1.24	245	304
	JP-4	6.5	4830	1.20	22	1.28	248	317
	Hydrazine	1.6	4690	1.22	19	1.25	263	329
Red fuming nitric acid (RFNA) (22% NO_2)	Ammonia	2.15	4220	1.24	21	1.12	237	265
	JP-4	4.1	5150	1.23	25	1.30	238	309
Oxygen (LOX)	Nitropropane	0.9	5620	1.23	23	1.05	257	270
	Ethyl alcohol, 92.5%	1.5	5370	1.21	23	0.98	256	251
	JP-4	2.2	5880	1.24	22	0.98	264	259
	Ammonia	1.3	4910	1.23	19	0.88	266	234
	Hydrazine	0.70	5370	1.25	18	1.06	282	299
	Hydrogen (max dI_s)	8.0	5870	1.22	16	0.43	317	136
	Hydrogen (max I_s)	3.5	4500	1.26	9	0.26	364	95
Ozone (100%)	JP-4	1.9	6380	1.25	21	1.17	283	331
	Hydrogen	3.2	4840	1.26	8.3	0.26	393	102

At present, the propellant combination most widely used for large rockets is liquid oxygen LOX and a specially selected petroleum naphtha fraction in the gasoline–kerosene boiling range termed JP-4. These materials have the advantage of being readily available at low cost and possess performance characteristics satisfactory for many purposes.

Much research is in progress to develop high-performance propellants having maximum values of I_s and dI_s. The data of Table 58 indicate the major improvements possible through use of fluorine as the oxidizer, together with nitrogen compounds such as ammonia or hydrazine. Other high-energy fuels receiving attention are the metals and their hydrides having low molecular weights such as lithium, beryllium, and boron.

The performance of rocket engines can be calculated by the principles of the foregoing sections if the composition of the products from the combustion chamber is known. Determining this composition requires consideration of chemical equilibria by methods discussed in Chapter 26.

Illustration 6. A rocket engine is operated at a combustion-chamber temperature of 4500° F at 270 psia by burning nitromethane with oxygen in a weight ratio of 1 to 0.05. The composition of the gases is as follows[8] in mole fraction:

CO_2	0.0720
CO	0.2593
H_2O	0.3157
H_2	0.1742
H	0.0068
HO	0.0063
N_2	0.1657
	1.0000

The gases are expanded through a nozzle having an energy efficiency of 0.85.

Average molecular weight of product	= 21.23
κ	= 1.232
Propellant flow rate (nitromethane plus oxygen)	= 1.23 lb per sec
A_j = nozzle throat area	= 0.462 sq in.

At an altitude of 68,000 ft ($T_0 = 393°$ R, $p_0 = 0.047$ atm $= 0.691$ psia) and a speed of 3000 miles per hour, calculate:

(*a*) Jet velocity.
(*b*) Specific impulse.
(*c*) Internal efficiency.
(*d*) Propulsion efficiency.
(*e*) Over-all efficiency.
(*f*) Horsepower per pound of propellant flow per second.
(*g*) Horsepower per square foot of nozzle discharge area.

[8] G. P. Sutton, *Rocket Propulsion Elements*, 2d ed., Table 4–4, p. 115, John Wiley & Sons (1956).

Solution: (*a*) **Jet velocity.**

$$u_j = \left\{\frac{\eta 2g_c R\kappa T_1}{M(\kappa - 1)}\left[1 - \left(\frac{p_o}{p_i}\right)^{(\kappa-1)/\kappa}\right]\right\}^{1/2}$$

$$= \left\{\frac{(0.85)(2)(32.2)(1.987)(778)(1.232)(4960)}{(21.23)(1.232 - 1.000)}\left[1 - \left(\frac{0.691}{270}\right)^{(1.232-1.000)/1.232}\right]\right\}^{1/2}$$

$$= 8418 \text{ ft per sec}$$

(*b*) **Specific thrust.**

$$\frac{F}{W} = \frac{u_j}{g_c} = \frac{8418}{32.2} = 261.4\ (\text{lb}_f)(\text{sec})/\text{lb}_m$$

(*c*) **Internal efficiency.** Calculation of this item requires the evaluation of $(\Delta H)_{298.16}$, based on the products developed in the rocket. The gross (higher) heat of combustion is used. Material-balance calculations show that 0.3312 g-mole of nitromethane produce 1 g-mole of products. The standard heat of combustion is calculated using heat of formation data as follows:

Heat of combustion per g-mole of products = [(0.0720)(−94,052)
+ (0.2593)(−26,416) + (0.3157)(−68,317) + (0.0068)(52,089) + (0.0063)(10,060)]
− [(0.3312)(−26,700)] = −25,929 g-cal

$$(\Delta H)_{298.16} = \frac{-25928.5}{21.23}(1.8) = -2198 \text{ Btu per lb of propellant (nitromethane and oxygen)}$$

$$\eta_I = \frac{u_j^2/2g_c}{-\Delta H_{298.16}}$$

$$= \frac{(70.863)10^6/2(32.2)}{-(-2198)778} = 0.643$$

(*d*) **Propulsion efficiency.**

$$\eta_P = \frac{2r_u}{1 + r_u^2}$$

$$r_u = \frac{u_p}{u_j} = \frac{3000(5280)/3600}{8418} = \frac{4400}{8418} = 0.5227$$

$$\eta_P = \frac{2(0.5227)}{1 + (0.5227)^2} = 0.821$$

(*e*) **Over-all efficiency.**

$$\eta_I = \frac{(u_j)(u_p)}{(-\Delta H)g_c + u_p^2/2}$$

$$= \frac{8418(4400)}{(2198)778(32.2) + (4400)^2/2} = 0.572$$

(*f*) **Horsepower per lb propellant flow per second.**

$$\frac{\text{Hp}}{W} = \frac{u_j u_p}{g_c(550)} = \frac{8418(4400)}{(32.2)550} = 2091\ (\text{hp})(\text{sec})/\text{lb}_m$$

(*g*) **Horsepower per square foot of nozzle area.**

$$= \frac{2091(1.23)}{0.462/144} = 8.02(10^5)\ \text{hp per sq ft}$$

Solid Propellants. Liquid-propellant combinations in general have the disadvantage of being difficult to store in the rocket ready for instant service. For this reason, much attention is being given to the development of solid propellants for large rockets.

The solid propellants long used in small rockets and guns are of the *homogeneous* type comprising nitrocellulose, nitroglycerin, or diethylene-glycol dinitrate, either singly in single-base or combined in double-base propellants. These materials are molded into shapes designed to give the desired burning rate, generally corresponding to burnout in less than 30 sec. Maximum specific impulses are of the order of 240 to 250 sec.

For large rockets much effort is being directed toward *composite* solid propellants, using separate oxidizers and fuels which are bonded together by a plastic binder. The oxidizers are generally perchlorates, particularly ammonium perchlorate AP, which is extensively used with polysulfide fuels. Research is being done on the use of lithium perchlorate, which has the advantage of high density and high oxygen content on a weight basis. However, the solid propellants are generally limited to specific impulses substantially below 300 sec.

Problems

1. Calculate the ideal performance characteristics of the engine of Illustration 1 at wide-open throttle if the compression ratio is increased to 11.0.

2. Repeat the calculations of problem 1 for the engine operated at part throttle with a manifold pressure of 15 in. Hg.

3. Repeat the calculations of problem 1 for operation at wide-open throttle and supercharged to a manifold pressure of 45 in. Hg.

4. It is desired to consider the desirability of producing the power output of the throttled engine of Illustration 1 with a smaller engine of the same design characteristics operating at wide-open throttle. Calculate the ideal piston displacement required, and compare the ideal performance characteristics with those of the engine of Illustration 1*b*.

5. A four-cycle compression-ignition engine has six cylinders with a total piston displacement of 1250 cu in. and a compression ratio of 16. It is operated at a speed of 900 rpm with sufficient excess air to correspond to an ideal Diesel-cycle temperature rise ($T_3 - T_2$) of 2800 F°. Air is supplied at 1 atm and 70° F. Assuming the air-standard Diesel cycle with $\kappa = 1.38$, calculate the performance characteristics listed below. The approximation equation may be used to evaluate n_1.

(a) r_t, the load ratio. For this purpose, assume $T_1 = T_a = 529.7°$ R.

(b) r_e, the expansion ratio.

(c) p_2, the compression pressure, in pounds per square inch.

(d) η_I, the indicated thermal efficiency.

(e) w_s, the indicated shaft work per cycle, in (ft)(lb_f).

(f) Imep, the indicated mean effective pressure, in pounds per square inch.

(g) Ihp, the indicated horsepower of the engine.

(h) Indicated torque of the engine.

6. For the Diesel cycle, the following equations may be developed:

$$r_t = 1 + \left(\frac{T_3 - T_2}{r_c^{\kappa-1}}\right)\frac{1}{T_1}$$

$$T_1 = T_a\left[\frac{1 - (1/r_t)(1/r_c)}{1 - (1/r_c)}\right]$$

Using the data given for problem 5, determine r_t and T_1 through the use of the foregoing two equations, employing a trial-and-error procedure. Then work out the various items called for in problem 5, using the values of r_t and T_1 obtained from the two equations given above. Avoid the use of the approximation equation for n_1.

7. A two-cycle compression-ignition engine has two cylinders with a total piston displacement of 300 cu in. It is supercharged by a mechanically driven blower to an inlet manifold pressure of 45 in. Hg. The compression ratio is 18. Assuming that the air-standard Diesel cycle is followed with $T_3 - T_2 = 2800°$ F and $\kappa = 1.38$, and atmospheric conditions of 70° F and 14.7 psia, calculate:

(a) Supercharger outlet temperature.

(b) Compression pressure and temperature.

(c) Expansion ratio r_e.

(d) Imep.

(e) Indicated horsepower at 2500 rpm.

(f) Power ideally required for supercharging at 2500 rpm.

(g) Over-all thermal efficiency.

(h) Indicated specific fuel consumption if the net heating value of the fuel is 18,300 Btu per lb.

Compare these results with those of Illustration 1.

8. Calculate the thermal efficiency of the Otto cycle when the compression and expansion ratios are each 10, assuming isentropic performance.

9. In an air-standard dual-cycle compression-ignition engine, air enters at 70° F and 14.6 psia. In isentropic compression, the volume is reduced from 3.0 to 0.2 cu ft.

In constant-volume combustion the temperature is increased to 2200° F, and in the constant-pressure combustion to 3000° F. Calculate:

(*a*) Cut-off ratio.

(*b*) Compression ratio.

(*c*) Heat added.

(*d*) Heat rejected.

(*e*) Thermal efficiency.

10. In an air-standard Otto cycle, air is supplied at 80° F and 15.0 psia. The heat added per cycle is 200 Btu with a cycle efficiency of 45%. On the basis of 1 lb of air, calculate:

(*a*) Heat rejected.

(*b*) Net work done.

(*c*) Temperature before expansion.

11. In an air-standard Diesel cycle, the compression ratio is 18; air is supplied at 70° F and 14.6 psia. A temperature of 3000° F is attained in combustion. Estimate the thermal efficiency of the cycle.

12. In an air-standard Brayton cycle, the efficiency of expansion is 0.84 and of compression 0.80 with a pressure ratio of 5. Air is supplied at 70° F, and the maximum temperature obtained is 1400° F. Calculate, on the basis of 1 lb of air:

(*a*) The work of compression.

(*b*) The work of expansion.

(*c*) The thermal efficiency.

13. It is desired to design a regenerative open-cycle gas turbine to develop 300 indicated horsepower with the following specifications:

Compressor efficiency		82%
Turbine efficiency		87%
Combustion efficiency		95%
Turbine inlet temperature		1600° F
Pressure ratio p_2/p_1,		3.5
Temperature difference in regenerator		100° F
Pressure drops:	compressor–turbine	1.5 psia
	turbine–atmosphere	0.7 psia
Ambient conditions		70° F, 1.0 atm.

Calculate:

(*a*) The temperatures and pressures at all points of the cycle.

(*b*) The thermal efficiency.

(*c*) The indicated specific fuel consumption in pounds per indicated horsepower-hour if the total heating value of the fuel is 19,900 Btu per lb.

(*d*) The air-flow rate in pounds per second.

(*e*) The regenerator area in the heat-transfer rate is 800 Btu/(hr)(sq. ft).

14. A free-piston gas turbine is to be designed for the same service as the gas turbine of Problem 13, using the following specifications:

Compressor pressure ratio	8.0	η_c	$= 82$
η_M	85%	r_c	$= 14$
W_E/W_F	25	q^*	$= 0.25$
$(p_2 - p_8)/p_2$	0.1	η_T	$= 0.87$

$q_F = 18{,}300$ Btu per lb (net)

Calculate the indicated specific fuel consumption, and compare with that of Problem 13.

15. The turbojet engine of Illustration 4 is provided with an afterburner which raises the temperature of the gases entering the nozzle to 1700° F. The nozzle is provided with a variable discharge area to approximate optimum expansion.

Calculate the results (*a*) through (*g*) of Illustration 4 when the afterburner is in use at an altitude of 40,000 ft.

16. Calculate the performance characteristics at 2000 mph of the ramjet engine of Illustration 5 if the temperature of the gases leaving the combustion chamber is increased to 2000° F.

17. A rocket engine is to be designed to produce a thrust of 10,000 lb at an altitude of 40,000 ft with the propellant and combustion-chamber conditions of Illustration 6. The nozzle is designed for optimum expansion at 40,000 ft where the atmospheric pressure is 0.509 psia.

As an approximation it may be assumed that equation 69 applies to overexpanding nozzles where $p_j < p_a$ resulting in a negative pressure thrust component. On this basis calculate:

(*a*) The thrust of the engine at sea level.

(*b*) The thrust at 80,000 ft where the atmospheric pressure is 0.206 psia.

19

Liquefaction of Gases, Refrigeration, and Evaporation

In chemical processing, several mechanical devices are of frequent occurrence such as those used in the liquefaction of gases, in mechanical and absorption refrigeration, and in evaporation and distillation by vapor recompression. The object of this chapter is to present the thermodynamic principles involved in these special operations.

Liquefaction of Gases by Self-Cooling

The objectives of the liquefaction of gases are reduction of storage space and transportation costs, production of extremely low temperatures for special processing, and the separation of gaseous mixtures into their pure components by fractional distillation of the liquefied mixture.

Where the critical temperature of a gas is below the temperature of the available cooling water, special methods of liquefaction are required. For example, oxygen must be cooled to at least its critical temperature, $-118.8°$ C, before it can be liquefied. Liquefaction of such gases is accomplished by compression and precooling, first by means of conventional heat exchange media or refrigeration, followed by self-cooling as shown in Fig. 213.

Liquefaction by Free Expansion. The gas to be liquefied is supplied to the system at temperature T_1 and pressure p_1 and is combined with recycle gas which lowers the temperature slightly to T_2. It is then compressed to pressure p_3, usually in a multistage compressor employing intercoolers. A cooler is also provided for the gas emerging from the last stage of the compressor. These coolers in the compressor system are supplied either with cooling water or with a cold fluid obtained from an auxiliary refrigeration system of the conventional type, not shown in Fig. 213. The compressed gas at temperature T_3 and pressure p_3 is passed to a heat exchanger, where it is cooled by the return flow of residual uncondensed recycle gas. The temperature drops from T_3 to T_4, with no appreciable drop in pressure. The cold compressed gas is then expanded through a throttle valve to pressure p_5, thus producing a drop in temperature sufficient to produce liquefaction. From the throttle valve, the mixture of saturated liquid and saturated gas goes to the separator, which is at temperature T_5 and pressure p_5. The uncondensed gas then passes through the heat exchanger, where its temperature rises from T_5 to T_7. There is no significant drop in pressure in the heat exchanger. Considering only that portion of the system including the heat exchanger,

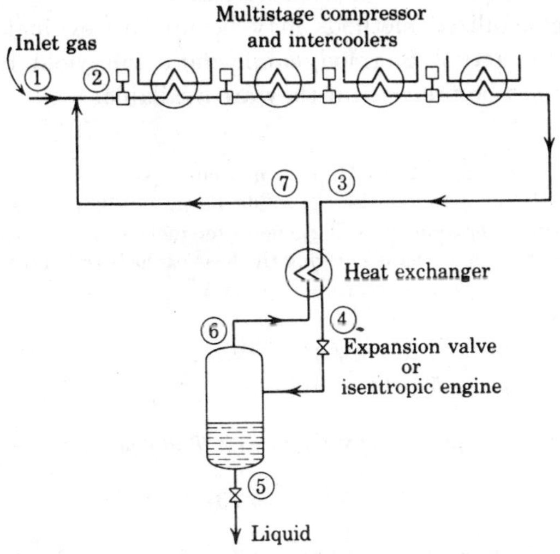

FIG. 213. Low-temperature liquefaction of gases

throttle valve, and separator, and assuming adiabatic operation, the following energy-balance equations may be written, in which y_5 represents the fraction of the gas entering at point 3 which is subsequently liquefied:

Over the heat exchanger:

$$\mathrm{H}_3 + (1 - y_5)\mathrm{H}_6 = \mathrm{H}_4 + (1 - y_5)\mathrm{H}_7 \tag{1}$$

Over the throttle valve and separator:

$$\mathrm{H}_4 = y_5\mathrm{H}_5 + (1 - y_5)\mathrm{H}_6 \tag{2}$$

Over-all balance:

$$\mathrm{H}_3 = y_5\mathrm{H}_5 + (1 - y_5)\mathrm{H}_7 \tag{3}$$

Since only two of these three equations are independent, they may be used to determine a maximum of two unknowns for the system. In a typical design problem, the unknowns are y_5 and T_1, with T_3, p_3, T_5, p_5, and T_7 being specified.

In such a problem, the fraction condensed y_5 is obtained from equation 3, rearranged as follows:

$$y_5 = \frac{\text{H}_7 - \text{H}_3}{\text{H}_7 - \text{H}_5} \tag{4}$$

The unknown temperature T_4 is determined through the use of either equation 1 or 2. By a trial-and-error procedure, the value of T_4 satisfying either equation 1 or 2 is determined.

Although generalized methods may be used to evaluate the various enthalpy items, actual thermodynamic data will yield more accurate results and are to be preferred if they are available.

Illustration 1. It is proposed to liquefy methane, using a system like that shown in Fig. 213, with $p_3 = 1500$ psia, $T_3 = 80°$ F, $p_5 = 150$ psia, and $T_7 = 75°$ F. Assuming adiabatic operation and using generalized methods, determine the following, on the basis of 1 lb-mole methane entering the heat exchanger at point 3: (*a*) fraction of methane condensed, (*b*) temperature at point 4, (*c*) heat-transfer duty of the heat exchanger.

$$T_c = 190.7° \text{ K} = 343.3° \text{ R}, \qquad z_c = 0.290$$

$$p_c = 45.8 \text{ atm} = 673.1 \text{ psia}$$

Molal heat-capacity equation, low range, with T in degrees Rankine:

$$c^*_p = 7.95 + (6.4)10^{-12}T^4$$

Saturation temperature corresponding to 150 psia $= -190°$ F $= 269.7°$ R.

TABLE A

	Point 3	Point 4	Point 5	Point 6	Point 7
Temperature, °R	539.7		269.7	269.7	534.7
Reduced temperature	1.572		0.786	0.786	1.558
Pressure, psia	1500	1500	150	150	150
Reduced pressure	2.229	2.229	0.2229	0.2229	0.2229
Enthalpy correction, Tables 47 and 50	1.799		9.53	0.66	0.120
Entropy correction, Tables 47 and 52			11.89	0.582	

In solving the problem through the use of the generalized tables, it is convenient to take the saturated liquid at point 5 as datum, and to express all enthalpies and entropies relative to that state. Accordingly, equations 2 and 3 where H_5 is zero, become, respectively,

$$\text{H}_4 = (1 - y_5)\text{H}_6 \tag{a}$$

$$\text{H}_3 = (1 - y_5)\text{H}_7 \tag{b}$$

Rearranging equation *b* gives

$$y_5 = \frac{H_7 - H_3}{H_7} \qquad (c)$$

$$H_3 = \left(\frac{H^* - H}{T_c}\right)_5 T_c + \int_{269.7}^{539.7} c^*_p \, dT - \left(\frac{H^* - H}{T_c}\right)_3 T_c$$

$$= (9.53)(343.3) + [7.95(539.7 - 269.7) + (1.28)10^{-12}(539.7^5 - 269.7^5)] - (1.799)(343.3)$$

$$= 3271 + 2203 - 618 = 4856 \text{ Btu per lb-mole}$$

Similarly,

$$H_7 = (9.53)(343.3) + [7.95(534.7 - 269.7) + 1.28(10^{-12})(534.7^5 - 269.7^5)] - (0.120)(343.3)$$

$$= 3271 + 2161 - 41 = 5391 \text{ Btu per lb-mole}$$

$$y_5 = \frac{5391 - 4856}{5391} = \frac{535}{5391} = 0.0992$$

To determine T_4, it is necessary to select a value of temperature that satisfies equation (*a*), which is as follows:

$$H_4 = (1 - y_5)H_6 = (1 - 0.0992)[(9.53 - 0.66)(343.3)]$$
$$= (0.9008)(3045) = 2743 \text{ Btu per lb-mole}$$

Equation 1 may equally well be used to calculate H_4.

By the method used in evaluating H_3 and H_7, a series of enthalpy values is developed, as given in Table B, all at a pressure of 1500 psia and relative to saturated liquid at 150 psia as datum.

TABLE B

Reduced Temperature	Temperature °R	Temperature °F	Enthalpy, Btu/lb-mole	Entropy, Btu/(lb-mole)(R°)
1.12	384.5	−75.2	2457	6.774
1.14	391.3	−68.4	2624	7.202
1.16	398.2	−61.5	2767	7.564
1.18	405.0	−54.7	2901	7.896
1.20	411.9	−47.8	3030	8.206
1.22	418.8	−40.9	3148	8.483
1.24	425.6	−34.1	3265	8.757

By interpolation in the table, $T_4 = 397.0°$ R $= -62.7°$ F.

The heat-transfer duty of the heat exchanger can be computed by considering either of the two streams of fluid passing through:

$q = H_3 - H_4 = 4856 - 2743 = 2113$ Btu per lb-mole methane entering at point 3

Liquefaction by Isentropic Expansion. Instead of the cooled compressed gas expanding through a throttle valve as indicated in Fig. 213, it may be expanded isentropically in an engine. The power thus developed may be used to supply in part the power requirements of the compressor, or it may be used for other purposes. This method of liquefying gases is termed the Claude process.

The Linde process, which involves expansion through a throttle valve as indicated in Fig. 213, is more commonly used than the Claude process, because of its freedom from moving parts and the avoidance of difficulties with condensate in an engine.

The Linde process is applicable only in the region where gases cool by free expansion. At high T_r values, the isotherms relating $(H^* - H)/T_c$ with p_r decrease progressively with increase of p_r. Hence, it is impossible in this region to cool by free expansion. This explains why hydrogen must be pre-cooled to low temperatures before the Linde process may be employed.

Energy-balance equations analogous to those for the Linde process may be set up, as follows, assuming as before that the system operates adiabatically.

Over the heat exchanger:

$$H_3 + (1 - y_5)H_6 = H_4 + (1 - y_5)H_7 \tag{5}$$

Over the engine and separator:

$$H_4 = w_s + y_5H_5 + (1 - y_5)H_6 \tag{6}$$

Over-all balance:

$$H_3 = w_s + y_5H_5 + (1 - y_5)H_7 \tag{7}$$

In addition, since the engine operates isentropically, the following entropy-balance equation is valid:

$$S_4 = y_5S_5 + (1 - y_5)S_6 \tag{8}$$

Since three independent equations are available, a maximum of three unknowns may be determined through their use. In a typical design problem, the unknowns are y_5, T_4, and w_s.

In a problem of this type, T_4 is determined through the use of the following equation, obtained by combining equations 5 and 8:

$$\frac{(H_3 - H_4) - (H_7 - H_6)}{H_7 - H_6} - \frac{S_4 - S_6}{S_6 - S_5} = 0 \tag{9}$$

A trial-and-error procedure is followed to determine the value of T_4 which satisfies equation 9.

The value of y_5 is then obtained through the use of either equation 5 or equation 8. Finally, shaft work is calculated from either equation 6 or equation 7.

Illustration 2. Using the same temperature and pressure conditions specified for Illustration 1, but with an isentropic engine substituted for the throttle valve, calculate the following items, assuming adiabatic operation and using generalized methods: (*a*) fraction of methane condensed, (*b*) temperature at point 4, (*c*) shaft work delivered by the isentropic engine, (*d*) heat-transfer duty of the heat exchanger. Base calculations on 1 lb-mole methane entering the heat exchanger at point 3, and choose the saturated liquid at point 5 as the reference state for enthalpy and entropy values.

A series of entropy values for superheated gas at 1500 psia and various temperatures is developed, using the following equation:

$$\mathrm{s}_T = (\mathrm{s}^* - \mathrm{s})_5 + \int_{269.7}^{T} \frac{c^*_p}{T}\, dT - R \ln \frac{1500}{150} - (\mathrm{s}^* - \mathrm{s})_T \qquad (a)$$

The results are summarized in the Table B of Illustration 1.

Equation 9 simplifies to the following:

$$\frac{\mathrm{H}_3 - \mathrm{H}_4}{\mathrm{H}_7 - \mathrm{H}_6} - \frac{\mathrm{s}_4}{\mathrm{s}_6} = 0 \qquad (b)$$

The values of H_6 and s_6 for use in equation *b* are determined as follows:

$$\mathrm{H}_6 = \left(\frac{\mathrm{H}^* - \mathrm{H}}{T_c}\right)_5 T_c - \left(\frac{\mathrm{H}^* - \mathrm{H}}{T_c}\right)_6 T_c = (9.53)(343.3) - (0.66)(343.3)$$
$$= 3271.6 - 226.6 = 3045 \text{ Btu per lb-mole}$$

$$\mathrm{s}_6 = (\mathrm{s}^* - \mathrm{s})_5 - (\mathrm{s}^* - \mathrm{s})_6 = 11.89 - 0.582 = 11.31 \text{ Btu/(lb-mole)(R°)}$$

From Illustration 1,
$\mathrm{H}_3 = 4856$ Btu per lb-mole, and $\mathrm{H}_7 = 5391$ Btu per lb-mole.
By the use of the foregoing values of H_3, H_6, H_7, and s_6, and the enthalpy and entropy values in the table of Illustration 1, it is found that equation *b* is satisfied when $T_4 = 416.8°$ R $= -42.9°$ F. At this temperature, $\mathrm{H}_4 = 3114$ Btu per lb mole and $\mathrm{s}_4 = 8.403$ Btu/(lb-mole)(R°).

Equation 8 is now used to evaluate y_5.

$$y_5 = \frac{\mathrm{s}_6 - \mathrm{s}_4}{\mathrm{s}_6} = \frac{11.31 - 8.403}{11.31} = 0.2570$$

Equation 5 may equally well be used to evaluate y_5.

Using equation 6 to evaluate shaft work,

$$w_s = \mathrm{H}_4 - (1 - y_5)\mathrm{H}_6 = 3114 - (1 - 0.2570)3045 = 3114 - 2262$$
$$= 852 \text{ Btu per lb-mole}$$

Equation 7 may equally well be used to evaluate shaft work.

The heat-transfer duty of the heat exchanger may be calculated by considering either of the two streams of fluid passing through.

$$q = H_3 - H_4 = 4856 - 3114 = 1742 \text{ Btu per lb-mole methane entering at point 3}$$

The chief source of error involved in the use of the generalized tables resides in the enthalpy and entropy correction terms for the saturated liquid at point 5. Thermodynamic tables for methane indicate the latent heat at −190° F to be 2855 Btu per lb-mole.

Using data from the enthalpy table,

$$\lambda = \left[\left(\frac{H^* - H}{T_c}\right)_{sL} - \left(\frac{H^* - H}{T_c}\right)_{sG}\right] T_c = (9.53 - 0.66)343.3$$

$$= (8.87)(343.3) = 3045 \text{ Btu per lb-mole}$$

Using data from Table 47,

$$\lambda = [(S^* - S)_{sL} - (S^* - S)_{sG}]T_s = (11.89 - 0.582)269.7$$

$$= (11.31)(269.7) = 3050 \text{ Btu per lb-mole}$$

Refrigeration

The purpose of refrigeration is to produce a region of temperature below that of the atmosphere. The net result is the extraction of heat from the low-temperature region and its rejection at the higher temperature of the surroundings or of the available cooling medium. Thus, in effect a refrigeration machine is a heat pump performing the reverse function of an engine. Refrigeration is of importance in the preservation of foods, in air conditioning, and in chemical engineering operations where low temperatures are required for the control of reactions, the condensation of vapors, or the crystallization of solids.

Two general types of refrigeration cycles are in common use. Both depend on the attainment of a cold region by the vaporization of a refrigerant fluid from an evaporator at a low temperature and pressure. In *compression refrigeration*, shown in Fig. 215, mechanical energy is used to compress the vapor leaving the evaporator, so that the heat absorbed at the low temperature of the evaporator may be rejected at the temperature level of the condenser. A refrigerant fluid is employed to absorb heat at low pressure and temperature, and is then compressed to a higher pressure and temperature, where it gives off heat, usually with condensation. Expansion to the lower pressure and temperature completes the cycle. This is usually accomplished by free expansion; isentropic expansion has rarely proved feasible.

In *absorption refrigeration*, shown in Fig. 224, heat that is absorbed at a low temperature and pressure is rejected at an intermediate temperature and high pressure, after its temperature level has been increased by addition of heat from a high-temperature source such as the combustion of a gas. In this scheme no mechanical energy is required except in some systems for pumping the liquid absorbent, and the refrigerant is compressed from the low pressure at which it absorbs heat to the high pressure at which it rejects heat, through intermediate absorption or

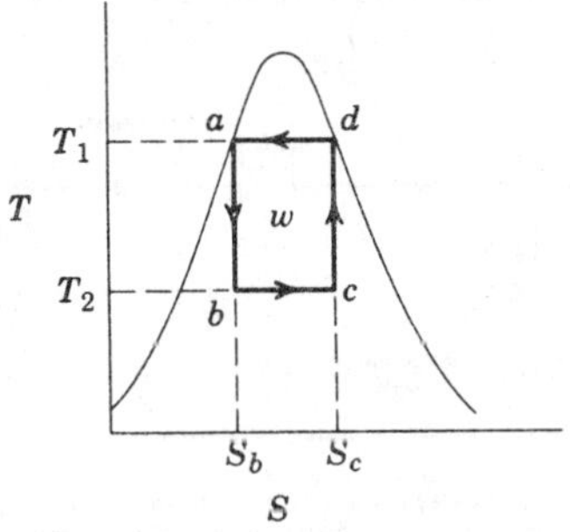

FIG. 214. Reverse Carnot cycle

adsorption at a low-pressure level. The refrigerant is released from the absorbent or adsorbent at a high pressure through the addition of heat from the high-temperature source. This method has the advantage of minimizing or eliminating moving mechanical parts, and may effectively utilize low-pressure exhaust steam or fuel gas as the source of heat.

Compression Refrigeration. A Carnot engine operating in reverse represents an ideal reversible refrigeration cycle. The *TS* diagram of such a cycle operating with saturated vapor is shown in Fig. 214. A quantity of heat q_2 is absorbed in increasing the quality of the vapor at temperature T_2 along *bc*. Isentropic compression along *cd* raises the temperature to T_1, where a quantity of heat $-q_1$ is rejected through condensation of the vapor along *da*. The liquid is isentropically expanded and partially vaporized along *ab* to temperature T_2, thus completing the cycle.

As developed in consideration of the Carnot engine, the net work done by the system is the difference between the heats absorbed at the higher- and lower-temperature levels, and equation 17–6 is directly applicable to the reverse cycle operating as a heat pump. However, in this case both w_s and $S_b - S_c$ are negative. It follows that the amount of heat rejected at the high temperature is greater than that absorbed at the low temperature by the amount of the work done on the system.

For comparing the performance of refrigeration machines, a *coefficient of performance* is used as an index. This is defined as the ratio of the heat absorbed in the evaporator to the work done on the fluid. Thus, the coefficient of performance for a Carnot cycle is

$$\frac{q_2}{-w_s} = \frac{T_2\,\Delta S_2}{\Delta S_2(T_1 - T_2)} = \frac{T_2}{T_1 - T_2} \tag{10}$$

where $\Delta S_2 = S_c - S_b$. According to the Carnot principle, the coefficient of performance of any reversible refrigeration machine operating between

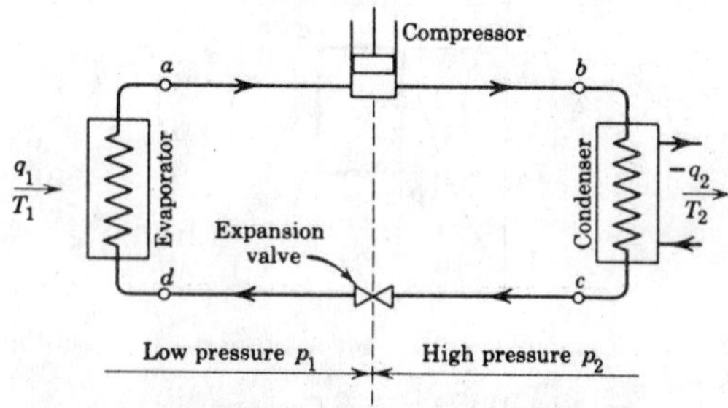

Fig. 215. One-stage compression–refrigeration cycle

temperatures T_1 and T_2 is expressed by equation 10, regardless of the nature of the refrigerant or the steps of the cycle.

The capacity of a refrigeration machine is ordinarily expressed in tons of ice-making capacity per 24 hours. *One ton of refrigeration capacity is capable of absorbing in 24 hours the heat evolved in freezing one ton of water at 32° F.* On this basis, 1 ton refrigeration capacity = 200 Btu per min = 4.71 hp. To define completely the characteristics of a refrigeration operation, it is necessary to specify the coefficient of performance, the refrigeration capacity, and the two temperature and pressure levels of heat absorption and rejection.

A commonly used refrigeration cycle with one-stage compression is shown in Fig. 215. Liquid refrigerant at pressure p_2 is expanded through a throttle valve into an evaporator at pressure p_1 and temperature T_1. As a result of this throttling, part of the liquid is vaporized, and the material entering the evaporator is a mixture of liquid and vapor at pressure p_1. In the evaporator, heat is absorbed from the region being cooled, and the liquid that enters the evaporator undergoes vaporization at temperature T_1. The vapors are then compressed to a pressure p_2 and condensed at the higher temperature T_2 with rejection of the heat of condensation $-q_2$.

The TS diagram of the cycle is shown in Fig. 216. This cycle deviates from the reversed Carnot cycle in the following particulars: (1) The saturated liquid at c is passed through a throttle valve rather than expanded isentropically in a reversible engine, and (2) saturated vapor at a is compressed to pressure p_2, with the result that the gas is superheated above the saturation temperature T_2.

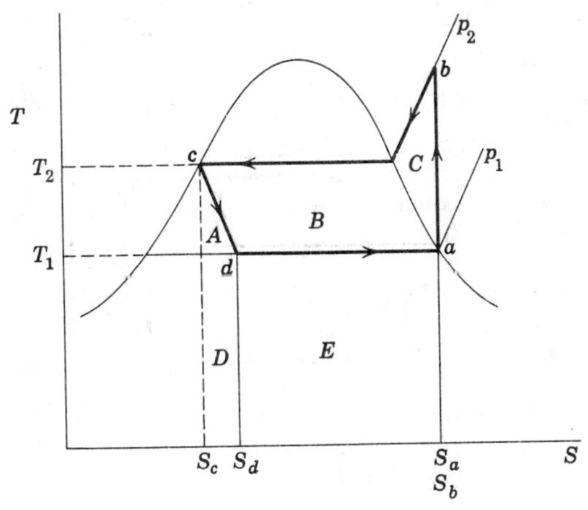

FIG. 216. TS diagram for compression–refrigeration cycle

Though throttling introduces irreversibility into the cycle, the use of a throttle valve instead of an engine permits the construction of cheaper and more robust equipment with lower maintenance costs.

Fig. 216 shows the TS diagram for the compression–refrigeration cycle of Fig. 215, while Fig. 217 shows the corresponding pH diagram with pressure plotted on a logarithmic scale. The latter is particularly convenient in carrying out refrigeration calculations.

The various relationships of the compression–refrigeration cycle are as follows:

$$\text{Heat absorbed in evaporator} = q_1 = E \text{ (Fig. 216)} = T_1(S_a - S_d) = H_a - H_d \tag{11}$$

$$\text{Heat rejected in condenser} = -q_2 = A + B + C + D + E \text{ (Fig. 216)} = H_b - H_c \tag{12}$$

Work done by compressor $= -w_s = -q_1 - q_2$

$= A + B + C + D$ (Fig. 216)

$= (H_b - H_c) - (H_a - H_d)$

$$= (H_b - H_d) - (H_a - H_d) = H_b - H_a \tag{13}$$

$$\text{Coefficient of performance} = \frac{q_1}{-w_s} = \frac{H_a - H_d}{H_b - H_a} \tag{14}$$

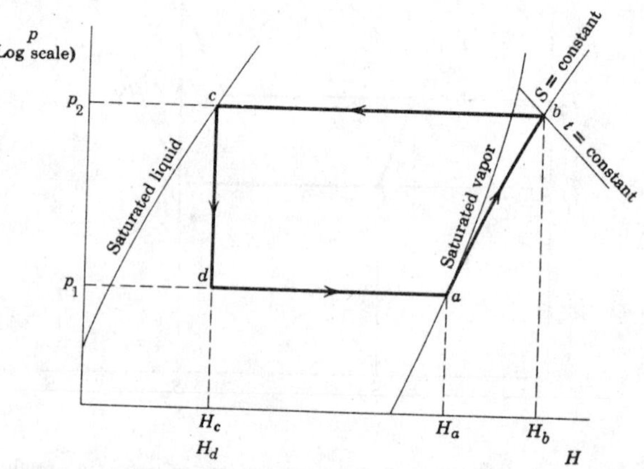

FIG. 217. *pH* diagram for compression–refrigeration cycle

Equation 13 for the shaft-work requirement of the compressor is based on the assumption of reversible operation; the actual work requirements are greater.

Both the isentropic compression efficiency and the volumetric efficiency decrease with increase in the ratio of p_2/p_1. For a noncooled single-stage compressor the efficiencies are expressed approximately by the following equations:[1]

$$\eta_S = 0.86 - 0.038(p_2/p_1) \tag{15}$$

$$\eta_V = 1.00 - 0.050(p_2/p_1) \tag{16}$$

Illustration 3. A single-stage compression–refrigeration plant using propane is to be designed for a capacity of 50 tons of refrigeration per day at an evaporator temperature of 0° F. Cooling water is available at 80° F, and the condenser is to be designed for a condensation temperature of 100° F. Isentropic compression

[1] G. Lorentzen, *Volumetric Efficiency and Compression Efficiency of Refrigeration Compressors*, Doctor's thesis, Fiskeridirektoratet, Bergen, Norway (1949).

efficiency = 0.75; volumetric efficiency = 0.81. Figure 218 shows the pH chart for propane.

Calculate:

(a) The pounds of propane circulating per hour.

(b) The theoretical power required to drive the compressor.

(c) The coefficients of performance at 0° and 100° F, for a Carnot cycle for the theoretical compression cycle, and for the real cycle.

(d) Volumetric displacement of the compressor.

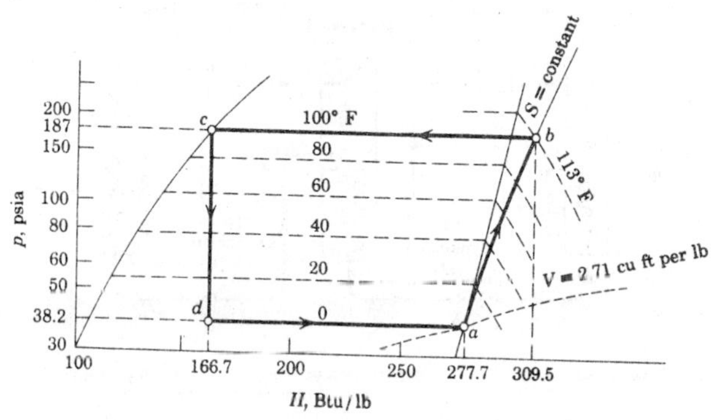

FIG. 218. pH diagram for propane (Illustration 3)

Solutions: (a) **Pounds of propane circulated per hour.**

Enthalpy after evaporator, H_a	= 277.7 Btu per lb
Enthalpy before evaporator, H_d	= 166.7 Btu per lb
Heat absorbed in evaporator, $H_a - H_d$	= 111.0 Btu per lb
Circulating propane, W = (50)(60)(200)/(111.0)	= 5400 lb per hour

(b) **Theoretical power required.**

Enthalpy after isentropic compression, H_b	= 309.5 Btu per lb
Enthalpy before isentropic compression, H_a	= 277.7 Btu per lb
Work required for isentropic compression, $H_b - H_a$	= 31.8 Btu per lb
Power required (5400)(31.8)/(3412)	= 50.3 kw

(c) **Coefficients of performance.**

For Carnot cycle,
$$\frac{460}{560 - 460} = 4.60$$

For theoretical compression cycle,
$$\frac{H_a - H_d}{H_b - H_a} = \frac{111}{31.8} = 3.49$$

For actual cycle,
$$\frac{(H_a \quad H_d)}{(H_b - H_a)/\eta_S} = \frac{111}{31.8/0.75} = 2.62$$

(*d*) **Volumetric displacement.**

Specific volume of saturated vapor in suction line (Fig. 218), v_a = 2.71 cu ft per lb

Suction volume, $Wv_1 = 5400(2.71)/60$ = 244 cu ft per min

Volumetric efficiency, η_v = 0.81

Volumetric displacement of the compressor 244/0.81 = 300 cu ft per min

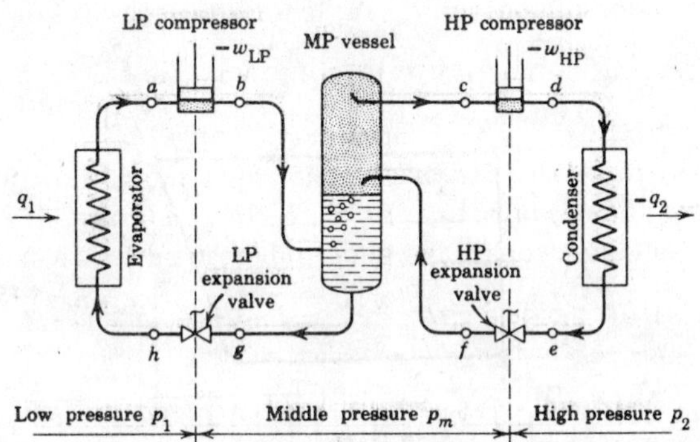

Fig. 219. Compression–refrigeration cycle with two-stage compression and expansion

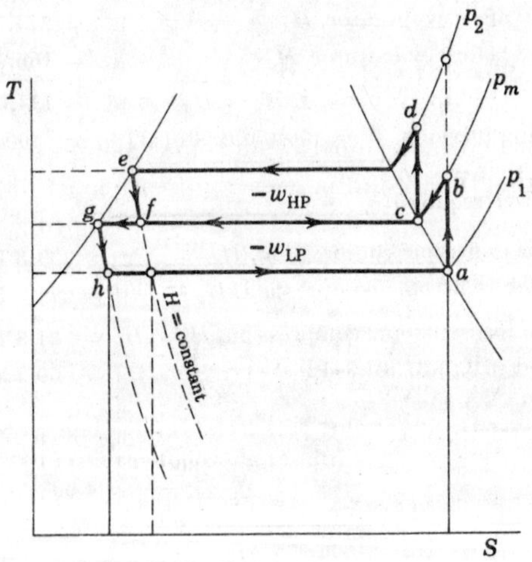

Fig. 220. *TS* diagram for two-stage compression-refrigeration cycle. Note: $-w_{HP}$ and $-w_{LP}$ are work per pound of refrigerant passing through the high-pressure and the low-pressure compressors, respectively

For high-pressure ratios, multistage compression is used to minimize lubrication difficulties, to increase the volumetric efficiency, and to reduce the work. Multistage compression is usually combined with multistage expansion. Intercooling is then often carried out by evaporation of the same liquid at the intermediate pressures. A commonly used scheme for a two-stage refrigeration plant is shown in Fig. 219. The corresponding *TS* diagram for isentropic compression is shown in Fig. 220, and the corresponding *pH* diagram in Fig. 221.

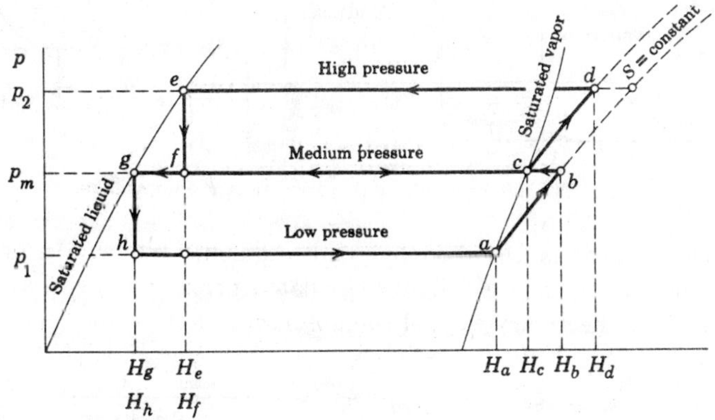

FIG. 221. *pH* diagram for two-stage compression–refrigeration cycle

The most widely used refrigerants in reciprocating and rotary compressors are ammonia and the chlorofluoroparaffins, of which Freon-12, difluorodichloromethane, is most commonly used. Sulfur dioxide, methyl chloride, ethane, and butane are used for special purposes. With turbocompressors, Freon-12, Freon-11, Freon-114, ammonia, and water vapor are used.

Vapor-Ejection Refrigeration. A vapor-ejector refrigeration system is shown in Fig. 222. The main difference from the previously described systems is that the compressor is replaced by a vapor ejector. This system is limited to the case where the vapors from the evaporator and the boiler are the same.

This system has been improved by multistage evaporation as shown in Fig. 223. The effect of the multistage evaporation is equivalent to regenerative heating of feed water in power plants.

Absorption Refrigeration. The most widely applied absorption refrigeration cycle uses ammonia as the refrigerant and water as the absorbent. A flow diagram of such a process is shown in Fig. 224. This flow diagram indicates a type of operation that may be used for large-scale process refrigeration. Liquid ammonia from the accumulator is throttled through an expansion valve, where it is partly vaporized, and then passes to the evaporator. The ammonia vapor passes from the evaporator to an

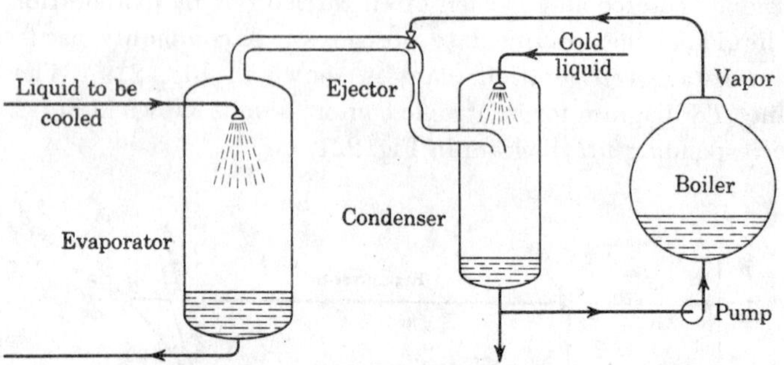

Fig. 222. Refrigeration with vapor-ejector compression

absorber where it is absorbed by weak ammonia liquor. In order to reduce the amount of weak liquor circulated per unit mass of ammonia absorbed, it is necessary to cool the absorbent close to the temperature

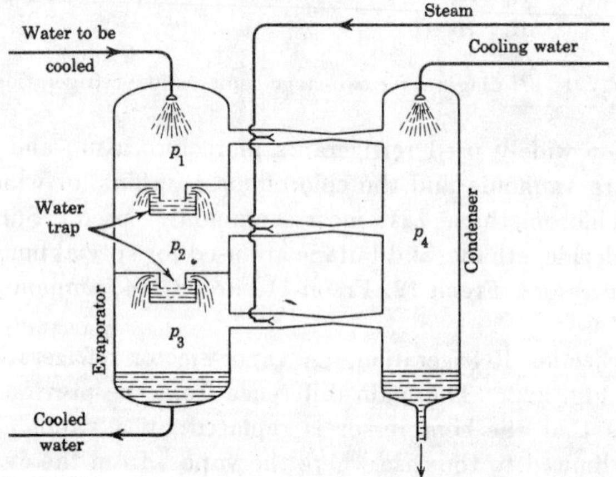

Fig. 223. Refrigeration with vapor ejectors and three-stage evaporation

of the available cooling water, and to provide intermediate cooling surface in the absorber.

The strong ammonia liquor on its way from the absorber to the stripper is passed through a heat exchanger, where it is heated countercurrently by the hot weak liquor on its way from the stripper to the absorber.

The weak liquor, after cooling, enters the top of the absorber. The strong liquor is fed to the stripping column, where heat is added by means of a reboiler to evaporate the ammonia. The vapors from the top of the stripping column are condensed, and a portion of the liquid is returned as reflux to the top of the stripping column, which contains bubble decks or packing to effect fractional distillation. It is evident that the stripper must be operated at a pressure high enough to produce

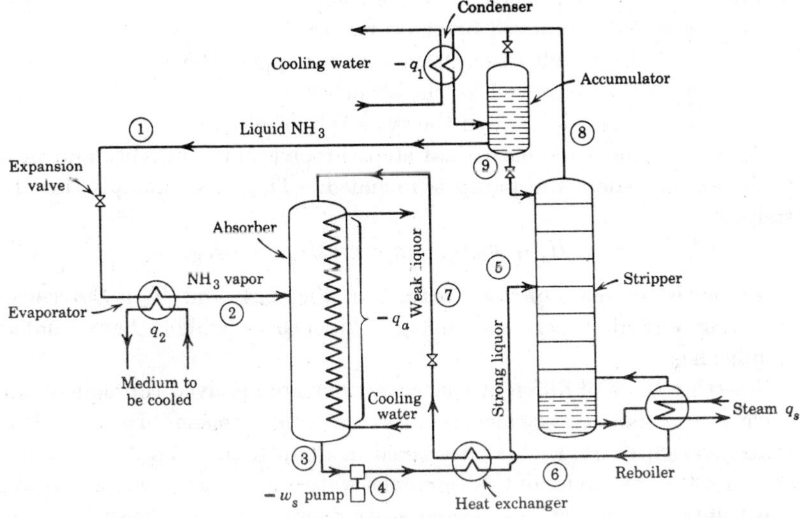

FIG. 224. Ammonia-absorption refrigeration

condensation of the ammonia at the temperature attainable with the available cooling water.

Absorption refrigeration cycles are also in common use, in which a simple still, with or without a partial condenser, replaces the stripping column. In other modifications used in household refrigeration, even the pump used for pumping liquid to the still is eliminated. This is sometimes accomplished by mixing with the refrigerant an inert gas of low density such as hydrogen, which permits the entire system to be kept at a constant total pressure. However, different partial pressures of the refrigerant are maintained for condensation and for evaporation. Circulation is induced by thermal convection, and absorption proceeds by the partial-pressure gradient of the refrigerant from vapor to liquid phase. Such units are described by Keenan.[2] In other systems water vapor is the solute and

[2] J. H. Keenan, *Thermodynamics*, John Wiley & Sons (1941).

aqueous lithium bromide is the absorbent. The pressure differences are less than the gravity head, thus obviating the use of a pump.

The problems of design and analysis of absorption refrigeration systems involve a series of energy balances embracing the entire cycle and each of its parts. Thus, for the over-all process, with reference to Fig. 224, if changes in kinetic, potential, and surface energies are neglected,

$$q_2 + q_s + q_1 + q_a - w_s = 0 \tag{17}$$

where q_2 = heat absorbed in evaporation
$-q_1$ = heat removed by the condenser
q_s = heat added to system in the stripper reboiler
$-q_a$ = heat removed by the absorber
$-w_s$ = work done upon the system by the pump

Energy balances for individual steps involve only enthalpy and heat terms except where the pump is included. Thus, for example, for the stripper,

$$m_5H_5 + m_9H_9 + q_s = m_8H_8 + m_6H_6 \tag{18}$$

where m_5 is the mass of the stream 5 in Fig. 224, and H_5 is the corresponding enthalpy per unit mass. The other symbols have similar significance.

Since mixtures of different concentrations are involved throughout the cycle, energy balances are most readily evaluated by means of an enthalpy-concentration chart, such as described in Chapter 9. Fig. 225 is such a chart for the enthalpies of the ammonia–water system at various pressures and temperatures and for all compositions and phases. Enthalpies are based on 1 lb of the fluid and are referred to saturated liquid ammonia at −40° F and to saturated water at 32° F. This chart is a modification of one prepared by Bošnjakovic[3] by the methods described in Chapter 9.

The lowest group of lines on the chart represents the enthalpies of various solid phases of ice and ammonia with the indicated eutectic and congruent points. The next lowest group of curves represents the enthalpies of saturated solutions over the entire range of composition for various temperatures and pressures. The third group from the bottom represents construction lines for obtaining vapor–liquid equilibrium relations. The next higher group of lines represents the enthalpies of saturated vapors covering the entire range of composition at various temperatures and pressures.

Conditions of vapor–liquid equilibria may be established by selection of points of equal temperature and pressure in the saturated liquid and

[3] F. Bošnjakovic, *Technische Thermodynamik II*, Theodor Steinkopff, Dresden and Leipzig (1935).

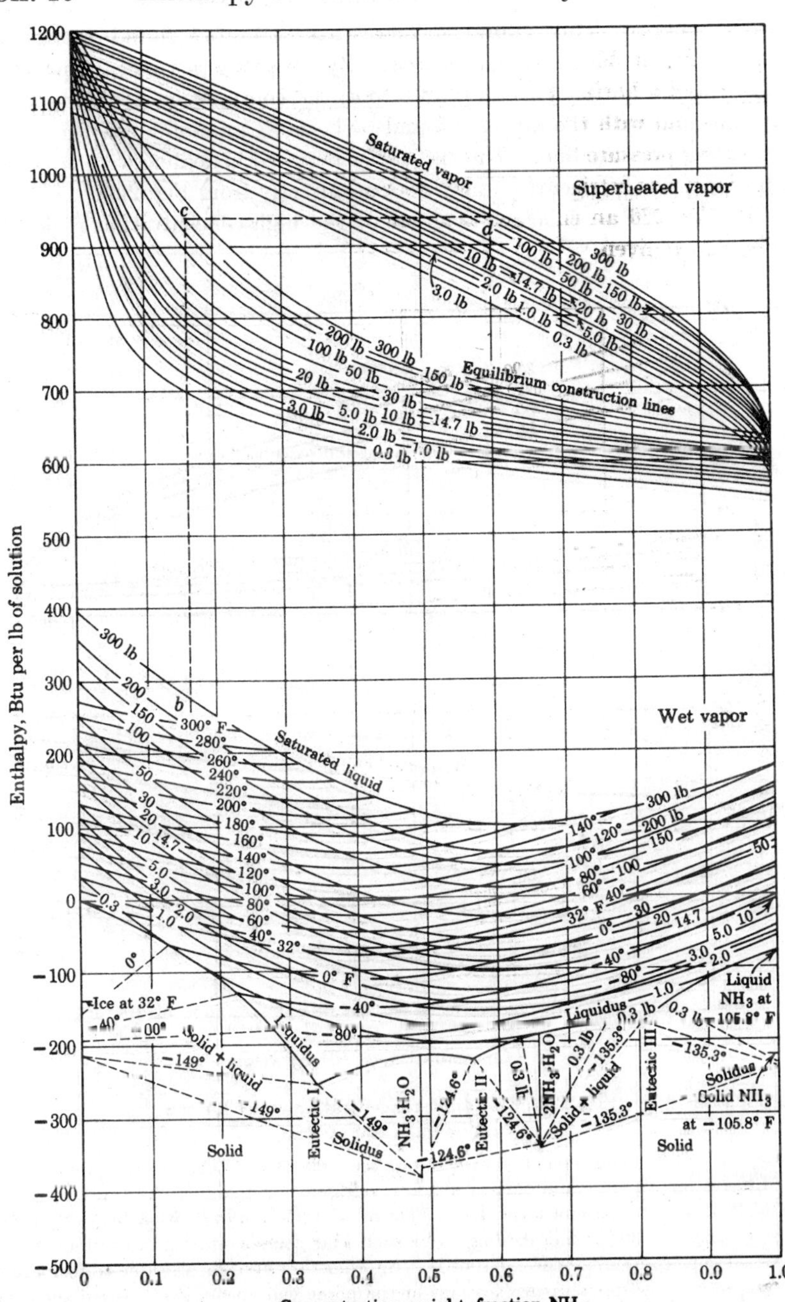

FIG. 225. Enthalpy concentration of the ammonia–water system
Reference states: Water at 32° F, liquid ammonia at −40° F

vapor curves. Thus, point b represents saturated liquid containing 16% NH_3 at 300° F and 200 psia. By following a vertical line from b to c and a horizontal line from c to d, the composition of the vapor in equilibrium with the liquid is found to be 60% ammonia, on the corresponding pressure line. The enthalpies required in dealing with problems in absorption refrigeration may be read directly from Fig. 225.

In Fig. 226 an enlarged diagram of the high-concentration region of Fig. 225 is given.

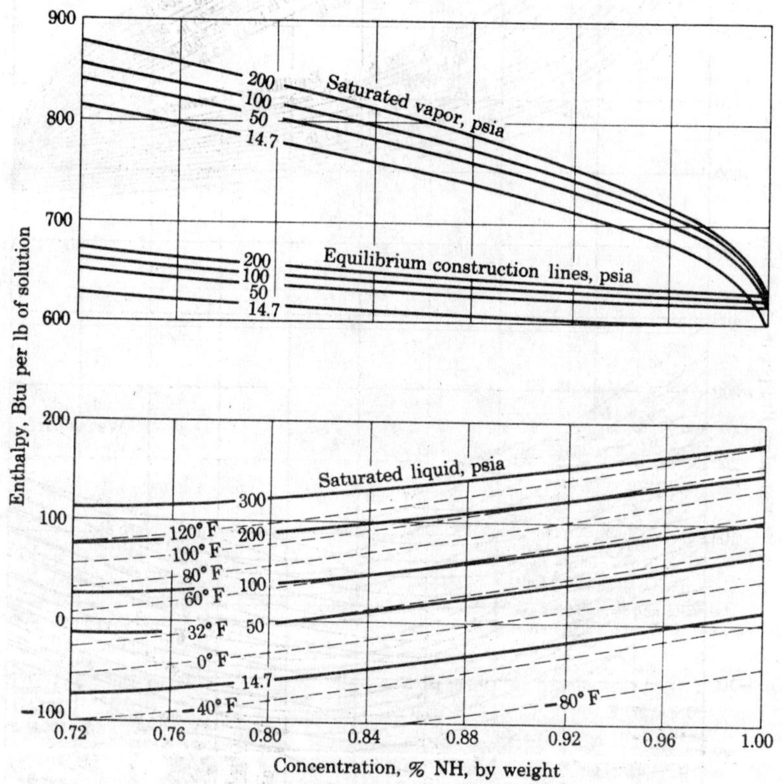

Fig. 226. Enthalpy–concentration chart for the ammonia–water system

Illustration 4. An ammonia-absorption refrigeration plant of the type shown in Fig. 224 is to be designed for a capacity of 50 tons per day, with an evaporator temperature of 30° F and cooling water such that condensation and cooling can be carried out at a temperature of 100° F. In order that no difficulty due to formation of ice in the low-temperature sections of the plant will ensue, it is desired that the ammonia from the stripper shall contain only 0.5% by weight of water. It may be assumed that the absorber produces contacting of such effectiveness that the ammonia concentration of the strong liquor is 95% of its equilibrium value. It also may be assumed that vaporization is completed in the evaporator under conditions such that the total vapor formed is at all times in equilibrium with the residual liquid.

The design study is to be based on the following arbitrary specifications:

1. Temperature of weak liquor leaving the heat exchanger, 120° F.
2. Reflux ratio of stripper (weight of stream 9/weight of stream 1) = 0.50.
3. Ammonia content of weak liquor, 20% by weight.
4. Temperature of strong liquor leaving absorber, 100° F.

These specifications might be varied over wide ranges, with resulting changes in the design of the plant for the given duty. These changes would affect both capital and direct operating costs. A series of design studies in which such specifications are varied is necessary in order to arrive at an optimum design which results in minimum over-all operating costs.

In this design study it is permissible to neglect:

1. Heat exchange between the equipment and its surroundings.
2. Variations in pressure in both the high- and low-pressure sections of the plant.
3. Variation in temperature of the ammonia–water mixture in the evaporator.

The evaporator pressure may be conservatively established as corresponding to complete vaporization at the specified temperature.

On this basis, calculate complete material and energy balances of the plant together with the temperatures and pressures at all points. Also calculate:

(*a*) The heat-transfer duties of the condenser, absorber cooler, heat exchanger, and reboiler, in Btu per hour.

(*b*) The horsepower of the pump.

(*c*) The pressure and quantity of dry saturated steam required for heating the reboiler with a 10° F temperature difference.

Solution: Material and energy balances are established, starting with the evaporator, and using data from Fig. 225. These calculations are summarized in Table A where the streams are identified by the numbers of Fig. 224. The values for m and mH in this table are reported in more significant figures than are justified by the accuracy of the data, in order that the numerical accuracy of the calculations may be readily verified by over-all balances.

Figure 227 shows how the enthalpy–concentration diagram of Fig. 225 is used to determine the enthalpies of various streams.

1. The pressure of stream 1 is read from Fig. 225 as 210 psia, the equilibrium pressure of liquid containing 99.5% NH_3 at 100° F.

2. The pressure of stream 2 is the equilibrium pressure of saturated vapor containing 99.5% NH_3 at 30° F, which is read from Fig. 225 by a trial-and-error procedure. A trial value of the equilibrium pressure is assumed, and a horizontal line is followed from the point representing saturated vapor at this pressure and 99.5% NH_3 over to the corresponding construction line. A vertical line is followed downward from this point to the 30° F line of the saturated liquid curves, and the corresponding equilibrium pressure is read. This second approximation pressure is used to repeat the operation until consistent values are obtained. In this manner it is found that the evaporator pressure is approximately 45 psia, and that the liquid in equilibrium with the vapor leaving the evaporator contains 77% NH_3.

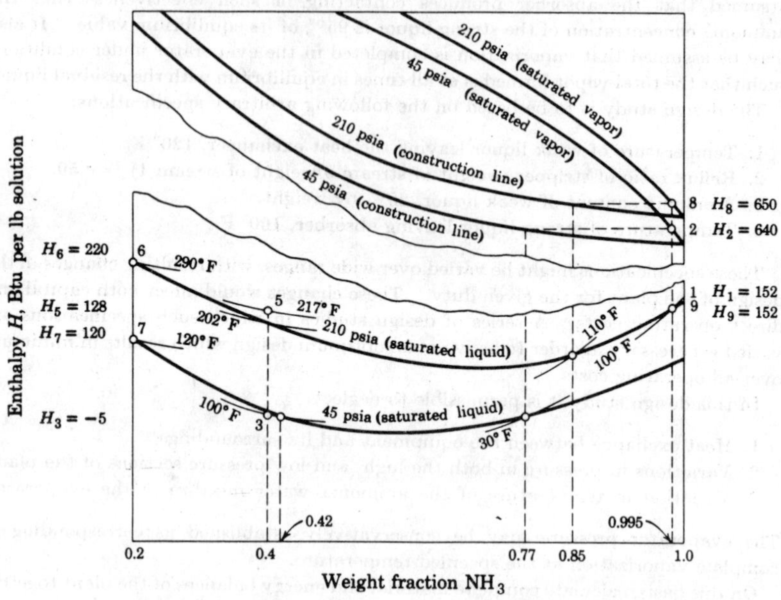

FIG. 227. Method of evaluating enthalpies of various streams in Illustration 4 by use of enthalpy–concentration chart

TABLE A. MATERIAL AND ENERGY BALANCES OF AMMONIA ABSORPTION REFRIGERATION PLANT

Stream	NH_3 % by Weight	t, °F	Absolute Pressure, psia	H, Btu/lb	m, lb/hr	mH, Btu/hr
1	99.5	100	210	152	1229.5	186,880
2	99.5	30	45	640	1229.5	786,880
3	40.0	100	45	−5	4887.3	−24,440
4	40.0	100+	210	−4.4	4887.3	−21,660
5	40.0	217	210	128.8	4887.3	629,432
6	20.0	290	210	220	3657.8	804,720
7	20.0	120	45	42	3657.8	153,628
8	99.5	110	210	650	1844.3	1,198,800
9	99.5	100	210	152	614.8	93,450

The rate of stream 2 is fixed by the heat-transfer duty of the evaporator and the differences in specific enthalpies of streams 1 and 2. Thus,

$$m_1 = 50(200)60/(640 - 152) = 1229.5 \text{ lb per hr}$$

3. The ammonia content of the strong liquor is 95% of the equilibrium concentration at 100° F and 45 psia which is read from Fig. 225. Thus, NH_3 in stream 3 = $0.95 \times 42 = 40\%$. It may be assumed that the enthalpy of a liquid which is not

completely saturated with NH_3 is the same as that of a saturated solution of the same NH_3 content and at the same temperature. On this basis $H_3 = -5$ Btu per lb.

The rate of stream 3 is determined by the over-all material balance and the ammonia balance of the absorber. Since, from the over-all material balance, $m_7 = m_3 - 1229.5$,

$$0.40m_3 = (m_3 - 1229.5)0.20 + (1229.5)(0.995)$$

$$m_3 = 4887.3 \qquad m_7 = 3657.8$$

4. The work done by the pump is calculated from equation 17-17. The specific gravity of the solution is 0.86.

$$-w_s = \frac{(210 - 45)144(4887.3)}{(62.4)(0.86)} = 2{,}160{,}000 \text{ ft-lb per hr}$$

$$= 2780 \text{ Btu per hr}$$

$$\text{Power input to fluid} = \frac{2{,}160{,}000}{60(33{,}000)} = 1.09 \text{ hp}$$

In order to determine the power required for the pump, the horsepower must be divided by the efficiency of the pump.

Enthalpy increase of the fluid in the pump 2780/4887.3 = 0.57 Btu per lb

5. The temperature of the feed to the stripper is determined by an energy balance around the heat exchanger. The quantity of stream 7 is determined, and its enthalpy may be evaluated from the specified temperature of 120° F. The temperature of stream 6 is the equilibrium temperature of a solution containing 20% NH_3 at a pressure of 210 psia, which is read from Fig. 225 at 290° F with an enthalpy of 220 Btu per lb. The enthalpy of stream 5 is then calculated from the specific enthalpies of Table A and the energy-balance equation $m_5H_5 = m_6H_6 + m_4H_4 - m_7H_7$, or

$$m_5H_5 = 804{,}720 - 21{,}660 - 153{,}628 = 629{,}432 \text{ Btu per hr}$$

$$H_5 = 629{,}432/4887.3 = 128.8 \text{ Btu per lb}$$

This specific enthalpy corresponds to a temperature of approximately 217° F on the saturated liquid lines of Fig. 225 which is above the saturation temperature of 202° F for a 40% solution at 210 psia. Thus, the feed to the stripper is partially vaporized by the heat exchanger. The heat-transfer duty of the exchanger is 804,720 − 153,628 = 651,092 Btu per hour.

6. The heat input to the reboiler is determined by an over-all energy balance around the stripping column. The rate of stream 9 is fixed by the specified reflux ratio and $m_8 = m_9 + m_1$. The corresponding enthalpies are read from Fig. 225 as listed in Table A. The temperature of stream 8 is the temperature of the saturated liquid which is in equilibrium with vapors containing 99.5% NH_3 at 210 psia. From Fig. 225, it is seen that this dew-point liquid has a composition of approximately 85% NH_3 and a temperature of 110° F. From Table A,

$$q_s = m_8H_8 + m_6H_6 - m_5H_5 - m_9H_9 = 1{,}198{,}800 + 804{,}720 - 629{,}432 - 93{,}450$$

$$= 1{,}280{,}638 \text{ Btu per hr}$$

The heat removed by the condenser is expressed by

$$-q_1 = m_8H_8 - m_1H_1 - m_9H_9 = 1{,}198{,}800 - 186{,}880 - 93{,}450$$

$$= 918{,}470 \text{ Btu per hr}$$

The heat removed in the absorber is expressed by

$$-q_a = m_7H_7 + m_2H_2 - m_3H_3 = 153{,}628 + 786{,}880 - (-24{,}440)$$
$$= 964{,}948 \text{ Btu per hr}$$

These results may be substituted in equation 17 as a verification of the calculation. Thus,

$$600{,}000 + 1{,}280{,}638 - 918{,}470 - 964{,}948 + 2780 = 0$$

7. The steam used for heating the reboiler must be at a temperature of 300° F which corresponds to a pressure of 67 psia. If it is assumed that the condensate is withdrawn at 300° F, heat to the reboiler per pound of steam is the heat of vaporization which from the steam table is 910 Btu per lb.

$$\text{Reboiler steam rate} = \frac{1{,}280{,}638}{910} = 1407 \text{ lb per hr}$$

The relative desirability of compression and absorption refrigeration cycles is largely dependent on local conditions. The absorption operation is favored by the availability of low-cost exhaust steam for heating the reboiler. The choice of a refrigeration cycle thus involves consideration of relative costs of power and steam. As previously mentioned, many modifications of the absorption cycle are possible in order to adapt it to local conditions.

Refrigeration cycles based on the adsorption of refrigerant vapors on solid adsorbents have been used to a limited extent and may develop to greater importance. Such schemes operate on an intermittent cycle, adsorbing vapors at low temperature and pressure in one chamber, while another chamber is evolving previously adsorbed vapors at a higher temperature and pressure. Adsorbents of high capacity such as silica gel, activated alumina, or charcoal may be used with a variety of fluids. The disadvantages of these methods are the intermittent nature of the operation and the difficulty of transferring heat to a granular solid. The thermodynamic problems of adsorption refrigeration may be handled by the same principles as demonstrated here for the absorption systems. It is desirable to develop an enthalpy–composition chart for the adsorbed system to facilitate calculations. Such a chart is shown in Fig. 104, page 390, for water vapor on silica gel.

Selection of a Working Fluid. In the selection of a working fluid for use in mechanical refrigeration, thermodynamic principles are of value in calculating the work of reversible expansion and in securing the desired pressure range in operating between two temperature levels. The final selection of the working fluid, however, is limited by such properties as toxicity, stability, and flammability, even at the expense of favorable thermodynamic properties.

In the selection of a binary system for use in absorption refrigeration, thermodynamic principles are of value in securing a system possessing high mutual solubility, ease of separation by a change of pressure, and the desired pressure range in operating between two temperature levels. The conditions imposed of high chemical stability over years of operation through millions of cycles are extremely difficult to meet with organic compounds. For a given binary system, the desired pressure range can be secured by operating between two concentration levels. In some systems, the use of pumps has been eliminated by limiting the pressure drop to the gravimetric head available in the apparatus. The advantage of mechanical simplicity is partially offset by increased circulation rates. The practical design of an absorption refrigeration plant involves many problems of economic analysis and manufacture to obtain an optimum balance of equipment and operating costs.

Heat Pump

A heat pump is a refrigeration machine in which application is centered on both the delivery of heat at the higher-temperature level and the removal of heat at the lower-temperature level. Its importance is rapidly increasing as a means of obtaining space heating in winter and cooling in summer. With a heat pump operating on the reverse Carnot cycle, the fluid is expanded isothermally at the low-temperature level with input of heat q_1, compressed isentropically from T_1 to T_2, compressed isothermally at the high-temperature level T_2 with rejection of heat $-q_2$, and expanded isentropically from T_2 to T_1. The net work done on the working medium is $-w_s$, the heat absorbed is q_1, and the heat rejected is $-q_2$. Accordingly,

$$-q_2 = q_1 - w_s \tag{19}$$

The coefficient of performance η of a heat pump in its heating cycle is defined as the ratio of heat rejected at the higher-temperature level to the net mechanical work input to the pump, or

$$\eta = \frac{q_1 - w_s}{-w_s} = \frac{-q_2}{-w_s} \tag{20}$$

For a reversed Carnot cycle it is evident from Fig. 228 that

$$\eta = \frac{T_2}{T_2 - T_1} \tag{21}$$

From equation 21 it follows that the coefficient of performance of a heat pump increases with decreasing temperature difference $T_2 - T_1$ and with increasing temperature level T_2. Hence, the most successful industrial applications of heat pumps are in fields where the temperature difference between the heat source and the heat delivery is small, as in distillation and evaporation.

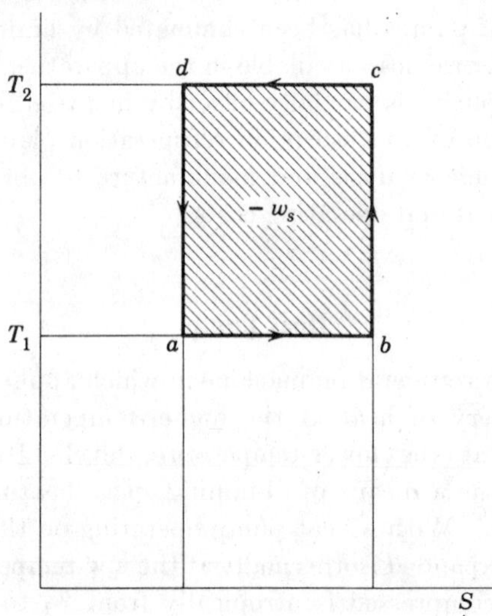

Fig. 228. Reversed Carnot cycle for heat pump

Vapor-Recompression Evaporation. The evaporation of water from nonvolatile solutes or suspensions is generally accomplished by multiple-effect evaporators, such as illustrated in Fig. 51, page 211. By this method several pounds of water may be evaporated by 1 lb of steam, depending on the number of effects used. A similar result can be produced in more compact and simple equipment by the use of a heat pump, as shown in Fig. 229.

A dilute solution at a temperature T_1 is charged to the evaporator containing the boiling concentrated solution at temperature T_2 and pressure p_2. As a result of the elevation of the boiling point of the solution over that of pure water, the vapors evolved are superheated. These vapors are compressed to p_4. The compressed steam is combined with throttled supply steam at pressure p_4, and the mixed steam is then delivered to the heat exchanger of the evaporator. Condensate is withdrawn substantially at the saturation temperature of the steam in the heat exchanger. Concentrated solution is continuously withdrawn from the bottom of the evaporator. The supply of fresh feed maintains a constant liquid level and concentration in the evaporator.

By this method it is possible to evaporate several pounds of water per pound of steam through the addition of a relatively small amount of energy as shaft work in the compressor; in fact, in some cases, steam may be rejected rather than supplied. However, the economy of the operation depends on the relative cost of energy supplied as shaft work and that

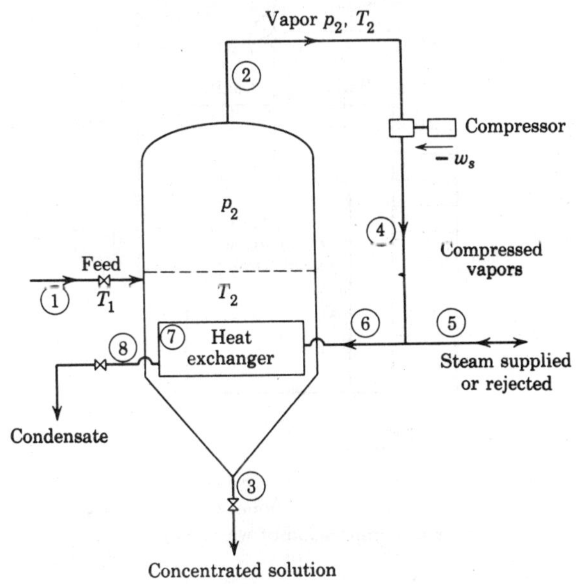

FIG. 229. Vapor-recompression evaporation

supplied as heating steam. Neglecting heat losses, the following equation expresses the energy balance of an evaporator in which m_1 is the feed rate in pounds per hour, m_2 the rate of evaporation, and m_5 the rate of supply of fresh steam:

$$-w_s + m_1H_1 + m_5H_5 = (m_2 + m_5)H_8 + (m_1 - m_2)H_3 \quad (22)$$

where the H values represent enthalpies per pound corresponding to the designations on Fig. 229. For specified values of m_1, m_2, p_2, and p_4, the work done on the vapors in the compressor is calculated by the methods previously demonstrated. The heating steam m_5 is then obtained directly from equation 22, and the ratio of evaporation to steam m_2/m_5 may be calculated.

This method of evaporation is of interest where power costs are low and fuel costs high. Such a situation might exist in the vicinity of large hydroelectric-power developments.

Another application of the heat pump is in fractional distillation as shown in Fig, 230, where the heating of the column reboiler is conducted by the compression of water vapor that is produced by the condensation of overhead vapors in the reflux condenser at the top of the column.

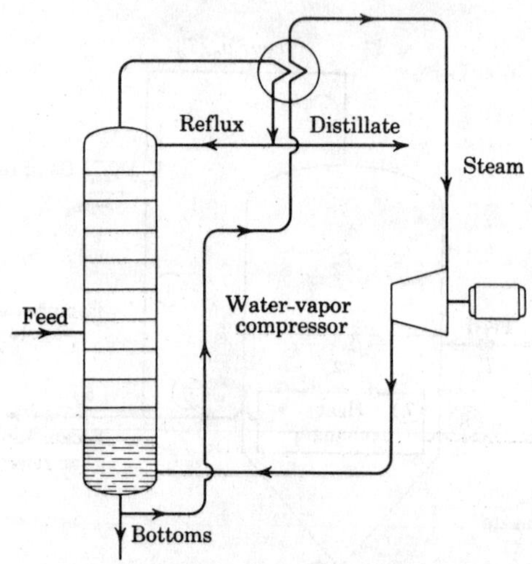

Fig. 230. Heat pump for fractional distillation of alcohol with compression of water vapor

The mechanical compressors indicated in Figs. 229 and 230 may be replaced by steam ejectors, which have the advantage of being less expensive.

Problems

1. It is desired to design a plant for the liquefaction of 2000 lb ethylene per day, using the method shown in Fig. 213. The ethylene is supplied to the system at 1 atm and 90° F (point 1). A four-stage compressor compresses the gas to 50 atm, with intercooling between stages to 90° F. The gas leaving the compressor at 50 atm is cooled to 90° F before entering the heat exchanger at point 3. The gas leaving the heat exchanger at point 4 is passed through an expansion valve where the pressure drops from 50 atm to 1 atm. The ethylene that is liquefied is drawn off at point 5 at 1 atm pressure. The recycle gas is passed into the heat exchanger and emerges at 1 atm and 70° F (point 7). Gain or loss of heat by the equipment may be neglected, and isentropic compression may be assumed in each stage of compression. Equation 15-81 may be used to establish the intermediate pressures in multistage compression.

Calculate:

(a) The fraction of the compressor discharge that is liquefied.

(b) The temperatures at all points of the cycle.

(c) The horsepower input to the gas in each stage of the compressor.

(d) The heat-transfer duties of the coolers and the heat exchanger in Btu per hour.

(e) Repeat the calculations for a plant in which the expansion of the gas is conducted by means of an isentropic engine rather than by means of a throttle valve.

2. An ammonia compression refrigerator operates between the temperature levels of $-20°$ F and 80° F. The vapor leaves the evaporator saturated. Under isentropic conditions calculate on the basis of one ton of refrigeration:

(a) The coefficient of performance.

(b) The horsepower required.

3. A vapor compression heat pump operated with Freon-12 as the refrigerant delivers 200,000 Btu per hour to a heat exchanger at 100° F with outside air at 40%. Estimate:

(a) The coefficient of performance of the heat pump.

(b) The power required.

4. In an ammonia–water absorption refrigerator, the temperature in the evaporator is 40° F and in the absorber 100° F. Saturated liquid containing 97% ammonia by weight leaves the condenser at 100° F. The liquid ammonia solution enters and leaves the absorber at 100° F and at the pressure of the evaporator. Calculate the following:

(a) The refrigeration capacity in Btu per pound of vapor leaving the evaporator at 40° F, saturated.

(b) The pounds of liquid solution leaving the absorber per pound of vapor entering, assuming that the ammonia content on a weight basis in the entering liquid is 45% of its saturation value, and in the leaving liquid is 90% of its saturation value at 100° F and the pressure of the absorber.

5. In an ammonia compression refrigeration cycle the condenser operates at 85° F. For a rated refrigeration capacity of 10 tons calculate the theoretical horsepower required to drive the compressor with:

(a) The evaporator at 25° F.

(b) The evaporator at 5° F.

6. In an ammonia absorption refrigerator 150,000 Btu per min are removed from the evaporator maintained at a temperature of $-20°$ F. Heat is supplied to the reboiler from saturated steam at 30 psia. Heat is rejected to the air at 80° F. Calculate:

(a) The heat supplied to the refrigerator in Btu per minute.

(b) The coefficient of performance.

(c) The steam supply in pounds per ton of refrigeration.

7. The refrigeration duty of Illustration 3 is to be performed by a plant using ammonia as the refrigerant, and with the same temperature levels. Assuming the same compressor efficiencies, calculate the power input, piston displacement, coefficient of performance, and heat input to the condenser for the ammonia machine. Compare these values and also the operating temperatures and pressures with those of the propane system.

8. Design an ammonia–water absorption refrigeration plant for the conditions specified in Illustration 4 except that the evaporator temperature is to be maintained at 0° F and the reflux ratio increased to 0.9 because of the more dilute solutions handled. Compare the heat-transfer duties, circulation rates, and steam requirement with those of Illustration 4.

9. Compare the relative costs of heating a house by use of a heat pump and by burning oil, where the outside temperature is 30° F and the house is to be kept at 72° F with electric energy at 6.0 cents per kilowatt-hour and fuel oil at 2.0 cents per pound. The oil has a net heating value of 14,500 Btu per lb and is burned with 20% heat loss in the stack gas.

10. A dilute aqueous solution of calcium chloride is to be evaporated in a vapor recompression evaporator like that shown in Fig. 229. Data are as follows:

Feed (point 1): 1000 lb per hr, 10% $CaCl_2$, 100° F, 1 atm
Concentrated liquor (point 3): 40% $CaCl_2$, 1 atm
Evaporator pressure (point 2): 1 atm
Outside steam supply: Furnished at 100 psia, saturated, and dry, throttled down to the delivery pressure of the compressor p_4
Isentropic compression efficiency: 75%
Temperature difference between boiling liquor in the evaporator and condensing steam in heat exchanger: 20 F°

Neglect heat losses from the entire system, including the compressor. Assume that the condensate leaves at point 8 as saturated liquid; that is, assume that it does not cool below the condensation temperature to the temperature of the boiling liquor inside the evaporator. Enthalpy and boiling-point data for $CaCl_2$–H_2O solutions are given in Fig. 82, p. 326. Enthalpy and entropy data for steam may conveniently be read from a Mollier chart.

20

Properties of Solutions

The term *solution* includes homogeneous mixtures of two or more components whether in the gas, liquid, or solid phase. Thermodynamic principles, as developed in previous chapters for pure substances, are equally applicable to solutions of fixed composition. Where variations in composition occur, the functional relations for all thermodynamic properties require one additional term for each additional component present.

Partial Molal Properties. As stated in Chapter 9, the properties of a solution are, in general, not additive properties of the pure components; the actual contribution to any extensive property is designated as its partial property. The term *partial property* is used to designate the property of a component when it is in admixture with one or more other components. In Chapter 9, the partial molal property of a given component in solution is defined as the differential change in that property with respect to a differential change in the amount of a given component under conditions of constant pressure and temperature, and constant number of moles of all components other than the one under consideration. Thus, for partial molal volume $\bar{v}_i$ and partial molal enthalpy $\bar{H}_i$ of component i,

$$\bar{v}_i = \left(\frac{\partial V}{\partial n_i}\right)_{pTn_j} \tag{1}$$

$$\bar{H}_i = \left(\frac{\partial H}{\partial n_i}\right)_{pTn_j} \tag{2}$$

where j designates all components other than i. Similar expressions may be written for partial molal properties of all other extensive properties.

The partial molal energy functions of any component in solution are defined in the same manner as the energy functions of the pure component, with the restriction that the composition must be stated; thus, for a solution of given composition,

$$\bar{H}_i = \bar{U}_i + p\bar{v}_i \tag{3}$$

$$\bar{A}_i = \bar{U}_i - T\bar{S}_i \tag{4}$$

$$\bar{G}_i = \bar{H}_i - T\bar{S}_i \tag{5}$$

The superscript bars indicate partial molal quantities.

Chemical Potentials. To define the thermodynamic properties of a single-component system of fixed mass,. it is necessary to specify mass and two other independent variables such as temperature, pressure, volume, and entropy. To define the thermodynamic properties of a solution of given mass, one additional composition variable must be introduced for each component.

Any infinitesimal change taking place in a system can be expressed in terms of all the infinitesimal changes in its independent properties in agreement with the properties of a continuous function represented by equation 13–57, page 529. Thus, for an infinitesimal change in internal energy, where volume, entropy, and composition are selected as the independent variables, the general differential relation is as follows:

$$dU = \left(\frac{\partial U}{\partial S}\right)_{Vn_1n_2\ldots} dS + \left(\frac{\partial U}{\partial V}\right)_{Sn_1n_2\ldots} dV + \left(\frac{\partial U}{\partial n_1}\right)_{VSn_2\ldots} dn_1 + \left(\frac{\partial U}{\partial n_2}\right)_{SVn_1\ldots} dn_2 + \cdots \quad (6)$$

where the subscripts 1,2, $\cdots$ refer to the individual components of the phase which undergoes change in mass.

Where composition is constant, all except the first three terms of equation 6 becomes zero, and in agreement with the relation $dU = T\,dS - p\,dV$, for fixed mass and composition, the following equations result:

$$\left(\frac{\partial U}{\partial S}\right)_{Vn_1n_2\ldots} = T \quad (7) \qquad \text{and} \qquad \left(\frac{\partial U}{\partial V}\right)_{Sn_1n_2\ldots} = -p \quad (8)$$

Equation 6 may be written as

$$dU = T\,dS - p\,dV + (\mu_1)_{VSn_2\ldots}\,dn_1 + (\mu_2)_{VSn_1\ldots}\,dn_2 + \cdots \quad (9)$$

where

$$(\mu_1)_{VS} = \left(\frac{\partial U}{\partial n_1}\right)_{VSn_2\ldots} \quad (10) \qquad (\mu_2)_{VS} = \left(\frac{\partial U}{\partial n_2}\right)_{VSn_1\ldots} \quad (11)$$

Thus, according to equation 9, the change in the total internal energy of a system is the sum of the changes of a number of energy terms, each of which comprises an intensive and an extensive factor. For heat energy, temperature is the intensive factor, and entropy is its extensive factor; for work of expansion, pressure is the intensive and volume the extensive factor. Similarly, n_1 and n_2 are extensive factors, and $(\mu_1)_{VS}$ and $(\mu_2)_{VS}$ are the intensive factors of internal energy associated with components 1,2, $\cdots$ in solution. These intensive factors of composition are designated *chemical potentials.*

Expressions similar to equation 9 may be developed for the other three energy functions; thus:

$$dH = T\,dS + V\,dp + (\mu_1)_{Sp}\,dn_1 + (\mu_2)_{Sp}\,dn_2 + \cdots \tag{12}$$

$$dA = -S\,dT - p\,dV + (\mu_1)_{TV}\,dn_1 + (\mu_2)_{TV}\,dn_2 + \cdots \tag{13}$$

$$dG = -S\,dT + V\,dp + (\mu_1)_{Tp}\,dn_1 + (\mu_2)_{Tp}\,dn_2 + \cdots \tag{14}$$

From the definitions of the energy functions, for a given mass and given composition,

$$U = H - pV = A + TS = G + TS - pV \tag{15}$$

The total derivative of equation 15 gives

$$\begin{aligned} dU &= dH - p\,dV - V\,dp = dA + T\,dS + S\,dT \\ &= dG + T\,dS + S\,dT - p\,dV - V\,dp \end{aligned} \tag{16}$$

Adding $(-T\,dS + p\,dV)$ to each of the equalities, gives

$$\begin{aligned} dU - T\,dS + p\,dV = dH - T\,dS - V\,dp &= dA + p\,dV + S\,dT \\ &= dG + S\,dT - V\,dp \end{aligned} \tag{17}$$

Subtracting the equalities of equation 17 from equations 9 through 14 and equating like coefficients, there results

$$(\mu_1)_{VS} = (\mu_1)_{Sp} = (\mu_1)_{TV} = (\mu_1)_{Tp} = \mu_1 \tag{18}$$

From the properties of an exact differential equation written for dU, dH, dA, and dG, as in equation 6, it follows that

$$\mu_1 = \left(\frac{\partial U}{\partial n_1}\right)_{VSn_2\cdots} = \left(\frac{\partial H}{\partial n_1}\right)_{Spn_2\cdots} = \left(\frac{\partial A}{\partial n_1}\right)_{TVn_2\cdots} = \left(\frac{\partial G}{\partial n_1}\right)_{Tpn_2\cdots} = \bar{G}_1 \tag{19}$$

The chemical potential μ_1 is thus the intensive factor of internal energy associated with component 1 in solution. It may also be defined in terms of any of the other three energy functions by equation 19. From the definition of partial molal quantities (page 331) and equation 19, it is evident that *the chemical potential of a component is equal to the partial molal free energy*, $\mu_i = \bar{G}_i$. The other partial derivatives in equation 19 are also chemical potentials but are not partial molal quantities since they are not restricted to constant temperature and pressure.

Effects of Pressure and Temperature on Chemical Potentials and Fugacities. The fugacity of component i in solution is defined in terms of partial molal free energy, or chemical potential, as for pure components (equations 14–61) as follows:

$$(d\bar{G}_i = d\mu_i = RTd\ln f_i)_T \tag{20}$$

For component i in solution, equations 13–42 may be written

$$d\bar{G}_i = d\mu_i = -\bar{S}_i\, dT + \bar{v}_i\, dp \tag{21}$$

At constant temperature

$$\left(\frac{\partial\bar{G}_i}{\partial p}\right)_T = \left(\frac{\partial\mu_i}{\partial p}\right)_T = \bar{v}_i \tag{22}$$

Combining equations 20 and 22 gives the effect of pressure on the fugacity of component i in solution at constant temperature as

$$\left(\frac{\partial\ln f_i}{\partial p}\right)_T = \frac{\bar{v}_i}{RT} \tag{23}$$

At a low pressure, the fugacity f^* of a pure substance becomes equal to pressure. Where G^* is the molal free energy corresponding to fugacity f^*, integration of equations 14–59 gives

$$G^* - G = RT(\ln f^* - \ln f)_T \tag{24}$$

Differentiation of equation 24 with respect to temperature at constant pressure gives

$$\left(\frac{\partial G^*}{\partial T}\right)_p - \left(\frac{\partial G}{\partial T}\right)_p = R\ln\frac{f^*}{f} + RT\left(\frac{\partial\ln f^*}{\partial T}\right)_p - RT\left(\frac{\partial\ln f}{\partial T}\right)_p \tag{25}$$

But since f^* is equal to p,

$$\left(\frac{\partial\ln f^*}{\partial T}\right)_p = 0 \tag{26}$$

Combining equations 24, 25, and 26 gives

$$\left(\frac{\partial G^*}{\partial T}\right)_p - \left(\frac{\partial G}{\partial T}\right)_p = \frac{G^*}{T} - \frac{G}{T} - RT\left(\frac{\partial\ln f}{\partial T}\right)_p \tag{27}$$

Combining equations 27 and 13–26 gives

$$-S^* + S = \frac{H^*}{T} - \frac{H}{T} - S^* + S - RT\left(\frac{\partial\ln f}{\partial T}\right)_p \tag{28}$$

or

$$\left(\frac{\partial\ln f}{\partial T}\right)_p = \frac{H^* - H}{RT^2} \tag{29}$$

For component i in solution, equation 29 can be written

$$\left(\frac{\partial \ln f_i}{\partial T}\right)_p = \frac{\text{H}^*_i - \bar{\text{H}}_i}{RT^2} \tag{30}$$

Properties of an Ideal Solution. An *ideal solution* is defined as a solution in which the fugacity of each component i is equal to the product of its mole fraction and the fugacity f°_i of the pure component at the same temperature, pressure, and state of aggregation as of the solution; then, for any phase,

$$(f_i = N_i f^\circ_i)_{pT} \tag{31}$$

The fugacity requirement of equation 31 is in agreement with Raoult's law for liquid solutions and with Dalton's law for gaseous mixtures. These latter laws are less general statements of ideal-solution behavior than equation 31.

An ideal liquid solution presupposes that, when one component is mixed with another, mutual solubility results, that no chemical interaction occurs, that molecular diameters are the same, and that the intermolecular forces of attraction and repulsion are the same between unlike as between like molecules. At low pressures, gaseous mixtures exhibit nearly ideal-solution behavior.

Applying equation 23 to both f_i and f°_i, there results

$$\left[\frac{\partial \ln (f_i/f^\circ_i)}{\partial p}\right]_T = \frac{\bar{v}_i - v^\circ_i}{RT} \tag{32}$$

For an ideal solution, since $f_i = N_i f^\circ_i$, it follows that equations 31 and 32 reduce to zero at constant composition N_i, and hence

$$\bar{v}_i - v^\circ_i = 0 \tag{33}$$

The total volume change per mole of solution in mixing at constant temperature and pressure then becomes

$$\left[\Delta v^M - \sum^i N_i(\bar{v}_i \quad v^\circ_i) - 0\right]_{pT} \tag{34}$$

Thus, the total volume change in mixing Δv^M to form an ideal solution from the pure components at the temperature, pressure, and state of aggregation of the solution is zero.

Applying equation 30 to component i in solution and at its reference state f°_i gives

$$\left[\frac{\partial \ln (f_i/f^\circ_i)}{\partial T}\right]_p = \frac{\text{H}^\circ_i - \bar{\text{H}}_i}{RT^2} \tag{35}$$

For an ideal solution, since $f_i = N_i f^\circ_i$, equation 35 reduces to zero, and hence

$$\bar{\text{H}}_i - \text{H}^\circ_i = 0 \tag{36}$$

and

$$\Delta \text{H}^M = \sum N_i(\bar{\text{H}}_i - \text{H}^\circ_i) = 0 \tag{37}$$

For mixing at constant pressure and temperature,

$$\Delta H^M = \Delta U^M + p\,\Delta V^M \tag{38}$$

where superscript M refers to the property changes in mixing at constant pressure and temperature. Since for an ideal solution $\Delta H^M = 0$ and $\Delta V^M = 0$, it follows that the internal energy of mixing to form an ideal solution from the pure components at the temperature, pressure, and state of aggregation of the solution, is zero.

$$\Delta U^M = 0 \tag{39}$$

Since

$$\Delta G_i{}^M = \Delta H_i{}^M - T\,\Delta S_i{}^M \tag{40}$$

and

$$\Delta A_i{}^M = \Delta U_i{}^M - T\,\Delta S_i{}^M \tag{41}$$

it follows that, for one mole of component i in an ideal solution,

$$\Delta \text{G}_i{}^M = \Delta \text{A}_i{}^M = RT \ln N_i \tag{42}$$

$$(\Delta \text{S}_i{}^M = -R \ln N_i)_{pT} \tag{43}$$

From equation 43 it follows that the entropy of a solution is not an additive property of the separate components even for ideal solutions; the same statement applies to extensive properties based on entropy. Thus, for ideal solutions, volume, enthalpy, and internal energy are additive properties but entropy, free energy, and work function include a term for the entropy of mixing:

$$S = \sum n_i \text{S}_i - \sum n_i R \ln N_i \tag{44}$$

$$G = \sum n_i \text{H}_i - T \sum n_i \text{S}_i + T \sum n_i R \ln N_i \tag{45}$$

Nearly ideal solutions are formed by closely related homologs of a series of organic compounds having nearly equal molal volumes such as heptane and octane, or benzene and toluene, at low temperatures. However, because of differences in molecular diameter, even these liquid solutions deviate from ideality. At conditions where a volatile component is in a liquid solution at temperatures above its own critical temperature, nonideal-solution behavior is always manifest. For gas mixtures, the pressure and temperature range of ideal-solution behavior is much greater than for liquid solutions. Usually ideal-solution behavior exists in nonchemically reacting gaseous mixtures at atmospheric pressure.

Nonideal Liquid Solutions. Hildebrand[1] classifies liquid solutions on the basis of molecular structure and intermolecular forces, as ideal, athermal semi-ideal, regular, associated, and solvated.

Athermal semi-ideal solutions are those showing no heat effect and no volume change on mixing the liquid components at the temperature and pressure of the system: $\Delta H^M = 0$ and $\Delta V^M = 0$. Hence,

$$\Delta G^M = -T\,\Delta S^M \tag{46}$$

Athermal solutions show nonideality with respect to a change in entropy. For athermal semi-ideal solutions, entropy deviations are due only to differences in the free volumes of the separate components, and in this respect are not completely random. Under these circumstances, the molal entropy of mixing becomes

$$\Delta s^M = -R(\sum N_i \ln \phi_i) \tag{47}$$

where

$$\phi_i = \frac{N_i v_i^f}{\sum\limits_i N_i v_i^f} \tag{48}$$

$$= \begin{array}{l}\text{fraction of molal free volume}\\ \text{occupied by component } i\end{array}$$

The free volume v_i^f is defined as the effective space occupied by individual molecules per mole. Hildebrand[1] has shown that, where the ratio of liquid molal volumes v_2^f/v_1^f is 2, the deviation of entropy from ideality for component 1 varies from 0 to 16%, and for component 2 from 0 to 33%. For athermal solutions the entropy of mixing is always greater than that for ideal solutions. In all cases athermal solutions show negative deviations from Raoult's law.

Regular solutions are those that are endothermic in formation but ideal in entropy change; thus, for each component,

$$\bar{\text{H}}_i - \text{H}^\circ_i > 0 \tag{49}$$

$$\bar{\text{s}}_i - \text{s}^\circ_i = -R \ln N_i \tag{50}$$

In regular solutions complete randomness exists as in ideal solutions, but the heat of mixing is positive instead of zero: $\Delta H^M > 0$. For regular

[1] J. H. Hildebrand and R. L. Scott, *The Solubility of Nonelectrolytes*, 3d ed., Reinhold Publishing Corp. (1950).

solutions, evaluation of the heat of mixing gives a measure of free energy of mixing. Thus,

$$\bar{G}_i - G^\circ_i = \bar{H}_i - H^\circ_i - RT \ln N_i \tag{51}$$

Solutions that depart widely from ideality with respect to both entropy and heat of solution are classified as *associated* and as *solvated*. When associated solutes are dissolved, dissociation takes place with absorption of heat and with an entropy increase exceeding that of ideal behavior. Thus, in *associated solutions*,

$$\bar{H}_i - H^\circ_i > 0 \quad \text{(endothermic)} \tag{52}$$

$$\bar{S}_i - S^\circ_i > -R \ln N_i \tag{53}$$

Solvation of a solute takes place with evolution of heat and with an entropy change less than for ideal behavior. Thus, in *solvated solutions*,

$$\bar{H}_i - H^\circ_i < 0 \quad \text{(exothermic)} \tag{54}$$

$$\bar{S}_i - S^\circ_i < -R \ln N_i \tag{55}$$

Equations of State for Mixtures

The laws of additive pressures and of additive volumes discussed in Chapter 3, page 57, when considered separately, do not necessarily imply ideal-gas behavior. For example, additivity of volumes can apply to liquids, but the relation $pV = RT$ does not apply. In general, these laws apply with much greater accuracy than the ideal-gas law and are reasonably satisfactory, even where pressures are moderately high. Either Dalton's or Amagat's law may hold for actual gases, but together they hold only for ideal gases. Which of these laws gives the better approximation depends on the conditions and the nature of the gaseous mixture. For example, for mixtures of argon and ethylene, Dalton's law of additive pressures is the more nearly accurate; for mixtures of nitrogen and hydrogen, the law of additive volumes holds better. The various equations of state developed for pure gases may be extended to gaseous mixtures by combinations of the constants for the separate pure components. For the van der Waals equation (page 560) the following combinations of constants are in common use:

$$\begin{aligned} a_m &= (\sum N_i a_i^{1/2})^2 \\ b_m &= \sum N_i b_i \end{aligned} \tag{56}$$

where a_m, b_m = average constants for the mixture.

For the Beattie–Bridgeman[2] equation (page 563), Beattie proposed the following combinations of constants;

$$
\begin{aligned}
(A_{0m})^{1/2} &= \sum N_i (A_{0i})^{1/2} \\
a_m &= \sum N_i a_i \\
B_{0m} &= \sum N_i B_{0i} \\
b_m &= \sum N_i b_i \\
c_m &= \sum N_i c_i
\end{aligned}
\tag{57}
$$

where the subscript m refers to the average constants for the mixture. Beattie found that volumes calculated from these average constants agreed with experimental data, with a maximum deviation of 0.55% for the systems and ranges investigated.

For the Benedict–Webb–Rubin[3] equation 14–14, which holds for both the gaseous and liquid states, the following combinations of constants are given.

$$
\begin{aligned}
B_{0m} &= \sum N_i B_{0i} & a_m &= [\sum (N_i a_i^{1/3})]^3 \\
A_{0m} &= (\sum N_i A_{0i}^{1/2})^2 & c_m &= (\sum N_i c_i^{1/3})^3 \\
C_{0m} &= [\sum (N_i C_{0i}^{1/2})]^2 & \alpha_m &= (\sum N_i \alpha_i^{1/3})^3 \\
b_m &= [\sum (N_i b_i^{1/3})]^3 & \gamma_m &= (\sum N_i \gamma_i^{1/2})^2
\end{aligned}
\tag{58}
$$

Mean Compressibility Factors. The equation of state for a mixture may be written in terms of a mean compressibility factor z_m; thus,

$$pv = z_m RT \tag{59}$$

The mean compressibility factor is a function of the pressure, temperature, and composition of the mixture. The compressibility factors of a solution of fixed composition are a function of pressure. Bartlett[4] has determined experimentally the compressibility factors of mixtures of hydrogen and nitrogen at 0° C. In Fig. 231 is shown the variation of the compressibility factor with composition at various pressures. Where volumes are additive at constant pressure and temperature, the mean compressibility factor is

[2] J. A. Beattie, *J. Am. Chem. Soc.*, **51**, 19 (1929).
[3] M. Benedict, G. B. Webb, and L. C. Rubin, *Chem. Eng. Prog.*, **47**, 419 (1951).
[4] E. P. Bartlett, *J. Am. Chem. Soc.*, **49**, 687, 1955 (1927).

an additive property of the compressibility factors of the separate components; thus,

$$z_m = \sum N_i z_i \tag{60}$$

z_i = compressibility factor of component i at the temperature and the total pressure of the mixture

The dotted lines on Fig. 231 represent the results obtained by the application of equation 60 to mixtures of hydrogen and nitrogen. For

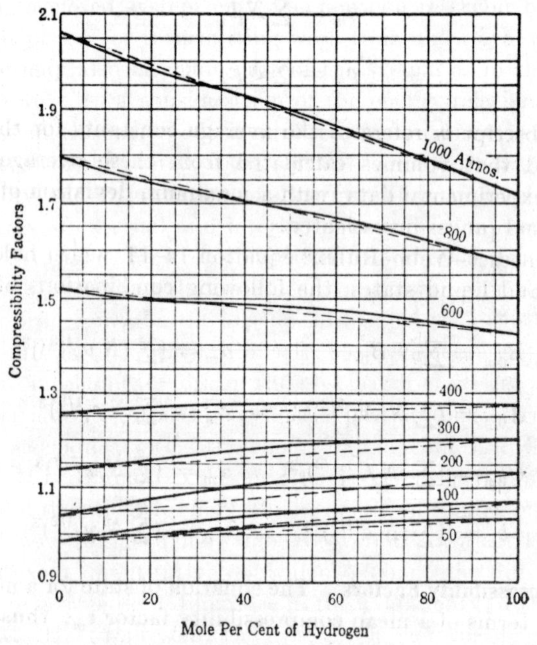

Fig. 231. Compressibility factors of mixtures of hydrogen and nitrogen at 0° C

this system, the validity of Amagat's law is satisfactory, the maximum error being 2.5%.

This additivity of volumes and compressibility factors is unsatisfactory for nonideal-solution behavior, especially at conditions close to the critical points of any of the components.

Pseudocritical Properties. The reduced properties of mixtures based on the true critical properties of the mixture do not give the same functional relations for z and properties derived therefrom as for pure components. Many attempts have been made to establish values for the pseudocritical constants of mixtures in terms of the critical properties of the separate components, so that compressibility factors of mixtures would follow the same functional relationships in terms of reduced properties as do pure components. To this end Kay[5] used the following simple additive relations:

$$p'_c = \sum N_i p_{ci}, \qquad p_r = p/p'_c \tag{61}$$

$$T'_c = \sum N_i T_{ci}, \qquad T_r = T/T'_c \tag{62}$$

The third parameter z_c may be treated in a similar manner:

$$z'_c = \sum N_i z_{ci} \tag{63}$$

In examining 40 different binary mixtures, Tang[6] found that these simple additive relationships gave a fair approximation for critical temperatures, but in general the pseudocritical pressure p'_c deviated widely from linearity. This deviation was found to be unsymmetrical with respect to the molal composition of a binary solution and, in general, to give values lower than the arithmetic mean. In estimating the compressibility of mixtures, using equations 61 through 63 with generalized values (Table 49), Chao and Tang[7] found a root-mean-square deviation of 8.9% based on 1100 mixtures in a total of 40 different systems.

The critical properties of a pure substance can be expressed in terms of the constants of the equations of state. If the relationship of both mixtures and pure fluids are represented by the same form of equation of state, pseudocritical properties of the mixtures can be expressed in terms of the constants of the equation by expressions analogous to those for pure components. If, in addition, the constants of the equations of state of the mixtures are predictable as functions of composition, the pseudocritical properties also can be expressed as functions of composition.

On the basis of these considerations Joffe[7a] proposed the following relationships for predicting p'_c and T'_c based on the van der Waals equation of state:

$$A = \frac{T'_c}{(p'_c)^{1/2}} = \sum_{i=1}^{n} N_i \frac{T_{ci}}{(p_{ci})^{1/2}} \tag{64}$$

$$B = \frac{T'_c}{p'_c} = \frac{1}{8} \sum_{i=1}^{n} \sum_{j=1}^{n} N_i N_j \left[\left(\frac{T_{ci}}{p_{ci}} \right)^{1/3} + \left(\frac{T_{cj}}{p_{cj}} \right)^{1/3} \right]^3 \tag{65}$$

where the double summation contains one term for each possible permutation in pairs of like and unlike components of the mixture 1,2,3, · · · , *n*.

[5] W. B. Kay, *Ind. Eng. Chem.*, **28**, 1014 (1936).

[6] W. K. Tang, *M. S. thesis*, University of Wisconsin (1956).

[7a] J. Joffe, *Ind. Eng. Chem.* **39**, 837 (1947).

For example, in a three-component mixture 1,2,3, the terms 11, 12, 13, 21, 22, 23, 31, 32 and 33 must be evaluated and summed.

Recently two modifications of Joffe's equations have been proposed. Leland and Mueller[7b] arrived at the following:

$$T'_c = \left[\frac{\sum_{i=1}^{n} \sum_{j=1}^{n} x_i x_j \left(\frac{z_c T_c^{\alpha+1}}{p_c} \right)_i^{1/2} \left(\frac{z_c T_c^{\alpha+1}}{p_c} \right)_j^{1/2}}{\sum_{i=1}^{n} \sum_{j=1}^{n} x_i x_j \left[\frac{1}{2} \left(\frac{z_c T_c}{p_c} \right)_j^{1/3} + \frac{1}{2} \left(\frac{z_c T_c}{p_c} \right)_j^{1/3} \right]^3} \right]^{1/\alpha} \tag{66}$$

$$p'_c = \left[\frac{T'_c \sum_{i=1}^{n} x_i (z_c)_i}{\sum_{i=1}^{n} \sum_{j=1}^{n} x_i x_j \left[\frac{1}{2} \left(\frac{z_c T_c}{p_c} \right)_i^{1/3} + \frac{1}{2} \left(\frac{z_c T_c}{p_c} \right)_j^{1/3} \right]^3} \right] \tag{67}$$

where α is an empirically evaluated exponent which is expressed as a function of Tp'_{cK}/pT'_{cK} in Table 58a where p'_{cK} and T'_{cK} are the appropriate pseudocritical values calculated from equations 61 and 62.

Table 58a. Values of Exponent α in Equations 66 and 67 for Pseudocritical Properties

$(Tp'_{cK})/(pT'_{cK})$	α	$(Tp'_{cK})/(pT'_{cK})$	α
0.400	2.20	1.300	1.47
0.500	2.06	1.400	1.39
0.600	1.98	1.500	1.32
0.700	1.91	1.600	1.24
0.800	1.84	1.700	1.17
0.900	1.76	1.800	1.09
1.000	1.69	1.900	1.02
1.100	1.61	2.000	1.00
1.200	1.54		

In using the pseudocritical values derived from equations 66 and 67 it was suggested that a pseudocritical compressibility factor be defined as follows:

$$z'_c = \sum_{i}^{n} N_i z_{ci} \tag{67a}$$

Leland and Mueller also used the pseudocritical constants from equations 66 and 67 in conjunction with the Benedict–Webb–Rubin equation of state for a *reference substance* which was taken as the paraffin hydrocarbon

[7b] T. W. Leland and W. H. Mueller, "Application to the Theory of Corresponding States to Multicomponent Systems," AIChE meeting, Atlantic City, Mar. 17, 1959.

having a value of z_c nearest to the value of z'_c for the mixture from equation 67*a*.

It will be noted that the pseudocritical properties calculated from equations 66 and 67 are not constant for a mixture of constant composition but vary with both T and p.

Stewart, Burkhart, and Voo[7c] proposed the following simplification of equation 65:

$$B = \frac{T'_c}{p'_c} = \frac{1}{3}\sum_{i=1}^{n} N_i \frac{T_{ci}}{p_{ci}} + \frac{2}{3}\left[\sum_{i=1}^{n} N_i \left(\frac{T_{ci}}{p_{ci}}\right)^{1/2}\right]^2 \tag{67b}$$

Equations 64 and 67*b* are used to evaluate p'_c and T'_c in conjunction with the following relations:

$$T'_c = A^2/B \tag{67c}$$

$$p'_c = T'_c/B \tag{67d}$$

Preliminary comparisons indicate that both equations 66–67 and equations 64, 67*b*–67*d* reduce by more than 50% the average deviations in calculations of densities of gaseous and liquid mixtures over wide ranges of conditions as compared to equations 61 and 62. Near the two-phase region in the vicinity of the pseudocritical point much greater improvement is realized. For conditions somewhat removed from the two-phase region a standard deviation of 2 to 3% is indicated.

Densities of Liquid Solutions. For ideal solutions, specific volumes are additive at constant pressure and temperature. For nonideal solutions, specific molal volumes and molal densities are related to the partial molal properties; thus,

$$v = \sum N_i \bar{v}_i$$
$$\frac{1}{\rho} = \sum (N_i/\bar{\rho}_i) \tag{68}$$

Where the density of a liquid solution is known at one condition of temperature and pressure, its value at another temperature and pressure can be estimated by generalized methods from the reduced conditions of the solution at two conditions from the ratio

$$\rho_m = \rho_{m1}\left(\frac{\rho'_r}{\rho'_{r1}}\right) \tag{69}$$

This ratio method gives accuracy greater than the direct use of generalized methods.

[7c] W. E. Stewart, S. F. Burkhart, and D. Voo, "Prediction of Pseudocritical Constants for Mixtures," presented at the AIChE meeting, Kansas City, Mo., May 18, 1959.

Calculation of Fugacities

For ideal-solution behavior in gaseous mixtures, the fugacity of component i is obtained from the fugacity of the pure gaseous component at the temperature and pressure of the system:

$$(f_i = y_i f^\circ_i) \quad \text{(gas)} \tag{70}$$

In gaseous mixtures, ideal-solution behavior is approached at low pressures, even where great deviations occur in the corresponding liquid solutions.

Similarly, for ideal-liquid solutions the fugacity of component i is obtained from the fugacity of the pure liquid at the temperature and pressure of the system:

$$(f_i = x_i f^\circ_i) \quad \text{(liquid)} \tag{71}$$

The fugacity f°_i of pure component i as liquid or gas at the temperature and pressure of the system may be obtained from the generalized values of fugacity coefficients, Table 51.

The assumption that gases form ideal solutions is of a much higher order of accuracy than the ideal-gas law and may be applied with accuracy satisfactory for many purposes to gaseous mixtures where the reduced pressure of each component is less than 0.8.

Illustration 1. A gaseous mixture has the following composition expressed in mole fractions:

Methane	$N_1 = 0.17$
Ethane	$N_2 = 0.35$
Propane	$N_3 = 0.48$
Total	1.00

Assuming an ideal-gaseous solution, calculate from Table 51 the fugacity of each component when the mixture is at an absolute pressure of 300 psia and a temperature of 100° F.

Solution:

	Methane	Ethane	Propane
T_c °R	343.3	549.7	665.8
p_c psia	673.2	708.3	617.4
z_c	0.290	0.285	0.277
T_r	1.635	1.022	0.855
p_r	0.446	0.424	0.486
$\nu_i = f_i/p$	0.976	0.886	0.56
$f^\circ_i = 300\,\nu_i$	292	266	168
$f_i = N_i f^\circ_i$ psia	49.6	93.2	80.5
p_i psia	51	105	144

Fugacities from Partial Molal Volumes. Integration of equation 23 at constant temperature from some low pressure p^*, where ideal-solution behavior is exhibited to the existing pressure, gives

$$\left(\int_{f^*_i}^{f_i} d\ln f_i\right)_T = \left(\frac{1}{RT}\int_{p^*}^{p} \bar{v}_i\,dp\right)_T \tag{72}$$

In an ideal solution

$$f^*_i = N_i p^* \tag{73}$$

Integration of equation 72, combined with equation 73, gives

$$\ln f_i = \frac{1}{RT}\int_{p^*}^{p} \bar{v}_i\,dp + \ln N_i + \ln p^* \tag{74}$$

Equation 74 permits calculation of the fugacity of each component in a mixture from pvT composition data of the mixture. The evaluation of f_i is facilitated by using the residual molal volume α, where

$$\alpha = v^* - v = \frac{RT}{p} - v \tag{75}$$

and v^* = ideal molal volume of the gas mixture. Differentiation of equation 75, at constant p, t, and n_j, gives

$$\left(\frac{\partial \alpha}{\partial n_i}\right)_{pTn_j} = \left(\frac{\partial v^*}{\partial n_i}\right)_{pTn_j} - \left(\frac{\partial v}{\partial n_i}\right)_{pTn_j} \tag{76}$$

or

$$\bar{\alpha}_i = \bar{v}^*_i - \bar{v}_i \tag{77}$$

where j = components other than i. For ideal-gas behavior, the partial molal volume is equal to the molal volume of the mixture

$$\bar{v}^*_i = v^* = RT/p \tag{78}$$

or

$$\bar{\alpha}_i = \frac{RT}{p} - \bar{v}_i \tag{79}$$

where $\bar{\alpha}_i$ = partial residual molal volume of component i.

From equation 78,

$$\int_{p^*}^{p} \bar{v}_i\,dp = RT\ln\frac{p}{p^*} - \int_{p^*}^{p} \bar{\alpha}_i\,dp \tag{80}$$

Combining equations 80 and 74 gives

$$\ln \frac{f_i}{N_i p} = -\frac{1}{RT}\int_{p^*}^{p} \bar{\alpha}_i \, dp \tag{81}$$

Values of the partial residual molal volume $\bar{\alpha}_i$ are obtained by the method of tangent intercepts (page 337) as shown in Fig. 232, by plotting experimental values of residual molal volumes against mole fractions. The values of $\bar{\alpha}_i$ are plotted against pressure extrapolated to low pressures where ideal behavior exists, and the graphical integration is carried out in accordance with equation 81. This method is shown in the following illustration.

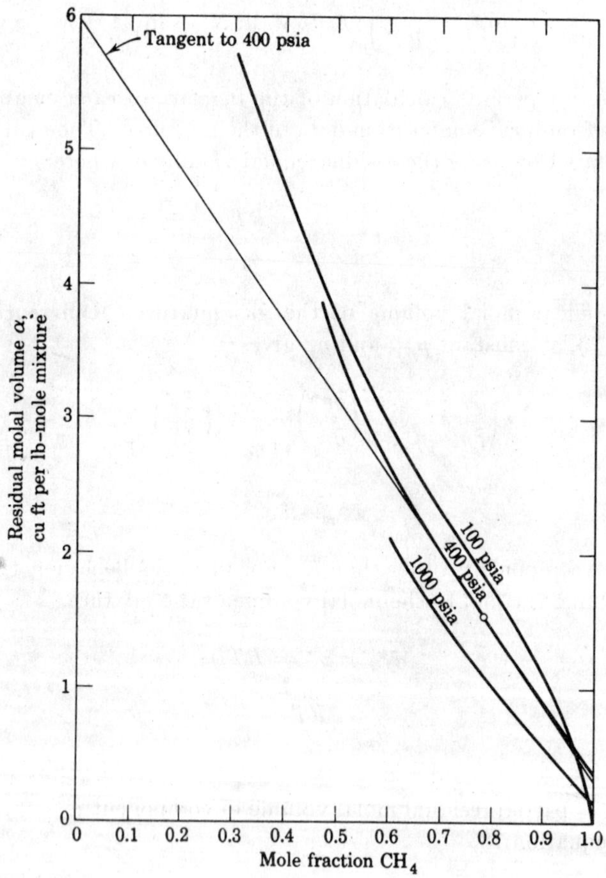

Fig. 232. Residual molal volume of a methane-n-butane gaseous mixture at a temperature of 220° F

Illustration 2. It is desired to calculate the fugacity of methane in a gaseous mixture composed of 78.4 mole per cent methane and 21.6 mole per cent n-butane at 220° F and 1000 psia. Experimental values of the compressibility factor for this system are reported by Sage, Budenholzer, and Lacey[8] in Table A.

TABLE A. COMPRESSIBILITY FACTORS OF THE METHANE-n-BUTANE MIXTURE AT 220° F

Pressure, psia	Mole % Methane								
	28.7	47.5	60.8	70.7	78.4	84.5	89.4	93.5	97.0
0	1.000	1.000	1.000	1.000	1.000	1.000	1.000	1.000	1.000
100	0.9240	0.9435	0.9590	0.9685	0.9745	0.9805	0.9854	0.9893	0.9925
200	0.8475	0.8900	0.9202	0.9390	0.9521	0.9630	0.9720	0.9795	0.9855
400	—	0.7870	0.8528	0.8895	0.9150	0.9339	0.9494	0.9618	0.9725
600	—	—	0.7985	0.8512	0.8860	0.9124	0.9318	0.9475	0.9620
800	—	—	0.7520	0.8192	0.8624	0.8958	0.9182	0.9365	0.9532
1000	—	—	0.7100	0.7920	0.8428	0.8820	0.9070	0.9275	0.9460

The residual molal volume α is obtained for various values of pressure and at a temperature of 680° R, from the values of compressibility z in Table A from the relation,

$$\alpha = (1 - z)RT/p \tag{82}$$

The calculated values of residual molal volume are tabulated in Table B.

TABLE B. CALCULATED VALUES OF RESIDUAL MOLAL VOLUMES α
TEMPERATURE = 220° F

Pressure, psia	100	200	400	600	800	1000
$\frac{RT}{p}$	72.9	36.4	13.2	12.13	9.10	7.29
$100N_i$						
28.7	5.55	5.55	—	—	—	—
47.5	4.12	4.00	3.88	—	—	—
60.8	2.99	2.91	2.68	2.44	2.26	2.12
70.7	2.30	2.22	2.01	1.81	1.645	1.52
78.4	1.86	1.745	1.55	1.383	1.252	1.147
84.5	1.42	1.340	1.202	1.002	0.950	0.800
89.4	1.065	1.020	0.921	0.826	0.745	0.679
93.5	0.780	0.746	0.695	0.637	0.579	0.529
97.0	0.546	0.528	0.500	0.461	0.426	0.394

To obtain values of partial residual molal volumes $\bar{\alpha}$, values of residual molal volume α are plotted against mole fraction of methane (Fig. 232). By the method of tangent intercepts, values of the partial residual molal volumes are obtained at 0.784 mole fraction methane and tabulated in Table C.

[8] B. H. Sage, R. A. Budenholzer, and W. N. Lacey, *Ind. Eng. Chem.*, **32**, 1262 (1940).

TABLE C. PARTIAL RESIDUAL MOLAL VOLUME OF METHANE AT $N_i = 0.784$; $t = 220°$ F

Pressure, psia	$\bar{\alpha}_i$, cu ft per lb-mole	$\bar{\alpha}_i$, cu ft per lb
0	0.288	0.018
100	0.288	0.018
200	0.273	0.017
400	0.240	0.015
600	0.208	0.013
800	0.176	0.011
1000	0.144	0.009

Values of $\ln (f_i/N_ip)$ are then obtained by graphical integration, in accordance with equation 81, using the values of Table C. The area obtained by graphical integration corresponds to 14.6 (cu ft/lb$_m$)(lb$_f$/sq in.). Substitution of this value into equation 81 gives

$$\ln \frac{f_2}{N_2p} = -\frac{(14.6)(16.04)}{(10.71)680} = -0.0322$$

$$\frac{f_2}{N_2p} = 0.9683$$

$$f_2 = (0.784)1000(0.9683) = 757 \text{ psia}$$

Where equations of state are known for mixtures in terms of composition, fugacities may be obtained by direct analytical methods. For example, applying equation 74 to the Benedict–Webb–Rubin[9] equation of state for mixtures (equations 58 and 14–14), the following relation for fugacity is obtained, applicable to both gaseous and liquid phases:

$$\begin{aligned} RT \ln f_i = {} & RT \ln(RTN_i\rho) \\ & + [(B_0 + B_{0i})RT - 2(A_0A_{0i})^{1/2} - 2(C_0C_{0i})^{1/2}/T^2]\rho \\ & + \tfrac{3}{2}[RT(b^2b_i)^{1/3} - (a^2a_i)^{1/3}]\rho^2 + \tfrac{3}{5}[a(\alpha^2\alpha_i)^{1/3} \\ & + \alpha(a^2a_i)^{1/3}]\rho^5 + \frac{3\rho^2(c^2c_i)^{1/3}}{T^2}\left[1 - \frac{e^{-\gamma\rho^2}}{\gamma\rho^2} - \frac{e^{-\gamma\rho^2}}{2}\right] \\ & - \frac{2\rho^2c}{T^2}\left(\frac{\gamma_i}{\gamma}\right)^{1/2}\left(\frac{1 - e^{-\gamma\rho^2}}{\gamma\rho^2} - e^{-\gamma\rho^2} - \frac{\gamma\rho^2e^{-\gamma\rho^2}}{2}\right) \end{aligned} \tag{83}$$

Activity

The ratio of fugacities of component i in any state compared to its fugacity in the standard state is designated as its *activity* a_i; thus,

$$(a_i)_T = (f_i/f^\circ_i)_T \tag{84}$$

[9] M. Benedict, G. B. Webb, and L. C. Rubin, *Chem. Eng. Prog.*, **47**, 419 (1951).

Integration of equation 20 at constant temperature gives

$$\bar{G}_i - G^\circ_i = \mu_i - \mu^\circ_i = RT \ln \frac{f_i}{f^\circ_i} \tag{85}$$

Combining equations 84 and 85 gives

$$(\Delta \bar{G}^\circ_i)_T = (\Delta \mu^\circ_i)_T = (RT \ln a_i)_T \tag{86}$$

The concept of activity is particularly useful when directly related to composition, and hence to a material balance. For example, if the standard state of a component in solution is taken as the pure component at the temperature and pressure of the solution, the activity of that component becomes a function of its mole fraction or concentration, and for ideal solutions becomes equivalent to mole fraction. In turn, isothermal changes in the chemical potential for a given component can be expressed in terms of changes in concentration or pressure. The term activity is a ratio without dimensions; its numerical value depends on the standard state chosen.

Standard States

The choice of standard state of fugacity f°_i is arbitrary and based on experimental convenience and reproducibility. One standard state is chosen for each component at each state of aggregation. The temperature of the standard state is the same as that of the system under study and is not a fixed reference value. The standard state with respect to pressure or concentration for each state of aggregation is arbitrarily chosen as some value that can be conveniently measured or accurately calculated. At certain temperatures, this standard state represents a hypothetical condition which cannot be attained experimentally but which should be capable of calculation with reproducible results. Different standard states may be selected for the same component for use in different systems, the choice being dictated by convenience. Throughout any one series of calculations it is important that the standard state of each component be kept the same.

The following standard states have been found convenient and are in more or less general use. Other standard states may be introduced as their convenience arises.

Gaseous Components. For gases, two standard states are in common use, namely; (*a*) the pure component gas in its ideal state at 1 atm pressure, and (*b*) the pure component gas at the pressure of the system; both standard states are at the temperature of the system.

(*a*) With the first standard state, the fugacity is unity when expressed in atmospheres, and the activity of a component gas in a mixture becomes numerically equal to its fugacity expressed in atmospheres, and in mixtures of gases showing ideal-gas behavior the activity of each component becomes equal to its partial pressure. This standard state is used in dealing with equilibria in chemical reactions.

(*b*) Where the standard state is taken as the pure gas at the pressure of the system, as in dealing with vapor–liquid equilibria, the activity of each component becomes equal to its mole fraction in ideal solutions. This standard state becomes hypothetical at temperatures where the total pressure exceeds the saturation pressure of the pure component gas.

Liquid Components. (*a*) The standard state of a pure component in the liquid state can be taken at a pressure of 1 atm. This standard state becomes hypothetical when the vapor pressure of the pure component exceeds 1 atm.

(*b*) The standard state may also be taken as the pure liquid at the pressure of the system, as in dealing with vapor–liquid equilibria. This state becomes hypothetical at temperatures above the critical or saturation temperatures of the pure liquid.

Solid Components. The standard state may be taken as the pure component in the solid state at a pressure of 1 atm. The assumption of unit activity for pure solid compounds should not be taken for granted. The activity of a solid may be greatly altered by small impurities, by lack of equilibrium in the lattice structure, and by crystal size.

Components in Solutions of Nonelectrolytes. In liquid solutions of nonelectrolytes, special standard states are convenient besides those mentioned above for pure components.

For components of limited solubility, it is convenient to define activity of a component by reference to its behavior at infinite dilution where its mole fraction approaches zero; thus, as x_i approaches 0,

$$\frac{a'_i}{x_i} = \frac{f_i}{f^{\circ\prime}{}_i x_i} = 1 \quad \text{or} \quad \left(\frac{\partial f_i}{\partial x_i}\right)_{x_i=0} = f^{\circ\prime}{}_i \tag{87}$$

As infinite dilution is approached the fugacity of a solute becomes proportional to mole fraction,

$$f_i = k_i x_i \quad \text{and} \quad \left(\frac{\partial f_i}{\partial x_i}\right)_{x_i=0} = k_i \tag{88}$$

From equations 87 and 88,

$$f^{\circ\prime}{}_i = k_i \tag{89}$$

This standard state is not the state of infinite dilution but a hypothetical liquid state which is defined by reference to the behavior of the component at infinite dilution. A state defined in this manner is termed a standard state referred to infinite dilution. The pressure of the system should also be designated in defining the above state. This pressure effect is usually negligible.

A similar standard state is represented as the behavior of a solute component at unit molality referred to its behavior at infinite dilution. Thus,

$$\frac{a''_i}{m_i} = \frac{f_i}{f^{\circ\prime\prime}_i m_i} = 1.0 \quad \text{(as } m \text{ approaches zero)} \tag{90}$$

As infinite dilution is approached,

$$f_i = k''_i m_i \tag{91}$$

or

$$\left(\frac{\partial f_i}{\partial m_i}\right)_{m_i=0} = k''_i \tag{92}$$

From equations 91 and 92, $k''_i = f^{\circ\prime\prime}_i$ (93)

From the definition of activity, the activities of a component when expressed on the basis of these three standard states are related to each other by constant factors, independent of concentration; thus,

$$f_i = a_i f^{\circ}_i = a'_i f^{\circ\prime}_i = a''_i f^{\circ\prime\prime}_i \tag{94}$$

The conversion of activities from one standard state to another for a given system is obtained by rearranging equation 94:

$$a_i = a'_i \left(\frac{f^{\circ\prime}_i}{f^{\circ}_i}\right) = a''_i \left(\frac{f^{\circ\prime\prime}_i}{f^{\circ}_i}\right) \tag{95}$$

Thus, the ratios of activities relative to different standard states are given by the ratios of the fugacities of the standard states according to equation 95, and these conversion factors are constant over the entire concentration range.

Activities in Nonelectrolytic Solutions. An evaluation of the activities of component 2 in an aqueous solution with reference to three standard states is shown in Fig. 234. The actual fugacity f_2 of component 2 as a function of concentration at constant temperature T is plotted in Fig. 233. At $x_2 = 1.0$, the fugacity of component 2 is 0.5. This value defines the standard-state fugacity $f^{\circ}_2 = 0.5$ and the activity is unity.

The value of $f^{\circ\prime}_2$ is obtained by drawing a tangent to the fugacity curve at zero mole fraction and extending it to $x_2 = 1.0$, giving a value of $f^{\circ\prime}_2 = 1.3$ and a corresponding activity a'_2 equal to $0.5/1.3 = 0.384$.

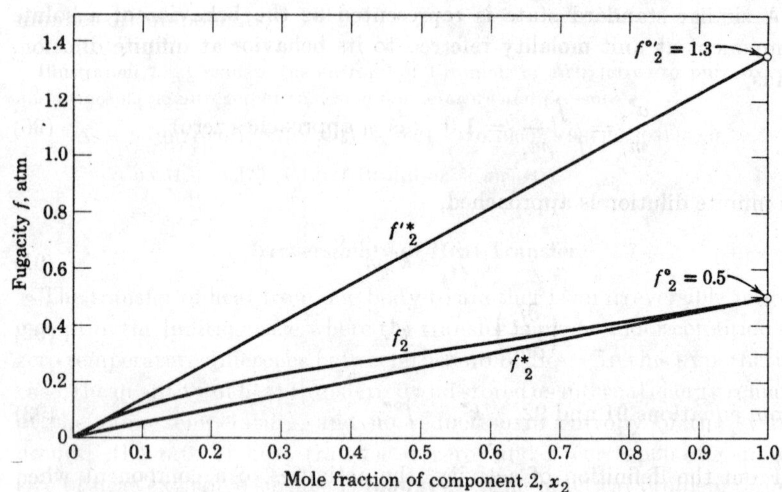

Fig. 233. Actual fugacity of component 2 as related to concentration; fugacity at different standard states

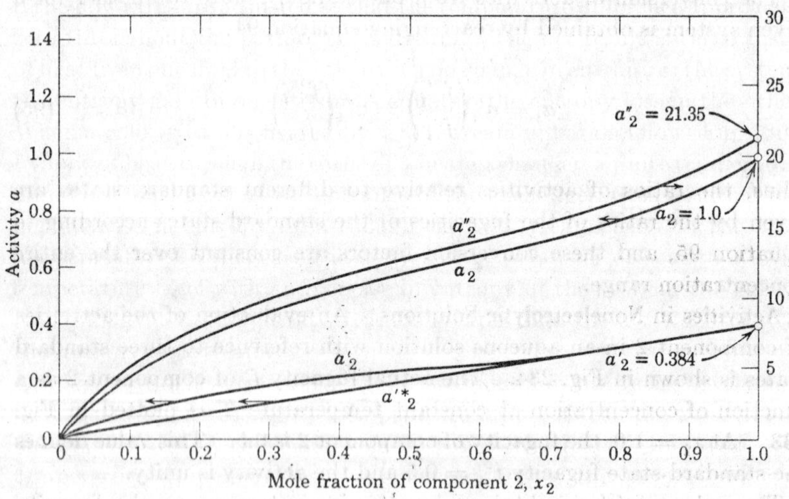

Fig. 234. Comparison of activities for different standard states

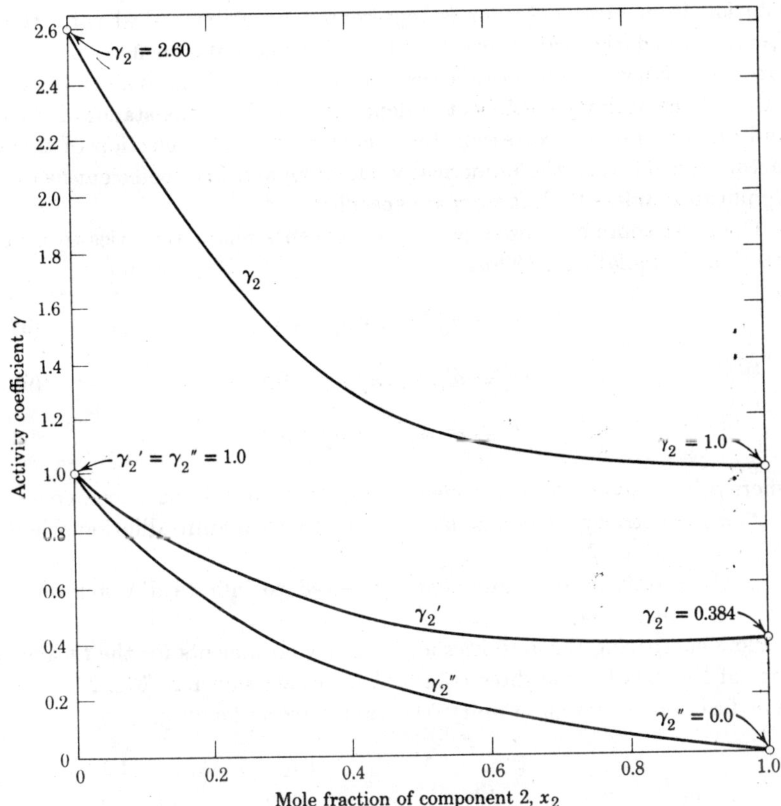

FIG. 235. Comparison of activity coefficients for different standard states

The value of $f^{\circ\prime\prime}_2$ is obtained by replotting f_2 against molality m_2, where at unit molality the corresponding mole fraction in an aqueous solution is 0.0177. A tangent to the fugacity–molality curve at zero molality extended to $m_2 = 1.0$ gives a value of $f^{\circ\prime\prime}_2 = 1.3/55.51 = 0.0234$. The ratios of activities referred to the different standard states for this particular solution are given by the following from equation 95:

$$a_2 = \frac{1.3}{0.5}\, a'_2 = \frac{0.0234}{0.5}\, a''_2 = 2.6a'_2 = 0.0468a''_2 \tag{96}$$

Hence, at $x_2 = 1.0$, $a_2 = 1.0$, $a'_2 = 0.384$, and $a''_2 = 21.35$.

Activity Coefficients. The concept of activity provides a thermodynamically defined quantity which with proper choice of standard state is directly related to composition in ideal solutions. The departure of a solution from ideal behavior is represented by an empirical correction factor termed the *activity coefficient* γ. The activity coefficient is the ratio of activity to a numerical expression of composition. The numerical value of the activity coefficient is dependent both on the standard state and on the units of expression for composition. The selection of these factors is arbitrary; the numerical value of an activity coefficient has no significance unless both factors are specified.

The most commonly used activity coefficients relate activities to mole fractions or molalities. Thus,

$$\gamma_i = a_i/N_i = f_i/(f^\circ{}_i N_i) \tag{97}$$

$$\gamma'_i = a'_i/N_i = f_i/(f^{\circ\prime}{}_i N_i) \tag{98}$$

$$\gamma''_i = a''_i/m_i = f_i/(f^{\circ\prime\prime}{}_i m_i) \tag{99}$$

where a_i = activity of component i referred to the pure component
a'_i = activity of component i referred to infinite dilution where $a'_i = N_i$
a''_i = activity of component i referred to infinite dilution where $a''_i = m_i$

Plots comparing the activities and activity coefficients for the fugacity data of Fig. 234 for the three standard states are shown in Fig. 235. At mole fractions of zero and unity, the values are as follows:

	At $x_2 = 0$	At $x_2 = 1.0$		At $x_2 = 0$	At $x_2 = 1.0$
a_2	0	1.0	γ_2	2.60	1.0
a'_2	0	0.384	γ'_2	1.0	0.384
a''_2	0	21.35	γ''_2	1.0	0

Calculation of Activities Where Fugacities Are Unknown. In solutions of components having low vapor pressures such as solids, fugacities may be unknown or difficult to evaluate. The ratio of activities based on the different standard states can then be evaluated from the activity coefficients at infinite dilution, where $\gamma_i = \gamma_{i0}$, $\gamma'_i = 1.0$, and $\gamma''_i = 1.0$. Applying equations 95 and 97 at infinite dilution,

$$\gamma_i x_i = x_i\left(\frac{f^{\circ\prime}{}_i}{f^\circ{}_i}\right) = m_i\left(\frac{f^{\circ\prime\prime}{}_i}{f^\circ{}_i}\right) \quad \text{and} \quad x_i = \left(\frac{m_i M}{1000}\right) \quad (\text{as } m_i \text{ and } x_i \to 0) \tag{100}$$

or

$$\frac{f^{\circ\prime}{}_i}{f^\circ{}_i} = \gamma_{io} \quad \text{and} \quad \frac{f^{\circ\prime\prime}{}_i}{f^\circ{}_i} = \frac{M}{1000}\gamma_{io} \tag{101}$$

Hence, from equation 95,

$$a_i = \gamma_{io} a'_i = \frac{\gamma_{io} a''_i M}{1000} \tag{102}$$

Effect of Pressure on Activity Coefficients. From equation 32 combined with equation 97 the effect of pressure on the activity coefficient for component i at constant composition and temperature gives

$$\left(\frac{\partial \ln \gamma_i}{\partial p}\right)_{TN_i} = \left(\frac{\bar{v}_i - v^\circ_i}{RT}\right)_{TN_i} \tag{103}$$

The molal volumes $\bar{v}_i$ and v°_i apply to the particular phase under consideration. For liquid solutions, the effect of pressure on activity coefficients is negligible at pressures below atmospheric. For gaseous mixtures, activity coefficients are nearly unity at reduced pressures below 0.8.

Effect of Temperature on Activity Coefficients. From equation 35 combined with equation 97, the effect of temperature on the activity coefficient of component i at constant pressure and composition is shown as follows.

$$\left(\frac{\partial \ln \gamma_i}{\partial T}\right)_{pN_i} = \left(\frac{\mathrm{H}^\circ_i - \bar{\mathrm{H}}_i}{RT^2}\right)_{pN_i} \tag{104}$$

The enthalpies H°_i and $\bar{\mathrm{H}}_i$ refer to the particular phase under consideration. The term $\bar{\mathrm{H}}_i - \mathrm{H}^\circ_i$ is the partial heat of mixing component i from its standard state to the solution of given composition both in the same state of aggregation and pressure. The value of H°_i may correspond to a hypothetical condition. For gaseous mixtures the term $\bar{\mathrm{H}}_i - \mathrm{H}^\circ_i$ is negligible at low pressures. For athermal solutions, $\bar{\mathrm{H}}_i - \mathrm{H}^\circ_i$ is zero.

For regular binary solutions, the activity coefficient of component 1 may be expressed as

$$\ln \gamma_1 = \frac{kN_2^2}{RT} \tag{105}$$

where k is independent of temperature and composition. Then

$$\left(\frac{\partial \ln \gamma_1}{\partial T}\right)_{pN_i} = -\frac{kN_2^2}{RT^2} = -\frac{(\bar{\mathrm{H}}_1 - \mathrm{H}^\circ_1)}{RT^2} \tag{106}$$

For regular solutions the differential heat of solution $\bar{\mathrm{H}}_i - \mathrm{H}^\circ_i$ is thus independent of temperature but dependent on composition only.

Gibbs–Duhem Equation

The relationship among the chemical potentials of the different components in a solution was first formulated by Gibbs.

In terms of partial molal properties, the total free energy of a solution at constant pressure and temperature is given by the relation

$$G_{pT} = (n_1\bar{G}_1 + n_2\bar{G}_2 + \cdots)_{pT} \tag{107}$$

or, since $\mu_i = \bar{G}_n$,

$$G_{pT} = (n_1\mu_1 + n_2\mu_2 + \cdots)_{pT} \tag{108}$$

The total derivative of G may then be written as

$$(dG)_{pT} = (n_1\, d\mu_1 + \mu_1\, dn_1 + n_2\, d\mu_2 + \mu_2\, dn_2 + \cdots)_{pT} \tag{109}$$

At constant pressure and temperature, equation 109 reduces to

$$(dG)_{pT} = (\mu_1\, dn_1 + \mu_2\, dn_2 + \cdots)_{pT} \tag{110}$$

Subtracting equation 110 from equation 109 gives

$$(n_1\, d\mu_1 + n_2\, d\mu_2 + \cdots = 0)_{pT} \tag{111}$$

In terms of mole fraction, equation 111 becomes

$$(N_1\, d\mu_1 + N_2\, d\mu_2 + \cdots = 0)_{pT} \tag{112}$$

Since $(d\mu_i = RT\, d \ln f_i)_T$, equation 112 becomes

$$\left[N_1\left(\frac{\partial \ln f_1}{\partial N_1}\right) + N_2\left(\frac{\partial \ln f_2}{\partial N_1}\right) + N_3\left(\frac{\partial \ln f_3}{\partial N_1}\right) + \cdots = 0\right]_{pT} \tag{113}$$

Equation 113 relates the fugacities of the various components in a solution with one another and to composition at constant temperature and pressure. This important relationship is termed the Gibbs–Duhem equation.

When applied to a binary solution, since $dN_1 = -dN_2$, equation 113 becomes

$$\left[N_1\left(\frac{\partial \ln f_1}{\partial N_1}\right)\right]_{pT} = \left[N_2\left(\frac{\partial \ln f_2}{\partial N_2}\right)\right]_{pT} \tag{114}$$

This relationship can also be expressed in terms of activities and activity coefficients. Where fugacities are unknown or difficult to measure, these latter forms become more useful.

Since $a_i = f_i/f^\circ_i$ and f°_i is a constant at a given pressure and temperature, equation 114 becomes

$$\left[N_1\left(\frac{\partial \ln a_1}{\partial N_1}\right)\right]_{pT} = \left[N_2\left(\frac{\partial \ln a_2}{\partial N_2}\right)\right]_{pT} \tag{115}$$

In terms of activity coefficients, $a_i = N_i\gamma_i$; hence,

$$\left(N_1\frac{\partial \ln \gamma_1 N_1}{\partial N_1}\right)_{pT} = \left(N_2\frac{\partial \ln \gamma_2 N_2}{\partial N_2}\right)_{pT} \tag{116}$$

$$N_1\left(\frac{\partial \ln \gamma_1}{\partial N_1}\right) + N_1\left(\frac{\partial \ln N_1}{\partial N_1}\right) = N_2\left(\frac{\partial \ln \gamma_2}{\partial N_2}\right) + N_2\left(\frac{\partial \ln N_2}{\partial N_2}\right) = 0 \tag{117}$$

but $N_1\left(\frac{\partial \ln N_1}{\partial N_1}\right) = \frac{N_1\,\partial N_1}{N_1\,\partial N_1} = 1$ and similarly $N_2\left(\frac{\partial \ln N_2}{\partial N_2}\right) = 1.0$ (118)

Hence

$$\left[N_1\left(\frac{\partial \ln \gamma_1}{\partial N_1}\right)\right]_{pT} = \left[N_2\left(\frac{\partial \ln \gamma_2}{\partial N_2}\right)\right]_{pT} \tag{119}$$

The Gibbs–Duhem equation involving activity coefficients, equation 119, is its most useful form, since activity coefficients measure directly departure from ideal-solution behavior.

The various forms of the Gibbs–Duhem equation are rigorous thermodynamic relations that are valid for conditions at constant temperature and pressure. They are of particular value in minimizing the number of experimental data necessary to evaluate the properties of a system and for detection of inconsistent or erroneous measurements.

Equation 119 in terms of activity coefficients is not applicable to aqueous solutions where ionization occurs.

Illustration 9. From the data for the vapor pressure of water above sugar ($C_{12}H_{22}O_{11}$) solutions taken from the *International Critical Tables* and summarized in Table A, calculate the corresponding:

(*a*) Fugacities of the water in solution.

(*b*) Activities a_1 and activity coefficients γ_1 of the water referred to pure liquid water.

(*c*) Activities a_2 and activity coefficients γ_2 of the sugar referred to the pure solid.

(*d*) Activities a'_2 and activity coefficients γ'_2 of the sugar referred to mole fractions at infinite dilution.

(*e*) Activities a''_2 and activity coefficients γ''_2 of the sugar referred to molality at infinite dilution.

TABLE A. ACTIVITIES OF WATER IN SUGAR SOLUTIONS AT 25° C

Molality, m	Mole Fractions $C_{12}H_{22}O_{11}$ N_2	H_2O N_1	Vapor pressure, mm Hg, $p_1 = f_1$	a_1	γ_1	$\log \gamma_1$	N_1/N_2
0	0	1.0	23.756	1.0	1.0	0	∞
0.1	0.00180	0.99820	23.714	0.99823	1.0000	0.0000	554.56
0.4	0.00715	0.99285	23.585	0.99280	0.99995	−0.00002	138.86
1.0	0.01770	0.98230	23.302	0.98089	0.99856	−0.00063	55.497
2.0	0.03478	0.96522	22.762	0.95816	0.99269	−0.00319	27.752
3.0	0.05127	0.94873	22.166	0.93307	0.98349	−0.00723	18.505
4.0	0.06722	0.93278	21.521	0.90592	0.97120	−0.01269	13.877
5.0	0.08263	0.91737	20.846	0.87750	0.95654	−0.01930	11.102
6.0	0.09755	0.90245	20.20	0.85031	0.94222	−0.02585	9.251
6.18 (sat)	0.10018	0.89982	20.08	0.84526	0.93937	−0.02716	8.9820

Solution: (*a*) At the low pressures involved it may be assumed that the fugacities of the water vapor f_1 are equal to partial pressures p_1. It is assumed that at equilibrium the fugacity of each component is the same in both phases, a condition that is proved in the next chapter.

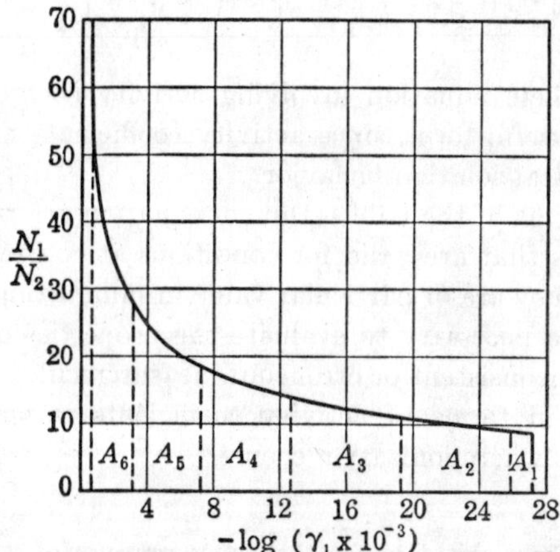

FIG. 236. Evaluation of the activity coefficients of sucrose

(*b*) The activities a_1 of the water in the solution are obtained by dividing p_1 by 23.756, the vapor pressure and fugacity of the pure water at 25° C, which is chosen as the standard state. At the low pressures involved, the pressure designation of the standard state is of no consequence, since pressures of this order have a negligible effect on the fugacity of the liquid. The activity coefficients γ_1 are equal to a_1/N_1.

(*c*) The activity coefficients of the sugar γ_2 are determined by graphical integration of equation 119. Values of $-\log \gamma_1$ and N_1/N_2 from Table A are plotted in Fig. 236. Since the fugacity of the sugar in the saturated solution is equal to that of the pure solid, the activity a_2 of the sugar in the saturated solution is 1.0 referred to the pure solid. The corresponding activity coefficient $\gamma_2 = 1.0/N_2 = 9.9820$ where $N_1/N_2 = 8.9820$. These values fix the lower limit of the integral which may be written

$$\log \gamma_2 = -\int_{8.9820}^{N_1/N_2} \frac{N_1}{N_2}\, d\log \gamma_1 + \log 9.9820$$

The incremental evaluation of the integral is indicated in Fig. 236, and the resultant values of γ_2 are given in Table B. Corresponding values of a_2 are obtained by multiplying N_2 by γ_2.

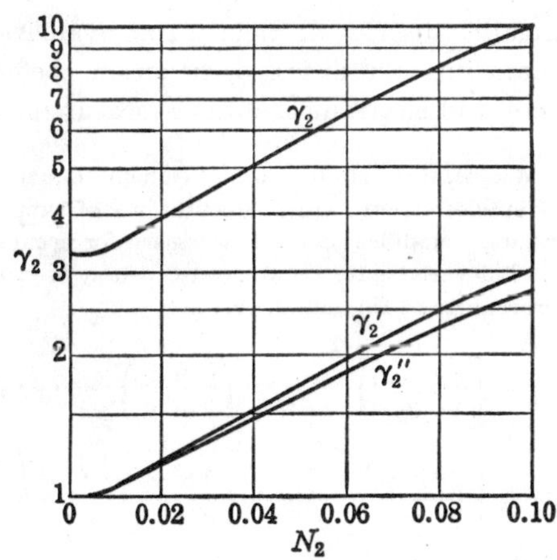

FIG. 237. Activity coefficients of sucrose in aqueous solutions at 25° C

(*d*) In Fig. 237, values of γ_2 are plotted as ordinates on a logarithmic scale against N_2, and the resulting curve is extrapolated to $N_2 = 0$. Values of a'_2 and a''_2 are then calculated from equation 95. The corresponding activity coefficients are by definition $\gamma'_2 = a'_2/N_2$ and $\gamma''_2 = a''_2/m$. These results are summarized in Table B and Fig. 237.

TABLE B. ACTIVITIES OF SUGAR IN AQUEOUS SOLUTIONS AT 25°C

m	N_2	γ_2	a_2	a'_2	γ'_2	a''_2	γ''_2
0	0.	3.2800	0	0	1.000	0	1.000
0.1	0.00180	3.2800	0.00502	0.00180	1.000	0.1000	1.000
0.4	0.00715	3.3696	0.02409	0.00734	1.027	0.4074	1.019
1.0	0.01770	3.8539	0.06821	0.02080	1.175	1.155	1.155
2.0	0.03478	4.7222	0.16424	0.0501	1.400	2.779	1.390
3.0	0.05127	5.7860	0.29665	0.0904	1.764	5.020	1.673
4.0	0.06722	7.0175	0.47172	0.1438	2.139	7.983	1.996
5.0	0.08263	8.3880	0.60310	0.2113	2.557	11.730	2.346
6.0	0.09755	9.7225	0.94843	0.2899	2.964	16.05	2.675
6.18	0.10018	9.9820	1.00000	0.3049	3.043	16.92	2.738

From inspection of Fig. 237, it is evident that the activity coefficients γ_2 and γ'_2 are widely different in numerical values but are related by a constant factor. The coefficient γ''_2 approaches γ'_2 at low concentrations but is not related to the other two coefficients by a constant factor.

It may be noted that the graphical integration shown in Fig. 236 becomes uncertain at low concentrations where N_1/N_2 approaches infinity. Special graphical methods have been developed by Lewis and Randall[10] to circumvent this difficulty and permit accurate integrations with zero concentration as one limit.

Although determination of activities and activity coefficients from vapor-pressure data is most direct, in principle the results are frequently less accurate than those obtained by other methods involving different types of equilibrium.

Lewis and Randall[10] discuss in detail the evaluation of activity coefficients from data on solubilities, distribution coefficients, freezing points, boiling points, and electromotive-force measurements.

Variable-Pressure Modification of the Gibbs–Duhem Equation. The Gibbs–Duhem equation 119 applies strictly only to conditions of simultaneous constant pressure and temperature. Modifications are necessary for accuracy at conditions of variable pressure and of variable temperature at constant pressure. These modifications follow the procedures of Ibl and Dodge.[11]

$$(d \ln \gamma_1)_T = \left[\left(\frac{\partial \ln \gamma_1}{\partial p}\right)_{TN_1} dp + \left(\frac{\partial \ln \gamma_1}{\partial \ln N_1}\right)_{pT} d \ln N_1\right]_T \tag{120}$$

$$(d \ln \gamma_2)_T = \left[\left(\frac{\partial \ln \gamma_2}{\partial p}\right)_{TN_2} dp + \left(\frac{\partial \ln \gamma_2}{\partial \ln N_2}\right)_{pT} d \ln N_2\right]_T \tag{121}$$

Combining equations 119, 103, and 120 gives

$$\left(\frac{d \ln \gamma_1}{d \ln N_1}\right)_T = \left(\frac{d \ln \gamma_2}{d \ln N_2} + X\right)_T \tag{122}$$

where

$$X = \left(\frac{\Delta v}{RT}\frac{dp}{dN_1}\right)_T \tag{123}$$

$$\Delta v = v - N_1 v^\circ_1 - N_2 v^\circ_2 \tag{124}$$

At the temperature of the system and the pressure corresponding to the pure components and composition of the solution, v, v°_1, and v°_2 = molal volumes of the solution and pure components, respectively, of the particular phase under consideration.

The term Δv is the volume change in mixing the separate pure components, all at the same temperature and pressure corresponding to the given composition of the mixture, to form 1 mole of solution.

Variable-Temperature Modification of the Gibbs–Duhem Equation. The modification of the Gibbs–Duhem equation may become significant for conditions of

[10] G. N. Lewis and M. Randall, *Thermodynamics*, McGraw-Hill Book Co. (1923).

[11] N. V. Ibl and B. F. Dodge, *Chem. Eng. Sci.*, **2**, 120 (1953).

constant pressure because of the sensitivity of activity coefficients to variations in temperature as composition changes. For a binary solution at constant pressure,

$$(d \ln \gamma_1)_p = \left[\left(\frac{\partial \ln \gamma_1}{\partial T}\right)_{pN_1} dT + \left(\frac{\partial \ln \gamma_1}{\partial \ln N_1}\right)_{pT} d \ln N_1\right]_p \tag{125}$$

$$(d \ln \gamma_2)_p = \left[\left(\frac{\partial \ln \gamma_2}{\partial T}\right)_{pN_2} dT + \left(\frac{\partial \ln \gamma_2}{\partial \ln N_2}\right)_{pT} d \ln N_2\right]_p \tag{126}$$

Combining equations 104, 119, and 125 gives

$$\left(\frac{d \ln \gamma_1}{d \ln N_1}\right)_p = \left(\frac{d \ln \gamma_2}{d \ln N_2} + Z\right)_p \tag{127}$$

where

$$Z = -\left(\frac{\Delta\text{H}}{RT^2}\frac{dT}{dN_1}\right)_p \tag{128}$$

$$\Delta\text{H} = \text{H} - N_1\text{H}^\circ_1 - N_2\text{H}^\circ_2 \tag{129}$$

At constant pressure and the temperature corresponding to the composition of the pure components and the solution, H, H°_1, and H°_2 = molal enthalpies of mixture and components, respectively, of the particular phase under consideration.

The term ΔH is the integral heat of solution per mole in mixing the pure components in their standard states to form a solution of a given composition at constant pressure and corresponding temperature. Equations 122 and 127 apply to any phase. The volumes and enthalpies refer to the particular phase under consideration. For the vapor phase at low pressures, the activity coefficients are nearly unity, and the Gibbs–Duhem equation is not required. In the liquid phase for pressures below 25 atm (reduced pressures below 0.8), the X term is negligible, and the unmodified Gibbs–Duhem equation is applicable, despite the variation of pressure with composition at constant temperature. The Z term may not be negligible where the variation of the bubble point with composition at constant pressure $(dT/dN_1)_p$ is considerable.

Enthalpy of Solutions

The enthalpy of a solution at a given temperature, pressure, and composition is usually given relative to the pure components at a reference temperature T_o and at reference pressures of p_{o1}, p_{o2} for saturated liquid components at temperature T_o. Where the normal state of aggregation of any component at p_o and T_o is a liquid, the reference pressure p_o is the saturation pressure. Where the normal state of aggregation of any component at T_o and atmospheric pressure is a gas the pressure p_o is taken as 1 atm.

Relative to these reference conditions, the molal enthalpy H of a binary liquid solution at temperature T and pressure p is given as follows where p is above the bubble-point pressure:

$$\text{H} = N_1\text{H}_1 + N_2\text{H}_2 + \Delta\text{H}^M + \int_{T_M}^{T} c_{pM}\, dT + \int_{p_M}^{p}\left(\frac{\partial \text{H}}{\partial p}\right)_T dp \tag{130}$$

where $p \gtreqless p_b$. The bubble-point pressure is the pressure at which the first bubble of vapor arises from a solution at a given equilibrium temperature.

H_1 = molal enthalpy of liquid component 1 at T_M and p_M relative to pure liquid at T_o and p_{o1} where p_{o1} is the saturation pressure corresponding to T_o

H_2 = is similarly defined where p_{o2} is the saturation pressure of component 2 at temperature T_o

ΔH^M = heat of mixing per mole of solution of given composition

p_M and T_M = pressure and temperature at which heat of solution ΔH^M is measured

c_{pM} = molal heat capacity of liquid solution at pressure p_M

$$\text{H}_1 = \int_{p_{o1}}^{p_M} \left(\frac{\partial \text{H}_1}{\partial p}\right)_{T_o} dp + \int_{T_o}^{T_M} c_{p1}\, dT \tag{131}$$

$$\text{H}_2 = \int_{p_{o2}}^{p_M} \left(\frac{\partial \text{H}_2}{\partial p}\right)_{T_o} dp + \int_{T_o}^{T_M} c_{p2}\, dT \tag{132}$$

where c_{p1}, c_{p2} = molal heat capacities of liquid components 1 and 2, respectively, at pressure p_M. Usually T_M is taken as = 298° K and p_M as 1 atm.

Where component 1 is a gas at T_o and atmospheric pressure, equation 131 is replaced by

$$\text{H}_1 = \int_{T_o}^{T_M} c_{p1}\, dT + \int_{1}^{p_M} \left(\frac{\partial \text{H}_1}{\partial p}\right)_{T_M} dp \tag{133}$$

where c_{p1} = molal heat capacity of gas at 1 atm.

The pressure term integrals in equations 130 through 133 for liquids are usually negligible.

For gaseous mixtures relative to the pure liquid components at temperature T_o and saturation pressures p_{o1} and p_{o2}, the enthalpy per mole of binary mixture is as follows:

$$\text{H} = N_1\lambda_1 + N_2\lambda_2 - N_1\int_{o}^{p_{o1}} \left(\frac{\partial \text{H}_1}{\partial p}\right)_{T_o} dp - N_2\int_{o}^{p_{o2}} \left(\frac{\partial \text{H}_2}{\partial p}\right)_{T_o} dp$$

$$+ N_1\int_{T_o}^{T} c^*_{p1}\, dT + N_2\int_{T_o}^{T} c^*_{p_2}\, dT + \Delta\text{H}^M + \int_{o}^{p} \left(\frac{\partial \text{H}}{\partial p}\right)_{T} dp \tag{134}$$

where ΔH^M = heat of mixing gases at zero pressure per mole of solution and at temperature T. (This term is usually negligible).

λ_1, λ_2 = molal heats of vaporization at pressures p_{o1} and p_{o2}, respectively

Where component 1 is a gas at T_o and atmospheric pressure, the term

$$\left[N_1\lambda_1 - N_1\int_o^{p_{o1}}\left(\frac{\partial \text{H}_1}{\partial p}\right)_{T_o} dp\right] \text{ is replaced by } -N_1\int_o^1\left(\frac{\partial \text{H}_1}{\partial p}\right)_{T_o} dp$$

Where generalized tables are used, the pressure integrals may be replaced by corresponding $(\text{H}^*-\text{H})/T_c$ and $(\text{H}^*-\text{H})T'_c$ terms, using pseudoreduced properties where mixtures are involved.

Illustration 4. Calculate the enthalpy of 1 lb of a 60 mole per cent solution of ethyl alcohol in water relative to the separate liquid components each saturated at 32° F:

(a) Of the liquid solution at atmospheric pressure and its bubble point.
(b) Of the gas mixture at atmospheric pressure and its dew point.
(c) Of the gas mixture at 500° F and 10 atm.

Component 1 = water, Component 2 = alcohol

At 32° F, $\Delta\text{H}^M = -212.4$ Btu per lb-mole solution where $N_2 = 0.60$.

Bubble point = 175° F, Dew point = 178° F

TABLE A

	Water	Alcohol
Vapor Pressure at 32° F	$p_{o1} = 0.0060$ atm	$p_{o2} = 0.0161$ atm
Heat of Vaporization, Btu/lb-mole at 32° F	$\lambda_1 = 19{,}400$	$\lambda_2 = 18{,}200$
z_c	0.23	0.248
T_c, °K	647.4	516.3
p_c, atm	218.3	63
M	18	46

Mean molal heat capacity of liquid, $N_2 = 0.60$ (32 to 175° F) = 23.9 Btu/(lb-mole)(R°).
Mean molecular weight = 34.8.

For $N_2 = 0.60$: $z'_c = 0.4(0.23) + 0.6(0.248) = 0.241$

$T'_c = 0.4(647.4) + 0.6(516.3) = 569°$ K (1024° R)

$p'_c = 0.4(218.3) + 0.6(63.0) = 125$ atm

For the gas mixture at $T = 500°$ F, $p = 10$ atm, $T_r = 0.94$, $p_r = 0.08$:

$(\text{H}^* - \text{H})/T'_c = 0.252$

$\text{H}^* - \text{H} = (0.252)1024 = 258$ Btu per lb-mole

For the gas mixture at 178° F, 1 atm, $T_r = 0.624$, $p_r = 0.008$.

$(\text{H}^* - \text{H})/T'_c = 0.066$; $\text{H}^* - \text{H} = (0.066)1024 = 68$ Btu per lb-mole

(a) Enthalpy of the liquid solution at 1.0 atm and 175° F, from equation 130:
Since ΔH^M is given at 32° F, $\text{H}_1 = 0$, $\text{H}_2 = 0$. The pressure integral terms are negligible.

$$\text{Hence } H = \frac{-212.4 + 23.9(175 - 32)}{34.8} = 92 \text{ Btu per lb}$$

(*b*) Enthalpy of the gas mixture at 1.0 atm and 178° F (354° K) from equation 134: The pressure integral terms and ΔH^M are negligible. Molal heat capacity of gases (T = °K) cal per g-mole:

Water: $c^*_p = 7.700 + (4.594)10^{-4}T + (25.21)10^{-7}T^2 - (8.587)10^{-10}T^3$

Ethyl alcohol: $c^*_p = 4.75 + (5.006)10^{-2}T - (2.479)10^{-5}T^2 + (4.790)10^{-9}T^3$

For Water:

$$1.8\int_{273}^{354} c^*_p\, dT = 1179 \text{ Btu per lb-mole}$$

For Alcohol:

$$1.8\int_{273}^{354} c^*_p\, dT = 2655 \text{ Btu per lb-mole}$$

From equation 134:

$$H = \frac{0.4(19{,}400) + 0.6(18{,}200) + 0.4(1179) + 0.6(2655) - 68}{34.8}$$

$$= 594 \text{ Btu per lb}$$

(*c*) Enthalpy of the gas mixture at 500° F (533° K) and 10 atm:

$$1.8\int_{273}^{533} c^*_{p1}\, dT = 3861, \qquad 1.8\int_{273}^{533} c^*_{p2}\, dT = 9886 \text{ Btu per lb-mole}$$

$$H = \frac{0.4(19{,}400) + 0.6(18{,}200) + 0.4(3861) + 0.6(9886) - 258}{34.8}$$

$$= 744 \text{ Btu per lb}$$

An enthalpy–composition diagram for the ethyl alcohol–water system is shown in Fig. 83, page 327.

Entropy of Solutions

The same reference states for the components of a mixture may be used for entropies as for enthalpies, as stated on page 879. The molal entropy of a binary liquid solution at temperature T and pressure p where p is above the bubble-point pressure is given thus:

$$s = N_1 s_1 + N_2 s_2 + \Delta s^M + \int_{T_M}^{T} \frac{c_{pM}}{T}\, dT + \int_{p_M}^{p} \left(\frac{\partial s}{\partial p}\right)_T dp \qquad (135)$$

where $p \geqq p_b$

s_1 = molal entropy of liquid component 1 at T_M and p_M relative to liquid T_o and p_{o1}

s_2 is defined similarly to s_1 except that T_{o2} and p_{o2} are used

Δs^M = entropy of mixing per mole of solution at temperature T_M and pressure p_M

$$s_1 = \int_{p_{o1}}^{p_M} \left(\frac{\partial s_1}{\partial p}\right)_{T_o} dp + \int_{T_o}^{T_M} \left(\frac{c_{p1}}{T}\right) dT \qquad (136)$$

$$s_2 = \int_{p_{o2}}^{p_M} \left(\frac{\partial s_2}{\partial p}\right)_{T_o} dp + \int_{T_o}^{T_M} \left(\frac{c_{p2}}{T}\right) dT \qquad (137)$$

At low pressures the three pressure integrals are negligible for liquids.

Where component 1 is a gas at T_o and 1 atm, equation 136 is replaced by

$$\mathrm{s}_1 = \int_{T_o}^{T_M} \left(\frac{c_{p1}}{T}\right)_{p=1} dT + \int_1^{p_M} \left(\frac{\partial \mathrm{s}_1}{\partial p}\right)_{T_M} dp \tag{138}$$

where the pressure integral is equal to $-R \ln p$. Where activity coefficients are known, $\Delta\mathrm{s}^M$ may be obtained from the values of the free energy of mixing $\Delta\mathrm{G}^M$; thus,

$$\Delta\mathrm{G}^M = +RT(N_1 \ln \gamma_1 + N_2 \ln \gamma_2) + RT(N_1 \ln N_1 + N_2 \ln N_2) \tag{139}$$

$$\Delta\mathrm{s}^M = \frac{-\Delta\mathrm{G}^M + \Delta\mathrm{H}^M}{T} = -R(N_1 \ln \gamma_1 + N_2 \ln \gamma_2 + N_1 \ln N_1 + N_2 \ln N_2) + \frac{\Delta\mathrm{H}^M}{T} \tag{140}$$

For *gaseous mixtures* relative to pure liquid components at temperature T_o and saturation pressures p_{o1} and p_{o2}, the entropy per mole of a binary mixture is as follows:

$$\mathrm{s} = \frac{N_1\lambda_1}{T_o} + \frac{N_2\lambda_2}{T_o} + N_1\int_{T_o}^{T}\left(\frac{c_{p1}}{T}\right) dT + N_2\int_{T_o}^{T}\left(\frac{c_{p2}}{T}\right) dT + N_1\int_{p_{o1}}^{1}\left(\frac{\partial \mathrm{s}_1}{\partial p}\right)_T dp + N_2\int_{p_{o2}}^{1}\left(\frac{\partial \mathrm{s}_2}{\partial p}\right)_T dp + \Delta\mathrm{s}^M + \int_1^p \left(\frac{\partial \mathrm{s}}{\partial p}\right)_T dp \tag{141}$$

where c_{p1} and c_{p2} = molal heat capacities of gases at p_{o1} and p_{o2}, respectively. $\Delta\mathrm{s}^M$ is the heat of mixing at T and 1 atm.

$$\int_{p_{o1}}^{1}\left(\frac{\partial \mathrm{s}_1}{\partial p}\right)_T dp = R \ln p_{o1} + (\mathrm{s}^* - \mathrm{s})_{p_{o1}T} - (\mathrm{s}^* - \mathrm{s})_{1T} \tag{142}$$

where the $(\mathrm{s}^* - \mathrm{s})_{pT}$ terms may be obtained from generalized tables;

$$\int_1^p \left(\frac{\partial \mathrm{s}}{\partial p}\right) dp = -R \ln p + (\mathrm{s}^* - \mathrm{s})_{1T} - (\mathrm{s}^* - \mathrm{s})_{pT} \tag{143}$$

where the $(\mathrm{s}^*-\mathrm{s})_{pT}$ terms may be obtained from generalized tables at pseudoreduced properties of the gas mixture.

$\Delta\mathrm{s}^M$ is the entropy of mixing gases at 1 atm and $T°$ K per mole of gaseous mixture.

For regular solutions $\Delta\mathrm{s}^M$ is equal to $-R(N_1 \ln N_1 + N_2 \ln N_2)$.

Where heats of mixing gases are not negligible, and activity coefficients are not unity, the value of $\Delta\mathrm{s}^M$ may be computed from equation 140, where the activity coefficients are for the components in the gas mixture at 1 atm and $T°$ K.

Systematic procedures for calculating the molal entropy of a liquid solution are summarized in Tables 59 and 60.

TABLE 59. STEPS INVOLVED IN CALCULATING THE MOLAL ENTROPY OF A LIQUID SOLUTION

Reference State: Pure Saturated Liquid Components at Temperature T_o

T_M = temperature of mixing

p_M = pressure of mixing

c_{pM} = molal heat capacity of solution

ΔH^M = heat of mixing at T_M and p_M ($\Delta H^M = 0$ for ideal solution behavior)

Step	Temperature at End of Step	Pressure at End of Step	Entropy Contributions: Rigorous Procedure	Entropy Contributions: Generalized Procedure (Ideal Solution Behavior Only)
Reference: separate liquids	T_o	p_{oi}	0	0
(1) Compression of separate liquids	T_o	p_M	$-\sum N_i \int_{p_{oi}}^{p_M} \left(\frac{\partial v_i}{\partial T}\right)_p dp$ usually negligible	$+\sum N_i \left[-R \ln \frac{p_M}{p_{oi}} - (s^*_2 - s_2)_{iL} + (s^*_1 - s_1)_{iL} \right]$ usually negligible
(2) Heating of separate liquids	T_M	p_M	$+\sum N_i \int_{T_o}^{T_M} \frac{c_{pi}\, dT}{T}$	$+\sum N_i \int_{T_o}^{T_M} \frac{c_{pi}}{T} dT$
(3) Mixing liquids	T_M	p_M	$-R \sum N_i \ln \gamma_i - R \sum N_i \ln N_i + \frac{\Delta H^M}{T}$	$-R \sum N_i \ln N_i$
(4) Heating of liquid solution	T	p_M	$+\int_{T_M}^{T} \frac{c_{pM}}{T} dT$	$+\int_{T_M}^{T} \frac{c_{pM}}{T} dT$
(5) Compression of liquid solution	T	p	$-\int_{p_M}^{p} \left(\frac{\partial v}{\partial T}\right)_p dp$	$-R \ln \frac{p}{p_M} - (s^*_2 - s_2)_L + (s^*_1 - s_1)_L$

TABLE 60. STEPS INVOLVED IN CALCULATING THE MOLAL ENTROPY OF A GASEOUS MIXTURE
Reference State: Pure Saturated Liquid Components at Temperature T_o

Step	Temperature at End of Step	Pressure at End of Step	Entropy Contributions	
			Rigorous Procedure	Generalized Procedure (Ideal Solution Behavior Only)
Reference: separate liquids	T_o	p_{oi}	0	0
(1) Vaporization of separate liquids	T_o	p_{oi}	$+\sum N_i \frac{\lambda_i}{T_o}$	$+\sum N_i \frac{\lambda_i}{T_o}$
(2) Heating of separate vapors	T	p_{oi}	$+\sum N_i \int_{T_o}^{T} \frac{c_{pi}}{T}\, dT$	$+\sum N_i \int_{T_o}^{T} \frac{c_{pi}}{T}\, dT$
(3) Compression of separate vapors	T	p	$-\sum N_i \int_{p_{oi}}^{p} \left(\frac{\partial v_i}{\partial T}\right)_p dp$	$-\sum N_i R \ln \frac{p}{p_{oi}} - \sum N_i[(s^*_2 - s_2)_{iG} - (s^*_1 - s_1)_{iG}]$
(4) Mixing vapors	T	p	$-R[\sum N_i \ln \gamma_i + \sum N_i \ln N_i] + \frac{\Delta H^M}{T}$ In the gas phase usually $\gamma_i = 1$ and $\Delta H^M = 0$	$-R \sum N_i \ln N_i$
(5) Compression of gaseous mixture	T	p_2	$-\int_{p}^{p_2} \left(\frac{\partial v}{\partial T}\right)_p dp$	$-R \ln \frac{p_2}{p} - (s^*_2 - s_2)'_G + (s^*_1 - s_1)'_G$

Illustration 5. Calculate the entropy of 1 lb of a 60 mole per cent solution of ethyl alcohol in water for the same conditions as given in Illustration 4.

At 32° F for a 60 mole per cent solution:

$$\gamma_1 = 1.80, \qquad N_1 = 0.4$$
$$\gamma_2 = 1.45, \qquad N_2 = 0.6$$

(*a*) Entropies of liquid solution at 1 atm and 175° F: At $\text{T}_M = 32°$ F,

$$\text{s}_1 = 0 \qquad \text{s}_2 = 0$$

From equation 140,

$$\Delta \text{s}^M = -1.987(0.4 \ln 1.80 + 0.6 \ln 1.45 + 0.4 \ln 0.4 + 0.6 \ln 0.6)$$
$$+ \frac{(-212.4)}{492} = -\ 0.0045 \text{ Btu/(lb-mole)(R°)}$$

From equation 135,

$$\text{s} = \frac{-\ 0.0045 + 23.9 \ln (635/492)}{34.8} = 0.175 \text{ Btu/(lb-mole)(R°)}$$

(*b*) Entropy of gas mixture at 1.0 atm and 178° F: For the gaseous mixture,

$$\Delta \text{s}^M = -1.987(0.4 \ln 0.4 + 0.6 \ln 0.6) = 1.33 \text{ Btu/(lb-mole)(R°)}$$

This value for Δs^M is uncorrected for activity coefficients and for heat of mixing in the gas phase. These values are unknown but negligible.

$$\frac{\lambda_1}{T_o} = \frac{19{,}400}{492} = 39.43 \text{ Btu/(lb-mole)(R°)}$$

$$\frac{\lambda_2}{T_o} = \frac{18{,}200}{492} = 37.0 \text{ Btu/(lb-mole)(R°)}$$

$$\int_{p_{o1}}^{1} \left(\frac{\partial \text{s}}{\partial p}\right)_T dp = R \ln p_{o1} = 4.576(\log 0.0060) = -10.17 \text{ Btu/(lb-mole)(R°)}$$

$$\int_{p_{o2}}^{1} \left(\frac{\partial \text{s}}{\partial p}\right)_T dp = R \ln p_{o2} = 4.576(\log 0.0161) = -8.20 \text{ Btu/(lb-mole)(R°)}$$

$$\int_{492}^{638} \left(\frac{c^*_{p1}}{T}\right) dT = 2.098 \text{ Btu/(lb-mole)(R°)}$$

$$\int_{492}^{638} \left(\frac{c^*_{p2}}{T}\right) dT = 4.708 \text{ Btu/(lb-mole)(R°)}$$

$$\text{s} = 0.4(39.43) + 0.6(37.0) + 0.4(2.098) + 0.6(4.708) + 0.4(-10.17)$$
$$+ 0.6(-8.20) + 1.338 = 33.984$$

or

$$\frac{33.984}{34.8} = 0.977 \text{ Btu/(lb)(R°)}$$

(c) Entropy of gas mixture at 500° F and 10 atm:

$$\int_{492}^{960} \left(\frac{c^*_{p1}}{T}\right) dT = 5.496 \text{ Btu/(lb-mole)(R°)}$$

$$\int_{492}^{960} \frac{c^*_{p2}}{T} dT = 13.807 \text{ Btu/(lb-mole)(R°)}$$

At 10 atm, 500° F:

$z'_c = 0.241,\qquad p_r = 0.08$

$p'_c = 125,\qquad T_r = 0.94$

$T'_c = 569,\qquad s^* - s = 0.21$

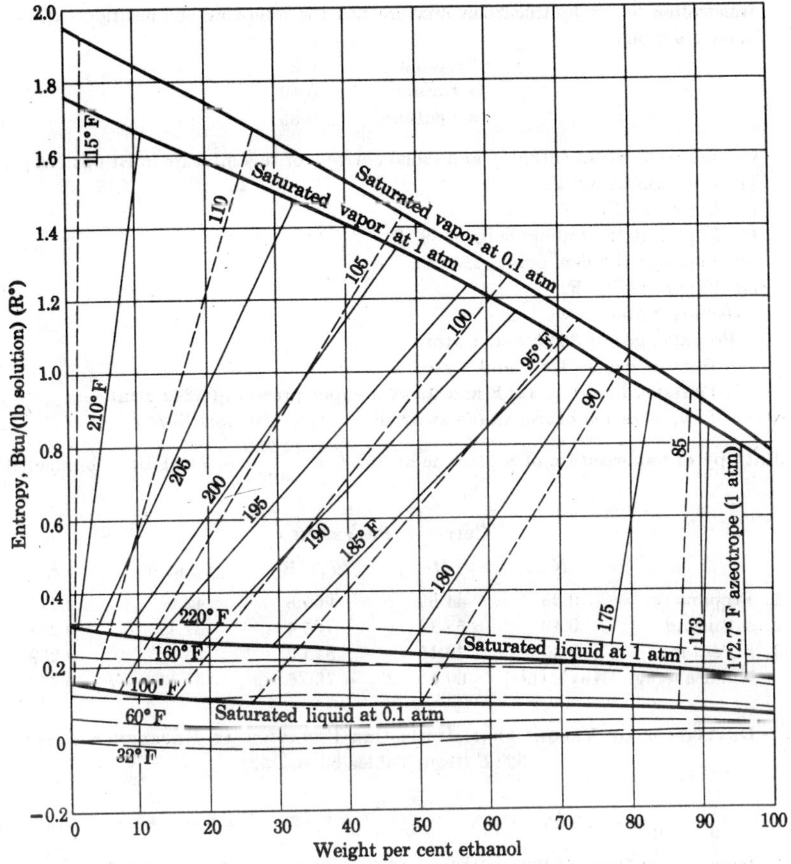

Fig. 238. Entropy–composition chart for the ethanol–water system Reference substances: pure saturated liquid C_2H_5OH and H_2O at 1 atm and 32° F

At 1 atm, 500° F, $(s^* - s) = 0.02$,

$$\int_1^p \left(\frac{\partial s_M}{\partial p}\right)_T dp = -R \ln p - (s^* - s)_{pT} + (s^* - s)_{1T}$$

$$= -R \ln 10 - 0.21 + 0.02 = -4.77 \text{ Btu/(lb-mole)(R°)}$$

$$\Delta s^M = 1.338$$

$$s = 0.4(39.43) + 0.6(37.0) + 0.4(5.496) + 0.6(13.807) + 0.4(-10.17)$$

$$+ 0.6(-8.20) + 1.338 - 4.77 = 36.034 \quad \text{or} \quad \frac{36.034}{34.8} = 1.035 \text{ Btu/(lb)(R°)}$$

An entropy–composition diagram for the ethyl alcohol–water system is shown in Fig. 238.

Illustration 6. A hydrocarbon mixture has the following composition expressed in mole fractions:

Propane	0.25
n-Butane	0.40
n-Pentane	0.35

Calculate the molal enthalpy and molal entropy of this mixture in Btu at 200 psia:
(a) Ideal gas at 32° F.
(b) Liquid at 100° F.
(c) Liquid at the bubble point (189° F).
(d) Vapor at the dew point (223° F).
(e) Vapor at 300° F.

Reference States:

Propane, gas at 32° F and 1 atm.
n-Butane, gas at 32° F and 1 atm.
n-Pentane, liquid at 32° F and its own vapor pressure (0.242 atm).

Heat of vaporization of n-pentane at 32° F = 11,960 Btu per lb-mole

Entropy of vaporization of n-pentane at 32° F $= \dfrac{11,960}{492} = 24.31$ Btu/(lb-mole)(R°)

CRITICAL PROPERTIES

	N	M	T_c, °R	p_c, atm	z_c
1. Propane	0.25	44.09	665.8	42.0	0.277
2. n-Butane	0.40	58.12	765.4	37.47	0.274
3. n-Pentane	0.35	72.15	845.6	33.3	0.269
Molal average (Kay's rule)		60.5	$T'_c = 768.5$	$p'_c = 37.14$	$z'_c = 0.273$

DEVIATIONS OF VAPORS FROM IDEAL-GAS BEHAVIOR AT REFERENCE STATE, 32° F (from Tables 50 and 52)

	T_r	p_r	z_c	$\frac{H^* - H}{T_c}$	$H^* - H$	$N(H^* - H)$	$s^* - s$	$N(s^* - s)$
1. Propane	0.739	0.0238	0.277	0.098	65.5	16.4	0.123	0.031
2. n-Butane	0.643	0.0267	0.274	0.077	59.0	23.6	0.119	0.048
3. n-Pentane	0.582	sat	0.269	0.035	29.6	10.4	0.063	0.022

$$\sum N\,(H^* - H) = 50.4; \quad \sum N_i(s^* - s) = 0.101$$

DEVIATIONS OF MIXTURE FROM IDEAL-GAS BEHAVIOR AT 200 PSIA, $z'_c = 0.273$, $p_r = 0.366$

	T_r	$\dfrac{H^* - H}{T'_c}$	$H^* - H$	$S^* - S$
Liquid at 100° F	0.728	11.88	9130	13.67
Liquid at 189° F (bubble point)	0.845	10.47	8046	12.05
Vapor at 223° F (dew point)	0.889	1.71	1310	1.41
Vapor at 300° F	0.988	0.93	715	0.70

Ideal molal heat capacities of gases in Btu per pound-mole are given by Kobe[12] where T is in degrees Rankine:

Propane: $c^*_p = -0.966 + (4.044)10^{-2}T - (1.159)10^{-5}T^2 + (1.300)10^{-9}T^3$

n-Butane: $c^*_p = 0.945 + (4.929)10^{-2}T - (1.352)10^{-5}T^2 + (1.433)10^{-9}T^3$

n-Pentane: $c^*_p = 1.618 + (6.029)10^{-2}T - (1.656)10^{-5}T^2 + (1.732)10^{-9}T^3$

mixture $c^*_{pM} = 0.703 + (5.093)10^{-2}T - (1.410)10^{-5}T^2 + (1.504)10^{-9}T^3$

t, °F	$\int_{492}^{T} c^*_{pM}\,dT$	$\int_{492}^{T} \frac{c^*_{pM}}{T}\,dT$
100	1614	3.08
189	3950	7.08
223	4900	8.50
300	7284	11.75

Molal Enthalpies at 200 psia. (*a*) **Ideal gas at 32° F.**

$$H^* = \sum N_i(H^* - H) + N_3\lambda_3$$

$$H^* = 50.4 + 0.35(11{,}960) = 4236 \text{ Btu per lb-mole}$$

(*b*) **Liquid at 100° F.**

Ideal gas: $H^* = 4236 + 1614 = 5850$ Btu per lb-mole

Liquid: $H = 5850 - 9130 = -3280$ Btu per lb-mole

(*c*) **Liquid at bubble point, 189° F.**

Ideal gas: $H^* = 4236 + 3950 = 8186$ Btu per lb-mole

Liquid (bubble point): $H = 8186 - 8046 = 140$ Btu per lb-mole

(*d*) **Vapor at dew point, 223° F.**

Ideal gas: $H^* = 4236 + 4900 = 9136$ Btu per lb-mole

Actual gas (dew point): $H = 9136 - 1310 = 7826$ Btu per lb-mole

(*e*) **Vapor at 300° F.**

Ideal gas: $H^* = 4236 + 7284 = 11{,}520$ Btu per lb-mole

Actual gas: $H = 11{,}520 - 715 = 10{,}805$ Btu per lb-mole

Molal Entropies at 200 psia. Entropy increase upon compression (ideal-gas behavior):

$$\Delta S = -R \sum N_i \ln (p_2/p_1)$$

$$= -1.987\left(0.25 \ln \frac{200}{14.7} + 0.40 \ln \frac{200}{14.7} + 0.35 \ln \frac{200}{3.55}\right)$$

$$= -6.18 \text{ Btu/(lb-mole)(R°)}$$

[12] K. A. Kobe, "Thermochemistry for the Petrochemical Industry", *Petroleum Refiner* (1949–1954).

Entropy of mixing (ideal-gas behavior):

$$\Delta s^M = -R \sum N_i \ln N_i$$
$$= -1.987(0.25 \ln 0.25 + 0.40 \ln 0.40 + 0.35 \ln 0.35)$$
$$= +2.147 \text{ Btu/(lb-mole)(R°)}$$

(*a*) **Ideal Gas at 32° F.**

$$s^* = 0.35 \frac{\lambda}{492} + \sum N_i(s^*_i - s_i) + \Delta s + \Delta s^M$$

$$s^* = 0.35(24.31) + 0.125 - 6.18 + 2.15 = 4.60 \text{ Btu/(lb-mole)(R°)}$$

(*b*) **Liquid at 100° F.**

Ideal gas: $s^* = 4.60 + 3.08 = 7.68$ Btu/(lb-mole)(R°)

Liquid: $s = 7.68 - 13.67 = -5.99$ Btu/(lb-mole)(R°)

(*c*) **Liquid at bubble point, 192° F.**

Ideal gas: $s^* = 4.60 + 7.08 = 11.68$ Btu/(lb-mole)(R°)

Liquid (bubble point): $s = 11.68 - 12.05 = -0.37$ Btu/(lb-mole)(R°)

(*d*) **Vapor at dew point, 231° F.**

Ideal gas: $s^* = 4.60 + 8.50 = 13.10$ Btu/(lb-mole)(R°)

Actual gas (dew point): $s = 13.10 - 1.41 = 11.69$ Btu/(lb-mole)(R°)

(*e*) **Vapor at 300° F.**

Ideal gas: $s^* = 4.60 + 11.75 = 16.35$ Btu/(lb-mole)(R°)

Actual gas: $s = 16.35 - 0.70 = 15.65$ Btu/(lb-mole)(R°)

Problems

1. A mixture of hydrocarbon gases has the following composition in mole per cent:

CH_4	38.38
C_2H_6	7.56
C_3H_8	7.05
n-C_4H_{10}	11.29
n-C_5H_{12}	35.72
	100.00

Calculate the density of this mixture in pounds per cubic foot in the gaseous state at a temperature of 200° F and an absolute pressure of 400 psia.

2. Calculate the density in grams per cubic centimeter of the mixture of problem 1 in the liquid state at a temperature of 100° F and an absolute pressure of 1300 psia.

3. Calculate the enthalpy in Btu per pound of the mixture of problem 1 in the gaseous state at the specified conditions. As the reference state of zero enthalpy, use the ideal gaseous state at 60° F for the propane and lighter constituents. For butane and the heavier constituents use the saturated liquid state at 60° F as the reference. It may be assumed that the enthalpies of the components are additive under these reference conditions. Doss[13] gives the following values of heats of vaporization in calories per gram at the normal boiling point: n-butane 92.0; n-pentane 85.5.

[13] M. P. Doss, *Physical Constants of the Principal Hydrocarbons*, 3d ed., Texas Co., Beacon, N.Y. (1942).

4. Calculate the enthalpy in Btu per pound of the mixture in the liquid state at the conditions of problem 2, using the reference states specified in problem 3.

5. Assuming that an ideal solution is formed, calculate the fugacities of the components of the mixture of problem 1 under the specified conditions.

6. The gas from the converter of a synthetic ammonia plant has the following composition in mole per cent:

$$N_2 = 20.2$$
$$H_2 = 60.8$$
$$NH_3 = 19.0$$

Calculate the fugacity of each component of the mixture at 800° F and an absolute pressure of 4500 psia, assuming that an ideal solution is formed.

7. The *International Critical Tables* give the following data for the lowering of the vapor pressure of water at 0° C by urea [$CO(NH_2)_2$]:

m	$100R$
1	1.52
2	1.49
4	1.46
6	1.45
10	1.43

$$R = \frac{(P_0 - p)}{mP_0}$$

where m = molality; P_0 = vapor pressure of the pure water; p = vapor pressure of H_2O above the solution. The solubility of urea at 0° C is 67.1 grams per 100 grams of H_2O.

Calculate and plot as functions of molality the activity of the water a_1 and of the urea a_2 referred to the solid at 0° C, and estimate the activities of the urea a'_2 and a''_2 referred to infinite dilution. Also calculate and plot the corresponding activity coefficients, γ_1, γ_2, γ'_2, and γ''_2.

8. The data in Table A from the *International Critical Tables* show the partial pressures in millimeters of Hg of toluene and acetic acid in solutions at a temperature of 69.94° C:

TABLE A

x_1, mole % Toluene	p_1, Toluene	p_2, Acetic Acid
0	0	136
12.50	54.8	120.5
23.10	84.8	110.8
31.21	101.9	103.0
40.19	117.8	95.7
48.60	130.7	88.2
53.49	137.6	83.7
59.12	145.2	78.2
66.20	155.7	69.3
75.97	107.3	57.8
82.80	170.2	46.5
90.58	186.1	30.5
95.65	193.5	17.2
100	202	0

Calculate the activity coefficients γ_1, γ_2 from these data. Plot the activity coefficients as ordinates on a logarithmic scale against the mole per cent toluene on a uniform scale, and check these curves for consistency with the Gibbs–Duhem equation at compositions of 10, 30, 50, 70, and 90% toluene.

21

Vapor–Liquid Equilibria at Low Pressures

The discussion on page 524 of equilibria in a closed system is restricted to the conditions attained by the system as a whole. When two phases are present they together constitute a closed system in which the phases reach equilibrium with each other. Each phase constitutes an open system, which can undergo changes in composition and mass. For complete equilibrium to exist, it is also necessary that equilibrium be maintained between all phases within the system. In this chapter the criteria of equilibria are extended to multiphase systems and the effects of composition thereon.

Equality of Chemical Potentials. In a closed system consisting of several phases, each of which is an open system, the properties as a whole must satisfy the equilibrium criteria established on page 524. If the conditions of restraint are constant temperature and pressure, then at equilibrium the free-energy change is zero if no means of performing work is present. Any change in the free energy of the entire system must equal the sum of the corresponding changes of free energy in all its parts. Representing the properties of the different individual phases of the system by primes and the different components of the various phases by subscripts, equation 20–14, page 851, becomes

$$\begin{aligned} dG = &(-S'\,dT + V'\,dp + \mu'_1\,dn'_1 + \mu'_2\,dn'_2 + \cdots) \\ &+ (-S''\,dT + V''\,dp + \mu''_1\,dn''_1 + \mu''_2\,dn''_2 + \cdots) \\ &+ (-S'''\,dT + V'''\,dp + \mu'''_1\,dn'''_1 + \mu'''_2\,dn'''_2 + \cdots) + \cdots \end{aligned} \tag{1}$$

For equilibrium at constant pressure and temperature among the different phases of a multiphase system, the free energy of the entire system is a minimum, and, for any change, the total free energy change is zero; hence,

$$dG = \begin{bmatrix} \mu'_1\,dn'_1 + \mu''_1\,dn''_1 + \mu'''_1\,dn'''_1 + \cdots \\ + \mu'_2\,dn'_2 + \mu''_2\,dn''_2 + \mu'''_2\,dn'''_2 + \cdots \\ + \mu'_3\,dn'_3 + \mu''_3\,dn''_3 + \mu'''_3\,dn'''_3 + \cdots \end{bmatrix}_{pT} = 0 \tag{2}$$

The total number of moles of each component in the closed system remains constant; hence

$$\begin{aligned} dn'_1 &= -dn''_1 - dn'''_1 \cdots \\ dn'_2 &= -dn''_2 - dn'''_2 \cdots \\ dn'_3 &= -dn''_3 - dn'''_3 \cdots \end{aligned} \tag{3}$$

Substitution of equation 3 into equation 2 gives

$$\begin{bmatrix} (\mu''_1 - \mu'_1)\, dn''_1 + (\mu'''_1 - \mu'_1)\, dn'''_1 + \cdots \\ + (\mu''_2 - \mu'_2)\, dn''_2 + (\mu'''_2 - \mu'_2)\, dn'''_2 + \cdots \\ + (\mu''_3 - \mu'_3)\, dn''_3 + (\mu'''_3 - \mu'_3)\, dn'''_3 + \cdots \end{bmatrix}_{pT} = 0 \tag{4}$$

All the dn terms in equation 4 are independent, since the dependent terms of one phase have been eliminated. Equation 4 thus holds for all and any values of n. Where all values of dn in equation 4 are zero except dn''_1,

$$(\mu''_1 - \mu'_1)\, dn''_1 = 0 \tag{5}$$

or

$$\mu''_1 = \mu'_1 \tag{6}$$

Similarly, all values of dn except dn'''_1 may be zero; hence,

$$(\mu'''_1 - \mu'_1)\, dn'''_1 = 0 \tag{7}$$

or

$$\mu'''_1 = \mu'_1 \tag{8}$$

Combining equations 6 and 8 gives

$$\mu'_1 = \mu''_1 = \mu'''_1 = \cdots \tag{9}$$

Similarly, it can be shown that

$$\begin{aligned} \mu'_2 &= \mu''_2 = \mu'''_2 = \cdots \\ \mu'_3 &= \mu''_3 = \mu'''_3 = \cdots \end{aligned} \tag{10}$$

In general, *the chemical potentials of any given component in all phases of a multicomponent system are identical at equilibrium.* Equation 10 is a necessary condition for equilibrium in a multiphase system.

Equation 10 can also be shown to be sufficient. For example, if in equation 4 it were assumed that $\mu''_1 - \mu'_1$ were not zero, then equation 4 would no longer be zero for all values of dn contrary to its statement. It is concluded that the equality of chemical potentials of a given component is a necessary condition for expressing the state of equilibrium in a multiphase system at constant pressure and temperature.

Gibbs Phase Rule. In any system comprising several phases existing at equilibrium, there are a limited number of intensive properties which can be varied freely without causing a change in the number of phases or the number of components present. For example, in a system comprising a pure liquid and its vapor at equilibrium, temperature can be varied over a wide range without changing the number of phases. However, it is impossible to vary both temperature and pressure freely without causing the disappearance of one phase or the other. The number of intensive properties that can be varied without changing the number of phases or the number of components in any phase is termed the *degrees of freedom* of the system.

Since, in addition to temperature and pressure, the chemical potential of each component is an intensive property, the total number of intensive properties subject to variation in a single phase containing C components is $C+2$. If this phase is a part of a system in equilibrium at constant temperature and pressure, a differential expression for the free-energy change may be written in the form of equation 1 for each phase. Thus

$$\begin{aligned} dG' &= f'(T, p, \mu_1, \mu_2, \cdots) = 0 \\ dG'' &= f''(T, p, \mu_1, \mu_2, \cdots) = 0 \\ dG''' &= f'''(T, p, \mu_1, \mu_2, \cdots) = 0 \end{aligned} \tag{11}$$

where the functions f', f'', f''' are characteristic of the respective phases. In this manner, if ϕ phases are present simultaneous equations may be written in terms of $C+2$ variables. Since for the numerical evaluation of the variables in simultaneous equations one equation is required for each variable, it follows that the number of variables not fixed by the equations is $C+2-\phi$ or

$$F = C + 2 - \phi \tag{12}$$

where F = degrees of freedom

C = number of components

ϕ = number of phases

Equation 12 is the famous phase rule of Gibbs, first developed in 1875. The foregoing derivation follows that presented by Keenan:[1]

As an example of the application of this rule, the equilibrium among the liquid, vapor, and solid states is considered. Thus, if only a pure liquid, for example water, is in equilibrium with its vapor, $C = 1$, $\phi = 2$, and $F = 1$; one intensive property, either temperature or pressure, but not both, may be varied freely. If, however, ice is also present in the system, $C = 1$, $\phi = 3$, and $F = 0$. Under such conditions, no intensive property can be varied, and the specification of three phases fixes both temperature and pressure. If, instead of pure water, a binary solution of

[1] J. H. Keenan, *Thermodynamics*, John Wiley & Sons (1941).

water and alcohol is in equilibrium with its vapor, $C = 2$, $\phi = 2$, and $F = 2$; hence, two intensive properties may be freely varied. For example, both temperature and pressure may be varied freely over restricted ranges with corresponding changes in composition of the phases, but all three variables cannot be independently varied. A specified composition and temperature fixes the pressure.

When used in connection with the phase rule, the number of components is the least number of independently variable chemical substances from which the system in all its variations can be produced. Elsewhere in this text the term component is not used in this restricted sense.

If some intensive property besides temperature, pressure, and chemical potential is significant, the constant 2 should be increased by one for each additional intensive property. For example, if surface tension has a significant effect on the pvT-composition relationships of a multiphase system, the phase rule becomes $F = C + 3 - \phi$.

Fugacity as a Criterion of Equilibrium. For equilibrium between phases of a multicomponent system, it was previously demonstrated (page 893) that the chemical potential of any component i must be the same in all phases. Since the fugacity of component i is directly related to its chemical potential by the relation $(d\mu_i = RT\, d \ln f_i)_T$, fugacity can also be used as a criterion of equilibrium between phases. This is a more useful property than chemical potential for defining equilibrium, since fugacities can be expressed in absolute values, whereas chemical potentials can be expressed only relative to some arbitrary reference state. For equilibrium of component i among various phases in equilibrium,

$$f_{iL} = f_{iL'} = f_{iV} = f_{is} \tag{13}$$

where the subscripts refer to the different phases.

Since fugacity is equal to the product of activity times the fugacity of the standard state, equation 13 may be written as

$$a_{iL}f^\circ_{iL} = a_{iL'}f^\circ_{iL'} = a_{iv}f^\circ_{iv} = a_{is}f^\circ_{is} = \cdots \tag{14}$$

where $a_{iL}, a_{iL'}, \cdots =$ activities of component 1 in the phases $L, L', \cdots$

$f^\circ_{iL}, f^\circ_{iL'}, \cdots =$ fugacities in the standard states of component 1 in the phases $L, L', \cdots$

Vapor–Liquid Equilibrium

For a multicomponent system comprising a single liquid phase L and a vapor phase v, equation 14 may be written

$$\frac{a_{iv}}{a_{iL}} = \frac{f^\circ_{iL}}{f^\circ_{iv}} = K_i \tag{15}$$

where K_i is termed the *vaporization equilibrium constant* of component i.

It is evident that the numerical value of the constant K_i depends on the definitions given to the standard states for both phases. It is convenient to choose as the standard state in the vapor phase the pure component vapor at the temperature and pressure of the system. Similarly, the standard state for each component in the liquid phase is chosen as the pure component in the liquid state at the temperature and pressure of the system. This selection of standard states has the advantages that, where the liquid and vapor possess ideal-solution behavior, activities are numerically equal to mole fractions. It follows from these definitions that K_i is a function of both the temperature and the pressure of the system, as well as of the nature of the component.

For nonideal solutions, the activity of component i is related to its mole fraction by means of an arbitrary factor called the activity coefficient. Thus, in the vapor phase,

$$a_{iv} = \gamma_{iv} y_i = \frac{f_i}{f^{\circ}_{iv}} \tag{16}$$

where y_i = mole fraction of component i in the vapor phase. Similarly, in the liquid phase,

$$a_{iL} = \gamma_{iL}\, x_i = \frac{f_i}{f^{\circ}_{iL}} \tag{17}$$

where x_i = mole fraction of component i in the liquid phase. The activity coefficients γ_i are dependent on the properties of component i and the other components with which it is mixed, as well as on the temperature and pressure. At atmospheric pressure or below the coefficients of the components in the gas phase are nearly unity. Under these conditions, $f_i = p_i = y_1 \pi$ and $f^{\circ}_{iL} = P_i$ and equation 17 becomes

$$\gamma_i = \frac{y_i \pi}{x_i P_i} \tag{18}$$

Binary Solutions. Four types of behavior of binary solutions are represented in Figs. 239 through 242, wherein pressures, activity coefficients, vapor compositions, and boiling points are plotted against mole fraction of the more volatile component in the liquid phase. These types are represented by ideal solutions, and those showing azeotropes with minimum boiling points, with maximum boiling points, and with separation of the liquid into two phases. Solutions that have maximum or minimum boiling points and that evolve vapors of the same composition as the liquid are termed *azeotropes*.

In these figures, total and partial pressures are plotted against composition at constant temperature, boiling points are plotted against composition at constant total pressure, activity coefficients are plotted against compositions at constant pressure, and vapor compositions are plotted against liquid compositions at constant pressure. Curves similar

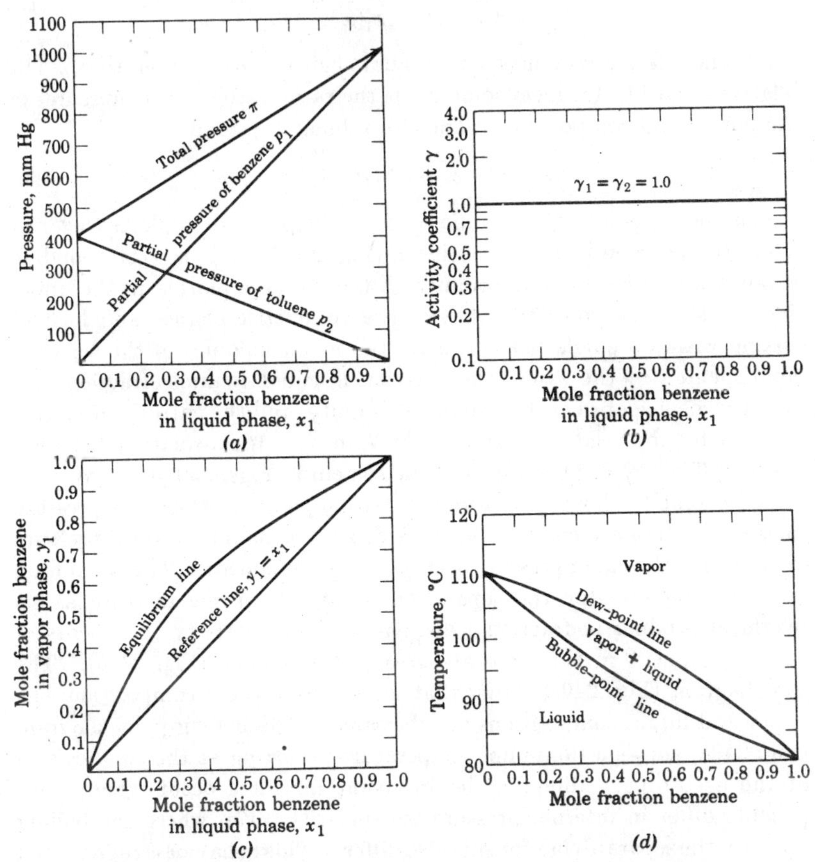

FIG. 239. Type I. Ideal solution behavior, benzene–toluene system. (a) Partial and total pressures at 90° C. (b) Activity coefficients. (c) Vapor–liquid equilibrium. (d) Phase diagram at 1 atm pressure

to the last three could also be plotted at conditions of constant temperature; the former presentation is more useful since separation processes are conducted more often at constant pressure than at constant temperature. Furthermore, vapor–liquid composition relations and activity coefficients are much more sensitive to temperature than to pressure.

For ideal-solution behavior (Fig. 239), the partial and total pressures at constant temperature show straight-line relationships with composition, the activity coefficients of each component are unity (Fig. 239*b*), and the y–x relationship is given as

$$y_1 = \frac{\alpha x_1}{1 + x_1(\alpha - 1)} \tag{19}$$

where the relative volatility α varies but slightly with temperature. The relative volatility for ideal solutions is the ratio of the vapor pressures of the pure liquid components. Thus for a binary system

$$\alpha = P_1/P_2 \tag{19a}$$

The boiling point of the liquid at constant pressure is plotted against liquid composition in the lower line L (Fig. 239*d*), and the corresponding vapor composition in equilibrium with the liquid is plotted in the upper line v. For ideal-solution behavior the volumetric change and heat of mixing within a single phase is zero, the molal volumes of the separate components are the same, the activity is equal to mole fraction, the activity coefficient of each component is unity, and the entropy of mixing is given by the relation $\Delta s^M = R \sum N_i \ln N_i$. Ideal-solution behavior is exemplified by the benzene–toluene system in Figs. 239*a* to 239*d*.

For a solution showing a *minimum boiling-point azeotrope*, the pressure lines (Fig. 240*a*) are concave downward, with a maximum total pressure greater than the vapor pressure of either pure component. This maximum pressure accounts for the separation of an azeotropic mixture as an overhead product in fractional distillation. The activity coefficients of each component in this case are greater than unity (Fig. 240*b*). The y–x diagram (Fig. 240*c*) shows that, at the azeotropic composition, the vapor and liquid compositions are the same. The boiling-point diagram (Fig. 240*d*) shows a minimum temperature occurring at the composition of the azeotrope. Minimum boiling-point mixtures occur where components differ in internal pressure and polarity, even where the boiling points of the separate components also differ. This behavior is represented by the isopropyl alcohol–isopropyl ether system which forms an azeotrope at 78 mole per cent isopropyl alcohol. Of known azeotropic solutions, the number of minimum boiling azeotropes is about ten times that of the maximum boiling type.

For a solution showing a *maximum boiling-point azeotrope*, the pressure lines are concave upward, with a minimum total pressure below the vapor pressure of either pure component. This accounts for an azeotrope residue formed during fractional distillation having the same final liquid and vapor compositions. The activity coefficients of the two components in this case (Fig. 241*b*) are each less than unity. The y–x diagram

(Fig. 241*c*) shows an azeotropic mixture where the values of y and x are identical. The boiling-point diagram (Fig. 241*d*) shows a maximum point corresponding to the azeotropic composition. Azeotropes having maximum boiling points are to be found in solutions that show a negative

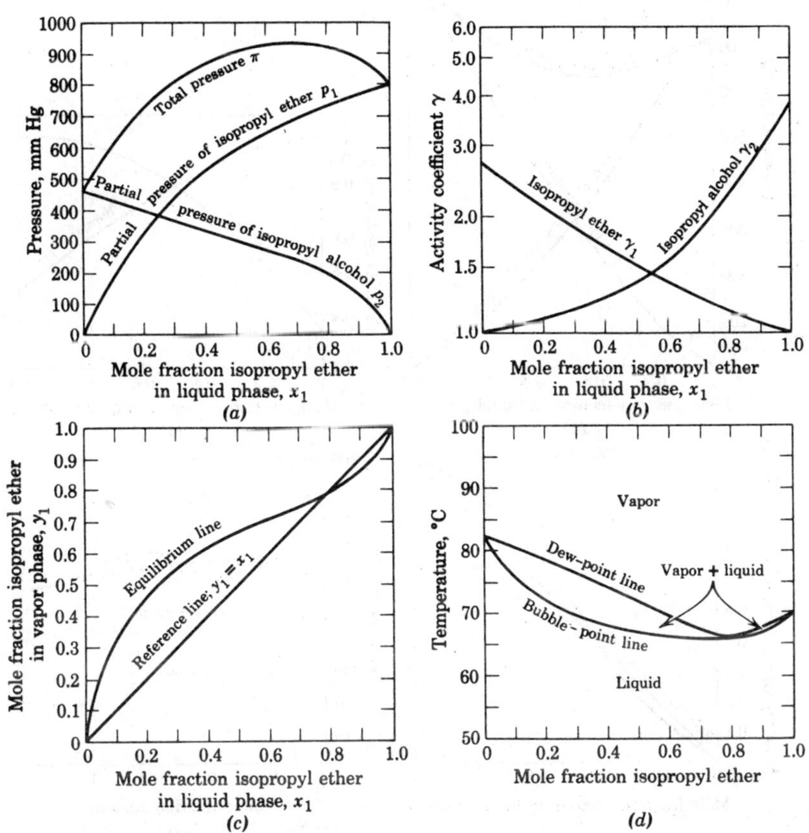

FIG. 240. Type II. Minimum boiling azeotrope, isopropyl ether–isopropyl alcohol system.
(*a*) Partial and total pressures at 70° C. (*b*) Activity coefficients.
(*c*) Vapor–liquid equilibria. (*d*) Phase diagram at 1 atm pressure

deviation from Raoult's law: that is, where solvation occurs with evolution of heat as evidence of chemical union. The components of such solutions occur where the separate components show wide differences in polarity, or in acidic and basic properties. This behavior is represented in Fig. 241 by an acetone–chloroform solution, having a maximum boiling azeotrope at 40 mole per cent acetone.

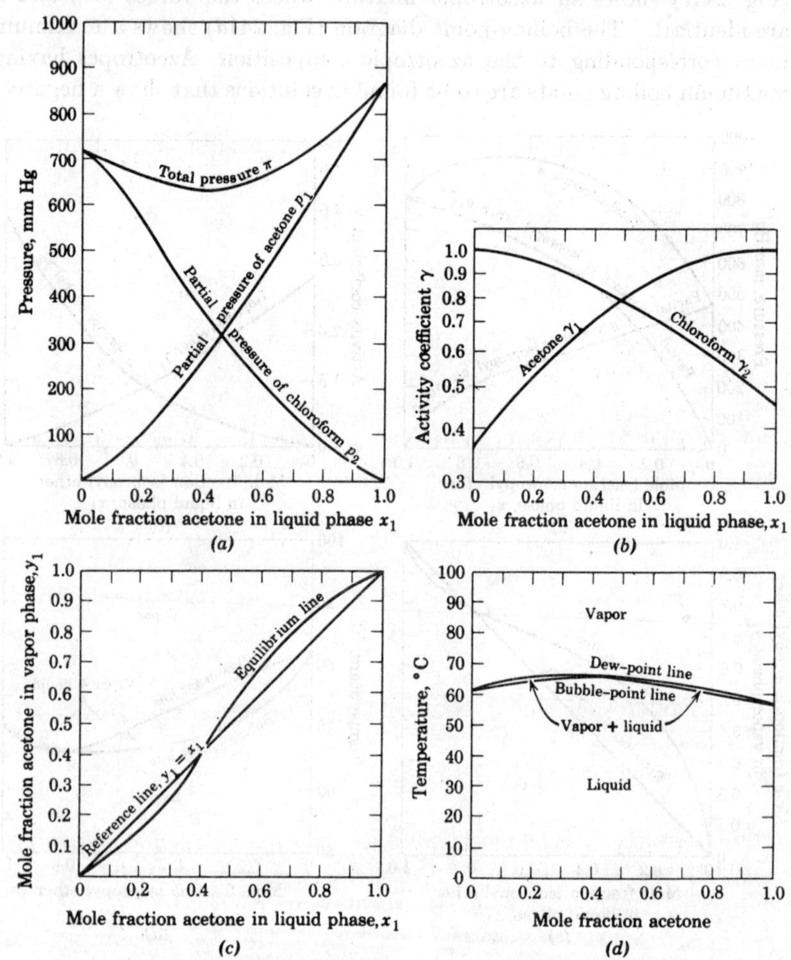

FIG. 241. Type III. Maximum boiling-point azeotrope, acetone–chloroform system.
(*a*) Partial and total pressures at 60° C. (*b*) Activity coefficients. (*c*) Vapor–liquid equilibria. (*d*) Phase diagram at 1 atm pressure

In the fourth type of solution behavior, solubility in the liquid phase is limited; separation of the liquid into two liquid phases occurs over the concentration range marked by a constant total pressure (Fig. 242*a*). In the two liquid-phase regions, the partial and total pressures remain constant as the relative amounts of the two phases change. The excessively high activity coefficients encountered results in the separation into two phases. The y–x diagram (Fig. 242c) shows a horizontal line over the immiscible region, and the boiling-point diagram (Fig. 242d)

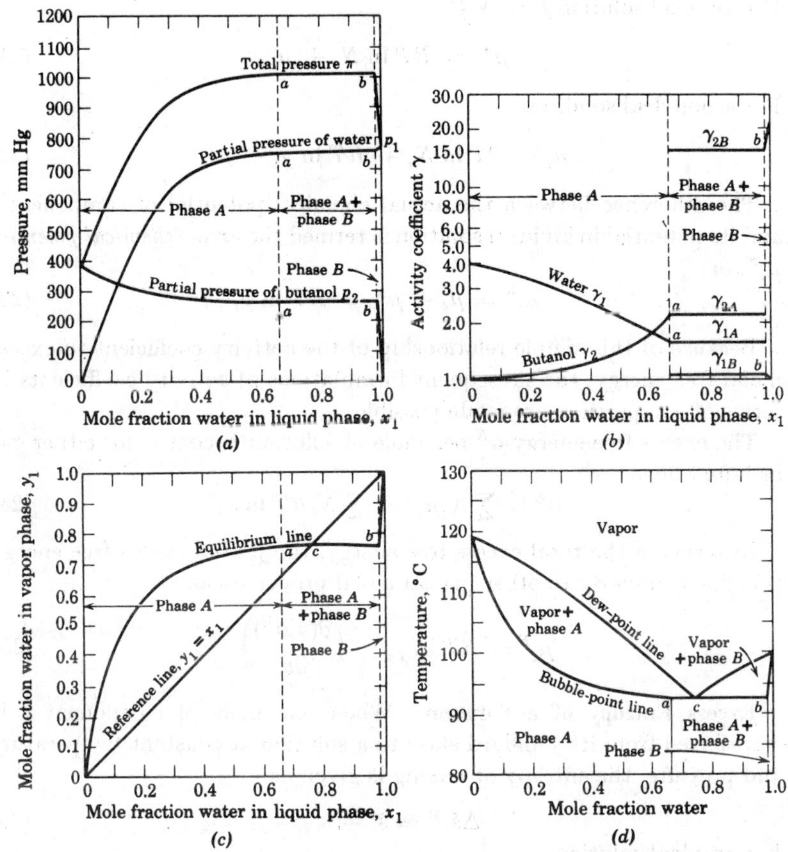

FIG. 242. Type IV. Minimum-boiling-point (two liquid phases), water–n–butanol system.
(a) Partial and total pressures at 100° C. (b) Activity coefficients.
(c) Vapor–liquid equilibria. (d) Phase diagram at 1 atm pressure

shows a minimum constant value over the same range. This system is represented by water–butanol, showing a minimum boiling point at 75 mole per cent water in the vapor phase.

Excess Free Energy of a Solution. When 1 mole of component i is transferred from its standard state of unit fugacity to a solution at the same temperature and pressure, the change in chemical potential $\Delta\mu_i^M$ of component i is expressed in terms of fugacity; thus,

$$(\Delta\mu_i^M)_T = (\mu_i - \mu^\circ_i)_T = [RT \ln (f_i/f^\circ_i)]_T \tag{20}$$

For an ideal solution $f_i = N_i f^\circ_i$:

$$\mu^*_i = RT \ln N_i + \mu^\circ_i \tag{21}$$

For a nonideal solution:

$$\mu_i = RT \ln N_i + RT \ln \gamma_i + \mu^\circ_i \tag{22}$$

The difference between the actual chemical potential of component i and its potential in an ideal solution is termed the *excess chemical potential* μ_i^E, or

$$\mu_i^E = \mu_i - \mu^*_i = RT \ln \gamma_i \tag{23}$$

Because of this simple relationship of the activity coefficient to excess molal free energy, the subsequent formulations of activity coefficients in terms of composition are made possible.

The excess free energy G^E per mole of solution becomes, for either gas or liquid phase,

$$\mathrm{G}^E = \sum N_i \mu_i^E = \sum N_i RT \ln \gamma_i \tag{24}$$

In terms of the total excess free energy, the partial excess free energy is defined similarly to other partial molal properties as

$$\mu_i^E = RT \ln \gamma_i = \left(\frac{\partial(n_t \mathrm{G}^E)}{\partial n_i}\right)_{pT} \tag{25}$$

Excess Entropy of a Solution. When one mole of component i is transferred from its standard state to a solution at constant temperature and pressure, the entropy of mixing is given as

$$\Delta\bar{\mathrm{s}}_i^M = \bar{\mathrm{s}}_i - \mathrm{s}^\circ_i \tag{26}$$

For an ideal solution

$$\Delta\bar{\mathrm{s}}_i^{M*} = -R \ln N_i \tag{27}$$

and the excess entropy $\bar{\mathrm{s}}_i^E$ of component i is $\Delta\bar{\mathrm{s}}_i^M - \Delta\bar{\mathrm{s}}_i^{M*}$, or

$$\bar{\mathrm{s}}_i^E = \Delta\bar{\mathrm{s}}_i^M + R \ln N_i \tag{28}$$

For an ideal solution the heat of mixing is zero, $\Delta\bar{\mathrm{H}}_i^{M*} = 0$; hence

$$\bar{\mathrm{H}}_i^E = \Delta\bar{\mathrm{H}}_i^M \tag{29}$$

$$\bar{\mathrm{s}}_i^E = \frac{\bar{\mathrm{H}}_i^E - \mu_i^E}{T} = \frac{\Delta\bar{\mathrm{H}}_i^M}{T} - R \ln \gamma_i \tag{30}$$

The excess entropy s^E per mole of solution becomes, for either gas or liquid,

$$s^E = \frac{\Delta H^M}{T} - R \sum N_i \ln \gamma_i \tag{31}$$

Wohl's Equation for Excess Free Energy. Inasmuch as all the effects of differences in intermolecular forces, polarity, chemical structure, and molecular size causing nonideal solution behavior are included in the concept of excess free energy, it becomes convenient to express this property empirically as a function of composition.

A general equation for expressing excess free energy was developed statistically by Wohl[2] in terms of composition, effective molal volume, and effective volumetric fraction z_i of the separate components. Thus

$$\frac{G^E}{RT(\sum q_i x_i)} = \sum_{ih} z_i z_h a_{ih} + \sum_{ihj} z_i z_h z_j a_{ihj} + \sum_{ihjl} z_i z_h z_j z_l a_{ihjl} + \cdots \tag{32}$$

where, for component i,

x_i = mole fraction i

q_i = effective molal volume

z_i = effective volume fraction

Subscripts i, h, j, l correspond to the separate components of the mixture.

$a_{ih} a_{ihj} \cdots$ = empirical constants.

$$z_i = \frac{x_i}{x_i + \sum \dfrac{q_j x_j}{q_i}} \tag{32a}$$

where j = any component other than i.

The term $\sum_{ih} z_i z_h a_{ih}$ represents the contribution due to the interactions of unlike components in groups of two and is designated as a *two-suffix term*. The term $\sum_{ihj} z_i z_h z_j a_{ihj}$ represents contributions due to the interaction of unlike components in groups of three. The combination of these two terms represents a *three-suffix* equation. A similar interpretation is given to the remaining terms. The number of terms actually used depends on the complexity of the solution and the precision of experimental data. The advantage of this equation is that each succeeding term reflects only the deviations unaccounted for by the preceding terms, and no advantage is gained in using more terms than is justified by experimental precision.

[2] K. Wohl, *Trans. Am. Inst. Chem. Engrs.*, **42**, 215 (1946).

Equation 32 is a four-suffix equation as characterized by the last term. If the last summation is omitted, the remainder is a three-suffix equation, whereas, if the last two terms are omitted, it becomes a two-suffix equation.

Activity Coefficients in Binary Solutions

Wohl's Three-Suffix Equation. For a binary system of components 1 and 2, a three-suffix equation of the form of equation 32 may be expanded to give

$$\frac{G^E}{RT(q_1x_1 + q_2x_2)} = z_1z_2 2a_{12} + z_1^2z_2 3a_{112} + z_1z_2^2 3a_{122} \tag{33}$$

or, since $z_1 + z_2 = 1.0$,

$$\frac{G^E}{RT} = \left(x_1 + \frac{q_2}{q_1}x_2\right)z_1z_2[z_1q_1(2a_{12} + 3a_{112}) + z_2q_1(2a_{12} + 3a_{122})] \tag{34}$$

By letting

$$A' = q_1(2a_{12} + 3a_{122}) \tag{35}$$

$$B' = q_2(2a_{12} + 3a_{112}) \tag{36}$$

equation 34 becomes

$$\frac{G^E}{RT} = \left(x_1 + \frac{q_2}{q_1}x_2\right)z_1z_2\left(z_1\frac{q_1}{q_2}B' + z_2A'\right) = x_1 \ln \gamma_1 + x_2 \ln \gamma_2 \tag{37}$$

From equation 25 and expressing x_i in terms of z_i, the following equations result:

$$\begin{aligned} \log \gamma_1 &= z_2^2\left[A + 2\left(B\frac{q_1}{q_2} - A\right)z_1\right] \\ \log \gamma_2 &= z_1^2\left[B + 2\left(A\frac{q_2}{q_1} - B\right)z_2\right] \end{aligned} \tag{38}$$

where $A = 2.303A'$ and $B = 2.303B'$.

Equation 38 involves three constants A, B, and q_1/q_2, which are characteristic of each binary system.

Margules Equations. Where the term q_2/q_1 is unity, equation 38 may be reduced to the Margules[3] three-suffix equation as modified by Carlson and Colburn;[4] thus,

$$\begin{aligned} \log \gamma_1 &= x_2^2[A + 2(B - A)x_1] = (2B - A)x_2^2 + 2(A - B)x_2^3 \\ \log \gamma_2 &= x_1^2[B + 2(A - B)x_2] = (2A - B)x_1^2 + 2(B - A)x_1^3 \end{aligned} \tag{39}$$

[3] M. Margules, *Sitzber. Akad. Wiss. Wien. Math. naturw. Klasse,* (II) **104,** 1243 (1895).

[4] H. C. Carlson and A. P. Colburn, *Ind. Eng. Chem.,* **34,** 581 (1942).

Where equation 39 holds, A is the terminal value of $\log \gamma_1$ at $x_1 = 0$, and B is the terminal value of $\log \gamma_2$ at $x_2 = 0$.

In the simplest case, where A and B are equal, the Margules two-suffix equation results:

$$\begin{aligned} \log \gamma_1 &= A x_2^2 \\ \log \gamma_2 &= A x_1^2 \end{aligned} \tag{40}$$

van Laar Equation. Where $q_1/q_2 = A/B$, equation 38 reduces to the form of the two-suffix equation as developed by van Laar[5,6] and rearranged by Carlson and Colburn:[4]

$$\begin{aligned} \log \gamma_1 &= A z_2^2 = \frac{A x_2^2}{\left(\dfrac{A}{B} x_1 + x_2\right)^2} \\ \log \gamma_2 &= B z_1^2 = \frac{B x_1^2}{\left(x_1 + \dfrac{B}{A} x_2\right)^2} \end{aligned} \tag{41}$$

Where the van Laar equation applies, A is the terminal value of $\log \gamma_1$ at $x_1 = 0$, and B is the terminal value of $\log \gamma_2$ at $x_2 = 0$.

It will be noted that, for both the Margules and the van Laar equations, A is the terminal value of $\log \gamma_1$ at $x_1 = 0$, and B is the terminal value of $\log \gamma_2$ at $x_2 = 0$.

The selection of the proper suffix equation depends on the molecular complexity of the system and the precision of experimental data. Where an equation is selected that fits the experimental data, the constants will be different for conditions of constant pressure or constant temperature.

The Margules three-suffix equation is suited for symmetrical systems: that is, where the constants A and B are nearly the same The van Laar equation can be used for unsymmetrical solutions where the ratio of A/B does not exceed 2, but is not applicable where maxima or minima values of $\log \gamma$ are shown.

In equations having only two constants, determination of γ_1 and γ_2 at a single known composition permits evaluation of these constants and the complete γ curves. Measurements of a single set of equilibrium liquid and vapor compositions, together with a knowledge of the vapor pressures of the pure components, suffice for calculations of γ_1 and γ_2.

van Laar Constants from Azeotropic Conditions. Where an azeotrope is formed, the composition of only one phase need be measured since the

[5] J. J. van Laar, *Z. Phys. Chem.*, **72**, 723 (1910).

[6] J. J. van Laar, *Z. Phys. Chem.*, **185**, 35 (1929).

liquid and vapor compositions are the same. By rearrangement of equation 41 the constants A and B are obtained directly:

$$A = \log \gamma_1 \left(1 + \frac{x_2 \log \gamma_2}{x_1 \log \gamma_1}\right)^2$$
$$B = \log \gamma_2 \left(1 + \frac{x_1 \log \gamma_1}{x_2 \log \gamma_2}\right)^2 \tag{42}$$

Illustration 1. The azeotrope of the ethanol–benzene system has a composition of 44.8 mole per cent ethanol with a boiling point of 68.24° C at 760 mm Hg: At 68.24° C, the vapor pressure of pure benzene is 517 mm Hg, and that of ethanol is 506 mm Hg. Calculate the van Laar constants for the system, and evaluate the activity coefficients for a solution containing 10 mole per cent ethanol.

Solution: Ethanol is designated as component 1. At the azeotropic composition, from equation 18, since $y_i = x_i$,

$$\gamma_1 = \frac{760}{506} = 1.502, \qquad \log \gamma_1 = 0.177$$

$$\gamma_2 = \frac{760}{517} = 1.47, \qquad \log \gamma_2 = 0.167$$

From equation 42,

$$A = 0.177 \left[1 + \frac{(0.552)(0.167)}{(0.448)(0.177)}\right]^2 = 0.830$$

$$B = 0.577$$

For a solution containing 10% ethanol, from equation 41,

$$\log \gamma_1 = \frac{(0.830)(0.90)^2}{\left[\dfrac{(0.830)(0.10)}{0.577} + 0.90\right]^2} = 0.6168, \qquad \gamma_1 = 4.14$$

$$\log \gamma_2 = 0.0109, \qquad \gamma_2 = 1.026$$

By this procedure the activity coefficients over the entire range of compositions may be calculated.

Redlich–Kister Equation. A commonly used equation for relating activity coefficients was developed by Redlich and Kister.[7] In this development, at conditions of constant temperature and pressure, excess free energy is related to composition by a series function using terms sufficient to fit the experimental data; thus:

$$G^E = RTx_1x_2[B + C(x_1 - x_2) + D(x_1 - x_2)^2 + \cdots]_{pT} \tag{43}$$

[7] O. Redlich and A. T. Kister, *Ind. Eng. Chem.*, **40**, 345 (1948).

The term x_1x_2 provides for the zero value of G^E at compositions corresponding to the separate pure components.

Formulation of individual activity coefficients is obtained by differentiation of equation 43 according to the definition of a partial molal property (page 331) to give

$$\ln \gamma_1 = x_1x_2[B + C(x_1 - x_2) + D(x_1 - x_2)^2 + \cdots] + x_2[B(x_2 - x_1) + C(6x_1x_2 - 1) + D(x_1 - x_2)(8x_1x_2 - 1) + \cdots] \tag{44}$$

$$\ln \gamma_2 = x_1x_2[B + C(x_1 - x_2) + D(x_1 - x_2)^2 + \cdots] - x_1[B(x_2 - x_1) + C(6x_1x_2 - 1) + D(x_1 - x_2)(8x_1x_2 - 1) + \cdots]$$

Subtracting the two parts of equation 44 gives

$$\ln \frac{\gamma_1}{\gamma_2} = B(x_2 - x_1) + C(6x_1x_2 - 1) + D(x_2 - x_1)(1 - 8x_1x_2) + \cdots \tag{45}$$

The same relationship for $\ln(\gamma_1/\gamma_2)$ may be obtained directly by differentiation of equation 43 with respect to x_1:

$$\left[\frac{\partial}{\partial x_1}\left(\frac{G^E}{RT}\right)\right]_{pT} = \left(\ln \frac{\gamma_1}{\gamma_2}\right)_{pT} \tag{46}$$

Experimental data on activity coefficients for binary systems are thus correlated by a single equation without separate expressions being used for the individual coefficients.

In calculating vapor compositions from liquid compositions, temperature sensitivity is minimized by using ratios of activity coefficients and of vapor pressures by applying equation 18 for the two components; thus,

$$y_1 = \frac{\gamma_1 x_1 P_1}{E\gamma_2 x_2 P_2}$$

$$y_2 = 1/E \tag{47}$$

$$E = \frac{\gamma_1 x_1 P_1}{\gamma_2 x_2 P_2} + 1$$

Evaluation of Constants of the Redlich–Kister Equation. The constants in equation 45 are usually evaluated by the method of least squares. Tierney[8] has shown that more weight should be assigned to data in the

[8] J. W. Tierney, *Ind. Eng. Chem.*, **50**, 707 (1958).

middle composition range than is given by the method of least squares. Without using the method of least squares computations, the constants may be obtained from a smoothed curve of ln (γ_1/γ_2) at the particular values of x_1 shown in Table 61.

Table 61. Evaluating Redlich–Kister Constants

Point	x_1	ln (γ_1/γ_2)
1	0	$B-C+D$
2	0.1464	$0.7071B-C/4$
3	0.2113	$0.5773(B-D/3)$
4	0.2959	$0.4082(B-2D/3)+C/4$
5	0.5	$C/2$
6	0.7041	$-0.4082(B-2D/3)+C/4$
7	0.7887	$-0.5773(B-D/3)$
8	0.8536	$-0.7071B-C/4$
9	1.000	$-B-C-D$

Redlich and Kister classify the complexity of binary solutions into the following five types:

Type 1. $B = 0, C = 0, D \cdots = 0$. Ideal solutions, ln $(\gamma_1/\gamma_2) = 0$.

Type 2. $B \neq 0$, $C = 0$, $D \cdots = 0$. Ln (γ_1/γ_2) versus x plots as a straight line passing through 0 at $x = 0.5$.

Type 3. $B \neq 0, C \neq 0, D \cdots = 0$. Only points 1, 3, 5, 7, and 9 need be considered. Values of ln (γ_1/γ_2) are equal at points 3 and 7. This type corresponds to components that are not associated, interact only moderately, and have nearly equal molal volumes.

Type 4. $B \neq 0$, $C \cdots = 0$, $D \neq 0$. Ln (γ_1/γ_2) versus x gives an S-shaped plot. Values are equal at $x = 0$ and $x = 1.0$. The curve passes through 0 at $x = 0.5$. Values of B and D are derived from points 2, 4, 6, and 8. Association of one of the components is reflected in this type.

Type 5. $B \neq 0$, $C \neq 0$, $D \neq 0$. Calculate C from point 5, B from points 2 and 8, D from points 4 and 6.

Chao's Modification of the Redlich–Kister Equation.[9] For vapor–liquid equilibrium in a binary system at constant pressure, temperature is fixed by composition and is not an independent variable; however, the same functional form of equation 43 may be used to express excess free energy under isobaric conditions. The numerical values of the constants will be slightly different and will be dependent on pressure only. Thus, at constant pressure,

$$G^E = RTx_1x_2[B' + C'(x_1 - x_2) + D'(x_1 - x_2)^2 + \cdots]_p \tag{48}$$

[9] K. C. Chao and O. A. Hougen, *Chem. Eng. Sci.*, **7**, 246 (1958).

The correct form of the Gibbs–Duhem relation for isobaric conditions has been shown by Ibl and Dodge[10] to be

$$x_1\, d\ln\gamma_1 + x_2\, d\ln\gamma_2 = Z\, dx_1 \tag{49}$$

where

$$Z = -\frac{\Delta\text{H}}{RT^2}\left(\frac{dT}{dx_1}\right)_p \tag{50}$$

$$\Delta\text{H} = \text{H} - x_1\text{H}^\circ_1 - x_2\text{H}^\circ_2$$

$\text{H}, \text{H}^\circ_1, \text{H}^\circ_2$ = molal enthalpies of mixture, and components respectively, each in the liquid phase

Differentiating equation 43 and combining with equation 49 gives

$$\ln\frac{\gamma_1}{\gamma_2} = \left[\frac{d}{dx_1}\left(\frac{\text{G}^E}{RT}\right)\right]_p - Z \tag{51}$$

Where Z is negligible, substitution of equation 48 into equation 51 gives the Redlich and Kister equation 45.

From equation 51,

$$\int_0^1 \ln\frac{\gamma_1}{\gamma_2}\, dx_1 = \int_0^1 \frac{\partial}{\partial x_1}\left(\frac{\text{G}^E}{RT}\right) dx_1 - \int_0^1 Z\, dx_1 \tag{52}$$

$$\int_0^1 \frac{\partial}{\partial x_1}\left(\frac{\text{G}^E}{RT}\right) dx_1 = \frac{\text{G}^E}{RT}\Bigg|_{x_1=0}^{x_1=1} = 0 \tag{53}$$

Combining equations 52 and 53

$$\int_0^1 \ln\frac{\gamma_1}{\gamma_2}\, dx_1 = -\int_0^1 Z\, dx_1 = -\int_0^1 \frac{\Delta\text{H}}{RT^2}\left(\frac{dT}{dx_1}\right) dx_1 = a \tag{54}$$

where a is a function of pressure.

The values of Z are not obtainable by direct experimentation, since they involve measurements of the variation of temperature with compositions at constant pressure and the heats of solution at hypothetical standard states. These measurements are circumvented by including the effect of Z in the empirical equation for the ratio of activity coefficients, thus retaining the same form as for isothermal conditions but with constants having slightly different numerical values, depending on the magnitude of Z.

The following modification of equation 45 was made by Chao[9] for isobaric conditions, allowing for the effect of the Z term and the requirements of equation 54:

$$\ln\frac{\gamma_1}{\gamma_2} = a + b(x_2 - x_1) + c(6x_1x_2 - 1) + d(x_2 - x_1)(1 - 8x_1x_2) + \cdots \tag{55}$$

[10] N. V. Ibl and B. F. Dodge, *Chem. Eng. Sci.*, **2**, 120 (1953).

The coefficients $b, c, d \cdots$ are not the same as $B, C, D \cdots$ in equation 43.

The activity coefficients of the individual components may be obtained by combining equations 49 and 55 to give

$$\begin{aligned} \ln \gamma_1 &= x_1x_2[B' + C'(x_1 - x_2) + D'(x_1 - x_2)^2 + \cdots] \\ &\quad + x_2[a + b(x_2 - x_1) + c(6x_1x_2 - 1) \\ &\quad + d(x_2 - x_1)(1 - 8x_1x_2) + \cdots] \\ \ln \gamma_2 &= x_1x_2[B' + C'(x_1 - x_2) + D'(x_1 - x_2)^2 + \cdots] \\ &\quad - x_1[a + b(x_2 - x_1) + c(6x_1x_2 - 1) \\ &\quad + d(x_2 - x_1)(1 - 8x_1x_2) + \cdots] \end{aligned} \tag{56}$$

For calculation of temperature under isobaric conditions, it is necessary to have equations for the separate activity coefficients and by iteration establish the temperature that satisfies the relation $y_1\pi = x_1\gamma_1P_1$ in agreement with the correct value of vapor pressure of the pure component.

Illustration 2. For a liquid solution having a molal composition of ethyl acetate $x_A = 0.836$ and cyclohexane $x_C = 0.164$, calculate (*a*) the composition of the vapor and (*b*) the bubble-point temperature at 760 mm Hg pressure. The fugacity coefficients in the vapor state at 1 atm are unity for each component.

$$\log P_A = 7.85401 - \frac{1742.32}{T}, \qquad (\text{nbp})_A = 77.2° \text{ C} \tag{a}$$

$$\log P_C = 6.84498 - \frac{1203.53}{t + 222.86}, \qquad (\text{nbp})_C = 80.7° \text{ C} \tag{b}$$

where P is in millimeters of mercury and t is in degrees centigrade.

From the experimental data of Chao[11] at 1 atm,

$$\log \frac{\gamma_C}{\gamma_A} = 0.3849(x_A - x_C) + 0.0321(6x_Ax_C - 1)$$

From this equation $\gamma_A/\gamma_C = 0.5585$. Assuming a temperature of 75° C, $P_A =$ 708.3 mm, $P_B = 673.3$ mm, $P_A/P_C = 1.1114$.

$$\frac{x_A}{x_C} = \frac{0.836}{0.164} = 5.097$$

Substituting these values into equation 47 gives

$$E = 1.3159$$

$$y_C = 0.2408$$

$$y_A = 0.7592$$

[11] K. C. Chao, Ph.D. thesis, University of Wisconsin (1956).

In order to verify the temperature, it is necessary to have values for individual coefficients. From equation 44 uncorrected for the effect of Z,

$$\log \gamma_A = x_A x_C[0.3849 + 0.0321(x_C - x_A)] - x_C[0.3849(x_A - x_C) + 0.0321(6x_A x_C - 1)]$$

$$\log \gamma_C = x_A x_C[0.3849 + 0.0321(x_C - x_A)] + x_A[0.3849(x_A - x_C) + 0.0321(6x_A x_C - 1)]$$

$$\gamma_A = 1.0194$$

$$\gamma_C = 1.8253$$

$$P_C = \frac{\pi y_C}{x_C \gamma_C} = 609.7 \text{ mm}$$

The value of $P_C = 609.7$ mm corresponds to a temperature of 73.6° C from equation (*b*) which is not in agreement with the assumed value of 75° C. When the calculation was repeated at 73.7° C, agreement was obtained with results shown in the first column. Considering the effect of Z for isobaric conditions the four constants in equation 56 are $B' = 0.4190$, $C' = 0.0149$, $b = 0.3849$, and $c = 0.0321$. The calculated results are given in Table A.

TABLE A

	Neglecting Z	Considering Z
γ_A	= 1.0194	1.0341
γ_C	= 1.8253	1.8517
γ_A/γ_C	= 0.5585	0.5585
y_A	= 0.759	0.759
y_C	= 0.241	0.241
t	= 73.7° C	73.3° C

In this system where the difference in normal boiling points is only 3.5° C, neglect of the Z term has no effect on vapor composition, and the calculated temperature is only 0.4° C high.

Testing Consistency of Activity Coefficients. For testing the consistency of experimental data on activity coefficients in binary systems where such data are available over the entire concentration range, Redlich and Kister[7] have developed the following convenient device.

From equation 46, at constant temperature,

$$\frac{dG^E}{dx_1} = RT \ln \left(\frac{\gamma_1}{\gamma_2}\right) \tag{57}$$

Since G^E is equal to zero at $x_1 = 0$ and $x_1 = 1$, it follows that

$$RT \int_0^1 \ln \left(\frac{\gamma_1}{\gamma_2}\right) dx_1 = G^E_{x=1} - G^E_{x=0} = 0 \tag{58}$$

This equation holds strictly only at constant temperature and pressure.

By plotting experimental data of $\ln(\gamma_1/\gamma_2)$ against x_1 from 0 to 1.0, the net area of the diagram for isothermal conditions should equal zero if the data are accurate and consistent. This relation does not necessarily hold under isobaric conditions. For example, in precise measurements by

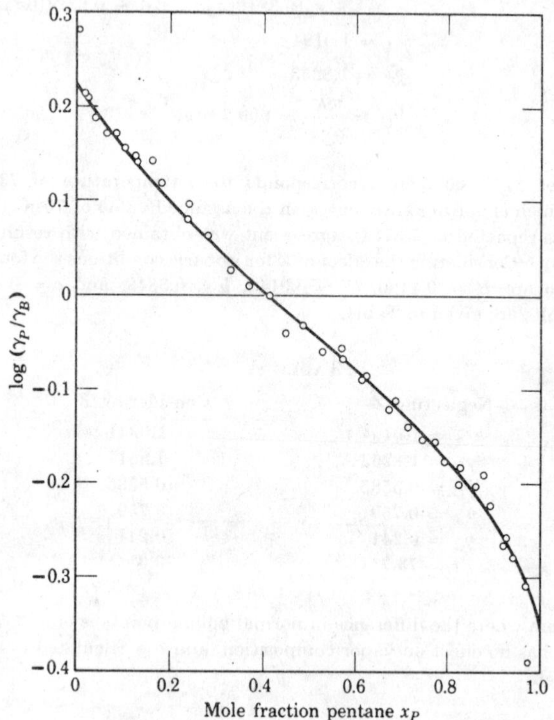

FIG. 243. Ratio of activity coefficients in the system pentane–benzene, $\pi = 760$ mm Hg

Myers[12] on the pentane–benzene system at constant pressure, the area below the zero ordinate line of such a plot is twice as large as the area above the zero ordinate line as shown in Fig. 243. In this case the difference is due to the Z effect and not to inconsistent measurements.

Activity Coefficients by Method of Carlson and Colburn. It was pointed out by Carlson and Colburn[4] that the constants in various equations for activity coefficients may be evaluated without vapor–liquid equilibrium composition data if a series of isothermal vapor-pressure or isobaric

[12] H. S. Myers, *Ind. Eng. Chem.*, **47**, 2215 (1955).

boiling-point measurements are available for solutions of known composition over the entire composition range. It is desirable to work from such data because of the difficulty of securing a reliable analysis of vapor mixtures and the ease of making up liquid solutions of a desired composition. Since, for ideal-gas behavior,

$$p_1 = \pi - p_2 = \pi - \gamma_2 x_2 P_2 \tag{59}$$

as x_2 approaches 1.0 where γ_2 is 1.0,

$$\lim_{x_1=0} \gamma_1 = \frac{\pi - x_2 P_2}{x_1 P_1} \tag{60}$$

Similarly,

$$\lim_{x_2=0} \gamma_2 = \frac{\pi - x_1 P_1}{x_2 P_2} \tag{61}$$

In working from isothermal total vapor-pressure data, the observed values of π are substituted in equation 60 with the corresponding mole fractions, and vapor pressures and apparent values of γ_1 are calculated. By plotting log γ_1 against x_1 and extrapolating to $x_1 = 0$, the value of A is determined. Similarly, when working from isobaric boiling-point data, values of the vapor pressures of the pure components at the observed temperatures are substituted in equation 60 with the corresponding total pressure and mole fractions. The calculated apparent values of γ_1 are extrapolated to $x_1 = 0$ as before. This latter procedure involves the assumption that the activity coefficients are independent of temperature over the range covered.

Illustration 3. In Table A are data for the boiling points of ethanol–benzene solutions at 750 mm Hg and the vapor pressures of pure ethanol and benzene at

TABLE A. BOILING POINTS OF ETHANOL–BENZENE SOLUTIONS AT 750 mm Hg

Mole Fractions		Boiling Point	Vapor Pressure of Pure Components, mm Hg at t_B° C		Apparent	Apparent	True	True
Ethanol x_1	Benzene x_2	t_B, °C	Ethanol, P_1	Benzene, P_2	γ_1	γ_2	γ_1	γ_2
0.0	1.0	79.7	804 mm	750 mm		1.00	5.70	1.00
0.04	0.96	75.2	671	648	4.74	1.16	4.20	1.00
0.11	0.89	70.8	560	502	4.07	1.38	3.66	1.05
0.28	0.72	68.3	507	518	2.66	1.63	2.11	1.16
0.43	0.57	67.8	497	509	2.15	1.84	1.58	1.34
0.61	0.39	68.3	507	518	1.77	2.18	1.22	1.84
0.80	0.20	70.1	545	549	1.47	2.80	1.05	2.64
0.89	0.11	72.4	598	592	1.285	3.33	1.02	3.14
0.94	0.06	74.4	650	632	1.165	3.64	1.00	3.58
1.00	0.0	78.1	750	711	1.00	—	1.00	4.06

Data from the *International Critical Tables*, III, pp. 217–221, 313.

these temperatures. Calculate the van Laar constants from these data, assuming the activity coefficients to be independent of temperature.

Solution: The values of apparent γ_1 and γ_2 shown in Table A are calculated from equations 60 and 61 and plotted in Fig. 244 against x_1. Extrapolating the γ_1 curve to $x_1 = 0$ gives the value of $\gamma_1 = 5.7$ or $A = \log \gamma_1 = 0.756$. Similarly, γ_2 is determined to be 4.06 by extrapolating the γ_2 curve to $x_1 = 1.0$ and $B = \log \gamma_2 = 0.608$. These constants are in fair agreement with those derived from the azeotrope composition.

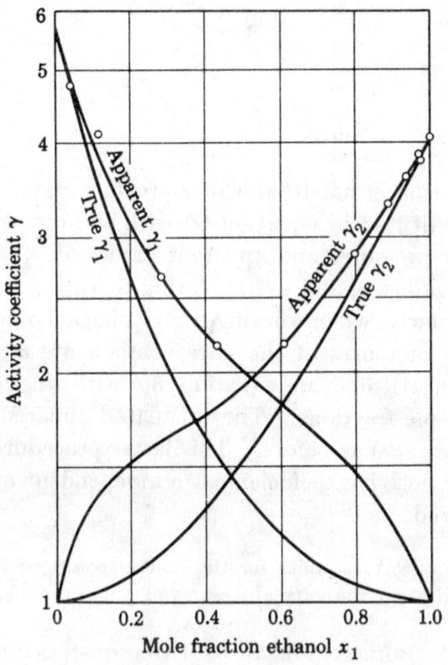

FIG. 244. Apparent and true activity coefficients of the ethanol–benzene system at 750 mm Hg

Improved accuracy in the determination of van Laar constants from boiling-point or total-pressure data is obtained by first calculating the constants from equations 59 and 60 as in Illustration 1. Values of γ_2 are then calculated from the van Laar equation with these approximate values of A and B substituted into equation 59 instead of $\gamma_2 = 1.0$ being assumed as in the first approximation. Similarly, calculated values of γ_1 are substituted into equation 60, and the extrapolation of the two equations to zero concentration is repeated to obtain second approximations of A and B. In the second approximation of Illustration 1, the same values of A and B are obtained. The corrected values of γ_1 and γ_2 are plotted in Fig. 244.

The preferred method for evaluating the constants of the van Laar equation depends on the nature of the system and the data available. Reliable vapor–liquid composition data in the dilute ranges are most desirable. A generally applicable method for obtaining such data by equilibrium condensation has been developed by Colburn, Schoenborn, and Shilling.[13] Where such data are not available, the composition of the azeotrope generally furnishes the best basis if it is in the middle of the composition range, between $x_1 = 0.25$ and $x_1 = 0.75$. If the azeotrope composition is outside this range, boiling-point or vapor-pressure data provide a better basis.

Once the activity coefficients of a system are established, equation 18 permits calculation of equilibrium compositions and vapor-pressure and boiling-point curves at low pressures where the ideal-gas law may be assumed. For a binary system of components 1 and 2,

$$\pi = p_1 + p_2 = y_1\pi + y_2\pi \tag{62}$$

Combining equations 59 and 62, since $x_2 = 1.0 - x_1$, gives

$$\pi = x_1 P_1 \gamma_1 + (1 - x_1) P_2 \gamma_2 \tag{63}$$

$$y_1 = \frac{x_1 P_1 \gamma_1}{x_1 P_1 \gamma_1 + (1 - x_1) P_2 \gamma_2} = \frac{\dfrac{P_1}{P_2}\dfrac{\gamma_1}{\gamma_2}\left(\dfrac{x_1}{1 - x_1}\right)}{1 + \dfrac{P_1}{P_2}\dfrac{\gamma_1}{\gamma_2}\left(\dfrac{x_1}{1 - x_1}\right)} \tag{64}$$

Equations 63 and 64 permit direct calculation of isothermal vapor-pressure and vapor–liquid composition curves. For deriving the more valuable isobaric boiling-point curves and the corresponding isobaric vapor–liquid composition curves, it is necessary that P_1 and P_2 be known as functions of temperature. Equation 64 may then be solved by a trial-and-error method for the boiling points and vapor compositions corresponding to selected liquid compositions. As a first approximation, the boiling point is assumed; this fixes a trial value of the *relative volatility* P_1/P_2 which is substituted in equation 64, and a first approximation y_1 is calculated. Then, P_1 is equal to $y_1\pi/x_1\gamma_1$, and a corrected boiling point is obtained as the temperature corresponding to this value of P_1. Since P_1/P_2 varies but little with temperature, the value of y_1 calculated from equation 64, using the corrected temperature is generally satisfactory.

[13] A. P. Colburn, E. M. Schoenborn, and C. D. Shilling, *Ind. Eng. Chem.*, **35**, 1250 (1943).

Illustration 4. From the van Laar constants of Illustration 1, calculate the boiling point at 750 mm Hg of a solution containing 28 mole per cent ethanol in benzene and the equilibrium composition of the vapor evolved. The vapor pressures of the pure components may be estimated from the data of Illustration 3.

Solution: As a first approximation assume that $t_B = 70°$ C. Then,

$$\frac{P_1}{P_2} = \frac{543}{548} = 0.99$$

From equation 41, the values of A and B from Illustration 1 give

$$\gamma_1 = 2.19; \qquad \gamma_2 = 1.18$$

By substitution in equation 64, a first approximation of y_1 is obtained:

$$y_1 = \frac{(0.99)\left(\frac{2.19}{1.18}\right)\left(\frac{0.28}{0.72}\right)}{1 + (0.99)\left(\frac{2.19}{1.18}\right)\left(\frac{0.28}{0.72}\right)} = \frac{0.715}{1.715} = 0.416$$

Substitution of this value in equation 18 gives

$$P_1 = (y_1\pi)/x_1\gamma_1 = \frac{(0.416)(750)}{(0.28)(2.19)} = 509 \text{ mm Hg}$$

This vapor pressure corresponds to a temperature of 68.4° C, the corrected boiling point. At this temperature $P_2 = 520$ mm, $P_1/P_2 = \frac{509}{520} = 0.978$. Substitution in equation 64 gives

$$y_1 = \frac{0.706}{1.706} = 0.413$$

In this manner complete boiling-point and vapor–liquid composition curves may be derived. The differences between the values calculated in this manner from the data of Table A are attributable to the inadequacy of the van Laar equation and the uncertainty of experimental data of this type.

Effects of Pressure and Temperature on the Composition of an Azeotrope. Because of its industrial importance in the separation of the components of a solution by distillation, considerable interest is attached to the effect of pressure in changing the temperature and composition of the azeotrope. The variation of the temperature of azeotropic vaporization with pressure can be obtained directly from the Clapeyron equation to give

$$\frac{dT}{dp} = \frac{T\,\Delta V_v}{\Delta H_v} \tag{65}$$

where ΔH_v = heat of vaporization of the azeotrope

ΔV_v = volume change in vaporization of the azeotrope

The variation of composition of the azeotrope with pressure is obtained from the variation of the fugacities of the azeotrope.

For the liquid phase,

$$d \ln f_1 = \left(\frac{\partial \ln f_1}{\partial T}\right)_{px} dT + \left(\frac{\partial \ln f_1}{\partial p}\right)_{Tx} dp + \left(\frac{\partial \ln f_1}{\partial x_1}\right)_{Tp} dx_1 \tag{66}$$

For the vapor phase,

$$d \ln f'_1 = \left(\frac{\partial \ln f'_1}{\partial T}\right)_{py} dT + \left(\frac{\partial \ln f'_1}{\partial p}\right)_{Ty} dp + \left(\frac{\partial \ln f'_1}{\partial y_1}\right)_{Tp} dy_1 \tag{67}$$

If equations 66 and 67 are applied under the conditions of azeotropic vaporization,

$$f_1 = f'_1 \quad \text{and} \quad x_1 = y_1$$

Combining equations 66 and 67 with 20–23 and 20–30 for component 1,

$$\left(\frac{\bar{H}'_1 - \bar{H}_1}{RT^2}\right) dT + \left(\frac{\bar{v}_1 - \bar{v}'_1}{RT}\right) dp + \left(\frac{\partial \ln f_1}{\partial x_1}\right)_{Tp} dx_1 - \left(\frac{\partial \ln f'_1}{\partial y_1}\right)_{Tp} dy_1 = 0 \tag{68}$$

Similarly, for component 2,

$$\left(\frac{\bar{H}'_2 - \bar{H}_2}{RT^2}\right) dT + \left(\frac{\bar{v}_2 - \bar{v}'_2}{RT}\right) dp + \left(\frac{\partial \ln f_2}{\partial x_2}\right)_{Tp} dx_2 - \left(\frac{\partial \ln f'_2}{\partial y_2}\right)_{Tp} dy_2 = 0 \tag{69}$$

Assuming activity coefficients in the vapor phase as unity and eliminating dT from equations 68 and 69 results in

$$\frac{dx_1}{dp} = \frac{(\Delta\bar{v}_1/\Delta\bar{H}_1) - (\Delta\bar{v}_2/\Delta\bar{H}_2)}{RT\left[\dfrac{(\partial \ln \gamma_1/\partial x_1)_{Tp}}{\Delta\bar{H}_1} + \dfrac{(\partial \ln \gamma_2/\partial x_2)_{Tp}}{\Delta\bar{H}_2}\right]} \tag{70}$$

where $\Delta\bar{H}_1 = \bar{H}'_1 - \bar{H}_1 =$ partial heat of vaporization of component 1
$\Delta\bar{H}_2 = \bar{H}'_2 - \bar{H}_2 =$ partial heat of vaporization of component 2
$\Delta\bar{v}_1 = \bar{v}'_1 - \bar{v}_1 =$ partial volume change in vaporization of component 1
$\Delta\bar{v}_2 = \bar{v}'_2 - \bar{v}_2 =$ partial volume change in vaporization of component 2

Neglecting the volume of the liquid and assuming ideal-gas behavior, equation 70 becomes

$$\frac{dx_1}{d \ln p} = \frac{1/\Delta\bar{H}_1 - 1/\Delta\bar{H}_2}{(\partial \ln \gamma_1/\partial x_1)_{pT}(\Delta H/x_2\, \Delta\bar{H}_1\, \Delta\bar{H}_2)} \tag{71}$$

where $\Delta H = x_1\, \Delta\bar{H}_1 + x_2\, \Delta\bar{H}_2$

Where activity coefficients are greater than unity, the denominator becomes negative, and the mole fraction of component 1 in the azeotrope increases with pressure, provided $\Delta\bar{H}_1$ is greater than $\Delta\bar{H}_2$.

An approximate method of estimating the effect of temperature on the composition of an azeotrope was suggested by Carlson and Colburn,[4] where the ratio of activity coefficients in a solution is assumed to be independent of temperature.

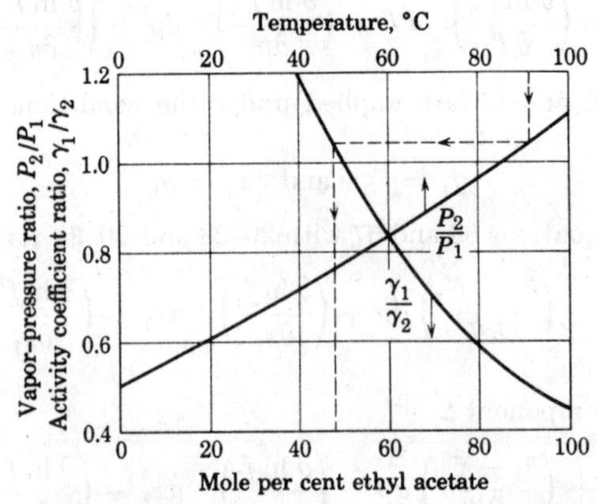

Fig. 245. Effect of temperature on the composition of the ethyl acetate–ethyl alcohol azeotrope

For ideal-gas behavior,

$$\gamma_1 = \frac{y_1 \pi}{x_1 P_1} \tag{72}$$

For an azeotrope,

$$\frac{\gamma_1}{\gamma_2} = \frac{P_2}{P_1} \tag{73}$$

Values of γ_1/γ_2 are plotted against x_1 at a known temperature. The azeotrope composition at any temperature is then the composition where γ_1/γ_2 is equal to P_2/P_1. This procedure as developed by Carlson and Colburn is shown in Fig. 245 for the ethyl acetate–ethyl alcohol system. The γ_1/γ_2 curve was derived from the composition of the azeotrope at atmospheric pressure and 71.8° C. To estimate the composition of the azeotrope at 91.4° C, the broken lines on the diagram are followed from this temperature vertically to the P_2/P_1 curve, then horizontally to the γ_1/γ_2 curve, and vertically to the composition axis, giving a value of 48 mole per cent ethyl acetate. Table 62 shows values estimated in this manner compared with the experimental measurements of Merriman.[14]

The total pressure P_z corresponding to the various azeotrope compositions and temperatures may be approximated by the following empirical equation,

$$P_z = \frac{P'_z(x_1P_1 + x_2P_2)}{(x'_1P'_1 + x'_2P'_2)} \tag{74}$$

where P_z, P_1, P_2 = vapor pressures of the azeotrope, component 1, and component 2 at temperature t

P'_z, P'_1, P'_2 = vapor pressures of the azeotrope, component 1, and component 2 at the reference temperature t'

x_1, x_2 = mole fractions of components 1 and 2 in the azeotrope at temperature t

x'_1, x'_2 = mole fractions of components 1 and 2 in the azeotrope at temperature t'

This method has not been sufficiently explored to permit estimation of the errors that may be encountered.

TABLE 62. EFFECT OF PRESSURE ON THE ETHYL ACETATE–ETHYL ALCOHOL AZEOTROPE[14]

Temperature, °C	P_2/P_1	Mole Fraction of Ethyl Acetate Calculated	Mole Fraction of Ethyl Acetate Measured	Total Pressure, mm Hg
18.7	0.598	0.787	0.734	77.4
40.5	0.725	0.677	0.660	220
56.3	0.827	0.602	0.601	423
71.8	0.921	0.539	0.539	760
83.1	0.994	0.498	0.490	1121
91.4	1.049	0.480	0.451	1476

Ternary Systems under Isobaric Conditions

The excess free energy of a ternary mixture is related to the activity coefficient by

$$G^E_{123} = x_1\mu^E{}_1 + x_2\mu^E{}_2 + x_3\mu^E{}_3 = RT(x_1 \ln \gamma_1 + x_2 \ln \gamma_2 + x_3 \ln \gamma_3) \tag{75}$$

Under conditions of constant pressure,

$$G^E_{123} = G^E{}_{12} + G^E{}_{23} + G^E{}_{31} + RTx_1x_2x_3[B_1 + C_1(x_2 - x_3) + C_2(x_3 - x_1) + C_3(x_1 - x_2) + \cdots] \tag{76}$$

where the first three terms on the right-hand side represent contributions by the separate binaries and the last term represents ternary effects. The binary contributions are related by assigning appropriate constants to equation 48, thus, for $G^E{}_{12}$,

$$G^E{}_{12} = RTx_1x_2[B'_{12} + C'_{12}(x_1 - x_2) + D'_{12}(x_1 - x_2)^2 + \cdots] \tag{77}$$

[14] R. W. Merriman, *J. Chem. Soc.*, **103**, 1801 (1913).

Since the temperature is fixed by composition the constants in equations 76 and 77 are dependent on pressure only.

The modified Gibbs–Duhem equation for a ternary system under isobaric conditions is given by an expansion of equation 49; thus

$$x_1\, d \ln \gamma_1 + x_2\, d \ln \gamma_2 + x_3\, d \ln \gamma_3 = Z\, dx_1 \tag{78}$$

Upon differentiating equation 75 and combining with equation 78, the individual activity coefficients may be expressed in terms of G^E,

$$\begin{aligned}\ln \gamma_1 = \frac{\text{G}^E}{RT} &+ \left[\frac{\partial}{\partial x_1}\left(\frac{\text{G}^E}{RT}\right)\right]_{x_2} + \frac{\Delta\text{H}}{RT^2}\left(\frac{\partial T}{\partial x_1}\right)_{x_2} \\ &- x_1\left[\frac{\partial}{\partial x_1}\left(\frac{\text{G}^E}{RT}\right)_{x_2} + \frac{\Delta\text{H}}{RT^2}\left(\frac{\partial T}{\partial x_1}\right)_{x_2}\right] \\ &- x_2\left[\frac{\partial}{\partial x_2}\left(\frac{\text{G}^E}{RT}\right)_{x_1} + \frac{\Delta\text{H}}{RT^2}\left(\frac{\partial T}{\partial x_2}\right)_{x_1}\right]\end{aligned} \tag{79}$$

and similarly for the other components. The ratio of the activity coefficients follows:

$$\ln\left(\frac{\gamma_1}{\gamma_2}\right) = \frac{\partial}{\partial x_1}\left(\frac{\text{G}^E}{RT}\right)_{x_3} + \frac{\Delta\text{H}}{RT^2}\left(\frac{\partial T}{\partial x_1}\right)_{x_3} \tag{80}$$

and similarly for the other ratios.

Expressions of activity coefficients and their ratios explicitly in terms of compositions result upon substituting equation 76 into equations 79 and 80, respectively. The difficulties with the correction term $(\Delta\text{H}/RT)(\partial T/\partial x)$ are circumvented by modifying the constants as is done with the binary system. Thus,

$$\begin{aligned}\ln \gamma_1 = (\text{G}^E_{123}/RT) &+ [x_2(x_2 + x_3) - x_1x_2] \\ &\quad [b_{12} + c_{12}(x_1 - x_2) + \cdots] \\ &- 2x_2x_3[b_{23} + c_{23}(x_2 - x_3) + \cdots] \\ &+ [(x_3 - x_1)(x_2 + x_3) + x_1x_2] \\ &\quad [b_{31} + c_{31}(x_3 - x_1) + \cdots] \\ &+ [x_1x_2(x_2 + x_3) + x_1x_2^2](c_{12} + \cdots) \\ &+ [x_2x_3(x_2 + x_3) - 2x_2^2x_3](c_{23} + \cdots) \\ &+ [-2x_3x_1(x_2 + x_3) + x_1x_2x_3](c_{31} + \cdots) \\ &+ [(x_2 + x_3)(x_2x_3 - x_1x_2) - x_2(x_1x_3 - x_1x_2)] \\ &\quad [b + c_1(x_2 - x_3) \\ &+ c_2(x_3 - x_1) + c_3(x_1 - x_2) + \cdots] \\ &+ x_1x_3x_3[(x_2 + x_3)(c_1 - 2c_2 + c_3) \\ &- x_2(2c_1 - c_2 - c_3) + \cdots]\end{aligned} \tag{81}$$

where G^E_{123} is given by equation 76.

$$\begin{aligned}\ln(\gamma_1/\gamma_2) = &-b_{12}(x_1 - x_2)\\ &+ c_{12}[2x_1x_2 - (x_1 - x_2)^2] + \cdots\\ &+ x_3\{b_{31} - b_{23} - c_{23}(2x_2 - x_3)\\ &+ c_{31}(x_3 - 2x_1) + \cdots\\ &- b_1(x_1 - x_2) - c_1[x_1(2x_2 - x_3) + x_2(x_3 - x_2)]\\ &- c_2[x_1(2x_2 - x_1) + x_3(x_1 - x_2)]\\ &+ c_3[2x_1x_2 - (x_1 - x_2)^2] + \cdots\}\end{aligned} \tag{82}$$

Similarly, expressions for the other components are obtained by cyclic advancement of the subscripts in the order 1, 2, 3, 1.

Illustration 5. For a liquid having a molal composition ethyl acetate $x_A = 0.192$, benzene $x_B = 0.099$, cyclohexane $x_C = 0.709$, calculate (*a*) the composition of the vapor and (*b*) the bubble-point temperature at a pressure of 760 mm Hg. Values of P_A and P_C in terms of temperature are recorded in Illustration 2; for P_B the corresponding equation is $\log P_B = 6.9056 - 1211.03/(t + 220.79)$; nbp = 80.1° C.

From experimental data at 760 mm Hg, Chao[11] reports the following values for use in equation 81, where log is substituted for ln and where A, B, C are substituted for 1, 2, 3:

$$\begin{aligned}b_{AB} &= 0.0372 & B'_{AB} &= 0.0359\\ b_{BC} &= 0.1443 & B'_{BC} &= 0.1507\\ b_{CA} &= 0.3849 & B'_{CA} &= 0.4190\\ c_{AB} &= 0 & C'_{AB} &= 0\\ c_{BC} &= 0.0100 & C'_{BC} &= 0.0100\\ c_{CA} &= 0.0321 & C'_{CA} &= 0.0149\end{aligned}$$

From equation 82,

$$\gamma_A/\gamma_B = 1.5072$$
$$\gamma_B/\gamma_C = 1.0266$$
$$\gamma_C/\gamma_A = 0.64631$$

$$\frac{x_A}{x_B} = \frac{0.192}{0.099} = 1.939, \qquad \frac{x_B}{x_C} = \frac{0.099}{0.709} = 0.1396, \qquad \frac{x_C}{x_A} = \frac{0.709}{0.192} = 3.693$$

Assuming a boiling point of 76.0° C,

$$P_A = 729.7 \qquad P_B = 668.7 \qquad P_C = 657.5$$

$$\frac{P_A}{P_B} = \frac{729.7}{668.7} = 1.091, \qquad \frac{P_B}{P_C} = \frac{668.7}{657.5} = 1.0170, \qquad \frac{P_C}{P_A} = \frac{657.5}{729.7} = 0.9011$$

Substitution of these values into equation 47 gives

$$E = 3.4642$$
$$y_A = 0.2887$$
$$y_B = 0.0905$$
$$y_C = 0.6208$$

In order to verify the temperature it is necessary to have values for at least one activity coefficient.

Substitution in equation 81 for isobaric conditions corrected for Z gives $\gamma_A = 1.6632$.

$$P_A = \frac{\pi y_A}{x_A \gamma_A} = \frac{(760)(0.2887)}{(0.192)(1.6632)} = 687 \text{ mm Hg}$$

with a corresponding temperature of 74.2° C.

The calculations are then repeated at 74.2° C to obtain the following final values:

	Experimental Values
$y_A = 0.288$	0.287
$y_B = 0.090$	0.092
$y_C = 0.622$	0.621
$t = 74.1°$ C	74.0° C

Using small-letter constants throughout gives a bubble-point temperature of 74.4° C and a vapor composition that is the same as above.

Separation of an Azeotrope by a Third Component. The separation of a binary azeotrope can in some cases be accomplished by a shift in the pressure of distillation as previously shown (page 916) and in some cases by the addition of a third component.

If a solution of components B and C when subject to fractional distillation at constant pressure produces an azeotropic mixture as a limiting overhead product, then the limiting bottom product will consist of pure component B or C, depending on which component is present in excess of the composition requirements of the azeotrope. For the azeotrope, the following conditions prevail:

$$x_B = y_B, \qquad x_C = y_C, \qquad \frac{x_B}{x_B + x_C} = \frac{y_C}{y_B + y_C} \tag{83}$$

and, from equation 18 for the two components,

$$\frac{\gamma_B}{\gamma_C} = \frac{P_C}{P_B} \tag{84}$$

If the addition of a third component A to a BC azeotrope depresses the ratio γ_B/γ_C the azeotropic ratios as defined by equation 83 will diminish. In fractional distillation of such a ternary mixture, the limiting overhead product will consist, in place of the azeotrope BC, of either pure component A or the azeotrope CA. Where the azeotrope CA is the limiting overhead product, the limiting bottom product will consist of pure component B.

When component C is added to the liquid BC azeotrope, the azeotropic ratios of equation 83 decrease along the 45° line as shown in Fig. 246 for the ternary system ethyl acetate A–benzene B–cyclohexane C. For the binary system of benzene and cyclohexane, the azeotrope contains 0.537 mole fraction of benzene ($x_B = 0.537$ and $x_C = 0.463$) and has a normal boiling point of 77.6° C. For the ternary system containing 0.15 mole fraction of ethyl acetate, the azeotropic condition of equation 83 is

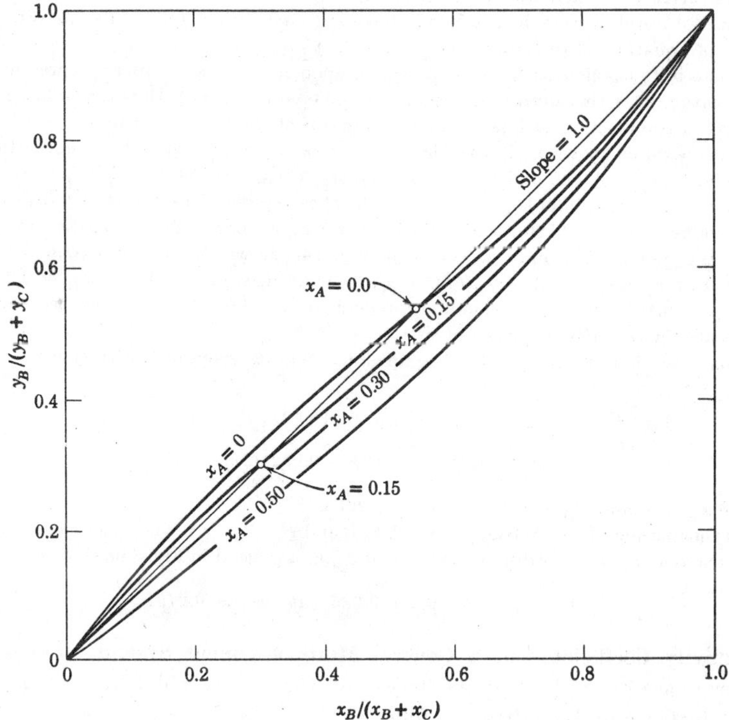

FIG. 246. Effect of ethyl acetate on the equilibrium ratios of benzene to benzene–cyclohexane in the vapor and liquid phases

satisfied when the two ratios equal 0.30, corresponding to $x_A = 0.15$, $x_B = 0.255$, and $x_C = 0.595$. The azeotropic conditions of equation 83 can no longer be satisfied when x exceeds a value of 0.275 mole fraction corresponding to $x_B = 0$ and $x_C = 0.725$ and where $\gamma_B/\gamma_C = P_C/P_B = 0.985$.

In fractional distillation of the binary system benzene–cyclohexane, the limiting overhead product is the corresponding azeotrope. When ethyl acetate is added and subjected to fractional distillation, the limiting overhead product is an ethyl acetate–cyclohexane azeotrope with ethyl acetate or benzene as the limiting bottom product, depending on which one is present in sufficient amount to exceed the composition requirements of the azeotrope. The advantage of forming the ethyl acetate–cyclohexane azeotrope is that it can be separated by the addition of water, whereas this is not possible with the benzene–cyclohexane azeotrope.

Illustration 6. Calculate the lowest concentration of ethyl acetate in a ternary mixture of ethyl acetate–benzene–cyclohexane that will just reduce the azeotropic ratios of equation 83 to zero, $x_B/(x_B + x_C) = y_B/(y_B + y_C) = 0$.

The activity coefficients for the separate components of this system are represented by equation 82 with values of the significant constants given in Illustration 5. For any given composition of liquid, the composition of the vapor and its temperature are calculated as shown in Illustration 5. For example, for a liquid of composition $x_A = 0.150$, $x_B = 0.050$, and $x_C = 0.800$; $x_B/(x_B + x_C) = 0.059$, and $y_B/(y_B + y_C) = 0.066$ with a normal bubble point of 74.8° C. These calculations are repeated at the same value of $x_A = 0.150$ for a series of values x_B covering the composition range. From the plot of such data as shown in Fig. 246, it will be seen that equality is reached at $x_B/(x_B + x_C) = y_B/(y_B + y_C) = 0.30$, where $x_A = 0.150$, $x_B = 0.255$, and $x_C = 0.595$. The corresponding temperature is 75.0° C. The same procedure is repeated with different values of x_A.

From equation 82 and values of significant constants given in Illustration 5,

$$\log \frac{\gamma_B}{\gamma_C} = -0.1443(x_B - x_C) + 0.0100[2x_B x_C - (x_B - x_C)^2] + x_A[0.0372 - 0.3849 - 0.0321(2x_C - x_A)]$$

where $x_B = 0$ and $x_A = 0.1$; $\gamma_B/\gamma_C = 1.2068$.

By the method of Illustration 5, $t = 73.8°$ C and $P_C/P_B = 0.985$. This calculation is repeated at $x_B = 0$ for different values of x_A and plotted in Fig. 246, showing that

$$P_C/P_B = \gamma_B/\gamma_C = 0.985 \quad \text{at} \quad x_A = 0.275$$

Activity Coefficients in a General Multicomponent System. For the general case of an N-component system, the excess molal free energy of the solution may be written as follows:

$$\frac{G^E_{1-N}}{RT} = \frac{1}{2}\sum_{i=1}^{N}\sum_{j=1}^{N}\frac{G^E_{ij}}{RT} \tag{85}$$

where $i \neq j$.

This summation considers the effect of all binary pairs. Upon expansion for a ternary system, equation 85 becomes

$$G^E_{123} = G^E_{12} + G^E_{13} + G^E_{23} \tag{86}$$

where $G^E_{ij} = G^E_{ji}$.

In terms of Redlich–Kister constants for an N-component system,

$$\frac{G^E_{1-N}}{RT} = \frac{1}{2}\sum_{i=1}^{N}\sum_{i=j}^{N} x_{ij}[B_{ij} + C_{ij}(x_i - x_j) + D_{ij}(x_i - x_j)^2 + \cdots] \tag{87}$$

where $i \neq j$ and where ternary and higher constants are neglected. For example, for a ternary system, equation 87 is expanded to give

$$\frac{\text{G}_{123}^{E}}{RT} = x_1x_2[B_{12} + C_{12}(x_1 - x_2) + D_{12}(x_1 - x_2)^2 + \cdots]$$
$$+ x_1x_3[B_{13} + C_{13}(x_1 - x_3) + D_{13}(x_1 - x_3)^2 + \cdots]$$
$$+ x_2x_3[B_{23} + C_{23}(x_2 - x_3) + D_{23}(x_2 - x_3)^2 + \cdots] \quad (88)$$

where $B_{ij} = B_{ji}$, $D_{ij} = D_{ji}$, but $C_{ij} = -C_{ji}$

Individual activity coefficients are obtained by differentiation of equation 87 according to definition; thus, for component i,

$$\ln \gamma_i = \left(\frac{\partial n_t \text{G}_{1-N}^{E}}{\partial n_i}\right)_{pT} \quad (89)$$

In this development, ternary and higher constants are omitted as well as the small effect due to the term $(\Delta\text{H}/RT^2)(\partial T/\partial x)_p$.

Expansion of Vapor Mixtures with Resultant Partial Condensation

The work and heat effects in the expansion and compression of vapor mixtures are calculated in the same manner as for pure gases, except that, in partial condensation, the compositions of the condensate and the residual vapor differ from that of the original vapor mixture and from each other. Three cases of expansion under flow conditions will be considered: namely, reversible isothermal, reversible isentropic, and free expansion in each case, from pressure p_1 to pressure p_2, neglecting changes in external kinetic and potential energies.

For *isothermal expansion*, reversible flow conditions,

$$w = -\Delta G = -\Delta H + T\,\Delta S \quad (90)$$

Where partial condensation occurs

$$w = -(1 - m)(\text{H}_{sG} - \text{H}_1) + m(\text{H}_{sL} - \text{H}_1) - (1 - m)T(\text{S}_{sG} - \text{S}_1) - mT(\text{S}_{sL} - \text{S}_1) \quad (91)$$

where the subscript s refers to saturated conditions at the final pressure p_2, and

m = mole fraction of vapor condensed on the basis of one mole of original gas mixture

H_1 = molal enthalpy of gas mixture at initial pressure p_1

S_1 = molal entropy of gas mixture at initial pressure p_2

Since the initial and final temperatures and pressures are known, the compositions of the products will also be fixed but unknown. The value of m must be such as to satisfy the material balance of all components as well as the final vapor–liquid equilibrium.

For component i per mole of original mixture of composition N_i,

$$N_i = mx_i + (1 - m)y_i \tag{92}$$

According to equilibrium conditions y_i is a known function of x_i.

A negative value for m indicates that no condensation takes place and that the vapor composition remains unchanged.

For isentropic expansion, reversible flow conditions, $w = -\Delta H$.

On the basis of one mole of vapor mixture with partial condensation,

$$w = -(1 - m)(\mathrm{H}_{sG} - \mathrm{H}_1) - m(\mathrm{H}_{sL} - \mathrm{H}_1) \tag{93}$$

$$(1 - m)(\mathrm{S}_{sG}) + m\mathrm{S}_{sL} - \mathrm{S}_1 = 0 \tag{94}$$

The final temperature T_2 is obtained by equating the entropy change to zero. This requires a trial-and-error procedure, starting with an assumed value of T_2.

For *free-expansion* flow conditions, the change of enthalpy is zero, and no work is done. On the basis of one mole of vapor mixture with partial condensation

$$w = -(1 - m)(\mathrm{H}_{sG} - \mathrm{H}_1) - m(\mathrm{H}_{sL} - \mathrm{H}_1) = 0 \tag{95}$$

The final temperature T_2 is obtained by a trial-and-error procedure, starting with an assumed value.

In all three cases, a negative value of m indicates that no condensation occurs, and the vapor composition remains unchanged, in which case the problems are solved exactly as for pure gases (Chapter 15).

Where a range in values of m satisfies these equations, a value must be selected that also satisfies the material balance and equilibrium requirements of each component.

Illustration 7. A saturated alcohol–water vapor mixture containing 47% alcohol by weight at 1 atm pressure and 200° F is expanded isentropically under flow conditions to a pressure of 0.1 atm. Calculate:

(*a*) The work done in Btu per pound-mole of original gas mixture.

(*b*) The final temperature.

(*c*) The fraction of original vapor condensed.

(*d*) The compositions of the final vapor and condensate.

This problem is solved through the use of the enthalpy and entropy–composition diagrams, pages 327 and 887.

Under initial conditions,

$$H_1 = 830 \text{ Btu per lb mixture}$$

$$S_1 = 1.33 \text{ Btu/(lb mixture)(R}^\circ\text{)}$$

Using equations 93 and 94 *on a weight basis* in pounds and composition in per cent by weight,

$$w = -(1 - m)(H_{sG} - H_1) - m(H_{sL} - H_1) \qquad (a)$$

$$(1 - m)S_{sG} + mS_{sL} = S_1$$

Assume that $t_2 = 105°$ F.

$$H_{sG} = 775, \qquad S_{sG} = 1.43$$

$$H_{sL} = 59, \qquad S_{sL} = 0.14$$

Substitution in equation (*a*) gives $m = 0.078$ lb, $w = 111$ Btu per lb
From vapor–liquid equilibria, $x = 0.075$ and $y = 0.485$. From a material balance,

	Liquid Phase	Vapor Phase	Total
Alcohol	0.006	0.447	0.453
Water	0.072	0.475	0.547
	0.078	0.922	1.000

The composition 0.453 is below the original value of 0.47.
Upon repetition at 104° F, the original composition of alcohol is 0.499.
Upon interpolation the following values are obtained:

(*a*) $w = 125$ Btu
(*b*) $t = 104.5°$ F
(*c*) $m = 0.063$ lb vapor condensed
(*d*) $x = 0.082$ weight fraction of alcohol in condensate
$y = 0.502$ weight fraction of alcohol in residual vapor

The condensate contains 8.2% of alcohol as compared with 47% in the original vapor and 50.2% in the final vapor. Under isothermal or free expansion at the given conditions, no condensation occurs.

Problems

1. The system toluene–acetic acid forms an azeotrope containing 62.7 mole per cent toluene and having a minimum boiling point of 105.4° C at 760 mm Hg. The vapor-pressure data from the *International Critical Tables* are given in Table A.

TABLE A

t, °C	Toluene	Acetic Acid
70	202.4	136.0
80	289.7	202.3
90	404.6	293.7
100	557.2	417.1
110	—	580.8
120	—	794.0
Normal boiling point, °C	110.7	118.5

Calculate the van Laar constants A and B for this system, and plot γ_1 and γ_2 as ordinates on a logarithmic scale against the mole fraction of toluene (component 1) on a uniform scale, neglecting the association of acetic acid in the vapor phase.

2. The data in Table B on the vapor pressures in millimeters of Hg of carbon tetrachloride–ethyl alcohol solutions are from the *International Critical Tables*:

Table B

Weight % CCl_4	34.8° C	66° C	Weight % CCl_4	34.8° C	66° C
0	103	462	58.25	206	752
10.23	122	520	71.68	220	782
20.02	142	576	76.69	223	789
27.13	156	614	84.25	226	788
39.06	179	677	92.98	225	780
43.87	187	700	97.43	221	741
48.86	193	716	*98.6*	*213*	*717*
			99.9	*187*	*593*
			100.0	173	544

Numbers in italics are graphical interpolations.

Evaluate the van Laar constants for the system at 34.8° C and 66° C.

3. From the results of problem 2, calculate vapor compositions corresponding to liquid compositions of 10, 20, 30, 40, 50, 60, 70, 80, and 90 mole per cent CCl_4 at pressures of 760 and 200 mm Hg. Plot these results as y–x diagrams. Assume that the activity coefficients are independent of temperature over the limited ranges involved. The vapor pressures of the pure component are given in Table C.

Table C. Vapor Pressures, mm Hg

t, °C	CCl_4	C_2H_5OH	t, °C	CCl_4	C_2H_5OH
20	91	43.9	55	379.3	280.6
25	114.5	59.0	60	450.8	352.7
30	143.0	78.8	65	530.9	448.8
35	176.2	103.7	70	622.3	542.5
40	215.8	135.3	75	—	666.1
45	262.5	174.0	80	843	812.6
50	317.1	222.2	Normal boiling point, °C	76.75	78.32

4. The composition of the azeotrope formed by carbon tetrachloride and ethyl alcohol at 760 mm Hg is given in the *International Critical Tables* as 61.3 mole per cent CCl_4 and the boiling point as 64.95° C. Using the vapor-pressure data of problem 3, calculate the composition of the azeotrope at a temperature of 34.8° C and its total vapor pressure, using equations 73 and 74. Evaluate the van Laar constants for the system at 34.8° C, and compare these results with those of problem 2.

5. Methylcyclohexane and n-heptane are each partially miscible with aniline at 25° C. The equilibrium compositions in mole per cent are as follows:[15]

	Methylcyclohexane–Aniline	Heptane–Aniline
Hydrocarbon layer	11.38% aniline	7.55% aniline
Aniline layer	82.35% aniline	93.91% aniline

[15] K. A. Varteressian and M. R. Fenske, *Ind. Eng. Chem.*, **29**, 270 (1937).

Calculate the van Laar constants for the methylcyclohexane–aniline system and for the *n*-heptane–aniline system at 25° C. Plot curves relating the activity coefficients of each of the respective hydrocarbons to its mole fraction in aniline.

6. (*a*) Using the results of problem 5, calculate the activity coefficients of methylcyclohexane and *n*-heptane in a solution of the following composition at 25° C, using the indicated liquid densities of the pure components:

	Mole %	Liquid Density (Pure), g per cc
n-Heptane	3.6	0.68
Methylcyclohexane	5.4	0.76
Aniline	91.0	1.02

It may be assumed that the *n*-heptane and methylcyclohexane form ideal binary solutions with each other.

(*b*) The vapor pressures at 25° C of *n*-heptane and methylcyclohexane may be taken as 46 and 37 mm Hg, respectively. Neglecting the vapor pressure of aniline, calculate the composition of the vapors and the total vapor pressure in equilibrium with the solution of part (*a*). Compare the effective relative volatilities ($\alpha = P_1\gamma_1/P_2\gamma_2$) of the heptane and methylcyclohexane in the presence and absence of the aniline.

7. In the system methanol–heptane Benedict et al.[16] give the mole percentage of methanol in the liquid and vapor phases as shown in Table D. At 58.8° C the azeotrope has a mole fraction of 74.7% ethanol.

Table D

°C	Liquid	Vapor
98.43	0	0
71.4	—	60.2
60.6	13.8	72.0
59.47	17.8	73.3
59.13	24.8	—
58.93	39.0	73.9
58.82	66.8	74.6
58.81	81.0	74.8
59.01	88.5	76.5
59.90	94.6	80.9
64.51	100.0	100.0

The activity coefficient for ethanol is given by the equation

$$RT \ln \gamma_1 = (12x_1^2x_2 - 12x_1^2x_2)A_1 + 12(x_1x_2^2 - 18x_1x_2^2)A_2 + (4x_2^3 - 12x_1x_2^3)A_3$$

where x_1 = mole fraction

$A_1 = 480.804$
$A_2 = 411.270$
$A_3 = 494.977$

Develop a similar expression for the activity coefficient of heptane.

[16] M. Benedict, C. A. Johnson, E. Solomon, and L. C. Rubin, *Trans. Am. Inst. Chem. Engrs.*, **41**, 371 (1945).

8. For the system benzene–cyclohexane at 760 mm Hg pressure, Chao[11] obtained the vapor–liquid equilibria data given in Table E, where component 1 is benzene and 2 is cyclohexane.

Table E

x_1	y_1	t °C	γ_1	γ_2	$\log\left(\frac{\gamma_1}{\gamma_2}\right)$
0.088	0.113	79.7	1.300	1.003	0.11264
0.156	0.190	79.1	1.256	1.008	0.09553
0.231	0.268	78.5	1.219	1.019	0.07783
0.308	0.343	78.0	1.189	1.032	0.06150
0.400	0.422	77.7	1.136	1.056	0.03172
0.470	0.482	77.6	1.108	1.075	0.01313
0.545	0.544	77.6	1.079	1.102	−0.00916
0.625	0.612	77.6	1.058	1.138	−0.03165
0.701	0.678	77.8	1.039	1.178	−0.05453
0.757	0.727	78.0	1.025	1.221	−0.07600
0.822	0.791	78.3	1.018	1.263	−0.09365
0.891	0.863	78.9	1.005	1.328	−0.12103
0.953	0.938	79.5	1.003	1.369	−0.13510

From these data, calculate the constants of the van Laar, Margules, and Redlich–Kister equations. Plot the activity coefficients calculated by each, and compare with experimental values.

22

Vapor–Liquid Equilibria at High Pressures

In the vaporization of mixtures at high pressures, unusual difficulties are experienced in calculating vapor–liquid equilibria at temperatures exceeding the normal critical temperature of any component present and where nonideal solution behavior occurs. The situation is further complicated in petroleum systems by the presence of a multiplicity of components which are usually not individually identified except in terms of boiling-point ranges.

Critical Phenomena of Mixtures. The behavior of a mixture including the region of the critical point is explained by reference to the pressure–temperature diagram of Fig. 247.

Curve AC' represents the vapor pressure of a pure compound having the critical temperature and pressure corresponding to point C'. For a single-component system, the area to the left of this curve represents the region of the liquid phase, and the area below it represents the region of the vapor phase.

Curve $BDECFGH$ is termed the *border line* of a mixture having the same average volatility as that of the pure compound represented by line AC', but made up of two substances having vapor pressures corresponding to lines B_1C_1 and B_2C_2. The area enclosed by the border-line curve represents a two-phase region in which both liquid and vapor are present in equilibrium. Line $BDEC$ is termed the *bubble-point line*. The area above line $BDEC$ is the region of complete liquefaction. Line $HGFC$ is termed the *dew-point line*. The area to the right of line $HGFC$ is the region of complete vaporization.

Point C represents the true critical point of the mixture. This critical point does not correspond to a maximum temperature at which the liquid phase can exist, as is the case with a pure component, but rather to the particular point on the border-line curve where the vapor and liquid phases become indistinguishable, where the bubble-point, dew-point, and quality lines meet. In general, both the critical temperature and the critical pressure of a mixture are higher than those of a pure compound having the same average volatility.

The dew-point line passes through a maximum temperature at F.

Thus, in the case of a mixture, liquid may exist at a temperature higher than the critical temperature. This maximum temperature F on the border-line curve is termed the *critical-condensation temperature*. Similarly, in the case of many mixtures the bubble-point line passes through a point of maximum pressure E, higher than the critical pressure. The areas $FJCM$ and $EKCL$ represent regions of *retrograde condensation*. If the mixture at the condition of point 1 is compressed at constant

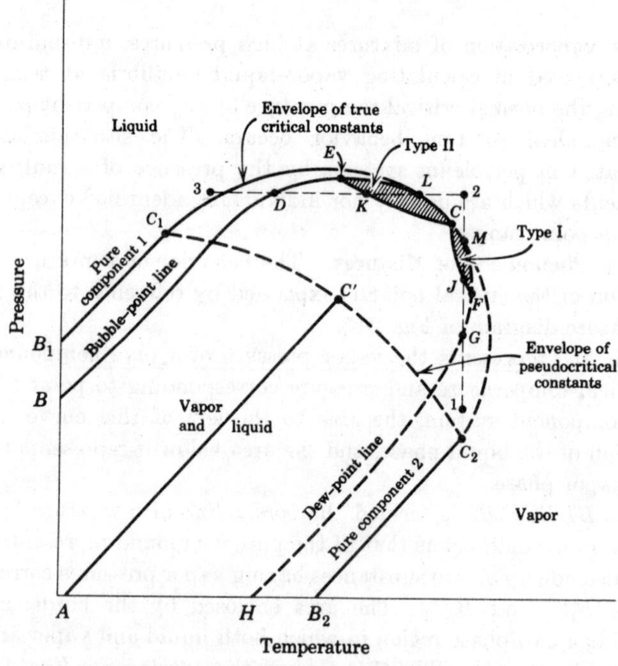

Fig. 247. Critical phenomena of mixtures

temperature, a denser phase appears at point G on the dew-point line. As the pressure is further increased, the quantity of this denser phase increases to a maximum at point J and then diminishes, disappearing entirely when point M is reached. This type of retrograde condensation occurring in area $FJCM$ is called the *first type*. If the liquid mixture at conditions of point 3 is heated at constant pressure, a less dense phase appears at point D on the bubble-point line, reaches a maximum at point K, and then diminishes, disappearing at point L. This type of retrograde condensation that occurs in area $EKCL$ is called the *second type*. Lines EKC and FJC are, respectively, defined by horizontal and vertical tangents to the quality curves in the two-phase region.

The area outside the border curve is a region of homogeneous fluid in which no phase separations occur. Moving clockwise in the region about the border-line curve, the liquid phase merges imperceptibly into the vapor phase.

Retrograde condensation can also occur by change of concentration alone at constant pressure and temperature. For example, at pressure p_2 (Fig. 248), the temperature–composition relationship for a binary solution

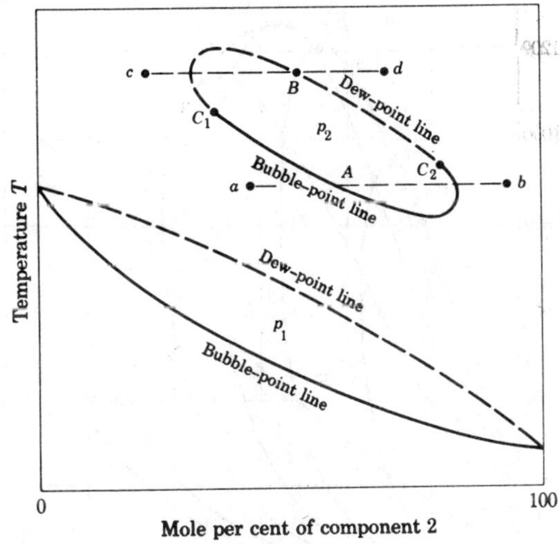

FIG. 248. Retrograde condensation with change in concentration

is given by curve C_1BC_2A where C_1BC_2 is the dew-point line and C_1AC_2 is the bubble-point line. It will be noted that, at pressure p_2, two critical points appear: namely, at C_1 and C_2, corresponding to different temperatures and compositions. By increasing the concentration of component 2 at constant temperature and pressure, a constant-temperature line from point a to point b twice crosses the bubble point line, indicating that the original liquid phase partly disappears and then re-forms; the latter corresponds to retrograde condensation of the second type. Similarly, by increasing the concentration of component 2 from point c to point d at constant pressure and temperature, the straight line twice crosses the dew-point line. The vapor phase begins to condense as concentration is increased and then re-forms. The initial condensation of vapor as the volatile component is added corresponds to retrograde condensation of the first type. These phenomena would not occur at the lower pressure p_1 which is below the critical pressure and temperature of each component.

Retrograde condensation is of particular interest in petroleum production. When natural gas is withdrawn from high-pressure wells, liquid gasoline condenses upon release of pressure. The residual gas may then be recompressed and recycled to the underground oil-bearing formation, where it again reaches equilibrium with the liquid. As a result, low-boiling fractions of liquid enter the gas phase, and are subsequently

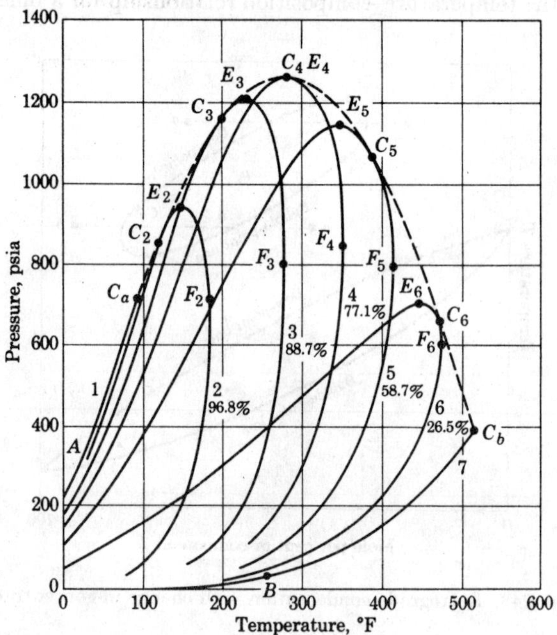

Fig. 249. Pressure–temperature diagrams of mixtures of ethane and n-heptane at various compositions

recovered by the retrograde condensation accompanying the expansion step. A discussion of this operation is given by Katz and Kurata.[1]

It may be noted from Fig. 247 that retrograde condensation of the first type occurs when a constant-temperature line twice crosses the dew-point line (broken line), and retrograde condensation of the second type occurs when a constant-pressure line twice crosses the bubble-point line (solid line). Both types of retrograde condensation occur in a given mixture when the critical point C lies on the border-line curve between the maximum pressure E and the maximum temperature F. When the border curves of a binary system are obtained for the entire range of

[1] D. L. Katz and F. Kurata, *Ind. Eng. Chem.*, **32**, 817 (1940).

compositions, it is found that the critical temperatures occur at different points on the border-line curves relative to E and F.

These effects are illustrated in Fig. 249 showing the border-line curves of various mixtures of ethane and heptane as determined by Kay.[2] Line AC_a is the vapor-pressure curve of pure ethane, and point C_a is its critical point. Line BC_b has the same significance for pure heptane. The lines 2–6 are the border-line curves for five different mixtures containing the

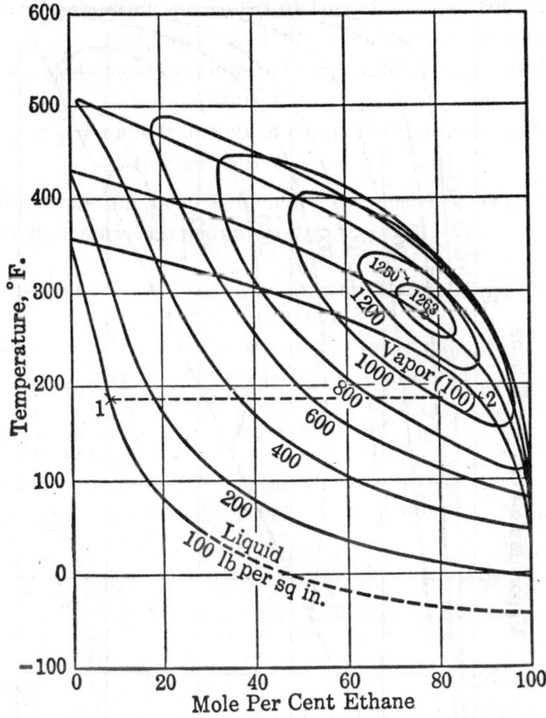

FIG. 250. Temperature–composition diagrams for mixtures of ethane and n-heptane at various pressures

indicated mole percentages of ethane. The corresponding critical points of the mixtures are designated as C_2–C_6, respectively, the corresponding maximum pressures as E_2–E_6, and the maximum temperatures as F_2–F_6. It may be noted that, for compositions 2 and 3, only retrograde condensation of the first type can occur, whereas, for compositions 5 and 6, both types can occur. Line $C_aC_2 \cdots C_6C_b$ represents the envelope of all critical points for all possible mixtures of ethane and heptane.

[2] W. B. Kay, *Ind. Eng. Chem.*, **30**, 459 (1938), reprinted with permission.

If the envelope curve were of such shape that it approached point C_b with a positive slope, it would indicate the existence of mixtures having critical temperatures higher than that of the higher-boiling component. These mixtures would have critical points lying in a clockwise direction beyond the corresponding points *E* and *F*, both of which would be on the bubble-point line. Such mixtures could exhibit only retrograde

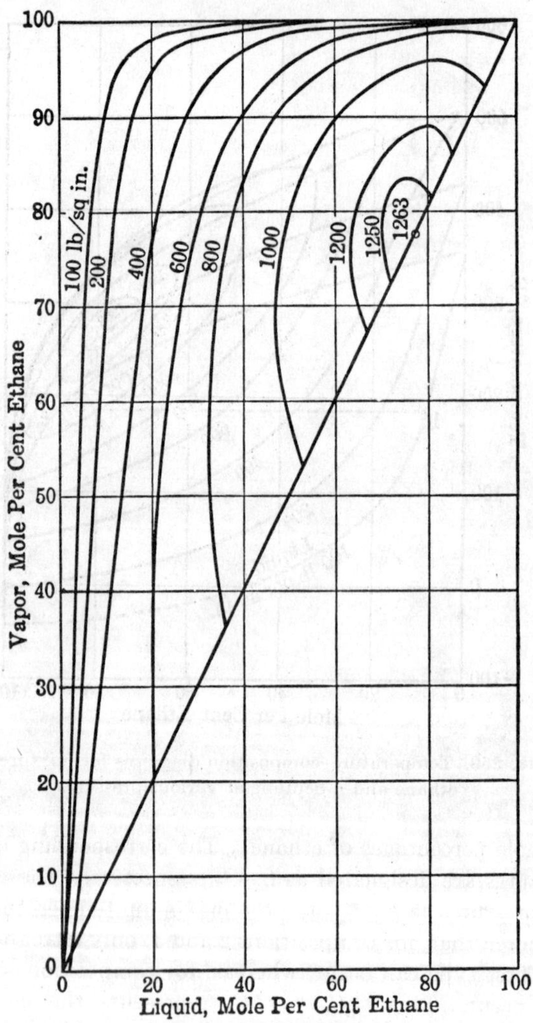

Fig. 251. Vapor–liquid mole fraction relations for ethane–*n*-heptane at various pressures

condensation of the second type. This situation is possible but unusual.

Isobaric boiling-point curves and liquid–vapor composition curves of the ethane–heptane system[2] at different pressures are plotted in Figs. 250 and 251.

Each envelope on Fig. 250 corresponds to the indicated constant pressure. The upper curve relates temperature to vapor composition and the lower to liquid composition. Thus, at a pressure of 100 psia and a temperature of 190° F, liquid of the composition corresponding to point 1 is in equilibrium with vapor of the composition of point 2. It may be noted that, as the pressure is increased, the range of the two-phase region is reduced. This behavior is of importance in the design of distillation equipment. At pressures higher than 396 psia, pure heptane cannot exist in the liquid phase in contact with ethane gas, and this represents the highest pressure at which pure heptane can be separated by distillation. At pressures between 396 and 712 psia, distillation can theoretically separate mixtures into pure ethane and liquid solutions of ethane in heptane. At pressures above 712 psia, pure ethane cannot be separated, and distillation can produce separation only into two mixtures, one rich and one poor in ethane. The maximum pressure at which a two-phase system can exist is 1263 psia, the maximum of the envelope curve of Fig. 249. This is the maximum pressure at which any type of distillation of the system can be conducted.

Vaporization Equilibrium Constants. Where the standard state of a pure component is taken as the temperature and pressure of the system, the free-energy change in going from the standard vapor state to the standard liquid state is given, as discussed in Chapter 20, as

$$-\Delta G^\circ_i = RT \ln \frac{a_{iv}}{a_{iL}} = RT \ln \frac{f^\circ_{iL}}{f^\circ_{iv}} = RT \ln K_i \tag{1}$$

where

$f^\circ_{iv}, f^\circ_{iL}$ = fugacities of component i in the vapor and liquid standard states

K_i = vaporization equilibrium constant of component i

From equations 21–15 through 17,

$$K_i = \frac{a_{iv}}{a_{iL}} = \frac{f^\circ_{iL}}{f^\circ_{iv}} = \frac{\gamma_{iv} y_i}{\gamma_{iL} x_i} \tag{2}$$

where γ_{iv}, γ_{iL} = activity coefficients of component i in the vapor and liquid phases.

The use of equilibrium constants for calculating vapor–liquid equilibria at elevated pressures was introduced by Brown[3] and by Lewis[4] and their co-workers.

For *ideal solutions* activity coefficients are unity and the equilibrium constant reduces to the ratio of mole fractions in the vapor and liquid phases

$$K_i = \frac{y_i}{x_i} \tag{3)*}$$

In terms of fugacity coefficients for either gas or liquid,

$$f^\circ{}_i = \nu_{i\pi}\pi \tag{4}$$

where π = total pressure.

Since, for liquids, density is nearly independent of pressure, the standard liquid-state fugacity can be obtained combining equations 14–59 and 60 and integrating to give

$$\left(\ln \frac{f^\circ{}_{iL}}{f_{is}}\right)_T = \left[\frac{v_{im}(\pi - p_s)}{RT}\right]_T \tag{5}$$

where f_{is} = fugacity of the pure liquid at saturation

v_{im} = mean molal volume of pure component i in the liquid state between pressure p_s and π, or

$$f^\circ{}_{iL} = p_{is}\nu_{is} \exp\left[\frac{v_{im}(\pi - p_{is})}{RT}\right] \tag{6}$$

In the use of equations 4, 5, and 6, the following hypothetical conditions are encountered which require arbitrary extrapolations in obtaining $f^\circ{}_i$ and $\nu^\circ{}_i$.

1. The liquid state is hypothetical at all temperatures above saturation: $T > T_s$.
2. The liquid state is hypothetical at all values of temperature greater than the critical temperature: $T > T_c$, $T_r > 1.0$.
3. The standard vapor state is hypothetical where the total pressure exceeds the vapor pressure: $\pi > p_s$.

The evaluation of vaporization equilibrium constants requires extrapolation procedures which in effect constitute arbitrary definitions of the thermodynamic properties of the vapors and liquids in the regions of the hypothetical states. Such arbitrary definitions are permissible, provided the corresponding values of the activity coefficients are used in calculating y/x relationships.

[3] M. Souders Jr., C. W. Selheimer, and G. G. Brown, *Ind. Eng. Chem.*, **24**, 517 (1932).
[4] W. K. Lewis and C. D. Luke, *Trans. Am. Soc. Mech. Engrs.*, **54**, 17 (1932).

It is desirable that logical and regular extrapolation procedures be employed so as to give smooth, continuous relationships of fugacities in the hypothetical standard state over the entire temperature and pressure ranges, and to produce activity coefficients equal to unity for ideal-solution behavior.

Early evaluations of vaporization equilibrium constants were based on experimentally observed values of y/x for hydrocarbon systems. The activity coefficients were assumed to be unity, and average values of K_i were arrived at empirically in the regions where K_i departs widely from 1.0. On this basis, extensive charts[5,6] were developed for the paraffin and olefin hydrocarbons.

In order to extend the vaporization equilibrium concept to systems other than hydrocarbons, Gamson and Watson[7] proposed generalized definitions of the properties of the standard states. These definitions were developed to produce substantial agreement with the empirical correlations of ideal vaporization equilibrium constants of hydrocarbons in the regions where they are supported by experimental data. In addition, the definitions permitted continuous extrapolations into the high-temperature and high-pressure regions to serve as a basis for correlation of activity coefficients where ideal-solution behavior cannot be assumed. These definitions are as follows:

1. Vapor pressures are extrapolated above the critical by the generalized reduced vapor–pressure equations 14–49.

2. The standard state of the liquid is taken as a hypothetical incompressible liquid having the vapor pressure of the real liquid and a constant thermal expansion coefficient equal to that of the real liquid extrapolated to 0° K.

3. The vapor-phase fugacity coefficient correlation was graphically extrapolated to high pressures at temperatures below the critical. This extrapolation was later modified[8] in the very high-pressure region to produce more consistent activity coefficients.

On the basis of these definitions, Smith and Smith[9] developed K charts for hydrocarbons covering wide ranges of temperature, pressure, and boiling points.

Introduction of the z_c parameter into the generalized thermodynamic

[5] J. H. Perry, *Chemical Engineers' Handbook*, 3d ed., p. 568, McGraw-Hill Book Co. (1950).

[6] Natural Gasoline Supply Men's Association, *Engineering Data Book*, Tulsa (1957).

[7] B. W. Gamson and K. M. Watson, *Nat. Petrol. News, Tech. Sec.*, **36**, R623 (Sept. 6, 1944).

[8] K. A. Smith and K. M. Watson, *Chem. Eng. Prog.*, **45**, 494 (1949).

[9] K. A. Smith and R. B. Smith, *Petroleum Processing*, **4** (Dec. 1949).

correlations[10] makes it possible to refine further the definitions of the ideal equilibrium constant and to express it as a general function of reduced temperature and pressure.

The molal volume of the hypothetical liquid standard state is represented by the relation

$$v_{im} = (0.25 + 0.132T_{ir})v_{ci} \tag{7}$$

where

$$\left(v_c = \frac{z_c R T_c}{p_c}\right)_i$$

Combining equations 2, 4, 6, and 7 gives

$$K_i = \frac{f^\circ_{iL}}{f^\circ_{iv}} = \frac{\nu^\circ_{iL}}{\nu^\circ_{iv}} = \frac{p_{irs}\nu_{is} \exp\left[z_c \dfrac{(0.25 + 0.132T_r)(p_r - p_{irs})}{T_r}\right]_i}{p_{ir}\nu_{i\pi}} \tag{8}$$

Thus, K_i is expressed as a unique function of z_{ci}, T_{ri} and p_{ri}.

Values of K_i have been calculated from equation 8 and tabulated in Table 63 and Fig. 252 as a function of p_r and T_r for a value of $z_c = 0.27$. For other values of z_c values of the ratio $K_{z_c}/K_{0.27}$ are nearly independent of pressure but highly dependent on temperature. The values of this ratio are represented by the relation

$$\frac{K_{z_c}}{K_{0.27}} = 10^{D(z_c - 0.27)} \tag{9}$$

Values of D as a function of temperature are given in Table 63.

The correlation is entirely different from that of Lydersen, Greenkorn, and Hougen,[10] which is in error.

Illustration 1. Determine the ideal vaporization equilibrium constant of ethyl chloride at 553° K and 26 atm. $T_c = 460.4°$ K, $p_c = 52$ atm, and $z_c = 0.274$.

$$T_r = 553/460.4 = 1.20$$
$$p_r = 26/52 = 0.50$$

From Table 63,

$$K_{0.27} = 3.08$$
$$D = -1.3$$

From equation 9,

$$K/K_{0.27} = 10^{(-1.3)(0.274-0.270)} = 0.988$$
$$K = (3.08)(0.988) = 3.04$$

For most purposes the activity coefficient ratio may be assumed to be unity for chemically similar materials in the range of Table 63.

[10] A. L. Lydersen, R. A. Greenkorn, and O. A. Hougen, *Univ. Wisconsin Eng. Exp. Sta. Rept.*, **4** (1955).

TABLE 63. IDEAL VAPORIZATION EQUILIBRIUM CONSTANTS, K FOR $z_c = 0.27$

T_r \ P_r	D	0.01	0.02	0.04	0.06	0.08	0.1	0.2	0.4	0.6	0.8	1.0	2.0
0.60	20.4	1.000	0.530	0.285	0.200	0.160	0.130	0.078	0.0520	0.0450	0.0450	0.0445	0.048
0.62	18.0	1.49	0.790	0.410	0.295	0.230	0.190	0.110	0.0743	0.0650	0.0640	0.0635	0.0665
0.64	15.8	2.20	1.12	0.620	0.425	0.330	0.270	0.150	0.102	0.0880	0.0875	0.0870	0.0890
0.66	13.8	3.13	1.59	0.850	0.583	0.467	0.370	0.206	0.138	0.118	0.114	0.113	0.116
0.68	11.9	4.30	2.15	1.10	0.770	0.600	0.490	0.275	0.180	0.152	0.146	0.145	0.149
0.70	10.3	5.65	2.85	1.43	1.00	0.790	0.640	0.360	0.230	0.193	0.184	0.180	0.186
0.72	8.8	7.25	3.60	1.87	1.29	0.980	0.840	0.455	0.281	0.239	0.227	0.220	0.230
0.74	7.6	9.20	4.55	2.40	1.62	1.21	1.02	0.560	0.341	0.289	0.274	0.264	0.274
0.76	6.5	11.4	5.70	2.98	2.00	1.48	1.26	0.690	0.408	0.350	0.321	0.313	0.322
0.78	5.5	14.0	7.00	3.59	2.41	1.80	1.49	0.835	0.480	0.414	0.373	0.364	0.370
0.80	4.7	16.7	8.50	4.25	2.84	2.15	1.70	1.00	0.600	0.480	0.435	0.420	0.420
0.85	3.0	25.0	12.8	6.43	4.17	3.34	2.70	1.43	0.820	0.660	0.600	0.581	0.570
0.90	1.7	35.0	17.9	8.90	5.93	4.80	3.70	1.98	1.15	0.860	0.785	0.720	0.680
0.95	0.7	47.0	23.4	11.8	7.85	6.30	4.80	2.56	1.48	1.07	0.965	0.883	0.831
1.00	0.0	62.0	30.0	15.0	10.1	7.90	6.07	3.20	1.82	1.31	1.16	1.00	0.870
1.10	−0.8	93.5	45.5	23.6	15.3	11.6	9.10	4.65	2.58	1.83	1.54	1.28	1.00
1.20	−1.3	134	57.0	33.0	21.8	16.4	13.3	6.90	3.65	2.50	2.00	1.67	1.15
1.30	−1.6	184	90.0	44.0	29.5	22.1	17.9	9.30	4.90	3.25	2.62	2.11	1.40
1.40	−1.7	242	1[illegible]8	57.0	38.2	28.8	23.4	11.9	6.30	4.13	3.38	2.80	1.67
1.50	−1.8	305	150	72.1	48.0	36.0	29.0	14.6	7.75	5.20	4.16	3.28	2.00
1.60	−1.95	380	183	89.0	59.0	44.7	34.8	17.4	9.35	6.50	5.00	4.00	2.35
1.70	−2.05	445	220	107	72.0	53.5	40.3	20.5	10.9	7.60	5.80	4.70	2.68
1.80	−2.20	517	254	126	84.0	62.5	46.3	23.9	12.4	8.62	6.58	5.36	3.03
1.90	−2.3	590	290	146	96.5	72.0	52.7	27.2	14.0	9.60	7.30	6.00	3.39
2.00	−2.4	670	330	167	110	82.0	60.0	31.0	15.6	10.6	8.05	6.60	3.74

For K at values of z_c other than 0.27,

$$\frac{K_{z_c}}{K_{0.27}} = 10^{D(z_c - 0.27)}$$

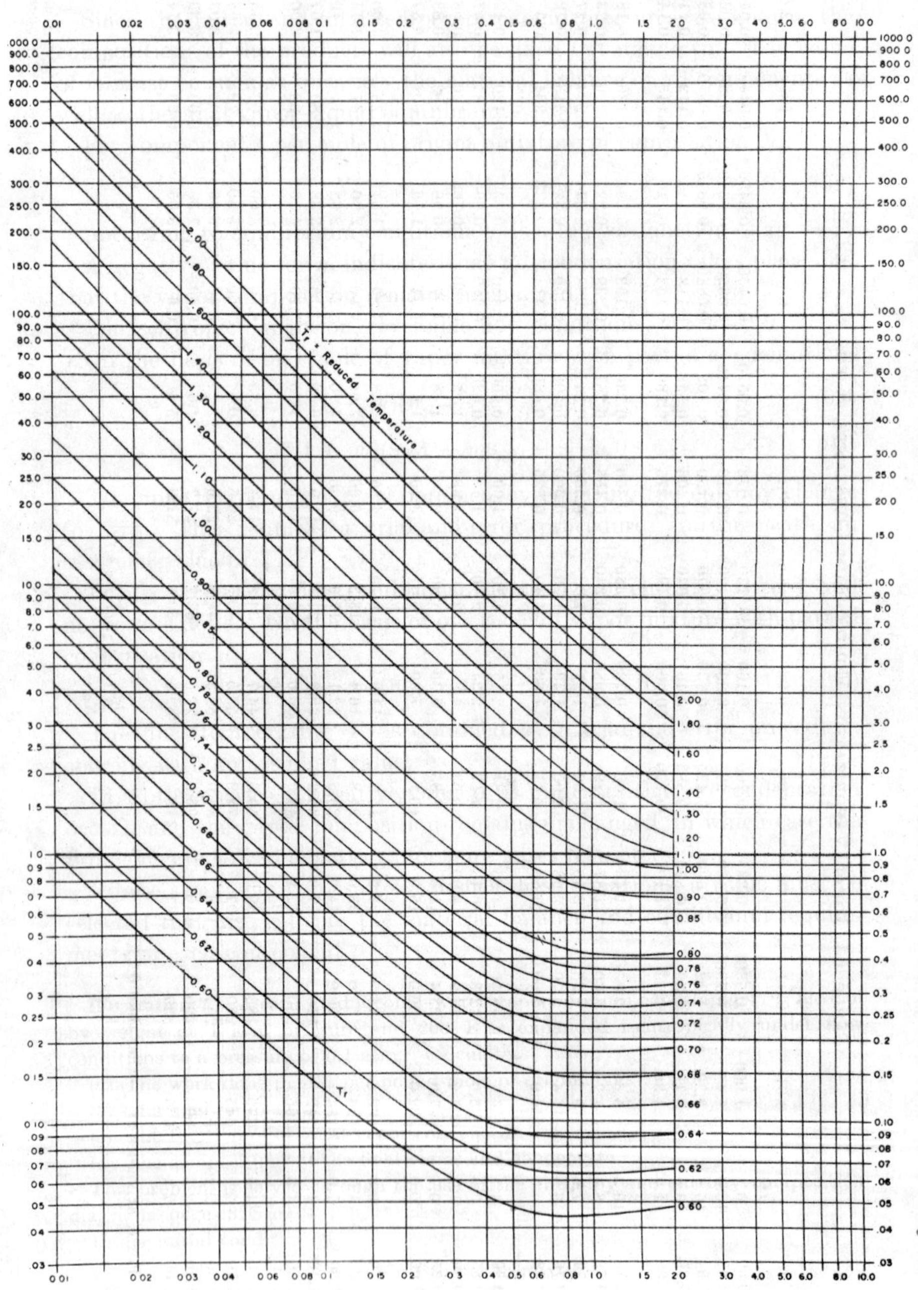

Fig. 25. aporization equilibrium constant $K(z_c = 0.27)$

Bubble-Point Equilibria. A liquid at its bubble-point temperature is in equilibrium with the first bubble of vapor that is formed upon heating the liquid at a given pressure.

Let N_i = mole fraction of component i in the original mixture

x_i = mole fraction of component i in the liquid phase

y_i = mole fraction of component i in the vapor phase

Under all conditions, $\sum N_i = 1.0$, $\sum x_i = 1.0$; $\sum y_i = 1.0$, and $y_i = K_i x_i$. At the bubble point, the composition of the liquid is the same as the entire system:

$$x_i = N_i$$

$$y_i = K_i x_i = K_i N_i \tag{10}$$

Hence, $$\sum K_i N_i = 1.0$$

If the total pressure is fixed, the bubble-point temperature is calculated by a graphical procedure, assuming various values of temperature until the value of $\sum K_i N_i = 1.0$. Since $x_i = N_i$, the composition of the vapor is then obtained from the ratio $y_i = K_i N_i$.

If the temperature is fixed, the bubble-point pressure is calculated by a similar graphical procedure, assuming various values of pressure until the value of $\sum K_i N_i = 1.0$.

Dew-Point Equilibria. A vapor is at its dew-point temperature when the first drop of liquid forms upon cooling the vapor at constant pressure and the composition of the vapor remaining is the same as the initial vapor mixture:

$$y_i = N_i = K_i x_i \quad \text{or} \quad x_i = N_i/K_i \tag{11}$$

$$\sum N_i/K_i = 1.0$$

The dew-point temperature or pressure is obtained by a similar graphical procedure to that used in calculating bubble-point equilibria.

Partial Vaporization. A solution comprising F moles is partially vaporized to form L moles of liquid and V moles of vapor. From an over-all material balance, following the method of Katz and Brown,[11]

$$V + L = F \tag{12}$$

Similarly, for component i,

$$Vy_i + Lx_i = FN_i \tag{13}$$

where N_i = mole fraction of component i in the F moles of total mixture. Combining equations 3 and 13 and solving for x_i, assuming equilibrium conditions,

$$x_i = \frac{FN_i}{VK_i + L} = \frac{F}{V}\left(\frac{N_i}{K_i + L/V}\right) \tag{14}$$

[11] D. L. Katz and G. G. Brown, *Ind. Eng. Chem.*, **25**, 1373 (1933).

Since $\sum x_i = 1$,

$$\sum \frac{N_i}{K_i + L/V} = \frac{V}{F} \tag{15}$$

Similarly, solving equation 13 for y_i gives

$$y_i = \frac{FN_i}{V + L/K_i} = \frac{F}{V}\left(\frac{N_i}{1 + L/VK_i}\right) \tag{16}$$

and

$$\sum \frac{N_i}{1 + L/(VK_i)} = \frac{V}{F} \tag{17}$$

Either equation 15 or 17 may be used in a graphical solution to establish the relationship among temperature, pressure, and percentage vaporization. Values of the quantity sought are assumed, and the corresponding summations are evaluated. The correct value is that producing a summation equal to V/F. The relationship between V/F and L/V is obtained by rearranging equation 12:

$$\frac{L}{V} = \frac{F}{V} - 1 \tag{18}$$

If equation 17 is used in the summation, values of vapor compositions are obtained directly:

$$y_i = \frac{N_i/(1 + L/VK_i)}{\sum[N_i/(1 + L/VK_i)]} \tag{19}$$

Equation 19 may be used with little error even if the summation does not exactly equal V/F. Corresponding values of x_i are then obtained from equation 13:

$$x_i = \frac{FN_i - Vy_i}{L} = \left(1 + \frac{V}{L}\right)N_i - \frac{V}{L}y_i \tag{20}$$

If equation 15 is used for the summation, values of x_i are obtained directly by an expression similar to equation 19. It is evident that, in a partially vaporized mixture at equilibrium, the vapor phase exists at its dew point while the liquid phase exists at its bubble point.

Illustration 2. A hydrocarbon mixture contains 25 mole per cent propane, 40% n-butane, and 35% n-pentane. Calculate the bubble-point temperature, the dew-point temperature, and the temperature of 45 mole per cent vaporization at an absolute pressure of 200 psia. Also, calculate compositions of the vapor formed at the bubble point, of the liquid formed at the dew point, and of the liquid and vapor resulting from 45 mole per cent vaporization. Ideal-solution behavior is assumed.

Table A

	T_c		p_c	z_c
Propane	369.9° K	665.8° R	42.01 atm	0.277
n-butane	425.2	765.4	37.47	0.274
n-pentane	469.8	845.6	33.31	0.260

Solution: (*a*) **Bubble-point temperature and vapor composition.** Values of K are obtained from Table 63 for substitution in equation 10. At various assumed temperatures at a pressure of 200 psia the values of K_i for each component are obtained from Table 63 at corresponding values of p_r, T_r, and z_c and presented in Table B.

TABLE B

Assumed Temperature				180° F (640° R)			200° F (660° R)		
	N_i	z_c	p_r	T_r	K_i	K_iN_i	T_r	K_i	K_iN_i
Propane	0.25	0.277	0.323	0.963	1.809	0.452	0.993	2.056	0.514
n-butane	0.40	0.274	0.362	0.837	0.856	0.342	0.862	0.925	0.390
n-pentane	0.35	0.269	0.408	0.756	0.389	0.136	0.780	0.520	0.182
$\sum K_iN_i$						0.930			1.086

By interpolation, the bubble-point temperature is found to be 189° F. At this temperature $\sum K_iN_i = 1.0$. The vapor composition y at the bubble-point is then obtained by re-evaluation of K_i at 189° F.

TABLE C

	N_i	z_c	T_r	p_r	K_i	$K_iN_i = y_i$
Propane	0.25	0.277	0.971	0.323	1.912	0.480
n-butane	0.40	0.274	0.848	0.362	0.925	0.370
n-pentane	0.35	0.269	0.767	0.408	0.430	0.150
$\sum K_iN_i$						1.000

(*b*) **Dew-point temperature and liquid composition.** At various assumed temperatures at a pressure of 200 psia the values of K_i for each component are obtained from Table 63 at the corresponding values of p_r, T_r, and z_c and presented in Table D.

TABLE D

Assumed Temperature				220° F (680° R)			240° F (700° R)		
Component	N_i	z_c	p_r	T_r	K_i	N_i/K_i	T_r	K_i	N_i/K_i
Propane	0.25	0.277	0.323	1.02	2.29	0.109	1.05	2.19	0.114
n-butane	0.40	0.274	0.362	0.890	1.17	0.342	0.916	1.37	0.292
n-pentane	0.35	0.269	0.408	0.805	0.610	0.505	0.829	0.714	0.490
$\sum N_i/K_i$						1.016			0.896

At $t = 223°$ F, $\sum N_i/K_i = 1.0$.

	N_i	K_i	$N_i/K_i = x_i$
Propane	0.25	2.29	0.109
n-butane	0.40	1.20	0.331
n-pentane	0.35	0.624	0.560
			1.000

(c) **Temperature and composition at 45 per cent vaporization.** $V/F = 0.45$

$$L/V = 1/0.45 - 1.0 = 1.22 \qquad V/L = 0.82$$

Table E

Assumed Temperature		200° F			220° F		
	N_i	K_i	$1 + L/VK_i$	$\frac{N_i}{1 + L/VK_i}$	K_i	$1 + L/VK_i$	$\frac{N_i}{1 + L/VK_i}$
Propane	0.25	2.056	1.593	0.157	2.29	1.533	0.163
n-butane	0.40	0.975	2.251	0.177	1.17	2.043	0.196
n-pentane	0.35	0.520	3.346	0.104	0.619	2.970	0.117
				0.438			0.476

$$\text{where} \sum \frac{N_i}{1 + L/VK_i} = 0.45, \qquad t = 206° \text{ F}$$

Values of y and x are obtained from equations 19 and 20, where $x_i = 1.82N_i - 0.82y_i$

Table F. At 206° F

	N_i	K_i	$1 + L/VK_i$	$\frac{N_i}{1 + L/VK_i}$	y	x
Propane	0.25	2.153	1.566	0.159	0.353	0.166
n-butane	0.40	1.053	2.158	0.185	0.412	0.390
n-pentane	0.35	5.288	3.307	0.106	0.235	0.444
			7.031	0.450	1.000	1.000

Equilibria near the Critical Region

If a multicomponent mixture is at its critical temperature, the variation with pressure of the vaporization equilibrium ratios y/x of its components must have the general form indicated by the solid lines of Fig. 253. At the critical point where the compositions of the liquid and vapor phases are identical, all equilibrium ratios equal 1.0. At pressures substantially below the critical, the equilibrium ratios are equal to the equilibrium constants or ideal fugacity ratios. In the intermediate region, the equilibrium ratios depart from the equilibrium constants, indicated by the broken lines, to converge at 1.0 at the critical point.

It is evident that the assumption that equilibrium ratios are equal to equilibrium constants leads to errors of increasing magnitude as the critical region is approached. The magnitude of these errors increases with increased difference in volatility between the lightest and heaviest component.

No generally satisfactory method is available for handling vaporization equilibrium calculations in or near the critical region. Three basically different approaches to this problem have been investigated, each of which has certain advantages.

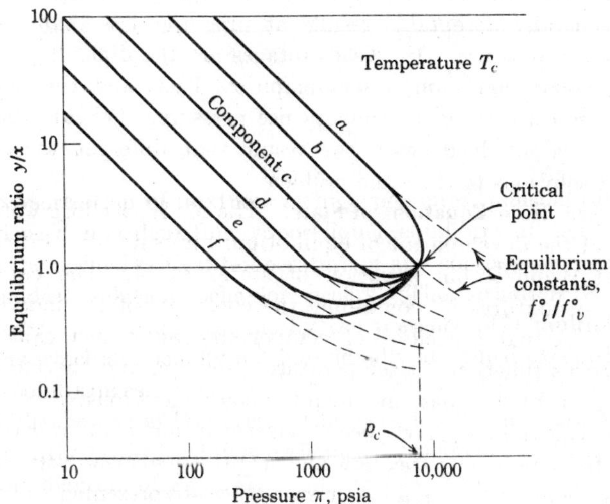

FIG. 253. Equilibrium ratios in a multicomponent system at its critical temperature

Convergence Pressures. It has been observed that the equilibrium ratios of components of mixtures which are not at their critical temperatures show a tendency to converge toward 1.0 at a single point in a manner similar to that indicated in Fig. 253. Actual convergence occurs only at the true critical point of the mixture, but, at temperatures other than the critical, the observed equilibrium ratio values deviate from the equilibrium constants along curves which if extrapolated appear to converge at a point termed the *convergence pressure* of the system at the temperature under consideration. Thus the convergence pressure equals the critical pressure only when the temperature is equal to the critical temperature. All other convergence pressures are in a hypothetical region.

The extensive literature on convergence pressures was reviewed by Hadden,[12] who developed methods for estimating convergence pressures for both binary and multicomponent systems.

The convergence pressure method has been extensively studied by a committee of the Natural Gasoline Association of America.[13] An elaborate set of charts has been developed and published, which presents equilibrium ratios for individual components as functions of temperature,

[12] S. T. Hadden, *Chem. Eng. Prog. Symp. Ser.* no. **49** (7), 53 (1953).

[13] Natural Gasoline Supply Men's Association, *Engineering Data Book*, Tulsa (1957).

pressure, and the convergence pressure of the mixture in which the component exists.

The convergence pressure method is relatively convenient to use and yields generally acceptable results at pressures less than 80% of the convergence pressure. Its disadvantages are the difficulty of preparing the necessary charts on a semi-empirical basis and the uncertainties involved in estimating the convergence pressure. Considerable personal judgment is involved, with the result that different workers obtain different solutions to the same problem.

Equilibria from Equations of State. The M. W. Kellogg Company has sponsored the development of equilibrium ratios and fugacities from the Benedict–Webb–Rubin equation of state, 14–14. The equation was applied to mixtures by combining the constants of the individual components. Actual fugacities of components were then calculated and expressed as functions of temperature, pressure, and the molal average boiling point of the phase in which the component exists.

These calculations have been performed for paraffin hydrocarbons ranging from methane to heptane and are presented in two sets of charts[14,15] for convenient use. Equilibrium ratios y/x are presented as functions of temperature and liquid and vapor phase molal average boiling points. Fugacities are presented as functions of temperature, mole fraction, and the molal average boiling points of the phase.

Although the Kellogg charts are reliable, their use is tedious because of the necessity of successive corrections for the compositions of both phases.

High-Pressure Activity Coefficients. It was proposed by Gamson and Watson[16] that the type of behavior indicated in Fig. 253 be taken into account by means of activity coefficients to express deviation from ideal-solution behavior in the liquid and vapor phases.

On the basis of generalized fugacity coefficients and the pseudocritical concept, Gamson and Watson developed equations for the activity coefficients. These relationships were simplified and modified empirically by Smith and Watson[17] as a basis for a general graphical correlation of activity coefficients, as functions of the pseudoreduced temperatures and pressures of the phases. The resulting charts were published in large scale by Smith and Smith.[18]

[14] M. W. Kellogg Co., *Liquid–Vapor Equilibrium in Mixtures of Light Hydrocarbons. Equilibrium Constants*, New York (1950).

[15] M. W. Kellogg Co., *Liquid–Vapor Equilibrium in Mixtures of Light Hydrocarbons. Fugacity Charts*, New York (1950).

[16] B. W. Gamson and K. M. Watson, *Nat. Petrol. News, Tech. Sec.*, **36**, R258 (May 3, 1944); R554 (Aug. 2, 1944).

[17] K. A. Smith and K. M. Watson, *Chem. Eng. Prog.*, **45**, 494 (1949).

[18] K. A. Smith and R. B. Smith, *Petroleum Processing*, **4** (Dec. 1949).

The Smith activity coefficients are probably the most reliable general method for estimating vaporization equilibrium ratios close to the critical point. At conditions removed from the critical, the agreement was less satisfactory. In this range, the convergence pressure charts and the Kellogg charts generally give more reliable results.

The difficulties in the generalized activity coefficient correlation originated from the inaccuracies of the generalized fugacity coefficient and pseudocritical correlations. The former has been improved by inclusion of the third parameter z_c in Table 51. Development of the improved pseudocritical relations of equations 64, 67*a* through *b* may open the way to reliable generalized activity coefficients.

Problems

1. Calculate the vaporization equilibrium constant of acetone at a temperature of 300° F and an absolute pressure of 250 psia. The density of acetone at 20° C is 0.7915 gram per cc. The vapor-pressure equation is given in Table 8, page 95.

2. A stream of gas in a natural gasoline plant has the following composition in mole per cent:

Ethane	10
Propane	14
Isobutane	19
n-Butane	54
Isopentane	3
	100

(*a*) Calculate the pressure necessary to condense this gas completely at a temperature of 100° F.

(*b*) For a condenser operating at the pressure of part (*a*), calculate the temperature at which condensation will start and the temperature of 50 mole per cent condensation. Also calculate the composition of the first liquid to condense and the compositions of the liquid and vapor phases at 50% condensation.

3. A fractionating column is to produce overhead and bottoms products having the following compositions in mole per cent:

	Overhead	Bottoms
Propane	23	—
Isobutane	67	2
n-Butane	10	46
Isopentane	—	15
n-Pentane	—	37
	100	100

(*a*) Calculate the pressure at which the column must operate in order to condense the overhead product completely at 120° F.

(*b*) Assuming that the overhead product vapors are in equilibrium with the liquid on the top plate of the column, calculate the temperature of the overhead vapors and the composition of the liquid on the top plate when operating at the pressure of part (*a*).

(*c*) Calculate the temperature of the liquid leaving the reboiler of the column under the pressure of part (*a*), assuming equilibrium conditions to exist in the reboiler. Also calculate the composition of the vapors leaving the reboiler.

4. A vapor mixture at 1 atm pressure and 100° C contains

Propane 50 mole %
n-Butane 30 mole %
n-Hexane 20 mole %

Estimate:

(*a*) The pressure required at a dew point of 100° C.
(*b*) The composition of the liquid at the dew point of 100° C.
(*c*) The pressure required for complete condensation at 100° C.
(*d*) The reversible work of compression for complete condensation at 100° C.

5. At a pressure of 500 psia and 100° F the composition of vapor in equilibrium with liquid hydrocarbon is

Methane 91.0 mole %
Ethane 7.0 mole %
Propane 2.0 mole %

Estimate the composition of the liquid in terms of these three components.

23

Solubility and Adsorption

In previous chapters the discussions of phase equilibria are limited to vapor–liquid systems. An elementary discussion with approximate procedures is presented here for other cases of phase equilibria. For comprehensive and satisfactory treatments, the reader is referred to the advanced texts of Hildebrand and Scott[1] and of Guggenheim.[2]

Solubility Parameter. In previous chapters lack of ideal-solution behavior is attributed to chemical reaction and to differences in molecular size and polarity among the different components present. A cause of nonideality common to all solutions results from differences in the van der Waals forces of attraction among the different species present. Hildebrand[1] has expressed this force in terms of a *solubility parameter*. For liquid solutions the solubility parameter of a particular component present is expressed in terms of internal energy of vaporization of the pure component saturated at the temperature of interest divided by its molal volume in the saturated liquid state. The internal energy of vaporization includes expansion to its ideal gaseous state. Thus

$$\delta_i = \left(\frac{\Delta E_i}{v_i}\right)_{sL}^{1/2} = \left(\frac{(U^* - U)_{isL}}{v_{isL}}\right)^{1/2} \tag{1}$$

where δ_i = solubility parameter

$\Delta E_i = (U^* - U)_{isL}$ = internal energy of vaporization of saturated liquid to the ideal gaseous state at temperature T

v_{isL} = molal volume of saturated liquid at temperature T

Regular Solutions. In Chapter 20, regular solutions are defined as those in which lack of ideality is due to differences in van der Waals forces of attraction, with an excess entropy of mixing equal to zero; hence, from equations 20-57 and 21-23

$$\Delta\mu_i^E = \Delta\bar{H}_i^M = RT \ln \gamma_i \tag{2}$$

[1] J. H. Hildebrand and R. L. Scott, *The Solubility of Nonelectrolytes*, Reinhold Publishing Corp. (1950).

[2] E. A. Guggenheim, *Mixtures*, Clarendon Press, London (1952).

where for component i,

$\Delta\mu_i^E$ = excess chemical potential

$\Delta\bar{H}_i$ = excess partial heat of mixing

γ_i = activity coefficient

In terms of its solubility parameters Hildebrand has expressed the excess molal free energy of a regular binary solution in terms of its solubility parameters and molal volumetric fractions of the liquid as follows:

$$\Delta G^E = \Delta H^M = v_m\phi_1\phi_2(\delta_1 - \delta_2)^2 \quad (2a)$$

where $\phi_1 = \dfrac{v_i x_i}{v_m}$

= molal volumetric fraction of component i in the liquid state (3)

$v_m = v_i x_i + v_2 x_2$

= mean molal volume of the liquid solution (4)

ΔH^M = molal heat of mixing in the liquid state

Regular solutions are endothermic, $\Delta H^M > 0$, excess entropies of mixing are zero, $\Delta S^E = 0$, entropies of mixing are regular, i.e., $\Delta S = -\sum x_i R \ln x_i$ and activity coefficients are all positive, $\gamma_i > 1.0$.

Partial molal heats of mixing $\Delta\bar{H}_1^M$ for component 1 are obtained by multiplying equation 1 by $(n_1 + n_2)$ and differentiating with respect to n_1, where $x_1 = n_1/(n_1 + n_2)$ and $x_1 + x_2 = 1$, to give

$$\Delta\bar{H}_1 = RT \ln \gamma_1 = v_1\phi_2^2(\delta_1 - \delta_2)^2 \quad (5)$$

or

$$\ln \gamma_1 = \frac{v_1}{RT}\phi_2^2(\delta_1 - \delta_2)^2 \quad (6)$$

In a multicomponent regular solution

$$\ln \gamma_i = \frac{v_i}{RT}(\delta_i - \bar{\delta})^2 \quad (7)$$

where

$$\bar{\delta} = \sum_1^n \phi_i\,\delta_i \quad (8)$$

Liquid–Liquid Systems. The mutual solubility of liquids is decreased by chemical dissimilarity of the components as reflected by an increase in their separate activity coefficients. Where the activity coefficients of the separate components become sufficiently great, the solution separates into two liquid phases. This progressive increase in activity coefficients and decrease in solubility is shown by a series of homologous alcohols of increasing molecular weights dissolved in water. The methanol–water system shows mutual solubility in all proportions with activity coefficients slightly greater than unity. The water–*n*-propanol solution has higher activity coefficients with complete miscibility over the temperature range existing at atmospheric pressure. The water–butanol system shows only partial miscibility with corresponding higher activity coefficients.

The properties of the water–butanol system at atmospheric pressure are shown in Fig. 242. Two liquid phases exist in the composition range from a to b corresponding to 66 to 98 mole per cent water. Between these limits the fugacity of each component is constant and equal in both phases. As a result, the boiling point is constant, and the vapor evolved is of constant composition c. The apparent activity coefficients in this range are inversely proportional to the mole fractions based on the combined phases. Although the vapor evolved from any liquid mixture of composition between a and b is of constant composition c, the liquid and vapor compositions are equal only at point c, which is termed the hetero-azeotropic composition. However, in each of the individual phases, the activity and composition are each constant, as indicated in Fig. 242. Since the fugacities in the two phases are equal at equilibrium, in the region of the two liquid phases,

$$a_1 f^\circ{}_1 = a'_1 f^{\circ\prime}{}_1 \tag{9}$$

where the primed quantities represent one liquid phase and the unprimed the other. Since the standard state is the pure liquid component, $f^\circ{}_1 = f^{\circ\prime}{}_1$ and

$$x_1\gamma_1 = x'_1\gamma'_1 \quad \text{or} \quad \frac{x_1}{x'_1} = \frac{\gamma'_1}{\gamma_1} \tag{10}$$

The ratio x_1/x'_1 is the distribution coefficient of component 1.

As the chemical dissimilarity of the components of a system increases, the deviation of the activity coefficients from unity increases, and the range of mutual solubility decreases. This effect corresponds to lengthening the horizontal range between a and b on the curves of Fig. 242. As complete immiscibility is approached, point a approaches zero, point b approaches 100% composition, and the activity coefficients become very large. As a rough guide to the relative magnitudes of the activity coefficients corresponding to various ranges of immiscibility, Colburn[3] prepared Table 64 for symmetrical systems.

[3] A. P. Colburn, private communication (1942).

Table 64. Range of Immiscibility in Symmetrical Systems

Immiscibility Range, mole fraction	Van Laar Constants, A and B	γ At $x = 0$	γ At $x = x_s$	
0	0.875	7.5	1.68	completely miscible
0.2 to 0.8	1.0	10.0	5.0	
0.1 to 0.9	1.2	15.8	9.8	
0.05 to 0.95	1.43	27	20	
0.02 to 0.98	1.77	59	50	
0.01 to 0.99	2.03	106	94	nearly immiscible

The constants in the Redlich–Kister equation can be evaluated from the mutual solubility of two partially miscible liquids. Thus, for a binary system, from equation 21–45, employing B and C for one liquid phase and B' and C' for the other.

$$\ln \frac{\gamma_1 \gamma'_2}{\gamma_2 \gamma'_1} = B(x_2 - x_1) + C(6x_2x_1 - 1) - B'(x'_2 - x'_1) - C'(6x'_1x'_2 - 1) + \cdots \quad (11)$$

By measurements of equilibrium solubility in the two liquid phases at two different temperatures, four equations will be available for establishing the four constants B, C, B', and C' applicable to vapor–liquid equilibria as well as to liquid–liquid solubility. This procedure neglects the small effect of temperature on the ratios of activity coefficients.

Illustration 1. Estimate the solubility of benzene in water at 20° C and atmospheric pressure.

	Water, Component 1	Benzene, Component 2
v	18	89 cc per g-mole
z_c	0.23	0.274
T_r	0.459	0.529
p_r	0.0046	0.0206
$(\text{U}^* - \text{U})/T_c$	15.34	12.776 g-cal/(g-mole)(K°)
$\text{U}^* - \text{U}$	9920	7188 g-cal per g-mole
$(\text{U}^* - \text{U})/v$	551	80.7 g-cal per cc
δ	23.5	8.98 (g-cal per cc)$^{1/2}$

Water phase (unprimed values): From equation 6,

$$\ln \gamma_2 = \frac{89}{(1.989)}(1)(23.5 - 8.98)^2 = 32.4$$

$$\gamma_2 = (1.3)10^{14}$$

at $x_1 = 1$, $\gamma_1 = 1.0$, $\phi_1 = 1.0$, and $\gamma_2 = (1.3)10^{14}$.

From equation 10, $x_2\gamma_2 = x'_2\gamma'_2$.

Since water is nearly insoluble in the benzene phase, $x'_2 = 1.0$, $\gamma'_2 = 1.0$, and $x'_2\gamma'_2 = 1.0$.

Hence,

$$x_2 = \frac{1}{\gamma_2} = (0.77)10^{-14}$$

The great difference in the solubility parameters of water and benzene accounts for the near insolubility of these two liquids as reflected by the high values of their activity coefficients. However, water and benzene do not form regular solutions, and nonideal-solution behavior is due to the polarity of water as well as to the difference in solubility parameters. For this reason the solubility at x_2 as calculated from the solubility parameter is too low. The actual experimental value of x_2 is $(1.6)10^{-4}$.

Critical Solution Temperature. As the temperature of two liquid phases of partially miscible liquids increases, the temperature at which complete miscibility occurs is termed the *critical solution temperature*. To attain this temperature, a corresponding pressure increase is required to prevent the boiling of either phase. In the region of immiscibility, the activity of a given component remains constant.

Where both phases merge at the critical solution temperature, the first and second derivatives of activity are zero:

$$\frac{\partial \ln \gamma_1 x_1}{\partial x_1} = 0 \quad \text{and} \quad \frac{\partial^2 \ln \gamma_1 x_1}{\partial x_1^{\,2}} = 0 \tag{12}$$

For a symmetrical regular solution where $\ln \gamma_1 = Ax_2^{\,2}$ (equation 21–40), the first and second derivatives of activity are zero when $x_1 = 0.5$. Since $a_1 = x_1\gamma_1$,

$$\frac{d \ln a_1}{dx_1} = -2A(1 - x_1) + \frac{1}{x_1} = 0 \tag{13}$$

Where $x_1 = 0.5$, the critical solution temperature T_{cs} occurs at a temperature where $A = 2$.

Hence, from equation 20–106, $k = 2RT_{cs}$.

Solubility of Gases. The solubility of gases is discussed extensively in Chapters 21 and 22 dealing with vapor–liquid equilibria. A special consideration is more convenient in dealing with gases dissolved in liquids above their critical temperatures. If the temperature of the system is below or not greatly above the critical temperature of the gas, the standard state for the gaseous component in the liquid phase may be taken as the pure component in a hypothetical liquid state at the tempera ture and pressure of the system.

For the soluble gas, component 2, unit fugacity is arbitrarily taken as the standard state in the gaseous phase, $f^\circ_{2\pi} = 1.0$. For ideal-solution behavior in the gaseous phase $\gamma_{2v} = 1.0$; hence, $a_{2v} = f_2 = \nu_{2\pi} y_2 \pi$.

In the liquid phase, $a_2 = \gamma_2 x_2 = f_2/f^\circ_{2L}$. By definition, for the equilibrium constant for gas solubility, $K_2 = a_{2v}/a_{2L}$; hence,

$$K_2 = \frac{\nu_{2\pi} y_2 \pi}{\gamma_{2L} x_2} = f^\circ_{2L} \tag{14}$$

where π = total pressure of the system

$\nu_{2\pi}$ = fugacity coefficient of pure gaseous component 2 at the temperature and pressure of the system

γ_2 = activity coefficient in the liquid state

Comparison of equation 14 with 6–26, page 181, written for component 2, in which the partial pressure p_2 is equal to πy_2, results in the following thermodynamic expression for Henry's constant H_2:

$$H_2 = \frac{K_2 \gamma_2}{\nu_{2\pi}} \tag{15}$$

Equations 14 and 15 may be used to predict solubilities where fugacity coefficients are obtained from Table 51 and where γ_2 is taken as unity.

For the effects of molecular size and van der Waals forces, the activity coefficients in the liquid phase may be calculated from equation 8, which gives good results in systems not involving highly polar compounds and for temperatures below the critical temperature of the gas. At temperatures above the critical temperature of the gas, values of P°_2 are obtained by extrapolation.

Illustration 2. Estimate the solubility of chlorine in carbon tetrachloride at a pressure of 1.0 atm and 0° C.

	Cl_2, Component 1	CCl_4, Component 2
Molecular weight	70.9	154
Density at −34.6° C, grams per cc	1.561 $\left\{ \begin{array}{l} T_r = 0.57 \\ \rho_r = 2.754 \end{array} \right.$	
Density at 20° C, grams per cc		1.595 $\left\{ \begin{array}{l} T_r = 0.526 \\ \rho_r = 2.871 \end{array} \right.$
T_c	417	556.4 °K
p_c	76.1	45.0 atm
z_c	0.276	0.272
At 273° K, 1 atm (by use of generalized tables):		
T_r	0.654	0.492
p_r	0.0131	0.0222
$(\mathrm{U}^* - \mathrm{U})/T_c$ (liquid)	11.03	13.44 g-cal/(g-mole)(K°)
$(\mathrm{U}^* - \mathrm{U})$ (liquid)	4600	7480 g-cal per g-mole
ρ_r (liquid)	2.584	2.933
ρ	1.464	1.630 grams per cc
v	48.4	94.5 cc per g-mole
$(\mathrm{U}^* - \mathrm{U})/v$	95.0	79.5 g-cal per cc
δ	9.75	8.92 (g-cal per cc)$^{1/2}$

At 0° C, the vapor pressure of liquid chlorine is 3.65 atm.

$$T_{rs} = 0.654, \quad p_{rs} = 0.047, \quad z_c = 0.276, \quad \nu_p = 0.971$$

At 0° C, 1 atm for chlorine gas,

$$T_r = 0.654, \quad p_r = 0.0131, \quad z_c = 0.276, \quad \nu_\pi = 0.992$$

From Raoult's law, the solubility of chlorine is

$$x_1 = \frac{y_1}{P^\circ_1} = \frac{1}{3.65} = 0.274$$

From equation 14, for component 1, where $\pi = 1.0$, and $y_1 = 1.0$

$$\gamma_1 x_1 = \frac{y_1 \nu_\pi}{P^\circ_1 \nu_p} \pi = \frac{0.992}{(3.65)(0.971)} = 0.280$$

Since chlorine and carbon tetrachloride form regular solutions, their activity coefficients may be calculated from solubility parameters from equation 6. Since ϕ_1, ϕ_2 and v_m are functions of x, a trial-and-error procedure is followed. As a first approximation, let $x_1 = 0.27$.

$$x_1 v_1 = (0.27)(48.4) = 13.1$$

$$x_2 v_2 = (0.73)(94.5) = 69.0$$

$$v_m = 82.1$$

$$\phi_2 = \frac{69.0}{82.1} = 0.84$$

$$\phi_1 = 0.16$$

From equation 6,

$$\ln \gamma_1 = \frac{48.4(0.84)^2(9.75 - 8.92)^2}{(1.987)273} = 0.043$$

$$\gamma_1 = 0.04$$

$$\text{Hence, } x_1 = \frac{0.280}{1.04} = 0.270.$$

In this illustration, a value of unity has been assumed for y_1. A corrected value may be obtained by considering also the vapor pressure of carbon tetrachloride.

A standard state for chlorine has been taken as unit fugacity. This illustration may also be solved by the generalized methods described in Chapter 22 where standard states are taken as the pure components at the temperature and pressure of the system. Vaporization equilibrium constants as defined in Chapter 22 may then be obtained directly from Table 63, where $K_1 = 2.65$ and $K_2 = 0.002$.

Estimation of the Solubility Parameter from Solubility Data. For a pure gas above its critical temperature hypothetical internal energies of vaporization and liquid volumes cannot be obtained accurately by extrapolation procedures. Hildebrand suggests calculating the solubility parameter of such a gas from its solubility data. In the following illustration the solubility parameter of methane is obtained from its solubility in chlorobenzene and then applied to obtain the solubility of methane in benzene. The generalized methods of obtaining vaporization equilibrium constants will be followed as developed in Chapter 22.

Illustration 3. Estimate the solubility parameter of methane (component 1) from its solubility in chlorobenzene at 25° C and 1.0 atm ($x_1 = 0.0021$ mole fraction), and from this value estimate the solubility of methane in benzene at 25° C and 1.0 atm. Both liquid solutions are regular with no polar components.

	Benzene	Chlorobenzene	Methane
T_c	562.1	632.4	190.7° K
p_c	48.6	44.6	45.8 atm
z_c	0.274	0.265	0.290
p_r	0.0206	0.0224	0.0218
T_r	0.530	0.470	1.60
$\frac{U^* - U}{T_c}$	12.73	14.37	g-cal/(g-mole)(K°)
$U^* - U$	7150	9100	g-cal per g-mole
ρ_r	2.843	3.005	
v	89.5	102.5	15 cc/per g-mole
$\frac{U^* - U}{v}$	79.9	88.9	g-cal per cc
δ	8.94	9.48	(g-cal per cc)$^{1/2}$
M	78.1	112.6	16
K	0.19	0.032	139

Values of K were obtained from Table 63, requiring graphical extension to obtain values of K for benzene and chlorobenzene by extrapolation.

Methane in Chlorobenzene. For ideal-solution behavior:

$$y_1 = K_1 x_1, \qquad y_1 = 139 x_1$$

$$y_2 = K_2 x_2, \qquad (1 - y_1) = 0.032 x_1$$

$$x_1 = 0.0070, \qquad y_1 = 0.965$$

$$x_2 = 0.9930, \qquad y_2 = 0.035$$

For nonideal-solution behavior, $x_1 = 0.0021$.

$$\gamma_1 = \frac{y_1}{K_1 x_1} = \frac{0.965}{139(0.0021)} = 3.30$$

$$\ln \gamma_1 = 1.194$$

Evaluation of solubility parameter for methane: $\phi_2 = 1.0$, $RT = 592$.

From equation 6,

$$\ln \gamma_1 = \frac{v_1}{RT} \phi_2^2 (\delta_1 - \delta_2)^2$$

$$1.194 = \frac{15}{592} (\delta_1 - 9.48)^2$$

$$(\delta_1 - 9.48)^2 = 47.2$$

$$\delta_1 = 16.38 = \text{solubility parameter of methane}$$

Methane in Benzene. For ideal-solution behavior:

$$y_1 = K_1x_1, \qquad y_1 = 139x_1$$
$$1 - y_1 = K_2(1 - x_1), \qquad 1 - y_1 = 0.19(1 - x_1)$$
$$x_1 = 0.0058, \qquad y_1 = 0.810$$
$$x_2 = 0.994, \qquad y_2 = 0.190$$

For nonideal-solution behavior: $\phi_1 = 1.0$, $RT = 592.$

$$\ln \gamma_1 = \frac{15}{592}(16.38 - 8.94)^2 = (0.0253)(55.5) = 1.40$$

$$\gamma_1 = 4.05$$

$$x_1 = \frac{y_1}{\gamma_1 K_1} = \frac{0.810}{(4.05)(139)} = 0.00145 = \text{solubility of methane in benzene}$$

Standard States Referred to Behavior at Infinite Dilution

In dealing with slightly soluble gases for experimental reasons it is convenient to express the activity of the solute gas in the liquid solution with reference to its behavior at infinite dilution with a corresponding standard-state fugactiy f°_{2L}. With this convention, equation 14 becomes

$$K'_2 = f^\circ_{2L} = \frac{y_2 \pi \nu_{2\pi}}{x_2 \gamma_{2L}} \tag{16}$$

Where γ'_{2L} is unity, values of K'_2 can be obtained from one experimental measurement of y_2 and x_2.

The variations of the solubility constant K'_2 with temperature follows directly from equation 20–30:

$$\left(\frac{\partial \ln K'_2}{\partial T}\right)_p = \frac{\text{H}^*_{2G} - \bar{\text{H}}^\circ_{2L}}{RT^2} \tag{17}$$

where

H^*_{2G} = enthalpy of component 2 in the ideal gaseous state at temperature T

$\bar{\text{H}}^\circ_{2L}$ = partial molal enthalpy of component 2 in solution at infinite dilution at the temperature and pressure of the system

The effect of pressure on K'_2 is derived from equation 20–23,

$$\left(\frac{\partial \ln K'_2}{\partial p}\right)_T = \frac{\bar{v}^\circ_{2L}}{RT} \tag{18}$$

where $\bar{v}^\circ_{2L}$ = partial molal volume of component 2 in the liquid phase at infinite dilution

Equation 17 may be used to calculate the effect of temperature on solubility from heat-of-dissolution data or, conversely, to calculate the differential heat of dissolution at infinite dilution $-(\text{H}^*_{2G} - \bar{\text{H}}^\circ_{2L})$ from solubility data at two temperatures if the ideal-solution behavior is approximated.

Illustration 4. At a partial pressure of 1.0 atm, pure CO_2 is in equilibrium with aqueous solutions containing 0.096 mole per cent CO_2 at 10° C and 0.0294 mole per cent CO_2 at 60° C.

(*a*) Assuming the solution to behave ideally, calculate the mole fraction of CO_2 in an aqueous solution saturated with CO_2 at 60° C and 100 atm. The effect of pressure on K'_2 may be neglected.

(*b*) Calculate the average differential heat of dissolution of CO_2 in water at infinite dilution in the range of 10 to 60° C.

Solution: (*a*) For CO_2,

$$T_c = 304.3° \text{ K}, \qquad p_c = 73.0 \text{ atm}$$

At 60° C, 1 atm, $\quad T_r = 1.10, \quad p_r = 0.014$

From Table 51, $\quad \nu_\pi = 0.995$

Substitution in equation 16, if it is assumed that $\gamma'_{2L} = 1.0$, gives

$$K'_2 = (1.0)(0.995)/(0.000294) = 3380$$

At 60° C, 100 atm, $\quad T_r = 1.10, \quad p_r = 1.37, \quad \nu_\pi = 0.63$

From equation 16,

$$x_2 = (100)(0.63)/3380 = 0.0186 \quad \text{or} \quad 1.86\%$$

(*b*) At 10° C, and 1.0 atm, $\quad T_r = 0.93, \quad p_r = 0.014, \quad \nu_\pi = 0.995$

$$K'_2 = (1.0)(0.995)/(0.00096) = 1036$$

Substitution in equation 17, after integration, gives

$$(\text{H}^*_{2G} - \bar{\text{H}}^\circ_{2L}) = -\left[(1.99)(2.303)\log\frac{3380}{1036}\right]\Big/\left(\frac{1}{333.2} - \frac{1}{283.2}\right)$$

$$= 4431 \text{ cal per g-mole}$$

The differential heat of dissolution of the ideal gas at infinite dilution is, therefore, −4431 cal per g-mole.

The experimentally observed mole percentage of CO_2 in water at 60° and 100 atm is approximately 1.8%. Assumption of Henry's law in this case would predict a value of 2.94%, an error of over 60% as compared to an error of less than 5% for equation 16. The good agreement with equation 16 indicates that the activity coefficients in the liquid phase are substantially unity as assumed, in spite of the chemical reaction occurring between carbon dioxide and water and the accompanying ionization. This may be interpreted as indicating that only a small fraction of the dissolved CO_2 reacts with the water. Where higher solute concentrations or more extensive reactions are involved, the assumption of ideal behavior may lead to large errors which require evaluation of the activity coefficients. The solubility of gases with accompanying chemical reaction is discussed in Chapter 27.

Solubility of Solids in Liquids

The distribution of a component among the phases of any system at equilibrium at a constant temperature is expressed by equation 21–15. This general relationship may be used to estimate activity coefficients and vapor–liquid equilibria from solubility or distribution data. Conversely, these same methods may be used to estimate distribution and solubility data from vapor–pressure or vapor–liquid equilibrium measurements.

For solutions of nonelectrolytes Hildebrand[1] recommends that the activity in the liquid phase of a solid solute be referred to the pure solute in a hypothetical liquid state at the temperature and pressure of the system. The activity in the solid state is taken as unity for the pure solid at the temperature and pressure of the system.

The equilibrium constant for the solubility of a solid in a liquid can be expressed in terms of activities, standard-state fugacities, and concentration in a manner analogous to vapor–liquid equilibria; hence, modification of equation 14 gives

$$K_2 = \frac{a_{2L}}{a_{2s}} = \frac{f^\circ_{2s}}{f^\circ_{2L}} = \frac{\gamma_{2L}x_2}{\gamma_2 s_2} \tag{19}$$

where

s_2, γ_{2s} = mole fraction and activity coefficient of component 2 in the solid phase

x_2, γ_{2L} = mole fraction and activity coefficient of component 2 in the liquid phase

f°_{2s} = fugacity of pure component 2 in the solid state at the temperature and pressure of the system

f°_{2L} = fugacity of pure component 2 in the hypothetical liquid state at the temperature and pressure of the system

K_2 = solubility constant of component 2, referred to the pure component

An expression for the solubility constant as a function of temperature is obtained from equation 20–30 applied to both phases:

$$\left(\frac{\partial \ln K_2}{\partial T}\right)_p = \frac{H^\circ_{2L} - H^\circ_{2s}}{RT^2} \tag{20}$$

where H°_{2L} and H°_{2s} are the molal enthalpies of pure component 2 in the hypothetical liquid and solid states, respectively, at the temperature and pressure of the system. If it is assumed that the heat of fusion is independent of temperature $H^\circ_{2L} - H^\circ_{2s}$ may be taken as the normal heat of fusion and equation 20 integrated with $K_2 = 1.0$ at the melting point T_m where $f^\circ_{2s} = f^\circ_{2L}$. Then

$$\ln K_2 = -\frac{\lambda_f}{R}\left(\frac{1}{T} - \frac{1}{T_m}\right) = -\frac{\lambda_f(T_m - T)}{RTT_m} \tag{21}$$

where λ_f = molal heat of fusion of component 2
T = temperature of the system
T_m = melting point of component 2

Similarly, the effect of pressure on the solubility equilibrium constant is obtained from equations 20–23 applied to both phases to give

$$\left(\frac{\partial \ln K_2}{\partial p}\right)_T = -\frac{(v^\circ_{L2} - v^\circ_{s2})}{RT} \tag{22}$$

or

$$\ln K_2 = -\frac{(v^\circ_{L2} - v^\circ_{s2})_{\text{avg}}}{RT}(\pi - P_m) \tag{23}$$

where

v°_{s2} = molal volume of the pure solid solute at the temperature and pressure of the system
v°_{L2} = molal volume of the hypothetical liquid solute at the temperature and pressure of the system
P_m = equilibrium pressure corresponding to melting point T_m

It may be noted from equation 23 that, for an ideal solution where $K_2 = x_2$, the solubility of a solid decreases with increase in molal heat of fusion, increases when the solid contracts upon melting, and when expressed as mole fraction is independent of the nature of the solvent; on a weight basis, the solubility decreases with increase in molecular weight of the solvent.

Equation 19 furnishes a basis for the evaluation of the corresponding activity coefficients. These coefficients approach unity in systems that do not involve polar components, large differences in molecular size, or large differences between T_m and T. If the activity coefficients are assumed to be unity, the *ideal solubility* of a pure solid is equal to K_2, or $x_2 = K_2$.

Equation 21 may be used to predict the solubility of a solid from its melting-point and heat-of-fusion data or, conversely, for calculating heats of fusion from solubility data where conditions are such that ideal solutions are approximated. The usefulness of equation 19 is restricted by the extensive experimental data required for evaluation of the activity coefficients.

Illustration 5. The solubility of naphthalene in benzene is 60% by weight at 45° C and 21% at 0° C. Assuming the solution to be ideal, calculate heats of fusion of naphthalene from these two solubility values. The melting point of naphthalene is 80.1° C.

Solution:

Basis: 100 grams of solution (component 2 is naphthalene).

		0° C			45° C		
	Molecular Weight	grams	g-moles	Mole Fraction	grams	g-moles	Mole Fraction
Naphthalene	128	21	0.164	0.14	60	0.469	0.478
Benzene	78	79	1.013	0.86	40	0.512	0.522
		100	1.177	1.00	100	0.981	1.000

$\log K_2 = \log x_2$ −0.854 −0.321

At 0° C, $\lambda_f = \dfrac{(2.303)(0.854)(1.99)(273.1)(353.2)}{(353.2 - 273.1)} = 4710$ cal per g-mole

At 45° C, $\lambda_f = \dfrac{(2.303)(0.321)(1.99)(318.1)(353.2)}{(353.2 - 273.1)} = 4710$ cal per g-mole

These results are in fair agreement with the observed value of 4550.

The fair agreement between the heats of fusion, as shown in Illustration 5, indicates that the solutions of benzene in naphthalene closely approximate ideal behavior. In other systems involving less similar components, much larger discrepancies can be expected, and the activity coefficients of equation 19 must be evaluated for satisfactory results. These deviations are discussed by Hildebrand.[1]

For solutes having melting points far removed from the temperature of the system or for electrolytes that dissociate into ions in solution, it is not advantageous to use the hypothetical standard state of equation 19 for the liquid phase. A more convenient convention is to express the activity of the solute in the liquid referred to the infinitely dilute solution at the pressure of the system. As mentioned on page 869, this convention corresponds to the selection of a hypothetical standard state, the fugacity of which is the value that would exist if the solution behaved ideally at unit mole fraction (or unit molality) as it does at zero concentration. The enthalpy and volume of this hypothetical standard state are therefore equal to the partial molal enthalpy and volume, respectively, at zero concentration. The activities in the solid phase are best referred to the pure solid at the pressure of the system as the standard state. On this basis, the solubility of a solid is expressed by

$$\frac{a'_{2L}}{a_{2s}} = \frac{f^\circ_{2s}}{f^{\circ\prime}_{2L}} = \frac{x_2\gamma'_{2L}}{s_2\gamma_{2s}} = K'_2 \tag{24}$$

where

s_2, γ_{2s} = mole fraction and activity coefficient of component 2 in the solid phase

x_2, γ'_{2L} = mole fraction and activity coefficient of component 2 in the given solution

f°_{2s} = fugacity of the pure solid at the temperature and pressure of the system

$f^{\circ\prime}_{2L}$ = fugacity of the hypothetical standard state for the liquid phase referred to infinite dilution

K'_2 = solubility equilibrium constant of component 2 referred to infinite dilution

The effect of temperature on the solubility constant is given by an equation expressing the effect of temperature on the fugacities of the standard states. From equation 20,

$$\left(\frac{\partial \ln K'_2}{\partial T}\right)_p = \frac{(\bar{H}^{\circ\prime}_{2L} - H^\circ_{2s})}{RT^2} \tag{25}$$

where

$\bar{H}^{\circ\prime}_{2L}$ = partial molal enthalpy of component 2 at infinite dilution at the temperature and pressure of the system

H°_{2s} = molal enthalpy of component 2 in the solid state at the temperature and pressure of the system

The quantity $\bar{H}^{\circ\prime}_{2L} - H^\circ_{2s}$ is the differential heat of solution at infinite dilution.

Similarly, by modifying equation 22 to this standard state, the effect of pressure on solubility is obtained:

$$\left(\frac{\partial \ln K'_2}{\partial p}\right)_T = -\frac{(\bar{v}^{\circ\prime}_{2L} - v^\circ_{2s})}{RT} \tag{26}$$

where

$\bar{v}^{\circ\prime}_{2L}$ = partial molal volume at infinite dilution at the temperature and pressure of the solution

v°_{2s} = molal volume of the pure solid at the temperature and pressure of the system

It may be noted from equation 25 that the solubility of a solid is increased by increased temperature if its differential heat of solution is positive. Similarly, the solubility is increased by increase in pressure if the partial molal volume in dilute solution is less than the volume of the solid.

The practical value of equation 24–26 is restricted by the great variations with concentration of the activity coefficients in liquid solutions. Except in dilute solutions, these coefficients must be evaluated from experimental data.

Gas–Solid Adsorption Equilibria. Equilibrium conditions in adsorption of gases on solids are discussed in Chapter 10, and the enthalpy of such systems in Chapter 9, pages 386–393. From equation 10–10

$$(w_2 + 1)H = H_1 + H_2 w_2 + \Delta H = \bar{H}_1 + w_2 \bar{H}_2 \tag{27}$$

or

$$\Delta H = \bar{H}_1 - H_1 + w_2(\bar{H}_2 - H_2) = \Delta\bar{H}_1 + w_2 \Delta\bar{H}_2 \tag{28}$$

where

$\Delta\bar{H}_1$ = the differential heat of adsorption per unit mass of adsorbent
$\Delta\bar{H}_2$ = the differential heat of adsorption per unit mass of adsorbate
ΔH = the integral heat of adsorption per unit mass of adsorbent
w_2 = mass of adsorbate per unit mass of adsorbent

From equation 28, at constant mass of adsorbent

$$\Delta\bar{H}_2 = \left(\frac{\partial \Delta H}{\partial w_2}\right)_{w_1=1,T} \tag{29}$$

Since the differential heat of adsorption varies with the mass adsorbed, the total heat of adsorption in adsorbing a mass w_2 of gas is obtained by integration; thus,

$$\Delta H = \int_0^{w_2} \Delta\bar{H}_2\, dw_2 \tag{30}$$

From equation 13–74, page 531, applied to gas adsorption under conditions of constant composition,

$$\left(\frac{\partial \Delta\bar{S}}{\partial \Delta\bar{V}}\right)_{T,w_2} = \left(\frac{\partial p}{\partial T}\right)_{w_2} \tag{31}$$

Since adsorption at constant temperature and composition also takes place at constant pressure, $T\,\Delta S = \Delta H$, equation 31 may be directly integrated:

$$\bar{S}_2 - S_{2g} = \left(\frac{\partial p}{\partial T}\right)_{w_2} (\bar{V}_2 - V_{2g}) = \frac{\bar{H}_2 - H_{2g}}{T} \tag{32}$$

If the partial molal volume of the adsorbed gas is neglected and ideal-gas behavior is assumed,

$$\bar{H}_2 - H_{2g} = -RT^2\left(\frac{\partial \ln p}{\partial T}\right)_{w_2} = R\left(\frac{\partial \ln p}{\partial \frac{1}{T}}\right)_{w_2} = \Delta\bar{H}_2 \tag{33}$$

and

$$\Delta H = \frac{R}{M_2}\int_0^{w_2} \left(\frac{\partial \ln p}{\partial \frac{1}{T}}\right)_{w_2} dw_2 \tag{34}$$

Illustration 6. In the adsorption of NO_2 gas by silica gel, the following equilibrium isotherm equations were derived from the experimental work of Foster:[4]

At 15° C, $w_2 = 0.330\ p_2^{1.333}$

At 25° C, $w_2 = 0.115\ p_2^{1.490}$

At 35° C, $w_2 = 0.022\ p_2^{1.870}$

where $\dot{p}_2$ = partial pressure of NO_2 in millimeters of mercury
w_2 = grams of NO_2 adsorbed per 100 grams of silica gel

Calculate the differential heat of adsorption per gram-mole of NO_2 when $w_2 = 5.0$.

From a cross-plot of ln p against $1/T$ at $w_2 = 5.0$,

$$\left(\frac{\partial \ln p}{\partial(1/T)}\right)_{w_2=5} = -3760° \text{ K}$$

From equation 33,

$$\Delta\bar{H}_2 = \bar{H}_2 - H_{2g} = -3760(1.987) = -7490 \text{ cal per g-mole } NO_2$$

To obtain the integral heat of solution of 5 grams of NO_2 on 100 grams of silica gel, it is necessary to obtain the differential heat of solution for different values of w_2 from 0 to 5 and then integrate according to equation 34.

Problems

1. The heat of fusion of α-naphthylamine ($C_{10}H_9N$) at its melting point of 49° C is 3150 cal per g-mole. Calculate the solubility at 30° C, assuming ideal behavior.

2. Calculate the ideal solubility of hydrogen cyanide gas in acetonitrile at a temperature of 100° C and a partial pressure of 5 atm. The density of liquid hydrogen cyanide is 0.699 at 20° C. The vapor pressures at 100° C are 9.2 and 1.89 atm, respectively.

3. Calculate the ideal solubility of chlorine in carbon tetrachloride at a partial pressure of 10 atm and a temperature of 100° C. The vapor pressures at 100° C are 37.6 and 1.92 atm, respectively.

4. The solubility of carbon monoxide in benzene at 20° C under a partial pressure of 1 atm is reported as 0.1533 cc of gas measured at 0° C and 1 atm per cc of liquid benzene. Assuming that the activity coefficients are unity, calculate the solubility in mole per cent at 20° C and 50 atm. Neglect the vapor pressure of benzene.

5. The following data for the solubility of carbon dioxide at a partial pressure of 1 atm in methyl alcohol are reported in cubic centimeters of gas measured at 0° C and 1 atm per cc of liquid.

t°, C	Solubility
15	4.366
20	3.918
25	3.515

(*a*) Assuming that ideal solutions are formed, calculate the average differential heats of dissolution of carbon dioxide in methyl alcohol in the temperature ranges of 15 to 20° C and 20 to 25° C. Neglect the vapor pressure of alcohol.

(*b*) From the results of part (*a*), calculate the solubility at a temperature of 35° C and a pressure of 100 atm.

[4] G. Foster, Ph.D. thesis, University of Wisconsin (1944).

6. At 300° K the equilibrium adsorption pressure exerted by water vapor adsorbed on a solid is represented by the following equation:

$$\frac{d \ln p}{dT} = 0.89w$$

where w = grams of water vapor adsorbed per gram of adsorbent and p = millimeters of mercury.

Calculate the integral heat of adsorption in calories per gram of adsorbent when 0.2 g of water vapor is adsorbed by 1 g of adsorbent, starting with water-free adsorbent.

24

Separation Processes

The methods of separating solutions and mixtures into their components are among the most important objectives in chemical processing. These separations include such unit operations as distillation, gas absorption, and solvent extraction. The minimum degradation of energy results where these separations are conducted reversibly with no net change in the entropy of the system and its surroundings. However, the usual and practical methods of separation are accomplished by methods that are inherently irreversible such as diffusion, turbulence, and the transfer of heat.

Reversible Separation. Reversible separation implies that each component must be transferred from a lower to a higher state of concentration with simultaneous separation of components. Such reversible separation involves the use of semipermeable membranes, selective for each of the separate components. Semipermeable membranes occur in living cells; biological processes depend thereon. In industrial processes such membranes appear in dialysis cells and in exchange resins, but otherwise have at present no extensive use.

For the isothermal reversible separation of a gas of components A and B from a given initial pressure π to the same final pressure for each component, the separation may be conceived as taking place mechanically by forcing the two gases to flow through separate membranes A and B, membrane A being permeable to component A only and membrane B to component B only (Fig. 254). Component A leaves membrane A at pressure p_A and is then compressed isothermally to the pressure π of the original mixture. Similarly, component B leaves membrane B at a pressure p_B and is compressed isothermally to pressure π of the original mixture.

For 1 mole of mixture, the isothermal reversible work of compression, assuming ideal-gas behavior, is, from equation 15–21,

$$-w = -RT(y_A \ln y_A + y_B \ln y_B)_T \tag{1}$$

where y_A, y_B = mole fractions of components A and B in the original mixture, or in terms of free energies,

$$-w = y_A \, \Delta\text{G}_A + y_B \, \Delta\text{G}_B \tag{2}$$

For ideal-gas behavior, $-w = -q$, and, since $\Delta s = q/T$ (reversible),

$$\Delta s = R(y_A \ln y_A + y_B \ln y_B) \tag{3}$$

In general, for the reversible isothermal separation of n components of 1 mole of an ideal mixture,

$$-w = \Delta G = -\sum_1^n y_i \, \Delta \bar{G}_i = -RT \sum_1^n y_i \ln y_i \tag{4}$$

For nonideal solution behavior, equation 4 becomes

$$-w = \Delta G = -\sum_1^n \Delta \bar{G}_i = +RT \sum_1^n y_i \ln \gamma_i y_i \tag{5}$$

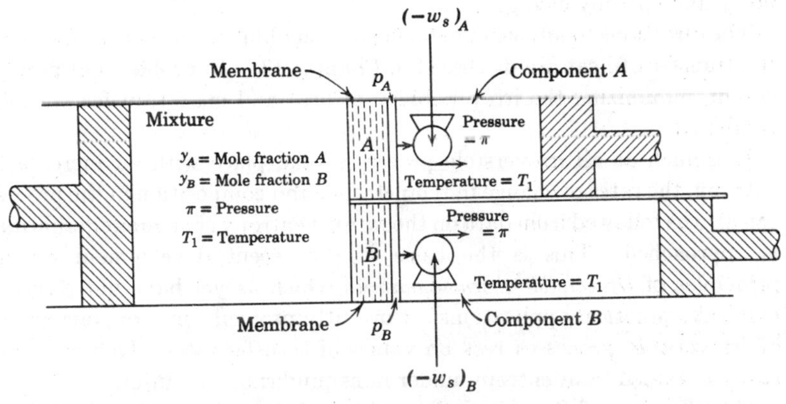

FIG. 254. Isothermal reversible separation

These same relationships apply to liquid solutions as well as to gases, since the change from liquid phase to gas phase of each pure component takes place at its boiling point with no change in free energy.

Separation by Heat and Mass Transfer. All processes involving heat and mass transfer and chemical reactions are inherently irreversible except in the limiting case of complete equilibrium where only infinitesimal temperature and concentration gradients exist, where reactants and products are at equilibrium, and where no turbulence is present in the fluid system. Reversibility is approached only as these gradients of temperature, pressure, concentration, and turbulence within each phase and between phases are reduced to zero and where chemical equilibrium also is approached. Thermodynamic principles are essential in determining conditions of solubility, chemical equilibria, vapor–liquid equilibria, the work of reversible expansion. These principles also indicate the direction to follow in order to minimize irreversibility. The displacement of a system from equilibrium establishes the driving force essential for the spontaneous occurrence of the process.

The major problems in process design are in establishing data on the rates of heat and mass transfer, on rates of chemical reaction and on pressure drops in fluid flow combined with cost data on energy, material, and equipment requirements. Minimizing production costs of power and chemical products is the major objective in process and equipment design. These problems go beyond the realm of chemical thermodynamics for their solution. Successful process design depends on minimizing costs, not on minimizing entropy increase. However, other factors being equal, economy of separation is improved by any change that reduces irreversibility and entropy change.

The directions to pursue in reducing irreversibility in problems involving the transfer of heat are discussed in Chapter 13. In problems of process design, minimizing the irreversibility of heat and mass transfer must be combined.

Inasmuch as all irreversible processes take place with an increase in entropy, the rates of change in temperature and composition in the process can also be followed from data on the rates of entropy change as equilibrium is approached. This is the basis of the recent development of the principles of *irreversible thermodynamics* which as yet have not received extensive practical applications. Currently problems in economic design of irreversible processes rest on values of transfer rates which are more easily obtained than entropy under nonequilibrium conditions.

Classification of Separation Processes. Separation processes may be classified on the basis of differential and stage procedures. In a *differential process,* the two fluids are in continuous contact and are usually in countercurrent flow. The change in concentration of each fluid is gradual in the direction of the flow of each. In a *stage* process, the two fluids are intermittently mixed at various points in the path of flow with abrupt changes in concentration of either stream occurring at each point of flow interruption. The assembly of stages in series is designated as a *cascade.* In gas absorption, differential separation is exemplified by the continuity of absorption in a packed bed, and stage separation by the discontinuity of absorption in a bubble-plate column.

On the basis of irreversibility Benedict[1] has classified the separation processes as shown in Table 65.

The following discussion is limited to gas absorption and fractional distillation, taking place in columns.

[1] M. Benedict, *Trans. Am. Inst. Chem. Engrs.*, **43**, 41 (1947), with permission.

TABLE 65. CLASSIFICATION OF SEPARATION PROCESSES

Irreversible processes
- *In columns*
  - Thermal diffusion
  - Countercurrent gas centrifuge
- *In separate stages*
  - Gaseous diffusion
  - Mass diffusion
  - Electrolysis

Partially reversible processes
- *In columns*
  - Gas absorption
  - Solvent extraction
  - Azeotropic distillation
  - Extractive distillation
  - Chemical exchange
  - Ion exchange
- *In separate stages*
  - Solvent extraction in units

Potentially reversible processes
- *In columns*
  - Distillation
  - Reversible absorption

Gas Absorption

Gas absorption may be carried out by passing a gaseous mixture countercurrent to a stream of liquid in a packed column. The solubility of the solute gas in the liquid as a function of temperature and partial pressure comprises the most essential thermodynamic data needed. For the sake of simplicity, isothermal conditions will be assumed, without consideration of the attendant problems involved in heat transfer. For any given ratio of the rate of liquid flow to the rate of gas flow, the relation of solute concentration y in the gas phase to solute concentration x in the liquid phase is represented by the so-called *operating* line as shown in Fig. 255. For dilute systems, the operating line is straight. The yx operating line is obtained from experimental data.

The equilibrium line y^*x represents the equilibrium concentration of the solute in the two phases in equilibrium with each other and is independent of transfer rates. Here y^* is the mole fraction of solute in the gas phase in equilibrium with a liquid of composition x. From the standpoint of removing the solute from the gas mixture, it is desirable to select a solvent that will render the equilibrium line as nearly horizontal as possible: that is, to obtain a high solubility of the gas at a given temperature and partial pressure. This condition may be undesirable for the subsequent recovery of the solute from the liquid. The factors influencing gas solubility are known only in part as given in Chapters 20 through 23.

The distance between the operating and equilibrium line y–y^* is a measure of the driving force bringing about solution of the gas. This

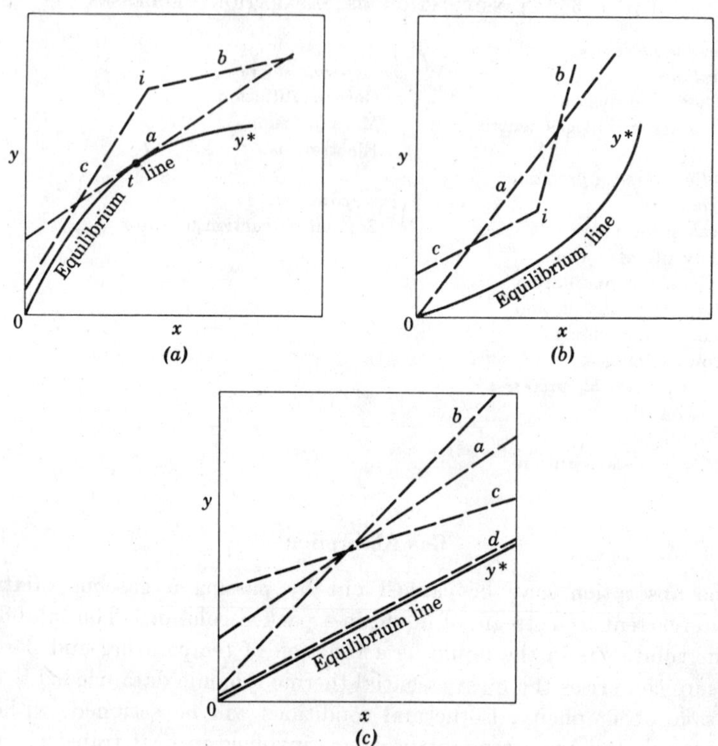

Fig. 255. Irreversibility in gas absorption
——— Equilibrium lines
– – – – Operating lines
(a) Very soluble gas. (b) Slightly soluble gas. (c) Henry's law

difference is also evidence of the irreversibility of the process as far as mass transfer is concerned.

Three types of equilibrium lines for gas solubility are shown in Fig. 255: for gases of high solubility Fig. 255*a*, for gases of low solubility Fig. 255*b*, and for gases following Henry's law Fig. 255*c*.

In general, $$y^* = \frac{\gamma x P_1}{\pi} \tag{6}$$

where the activity coefficient γ is small for highly soluble gases, and is large for gases of low solubility. Where Henry's law applies, $y^* = Hx$ and $H = P_1/\pi$.

For the sake of illustration, the operating lines are assumed to be straight. Actually the operating lines are straight only when dealing with dilute systems in both gas and liquid phases; they will be slightly concave upward when dealing with slightly soluble gases at higher concentrations and concave downward when dealing with highly soluble gases at high liquid concentrations. Reversibility is attained only when the operating and equilibrium lines coincide throughout. For a constant flow rate, this can be approached only in case *c*, where the dotted and full lines coincide. Under actual operation, a departure between the two lines is necessary, preferably keeping a fairly uniform distance between them to maintain an even performance load over the length of column (line *a*). The actual slope to be employed depends on whether a high removal of solute gas is desired (line *b*) or a liquid product of high concentration is desired (line *c*).

For the dissolution of a slightly soluble gas as shown in Fig. 255*b*, reversibility can be attained for a constant flow rate only at terminal conditions (line *a*0). To attain fairly uniform performance throughout, it becomes desirable to change the slope of the operating line by changing from line *bi* to line *ci* at point *i*. This requires increasing the rate of liquid flow in the bottom half of the column, thus obtaining a high concentration of liquid leaving the bottom and a high recovery of gas leaving the top. This is accomplished by employing a high ratio of liquid flow to gas flow L/V in the bottom section of the column (line *bi*) and a low ratio in the top section (line *ci*).

For the dissolution of a highly soluble gas as shown in Fig. 255*c*, reversibility can be attained only at point *t*. To attain nearly uniform performance throughout the length of columns, it becomes desirable to operate with two liquid flow rates for a given gas flow rate as represented by lines *ci* and *bi*, to obtain a favorable recovery of gas and a liquid of high concentration. In the bottom section of the column, the ratio of liquid flow to gas flow L/V is small; in the top section (line *ci*) this ratio is large. The situation is the reverse of that shown in Fig. 255*b*.

The operation of columns with the same flow rate of inert gas but with a change in liquid flow rate presents the problem of producing two liquid products of different concentrations. Only part of the liquid may be passed through the bottom part of the column.

In all cases of gas absorption, the position of the operating line becomes one of the most important considerations. This becomes a problem of economic studies involving operating temperatures and pressures, rates of mass transfer, pressure drops, and costs of materials, power, and equipment.

Gas Absorption under Pressure. To increase the recovery of a solute gas and to produce liquid solutions of higher concentration, it becomes

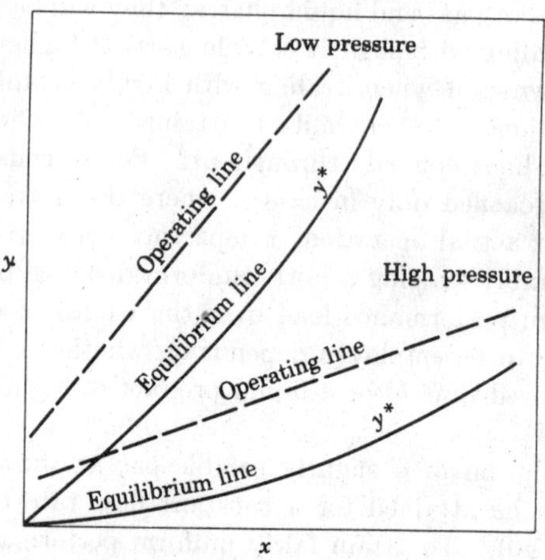

Fig. 256. Effect of pressure on equilibrium line

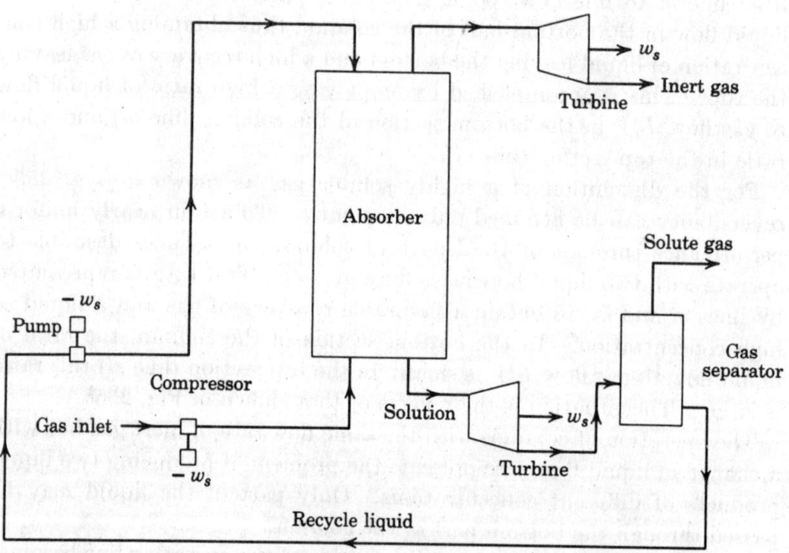

Fig. 257. Gas absorption under pressure with reversible compression and expansion

desirable to operate at high pressures whereby the y^*x equilibrium line becomes more favorable as shown in Fig. 256.

High-pressure absorption requires compression of the entering gas and pumping the absorbent liquid under greater pressure (Fig. 257). If the gas leaving the absorber is expanded freely to the atmosphere, the over-all result is a greatly increased irreversibility of the process over that of low-pressure absorption. This condition can be improved by expanding the spent gases in an engine to generate power for compressing the incoming gas. Also the liquid leaving the column could be expanded through a turbine whereby the reduction in pressure would result in the separation of the component solute gas in a nearly pure state. This expansion and separation could be conducted in stages to produce additional power. The stripped liquid is recycled for further absorption. In high-pressure absorption under these conditions of expansion and compression, the solvent liquid virtually behaves as a semipermeable membrane.

The irreversibility occurring in expansion and compression limits the general application of high-pressure absorption. In this case of gas absorption, thermodynamics becomes of additional importance in calculating the reversible work of compression and expansion.

Benedict[1] gives a more detailed account of high-pressure gas absorption applied to the separation of methane from hydrogen by absorption of the methane in a relatively nonvolatile liquid hydrocarbon at a high pressure, wherein stagewise expansion and compression steps are employed to reduce irreversibility.

Fractional Distillation

Separation of a liquid solution by fractional distillation approaches reversibility when the temperatures of the vapor and liquid in contact with each other are in equilibrium and at the same temperature and pressure throughout the column. An ideal binary solution of close-boiling components will be considered with complete separation into its pure liquid components, using a bubble-cap column (Fig. 258). Under normal conditions, heat is supplied to the reboiler at the bottom of the column at temperature T_B and removed in the condenser at the top of the column at a lower temperature T_D. The feed enters at some intermediate plate in the column and at an intermediate temperature. The temperature and vapor–liquid ratio of the feed can be adjusted for any given total composition so that the total enthalpy of the entering feed is equal to the combined enthalpies of the liquid overhead and bottom product; thus, for 1 mole of feed,

$$\mathrm{H}_F = z_F \mathrm{H}_D + (1 - z_F)\mathrm{H}_B \tag{7}$$

where z_F is the mole fraction of the more volatile component in the combined feed. Under these circumstances, for an adiabatic column the heat q_B added to the reboiler at temperature T_B is equal to the heat $-q_D$ removed from the condenser at temperature T_D: $q = q_B = -q_D$.

The entropy increase in the liquid used for condensing is equal to $-q/T_D$, and the entropy loss from the fluid used for boiling is q/T_B. The net increase in entropy of the two heat-exchange media is, hence,

$$\Delta S = q\left(\frac{1}{T_D} - \frac{1}{T_B}\right) \tag{8}$$

The vapor–liquid equilibrium during separation is shown by the y–x diagram (Fig. 259).

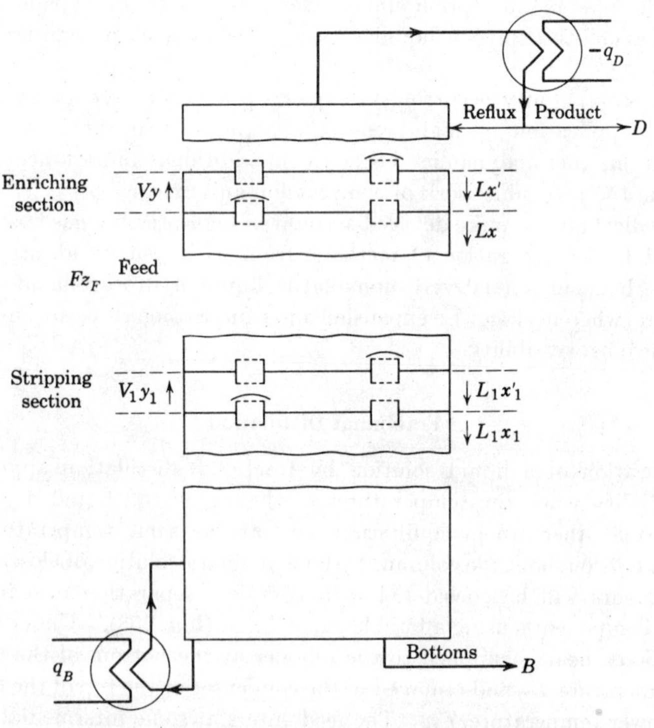

Fig. 258. Conventional fractional distillation

With heat being supplied only in the reboiler and removed only in the condenser, the minimum amount of heat will be obtained when the operating lines above and below the feed plate intersect at point y_F. Even greater heat economy can be theoretically secured by the progressive addition of heat throughout the length of column below the feed plate, instead of in the reboiler only, and the progressive removal of heat throughout the length of column above the feed plate, instead of only in the condenser at the top of the column, as shown in Fig. 260. Under

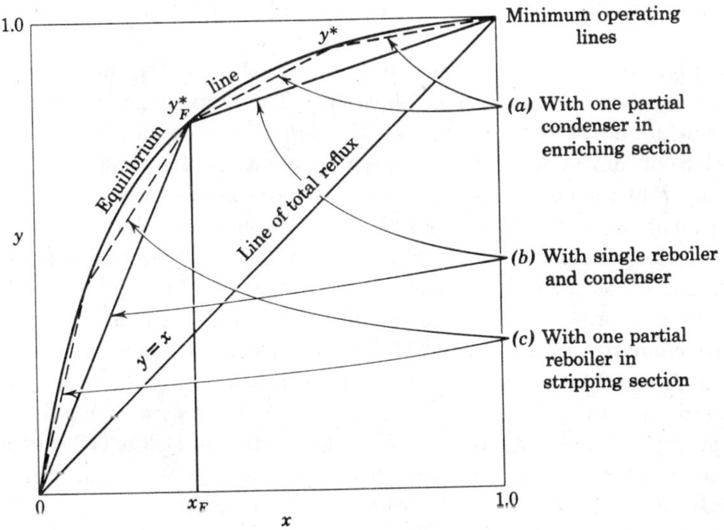

FIG. 259. Equilibrium and operating lines in fractional distillation

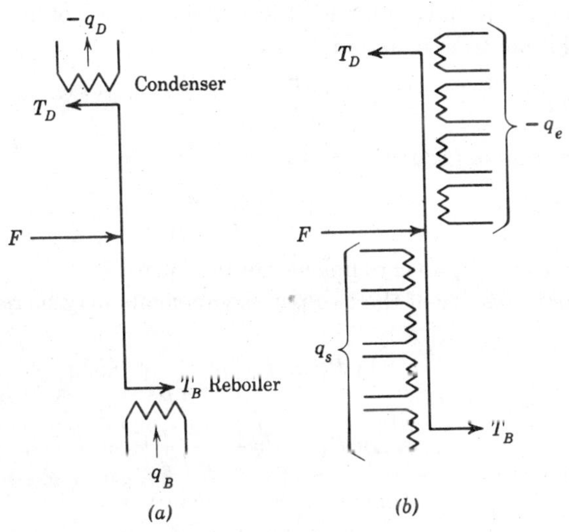

FIG. 260. Distribution of heat-exchange surface in fractional distillation

such circumstances, the operating line can be theoretically made to coincide with the equilibrium line throughout, and result in reducing the heat load to that required by reversible separation.

In Fig. 258, F moles of feed of composition z_F enter the column at some plate between the top and bottom. The more volatile component appears in the product D, leaving the top of the column with a mole fraction of unity for complete separation, $x_D = 1.0$, and the liquid product B at the bottom of the column contains none of the more volatile component, $x_B = 0$. The vapor phase is gradually enriched in the more volatile component toward the top of the column and gradually depleted in the more volatile component toward the bottom of the column.

In order to procure enrichment, part of the condensate must be returned to the column. In conventional operation, with a single heat source, the minimum heat is required when the operating line, having a slope equal to the ratio of liquid and vapor rates L/V, intersects the equilibrium line at y^*_F. The section of the column above the feed plate is designated as the *enriching section*, and that below the feed plate as the *stripping section.* In order to maintain steady operation of the column, the same upward flow D of the more volatile product must be maintained throughout the enriching section, and the same downward flow B of the less volatile product must be maintained throughout the stripping section.

Under equilibrium conditions at any point in the column for an ideal binary solution, from equation 21–19.

$$\alpha = \frac{y(1-x)}{x(1-y)} = \frac{P_1}{P_2} \tag{9}$$

and the *separation factor* $\alpha - 1$ is given as

$$\alpha - 1 = \frac{y-x}{x(1-y)} \tag{10}$$

Where $\alpha - 1$ is zero, no enrichment can take place.

The vapor pressure of the two pure components may be represented as follows

$$\begin{aligned} \ln P_1 &= A_1 - \frac{\lambda_1}{RT_D} \\ \ln P_2 &= A_2 - \frac{\lambda_2}{RT_B} \end{aligned} \tag{11}$$

For close-boiling components $\lambda_1 = \lambda_2$; hence,

$$A_1 - A_2 = \frac{\lambda}{R}\left(\frac{1}{T_D} - \frac{1}{T_B}\right) = \ln\frac{P_1}{P_2} = \ln \alpha \tag{12}$$

Where P_1 and P_2 are nearly equal as for difficult cases of separation,

$$\ln \alpha = \alpha - 1. \tag{13}$$

Hence
$$\frac{\lambda}{R}\left(\frac{1}{T_D} - \frac{1}{T_B}\right) = \alpha - 1 \tag{14}$$

A material balance around the enriching section for any plate (Fig. 258) gives, for the more volatile component,

$$yV = Lx' + D \tag{15}$$

The prime designates liquid coming from the plate above the plate under consideration. For both components,

$$V = L + D \tag{16}$$

Combining equations 15 and 16 gives

$$y - x' = (1 - y)\frac{D}{L} \tag{17}$$

Corresponding material balances in the stripping section for any given plate yield

$$L_1x'_1 = V_1y_1 \tag{18}$$

$$L_1 = V_1 + B \tag{19}$$

$$y_1 - x'_1 = \frac{B}{L_1}y_1 \tag{20}$$

The subscript 1 refers to conditions in the stripping section.

Combining equations 17 and 10 gives for the enriching section

$$x' - x = (\alpha - 1)x(1 - y) - (1 - y)D/L \tag{21}$$

Combining equations 20 and 10 gives for the stripping section

$$x' - x_1 = (\alpha - 1)x_1(1 - y_1) - y_1B/L_1 \tag{22}$$

In the enriching section, the difference in composition between the two streams leaving the same stage is given as

$$y - x = (\alpha - 1)x(1 - y) \tag{23}$$

Thus the excess of enrichment in the liquid stream from stage to stage is given by the term $(1 - y)D/L$, or is inversely proportional to the down-flow rate L. The minimum value of L is secured where the enrichment $x' - x$ becomes zero; thus, from equation 21, where $x_F = z_F$:

$$L_m = \frac{D}{(\alpha - 1)z_F} \tag{24}$$

Since $F = D + B$, and $z_FF = D$, equation 24 may be written,

$$L_m = \frac{F}{\alpha - 1} \tag{25}$$

Similarly, for the stripping section,

$$L_m = \frac{F(1 - z_F)y_1}{(\alpha - 1)x_1(1 - y_1)} \tag{26}$$

The minimum heat required is, hence,

$$q_{\min} = L_m\lambda \tag{27}$$

or

$$\Delta S_m = L_m\lambda\left(\frac{1}{T_D} - \frac{1}{T_B}\right) \tag{28}$$

Combining equations 28, 25, and 14 gives

$$\Delta S_m = RF \tag{29}$$

The minimum entropy increase in conventional fractional distillation of ideal solutions is hence independent of the enrichment factor or the heat of vaporization. For components boiling close together, the temperature difference from reboiler to condenser is small, but the amount of heat required for separation is large so that the minimum entropy increase tends to remain constant. For complete separation of 1 mole of a binary ideal solution of close-boiling components, a minimum net entropy increase of the heat-exchange media of R units in conventional fractional distillation is required independent of feed composition.

The ratio of the minimum change in entropy in reversible isothermal separation to the minimum change in conventional fractional distillation with minimum reflux gives the thermodynamic efficiency as shown in Table 66.

TABLE 66. EFFICIENCY OF CONVENTIONAL DISTILLATION

Feed Composition, mole %	Thermodynamic Efficiency of Conventional Distillation
0 or 100	0.000
1 or 99	0.056
5 or 95	0.189
10 or 90	0.325
20 or 80	0.500
30 or 70	0.611
50	0.693

The efficiency of separation by conventional fractional distillation at minimum reflux is highest for a feed composition of 0.5 mole fraction, but becomes progressively poorer as the mole fraction in either component is diminished and approaches zero at zero concentration for either component.

Improvement over Conventional Distillation. As previously discussed, the thermodynamic efficiency of separation by conventional fractional distillation can be improved by approaching a minimum reflux ratio not only at the feed plate but also at every plate in the column: that is, by approaching a zero temperature difference for heat exchange at each plate. This is conceivable by employing a partial reboiler at each plate in the stripping section and a partial condenser at each plate in the enriching section.

When a distillation column is run in this manner, only a small fraction of the heat added to the column flows from the bottom to the top, so that the entropy increase is less than in conventional distillation. This is especially true for feed compositions containing small amounts of either component.

This type of distillation is impracticable but indicates the direction of possible improvement. The addition or removal of heat should be tapered toward each end of the column.

Based on the example given by Benedict for the separation of a feed composition of 0.1 mole fraction into pure components, the values of entropy increase and heat input per mole of feed are recorded in Table 67 for the indicated methods of separation. The results in the last column are for a temperature of 350° K, where $\lambda = 8000$ cal per g-mole and $\alpha = 2$.

TABLE 67. COMPARISON OF VARIOUS METHODS OF SEPARATION

	ΔS, Entropy Increase	Heat or Work Added	
1. Reversible separation	$0.325R$	$-w = 0.325RT$	$-w = 228$ cal
2. Conventional distillation			
(*a*) Minimum reflux	R	$q = \lambda/(\alpha - 1)$	$q = 8000$ cal
(*b*) 1.25 minimum reflux	$1.25R$	$q = \dfrac{1.25\lambda}{\alpha - 1}$	$q = 10{,}000$ cal
3. Addition of 1 partial condenser in the stripping section at $x = 0.3$ with 1.25 minimum reflux	$0.655R$	$q = \dfrac{0.53\lambda}{\alpha - 1}$	$q = 5300$ cal

Problems

1. Calculate the minimum total work in Btu in separating 100 lb of a solution at 80° F containing 30 mole per cent benzene, 25% toluene, and 45% xylene into the three pure components. Also calculate the entropy change per pound of solution.

2. Calculate the minimum work and the entropy change in separating 100 lb of a 2-molal aqueous solution of sucrose into its components at 25° C. Compare this result with the heat required to vaporize the water at the same temperature.

3. Verify the values of ΔS, q, and w recorded in Table 67.

25
Chemical Equilibrium Constants

As explained in Chapter 13, the criterion of equilibrium in a chemically reacting system at constant temperature and pressure is that the change in free energy of any possible reaction shall be zero. A negative free-energy change may be looked on as the driving potential which is directing the reaction towards a state of equilibrium, and as a direct measure of the departure of the reacting system from its equilibrium state. From a kinetic viewpoint the reaction at equilibrium may still be considered as proceeding reversibly, but with equal rates in opposite directions and with no net change in composition.

The Equilibrium Constant

For developing the equations of chemical equilibrium the general reaction represented by the following stoichiometric equation is considered:

$$bB + cC + \cdots = rR + sS + \cdots \tag{1}$$

where b, c, r, s are the number of moles of reactants and products B, C, R, and S, respectively.

When this reaction proceeds isothermally at any temperature T, starting with each reactant in its standard state of unit activity and ending with products each at unit activity, the accompanying change of free energy is represented by the symbol $\Delta G°$. The corresponding activity $a°_B$, $a°_C$, $a°_R$, $a°_S$ of each component in the standard state is unity.

When this reaction proceeds isothermally until equilibrium is established, all the activities adjust themselves to new values at which the change in free energy is zero. At equilibrium the activities of the separate components are represented by symbols, a_B, a_C, a_R, and a_S.

Since the free-energy change is the difference between the free energies of the products and reactants,

$$\Delta G° = r\bar{G}°_R + s\bar{G}°_S + \cdots - b\bar{G}°_B - c\bar{G}°_C - \cdots \tag{2}$$

$$\Delta G = r\bar{G}_R + s\bar{G}_S + \cdots - b\bar{G}_B - c\bar{G}_C - \cdots \tag{3}$$

where $\bar{G}$ = partial molal free energy.

Combining equations 2 and 3 gives

$$\Delta G - \Delta G^\circ = r(\bar{G}_R - \bar{G}^\circ_R) + s(\bar{G}_S - \bar{G}^\circ_S) + \cdots - b(\bar{G}_B - \bar{G}^\circ_B) - c(\bar{G}_C - \bar{G}^\circ_C) \quad (4)$$

Equation 4 may be written in terms of activities by combination with equations 20–84 and 85. Thus,

$$\Delta G - \Delta G^\circ = rRT \ln a_R + sRT \ln a_S + \cdots - bRT \ln a_B - cRT \ln a_C - \cdots$$

or

$$\Delta G - \Delta G^\circ = RT \ln \left(\frac{a_R{}^r a_S{}^s \cdots}{a_B{}^b a_C{}^c \cdots} \right) \quad (5)$$

At equilibrium, $\Delta G = 0$, and

$$-\frac{\Delta G^\circ}{T} = R \ln \left(\frac{a_R{}^r a_S{}^s \cdots}{a_B{}^b a_C{}^c \cdots} \right) = R \ln K \quad (6)$$

where K is the *equilibrium constant* at temperature T. Thus,

$$K = \frac{a_R{}^r a_S{}^s \cdots}{a_B{}^b a_C{}^c \cdots} = \exp\left(-\frac{\Delta G^\circ}{RT}\right) = \exp\left(-\frac{\Delta H^\circ}{RT} + \frac{\Delta S^\circ}{R}\right) \quad (7)$$

where ΔG°, ΔH°, and ΔS° are the changes in free energy, enthalpy, and entropy, respectively, that accompany the stoichiometric reaction with all reactants and products in their standard states.

From equation 7 it is evident that the equilibrium constant is determined by the temperature and the free-energy change which would accompany the reaction of the indicated numbers of moles if each reactant were initially in its standard state and each product finally in its standard state of unit activity at the temperature of the system. This free-energy change ΔG° is termed the *standard free-energy change* of the reaction. The standard free-energy change depends on the temperature, the definition of the standard state of each component, and the number of moles entering into the stoichiometric equation under consideration. Accordingly, the numerical value of an equilibrium constant is without significance unless it is accompanied by specifying these three factors.

The effect on the equilibrium constant of the form of the stoichiometric equation is illustrated by consideration of the synthesis of ammonia. This reaction may be designated as

$$N_2 + 3H_2 = 2NH_3$$

The corresponding equilibrium constant is then

$$K' = \frac{(a_{NH_3})^2}{a_{N_2}(a_{H_2})^3} \tag{8}$$

The same reaction may be written

$$\tfrac{1}{2}N_2 + \tfrac{3}{2}H_2 = NH_3$$

and

$$K = \frac{a_{NH_3}}{(a_{N_2})^{1/2}(a_{H_2})^{3/2}} \tag{9}$$

It is evident that, in this case, $K' = K^2$, and specification of the number of moles involved in the stoichiometric equation is essential.

When the standard state of each component of a reacting system is taken at 1 atm pressure, whether gaseous, liquid, or solid, the equilibrium constant is taken at 1 atm pressure. As so defined, there is no change in pressure of the standard state, and the *reaction equilibrium constant is independent of the pressure of the system.* This is the standard state customarily employed in dealing with equilibria in chemical reactions.

In dealing with physical equilibrium, it is customary to select as the standard state the pure components at the temperature and pressure of the system. As shown in Chapter 20, this basis has the advantage of permitting useful correlations between mole fractions and activity coefficients. Where the standard state is referred to behavior at infinite dilution, the pressure of the system is also taken as the standard state. With the standard states defined in this manner, the equilibrium constants vary as a function of pressure.

$$\left(\frac{\partial \ln K}{\partial p}\right)_T = -\frac{\Delta V^\circ}{RT} \tag{10}$$

where $\Delta V^\circ = \sum V^\circ{}_P - \sum V^\circ{}_R$, and V° = volume at standard state.

Standard Free-Energy Changes

It is pointed out in Chapter 13 that free energy is an extensive property and that the free-energy change in any process is determined by the final and initial conditions and not by the intermediate path. Thus, free-energy changes may be treated in the same manner as enthalpy changes and are functions of temperature, pressure, and composition at the beginning and end of the change.

The free-energy change accompanying a reaction as it actually proceeds under the conditions of the reaction process is of little interest except in the consideration of electrochemical reactions, where the decrease in free energy of the system represents the electric energy released under reversible conditions at constant pressure. Also, it has been pointed out in Chapter 15 that the decrease in free energy of a flow process proceeding reversibly at constant temperature without generation of electric energy is equal to the mechanical work done. Usually chemical reactions proceed in a highly irreversible manner without generation of either electric energy or other useful work, and the loss in free energy is not accompanied by a corresponding release of useful work.

The greatest value of the free-energy concept is in the calculation of compositions of reacting systems at equilibrium under which conditions the actual free-energy changes are zero. These compositions are, however, related by equation 6 to the standard free-energy changes which are not equal to zero at the standard states. In considering any changes that take place in a system, actual and standard free-energy changes must not be confused.

Standard Free-Energy Change at 25° C. The standard free-energy change and the corresponding standard enthalpy and entropy changes which accompany a reaction at an arbitrary reference temperature may be calculated by the summation principles demonstrated in Chapter 9 for heats of reaction. The accepted reference temperature extensively used for thermodynamic data is 25° C (298.16° K).

It has been common practice in the past to present standard free-energy data in the form of tables of standard free energies of formation at 25° C. From such tables a standard free energy of reaction is calculated as the algebraic sum of the standard free energies of formation of the products less the algebraic sum of the standard free energies of formation of the reactants. When an element enters into a reaction, its standard free energy of formation is zero if its state of aggregation is that selected as the basis for the standard free energy of its compounds.

Illustration 1. Calculate the standard change of free energy at 298.16° K in the gas-phase alkylation of isobutane with ethylene to form neohexane. Free energies of formation at 1 atm, 298.16° K, and the ideal-gaseous state are as follows:

C_4H_{10}, isobutane (gas): $\Delta G^\circ_f = -5000.0$ g-cal/(g-mole)(K°)

C_2H_4, ethylene (gas): $\Delta G^\circ_f = +16{,}282.0$ g-cal/(g-mole)(K°)

C_6H_{14}, neohexane (gas): $\Delta G^\circ_f = -2370.0$ g-cal/(g-mole)(K°)

For the reaction

$$C_4H_{10}(g) + C_2H_4(g) \rightarrow C_6H_{14}(g)$$

$$\Delta G^\circ_{298.16} = \Sigma(\Delta G^\circ_f)_P - \Sigma(\Delta G^\circ_f)_R = (-2370) - (-5000 + 16{,}282)$$

$$\Delta G^\circ_{298.16^\circ} = -13{,}652 \text{ g-cal}$$

Effect of Temperature on $\Delta G°$ and K. A general differential relationship between $\Delta G°$ and temperature is expressed by equation 13–154. Combination of this with equation 6 yields a similar expression for the effect of temperature on the equilibrium constant. Thus

$$R\left(\frac{\partial \ln K}{\partial T}\right)_p = \frac{\Delta H°}{T^2}$$

or

$$R\left(\frac{\partial \ln K}{\partial(1/T)}\right)_p = -\Delta H° \tag{11}$$

Equation 11 may be used to determine the standard heat of reaction from equilibrium constants known at different temperatures from the slope of a curve relating $R \ln K$ to $1/T$.

An integrated relationship between the standard free-energy change and temperature is obtained by expressing the standard heat of reaction and entropy change in terms of empirical heat-capacity equations. If heat capacities are represented by the general empirical expression $c°_p = a + bT + cT^2 + dT^3 + e/T^2$, equation 9–71, page 347, may be written

$$\Delta H°_T = I_H + \Delta aT + \tfrac{1}{2}\Delta bT^2 + \tfrac{1}{3}\Delta cT^3 + \tfrac{1}{4}\Delta dT^4 - \Delta e/T \tag{12}$$

The integration constant I_H is determined from a single value of $\Delta H°$ at a temperature within the range of applicability of the heat-capacity equations.

Since at constant pressure $dS = c_p\, dT/T$, it follows that

$$d(\Delta S)° = \Delta C°_p\, dT/T$$

or

$$\Delta S°_T = I_S + \Delta a \ln T + \Delta bT + \tfrac{1}{2}\Delta cT^2 + \tfrac{1}{3}\Delta dT^3 - \frac{\Delta e}{2T^2} \tag{13}$$

The integration constant I_S is determined from a single value of $\Delta S°$ at any temperature within the range of the accuracy of the heat-capacity equations.

Since $\Delta G° = \Delta H° - T\Delta S°$, from equations 12 and 13,

$$\frac{\Delta G°}{T} = \frac{I_H}{T} + (\Delta a - I_S) - \Delta a \ln T - \frac{\Delta bT}{2} - \frac{\Delta cT^2}{6} - \frac{\Delta dT^3}{12} - \frac{\Delta e}{2T^2} \tag{14}$$

The integration constants I_H and I_S are determined from a single value of $\Delta H°$ together with a single value of either $\Delta S°$ or $\Delta G°$. These reference values may be at different temperatures.

The forms of equations 12 ,13, and 14 depend on the empirical equations used for expressing heat capacities.

Standard Free Energy of Formation. The standard free energy of formation of a compound from its elements at any temperature level is a special modification of equation 14 as illustrated in the following:

Illustration 2. Derive a general equation for the standard free energy of formation of $NH_3(g)$ at temperature $T°$ K with the standard state of each component being the ideal gas at 1 atm. The absolute entropies in the ideal-gaseous state at 298.16° K and 1 atm are as follows:

$$NH_3(g) = 46.010 \text{ g-cal/(g-mole)(K°)}$$
$$N_2(g) = 45.767 \text{ g-cal/(g-mole)(K°)}$$
$$H_2(g) = 31.211 \text{ g-cal/(g-mole)(K°)}$$

From Table C of the appendix,

$$NH_3\text{:}\quad c°_p = 6.5846 + (0.61251)10^{-2}T + (0.23663)10^{-5}T^2 - (1.5981)10^{-9}T^3$$
$$N_2\text{:}\quad c°_p = 6.903 - (0.03753)10^{-2}T + (0.1930)10^{-5}T^2 - (0.6861)10^{-9}T^3$$
$$H_2\text{:}\quad c°_p = 6.952 - (0.04576)10^{-2}T + (0.09563)10^{-5}T^2 - (0.2079)10^{-9}T^3$$

For the reaction:

$$\tfrac{1}{2}N_2(g) + \tfrac{3}{2}H_2(g) \rightarrow NH_3(g)$$
$$\Delta H°_{298.16} = -11{,}040 \text{ g-cal per g-mole}$$

$$\Delta a = 6.5846 - [\tfrac{1}{2}(6.903) + \tfrac{3}{2}(6.952)] = -7.2949$$
$$\Delta b = (0.61251)10^{-2} - [\tfrac{1}{2}(-0.03753)10^{-2} + \tfrac{3}{2}(-0.04576)10^{-2}] = (0.69992)10^{-2}$$
$$\Delta c = (0.23663)10^{-5} - [\tfrac{1}{2}(0.1930)10^{-5} + \tfrac{3}{2}(0.09563)10^{-5}] = -(0.003315)10^{-5}$$
$$\Delta d = (-1.5981)10^{-9} - [\tfrac{1}{2}(-0.6861)10^{-9} + \tfrac{3}{2}(-0.2079)10^{-9}] = -(0.9432)10^{-9}$$

$$I_H = \Delta H°_{298.16} - \Delta a(298.16) - \frac{\Delta b}{2}(298.16)^2 - \frac{\Delta c}{3}(298.16)^3 - \frac{\Delta d}{4}(298.16)^4$$

$$= -11{,}040 - (-7.2949)(298.16) - \frac{(0.69992)10^{-2}}{2}(8.8899)10^4$$

$$- \frac{(-0.003315)10^{-5}}{3}(2.6506)10^7 - \frac{(-0.9432)10^{-9}}{4}(7.9031)10^9 = -9173.9$$

$$\Delta S°_{298.16} = 46.010 - [\tfrac{1}{2}(45.767) + \tfrac{3}{2}(31.211)] = -23.69$$

$$I_S = \Delta S°_{298.16} - \Delta a(2.30259)\log 298.16 - \Delta b(298.16) - \frac{\Delta c}{2}(298.16)^2 - \frac{\Delta d}{3}(298.16)^3$$

$$= -23.69 - (-7.2949)(2.30259)(2.47445) - (0.69992)10^{-2}(298.16)$$

$$- \frac{(-0.003315)10^{-5}}{2}(8.8899)10^4 - \frac{(-0.9432)10^{-9}}{3}(2.6506)10^7 = 15.797$$

$$\frac{\Delta G°}{T} = \frac{I_H}{T} - \Delta a \ln T - \frac{\Delta b}{2}T - \frac{\Delta c}{6}T^2 - \frac{\Delta d}{12}T^3 + (\Delta a - I_S)$$

$$= -\frac{9173.9}{T} + 7.2949 \ln T - (0.34996)10^{-2}T + (0.0005525)10^{-5}T^2$$

$$+ (0.0786)10^{-9}T^3 - 23.092$$

The standard free-energy change of any reaction at any temperature level may be obtained by the method of Illustration 2 in which equation 14 is applied directly to the reaction combined with a knowledge of heat-capacity equations of the components and single values of ΔH° and ΔG° or ΔS° for the reaction, each at some known temperature. An alternative method is to evaluate the free energies of formation as functions of temperature, and calculate the free-energy change of the reaction by combining the standard free energies of formation of the separate components. There is little point in this latter procedure except where numerous different reactions that involve the same components are under consideration.

For reactions in which $\Delta C_p = 0$ both ΔH° and ΔS° are independent of temperature. Equation 14 then reduces to

$$\Delta G^\circ = \Delta H^\circ_1 - T\,\Delta S^\circ_1 \tag{15}$$

From equation 15 it is evident that the standard free-energy change of a reaction may vary widely with temperature, even though the standard heat of reaction and entropy change are constant.

Where the heat-capacity data for a reaction are not conveniently expressed by empirical equations, the standard free-energy change at any temperature may be calculated by correcting separately the standard enthalpy and entropy changes by graphical integrations. Thus, if ΔC_p is the difference between the heat capacities of the products and of the reactants, and it is desired to obtain the standard free-energy change at a temperature T_3 from values of standard enthalpy and entropy changes ΔH°_1 and ΔS°_2 at temperatures T_1 and T_2, respectively:

$$\Delta G^\circ_3 = \Delta H^\circ_1 + \int_{T_1}^{T_3} \Delta C^\circ_p\,dT - T_3\,\Delta S^\circ_2 - T_3 \int_{T_2}^{T_3} \frac{\Delta C^\circ_p}{T}\,dT \tag{16}$$

The integrations of equation 16 may be carried out graphically by plotting ΔC°_p and $\Delta C^\circ_p/T$ against T.

Standard Free Energies of Formation of Hydrocarbons. Standard free energies of formation of a few hydrocarbons as a function of temperature are shown in Fig. 261, modified from a similar figure of Parks and Huffman.[1] In this figure, the molal free energies of formation are expressed per gram-atom of carbon as graphite at unit activity, in order that the figure shall show visually the relative stabilities of the different compounds in their standard states with respect to the elements. For example, the fact that the acetylene curve C_2H_2 is positive indicates that, over the entire temperature range, this compound is thermodynamically unstable in its standard state and tends to decompose spontaneously

[1] G. S. Parks and H. M. Huffman, *The Free Energies of Some Organic Compounds*, Reinhold Publishing Corp. (1932).

to an equilibrium mixture containing large proportions of carbon and hydrogen. Similarly, since the curve for propane C_3H_8 is higher than that for ethane C_2H_6, the latter compound is the more stable. Propane becomes relatively unstable at temperatures above 450° K.

The standard free energies of formation of the hydrocarbons are obtained by multiplying the values from Fig. 261 by the number of

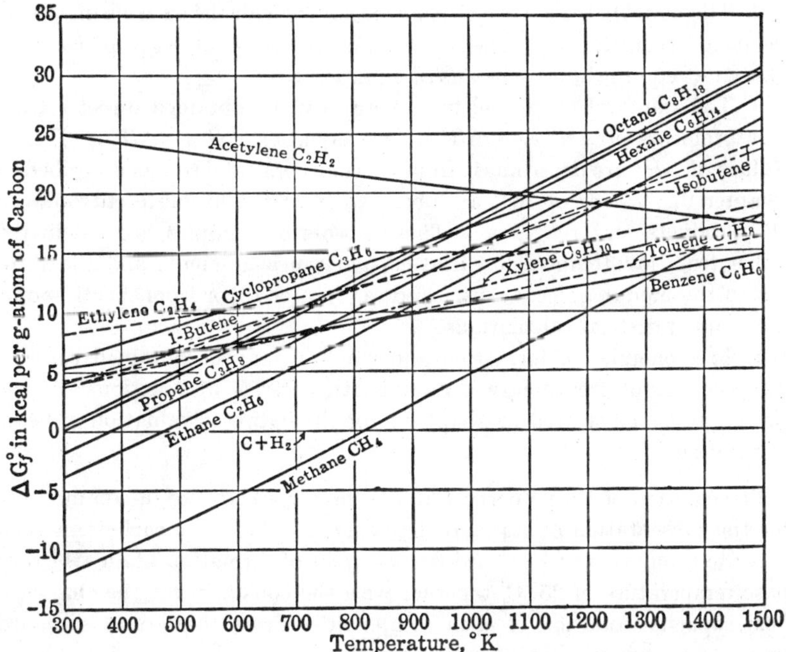

Fig. 261. Standard free energies of formation of hydrocarbons

(Revision of chart from Parks and Huffman, *Free Energies of Some Organic Compounds*, with permission of the Reinhold Publishing Corp.)

carbon atoms. The standard free-energy changes of reactions involving these hydrocarbons may then be calculated by summations of the standard free energies of formation.

Illustration 3. Calculate the standard free-energy change of the following reaction at 1000° K from the data of Fig. 261.

$$2C_2H_4 = iC_4H_8$$

$$iC_4H_8: \quad \Delta G^\circ_f = (4)(15{,}300) = 61{,}200 \text{ cal per g-mole}$$

$$C_2H_4: \quad \Delta G^\circ_f = (2)(14{,}000) = 28{,}000 \text{ cal per g-mole}$$

For the reaction at 1000° K,

$$\Delta G^\circ = 61{,}200 - 56{,}000 = 5200 \text{ cal}$$

Determination of Standard Free Energy. Four common methods are available for determining standard free energies of formation and the standard free-energy change of a chemical reaction.

1. For reactions that proceed reversibly in a galvanic cell at constant pressure and temperature, the free-energy change can be obtained directly from measurements of electric energy generated.

2. The standard free-energy change may be calculated from measurements of temperature, pressure, and composition of each phase in a system at equilibrium by means of equation 6.

3. The standard change in free energy can be obtained directly from measurements of ΔH° and ΔS° by the equation $\Delta G^\circ = \Delta H^\circ - T\,\Delta S^\circ$. Values of ΔH° are determined from thermochemical data, as described in Chapter 9. Values of ΔS° are obtained from the absolute entropies of the products and reactants. These absolute entropies are calculated either from low-temperature heat-capacity measurements and the third law of thermodynamics, as discussed in Chapter 13, or from spectroscopic data and statistical calculations.

4. Free energies of formation and reaction may be obtained through combination of free-energy data for other series of reactions by the procedure demonstrated in Chapter 9 for the indirect evaluation of heats of reaction.

Presentation of Free-Energy Data. Three methods are in common use for the presentation of standard free-energy data. The earliest method was the tabulation of standard free energies of formation at an arbitrary base temperature of 25° C, together with the constants for the empirical heat-capacity equation of each compound. From these data standard free-energy changes may be calculated at any desired temperature by the methods demonstrated in Illustration 2.

An alternative method is to present tables of heats of formation and absolute entropies at a selected base temperature. By combining these values, the standard free energies of reaction at the base temperature are obtained. Free energies at other temperatures are then calculated from the heat-capacity equations. This method has the advantage of separating the heat of formation and entropy data on which the free-energy calculations are based. Changes are constantly being made in the accepted values, and tabulation of separate heats of formation and entropies minimize the confusion resulting from such changes. Furthermore, the separate enthalpy and entropy values are useful for other calculations.

In Table 68 are values of absolute entropies at 25° C. The use of such data in conjunction with the heats of formation and reaction of Tables 29 and 30, pages 297 and 306, is demonstrated in Illustration 2.

With the extension of calorimetric measurements into the region of low temperatures and the development of statistical methods based on spectroscopic data, it has become possible to express values of enthalpy and free energy of many substances relative to their values at the absolute zero of temperature. A further improvement has been introduced by reporting values of $(H^\circ_T - H^\circ_0)$ and $(G^\circ_T - G^\circ_0)/T$ as direct tabular or graphical functions of temperature rather than formulating the results in terms of the constants of empirical heat-capacity equations. Consistent with this practice heats of formation ΔH°_{f0} at 0° K are now being reported relative to the elements at 0° K instead of at some arbitrary temperature level. By this scheme, the agreement of different values can be judged directly by numerical values rather than by the comparison of different equations which appear in as many forms as there are empirical heat-capacity equations for the separate components. The relative agreement of several forms of such equations can be judged only by repetition of calculations and usually not by inspection.

The standard heat of formation of a compound relative to its elements, all in their standard state of unit activities at any temperature level, is expressed by the following equation:

$$\Delta H^\circ_{fT} = [(H^\circ_T - H^\circ_0) + \Delta H^\circ_{f0}]_C - \sum (H^\circ_T - H^\circ_0)_E \tag{17}$$

where

ΔH°_{fT} = standard heat of formation at temperature T

H°_T = enthalpy of the compound or element at temperature T

H°_0 = enthalpy of the compound or element at 0° K

ΔH°_{f0} = standard heat of formation at 0° K

The subscripts C and E refer to the compound and to the element, respectively. At absolute zero temperature, the four energy properties become equal. Thus,

$$U^\circ_0 = H^\circ_0 = G^\circ_0 = A^\circ_0 \tag{18}$$

As a result, it is common practice to report data in terms of the exactly equivalent functions $(G^\circ_T - G^\circ_0)/T$, or $(G^\circ_T - H^\circ_0)/T$, or $(G^\circ_T - H^\circ_0)/T$, and ΔH°_{f0} or ΔG°_{f0}.

The standard free energy of formation of a compound at a temperature T from the elements at the same temperature is expressed by an equation similar to 17.

$$\left(\frac{\Delta G^\circ_f}{T}\right)_T = \left[\left(\frac{G^\circ_T - H^\circ_0}{T}\right) + \frac{\Delta H^\circ_{f0}}{T}\right]_C - \sum \left(\frac{G^\circ_T - H^\circ_0}{T}\right)_E \tag{19}$$

TABLE 68. STANDARD MOLAL ENTROPIES s° AT 25° C (298.16° K)
g-cal/(g-mole)(K°)

Standard States

Gases: The ideal-gaseous state at 1 atm pressure.
Liquids: The pure liquid at 1 atm pressure.
Solids: The pure solid at 1 atm pressure.
Aqueous Solutions: The hypothetical ideal 1.0 molal solution in which $a/m = 1.0$. The ionic entropies are relative to zero for the hydrogen ion H^+.

Abbreviations

c = crystalline state g = gaseous state
l = liquid state aq = dilute aqueous solution

INORGANIC COMPOUNDS

References

1. *Selected Values of Chemical Thermodynamic Properties*, as of July 1, 1953, edited by D. D. Wagman, National Bureau of Standards, with permission.
2. K. K. Kelley, "The Entropies of Inorganic Substances," *U.S. Bur. Mines Bull.*, **434** (1941), with permission.

	Aluminum		
$Al(c)$	6.769	Al_2O_3(corundum, c)	12.186
$Al(g)$	39.303	Al_2SiO_5(sillimanite, c)	27.0
$Al^{+++}(aq)$	−74.9		
	Antimony		
$Sb(c)$	10.5	$Sb_2O_4(c)$	30.3
$Sb(g)$	43.06	$Sb_2O_5(c)$	29.9
$Sb_2(g)$	60.9	$Sb_2S_3(c)$	39.6
$Sb_2O_3(c)$	29.4	$SbCl_3(g)$	80.8
$Sb_4O_6(g)$	102	$SbCl_3(c)$	44.5
	Argon		
$A(g)$	36.983		
	Arsenic		
$As(c)$	8.4	$As_2O_5(c)$	25.2
$As(g)$	33.22	$AsCl_3(g)$	78.2
$As_2(g)$	57.3	$AsCl_3(l)$	55.8
$As_4(g)$	69	$AsF_3(g)$	69.08
As_2O_3(octahedral, c)	25.6	$AsF_3(l)$	43.31
$As_4O_6(g)$	101		
	Barium		
$Ba(c)$	16	$BaCl_2 \cdot 2H_2O(c)$	48.5
$Ba(g)$	40.699	$BaF_2(c)$	23.0
$Ba^{++}(aq)$	3	$BaCO_3$(witherite, c)	26.8
$BaO(c)$	16.8	$BaNO_3(c)$	51.1
$BaO(g)$	55.70	$BaSO_4(c)$	31.6
$BaCl_2(c)$	30		
	Beryllium		
$Be(c)$	2.28	$BeO(c)$	3.37
$Be(g)$	32.545	$BeO(g)$	47.18

TABLE 68 (CONTINUED) INORGANIC COMPOUNDS

Substance	Entropy	Substance	Entropy
Bismuth			
$Bi(c)$	13.6	$Bi_2O_3(c)$	36.2
$Bi(g)$	44.67	$BiCl_3(g)$	85.3
$Bi_2(g)$	65.4	$BiCl_3(c)$	45.3
Boron			
$B(c)$	1.56	$BBr_3(g)$	77.49
$B(g)$	36.649	$BCl_3(g)$	69.29
$B_2O_3(c)$	12.91	$BF_3(g)$	60.70
$H_3BO_3(c)$	21.41	$B_4C(c)$	6.47
Bromine			
$Br_2(l)$	36.4	$Br^-(aq)$	19.29
$Br_2(g)$	58.639	$BrCl(g)$	57.34
$Br(g)$	41.8052	$BrO_3^-(aq)$	38.2
Cadmium			
$Cd(c)$	12.3	$CdCl_2(c)$	28.3
$Cd(g)$	40.067	$Cd(OH)_2(c)$	22.8
$Cd^{++}(aq)$	−14.6	$CdCO_3(c)$	25.2
$CdO(c)$	13.1	$CdSO_4(c)$	32.8
$CdO(g)$	46.9	$CdSO_4 \cdot H_2O(c)$	41.1
$CdS(c)$	17	$CdSO_4 \cdot \frac{8}{3}H_2O(c)$	57.9
Calcium			
$Ca(c)$	9.95	$Ca(OH)_2(c)$	18.2
$Ca(g)$	36.993	$CaCO_3$(calcite, c)	22.2
$Ca^{++}(aq)$	−13.2	$CaCO_3$(aragonite, c)	21.2
$CaO(c)$	9.5	$CaC_2O_4 \cdot H_2O(c)$	37.28
$CaO(g)$	52.3	$Ca_3(PO_4)_2(\alpha, c)$	57.6
$CaS(c)$	13.5	$CaSiO_3$(wallastonite, c)	19.6
$CaF_2(c)$	16.46	$CaSO_4$(anhydrite, c)	25.5
$CaH_2(c)$	10		
Carbon			
C(graphite)	1.3609	$CF_4(g)$	62.7
C(diamond)	0.5829	$COCl_2(g)$	69.13
$C(g)$	37.7611	$CH_4(g)$	44.50
$C_2(g)$	47.89	$CN(g)$	48.40
$CO(g)$	47.301	$C_2N_2(g)$	57.86
$CO_2(g)$	51.061	$CNBr(g)$	59.05
$CS(g)$	50.4	$CNCl(g)$	56.31
$CS_2(l)$	36.10	$CNI(g)$	61.26
$CS_2(g)$	56.84	$CO_3^{--}(aq)$	−12.7
$COS(g)$	55.34	$HCO_3^-(aq)$	22.7
$CBr_4(g)$	85.6	$H_2CO_3(aq)$	45.7
$CCl_4(g)$	73.95	$CN^-(aq)$	28.2
$CCl_4(l)$	51.25	$C_2O_4^{--}(aq)$	9.6
Chlorine			
$Cl_2(g)$	53.286	$ClO^-(aq)$	10.3
$Cl(g)$	39.4569	$ClO_2^-(aq)$	23.8
$Cl^-(aq)$	13.17	$ClO_3^-(aq)$	39.0
$Cl_2O(g)$	63.70	$ClO_4^-(aq)$	43.2
$ClO_2(g)$	59.6		

TABLE 68 (CONTINUED) INORGANIC COMPOUNDS

Chromium

$Cr(c)$	5.68	$CrCl_2(c)$	27.4
$Cr(g)$	41.637	$CrCl_3(c)$	30.0
$Cr_2O_3(c)$	19.4	$CrO_4^{--}(aq)$	9.2

Cobalt

$Co(c)$	6.8	$CoO(c)$	10.5
$Co(g)$	42.881	$CoCl_2(c)$	25.4

Copper

$Cu(c)$	7.96	$CuBr(c)$	21.9
$Cu(g)$	39.744	$CuBr(g)$	59.22
$Cu_2(g)$	58.9	$CuCl(c)$	20.2
$Cu^{++}(aq)$	−23.6	$CuCl(g)$	56.50
$Cu_2O(c)$	24.1	$CuF(g)$	54.9
$CuO(c)$	10.4	$CuI(c)$	23.1
$Cu_2S(c)$	28.9	$CuI(g)$	61.06
$CuS(c)$	15.9	$CuH(g)$	46.89
$CuCO_3(c)$	21	$CuSO_4(c)$	27.1

Fluorine

$F_2(g)$	48.6	$F^-(aq)$	−2.3
$F(g)$	37.917	$F_2O(g)$	58.95

Helium

$He(g)$	30.126		

Hydrogen

$H_2(g)$	31.211	$H_2S(g)$	49.15
$D_2(g)$	34.602	$HBr(g)$	47.437
$H(g)$	27.393	$HCl(g)$	44.617
$H^+(aq)$	zero by definition	$HF(g)$	41.47
$H_2O(l)$	16.716	$HI(g)$	49.314
$H_2O(g)$	45.106	$HCN(l)$	26.97
$D_2O(l)$	18.162	$HCN(g)$	48.23
$D_2O(g)$	47.379		

Iodine

$I_2(c)$	27.9	$IBr(c)$	33.0
$I_2(g)$	62.280	$ICl(g)$	59.12
$I(g)$	43.1841	$ICl_3(c)$	41.1
$I^-(aq)$	26.14	$IO_3^-(aq)$	27.7
$IBr(g)$	61.80		

Iron

$Fe(c)$	6.49	$Fe(OH)_2(c)$	19
$Fe(g)$	43.11	$FeS(c)$	16.1
$Fe^{++}(aq)$	−27.1	FeS_2(pyrites, c)	12.7
$Fe^{+++}(aq)$	−70.1	$FeCl_2(c)$	28.6
$FeO(c)$	13.4	$Fe_3C(c)$	25.7
$Fe_{0.95}O$(wustite, c)	12.9	$FeCO_3$(siderite, c)	22.2
$Fe_3O_4(c)$	35.0	$Fe_4N(c)$	37.3
$Fe_2O_3(c)$	21.5	$2FeO{\cdot}SiO_2(s)$	35.4

TABLE 68 (CONTINUED) INORGANIC COMPOUNDS

Substance	S^0	Substance	S^0
Lead			
Pb(*c*)	15.51	PbS(*c*)	21.8
Pb(*g*)	41.890	PbS(*g*)	61.2
Pb^{++}(*aq*)	5.1	$PbCl_2$(*c*)	32.6
PbO(yellow, *c*)	16.6	$PbCl_2$(*g*)	75.9
PbO(*g*)	57.4	$PbSO_4$(*c*)	35.2
PbO_2(*c*)	18.3	$PbCO_3$(*c*)	31.3
Pb_3O_4(*c*)	50.5		
Lithium			
Li(*c*)	6.70	LiF(*c*)	8.57
Li(*g*)	33.143	LiI(*g*)	55.68
Li_2(*g*)	47.06	LiH(*c*)	5.9
Li^+(*aq*)	3.4	LiH(*g*)	40.77
LiCl(*g*)	51.01	LiOH(*c*)	12
Li_2CO_3(*c*)	21.60		
Magnesium			
Mg(*c*)	7.77	$Mg(OH)_2$(*c*)	15.09
Mg(*g*)	35.504	$MgCl_2$(*c*)	21.4
Mg^{++}(*aq*)	−28.2	$MgCO_3$(*c*)	15.7
MgO(*c*)	6.4	$MgSiO_3$(*c*)	16.2
MgO(*g*)	50.7	$MgSO_4$(*c*)	21.9
Manganese			
Mn(*c*)	7.59	Mn_3O_4(*c*)	35.5
Mn(*g*)	41.493	MnS(*c*)	18.7
MnO(*c*)	14.4	$MnSO_4$(*c*)	26.8
MnO(*g*)	54.0	$MnCl_2$(*c*)	28.0
MnO_2(*c*)	12.7	$MnCO_3$(*c*)	20.5
MnO_4^-(*aq*)	45.4	Mn_3C(*c*)	23.6
Mn_2O_3(*c*)	22.9	$MnO \cdot SiO_2$(*c*)	21.3
Mercury			
Hg(*l*)	18.5	$HgBr_2$(*g*)	74.7
Hg(*g*)	41.80	$HgBr_2$(*c*)	38.9
Hg_2^{++}(*aq*)	17.7	$HgCl_2$(*g*)	70.4
HgO(red, *c*)	17.2	$HgCl_2$(*c*)	34.6
HgO(yellow, *c*)	17.5	HgI_2(*g*)	78.7
HgBr(*g*)	65.0	HgI_2(*c*)	42.2
Hg_2Br_2(*c*)	50.9	HgS(*c*)	18.6
Hg_2Cl_2(*c*)	40.8	Hg_2SO_4(*c*)	47.98
Molybdenum			
Mo(*c*)	6.83	MoS_2(*c*)	15.1
Mo(*g*)	43.462	MoS_3(*c*)	15.9
MoO_3(*c*)	18.68		
Neon			
Ne(*g*)	34.948		
Nickel			
Ni(*c*)	7.20	$NiCl_2$(*c*)	25.6
Ni(*g*)	43.592	$Ni(CO)_4$(*g*)	97
NiO(*c*)	9.22	$NiSO_4$(*c*)	18.6
$Ni(OH)_2$(*c*)	19		

TABLE 68 (CONTINUED) INORGANIC COMPOUNDS

		Nitrogen	
$N_2(g)$	45.767	$NH_3(g)$	46.01
$N(g)$	36.6145	$NH_4Cl(c)$	22.6
$N_2O(g)$	52.58	$NH_4HCO_3(c)$	28.3
$NO(g)$	50.339	$NH_4HS(c)$	27.1
$NO_2(g)$	57.47	$NH_4OH(aq)$	42.8
$N_2O_4(g)$	72.73	$NH_4^+(aq)$	26.97
$N_2O_5(c)$	36.6	$NO_2^-(aq)$	29.9
$N_2O_5(g)$	82	$NO_3^-(aq)$	35.0
$NOBr(g)$	65.16	$HNO_3(l)$	37.19
$NOCl(g)$	63.0		
		Oxygen	
$O_2(g)$	49.003	$OH(g)$	43.888
$O(g)$	38.469	$OH^-(aq)$	−2.519
$O_3(g)$	56.8		
		Phosphorus	
$P_2(g)$	52.13	$PF_3(g)$	64.13
$P_4(g)$	66.90	$PH_3(g)$	50.2
P(white, *s*)	10.6	$PN(g)$	50.45
$P(g)$	38.98	$H_3PO_4(aq)$	44.0
$PBr_3(g)$	83.11	$H_2PO_4^-(aq)$	21.3
$PCl_3(g)$	74.49	$HPO_4^{--}(aq)$	−8.6
$PCl_3(l)$	52.2	$PO_4^{---}(aq)$	−52
$PCl_5(g)$	84.3		
		Potassium	
$K(c)$	15.2	$KCl(g)$	57.24
$K(g)$	38.296	$KF(c)$	15.91
$K_2(g)$	59.69	$KI(c)$	24.94
$K^+(aq)$	24.5	$KH(g)$	47.3
$KBr(c)$	23.05	$KClO_3(c)$	34.17
$KBr(g)$	59.87	$KNO_3(c)$	31.77
$KCl(c)$	19.76	$K_2SO_4(c)$	42.0
		Silicon	
$Si(c)$	4.47	$SiCl_4(g)$	79.2
$Si(g)$	40.120	$SiCl_4(l)$	57.2
SiO_2(quartz, *c*)	10.0	$SiF_4(g)$	68.0
SiO_2(cristobalite, *c*)	10.10	$SiH_4(g)$	48.7
SiO_2(tridymite, *c*)	10.36	$SiC(c)$	3.935
SiO_2(glass)	11.2	$Si_3N_4(c)$	22.4
		Silver	
$Ag(c)$	10.206	$AgBr(g)$	62.1
$Ag(g)$	41.3221	$AgCl(g)$	58.5
$Ag^+(aq)$	17.67	$AgCl(c)$	22.97
$Ag_2O(c)$	29.09	$AgNO_3(c)$	33.68
$Ag_2S(\alpha, c)$	34.8	$Ag_2SO_4(c)$	47.8
$AgBr(c)$	25.60	$Ag(NH_3)_2^+(aq)$	57.8

TABLE 68 (CONTINUED) INORGANIC COMPOUNDS

		Sodium	
Na(*c*)	12.2	NaOH(*c*)	14.2
Na(*g*)	36.715	Na_2CO_3(*c*)	32.5
Na_2(*g*)	55.02	$NaHCO_3$(*c*)	24.4
Na^+(*aq*)	14.4	$NaNO_3$(*c*)	27.8
Na_2O(*c*)	17.4	Na_4SiO_4(*c*)	46.8
NaBr(*g*)	58.1	Na_2SiO_3(*c*)	27.2
NaCl(*c*)	17.30	$Na_2Si_2O_5$(*c*)	39.4
NaCl(*g*)	55.5	Na_2SO_3(*c*)	34.9
NaF(*c*)	14.0	Na_2SO_4(*c*)	35.73
NaF(*g*)	53.8	$Na_2SO_4 \cdot 10H_2O$(*c*)	141.7
NaH(*g*)	44.93	NaK(*g*)	58.6
NaI(*g*)	60.0		
		Sulfur	
S(rhombic)	7.62	SF_6(*g*)	69.5
S(monoclinic)	7.78	H_2S(*g*)	49.15
S(*g*)	40.085	H_2S(*aq*)	29.2
S_2(*g*)	54.41	HS^-(*aq*)	14.6
S_6(*g*)	92	H_2SO_3(*aq*)	54.7
S_8(*g*)	109	HSO_3^-(*aq*)	31.64
SO(*g*)	53.04	SO_3^{--}(*aq*)	10.4
SO_2(*g*)	59.40	HSO_4^-(*aq*)	30.32
SO_3(*g*)	61.24	SO_4^{--}(*aq*)	4.1
SO_3(*l*)	31.7		
		Tin	
Sn(white, *c*)	12.3	Sn(*g*)	40.245
Sn(gray, *c*)	10.7	Sn^{++}(*aq*)	−4.9
SnO(*c*)	13.5	$SnCl_4$(*g*)	87.2
SnO_2(*c*)	12.5	$SnCl_4$(*l*)	61.8
		Titanium	
Ti(*c*)	7.24	$TiCl_4$(*g*)	84.4
Ti(*g*)	43.069	$TiCl_4$(*l*)	60.4
TiO_2(rutile, *c*)	12.01		
		Vanadium	
V(*c*)	7.05	V_2O_4(*c*)	24.65
V(*g*)	43.546	V_2O_5(*c*)	31.3
V_2O_3(*c*)	23.58		
		Zinc	
Zn(*c*)	9.95	$ZnBr_2$(*c*)	32.84
Zn(*g*)	38.45	$ZnCl_2$(*c*)	25.9
Zn^{++}(*aq*)	25.45	ZnI_2(*c*)	38.0
ZnO(*c*)	10.5	$ZnCO_3$(*c*)	19.7
ZnO(*g*)	54.1	$ZnSO_4$(*c*)	29.8
ZnS(*c*)	13.8		
		Zirconium	
Zr(*c*)	9.18	Zr(*g*)	43.313

TABLE 68 (CONTINUED) ORGANIC COMPOUNDS

Hydrocarbons

Compound	C Atoms	$S^\circ_{298.16}$	Ref.
Methane(*g*)	1	44.50	1
Ethyne (acetylene)(*g*)	2	47.997	1
Ethene (ethylene)(*g*)	2	52.45	1
Ethane(*g*)	2	54.85	1
Propyne (allylene, methyl-acetylene)(*g*)	3	59.30	1
Propadiene (allene)(*g*)	3	58.30	1
Propene (propylene)(*g*)	3	63.80	1
Propane(*g*)	3	64.51	1
Cyclopropane(*g*)	3	56.79	10
2-Butyne (dimethyl-acetylene)(*g*)	4	67.71	1
2-Butyne (dimethyl-acetylene)(*l*)	4	46.63	18
2-Methylpropene (isobutene)(*g*)	4	70.17	1
cis-2-Butene(*g*)	4	71.90	1
trans-2-Butene(*g*)	4	70.86	1
1-Butene(*g*)	4	73.04	1
n-Butane(*g*)	4	74.12	1
n-Butane(*l*)	4	55.2	16
2-Methylpropane (isobutane)(*g*)	4	70.42	1
2-Methylpropane (isobutane)(*l*)	4	52.09	16
2-Methyl-2-butene(*g*)	5	80.92	1
3-Methyl-1-butene(*g*)	5	79.70	1
2-Methyl-1-butene(*g*)	5	81.73	1
n-Pentane(*g*)	5	83.40	1
2-Methylbutane (isopentane)(*g*)	5	82.12	1
2,2-Dimethyl-propane (neopentane)(*g*)	5	73.23	1

Compound	C Atoms	$S^\circ_{298.16}$	Ref.
Cyclopentane(*g*)	5	70.00	1
Cyclopentane(*l*)	5	48.82	1
n-Hexane(*g*)	6	92.83	1
2-Methylpentane(*g*)	6	90.95	1
2-Methylpentane(*l*)	6	69.51	1
2,2-Dimethyl-butane(*g*)	6	85.62	1
2,2-Dimethyl-butane(*l*)	6	65.08	1
2,3-Dimethyl-butane(*g*)	6	87.42	1
3-Methylpentane(*g*)	6	90.77	1
Benzene(*g*)	6	64.34	1
Benzene(*l*)	6	41.30	1
Cyclohexane(*g*)	6	71.28	1
Cyclohexane(*l*)	6	48.85	1
Methyl-cyclopentane(*l*)	6	59.26	1
n-Heptane(*g*)	7	102.24	1
2-Methylhexane(*g*)	7	100.35	1
Methyl-cyclohexane(*l*)	7	59.26	1
2,2,3-Trimethyl-butane(*g*)	7	91.60	1
Methylbenzene (toluene)(*g*)	7	76.42	1
Methylbenzene (toluene)(*l*)	7	52.48	1
Ethylcyclo-pentane(*l*)	7	67.00	1
n-Octane(*g*)	8	111.55	1
2,2,4-Trimethyl-pentane(*g*)	8	101.15	1
2,2,3,3-Tetra-methylbutane(*g*)	8	93.06	1
1,2-Dimethyl-benzene (*o*-xylene)(*g*)	8	84.31	1
1,2-Dimethyl-benzene (*o*-xylene)(*l*)	8	58.91	1
1,3-Dimethyl-benzene (*m*-xylene)(*g*)	8	85.49	1

TABLE 68 (CONTINUED) ORGANIC COMPOUNDS

Compound	C Atoms	$s^\circ_{298.16}$	Ref.	Compound	C Atoms	$s^\circ_{298.16}$	Ref.
1,3-Dimethylbenzene (*m*-xylene)(*l*)	8	60.27	1	*n*-Decane(*g*)	10	130.17	1
1,4-Dimethylbenzene (*p*-xylene)(*g*)	8	84.23	1	Naphthalene(*c*)	10	39.9	14
1,4-Dimethylbenzene (*p*-xylene)(*l*)	8	59.12	1	*n*-Butylbenzene(*g*)	10	105.04	1
Ethenylbenzene (styrene)(*g*)	8	82.48	1	β-Methylnaphthalene(*c*)	11	48.8	14
Ethenylbenzene (styrene)(*l*)	8	56.78	5	*n*-Dodecane(*g*)	12	148.78	1
Ethylbenzene(*l*)	8	60.99	1	Diphenyl(*c*)	12	49.2	14
n-Nonane(*g*)	9	120.86	1	*n*-Heptyl cyclohexane(*g*)	13	137.51	1
1,3,5-Trimethylbenzene (mesitylene)(*g*)	9	92.15	1	Anthracene(*c*)	14	49.6	14
				Phenanthrene(*c*)	14	50.6	14
				Pyrene(*c*)	16	51.4	14
				n-Dodecylcyclohexane(*g*)	18	184.05	1
				1,3,5-Triphenylbenzene(*c*)	24	87.9	14

Alcohols

Compound	$s^\circ_{298.16}$	Ref.	Compound	$s^\circ_{298.16}$	Ref.
Methyl alcohol(*g*)	56.8	12	Ethyl alcohol(*l*)	38.4	12
Methyl alcohol(*l*)	30.3	12	Isopropyl alcohol(*l*)	43.0	17
Ethyl alcohol(*g*)	67.4	12	Isopropyl alcohol(*g*)	73.4	17

Aldehydes, Ketones, Ethers

Compound	$s^\circ_{298.16}$	Ref.	Compound	$s^\circ_{298.16}$	Ref.
Formaldehyde(*g*)	52.26	12	Dimethyl ether(*g*)	63.72	12
Acetone(*g*)	72.7	17	Dimethyl ether(*l*)	44.98	9
Acetone(*l*)	47.9	17			

Acids

Compound	$s^\circ_{298.16}$	Ref.	Compound	$s^\circ_{298.16}$	Ref.
Formic acid(*g*) (monomer)	60.0	12	Acetic acid(*l*)	38.2	12
Formic acid(*l*)	30.82	12	Lactic acid(*c*)	34.30	7
Formic acid(*g*) (dimer)	83.1	12	Hippuric acid(*c*)	57.2	6

Carbohydrates

Compound	$s^\circ_{298.16}$	Ref.	Compound	$s^\circ_{298.16}$	Ref.
1-Sorbose(*c*)	52.8	8	β-Lactose(*c*)	92.3	2
β-Maltose monohydrate(*c*)	99.8	2	α-*d*-galactose(*c*)	49.1	8
α-Lactose monohydrate(*c*)	99.1	2			

Table 68 (continued) Organic Compounds

Compound	$s^\circ_{298.16}$	Ref.	Compound	$s^\circ_{298.16}$	Ref.
Halogen Compounds					
Carbon tetrachloride(*g*)	73.95	12	Ethylene dichloride(*g*)	72.74	15
Carbon tetrachloride(*l*)	51.25	12	Phosgene(*g*)	69.13	12
Chloroform(*g*)	70.86	12	Bromoform(*g*)	79.18	12
Methylene chloride(*g*)	64.68	12	Methylene bromide(*g*)	70.16	12
Methyl chloride(*g*)	55.97	12	Methyl bromide(*g*)	58.74	12
Methyl chloride(*l*)	36.74	11	Ethylene dibromide(*l*)	53.37	12
Fluorotrichloromethane(*g*)	74.06	12	Ethylene dibromide(*g*)	79.37	15
Fluorotrichloromethane(*l*)	53.92	13	Methyl fluoride(*g*)	53.30	12
Ethylene dichloride(*l*)	49.84	12	Methyl iodide(*g*)	60.85	12
Sulfur Compounds					
Methyl mercaptan(*g*)	60.90	12	Dimethyl sulfide(*l*)	46.94	12
Dimethyl sulfide(*g*)	68.28	12			
Nitrogen Compounds					
Methylamine(*g*)	57.73	12	Methyl cyanide(*g*)	58.18	12
Methylamine(*l*)	35.90	4	Methyl isocyanide(*g*)	58.78	12
Dimethylamine(*g*)	65.30	12	Urea(*c*)	25.00	12
Dimethylamine(*l*)	43.58	3	Cyanogen(*g*)	57.86	12

References for Organic Materials

1. "Selected Values of Properties of Hydrocarbons and Related Compounds," *Am. Petrol. Inst. Research Proj.*, **44**, as of July 1, 1953, edited by F. D. Rossini.
2. A. G. Anderson and G. Stegeman, *J. Am. Chem. Soc.*, **63**, 2120 (1941).
3. J. G. Aston, M. L. Eidinoff, and W. S. Forster, *J. Am. Chem. Soc.*, **61**, 1539 (1939).
4. J. G. Aston, C. W. Siller, and G. H. Messerly, *J. Am. Chem. Soc.*, **59**, 1743 (1937).
5. L. Guttman, E. F. Westrum, Jr., and K. S. Pitzer, *J. Am. Chem. Soc.*, **65**, 1246 (1943).
6. H. M. Huffman, *J. Am. Chem. Soc.*, **63**, 688 (1941).
7. H. M. Huffman, E. L. Ellis, and H. Borsook, *J. Am. Chem. Soc.*, **62**, 297 (1940).
8. G. W. Jack and G. Stegeman, *J. Am. Chem. Soc.*, **63**, 2120 (1941).
9. R. M. Kennedy, M. Sagenkahn, and J. G. Aston, *J. Am. Chem. Soc.*, **63**, 2268 (1941).
10. J. W. Linnett, *J. Chem. Phys.*, **6**, 700 (1938).
11. G. H. Messerly and J. G. Aston, *J. Am. Chem. Soc.*, **62**, 886 (1940).
12. National Bureau of Standards, *Selected Values of Chemical Thermodynamic Properties*, as of July 1, 1953, edited by D. D. Wagman.
13. D. W. Osborne, C. S. Garner, R. N. Doescher, and D. M. Yost, *J. Am. Chem. Soc.*, **63**, 3496 (1941).
14. G. S. Parks, *Chem. Rev.*, **27**, 75 (1940).
15. K. S. Pitzer, *J. Am. Chem. Soc.*, **62**, 331 (1940).
16. K. S. Pitzer, *Chem. Rev.*, **27**, 39 (1940).
17. S. C. Schumann and J. G. Aston, *J. Chem. Phys.*, **6**, 485 (1938).
18. D. M. Yost, D. W. Osborne, and C. S. Garner, *J. Am. Chem. Soc.*, **63**, 3492 (1941).

Similarly, the standard free-energy change and standard heat of reaction of any reaction at any temperature may be expressed in terms of heats of formation at the absolute zero. Thus,

$$\Delta H^\circ_T = \sum[(\mathrm{H}^\circ_T - \mathrm{H}^\circ_0) + \Delta \mathrm{H}^\circ_{f0}]_P - \sum[(\mathrm{H}^\circ_T - \mathrm{H}^\circ_0) + \Delta \mathrm{H}^\circ_{f0}]_R \quad (20)$$

$$\left(\frac{\Delta G^\circ}{T}\right)_T = \sum\left[\frac{\mathrm{G}^\circ_T - \mathrm{H}^\circ_0}{T} + \frac{\Delta \mathrm{H}^\circ_{f0}}{T}\right]_P - \sum\left[\frac{\mathrm{G}^\circ_T - \mathrm{H}^\circ_0}{T} + \frac{\Delta \mathrm{H}^\circ_{f0}}{T}\right]_R \quad (21)$$

where the subscript P indicates the products and R the reactants.

Standard enthalpy and free-energy changes are readily calculated by equations 20 and 21 from values of $\Delta \mathrm{H}^\circ_{f0}$ and tables or curves which express $(\mathrm{H}^\circ_T - \mathrm{H}^\circ_0)$ and $(\mathrm{G}^\circ_T - \mathrm{H}^\circ_0)/T$ as functions of temperature. It will be noted that, for any compound, the group $[(\mathrm{H}^\circ_T - \mathrm{H}^\circ_0) + \Delta \mathrm{H}^\circ_{f0}]$ represents its enthalpy in the standard state at temperature T relative to the elements in their standard states at 0° K. Similarly, $[(\mathrm{G}^\circ_T - \mathrm{H}^\circ_0) + \Delta \mathrm{H}^\circ_{f0}]$ is the standard free energy of the compound at temperature T relative to the elements at 0° K.

In Table 69 are values of $(\mathrm{G}^\circ_T - \mathrm{H}^\circ_0)/T$ and $\Delta \mathrm{H}^\circ_{f0}$ for a number of

TABLE 69. FREE-ENERGY FUNCTION AND STANDARD HEAT OF FORMATION AT 0° K

G°_T = molal free energy of the substance in its standard state, at temperature T, as g-cal/g-mole
H°_0 = molal enthalpy of the substance in its standard state, at 0° K, as g-cal/g-mole
$(\Delta \mathrm{H}^\circ_f)_0$ = standard molal heat of formation at 0° K, as kg-cal/g-mole

		$-(\mathrm{G}^\circ_T - \mathrm{H}^\circ_0)/T$							
	State	298.16° K	400° K	500° K	600° K	800° K	1000° K	1500° K	$(\Delta \mathrm{H}^\circ_f)_0$
Methane	*g*	36.46	38.86	40.75	42.39	45.21	47.65	52.84	−15.987
Ethane	*g*	45.27	48.24	50.77	53.08	57.29	61.11	69.46	−16.517
Ethene (ethylene)	*g*	43.98	46.61	48.74	50.70	54.19	57.29	63.94	+14.522
Ethyne (acetylene)	*g*	39.976	42.451	44.508	46.313	49.400	52.005	57.231	+54.329
Propane	*g*	52.73	56.48	59.81	62.93	68.74	74.10	85.86	−19.482
Propene (propylene)	*g*	52.95	56.39	59.32	62.05	67.04	71.57	81.43	+8.468
n-Butane	*g*	58.54	63.51	67.91	72.01	79.63	86.60	101.95	−23.67
2-Methylpropane (isobutane)	*g*	56.08	60.72	64.95	68.95	76.45	83.38	98.64	−25.30
1-Butene	*g*	59.25	63.64	67.52	71.14	77.82	83.93	97.27	+4.96
cis-2-Butene	*g*	58.67	62.89	66.51	69.94	76.30	82.17	95.12	+3.48
trans-2-Butene	*g*	56.80	61.31	65.19	68.84	75.53	81.62	94.91	+2.24
2-Methylpropene (isobutene)	*g*	56.47	60.90	64.77	68.42	75.15	81.29	94.66	+0.98
n-Pentane	*g*	64.52	70.57	75.94	80.96	90.31	98.87	117.72	−27.23
2-Methylbutane (isopentane)	*g*	64.36	70.67	75.28	80.21	89.44	97.96	116.78	−28.81
2,2-Dimethylpropane (neopentane)	*g*	56.36	61.93	67.04	71.96	81.27	89.90	108.91	−31.30
n-Hexane	*g*	70.62	77.75	84.11	90.06	101.14	111.31	133.64	−30.91
2-Methylpentane	*g*	70.50	77.2	83.3	89.1	100.1	110.3	132.5	−32.08
3-Methylpentane	*g*	68.56	75.69	82.05	88.0	99.08	109.3	131.6	−31.97
2,2-Dimethylbutane	*g*	65.79	72.3	78.3	84.1	95.0	105.1	127.4	−34.65
2,3-Dimethylbutane	*g*	67.58	74.06	80.05	85.77	96.54	106.57	128.70	−32 73
Graphite	*s*	0.5172	0.824	1.146	1.477	2.138	2.771	4.181	0
Hydrogen, H_2	*g*	24.423	26.422	27.950	29.203	31.186	32.738	35.590	0
H_2O	*g*	37.165	39.505	41.293	42.766	45.128	47.010	50.598	−57.107
CO	*g*	40.350	42.393	43.947	45.222	47.254	48.860	51.864	−27.2019
CO_2	*g*	43.555	45.828	47.667	49.238	51.895	54.109	58.481	−93.9686
O_2	*g*	42.061	44.112	45.675	46.968	49.044	50.697	53.808	0
N_2	*g*	38.817	40.801	42.415	43.688	45.711	47.306	50.284	0
NO	*g*	42.980	45.134	46.760	48.090	50.202	51.864	54.964	+21.477

Source: "Selected Values of Properties of Hydrocarbons and Related Compounds," *Am. Petrol. Inst. Research Proj.*, **44**, as of July 1, 1958, edited by F. D. Rossini, with permission.

Table 70. Enthalpy above 0° K

H°_T = molal enthalpy of the substance in its standard state, at temperature T
H°_0 = molal enthalpy of the substance in its standard state, at 0° K

		$(H^\circ_T - H^\circ_0)$, kg-cal/g-mole						
	State	298.16° K	400° K	500° K	600° K	800° K	1000° K	1500° K
Methane	*g*	2.397	3.323	4.365	5.549	8.321	11.560	21.130
Ethane	*g*	2.856	4.296	6.010	8.016	12.760	18.280	34.500
Ethene (ethylene)	*g*	2.525	3.711	5.117	6.732	10.480	14.760	27.100
Ethyne (acetylene)	*g*	2.3915	3.5412	4.7910	6.127	8.999	12.090	20.541
Propane	*g*	3.512	5.556	8.040	10.930	17.760	25.670	48.650
Propene (propylene)	*g*	3.237	4.990	7.076	9.492	15.150	21.690	40.570
n-Butane	*g*	4.645	7.340	10.595	14.376	23.264	33.540	63.270
2-Methylpropane (isobutane)	*g*	4.276	6.964	10.250	14.070	23.010	33.310	63.050
1-Butene	*g*	4.112	6.484	9.350	12.650	20.370	29.250	54.840
cis-2-Butene	*g*	3.981	6.144	8.839	12.010	19.510	28.230	53.620
trans-2-Butene	*g*	4.190	6.582	9.422	12.690	20.350	29.190	54.710
2-Methylpropene (isobutene)	*g*	4.082	6.522	9.414	12.750	20.490	29.370	55.000
n-Pentane	*g*	5.629	8.952	12.970	17.628	28.568	41.190	77.625
2-Methylbutane (isopentane)	*g*	5.295	8.596	12.620	17.300	28.300	41.010	77.740
2,2-Dimethylpropane (neopentane)	*g*	5.030	8.428	12.570	17.390	28.640	41.510	78.420
n-Hexane	*g*	6.622	10.580	15.360	20.892	33.880	48.850	92.010
2-Methylpentane	*g*	6.097	10.080	14.950	20.520	33.600	48.700	
3-Methylpentane	*g*	6.622	10.580	15.360	20.880	33.840	48.800	
2,2-Dimethylbutane	*g*	5.912	9.880	14.750	20.340	33.520	48.600	
2,3-Dimethylbutane	*g*	5.916	9.833	14.610	20.170	33.230	48.240	
Graphite	*s*	0.25156	0.5028	0.8210	1.1982	2.0816	3.0750	5.814
Hydrogen, H_2	*g*	2.0238	2.7310	3.4295	4.1295	5.5374	6.9658	10.6942
Water, H_2O	*g*	2.3677	3.1940	4.0255	4.8822	6.6896	8.6080	13.848
CO	*g*	2.0726	2.7836	3.4900	4.2096	5.7000	7.2570	11.3580
CO_2	*g*	2.2381	3.1948	4.2230	5.3226	7.6896	10.2220	17.004
O_2	*g*	2.0698	2.7924	3.5240	4.2792	5.8560	7.4970	11.7765
N_2	*g*	2.07227	2.7824	3.4850	4.1980	5.6686	7.2025	11.2536
NO	*g*	2.1942	2.9208	3.6440	4.3812	5.9096	7.5060	11.6940

Source: "Selected Values of Properties of Hydrocarbons and Related Compounds," *Am. Petrol. Inst. Research Proj.*, **44,** as of July 1, 1958, edited by F. D. Rossini, with permission.

compounds[2,3] and elements in the ideal-gaseous state at 1 atm for seven temperatures. In Table 70 is a corresponding summary of values of $(H^\circ_T - H^\circ_0)$. By interpolation standard enthalpy and free-energy changes at intermediate temperatures are obtained by equations 20 and 21 without the use of empirical heat-capacity equations. Values of $(\Delta G^\circ/T)_T$ calculated from equation 21 are directly related to the equilibrium constant. Thus, rearranging equation 6 gives

$$4.576 \log K = -\left(\frac{\Delta G^\circ}{T}\right)_T \tag{22}$$

where $(\Delta G^\circ/T)_T$ is expressed in g-cal/(g-mole)(K°) or Btu/(lb-mole)(R°).

In Fig. 262 are values of log K for several reactions plotted as functions of temperature. These values were calculated by equation 22, using the data of Tables 29, 30, 68, 69 and 70.

[2] K. S. Pitzer, *Chem. Rev.*, **27**, 39 (1940).

[3] F. D. Rossini, E. J. R. Prosen, and K. S. Pitzer, *J. Research Nat. Bur. Standards*, **27,** 529 (1941).

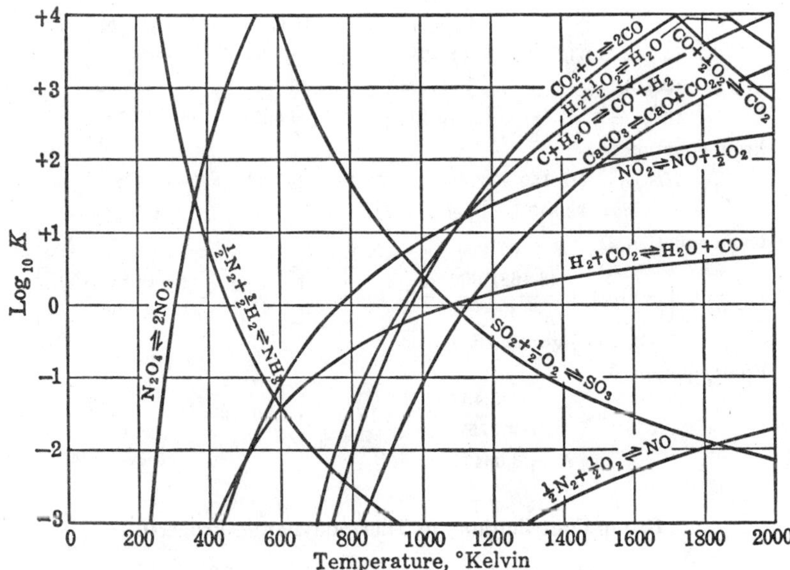

FIG. 262. Equilibrium constants of chemical reactions

Illustration 4. From the data of Tables 69 and 70, calculate the values of ΔH°_f and $\Delta G^\circ_f/T$ for the formation of isobutene from the elements at 298.16° K.

$$4C(s) + 4H_2(g) \rightarrow C_4H_8(g)$$

From Tables 69 and 70:

	$i\text{-}C_4H_8(g)$	$C(s)$	$H_2(g)$
$(G^\circ_T - H^\circ_0)/T$	−56.47	−0.5172	−24.423
$(\Delta H^\circ_f)_0$	0.980	0	0
$(H^\circ_T - H^\circ_0)_{298.16}$	4.082	0.25156	2.0238

From equation 17,

$$(\Delta H^\circ_f)_{298.16} = (4.082 + 0.980) - [4(0.25156) + 4(2.0238)]$$
$$= -4.0394 \text{ kcal per g-mole}$$

From equation 19,

$$\frac{\Delta G^\circ_f}{298.16} = \left[-56.47 + \frac{(0.980)1000}{298.16}\right] - [4(-0.5172) + 4(-24.423)]$$

$$= 46.578 \text{ g cal/(g mole)(K}^\circ)$$

Illustration 5. From the data of Tables 69 and 70, calculate the values of ΔH°, $\Delta G^\circ/T$, and the equilibrium constant K for the dehydrogenation of propane to propylene at 800° K.

$$C_3H_8(g) \rightarrow C_3H_6(g) + H_2(g)$$

From Tables 69 and 70 at 800° K,

	$C_3H_6(g)$	$H_2(g)$	$C_3H_8(g)$
$(G°_T - G°_0)/T$	−67.04	−31.186	−68.74
$(\Delta H°_f)_0$	8.468	0	−19.482
$(H°_T - H°_0)$	15.150	5.5374	17.760

From equation 20,

$$\Delta H°_{800} = (15.150 + 8.468 + 5.5374) - [17.760 + (-19.482)]$$
$$= 30.877 \text{ kcal per g-mole}$$

From equation 21,

$$\frac{\Delta G°}{800} = \left[-67.04 + \frac{(8.468)1000}{800} - 31.186\right] - \left[-68.74 + \frac{(-19.842)1000}{800}\right]$$
$$= 5.452 \text{ g-cal/(g-mole)(K°)}$$

From equation 6,

$$\log K = \frac{-5.452}{4.5757} = -1.19151 = 0.80849 - 2$$
$$K = 0.06434$$

Group Contributions to Thermodynamic Properties

Various empirical methods for correlating heats of formation, entropies, and heat capacities of organic compounds were reviewed by Andersen, Beyer, and Watson[4] who proposed a scheme whereby these properties for the ideal-gaseous state are resolved into contributions attributable to atomic groups. From the resulting tables, the properties of complex molecules are readily estimated by summation of the contributions of their component groups. The corresponding properties in the liquid and nonideal-gaseous states may then be calculated by the methods of Chapter 14.

Many of the values given in the original tables of Andersen, Beyer, and Watson were subsequently modified by Brown,[5] who made use of the extensive tables of data on hydrocarbons compiled by Rossini[6] and co-workers.

The molal values of heats of formation and entropies for the ideal-gaseous state at 25° C and 1 atm, and the constants a, b, and c of the three-term heat-capacity equation 8–20, page 252, were resolved into group contributions on the following basis.

[4] J. W. Andersen, G. H. Beyer, and K. M. Watson, *Nat. Petrol. News, Tech. Sec.*, **36,** *R*476 (July 5, 1944). Also *Process Engineering Data*, National Petroleum Publishing Co., Cleveland (1944).

[5] J. M. Brown, *Univ. Wisconsin Dept. Chem. Eng. Spec. Probs. Proj. Rept.* (June 1953).

[6] "Selected Values of Properties of Hydrocarbons and Related Compounds," *Am. Petrol. Inst. Research Proj.*, **44,** as of July 1, 1953, edited by F. D. Rossini.

Each compound is considered as composed of a basic group which is modified by the substitution of other groups for atoms comprising it. For example, all paraffin hydrocarbons may be considered as derived from methane by successive substitution of CH_3 groups for hydrogen atoms. Similarly, any secondary amine can be considered as derived from the base group $NH(CH_3)_2$. The contributions of ten base groups are given in Table 71.

The contributions resulting from the primary substitution of a methyl group for a hydrogen atom in any one of the base groups is given in Table 72. In the cases of benzene, naphthalene, cyclopentane, and cyclohexane, the base group contains several carbon atoms, and successive substitutions on different carbon atoms involve different contributions, depending on the number and position of the substituted groups.

Table 71. Base Group Properties

Base Group	$(\Delta H_f)^\circ_{298.16}(g)$, kcal/g-mole	$S^\circ_{298.16}(g)$, g-cal/(g-mole)(K°)	Constants of Heat-Capacity Equations a	$b(10^3)$	$c(10^6)$
Methane	−17.89	44.50	3.79	16.62	−3.24
Cyclopentane	−18.46	70.00	−9.02	109.28	−40.23
Cyclohexane	−29.43	71.28	−11.53	139.65	−52.02
Benzene	19.82	64.34	−4.20	91.30	−36.63
Naphthalene	35.4	80.7	3.15	109.40	−34.79
Methylamine	−7.1	57.7	4.02	30.72	−8.70
Dimethylamine	−7.8	65.2	3.92	48.31	−14.09
Trimethylamine	−10.9	—	3.93	65.85	−19.48
Dimethyl Ether	−46.0	63.7	6.42	39.64	−11.45
Formamide	−49.5	—	6.51	25.18	−7.47

The contributions resulting from the secondary substitution of methyl groups for hydrogen atoms are classified in Table 73, according to A, the type of the carbon atom on which the substitution is made, and B, the highest type number of an adjacent carbon atom. The carbon-atom types are defined on the basis of the number of hydrogen atoms attached. Thus,

Type	
1	$—CH_3$
2	$—CH_2$ (with one free bond below)
3	$—CH$ (with free bonds above and below)
4	$—C—$ (with free bonds above and below)
5	C atom in benzene or naphthalene ring

Table 72. Contribution of Primary CH_3 Substitution Groups Replacing Hydrogen

Base Group	$\Delta(\Delta H^\circ_f)_{298.16}(g)$, kcal/g-mole	$\Delta S^\circ_{298.16}(g)$, g-cal/(g-mole)(K°)	Constants of Heat-Capacity Equations Δa	$\Delta b(10^3)$	$\Delta c(10^6)$
1. Methane	−2.50	10.35	−2.00	23.20	−9.12
2. Cyclopentane					
(*a*) First primary substitution	−7.04	11.24	1.87	17.55	−6.68
(*b*) Second primary substitution					
To form 1,1	−7.55	4.63	−0.67	24.29	−10.21
To form 1,2(*cis*)	−5.46	6.27	−0.01	22.69	−9.46
To form 1,2(*trans*)	−7.17	6.43	0.28	21.97	−9.18
To form 1,3(*cis*)	−6.43	6.43	0.28	21.97	−9.18
To form 1,3(*trans*)	−6.97	6.43	0.28	21.97	−9.18
(*c*) Additional substitutions, each	−7.0	—	—	—	—
3. Cyclohexane					
(*a*) Enlargement of ring over 6 C, per carbon atom added to ring	−10.97	1.28	−2.51	30.37	−11.79
(*b*) First primary substitution on ring	−7.56	10.78	2.13	18.66	−5.71
(*c*) Second primary substitution on ring					
To form 1,1	−6.27	5.18	−2.14	25.69	−10.09
To form 1,2(*cis*)	−4.16	7.45	−0.65	22.19	−8.84
To form 1,2(*trans*)	−6.03	6.59	−0.06	22.59	−2.56
To form 1,3(*cis*)	−7.18	6.48	−0.34	21.49	−7.95
To form 1,3(*trans*)	−5.21	7.86	0.29	19.29	−7.23
To form 1,4(*cis*)	−5.23	6.48	0.29	19.29	−7.23
To form 1,4(*trans*)	−7.13	5.13	−0.72	23.79	−9.91
(*d*) Additional substitutions on ring, each	−7.0	—	—	—	—
4. Benzene					
(*a*) First substitution	−7.87	12.08	0.78	16.68	−5.41
(*b*) Second substitution					
To form 1,2	−7.41	7.89	4.27	9.72	−1.87
To form 1,3	−7.83	9.07	0.77	17.46	−6.19
To form 1,4	−7.66	7.81	1.76	13.45	−3.41
(*c*) Third substitution					
To form 1,2,3	−6.83	9.19	1.41	12.78	−2.71
To form 1,2,4	−7.87	10.42	1.61	12.72	−2.77
To form 1,3,5	−7.96	6.66	2.41	11.30	−1.90

TABLE 72. (CONTINUED)

Base Group	$\Delta(\Delta H^\circ_f)_{298.16}(g)$, kcal/g-mole	$\Delta S^\circ_{298.16}(g)$, g-cal/ (g-mole)(K°)	Constants of Heat-Capacity Equation Δa	$\Delta b(10^3)$	$\Delta c(10^6)$
5. Naphthalene					
(a) First substitution	−4.5	12.0	0.36	17.65	−5.88
(b) Second substitution					
To form 1,2	−6.3	8.1	5.20	6.02	1.18
To form 1,3	−6.5	9.2	1.72	14.18	−3.76
To form 1,4	−8.0	7.8	1.28	14.57	−3.98
6. Methylamine	−5.7	—	−0.10	17.52	−5.35
7. Dimethylamine	−6.3	—	−0.10	17.52	−5.35
8. Trimethylamine	−4.1	—	−0.10	17.52	−5.35
9. Formamide Substitution on C atom	−9.0	—	6.11	1.75	4.75

TABLE 73. CONTRIBUTION OF SECONDARY CH_3 SUBSTITUTIONS REPLACING HYDROGEN

A	*B*	$\Delta(\Delta H^\circ_f)_{298.16}(g)$, kcal/g-mole	$\Delta S^\circ_{298.16}(g)$, g-cal/ (g-mole)(K°)	Constants of Heat-Capacity Equations Δa	$\Delta b(10^3)$	$\Delta c(10^6)$
1	1	−4.75	10.00	0.49	22.04	−8.96
1	2	−4.92	9.18	1.09	17.79	−6.47
1	3	−4.42	9.72	1.00	19.88	−8.03
1	4	−5.0	11.0	1.39	17.12	−5.88
1	5	−4.68	10.76	1.09	18.71	−7.16
2	1	−6.31	5.57	−0.30	21.74	−8.77
2	2	−6.33	7.15	−0.64	23.38	−9.97
2	3	−5.25	6.53	0.80	19.27	−7.70
2	4	−3.83	7.46	2.52	16.11	−5.88
2	5	−6.18	6.72	0.37	19.25	−7.72
3	1	−8.22	2.81	−0.28	24.21	−10.49
3	2	−7.00	3.87	−0.93	24.73	−8.95
3	3	−5.19	3.99	−3.27	30.96	−14.06
3	4	−4.94	1.88	−0.14	27.57	−10.27
3	5	−9.2	1.3	0.42	16.20	1.68
1	—O— in ester or ether	−7.0	14.4	−0.01	17.58	−5.33
Substitution of H of OH group to form ester		0.5	16.7	0.44	10.03	−4.95

Two special secondary substitutions are defined in Table 73 for use in calculating the properties of esters and ethers. One is the substitution of a methyl group for the hydrogen of a carboxyl group to form a methyl ester. The other is the substitution of a methyl group for one of the hydrogens of a methyl ester or ether to form an ethyl ester or ether. If additional substitutions are made on this same carbon atom, the contributions are evaluated from the corresponding values of A and B based on the type numbers of the carbon atoms involved.

In Table 74 are the contributions resulting from the substitution of multiple bonds for single bonds between two carbon atoms of types A and B, respectively.

In Table 75 are given additional corrections to take account of the

Table 74. Multiple-Bond Contributions Replacing Single Bonds

Type of Bond	$\Delta(\Delta H^\circ_f)_{298.16}(g)$, kcal/g-mole	$\Delta S^\circ_{298.16}(g)$, g-cal/(g-mole)(K°)	Constants of Heat-Capacity Equations Δa	$\Delta b(10^3)$	$\Delta c(10^6)$
1=1	32.88	−2.40	0.21	−8.28	1.36
1=2	30.00	−0.21	1.12	−11.40	3.32
1=3	28.23	−0.11	2.18	−15.62	6.42
2=2 (*cis*)	28.39	−1.19	−3.57	0.14	1.08
2=2 (*trans*)	27.40	−2.16	1.27	−12.77	3.88
2=3	26.72	−0.28	−2.02	−10.42	3.83
3=3	25.70	−0.66	−0.41	−15.14	6.39
1≡1	74.58	−9.85	4.72	−24.36	6.29
1≡2	69.52	−4.19	3.16	−26.37	8.82
2≡2	65.50	−3.97	1.00	−25.70	−9.50

Table 75. Additional Corrections for Final Structure of Hydrocarbons

	$\Delta(\Delta H^\circ_f)_{298.16}(g)$, kcal/g-mole	$\Delta S^\circ_{298.16}(g)$, g-cal/(g-mole)(K°)	Constants of Heat-Capacity Equations Δa	$\Delta b(10^3)$	$\Delta c(10^6)$
Additional correction for length of each side chain on ring					
1. More than 2 C on cyclopentane side chain	−0.45	+0.12	−0.48	1.5	1.15
2. More than 2 C on cyclohexane side chain	0.32	−0.39	0.76	2.10	1.30
3. More than 4 C on benzene side chain	−0.70	−0.62	0.22	−0.20	0.08
Additional correction for double-bond arrangement					
1. Adjacent double bonds	13.16	−3.74	2.24	1.16	−0.25
2. Alternate double bonds	−4.28	−5.12	−0.94	3.88	−3.49
3. Double bond adjacent to aromatic ring					
(*a*) Less than 5 C in side chain	−2.0	−2.65	1.01	−3.24	1.31
(*b*) Over 4 C in side chain	−1.16	−2.65	1.01	−3.24	1.31

length of side chains of ring compounds and also to take account of the arrangement of double bonds.

In Table 76 are the incremental contributions resulting from the substitution of various groups for one or two methyl groups. Thus, if a methyl group is replaced by an OH group, the contribution to the heat of formation is —32.7 kcal per g-mole. If two methyl groups are replaced by an oxygen atom to form an aldehyde, the contribution is —12.9 kcal

TABLE 76. SUBSTITUTION GROUP CONTRIBUTIONS REPLACING CH_3 GROUP

Group	$\Delta(\Delta°H_f)_{298.16}(g)$, kcal per g-mole	$\Delta s°_{298.16}(g)$, cal/(g-mole)(°K)	Constants of Heat-Capacity Equations Δa	$\Delta b(10^3)$	$\Delta c(10^6)$
—OH (aliphatic, meta, para)	−32.7	2.6	3.17	−14.86	5.59
—OH ortho	−47.7	—	—	—	—
$—NO_2$	1.2	2.0	6.3	−19.53	10.36
—CN	39.0	4.0	3.64	−13.92	4.53
—Cl	0 for first Cl on a carbon; 4.5 for each additional	0	2.19	−18.85	6.26
—Br	10.0	3.0‡	2.81	−19.41	6.33
—F	−35.0	−1.0‡	2.24	−23.61	11.79
—I	24.8	5.0‡	2.73	−17.37	4.09
=O aldehyde	−12.9	−12.3	3.61	−55.72	22.72
=O ketone	−13.2	−2.4	5.02	−66.08	30.21
—COOH	−87.0	15.4	8.50	−15.07	7.94
—SH	15.8	5.2	4.07	−24.96	12.37
$—C_6H_5$	32.3	21.7	−0.79	53.63	−19.21
$—NH_2$	12.3	−4.8	1.26	−7.32	2.23

‡ Add 1.0 to the calculated entropy contributions of halides for methyl derivatives; for example, methyl chloride = 44.4 (base) + 10.4 (primary CH_3) − 0.0 (Cl substitution) + 1.0.

per g-mole. It may be noted that the phenyl group is included in Table 76, in addition to being designated as a base group in Table 71. This contribution is used in calculating the properties of complex compounds in which several base groups are combined, as, for example, in polybasic aromatic acids.

The contributions to heats of formation resulting from substitution of chlorine from methyl groups vary with the number of substitutions made on a single carbon atom. Corresponding variations were not found for the contributions to entropy or heat capacity or for the substitution of the other halogens. As noted in the table, a correction must be applied to the entropies calculated for the halogenated methanes. In general, the calculated results tend to be uncertain for single-carbon-atom compounds.

The suggested sequence of operations in estimating the properties of a complex compound is as follows:

1. Select the base group, and determine its properties from Table 71. Where a choice of base group is possible, select the group having the largest entropy.

2. In dealing with cyclic hydrocarbons having more than 6 carbon atoms, select cyclohexane as the base group, and take care of the ring enlargement effect through the use of the values given under 3(*a*) of Table 72.

3. Make all the primary substitutions before starting the secondary substitutions. In the case of methane, only one primary substitution is possible. With ring structures, one primary substitution is possible for each carbon atom of the base ring. If more than two primary substitutions are to be made on a base ring, the two substitutions closest together should be made first. If more than one substitution is to be made on a specific carbon atom in the base ring, such additional substitutions are treated as secondary substitutions and are made only after all the primary substitutions have been completed. Table 72 is used to evaluate the group contributions resulting from primary substitutions.

4. Complete the carbon skeleton of the compound by making a series of secondary substitutions, and use Table 73 to obtain the values of group contributions for such substitutions. In order to have the results conform with the procedures used in developing the tables, the following rules should be observed.

(*a*) The longest straight chain should be fully developed before any side chains are added.

(*b*) Side chains are added in the order of their length. However, if the carbon atom having the longest side chain is to have a second side chain, this second side chain should be added before additions are made to other carbon atoms of the straight chain.

(*c*) If more than one double bond is to be introduced into a chain, the double bond nearest the end of the chain should be introduced first.

5. Make additional CH_3 substitutions which subsequently will be removed for the groups listed in Table 76. Allowance for such CH_3 substitutions is made from Table 73.

6. Add multiple bonds, and obtain from Table 74 the contributions resulting from such additions.

7. If required, make the additional corrections called for by Table 75.

8. Replace CH_3 groups by the substitution groups listed in Table 76. Make allowance for such replacements by using the values given in Table 76.

Occasional instances may be encountered wherein different values result, dependent on the order in which the compound is built. In such instances, an average value should be taken. However, adherence to the rules regarding the addition of side chains and the introduction of double bonds as indicated under 4 above greatly reduce the possibility of such an occurrence.

Illustration 6. Approximate the standard heat of formation $\Delta H^\circ_{f298.16}$ (g) of 2,2,4-trimethylpentane at 25° C.

Base group (methane)	Table 71	−17.89
Primary CH_3	Table 72	−2.50

Secondary CH_3 substitutions by successive replacement of hydrogen (Table 73):

	A	B	
Secondary methyl groups in 5 carbon chain	1	1	−4.75
	1	2	−4.92
	1	2	−4.92
Side methyl groups	2	2	−6.33
	3	2	−7.00
	2	2	−6.33
			−54.64

The value given in the API tables[6] is −53.57 kcal per g-mole.

Illustration 7. Approximate the value of $\Delta H^\circ_{f298.16}$ (g) for dimethylphthalate.

```
   C     O
 /  \\   ||
C    C—C—OCH3
||   |
||   |
C    C—C—OCH3
 \  //   ||
   C     O
```

Base group (benzene)	Table 71	19.82
Primary CH_3 replacing H	Table 72	−7.87
Ortho CH_3 replacing H	Table 73	−7.41
—COOH replacing CH_3	Table 76	−87.0
—COOH replacing CH_3	Table 76	87.0
CH_3 replacing H of —COOH	Table 73	9.5
CH_3 replacing H of —COOH	Table 73	9.5
	$\Delta H^\circ_{f298.16}$ =	−150.5

The value of $\Delta H^\circ_{f298.16}$ calculated from the heat of combustion listed by Kharasch[7] is −147.1 kcal per g-mole.

[7] M. S. Kharasch, *Bur. Standards J. Research*, **2**, 359 (1929).

Illustration 8. Approximate the entropy of 2,2,3,3-tetramethylbutane.

	A	B		
Base group (methane)			Table 71	44.50
Primary CH_3			Table 72	10.35
Secondary CH_3 substitutions replacing hydrogen			Table 73	
To form 4-carbon chain	1	1		10.00
	1	2		9.18
	2	2		7.15
To form side chains	3	2		3.87
	2	4		7.46
	3	4		1.88
				94.39

The value given in the API tables[6] is 93.06 g-cal/(g-mole)(K°).

Illustration 9. Approximate the entropy of 1,2-dibromoethane.

	A	B		
Base group (methane)			Table 71	44.50
Primary CH_3 replacing hydrogen			Table 72	10.35
Secondary CH_3 substitutions replacing hydrogen:				
	1	1	Table 73	10.0
	1	2		9.18
Br substitution replacing CH_3			Table 76	3.0
Br substitution replacing CH_3			Table 76	3.0
				80.03

This result is in good agreement with the value reported by Pitzer.[8]

Illustration 10. Calculate the heat required to raise 1 mole of 2,3-dimethylpentane in the ideal-gaseous state from 298.16 to 1000° K.

	A	B	Δa	$\Delta b(10^3)$	$\Delta c(10^6)$
Base (Table 71)			3.79	16.62	−3.24
Primary (Table 72)			−2.00	23.30	−9.12
5 C chain (Table 73)	1	1	0.49	22.04	−8.96
	1	2	1.09	17.79	−6.47
	1	2	1.09	17.79	−6.47
Side methyl groups	2	2	−0.64	23.38	−9.97
	2	3	0.80	19.27	−7.70
			4.62	140.19	−51.93

$$\mathrm{H}_{10000} - \mathrm{H}_{298.16} = \int_{298.16}^{1000} \left[4.62 + (140.19)10^{-3}T - (51.93)10^{-6}T^2 \right] dT$$

$$= \left[4.62T + (70.095)10^{-3}T^2 - (17.31)10^{-6}T^3 \right]_{298.16}^{1000}$$

$$= 50{,}256 \text{ g-cal per g-mole}$$

The value obtained from the API tables[6] is 49,478 g-cal per g-mole.

[8] K. S. Pitzer, *Chem. Rev.*, **27**, 39 (1940).

Extensive comparisons with the data from the literature lead Andersen, Beyer, and Watson to conclude that, in general, molal heats of formation and entropies calculated by the group-contribution method differ from the better experimental values by less than 4.0 kcal per g-mole for heats of formation, and 2.0 g-cal/(g-mole)(K°) for entropy. Calculated heat capacities appear to be within 5% of the accepted values, although serious discrepancies exist among the experimental values of different investigators.

This method of calculation may be extended to other series of compounds if data are available for the base-group contribution. Properties of higher homologs or derivatives may then be estimated from group contributions.

Heats of formation and heat capacities estimated from group contributions are satisfactory for use in energy balances except where high accuracy is required. Equilibrium constants based on the estimated entropies and heats of formation are satisfactory for predicting the feasibility of a reaction but not for accurate calculation of equilibrium compositions.

Problems

1. Normal butane is isomerized to isobutane by the action of a catalyst at moderate temperatures. It is found that equilibrium is reached at the following compositions:

Temperature, °C	Mole % n-Butane	Mole % Isobutane
44	31	69
118	43	57

Assuming that activities are equal to mole fractions, calculate the standard free-energy change of the reaction at each temperature and the average values for the heat of reaction and entropy change over this temperature range.

2. From the data of Tables 17, 29, and 68, derive equations relating $\Delta G°/T$ to temperature for the following reactions:

(*a*) $SO_2(g) + \frac{1}{2}O_2(g) = SO_3(g)$

(*b*) $\frac{1}{2}N_2(g) + \frac{3}{2}H_2(g) = NH_3(g)$

(*c*) $C(s) + H_2O(g) = H_2(g) + CO(g)$

(*d*) $CO_2(g) + C(s) = 2CO(g)$

The atomic heat capacity of graphitic carbon, related to $T°$ K, is $2.673 + 0.002617T - 116{,}900/T^2$.

3. From the liquid-state values in Table 68, estimate the entropies in the ideal-gaseous state at 298.16° K for the following compounds:

		t_b, °C	Density, 20° C, g per cc
(*a*)	n-Decane	174	0.747
(*b*)	n-Butylbenzene	180	0.862
(*c*)	n-Heptylcyclohexane	223	0.801

In the absence of data on the physical properties of these compounds, use equations 4–16 and 4–3 for estimating vapor pressures at 298.16° K and heats of vaporization at the normal boiling points, and equation 8–40 for the effect of temperature on heat of vaporization. Critical temperatures should be estimated by equation 4–8, critical pressures by equation 4–9, and liquid densities by equation 14–53.

4. From the data of Tables 69 and 70, calculate the equilibrium constants and heats of reaction of the following reactions at temperatures of 298.16, 500, 800, and 1000° K:

(*a*) $C_2H_6(g) = C_2H_4(g) + H_2(g)$

(*b*) $2C_2H_4(g) = i\text{-}C_4H_8(g)$

(*c*) $C_3H_8(g) + C_2H_4(g) = C_5H_{12}(g)$ (tetramethylmethane)

5. From the data of Tables 29 and 70, calculate the values of ΔH°_{f0}, the heat of formation at 0° K in the ideal-gaseous state, for each of the following compounds.

	$(H^\circ_{298.16} - H^\circ_0)$ kcal per g-mole	λ_b, g-cal per gram	t_b, °C	t_c, °C
(*a*) Benzene	3.401	94.4	80.1	288.9
(*b*) Toluene	4.306	86.5	111	320.8
(*c*) Orthoxylene	5.576	82.8	144.5	358.4

6. From the data of Table 70, calculate the heats of reaction in the ideal-gaseous state of the following reactions at 600 and 1000° K:

(*a*) $C_3H_8 \rightarrow CH_4 + C_2H_4$

(*b*) $C_3H_8 \rightarrow C_3H_6 + H_2$

(*c*) $CO + \frac{1}{2}O_2 \rightarrow CO_2$

(*d*) $CH_4 + H_2O \rightarrow CO + 3H_2$

7. Estimate from the group contributions of Tables 71 through 75 the heats of formation, entropies, and heat-capacity equations for the following compounds in the ideal-gaseous state: (*a*) Isoprene (2-methyl-1,3-butadiene), (*b*) metadi-isopropyl benzene, (*c*) isobutyronitrile, (*d*) diphenyl, (*e*) *o*-nitrobenzoic acid.

8. The boiling point of isobutyronitrile is 108° C, and its liquid density at 20° C is 0.773 gram per cc. From the result of problem 7*c* and the generalizations of Chapters 4, 8, and 13, calculate the entropy and heat of formation of liquid isobutyronitrile at 25° C under its vapor pressure.

26

Equilibria in Chemical Reactions

By evaluation of the free-energy changes of the reactions of a chemical process, it is possible to calculate the composition of the equilibrium mixture and to determine the extent of conversion of the initial reactants. Such considerations are essential in determining the most favorable conditions of temperature, pressure, composition, and ratios of reactants to obtain the greatest conversion of reactants and the highest yield of products.

Equilibrium Compositions

Dodge[1] has aptly discussed the feasibility of a chemical reaction thus:

> The statement is sometimes made that a given reaction is thermodynamically impossible. This is a loose statement which has no meaning in the absence of qualifying statements. For example, any reaction, starting with pure reactants uncontaminated by any of the products, will have a tendency to proceed to some extent even though this may be only infinitesimal. Thus, the reaction
>
> $$H_2O(g) = H_2(g) + \tfrac{1}{2}O_2(g)$$
>
> proceeds to some extent at 25° C, and we can even calculate with considerable assurance the percentage of water vapor that would be decomposed. From the accurately known value of ΔG° of this reaction at 25° C, the equilibrium constant is about 1×10^{-40}, and the extent of decomposition is infinitesimally small but definite.
>
> From the value of the standard free-energy change for any reaction, we can form an opinion about the feasibility of the reaction without further calculation. Thus, if $\Delta G^\circ = 0$ at a given temperature, then $K = 1$, and it is obvious that the reaction must proceed to a considerable extent before equilibrium is reached. The situation becomes less favorable as ΔG° increases in the positive direction, but there is no definite value that one can choose as clearly indicating that the reaction is not feasible from the standpoint of industrial operation. At 600° K, the ΔG° for the methanol synthesis reaction is +11,000 cal per g-mole, and yet the reaction is certainly feasible at this temperature. In this case the unfavorable free-energy change for the standard

[1] B. F. Dodge, *Trans. Am. Inst. Chem. Engrs.*, **34**, 540 (1938), with permission.

state is partially overcome by utilizing high pressure to displace the equilibrium. Other means can also be used, such as changing the ratio of reactants or removing one of the reaction products.

For the purpose of ascertaining quickly and only approximately if any given reaction is promising at a given temperature, the following rough classification may be useful:

$\Delta G^\circ < 0$	Reaction is promising
$\Delta G^\circ > 0$ but $< +10{,}000$	Reaction is of doubtful promise but warrants further study
$\Delta G^\circ > +10{,}000$	Very unfavorable, would be feasible only under unusual circumstances

It should be understood that these are only approximate criteria that are useful in preliminary exploratory work.

From equation 25–7 compositions of systems at equilibrium can be calculated from a knowledge of the standard free-energy change and the relations of activities to stoichiometric compositions. These relationships involve the activity coefficients discussed in Chapter 20. If these coefficients are included, equation 25–7 becomes, for the general case,

$$K = \frac{(\gamma_R N_R)^r (\gamma_S N_S)^s \cdots}{(\gamma_B N_B)^b (\gamma_C N_C)^c \cdots}$$

$$\frac{N_R{}^r N_S{}^s \cdots}{N_B{}^b N_C{}^c \cdots} = K \frac{\gamma_B{}^b \gamma_C{}^c \cdots}{\gamma_R{}^r \gamma_S{}^s \cdots} \tag{1}$$

Gaseous Systems. As previously mentioned, in a system that involves only gaseous components, it is convenient to choose as the standard state the component gases, each in the ideal state at 1 atm pressure. This state is hypothetical at temperatures where saturation pressures are below unit fugacity. With this definition, the activity of component B in ideal solutions of gases is

$$a_B = f_B = N_B f_{B\pi} = N_B \nu_{B\pi} \pi \tag{2}$$

where $f_{B\pi}$, $\nu_{B\pi}$ = the fugacity and fugacity coefficient, respectively, of pure component B at the temperature and total pressure π of the system.

Combining equations 2 and 25–7 gives

$$K = \left(\frac{N_R{}^r N_S{}^s \cdots}{N_B{}^b N_C{}^c \cdots}\right) \left(\frac{\nu_R{}^r \nu_S{}^s \cdots}{\nu_B{}^b \nu_C{}^c \cdots}\right) \pi^{r+s+\cdots-b-c\cdots} \tag{3}$$

In a system containing n_I moles of inert gases not entering into the reaction,

$$N_R = \frac{n_R}{n_B + n_C + \cdots + n_R + n_S + \cdots + n_I} \tag{4}$$

where n_R = moles of R in the equilibrium mixture.

The ratio of the fugacity coefficients in equation 3 is constant for a given temperature and pressure and is designated by K_ν.

$$K_\nu = \frac{\nu_R^{\,r}\nu_S^{\,s}\cdots}{\nu_B^{\,r}\nu_C^{\,c}\cdots} \tag{5}$$

However, unlike the equilibrium constant K, the term K_ν is affected by changes in pressure as well as in temperature.

For reactions at low pressures, in the neighborhood of atmospheric or below, the fugacity coefficients may be taken as unity and the term K_ν neglected. At higher pressures, this term may have a marked effect on the equilibrium calculations and may be determined from Table 51, page 601.

Combining equations 3, 4, and 5 gives

$$\left(\frac{n_R^{\,r}n_S^{\,s}\cdots}{n_B^{\,b}n_C^{\,c}\cdots}\right) \times K_\nu\left(\frac{\pi}{n_B + n_C + \cdots + n_R + n_S + \cdots + n_I}\right)^{(r+s+\cdots)-(b+c+\cdots)} = K \tag{6}$$

Equation 6 is the most useful form of the equilibrium equation for calculating the composition of a reacting system at equilibrium. This calculation is best carried out by expressing the number of moles of each active material present at equilibrium in terms of the equilibrium conversion x of a given reactant and the numbers of moles of components in the original unreacted mixture. These values are substituted in equation 6, which is then solved for the equilibrium conversion. If the final equation is of a complicated form, graphical methods may offer the most convenient solution.

Illustration 1. The gases from the pyrites burner of a contact sulfuric acid plant have the following composition by volume:

SO_2	7.8%
O_2	10.8
N_2	81.4
	100.0%

This gaseous mixture is passed into a converter where in the presence of a catalyst the SO_2 is oxidized to SO_3. The temperature in the converter is maintained at 500° C, and the pressure at 760 mm Hg. Calculate the composition of the gases leaving the converter, assuming that equilibrium conditions are reached. It may be assumed that the fugacity coefficients and hence K_ν are equal to unity.

Solution:

Basis: 100 lb-moles of the original mixture.

$$SO_2(g) + \tfrac{1}{2}O_2(g) = SO_3(g)$$

From Fig. 262, at 500° C (773° K), $\log K = 1.93$, $K = 85$.

$\pi = 1.0$ atm

Limiting reactant = SO_2

Let x = equilibrium conversion of SO_2. Composition of equilibrium mixture:

$$SO_2 = 7.8 - 7.8x \text{ lb-moles} = n_{SO_2}$$

$$SO_3 = 7.8x = n_{SO_3}$$

$$O_2 = 10.8 - 7.8\left(\frac{x}{2}\right) \text{lb-moles} = n_{O_2}$$

$$N_2 = 81.4 \text{ lb-moles} = n_{N_2}$$

From equation 6,

$$\frac{n_{SO_3}}{n_{SO_2}(n_{O_2})^{1/2}}\left(\frac{\pi}{n_{SO_2} + n_{SO_3} + n_{O_2} + n_{N_2}}\right)^{-1/2} = K \qquad (a)$$

Substituting in this equation gives

$$\frac{7.8x}{(7.8 - 7.8x)(10.8 - 3.9x)^{1/2}}\left(\frac{\pi}{100 - 3.9x}\right)^{-1/2} = K \qquad (b)$$

$$\frac{x}{1.0 - x}\left(\frac{100 - 3.9x}{10.8 - 3.9x}\right)^{1/2} - 85 = \Delta \qquad (c)$$

The solution of (c) is obtained by assuming a series of values of x and calculating the corresponding values of Δ. The results may be plotted, the value of x corresponding to $\Delta = 0$ being the correct solution. Since the quantity in parentheses varies but little when x is close to 1.0, the correct solution also may be readily obtained by trial without a plot to give

$$x = 0.9585 \text{ or } 95.85\%$$

Composition of gases leaving the converter:

SO_2 = 7.8 − (7.8)(0.9585)	=	0.32 lb-mole	0.3%
SO_3 = (7.8)(0.9585)	=	7.49 lb-moles	7.8
O_2 = 10.8 − (3.9)(0.9585)	=	7.07 lb-moles	7.3
N_2 =		81.40 lb-moles	84.6
Total	=	96.28 lb-moles	100.0%

This same procedure may be followed to calculate the degree of completion in any system at conditions for which the necessary data are available. By repeating the calculations of Illustration 1 to correspond to other conditions of temperature and pressure, the effects of varying these conditions may be quantitatively predicted.

Under conditions of high pressure, the fugacity coefficients cannot be neglected, but the general method of solution is the same. Values of K_ν may be calculated directly from fugacity coefficients of the individual components obtained from Table 51. Where a number of calculations

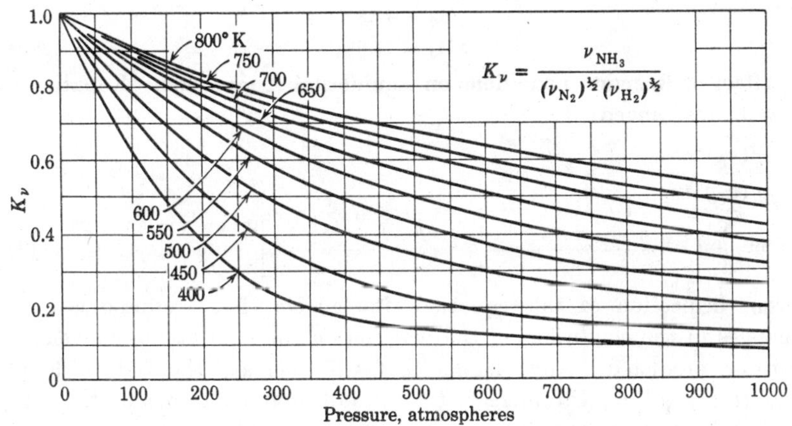

FIG. 263. Ratio of fugacity coefficients for ammonia synthesis

are to be made for a particular system of reactants and products, a chart may be prepared relating K_ν to temperature and total pressure. Figure 263 is such a chart for the ammonia-synthesis system, where

$$K_\nu = \frac{\nu_{NH_3}}{(\nu_{N_2})^{1/2}(\nu_{H_2})^{3/2}}$$

Illustration 2. Calculate the equilibrium percentage conversion of nitrogen to ammonia at 700° K and a pressure of 300 atm if the gas enters the converter with a composition of 75 mole per cent H_2 and 25 mole per cent N_2.

$$\tfrac{1}{2}N_2 + \tfrac{3}{2}H_2 = NH_3$$

At 300 atm, 700° K, $K = 0.0091$ (Fig. 262), and $K_\nu = 0.72$ (Fig. 263).

Let x represent the number of moles of ammonia formed at equilibrium, starting with $\frac{1}{2}$ mole N_2 and $\frac{3}{2}$ moles H_2. Then

$$\frac{n_{NH_3}}{(n_{N_2})^{1/2}(n_{H_2})^{3/2}} = \frac{x}{(\frac{1}{2} - \frac{1}{2}x)^{1/2}(\frac{3}{2} - \frac{3}{2}x)^{3/2}} = \frac{0.77x}{(1 - x)^2}$$

and $$2 - x = \text{total number of moles at equilibrium}$$

Substitution of the preceding values in equation 6 gives

$$\left[\frac{0.77x}{(1-x)^2}\right]0.72\left(\frac{300}{(2-x)}\right)^{1-2} = 0.0091$$

or

$$x = 0.589$$

The percentage conversion of N_2 to NH_3 is hence 58.9%, and the composition of the equilibrium mixture in mole per cent is

$$N_2 = 14.6\%$$
$$H_2 = 43.8\%$$
$$NH_3 = 41.6\%$$

Effect of Reaction Conditions on Equilibrium Conversion. Equation 6 may be rearranged:

$$\frac{(n_R)^r(n_S)^s\cdots}{(n_B)^b(n_C)^c\cdots} = \frac{K}{K_\nu}\left[\frac{n_B+n_C+\cdots+n_R+n_S+\cdots+n_I}{\pi}\right]^{(r+s+\cdots)-(b+c+\cdots)} \tag{7}$$

From inspection of this equation, the effects that are produced by changes in the conditions of the reaction on the equilibrium conversion can be predicted. Any change that will increase the right side of equation 7 will tend to increase the ratio of products to reactants in the equilibrium mixture and correspond to an increased conversion.

Effect of Temperature. From equation 7, it is apparent that an increased value of the equilibrium constant K must correspond to an increased conversion. The value of the equilibrium constant for the commonly chosen standard state depends only on temperature, as already discussed. It follows that the *equilibrium conversion is increased by a rise in temperature in an endothermic reaction and decreased in an exothermic reaction.*

Effect of Pressure. It has been pointed out that the equilibrium constant K is independent of pressure with the standard states for which equation 7 was derived. The effect of pressure on K_ν can be calculated from fugacity coefficients. When the compressibility of the products is greater than that of the reactants, an increase in pressure will decrease K_ν, and hence increase the conversion. In addition, from inspection of equation 7, it is apparent that the pressure under which a reaction proceeds will affect its equilibrium conversion in case the reaction produces a change in the total number of moles of gaseous components present in the system. If there is no change in the number of moles of gases, the exponent $(r+s+\cdots)-(b+c+\cdots)$ will equal zero, and the magnitude of the pressure π will have no effect on the extent of the reaction except as it affects K_ν.

However, *if a reaction produces a decrease in the total number of moles of gaseous components, the equilibrium degree of completion is increased by an increase in pressure.* If the total number of moles of gases is increased as a result of the reaction, an increase in pressure reduces the equilibrium degree of completion. This is in accord with the classical Le Chatelier–Braun principle.

Effect of Dilution with an Inert Gas. Dilution of a reacting system with an inert gas corresponds to an increase of n_I of equation 7. The effect produced is similar to that of a decrease in pressure. Hence, *if a reaction produces an increase in the number of moles of gaseous components, the equilibrium degree of completion is increased by dilution with an inert gas.* If no change in the total number of moles of gases accompanies a reaction, the presence of inert gases has no effect on the equilibrium conversion.

Effect of Excess Reactants. If component B of equation 7 is the limiting reactant, an increase in the number of moles of the other reactants $C \cdots$ increases the number of moles of products, $R, S, \cdots$, and also the degree of conversion of reactant B at equilibrium. Therefore, *the presence of excess of one reactant tends to increase the equilibrium conversion of the other reactant.*

Effect of Presence of Products in Initial Reacting System. It is apparent from equation 7 that the presence in the original unreacted system of any of the compounds that are products of a reaction reduces the amounts of these compounds that are formed by the reaction in proceeding to equilibrium conditions. Therefore, the *addition of reaction products to the initial reacting system reduces the equilibrium conversion of any reactant.*

Equilibrium-Conversion Charts. Where numerous calculations are to be made of equilibrium composition corresponding to various conditions of temperature and pressure for a reacting system of *constant initial composition*, it is frequently convenient to prepare a chart expressing the equilibrium degree of completion as a function of temperature and pressure. Such a chart obviates the necessity of frequent repetitions of tedious graphical solutions of the type of Illustration 1.

Even in a complicated system, the data for an equilibrium-conversion chart may be readily calculated by arbitrarily selecting a series of values of conversion. From equation 6, the equilibrium constant is calculated, corresponding to each of these selected degrees of completion and to a selected constant pressure. The temperature corresponding to each equilibrium constant may then be obtained from a diagram similar to Fig. 262. These results are plotted, temperatures being used as abscissas, and fractional conversion as ordinates. By repeating these calculations to correspond to other selected pressures, a complete chart may be constructed. In the following illustration this method is applied to the system discussed in Illustration 1.

Illustration 3. For the reacting system discussed in Illustration 1, calculate and plot curves relating the equilibrium conversion to temperature at pressures of 1.0 and 2.0 atm, respectively, assuming $K_\nu = 1.0$.

TABLE A

1	2	3	4	5	6	7	8
x	$1.0 - x$	$3.9x$	$10.8 - 3.9x$	$100 - 3.9x$	$[(5)/(4)]^{1/2}$	$(1)/(2)$	$K(\pi)^{1/2}$
0.50	0.50	1.95	8.85	98.0	3.33	1.0	3.33
0.70	0.30	2.73	8.07	97.3	3.47	2.33	8.09
0.90	0.10	3.51	7.29	96.5	3.64	9.0	32.8
0.95	0.05	3.70	7.10	96.3	3.68	19.0	70.1
0.98	0.02	3.82	6.98	96.2	3.71	49.0	181

Basis: 100 lb-moles of the original mixture.
Let x = fractional conversion of SO_2 at equilibrium.
From equation (*b*) of Illustration 1,

$$K(\pi)^{1/2} = \frac{(x)}{(1.0 - x)}\left(\frac{100 - 3.9x}{10.8 - 3.9x}\right)^{1/2}$$

A series of values of x is selected, and the corresponding values of K are calculated, first with π equal to 1.0 and then to 2.0 atm. The corresponding temperatures $T°$ K are then obtained from Fig. 262. The calculations are summarized in Tables A and B. In the column headings, the numbers in parentheses represent the results contained in the columns bearing these numbers.

TABLE B

	$\pi = 1.0$			$\pi = 2.0$		
x	K	log K	$T°$ K	K	log K	$T°$ K
0.5	3.33	0.523	978	2.35	0.371	1005
0.7	8.09	0.908	910	5.71	0.757	935
0.9	32.8	1.516	822	23.2	1.366	841
0.95	70.1	1.846	782	49.5	1.695	800
0.98	181	2.258	737	128	2.107	752

These calculations may be continued to correspond to other pressures. The results are plotted in curves E of Fig. 264, relating the equilibrium conversion as ordinates to temperature in degrees centigrade. This chart is applicable only to systems of the particular initial composition here considered. Change in relative proportions of any of the original reactants renders the chart inapplicable.

A chart of the type of curves E of Fig. 264 may be prepared for any reacting system, and from it, equilibrium conversion may be readily estimated by interpolation. For example, at a temperature of 600° C and a pressure of 1.5 atm, it is estimated from the E curves of Fig. 264 that the equilibrium conversion is 82%. If the initial composition of the reacting system undergoes change, a new chart must be prepared to correspond to each different composition.

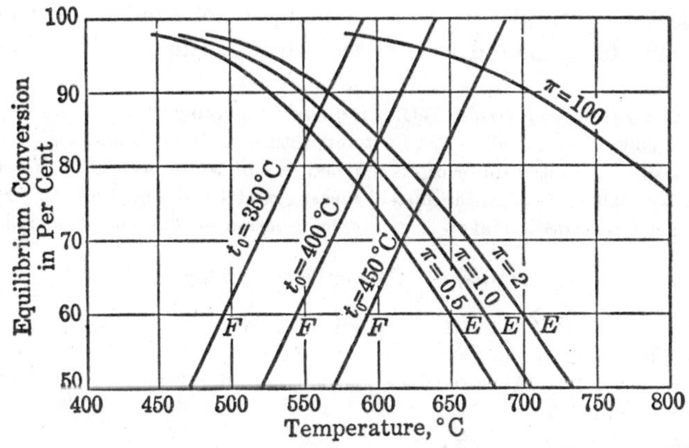

Fig. 264. Equilibrium conversion and adiabatic temperature relations for the oxidation of SO_2

Heterogeneous Reactions. Equation 6 may be applied to any gaseous system if the gas mixture behaves as an ideal solution. When a component of a reaction is involved in a heterogeneous reaction as a pure liquid or pure solid, its activity may be taken as unity, provided the pressure on the system does not differ much from the chosen standard state. The effect of pressure on the activity of a solid or liquid may be calculated from equation 20–32; this effect is negligible at moderate pressures.

If in the reaction represented by equations 25–1 and 3 there are, in addition to gaseous components B, C, R and S, d moles of a solid or liquid reactant D, and t moles of a solid or liquid product T, equation 3 becomes

$$K = \frac{a_T^{\,t}}{a_D^{\,d}}\left(\frac{N_R^{\,r}N_S^{\,s}}{N_B^{\,b}N^{c}}\right)\left(\frac{\nu_R^{\,r}\nu_S^{\,s}}{\nu_B^{\,b}\nu_C^{\,c}}\right)\pi^{r+s-b-c} \tag{8}$$

Where the standard state for solids and liquids is taken at atmospheric pressure or at low equilibrium vapor pressures, the activities of pure liquids and pure solids may be taken as unity at all moderate pressures, and the composition of the gaseous phase at equilibrium will not be affected by the presence of the solid or liquid. However, at high pressures, the activities of pure solids and liquids are affected by pressure, and the composition of the gaseous phase at equilibrium is affected by the presence of the liquid or solid. When solid or liquid solutions are formed, the activities of the components in solid or liquid solutions are no longer unity even at moderate pressures, and the equilibrium composition of the gas is greatly affected by the composition of the solid or liquid phase. In such cases, the activities in the solid or liquid solutions are expressed in terms of mole fractions and activity coefficients, as discussed in Chapter 20.

Illustration 4. Ferrous oxide, FeO, is reduced to metallic iron by passing over it a mixture of 20% CO and 80% N_2 at a temperature of 1000° C under a pressure of 1 atm. Assuming that equilibrium is reached, calculate the weight of metallic iron produced per 1000 cu ft of gas admitted, measured at 1 atm and 1000° C. Fugacity coefficients are assumed equal to unity. The reaction taking place is as follows:

$$\mathrm{FeO}(s) + \mathrm{CO}(g) \rightleftharpoons \mathrm{Fe}(s) + \mathrm{CO_2}(g)$$

At 1000° C, the value of K for this reaction is 0.403.

Basis: 1.0 lb-mole of entering gas.

Since the activities of FeO and Fe equal 1.0, equation 8 may be arranged in the form of equation 6. Then

$$\left(\frac{n_{\mathrm{CO_2}}}{n_{\mathrm{CO}}}\right)\left(\frac{\pi}{n_{\mathrm{CO}} + n_{\mathrm{CO_2}} + n_{\mathrm{N_2}}}\right)^{1-1} = K \quad \text{or} \quad \frac{n_{\mathrm{CO_2}}}{n_{\mathrm{CO}}} = K$$

Let x = fractional conversion of CO at equilibrium. At equilibrium:

$$n_{\mathrm{CO_2}} = 0.20x$$

$$n_{\mathrm{CO}} = 0.20(1 - x)$$

$$\frac{0.20x}{0.20(1 - x)} = 0.403$$

or

$$x = 0.287$$

CO_2 produced = (0.287)(0.20) = 0.0574 lb-mole

Fe produced = 0.0574 lb-atom = 3.2 lb

Volume of entering gas at 1000° C = $(359)(\frac{1273}{273})$ = 1673 cu ft

$$\text{Fe produced per 1000 cu ft of gas} = \frac{(3.2)(1000)}{1673} = 1.9 \text{ lb}$$

Pressures of Decomposition. Many solid compounds decompose to yield another solid and a gas as in the calcination of calcium carbonate to form lime and carbon dioxide. Such decompositions will proceed only when the activity of the gaseous product in contact with the solid is less than the equilibrium value determined by the temperature and the nature of the reaction.

A solid decomposition reaction may be represented by the following equation:

$$bB(s) \rightleftharpoons rR(s) + sS(g) \tag{9}$$

The activities of the solids are approximately unity at moderate pressures, so that, as long as any of compounds B and R are present, equation 7 may be applied to this reaction as follows:

$$K = a_S^{s} = \exp\left(-\frac{\Delta G^\circ}{RT}\right) = \exp\left(-\frac{\Delta H^\circ}{RT} + \frac{\Delta S^\circ}{R}\right) \tag{10}$$

At low pressures, where the activity of a gas may be taken as equal to its partial pressure, $a^*_S = p^*_S$ and

$$\ln p^*_S = -\frac{\Delta G^\circ}{sRT} = -\frac{\Delta H^\circ}{sRT} + \frac{\Delta S^\circ}{sR} \tag{11}$$

It may be noted that, if ΔH° and ΔS° are independent of temperature, the form of equation 11 is similar to that of the Clausius–Clapeyron vapor-pressure equation. In general, both ΔH° and ΔS° vary with temperature, but to a much smaller extent than ΔG°.

Illustration 5. Calculate the decomposition pressure of limestone at 1000° K and the temperature necessary to produce a decomposition pressure of 1.0 atm.

$$CaCO_3(s) \rightleftharpoons CaO(s) + CO_2(g)$$

From Table 08,

$$\Delta S^\circ_{298.16} = 9.5 + 51.1 - 22.2 = 38.4$$

From Table 29,

$$\Delta H^\circ_{298.16} = (-151.7) + (-94.03) - (-289.3) = 43.57 \text{ kcal per g-mole}$$

From Table 17, for CO_2:

$$c^\circ_p = 6.85 + (8.533)10^{-3}T - (2.475)10^{-6}T^2$$

From the data of Kelley,[2]

For CaO: $$c_p = 10.00 + (4.84)10^{-3}T - \frac{108{,}000}{T^2}$$

For $CaCO_3$: $$c_p = 19.68 + (11.89)10^{-3}T - \frac{307{,}600}{T^2}$$

Using the symbols of equation 24–12 yields

$$\Delta a = 6.85 + 10.00 - 19.68 = -2.83$$

$$\Delta b = (8.533 + 4.84 - 11.89)10^{-3} = (1.483)10^{-3}$$

$$\Delta c = -(2.475)10^{-6}$$

$$\Delta d = (-108{,}000) - (-307{,}600) = 199{,}600$$

Substitution in equation 25–12 gives

$$\Delta H_{298.16} = I_H + \Delta aT + \tfrac{1}{2}\Delta bT^2 + \tfrac{1}{3}\Delta cT^3 - \frac{\Delta d}{T}$$

$$43{,}570 = I_H + (-2.83)298.1 + \tfrac{1}{2}(1.483)10^{-3}(298.1)^2 + \tfrac{1}{3}(-2.475)10^{-6}(298.1)^3 - \frac{199{,}600}{298.1}$$

or $$I_H = 45.037 \quad \text{and} \quad \Delta H^\circ_{1000} = 41{,}924$$

[2] K. K. Kelley, "High-Temperature Specific Heat Equations for Inorganic Substances," *U.S. Bur. Mines Bull.*, 371 (1934).

Substitution in equation 25–13 gives

$$\Delta S^\circ_{298.16} = I_S + \Delta a \ln T + \Delta b T + \tfrac{1}{2}\Delta c T^2 - \frac{\Delta d}{2T^2}$$

$$38.4 = I_S + (-2.83)(5.697) + (1.483)10^{-3}(298.1)$$

$$+ \tfrac{1}{2}(-2.475)10^{-6}(298.1)^2 - \frac{1}{2}\left(\frac{199{,}600}{(298.1)^2}\right)$$

$$I_S = 55.31 \quad \text{and} \quad \Delta S^\circ_{1000} = 35.90$$

From equation 11,

$$\log p^*_S = \frac{1}{2.303}\left[\frac{-41{,}924}{(1.987)1000} + \frac{35.90}{1.987}\right] = -1.316$$

$$p_S = 0.0483 \text{ atm at } 1000^\circ \text{ K}$$

At a decomposition pressure of 1 atm, from equation 11, $\Delta G^\circ_T = 0$. The corresponding temperature T is calculated from equation 25–14:

$$\Delta G^\circ_T = I_H + (\Delta a - I_s)T - \Delta a T \ln T - \tfrac{1}{2}\Delta b T^2 - \tfrac{1}{6}\Delta c T^3 - \frac{\Delta d}{2T}$$

Substitution of known values gives

$$\Delta G^\circ_T = 45{,}037 - 58.14T + 6.517T \log T - (0.7415)10^{-3}T^2$$

$$+ (0.4125)10^{-6}T^3 - \frac{99{,}800}{T} = 0$$

This equation is solved graphically or by trial and error to give

$T = 1180^\circ$ K, the decomposition temperature of $CaCO_3$ at 1.0 atm

Vapor Pressure. Equations 25–22 and 11 are directly applicable to the equilibrium between a pure solid or liquid and its vapor. In this case $\Delta G^\circ, \Delta H^\circ, \Delta S^\circ$ represent the changes in free energy, enthalpy, and entropy, respectively, in the isothermal transformation of the solid or liquid to the ideal vapor at a pressure of 1.0 atm. In vaporization at low pressures, ΔH° becomes equal to the heat of vaporization at the given temperature.

Adiabatic-Reaction Temperatures

In Chapter 9, methods are demonstrated whereby the temperature attained by a reacting system may be calculated if the reaction goes to completion without loss of heat from the system. In a reaction reaching equilibrium conditions that do not approach 100 per cent conversion, the temperature attained depends on the degree of conversion actually produced, which in turn depends on temperature. The equilibrium temperature attained by such a reaction when proceeding adiabatically may best be determined by a graphical calculation, making use of an equilibrium-conversion chart for the system.

On an equilibrium-conversion chart of the type of curves E of Fig. 264, a curve is plotted relating calculated adiabatic reaction temperature to conversion. This curve may be established by selecting a series of values of degrees of conversion and calculating the reaction temperature corresponding to each by the method discussed in Chapter 9, page 351. The values on this curve are independent of pressure if it is assumed that enthalpies are independent of pressure. However, a new curve must be plotted to correspond to each change in the initial temperature of any of the reactants.

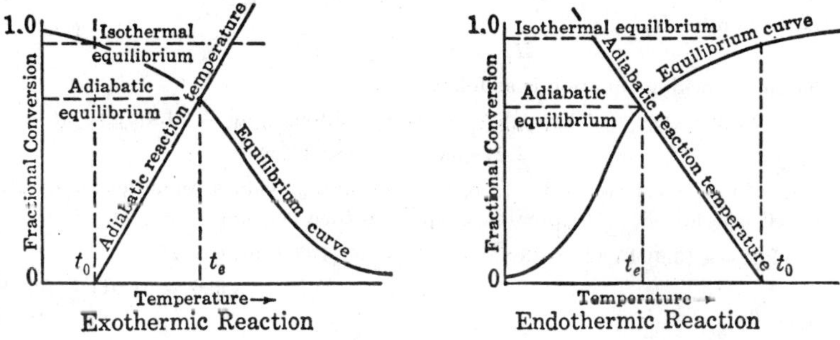

Fig. 265. Equilibrium conversion in exothermic and endothermic reactions for isothermal and adiabatic operations

In reactions proceeding adiabatically, the characteristic behaviors of exothermic and endothermic reactions are shown diagrammatically in Fig. 265. In exothermic reactions, the equilibrium conversion decreases with increase in temperature, whereas the temperature steadily rises as the reaction proceeds. The adiabatic reaction cannot proceed beyond the intersection of these two lines. In endothermic reactions, the reverse is true; the equilibrium degree of conversion is favored by a rise in temperature, whereas the temperature falls as the reaction proceeds. It may be noted that isothermal operation leads to a higher equilibrium conversion in either an endothermic or exothermic reaction.

Illustration 6. The mixture of 7.8% SO_2, 10.8% O_2, and 81.4% N_2 discussed in Illustrations 1 and 3 enters the converter at a temperature of 400° C. Calculate the equilibrium temperature attained in the converter, assuming that it is thermally insulated so that the heat loss is negligible.

Solution: Equation 9–74 is employed. Since the converter operates adiabatically, $\Delta H = 0$, and the equation reduces to the following form.

$$H_p = H_R - \Delta H_{25}$$

Basis: 100 g-moles of the original gaseous mixture.

Reference temperature: 25° C.

Expression for ΔH_{25}: From heat of formation data given in Table 29, page 301,

$$(\Delta H)_{25} = (-94{,}450) - (-70{,}960) = -23{,}490 \text{ g-cal per g-mole}$$

If x = degree of completion of the reaction, $7.8x$ g-moles of SO_2 are converted, and the total standard heat of reaction to be substituted in the above equation is $(7.8x)(-23{,}490)$ or $-183{,}222x$.

Determination of H_R: Mean heat capacity between 25 and 400° C is obtained from Table 19, page 258, and used to calculate H_R as follows:

$$SO_2 = (7.8)(10.94)375 = 32{,}000$$
$$O_2 = (10.8)(7.406)375 = 29{,}994$$
$$N_2 = (81.4)(7.089)375 = 216{,}392$$
$$H_R = 278{,}386 \text{ g-cal}$$

Expression for H_p:

$$H_p = \sum n_i c_{pmi}(t - 25)$$

The final working equation is as follows:

$$[(7.8 - 7.8x)(c_{pm})_{SO_2} + 7.8x(c_{pm})_{SO_3} + (10.8 - 3.9x)(c_{pm})_{O_2} + 81.4(c_{pm})_{N_2}](t - 25) = 278{,}386 - (-183{,}222x)$$

A value of $x = 0.5$ is assumed. In order to obtain c_{pm} values, an outlet temperature of 500° C is assumed. The preceding equation then becomes

$$[(3.9)(11.22) + (3.9)(15.82) + (8.85)(7.515) + (81.4)(7.159)](t - 25) = 278{,}386 + 91{,}611 = 369{,}997$$
$$t = 515° \text{ C}$$

A second evaluation of t is made, using c_{pm} values corresponding to the preliminary value of 515° C.

$$[(3.9)(11.255) + (3.9)(15.896) + (8.85)(7.530) + (81.4)(7.169)](t - 25) = 369{,}997$$
$$t = 514° \text{ C}$$

This procedure is repeated at several other assumed values of x. The results are plotted to give curve F of Fig. 264, corresponding to an initial temperature of 400° C. This curve crosses the E curve corresponding to a pressure of 1.0 atm at the point corresponding to a temperature of 592° C and a degree of completion of 81.5%. These are the equilibrium conditions of the reaction.

The F curves of Fig. 264 relate calculated reaction temperatures to conversion, each corresponding to a different initial temperature of the reactants entering the converter. These curves were established by the method demonstrated in Illustration 3. From Fig. 264, the equilibrium temperature and degree of completion corresponding to any selected adiabatic operating conditions may be readily estimated. For example, if the reacting gases enter the converter at 375° C and a pressure of 1.5 atm, an equilibrium temperature of 585° C and a degree of completion of 87% will be attained.

If a reaction cannot be assumed to proceed adiabatically, a chart of the type of Fig. 264 can be constructed, which takes into account the loss of heat from the reacting system. Such losses in no way affect the E curves. However, the heat losses, expressed as a function of the temperature of the system, must be included in the energy-balance equation from which the data for the F curves are calculated.

Equilibrium-Reaction Temperatures at High Pressures. When reactions proceed at high pressures, the method described for obtaining the equilibrium-reaction temperature should be modified to take into consideration the effect of pressure on enthalpies and on K_ν. For example, if a 1 : 3 mixture of nitrogen and hydrogen enters a catalytic reaction chamber at 400° C and reacts adiabatically to equilibrium at 300 atm, the temperature leaving will be 590° C, with a corresponding percentage conversion of nitrogen to ammonia of 25%. If the results were calculated with the effect of pressure on enthalpies and K_ν neglected, the calculated results would be 600° C and 21% conversion. The smaller temperature rise under actual conditions is due to the increased heat capacity of ammonia at 300 atm. The increased percentage conversion is due to the decrease in activities of ammonia as compared to nitrogen and hydrogen.

Equilibrium-reaction temperatures are readily calculated by the general methods already demonstrated if ideal solutions are formed. This condition is generally satisfactory where all reactants and products are well above their critical temperatures. For other cases involving reactions at conditions in the vicinity of the critical point of the mixture, the precise methods for estimating activities described in Chapter 21 must be used.

Illustration 7. It is desired to calculate the equilibrium percentage conversion of SO_2 to SO_3 and the corresponding temperature when the system described in Illustration 6 reacts adiabatically at 100 atm and the entering temperature is 400° C.

The equilibrium-temperature line E, corresponding to 100 atm, is constructed as in Illustration 3, except that the deviation of K_ν from unity is calculated from Table 51 as in Illustration 2. The adiabatic heating line F is constructed by the method employed in Illustration 6, except that allowance is made for the effect of the pressure of 100 atm on the enthalpies of the entering and reacting mixtures. These corrections are obtained from Table 50. The procedure is as follows:

The critical constants of the component gases are:

	T_c° K	p_c, atm
SO_2	430.7	77.8
N_2	126.2	33.5
O_2	154.4	49.7
SO_3	491.4	83.8

For the entering gas, which contains 7.8% SO_2, 81.4% N_2, 10.8% O_2:

Pseudocritical temperature = 0.078(430.7) + 0.814(126.2) + 0.108(154.4) = 152° K

Pseudocritical pressure = 0.078(77.8) + 0.814(33.5) + 0.108(49.7) = 38.6 atm

$$\text{Reduced temperature,}\quad T_r = \frac{673}{152} = 4.40$$

$$\text{Reduced pressure,}\quad p_r = \frac{100}{38.6} = 2.59$$

From Table 50, $\dfrac{H^* - H}{T_c} = 0$

The effect of the elevated pressure on the enthalpy at 400° C is negligible. When the gas has been heated adiabatically by reaction to equilibrium, the approximate conversion is 96% at 631° C.

The corresponding composition is

$SO_2 = 7.8(1 - 0.96)$	=	0.31 mole	0.3%
N_2	=	81.40 moles	84.6
$O_2 = 10.8 - \dfrac{(7.8 - 0.31)}{2}$	=	7.05 moles	7.3
$SO_3 = 7.8 - 0.31$	=	7.49 moles	7.8
		96.25 moles	100.0%

The correction of enthalpy for pressure may be shown by Table 50 to be negligible, and the adiabatic line at 100 atm coincides with the adiabatic line for 1 atm.

Electrochemical Reactions

In Chapter 13, it was indicated that some chemical reactions which ordinarily proceed spontaneously can be made to proceed reversibly in an electrolytic cell. In such instances, much of the energy is released as electric energy rather than as heat. Such a method of energy release depends on the building up of electric charges in the reacting system by the formation or discharge of ions, and the orderly flow of electrons through a closed conducting circuit. Energy conversion of this type is limited to systems that are capable of ionic dissociation, and requires the use of an electrolytic cell wherein opposing electrodes are employed to direct the inflow and outflow of electrons.

Relatively few chemical reactions can be carried out reversibly in an electrolytic cell. For example, no practical fuel cell has yet been developed wherein the combustion reaction, $C(s) + O_2(g) \rightarrow CO_2(g)$, is carried out.

The units most commonly used in calculations pertaining to electrolytic reactions are the following:

1 joule = 1 (watt)(second) = 1 (volt)(ampere)(second)
= 1 (volt) (coulomb)

1 (kw)(hr) = 3,600,000 (watt)(second)

1 National Bureau of Standards gram-calorie = 4.1840 international joules

1 faraday = 96,493.1 (abs. coulomb)/(gram-equivalent)

The faraday is the quantity of electricity, in coulombs, that will deposit or dissolve one gram-equivalent of material. For example, one faraday will deposit or dissolve (dependent on the direction of electron flow), 1.0080 grams of hydrogen, 35.457 grams of chlorine, 65.38/2 grams of zinc, 63.54/2 grams of copper from a solution containing cupric ions, 63.54 grams of copper from a solution containing cuprous ions, etc.

In some instances, an electrode reaction does not involve solution or deposition of material, but merely a change in the valence of an ion, as, for example, in the following reaction:

$$Fe^{+++} + e \rightarrow Fe^{++}$$

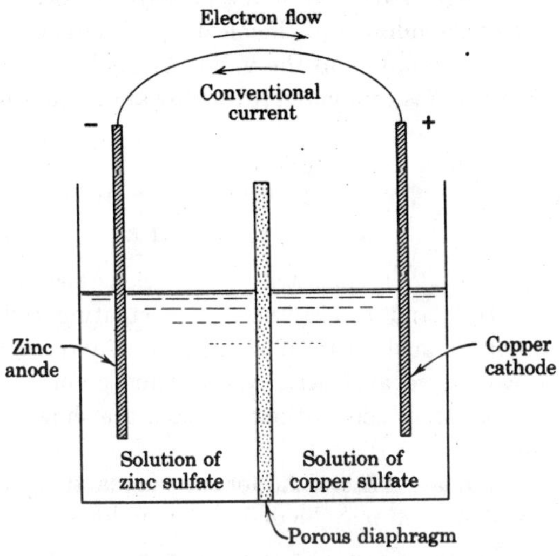

FIG. 266. The zinc–copper cell

The symbol e represents 96,493.1 coulombs of electricity per gram-ion. Thus, one faraday per gram-ion of material will produce a unit change of valence.

An electrolytic cell has two electrodes; the *anode* and the *cathode*. At the anode, a reaction occurs that involves a release of electrons, whereas, at the cathode, a reaction occurs that involves a combination of electrons. The anode reaction, involving release of electrons, is termed an oxidizing reaction, whereas the cathode reaction, involving the addition of electrons, is termed a reducing reaction.

These relations are indicated in Fig. 266, which shows a zinc–copper cell in schematic form. The anode reaction is

$$\tfrac{1}{2}Zn \rightarrow \tfrac{1}{2}Zn^{++} + e$$

The cathode reaction is

$$\tfrac{1}{2}Cu^{++} + e \rightarrow \tfrac{1}{2}Cu$$

The over-all cell reaction is:

$$\tfrac{1}{2}Zn + \tfrac{1}{2}Cu^{++} \rightarrow \tfrac{1}{2}Zn^{++} + \tfrac{1}{2}Cu$$

In the *external* circuit connecting the two electrodes, the flow of electrons is from anode to cathode. It should be noted that the direction of electron flow is opposite to the conventional designation of direction of flow of an electric current.

The energy change associated with an electrochemical reaction is equal to the product of the number of chemical equivalents undergoing change, the coulombs per faraday, and the voltage. As has been demonstrated earlier, if the reaction is conducted reversibly at constant temperature and pressure,

$$\Delta G = -w_f \tag{12}$$

Accordingly,

$$\Delta G = -w_f = -nFE \tag{13}$$

It is necessary to adopt consistent conventions regarding the sign of electrode potentials and the method of calculating cell voltages from electrode potentials such that, if a cell reaction tends to occur spontaneously, E will have a positive value, thus giving a negative value for ΔG. The consistent conventions adopted are shown in Table 76 and in Illustration 8.

Equations that relate ΔH and ΔS for the cell reaction with the electrical characteristics of the cell may be deduced as follows:

$$dG = -S\,dT + V\,dp \tag{14}$$

Applying this to the products and reactants of a reaction that proceeds at constant pressure and with terminal temperatures that are the same:

$$\left(\frac{\partial \Delta G}{\partial T}\right)_p = -\Delta S \tag{15}$$

Also,

$$\Delta G = \Delta H - T\,\Delta S \tag{16}$$

$$-\Delta S = \frac{\Delta G}{T} - \frac{\Delta H}{T} \tag{17}$$

Substituting for $-\Delta S$ in equation 15 gives

$$\left(\frac{\partial \Delta G}{\partial T}\right)_p = \frac{\Delta G}{T} - \frac{\Delta H}{T} \tag{18}$$

This is known as the Gibbs–Helmholtz equation.

From equation 13,

$$\left(\frac{\partial \Delta G}{\partial T}\right)_p = -nF\left(\frac{\partial E}{\partial T}\right)_p \tag{19}$$

Substituting in equation 15,

$$\Delta S = nF\left(\frac{\partial E}{\partial T}\right)_p \tag{20}$$

Substituting in equation 18,

$$\Delta H = -nFE + nFT\left(\frac{\partial E}{\partial T}\right)_p \tag{21}$$

Standard Electrode Potentials. In Table 76 are given a number of standard electrode potentials. These are the equilibrium potentials developed at 25° C, with the ions at unit activity. The potential values are all relative to the standard hydrogen electrode, which consists of platinized platinum saturated with hydrogen at 1 atm pressure, in contact with a solution containing hydrogen ions at unit activity.

The arrangement in Table 76 is such that, for any chosen pair of electrodes, the lower one in the table will function as cathode (+ terminal) and the upper one in the table will function as anode (— terminal) when the cell reaction is permitted to proceed in its normal direction, without being reversed by the application of an external voltage.

Illustration 8. Consider a cell at 25° C, having zinc and copper electrodes, each in contact with a solution of its own ions at unit activity, and with the copper ions in the cupric state.

(*a*) Which electrode will function as anode, and which will function as cathode?

(*b*) Write the individual electrode reactions and the over-all cell reaction.

(*c*) Calculate the cell voltage.

(*d*) Calculate ΔG° for the over-all cell reaction.

(*e*) Calculate the equilibrium constant for the over-all reaction and $a_{Zn^{++}}/a_{Cu^{++}}$ for an equilibrium system containing Cu, Cu^{++}, Zn, and Zn^{++}.

Solution: (*a*) Since copper is below zinc in the series, copper will function as cathode and zinc as anode.

(*b*) At the copper cathode, the reaction is as indicated Table 76, and is as follows:

$$\tfrac{1}{2}Cu^{++}\ (a = 1) + e \rightarrow \tfrac{1}{2}Cu$$

At the zinc anode, the reaction is the reverse of that written in Table 76, as follows:

$$\tfrac{1}{2}Zn \rightarrow \tfrac{1}{2}Zn^{++}\ (a = 1) + e$$

The over-all reaction equation is obtained by adding the electrode reaction equations:

$$\tfrac{1}{2}Cu^{++}\ (a = 1) + \tfrac{1}{2}Zn \rightarrow \tfrac{1}{2}Cu + \tfrac{1}{2}Zn^{++}\ (a = 1)$$

(*c*) Cell voltage $= E^\circ_{\text{cathode}} - E^\circ_{\text{anode}} = 0.345 - (-0.762) = 1.107$ volts.

(*d*) $\Delta G^\circ = -nFE = -1(96{,}493.1)(1.107) = -106{,}818$ joules $= -25{,}530$ g-cal.

(e)
$$\log K = \frac{-\Delta G^\circ}{4.57567T} = \frac{25{,}530}{(4.57567)(298.16)} = 18.7131$$

$$K = \left(\frac{a_{Cu}\, a_{Zn^{++}}}{a_{Zn}\, a_{Cu^{++}}}\right)^{1/2} = 5.16(10^{18})$$

Since the metals are at unit activity,

$$\frac{a_{Zn^{++}}}{a_{Cu^{++}}} = K^2 = 2.67(10^{37})$$

Table 76. Standard Electrode Potentials

E° = electrode potential at 25° C, in volts, relative to the standard hydrogen electrode; ions at unit activity.

Electrode	Reaction When Electrode is Cathode in Cell. Ions at Unit Activity	E°, volts
K; K^+	$K^+ + e = K$	−2.922
Ca; Ca^{++}	$\frac{1}{2}Ca^{++} + e = \frac{1}{2}Ca$	−2.87
Na; Na^+	$Na^+ + e = Na$	−2.712
Mg; Mg^{++}	$\frac{1}{2}Mg^{++} + e = \frac{1}{2}Mg$	−2.34
Be; Be^{++}	$\frac{1}{2}Be^{++} + e = \frac{1}{2}Be$	−1.70
Al; Al^{+++}	$\frac{1}{3}Al^{+++} + e = \frac{1}{3}Al$	−1.67
Mn; Mn^{++}	$\frac{1}{2}Mn^{++} + e = \frac{1}{2}Mn$	−1.05
Zn; Zn^{++}	$\frac{1}{2}Zn^{++} + e = \frac{1}{2}Zn$	−0.762
Cr; Cr^{+++}	$\frac{1}{3}Cr^{+++} + e = \frac{1}{3}Cr$	−0.71
Ga; Ga^{+++}	$\frac{1}{3}Ga^{+++} + e = \frac{1}{3}Ga$	−0.52
Fe; Fe^{++}	$\frac{1}{2}Fe^{++} + e = \frac{1}{2}Fe$	−0.440
Cd; Cd^{++}	$\frac{1}{2}Cd^{++} + e = \frac{1}{2}Cd$	−0.402
In; In^{+++}	$\frac{1}{3}In^{+++} + e = \frac{1}{3}In$	−0.340
Tl; Tl^+	$Tl^+ + e = Tl$	−0.336
Co; Co^{++}	$\frac{1}{2}Co^{++} + e = \frac{1}{2}Co$	−0.277
Ni; Ni^{++}	$\frac{1}{2}Ni^{++} + e = \frac{1}{2}Ni$	−0.250
Sn; Sn^{++}	$\frac{1}{2}Sn^{++} + e = \frac{1}{2}Sn$	−0.136
Pb; Pb^{++}	$\frac{1}{2}Pb^{++} + e = \frac{1}{2}Pb$	−0.126
Pt; $H_2(g)$; H^+	$H^+ + e = \frac{1}{2}H_2$	0.000
Normal calomel electrode	$\frac{1}{2}Hg_2Cl_2 + e = Hg + Cl^-$	0.2802
Cu; Cu^{++}	$\frac{1}{2}Cu^{++} + e = \frac{1}{2}Cu$	0.345
Cu; Cu^+	$Cu^+ + e = Cu$	0.522
Pt; Fe^{++}, Fe^{+++}	$Fe^{+++} + e = Fe^{++}$	0.771
Hg; Hg_2^{++}	$\frac{1}{2}Hg_2^{++} + e = Hg$	0.799
Ag; Ag^+	$Ag^+ + e = Ag$	0.800
Pd; Pd^{++}	$\frac{1}{2}Pd^{++} + e = \frac{1}{2}Pd$	0.83
Hg; Hg^{++}	$\frac{1}{2}Hg^{++} + e = \frac{1}{2}Hg$	0.854
Pt; $Cl_2(g)$; Cl^-	$\frac{1}{2}Cl_2(g) + e = Cl^-$	1.358
Pt; Pt^{++}	$\frac{1}{2}Pt^{++} + e = \frac{1}{2}Pt$	ca 1.2
Au; Au^{+++}	$\frac{1}{3}Au^{+++} + e = \frac{1}{3}Au$	1.42
Au; Au^+	$Au^+ + e = Au$	1.68

References: 1. H. H. Uhlig, *The Corrosion Handbook*, p. 1134, John Wiley & Sons, (1948), with permission.

2. Farrington Daniels, *Outline of Physical Chemistry*, p. 423, John Wiley & Sons (1955).

Efficiency of Electrochemical Reactions. If the external circuit of an electrolytic cell is closed and no external opposing voltage is applied, current will flow. As soon as flow of current occurs, heat is generated because of the electrical resistances offered by the electrolyte, by membranes separating the anode and cathode compartments, by gases discharged and accumulated on the electrodes, by the electrode metals, and by the external electric circuit. This generation of heat is irreversible; hence, under such conditions, the cell operates irreversibly. As the current is progressively decreased by the application of an external opposing voltage of progressively greater magnitude, the degree of irreversibility diminishes, and in the limit, at zero flow of current, the operation of the cell is reversible.

In industrial electrochemical reactions, it is necessary to use large currents in order to maintain reasonable production rates. Consequently, such operations are always irreversible in character, with a considerable loss of heat.

In many electrolytic reactions as conducted under industrial conditions, not all the current flowing forms the desired products; appreciable amounts may be used in the deposition of side products. In analyzing industrial electrochemical processes, it is, therefore, of interest to compute *current efficiencies* of the anode and cathode. In a given cell, these two current efficiencies are not necessarily equal. Current efficiency may be defined as follows:

$$\text{Current efficiency} = \left(\frac{\text{weight of desired product actually obtained}}{\text{weight of desired product if current had been used solely to deposit the desired product}}\right)(100)$$

$$= \left(\frac{\text{current theoretically required to deposit the desired material that is actually obtained}}{\text{total quantity of electricity passed}}\right)(100)$$

Current efficiencies of industrial electrochemical processes vary over wide limits. For example, in plating copper from an acid copper sulfate bath, cathode current efficiencies of 100% can be closely approached at low current densities. By way of contrast, in chromium plating, cathode current efficiencies approximate 10%, with most of the current being used to deposit hydrogen. Current efficiency is markedly influenced by current density.

In analyzing electrochemical processes, *energy efficiencies* are also frequently calculated. Energy efficiency may be defined as follows:

$$\text{Energy efficiency} = \left(\frac{\begin{array}{l}\text{energy theoretically required, based on the}\\ \text{decomposition voltage, and on the amount}\\ \text{of desired product actually obtained}\end{array}}{\text{actually energy input}}\right) \quad (100)$$

It will be noted that calculation of energy efficiency requires the use of a value for decomposition voltage. This is the minimum voltage, above

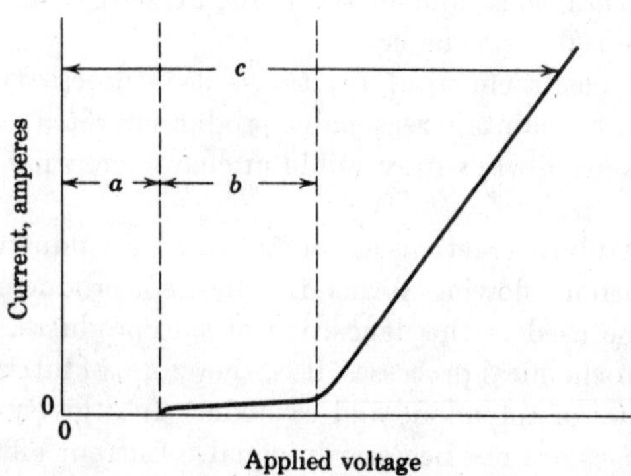

FIG. 267. Current–voltage relation in electrolysis
a = equilibrium cell voltage, no current flowing
$a + b$ = decomposition voltage
b = cell overvoltage
c = voltage of cell operating above decomposition voltage

which a large increase of current flow takes place with increase in cell voltage, as indicated in Fig. 267.[3]

If the electrolytic cell produces two desired products, as, for example, sodium hydroxide solution and chlorine gas, in the electrolysis of sodium chloride solution, the energy-efficiency calculation may be based on the anode product or on the cathode product. These energy efficiencies are not necessarily the same, unless current efficiencies for the anode and cathode are the same, in which case energy efficiency may be calculated as follows:

$$\text{Energy efficiency} = (\text{current efficiency})\left(\frac{\text{decomposition voltage}}{\text{actual cell voltage}}\right) \quad (100)$$

[3] W. A. Koehler, *Principles and Applications of Electrochemistry*, Vol. II, John Wiley & Sons (1944).

Illustration 9. A cell electrolyzing sodium chloride solution produces caustic soda solution, chlorine gas, and hydrogen gas. The actual operating cell voltage is 3.44 volts, whereas the decomposition voltage for NaCl at the concentration used in the cell is 2.3 volts. Production is 0.8944 lb NaOH and 0.7926 lb Cl_2 per kw-hr. The desired products are NaOH and Cl_2; the hydrogen is discarded. Calculate the following items:

(*a*) Current efficiencies based on both NaOH and Cl_2 productions.

(*b*) Energy efficiencies based on both NaOH and Cl_2 productions.

Basis; 1 kw-hr energy supplied.

(*a*) Faradays ker kw-hr $= \left[\dfrac{3,600,000}{(96,493.1)(\text{operating voltage})}\right]$

$$= \frac{3,600,000}{(96,493.1)(3.44)} = 10.845$$

Maximum possible production of NaOH, based on current actually supplied to cell

$$= (10.845)(40.0) = 433.8 \text{ grams} = 0.9564 \text{ lb}$$

Current efficiency based on NaOH $= (0.8944/0.9564)100 = 93.5\%$

Maximum possible production of Cl_2, based on current actually supplied to cell

$$= (10.845)(35.457) = 384.53 \text{ grams} = 0.8477 \text{ lb}$$

Current efficiency based on Cl_2 $= (0.7926/0.8477)100 = 93.5\%$

(*b*) If the cell were operated at the decomposition voltage of 2.3 volts,

$$\text{kw-hr per faraday} = \frac{(96,493.1)(2.3)}{3,600,000} = 0.06165$$

$$\text{Faradays required to form NaOH} = \frac{(0.8944)(453.6)}{40.00} = 10.143$$

Theoretical energy requirement for NaOH formation

$$= (0.06165)(10.143) = 0.625 \text{ kw-hr}$$

Energy efficiency based on NaOH production $= 62.5\%$

$$\text{Faradays required to form } Cl_2 = \frac{(0.7926)453{\cdot}6}{35.457} = 10.140$$

Theoretical energy requirement for Cl_2 formation $= (0.06165)(10.140) = 0.625$

Energy efficiency based on Cl_2 production $= 62.5\%$

In this particular instance, the current efficiencies at anode and cathode are the same; hence, energy efficiency may be calculated as follows:

$$\text{Energy efficiency} = (\text{current efficiency})\left(\frac{\text{decomposition voltage}}{\text{actual cell voltage}}\right)$$

$$= (93.5)\frac{2.3}{3.44} = 62.5\%$$

Problems

1. Nitrogen tetroxide dissociates into nitrogen peroxide according to the following reaction:

$$N_2O_4(g) = 2NO_2(g)$$

The standard free-energy change in calories per gram-mole of this reaction in the ideal state at 1 atm is represented by the following equation from the *International Critical Tables:*

$$-\Delta G^\circ = -13{,}600 + 41.6T^\circ \text{ K}$$

Calculate the equilibrium composition of the mixture formed from the dissociation of pure N_2O_4 under the following conditions, assuming $K_\nu = 1.0$:

(*a*) At a temperature of 273° K and 1 atm.

(*b*) At a temperature of 400° K and 1 atm.

2. Isobutane is alkylated by ethylene to form neohexane, 2, 2-dimethylbutane, at elevated temperatures and pressures. From the data of Table 69 calculate the composition of the equilibrium mixture resulting from reacting a mixture of 4 moles of isobutane and 1.0 mole of ethylene at 700° K and 100 atm.

3. Benzene may be produced by the catalytic dehydrogenation of *n*-hexane. From the results of problem 25-5 and the data of Tables 30 and 68, calculate the composition of the equilibrium mixture and the percentage conversion of hexane at a temperature of 1050° F and 1 atm. The following heat-capacity equations may be used where T is in degrees Kelvin.

Benzene[4] $\quad c^*_p = 0.23 + (77.83)10^{-3}T - (27.16)10^{-6}T^2$

n-Hexane[5] $\quad c^*_p = 4.296 + (118.661)10^{-3}T - (42.13)10^{-6}T^2$

4. Propane may be dehydrogenated catalytically to form propylene. From the data of Table 69, develop curves relating the percentage of propane dehydrogenated at equilibrium to temperature in the range of 400 to 1200° K and at pressures of 0.5, 1.0, and 2.0 atm.

5. Water and chlorine react at elevated temperatures according to the following equation:

$$\tfrac{1}{2}Cl_2(g) + \tfrac{1}{2}H_2O(g) = HCl(g) + \tfrac{1}{4}O_2(g)$$

Using the data of Tables 29, 68, and D, and the molal heat-capacity equation of chlorine,

$$c^*_p = 8.28 + 0.00031T(^\circ R):$$

(*a*) Calculate an equation for $\Delta G^\circ/T$.

(*b*) Calculate the composition of the equilibrium mixture at 500° C, starting with equal volumes of chlorine and water and assuming that $K_\nu = 1.0$.

(*c*) Repeat the calculation of part (*b*) for a pressure of 100 atm, evaluating K_ν from Table 51 and the critical data of Table 7A, page 92.

6. Sulfur dioxide is reduced by hydrogen according to the following equation:

$$3H_2(g) + SO_2(g) = H_2S(g) + 2H_2O(g)$$

(*a*) From the data of Tables 29, 68, and D, derive an equation for the standard free-energy change.

(*b*) Calculate the equilibrium composition of the mixture obtained at 1200° C and atmospheric pressure, starting with three parts of hydrogen and one part of SO_2.

(*c*) Repeat the calculation of part (*b*) at 100 atm, using Table 51 and the critical data from Table 7A, page 92.

7. Methyl alcohol is synthesized by passing a mixture of CO and H_2 over a catalyst according to the following equation:

$$CO(g) + 2H_2(g) = CH_3OH(g)$$

[4] J. W. Andersen, G. H. Beyer, and K. M. Watson, *Nat. Petrol. News, Tech. Sec.*, **36**, *R*476 (July 5, 1944). Also *Process Engineering Data*, National Petroleum Publishing Co., Cleveland (1944).

[5] H. M. Spencer and G. N. Flannagan, *J. Am. Chem. Soc.*, **64**, 2511 (1942).

From the data of Tables 30, 68, and D, calculate the composition of the equilibrium mixture obtained at a temperature of 300° C and a pressure of 240 atm, starting with two parts of H_2 and one part of CO, assuming that only this reaction takes place. The critical temperature of CH_3OH is 240° C, and the critical pressure 78.5 atm. The heat capacity is given by the following equation with T in degrees Kelvin:

$$c^*_p = 5.72 + (24.85)10^{-3}T - (8.167)10^{-6}T^2$$

8. Methyl alcohol is oxidized by air to formaldehyde at 550° C in a catalyst chamber at atmospheric pressure. Calculate the percentage yield of formaldehyde, using the theoretical air supply and assuming no further oxidation.

$$CH_3OH(g) + \tfrac{1}{2}O_2(g) \rightarrow HCHO(g) + H_2O(g)$$

Entropy and heat-of-reaction data are given in Tables 68 and 30. Heat capacities may be taken from Table D and the following equation for HCHO with T in degrees Kelvin:

$$c^*_p = 4.50 + (13.95)10^{-3}T - (3.73)10^{-6}T^2$$

9. In the Birkeland–Eyde process, the nitrogen of the atmosphere is oxidized in a long flaming electric arc:

$$\tfrac{1}{2}N_2 + \tfrac{1}{2}O_2 = NO$$

Assuming that only this reaction takes place, calculate the percentage conversion of nitrogen to NO in air of average atmospheric composition at a pressure of 1.0 atm and at temperatures of, respectively, 2000 and 3000° K.

10. Carbon dioxide is reduced by graphite according to the equation:

$$C(\text{graphite}) + CO_2(g) = 2CO(g)$$

Assuming that equilibrium is attained, calculate the degree of completion of the reduction of pure CO_2 under the following conditions, using the data of Table 69.

(*a*) A temperature of 1000° K and a pressure of 1.0 atm, assuming $K_\nu = 1.0$.

(*b*) A temperature of 1500° K and a pressure of 1.0 atm, assuming $K_\nu = 1.0$.

(*c*) A temperature of 1000° K and a pressure of 100 atm.

11. A mixture of 79% N_2 and 21% CO_2 by volume is passed over hot carbon (graphite) at a temperature of 1000° K and a pressure of 1.0 atm. Using the data of Table 69, calculate the equilibrium composition of the gases, and compare this result with that of part (*a*) of problem 10.

12. Carbon monoxide is burned with pure oxygen in the theoretically required proportions. Calculate the degrees of completion of the oxidation if equilibrium is attained at temperatures of, respectively, 1000 and 3000° K, under a pressure of 1.0 atm. Evaluate $\Delta G^\circ/T$ at 3000° K by extending the data of Tables 69 and 70 from 1500° K through graphical integration of the heat capacities of Table 17, page 255.

13. Water gas leaves a generator containing 51.1% H_2, 2.3% CO_2, and 46.6% CO by volume on the dry basis. Ten per cent of the steam which was introduced into the bottom of the generator passed through the bed of hot coke without decomposition and is present in the gases. This gaseous mixture is passed into a reaction chamber under a pressure of 1.0 atm in contact with a catalyst of chromium oxide and allowed to attain equilibrium at a temperature of 700° K (423° C). Calculate the equilibrium composition of the gaseous mixture, using the data of Fig. 262.

14. One volume of the initial wet water gas described in problem 13 is mixed with three volumes of additional water vapor. This mixture is passed into the reaction chamber operated at the conditions described in problem 13 and allowed to reach equilibrium.

(*a*) Calculate the equilibrium composition of the gaseous mixture.

(*b*) Calculate the composition of the residual gas if the CO_2 and H_2O are removed from the gaseous mixture of part (*a*) after equilibrium is attained.

15. The hydrate of sodium carbonate decomposes according to the following equation:

$$Na_2CO_3 \cdot H_2O(s) = Na_2CO_3(s) + H_2O(g)$$

The equilibrium pressure in atmospheres of water vapor in this reaction is given by the following equation:

$$\log p = 7.944 - 3000.0/T° \text{ K}$$

Derive an expression for the standard free-energy change of this reaction.

16. Zinc oxide is reduced with carbon monoxide under a pressure of 1.0 atm according to the following reaction:

$$ZnO(s) + CO(g) = Zn(g) + CO_2(g)$$

Calculate the degree of completion of the oxidation of CO at atmospheric pressure under the following different conditions, assuming that equilibrium conditions are attained and that ZnO is always present:

(*a*) At a temperature of 1000° C, using pure CO.

(*b*) At a temperature of 1500° C, using pure CO.

(*c*) At a temperature of 1000° C, using a mixture of 27.5% CO, 4.3% CO_2, and 68.2% N_2 by volume.

(*d*) At a temperature of 1500° C, using a mixture of 27.5% CO, 4.3% CO_2, and 68.2% N_2 by volume.

17. Sodium bicarbonate is calcined according to the following equation:

$$2NaHCO_3(s) = Na_2CO_3(s) + H_2O(g) + CO_2(g)$$

Calculate the pressure of the equimolecular mixture of H_2O and CO_2 in equilibrium with $NaHCO_3$ at a temperature of 100° C.

For Na_2CO_3, $c_p = 16.88 + (32.4)10^{-3}T$,

For $NaHCO_3$ $c_p = 10.73 + (34.4)10^{-3}T$,

where T is in degrees Kelvin.

18. (*a*) Calculate the boiling point of aluminum at atmospheric pressure from the data in Table A.

TABLE A

Temperature, °K	$\frac{G° - H°_{0g}}{T}$ g-cal per g-atom Gas	$\frac{G° - H°_{0s}}{T}$ g-cal per g-atom Solid or Liquid
298.16	−33.766	−3.13
1000	−40.217	−9.09
2000	−43.762	−14.64
2100	−44.009	−15.01
2200	−44.245	−15.37
2300	−44.470	−15.71
2400	−44.681	−16.04

The value of $\Delta H°_0$ for vaporization is 66.920 kcal per g-atom.

(*b*) Calculate the vapor pressure at 2000° K.

19. Considering each of the following reactions:

(1) $$SO_2(g) + \tfrac{1}{2}O_2(g) = SO_3(g)$$

(2) $$\tfrac{1}{2}N_2(g) + \tfrac{1}{2}O_2(g) = NO(g)$$

(3) $$C(s) + CO_2(g) = 2CO(g)$$

Tabulate the effects of the following changes on (*a*) the velocity of reaction (moles transformed per unit time per unit volume), (*b*) the equilibrium degree of completion, and (*c*) the actual degree of completion obtained in a specified time interval:

(1) Increase of temperature.
(2) Increase of pressure.
(3) Provision of a positive catalyst.
(4) Dilution with an inert gas.
(5) Agitation of the reacting system.

Tabulate each effect as an increase, decrease, no effect, or indeterminate.

20. In the American process of synthesizing ammonia a mixture of three volumes of hydrogen and one volume of nitrogen is passed into a reaction chamber in contact with a catalyst of granular iron oxide combined with oxides of potassium and aluminum. Using the data of Illustration 2 and Fig. 262:

(*a*) Plot curves relating the equilibrium degree of completion of this reaction to temperature at pressures of 1.0, 100, and 300 atm. The temperature range from 400 to 800° K should be included in the calculations.

(*b*) Calculate the equilibrium degree of completion of this same reaction at a pressure of 200 atm and a temperature of 750° K.

21. In the synthesis of ammonia described in problem 20, the mixture of N_2 and H_2 is introduced into the reaction chamber under a pressure of 300 atm and a temperature of 400° C. Assuming that heat loss from the reaction chamber is negligible, calculate the equilibrium-reaction temperature, using the curves of problem 20.

22. From the results of problem 4 and the data of Table 70, calculate the equilibrium temperatures and conversions reached when propane is dehydrogenated in an adiabatic reactor with an inlet temperature of 1200° F and pressures of 1.0 and 2.0 atm, respectively.

23. Calculate the values of $\Delta G°$, $\Delta H°$, and $\Delta S°$ for the reaction

$$Ag(s) + I(s) = AgI(s)$$

using electrochemical measurements made on two cells.

Cell 1 has a lead-plated platinum anode, immersed in a solution of a lead salt saturated with lead iodide. The cathode is platinum, covered with iodine and lead iodide, immersed in the same solution. The net reaction in the cell is

$$\tfrac{1}{2}Pb(s) + I(s) \rightarrow \tfrac{1}{2}PbI_2(s)$$

The cell voltage at 25° C is 0.89362 volt, and $(\partial E/\partial T)_p = (4.2)10^{-5}$ volts per K°.

Cell 2 has a lead-plated platinum anode surrounded by solid lead iodide, and immersed in a solution of potassium iodide. The cathode is platinum plated with silver, surrounded by solid silver iodide, and immersed in the potassium iodide solution. The net cell reaction is

$$\tfrac{1}{2}Pb(s) + AgI(s) \rightarrow \tfrac{1}{2}PbI_2(s) + Ag(s)$$

The cell voltage at 25° C is 0.2078 volt, and

$$(\partial E/\partial T)_p = -(18.8)10^{-5} \text{ volt per K°}$$

27

Equilibria in Complex Reactions

Where several reactions occur among a given group of reactants, the composition of the resultant products at equilibrium depends on the simultaneous equilibria of all the separate reactions. If large numbers of products are formed as in the pyrolysis of heavy hydrocarbons, or of cellulose, the resultant composition cannot ordinarily be calculated. However, even in such cases where many reactions are possible, restriction to a single course may be accomplished by the choice of a specific catalyst which promotes one selected reaction. This selection is dependent on kinetic considerations alone. It is common for equilibrium to be approached with respect to one or two reactions while many other reactions which are thermodynamically possible do not occur to any appreciable extent.

For example, in the reaction of carbon monoxide with hydrogen, the number of products theoretically possible is almost unlimited, and the free energy in the formation of many of these products may be extremely favorable; yet it is possible by the selection of a special catalyst to exclude consideration of nearly all possible products save methanol. An apparent equilibrium yield can be established by considering only the reaction $CO + 2H_2 \rightleftharpoons CH_3OH$, and ignoring all other possible reactions. However, such an equilibrium is valid only with respect to the one reaction. If the reacting system is permitted to remain at reaction conditions for an indefinite time, the other relatively slow reactions will take place with resultant changes in composition, and ultimately an equilibrium involving all possible products will be reached. The calculation of such complete equilibrium compositions for complex systems is of little practical value, and consideration is ordinarily limited to the relatively few rapid reactions.

Isomerization. The simplest case of simultaneous reactions is encountered in the isomerization of organic compounds, as, for example, the conversion of normal butane to isobutane. Similarly, normal pentane may be isomerized to either isopentane (2-methylbutane) or neopentane (2-2-dimethylpropane).

Calculation of equilibrium compositions in isomerization is simplified by the monomolecular nature of the reactions, and by the fact that the physical properties of the isomers are so similar that deviations from ideal behavior cancel. The composition of the equilibrium mixture of isomers is calculated on the basis of a unit quantity of the starting compound in the equilibrium mixture. The corresponding quantity of each isomer is then equal to the equilibrium constant of the reaction by which it is formed.

Illustration 1. From the data of Table 69, calculate the composition of the mixture obtained by isomerizing *n*-pentane to equilibrium at 400° K.

	n-Pentane	Isopentane	Neopentane
$(G°_T - H°_0)/400$	−70.57	−70.67	−61.93
$(\Delta H°_f)_0$	−27.23	−28.81	−31.30

For the reaction, *n*-pentane → isopentane:

$$\left(\frac{\Delta G°}{400}\right) = \left[-70.67 + \frac{(-28.81)1000}{400}\right] - \left[-70.57 + \frac{(-27.23)1000}{400}\right]$$

$$= 4.050$$

$$\log K = -\frac{\Delta G°/T}{4.5757} = -\frac{-4.050}{4.5757} = 0.88511$$

$$K = 7.676$$

For the reaction, *n*-pentane → neopentane:

$$\left(\frac{\Delta G°}{400}\right) = \left[-61.93 + \frac{(-31.30)1000}{400}\right] - \left[-70.57 + \frac{(-27.23)1000}{400}\right]$$

$$= -1.535$$

$$\log K = -\frac{\Delta G°/T}{4.5757} = -\frac{-1.535}{4.5757} = 0.33547$$

$$K = 2.165$$

On the basis of one mole of *n*-pentane in the equilibrium mixture, the following results are obtained by application of equation 26–6:

	Moles	Mole %
n-Pentane	1.000	9.2
Isopentane	7.676	70.8
Neopentane	2.165	20.0
	10.841	100.00

By repeating the calculations of Illustration 1 at other temperatures, the isomerization equilibrium diagram is developed as shown in Fig. 268. From this diagram, it is evident that low temperatures are required for maximum mole fraction of neopentane, while isopentane is at a maximum at approximately 500° K.

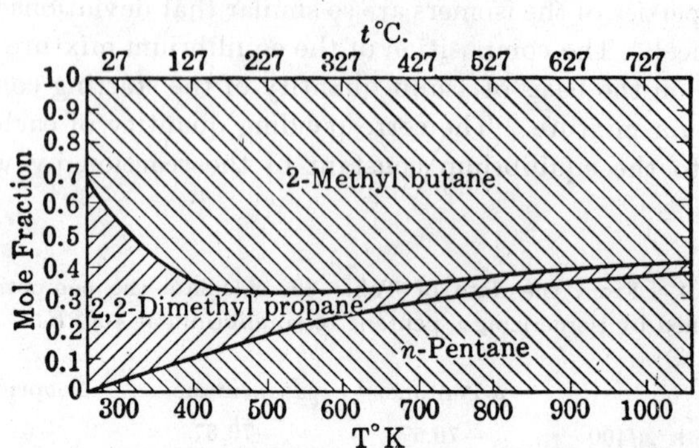

[F. D. Rossini, E. J. R. Pressen and K. S. Pitzer, *J. Nat. Bureau of Standards*, **27**, 529 (1941), with permission]

FIG. 268. Equilibrium concentrations of pentanes

Complex Reactions. Where a limited number of simultaneous reactions are known to proceed, a generalized procedure for equilibrium calculations may be illustrated by considering the following three reactions from initial reactants B and C:

(1) $$bB + cC \rightleftharpoons rR + sS$$

(2) $$bB + rR \rightleftharpoons tT + uU$$

(3) $$cC + sS \rightleftharpoons vV$$

If each reaction proceeded to completion, the over-all equation would be

(4) $$2bB + 2cC = tT + uU + vV$$

If the foregoing are all gaseous reactions, the respective equilibrium compositions are expressed by equation 26–6:

$$K_1 = \left(\frac{{n_R}^r {n_S}^s}{{n_B}^b {n_C}^c}\right) K_{\nu 1} \left(\frac{\pi}{\sum n}\right)^{(r+s)-(b+c)} \tag{1}$$

$$K_2 = \left(\frac{{n_T}^t {n_U}^u}{{n_B}^b {n_R}^r}\right) K_{\nu 2} \left(\frac{\pi}{\sum n}\right)^{(t+u)-(b+r)} \tag{2}$$

$$K_3 = \left(\frac{{n_V}^v}{{n_C}^c {n_S}^s}\right) K_{\nu 3} \left(\frac{\pi}{\sum n}\right)^{v-(c+s)} \tag{3}$$

where $$\sum n = n_B + n_C + n_R + n_S + n_T + n_U + n_V + n_I$$

From equation 25-6:

$$\Delta G^\circ_1 = -RT \ln K_1 \tag{4}$$
$$\Delta G^\circ_2 = -RT \ln K_2 \tag{5}$$
$$\Delta G^\circ_3 = -RT \ln K_3 \tag{6}$$

Combining equations 4, 5, and 6 gives

$$\Delta G^\circ_4 = \Delta G^\circ_1 + \Delta G^\circ_2 + \Delta G^\circ_3 = -RT \ln K_1K_2K_3 = -RT \ln K_4 \tag{7}$$

or

$$K_4 = K_1K_2K_3 \tag{8}$$

Thus, the equilibrium constant of the over-all reaction is the product of the equilibrium constants of the intermediate reactions. By substitution of equations 1, 2, and 3 in 8:

$$K_4 = \left(\frac{n_T{}^t n_U{}^u n_V{}^v}{n_B{}^{2b} n_C{}^{2c}}\right) K_{\nu 4} \left(\frac{\pi}{\sum n}\right)^{t+u+v-2b-2c} \tag{9}$$

Equation 9 is applicable to any equilibrium mixture of the system under consideration, though the intermediate products R and S are present but are not indicated by the equation. However, by itself it does not completely define the composition of the equilibrium mixture but fixes only the relative proportions of components B, C, T, U, and V. Thus, an over-all equilibrium equation representing the result of a sequence of reactions is useful for the calculation of complete equilibrium compositions *only when intermediate products are not present in significant quantities.*

In the general case where all intermediate and final products must be considered, it is necessary that the equilibrium equations of all reactions be satisfied by the composition of the system at equilibrium. Determination of the equilibrium composition therefore requires simultaneous solution of the independent equilibrium equations. The number of equations to be solved is equal to the number of independent equations involved. For example, in the system just considered, there are three independent reactions. Reaction 4 is not independent since it results from combination of the others. Thus, simultaneous solution of the equilibrium equations 1, 2, and 3 completely establishes the equilibrium composition, and this composition will of necessity satisfy equation 9. It is a general rule that *the number of independent reactions that must be considered is the least number that includes every reactant and product present to an appreciable extent in all phases of the equilibrium system, and accounts for the formation of each product from the original reactants.*

For simultaneous solution of the equilibrium equations, a definite quantity of the initial system is selected as a basis, and the number of moles converted by each reaction in proceeding to equilibrium is designated as an algebraic symbol. For example, in the system considered above, an equilibrium calculation may be based on 100 moles of the initial reactants. The number of moles of B converted by reaction 1 in going to equilibrium may be designated as x. Similarly, y may represent the moles of B converted by reaction 2, and z the moles of C converted by reaction 3. The number of moles of all components in equations 1, 2, and 3 may be expressed in terms of these three variables x, y, and z, which are then evaluated by simultaneous solution. The solution may require a graphical or trial-and-error procedure where the equations are complex.

These principles are well illustrated by consideration of the catalytic process for the production of hydrogen and carbon monoxide by the reaction of steam and methane at high temperatures. This process is extensively used for obtaining both pure hydrogen and mixtures of hydrogen and carbon monoxide. A thermodynamic analysis of the operation involved is presented by Dodge.[1] The kinetics of a catalytic reaction of this type may involve several intermediate steps, which are of importance in determining the rate of reaction as discussed in Part III. However, a sound thermodynamic treatment is possible without consideration or knowledge of reaction rates. It may be assumed that the principal reaction proceeds in two stages with intermediate formation and removal of carbon. Thus, these reactions with the corresponding equilibrium constants at 600° C from the data of Tables 30 and 68 are as follows:

$$(1) \qquad CH_4 = C + 2H_2, \qquad K_1 = \frac{(a_{H_2})^2}{a_{CH_4}} = 2.13$$

$$(2) \qquad C + H_2O = CO + H_2, \qquad K_2 = \frac{a_{CO}\, a_{H_2}}{a_{H_2O}} = 0.269$$

The over-all reaction is

$$(3) \qquad CH_4 + H_2O = CO + 3H_2, \qquad K_3 = \frac{(a_{H_2})^3\, a_{CO}}{a_{CH_4}\, a_{H_2O}} = 0.574$$

In order that no carbon may appear in the equilibrium mixture represented by these three reactions, it is necessary that sufficient steam be added so that the ratio $a^2_{H_2}/a_{CH_4}$ may be equal to or greater than K_1 and that the ratio $(a_{CO})(a_{H_2})/a_{H_2O}$ be equal to or less than K_2. If the first ratio is greater than K_1, then carbon added to such a system can react with hydrogen and form methane until the ratio is reduced to K_1. If the second ratio is less than K_2, then carbon added to such a system can react with steam until the ratio rises to K_2.

[1] B. F. Dodge, *Trans. Am. Inst. Chem. Engrs.*, **34**, 540 (1938).

Many other reactions are possible in this system. A few, including the more likely, follow, together with the corresponding equilibrium constants at 600° C.

		At 600° C
(4)	$CO + H_2O = CO_2 + H_2,$	$K_4 = 2.21$
(5)	$2CO = C + CO_2,$	$K_5 = 8.14$
(6)	$CO_2 = CO + \frac{1}{2}O_2,$	$K_6 = (4.95)10^{-13}$
(7)	$H_2O = H_2 + \frac{1}{2}O_2,$	$K_7 = (1.12)10^{-12}$
(8)	$2CH_4 = C_2H_6 + H_2,$	$K_8 = (5.51)10^{-5}$

Consideration of the values of these equilibrium constants indicates that, at 600° C, reactions 6, 7, and 8 can proceed only to negligible extents, and hence that O_2 and C_2H_6 cannot be appreciably present at equilibrium. When the ratio of steam to methane in the feed is sufficiently high so that carbon cannot be present at equilibrium, the equilibrium composition may be calculated from consideration of only reactions 3 and 4, which involve all the significant reactants in the absence of carbon. In order to determine the minimum steam ratio required for freedom from carbon, the equilibrium compositions corresponding to a series of steam ratios are calculated on the assumption of no carbon formation. The resultant ratios $(a_{H_2})^2/a_{CH_4}$ and $(a_{CO})(a_{H_2})/(a_{H_2O})$ are plotted against the steam–methane ratio. The minimum steam ratio is that where $(a_{H_2})^2/a_{CH_4} = K_1$ and $(a_{CO})(a_{H_2})/(a_{H_2O}) = K_2$. This procedure is demonstrated in the following illustration:

Illustration 2. Calculate the composition of the equilibrium mixture obtained when 5 moles of steam react with 1 mole of methane at 600° C and 1.0 atm, assuming that no carbon is present. Also determine the minimum ratio of steam required for freedom from carbon.

Basis: 1 mole CH_4, 5 moles H_2O.

Let x = moles CH_4 converted by reaction 3

y = moles CO converted by reaction 4

At equilibrium,

$$\text{Moles } CH_4 = 1 - x$$

$$\text{Moles } H_2O = 5 - x - y$$

$$\text{Moles } CO = x - y$$

$$\text{Moles } H_2 = 3x + y$$

$$\text{Moles } CO_2 = y$$

$$\text{Total moles} = 6 + 2x$$

Since the pressure is 1.0 atm, K_ν may be taken as 1.0. Substitution of the foregoing values in equation 26–6 gives

From reaction 3,

$$\left[\frac{(x-y)(3x+y)^3}{(1-x)(5-x-y)}\right]\left(\frac{1.0}{6+2x}\right)^2 = K_3 = 0.574$$

From reaction 4,

$$\left[\frac{(y)(3x+y)}{(x-y)(5-x-y)}\right] = K_4 = 2.21$$

These two equations are solved by assuming several values of x and calculating the corresponding values of y from each equation. Because of the complicated forms of the individual equations, the solutions are carried out graphically. The values of y from each equation are plotted against the assumed values of x. The intersection of these two curves gives the correct solution. Thus,

$$x = 0.9124$$
$$y = 0.633$$

Table A. Equilibrium Mixture

	Moles	Mole %	Mole % (Moisture-Free)
CH_4	$1 - x = 0.0876$	1.12	2.01
H_2O	$5 - x - y = 3.4546$	44.15	
CO	$x - y = 0.2794$	3.57	6.39
H_2	$3x + y = 3.3702$	43.07	77.11
CO_2	$y = 0.6330$	8.09	14.49
Total	$6 + 2x = 7.8248$	100.00	100.00

The corresponding activity ratios for reactions 1, 2, and 5 are

$$\frac{(a_{H_2})^2}{a_{CH_4}} = 16.5 > (K_1 = 2.13)$$

$$\frac{a_{CO}a_{H_2}}{a_{H_2O}} = 0.0348 < (K_2 = 0.269)$$

$$\frac{a_{CO_2}}{(a_{CO})^2} = 63.4 > (K_5 = 8.14)$$

It follows from these ratios that, with 5 moles of steam per mole of methane, no carbon is present at equilibrium. This calculation is repeated for steam ratios of 2 and 1.25. The corresponding values of $a_{CO}a_{H_2}/a_{H_2O}$ are plotted in Fig. 269, from which the minimum steam ratio is determined as 1.38. The limiting steam ratio can be calculated directly from reactions 1, 2, and 4, which include all components including carbon and shows the formation of CO.

It must be emphasized that the solution of Illustration 2 is based on the assumption that equilibrium is attained. In actual operations that do not reach equilibrium, quite different results might be obtained, depending on the rates of the various reactions. Thus, if reaction 1 were fast and reaction 2 were slow, carbon might form even with high steam ratios. Conversely, if reaction 3 were primarily a catalytic reaction and reactions 1 and 5 were relatively slow, operation might be possible

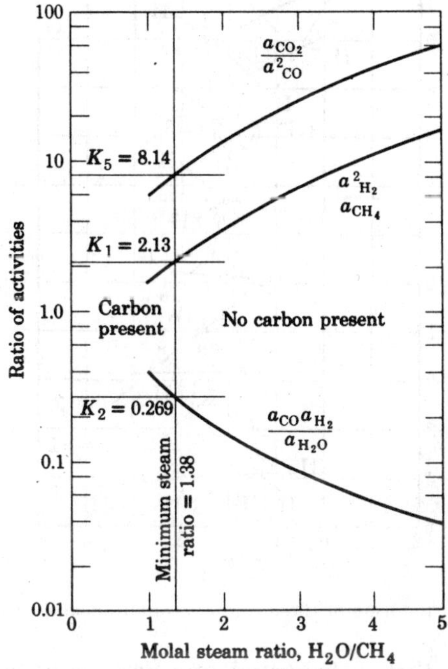

FIG. 269. Minimum molal steam ratio for absence of carbon

at low steam ratios without carbon formation. These effects are determined by reaction rates and the types of catalysts present.

Where three significant and independent reactions are involved, the solution of three simultaneous equations is required with the evaluation of three unknowns, x, y, and z. Where algebraic methods of solution fail or become too complex, a more general graphical solution involves finding the simultaneous intersection of three lines in space; this is equivalent to finding the point of intersection of two three-sided pyramids which meet at a common apex. This can be done by assuming various values of z and plotting the three equations corresponding to reactions I, II, and

III in terms of x and y as shown in Fig. 270. In this figure the intersection of the three lines gives a triangular area A at $z = 5.0$; at another value of $z = 4.0$ the area of intersection is reduced to B; and, by further decrease in z, the area of intersection diminishes to zero at $z = 2$ and increases again as values of z are further decreased.

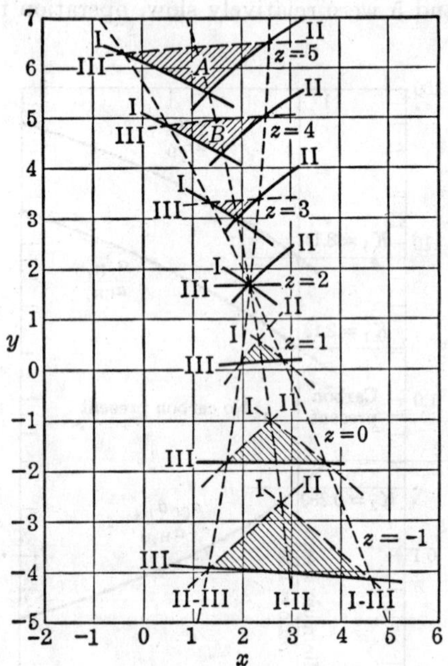

FIG. 270. Solution of three complex simultaneous equations

It is fortuitous that in industrial processes rarely more than three independent and significant reactions are involved under equilibrium conditions at a given temperature level.

Statistical methods of establishing equilibrium compositions of complex reacting systems are presented by Brinkley[2] and by Kobe and Leland.[3]

Illustration 3. Steam is passed through a bed of carbon briquettes in a steel reactor. The pressure is maintained at 1 atm. Calculate the equilibrium composition of the products at 600° C, assuming ideal gas behavior.

The following reactions and many more may be written. Reaction with the steel reactor itself is considered as a possibility.

[2] S. R. Brinkley, Jr., *J. Chem. Phys.*, **14**, 563 (1946); **15**, 107 (1947).

[3] K. A. Kobe and T. W. Leland, *Univ. Texas Bur. Eng. Research Spec. Pub.* **26**, (Apr. 1, 1954).

	Reactions	Equilibrium Constants at 600° C
(1)	$Fe(s) + H_2O(g) \rightleftharpoons FeO(s) + H_2(g)$,	$K_1 = 0.514$
(2)	$C(s) + H_2O(g) \rightleftharpoons CO(g) + H_2(g)$,	$K_2 = 0.269$
(3)	$C(s) + 2H_2O(g) \rightleftharpoons CO_2(g) + 2H_2(g)$,	$K_3 = 0.595$
(4)	$C(s) + CO_2(g) \rightleftharpoons 2CO(g)$,	$K_4 = 0.123$
(5)	$CO(g) + H_2O(g) \rightleftharpoons CO_2(g) + H_2(g)$,	$K_5 = 2.21$
(6)	$H_2O(g) \rightleftharpoons H_2(g) + \frac{1}{2}O_2(g)$,	$K_6 = (1.12)10^{-12}$
(7)	$CO_2(g) \rightleftharpoons CO(g) + \frac{1}{2}O_2(g)$,	$K_7 = (4.95)10^{-13}$
(8)	$C(s) + 2H_2(g) \rightleftharpoons CH_4(g)$,	$K_8 = 0.470$
(9)	$CH_4(g) + H_2O(g) \rightleftharpoons CO(g) + 3H_2(g)$,	$K_9 = 0.574$
(10)	$2CH_4(g) \rightleftharpoons C_2H_6(g) + H_2(g)$,	$K_{10} = (5.51)10^{-5}$
	Formation of the organic compounds	$K <<< 10^{-5}$
	Formation of free radicals and free atoms	$K <<< 10^{-5}$

From the inspection of the small equilibrium constants in equations 10, 7, and 6, $K_6 = (1.12)10^{-12}$, $K_7 = (4.95)10^{-13}$, and $K_{10} = (5.51)10^{-5}$, it is evident that the contents of ethane and molecular oxygen are negligible in the presence of the other components. Additional equations set up for the formation of hydrocarbons and organic compounds of complexity greater than methane (reaction 10) would prove their presence to be even less significant than for methane and need be given no further consideration. The presence of atomic gases or free radicals resulting from dissociation would ordinarily prove to be so small as not to affect the equilibrium composition of the principal components, and hence such reactions may be neglected as a first approximation. Similarly, equilibrium vaporization constants for gases present are so high as to indicate that such components are not present in the liquid state. From similar constants, the presence of iron, carbon, and iron oxide in the vapor states may be shown to be negligible. The presence of particular components in the solid or liquid states can be established from boiling-point and melting-point data.

The presence of solid FeO as a separate phase should be questioned.

By reaction 1,

$$Fe(s) + H_2O(g) \rightleftharpoons FeO(s) + H_2(g)$$

Let x = moles of H_2O decomposed per mole of H_2O supplied. At equilibrium,

$$\begin{aligned} \text{Moles } H_2O &= 1 - x \\ \text{Moles } H_2 &= x \\ \text{Total moles of gas} &= 1.0 \end{aligned}$$

Then $$K_1 = \frac{x}{1-x} = 0.514, \qquad x = 0.339, \qquad a_{H_2} = 0.339$$

$$1 - x = 0.661, \qquad a_{H_2O} = 0.661$$

$$\frac{a_{H_2}}{a_{H_2O}} = \frac{0.339}{0.661} = 0.514$$

If the ratio of a_{H_2}/a_{H_2O} exceeds 0.514 in the final mixture, then FeO does not exist at equilibrium. It will be initially assumed that FeO does not exist. The validity of this assumption can be approximated by assuming that reaction 2 is the principal reaction.

By equation 2, $C(s) + H_2O(g) \rightleftharpoons CO(g) + H_2(g)$

Let y = moles H_2O decomposed per mole H_2O supplied. At equilibrium,

$$H_2 = y \text{ moles} \qquad a_{H_2} = \frac{y}{1+y}$$

$$CO = y \text{ moles} \qquad a_{CO} = \frac{y}{1+y}$$

$$H_2O = 1 - y \text{ moles} \qquad a_{H_2O} = \frac{1-y}{1+y}$$

Total moles of gas = $(1 + y)$

$$K_2 = 0.269 = \frac{y^2}{(1+y)(1-y)}$$

$$y = 0.461$$

$$a_{H_2} = \frac{y}{1+y} = \frac{0.461}{1.461} = 0.314$$

$$a_{H_2O} = \frac{1-y}{1+y} = \frac{0.539}{1.461} = 0.368$$

$$\frac{a_{H_2}}{a_{H_2O}} = \frac{0.314}{0.368} = 0.854$$

The ratio of a_{H_2}/a_{H_2O} for reaction 1 is 0.512. This ratio is exceeded by reaction 2. Hence, the assumption that FeO is absent is reasonable and will be accepted in calculating the equilibrium composition. This assumption will then be tested for validity.

At equilibrium, it is therefore expected that the following reacting components will be present in appreciable quantities: $C(s)$, $H_2(g)$, $CO_2(g)$, $CO(g)$, $H_2O(g)$, $CH_4(g)$.

The scheme of procedure is to select the least number of independent equations that contain all the reacting components that are present to an appreciable extent in the equilibrium mixture including all phases, and that also account for the formation of all products from the original feed. It will be seen that reactions 2, 5, and 8 account for all reacting components in all phases, and, as independent reactions, do not fix the composition of the gaseous components to stoichiometric ratios. Since the iron is tentatively assumed unreactive in the presence of carbon, its presence will not affect the equilibrium composition, and an equation including Fe is not required. If an equation showing the formation of FeO were included, a negative content of either CO or CO_2 would result. This would be meaningless.

The possible selection of other reactions might be considered. For example, reactions 9 and 4 contain all the components, but do not account for the formation of CH_4. Reactions 8 and 5 contain all the components, but do not account for the formation of CO. A combination of reactions 8 and 2 gives

(11) $4C(s) + 3H_2O(g) \rightleftharpoons CH_4(g) + H_2(g) + 3CO(g)$

But reaction 11 is a dependent equation obtained from reactions 8 and 2. If reaction 11 were combined with reaction 3, these two reactions would account for all components, but the ratio of CO/CH_4 would be fixed to a value of 3 by reaction 11, whereas the actual ratio would be determined by the two reactions 8 and 2 proceeding independently.

Let x = moles of H_2O reacted by reaction 5

y = moles of CH_4 formed by reaction 8

z = moles of H_2O reacted by reaction 2

At equilibrium,

$$\begin{aligned} H_2O &= 1 - x - z \\ H_2 &= x - 2y + z \\ CO_2 &= x \\ CO &= -x + z \\ CH_4 &= y \end{aligned}$$

Total moles of gas $= n = 1 - y + z$

$$K_2 = \frac{(z - x)(x - 2y + z)}{(1 - x - z)(1 - y + z)} = 0.269$$

$$K_5 = \frac{x(x - 2y + z)}{(z - x)(1 - x - z)} = 2.21$$

$$K_8 = \frac{y(1 - y + z)}{(x - 2y + z)^2} = 0.470$$

By solution of the three simultaneous equations,

$$\begin{aligned} x &= 0.252 \\ y &= 0.0935 \\ z &= 0.457 \end{aligned}$$

At equilibrium:

		Mole %
H_2O	= 0.291 mole	21.3
H_2	= 0.522	38.3
CO_2	= 0.252	18.5
CO	= 0.205	15.0
CH_4	= 0.0935	6.9
	1.3635	100.0

The above molal composition not only satisfies reactions 2, 5, and 8, but also reactions 2, 3, and 9, or any equations resulting from the combinations of these. The ratio of a_{H_2}/a_{H_2O} in this final mixture is 1.79. This ratio exceeds the value 0.514 for reaction 1. This indicates that, in the presence of carbon, iron will not react with steam under equilibrium conditions at 600° C, and that equation 1 should be omitted. This indicates that the reaction of steam with carbon can be conducted at 600° C in a steel vessel without any reaction taking place with the steel under equilibrium conditions.

Calculations can now be extended to include the content of atoms, free radicals, and vapors which are present in insignificant amounts. This is done by considering the additional independent reactions which were previously eliminated because of their insignificant effect. The effect of these trace compounds and elements on total composition is so small that each equation can be taken separately to calculate the content of these traces without undertaking the difficult task of solving all equations simultaneously.

This discussion applies to conditions at equilibrium. The actual composition of the mixture before equilibrium is reached will depend on the relative rates of the various reactions, permitting any given component, including FeO, to exist temporarily at a composition greater or less than its equilibrium value.

Metallurgical Reactions. The methods previously developed are directly applicable to the calculation of equilibria in metallurgical processes such as calcining, roasting, and smelting. The majority of these reactions are heterogeneous and complex. Such problems may be handled by extension of the procedure demonstrated in Illustration 2.

An interesting example is encountered where the primary reactants are all solids while the products include substances in the gaseous and liquid phases. In such instances a reaction pressure is developed, analogous to the decomposition pressures previously discussed. The reaction pressure is a result of the nature of the reactants and the temperature, and it is not possible to vary independently temperature and pressure as is the case where a gaseous reactant is present.

When a reaction starts out in a heterogeneous system, any or all of the solid or liquid phases may be absent when equilibrium is reached.

A typical metallurgical reaction in which reactants are solids is the reduction of zinc oxide by carbon. This problem has been studied in detail by Maier.[4]

Illustration 4. Zinc oxide (ZnO) is reduced by roasting it with carbon in a closed retort from which the gaseous and liquid products of reaction may be continuously removed. Air is excluded from the retort so that the reaction is free to proceed under its own equilibrium pressure.

(*a*) Calculate the equilibrium reaction pressure as a function of the roasting temperature.

(*b*) Determine the temperature at which the operation must be conducted in order that the products may be withdrawn at a pressure of 1.0 atm.

(*c*) Calculate the minimum temperature and the corresponding pressure at which zinc is produced as a liquid.

(*d*) Calculate the temperature and pressure at which the operation must be conducted in order that 50 per cent of the zinc may be produced directly in the liquid state.

Solution: Although the over-all effect is the reaction of the two solids, it may be assumed, in order to account for all the products, that the actual reaction proceeds in two stages:

(1) $$ZnO(s) + CO(g) = Zn(g) + CO_2(g)$$

(2) $$C(s) + CO_2(g) = 2CO(g)$$

The over-all reaction becomes

(3) $$ZnO(s) + C(s) = Zn(g) + CO(g)$$

[4] C. G. Maier, "Zinc Smelting from a Chemical and Thermodynamic Viewpoint," *U.S. Bur. Mines Bull.*, **324** (1930).

In order to define completely the equilibrium composition, including the CO_2 present, reactions 1 and 2 must be considered simultaneously.

The standard free-energy changes of the two reactions may be expressed as functions of temperature by means of the data of Tables 16, 29, and 68, together with the data of Kelley[5,6] on the heat capacities and heats of vaporization.

For reaction 1,

$$\Delta G^\circ_1 = -RT \ln K_1 = 49{,}265 - 70.50T + 5.83T \ln T - (2.496)10^{-3}T^2 + (0.336)10^{-6}T^3 - \frac{(0.912)10^5}{T} \qquad (a)$$

For reaction 2,

$$\Delta G^\circ_2 = -RT \ln K_2 = 40{,}608 - 24.96T - 2.977T \ln T + (3.484)10^{-3}T^2 - (0.259)10^{-6}T^3 - 58{,}450/T \qquad (b)$$

Basis of Calculations. 1 lb-mole of ZnO reduced, which liberates 1 lb-atom of oxygen appearing as either CO or CO_2 in the products of the over-all reaction.

(*a*) Let x = lb-moles of CO produced. Then $(1 - x)$ lb-atoms of oxygen is present in CO_2 or

$$\tfrac{1}{2}(1 - x) = \text{lb moles of } CO_2 \text{ formed}$$

The total gas formed is the sum of the CO, Zn, and the CO_2, or

$$\tfrac{1}{2}(3 + x) = \text{total gaseous moles formed}$$

These values are substituted in equation 26–6, with the assumption that all zinc is vaporized and that activities are equal to partial pressures.

$$K_1 = \frac{\pi(1 - x)}{x(3 + x)} \qquad (c)$$

$$K_2 = \frac{4x^2\pi}{(3 + x)(1 - x)} \qquad (d)$$

Combining (*c*) and (*d*) by eliminating π gives

$$4K_1x^3 = K_2(1 - x)^2 \qquad (e)$$

For any given temperature, values of K_1 and K_2 are obtained from equations (*a*) and (*b*), and corresponding values of x are calculated from equation (*e*). The resultant pressures are then obtained from equations (*c*) or (*d*). These results are shown in the first four columns of Table A. In Fig. 271, the equilibrium pressure is plotted against temperature.

(*b*) From Fig. 271, where $\pi = 1.0$ atm, $T = 1170°$ K. From Table A, at 1.0 atm, $x = 0.978$. The composition of the gaseous products is obtained directly from this value. Thus, the total number of moles of products is $3.978/2 = 1.989$.

	Moles	Partial Pressure	Mole %
Zn	1.0	0.503 atm	50.3
CO	0.978	0.491	49.1
CO_2	0.011	0.006	0.6
	1.989	1.000	100.0

[5] K. K. Kelley, "High-Temperature Specific Heat Equations for Inorganic Substances," *U.S. Bur. Mines Bull.*, **371** (1934).

[6] K. K. Kelley, "The Free Energies of Vaporization and Vapor Pressures of Inorganic Substances," *U.S. Bur. Mines Bull.*, **383** (1935).

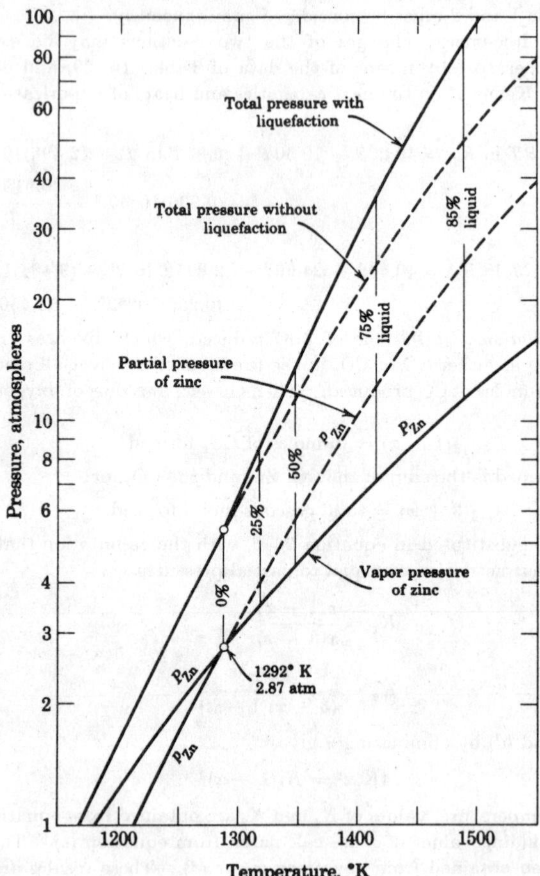

Fig. 271. Equilibrium conditions in the reduction of zinc oxide

For each lb-atom of carbon consumed, 0.503/0.497 or 1.012 lb-atoms of zinc are produced, or 5.51 lb of zinc are formed per lb of carbon consumed.

(c) *Minimum temperature for producing liquid zinc.* If operating conditions are such that liquid zinc may be produced directly at equilibrium conditions, the vaporization equilibrium of zinc must be considered as a simultaneous reaction. Thus,

(4) $$Zn(l) \rightleftharpoons Zn(g)$$

The equilibrium constant for this vaporization is equal to the vapor pressure of the liquid zinc if ideal behavior is assumed, or $K_4 = P_{Zn}$. In Fig. 271 are plotted vapor pressures as a function of temperature taken from the data of Maier.[4]

The minimum temperature at which liquid zinc can be produced occurs when the partial pressure of the zinc vapor is just equal to the vapor pressure of molten zinc.

This corresponds to the dew point of zinc vapor. The partial pressure of zinc vapor p_{Zn} at different reduction temperatures is calculated by the method of part (*b*) and tabulated in column 6 of Table A. In column 7 are the vapor pressures P_{Zn} of liquid zinc. The partial pressure of zinc vapor p_{Zn} and the vapor pressure of molten zinc P_{Zn} are plotted against temperature in Fig. 271. The intersection point of these two curves indicates that the minimum temperature at which molten zinc can occur is 1292° K, where the partial pressure of zinc vapor is 2.87 atm and the total pressure is 5.6 atm. However, under these conditions a negligible amount of zinc is formed, and higher temperatures and pressures must be employed to obtain appreciable yields of the liquid.

TABLE A. COMPLETE VAPORIZATION OF ZINC ASSUMED

Temperature, °K	K_1	K_2	x	π atm	p_{Zn} atm	P_{Zn} atm $= K_4$
1100	0.001666	15.35	0.980	0.322	0.162	0.43
1200	0.009014	70.8	0.978	1.607	0.808	1.20
1300	0.03853	255.3	0.976	6.31	3.174	3.06
1400	0.1300	760.4	0.975	20.01	10.07	6.33
1500	0.3740	1940	0.973	54.25	27.31	12.2

At temperatures below 1292° K, two solid phases and one gas phase exist. At temperatures above 1292° K, two solid phases, one liquid phase, and one gas phase exist. At all temperatures, the system is univariant; that is, fixing the temperature automatically fixes the pressure.

(*d*) *Yield of liquid zinc by direct reduction.* When liquid zinc is present at equilibrium conditions, the partial pressure of the zinc vapor must equal the vapor pressure of the liquid zinc at the equilibrium temperature.

Since equation 26–6 is applicable only where each component is present in only one phase, equation 26–8 must be used as the basis for developing equilibrium expressions. On the basis of 1 lb-mole of ZnO reduced, let x represent the pound-moles of CO formed and y the pound-moles of zinc vapor present at equilibrium. Then the pound-moles of CO_2 are $\frac{1}{2}(1-x)$, and the total moles of gaseous products are $\frac{1}{2}(1 + x + 2y)$.

Equilibrium partial pressures, if ideal behavior of the gases is assumed, are as follows:

$$p_{CO} = \frac{2x\pi}{1 + x + 2y} \qquad (f)$$

$$p_{CO_2} = \frac{(1 - x)\pi}{1 + x + 2y} \qquad (g)$$

$$P_{Zn} = \frac{2y\pi}{1 + x + 2y} = K_4 \qquad (h)$$

or, from (*h*),

$$\pi = \frac{P_{Zn}(1 + x + 2y)}{2y} \qquad (i)$$

These values are substituted in equation 26–8 with the assumption that partial pressures are equal to activities.

From reaction 1, $ZnO + CO \longrightarrow Zn + CO_2$

$$\frac{P_{Zn}(1 - x)}{2x} = K_1 \qquad (j)$$

From reaction 2, $C + CO_2 \longrightarrow 2CO$

$$\frac{4\pi x^2}{(1-x)(1+x+2y)} = K_2 \qquad (k)$$

Substitution of (*i*) in (*k*) gives

$$\frac{4P_{Zn}x^2}{(1-x)2y} = K_2 \qquad (l)$$

Solving (*j*) for x gives

$$x = \frac{P_{Zn}}{P_{Zn} + 2K_1} \qquad (m)$$

Substitution of (*m*) in (*l*) gives

$$\frac{P^3_{Zn}}{yK_1(P_{Zn} + 2K_1)} = K_2 \qquad (n)$$

or

$$y = \frac{P^3_{Zn}}{(P_{Zn} + 2K_1)K_1K_2} \qquad (o)$$

From equation (*o*), y is evaluated directly for any selected temperature that fixes the values of P_{Zn}, K_1, and K_2. The corresponding compositions and total pressures are evaluated from equations (*m*) and (*i*). These results are presented in Table B, together with the percentages of the total zinc formed that is present in the liquid state.

Table B. Presence of Liquid Zinc Considered

Temperature, °K	K_1	K_2	P_{Zn}	y	x	π	% Liquid Zn $100(1-y)$
1290	0.0339	226	2.80	1.00	0.976	5.555	0
1353	0.074	464	4.6	0.603	0.970	12.1	39.7
1400	0.1300	760.4	6.33	0.390	0.961	22.2	61.0
1500	0.3740	1940	12.2	0.192	0.941	83.9	80.8

The values of π in Table B are plotted in Fig. 271 for comparison with those of Table A, which were based on the assumption of complete vaporization. From inspection of Fig. 271, 50% liquefaction of zinc will occur at 1373° K, 16.3 atm pressure.

It is evident that the direct production of liquid zinc in yields approaching complete recovery would require operating conditions that cannot be attained with the structural materials at present available.

Chemical Equilibria in Internal-Combustion Engines

To simplify calculations in the cycle analysis of internal-combustion engines, air is generally taken as the working fluid, as described in Chapter 18. Under actual conditions, the products of combustion of a hydrocarbon fuel constitute the working fluid. Cycle analyses can be improved by considering the equilibrium products of combustion as the working fluid where equilibrium composition is progressively changing during the cycle. The assumption of equilibrium conditions is still an approximation. The actual composition changes are unknown since they involve the unknown chemical kinetics of combustion. Because of the laborious calculations involved and the uncertainties of reaction rates, cycle analyses based on products of combustion are rarely justified, and the air-standard cycle is generally accepted.

In the combustion of a hydrocarbon in air, high temperatures are obtained, resulting in a mixture of gases including carbon dioxide, water, nitrogen, and their dissociated products: carbon monoxide, hydrogen, nitric oxide, as well as the atomic forms of oxygen, nitrogen, and hydrogen, and free radicals such as the hydroxyl radical OH.

Illustration 5. The summary of a thermodynamic analysis is presented herewith, applied to the combustion of gaseous iso-octane C_8H_{18} with the minimum amount of air required for complete combustion of iso-octane to CO_2 and H_2O. On the basis of 1 g-mole of iso-octane, this requires 12.5 g-moles of O_2 accompanied by 47.0 g-moles of N_2, or 59.5 g-moles of dry air. The stages of the Otto cycle consist of the following:

1. Intake of the air–fuel mixture.
2. Isentropic compression of the mixture.
3. Adiabatic combustion at constant volume.
4. Isentropic expansion of the combustion products.
5. Expulsion of the combustion products.

It is assumed that no heat is transferred to the walls of the engine, that the compression and expansion strokes take place reversibly, and that clearance volume is negligible. The combustion at constant volume is highly irreversible. This irreversibility cannot be avoided by any known practical means. The expansion of the products is assumed to be isentropic. During this expansion, work is done, and the gaseous products undergo further chemical change. In this stage, it is assumed that both chemical reaction and expansion proceed reversibly and that chemical equilibrium is maintained during the changes in pressure and temperature.

For illustration, it is assumed that the feed enters at a pressure p_1 (1 atm), temperature T_1 (330° K), and at volume V_1 (1638.2 liters) corresponding to 60.5 g-moles of mixture. A compression ratio of 5 is assumed; hence, $V_2 = 327.6$ liters. The entropy, enthalpy, and internal energy of the total gas mixture are calculated, relative to the same properties of the separate gases at 0° K and a pressure of 1 atm.

All values of thermodynamic properties were taken from *Selected Values of Chemical Thermodynamic Properties*, National Bureau of Standards, Washington, D.C. (1950).

At the end of the intake stroke, the conditions are recorded in column I (Table A). In isentropic compression, the pressure is increased to 8.455 atm, and the temperature to 558° K, with no change in composition or entropy, but with increase in internal energy and enthalpy as recorded in column II.

At the end of the compression stroke, the fuel burns at constant volume to attain an equilibrium conversion at temperature T_3 and pressure p_3. The combustion process is inherently irreversible. At the high temperatures obtained, the products will not be entirely CO_2, H_2O, and N_2, but all the possible equilibrium products resulting therefrom; namely CO, H_2, O_2, H, O, NO, and OH.

To establish the composition of the equilibrium mixture, it is necessary to consider only the least number of independent reaction equations that account for all the possible products resulting from the original mixture of iso-octane and air. The gaseous phase is the only phase present. The formation of any organic compounds can be eliminated from consideration because of the extremely low value of their equilibrium constants at the high temperatures of combustion.

The following five independent chemical reactions, following the combustion of iso-octane, are the only ones that are significant:

(1) $CO_2 \rightleftharpoons CO + \frac{1}{2}O_2 \qquad K_1 = \frac{n_{CO}\sqrt{n_{O_2}}}{n_{CO_2}}\left(\frac{p_3}{n_t}\right)^{1/2}$

(2) $H_2O \rightleftharpoons \frac{1}{2}H_2 + OH \qquad K_2 = \frac{\sqrt{n_{H_2}}n_{OH}}{n_{H_2O}}\left(\frac{p_2}{n_t}\right)^{1/2}$

(3) $\frac{1}{2}N_2 + \frac{1}{2}O_2 \rightleftharpoons NO \qquad K_3 = \frac{n_{NO}}{\sqrt{n_{N_2}n_{O_2}}}$

(4) $\frac{1}{2}H_2 \rightleftharpoons H \qquad K_4 = \frac{n_H}{\sqrt{n_{H_2}}}\left(\frac{p_3}{n_t}\right)^{1/2}$

(5) $\frac{1}{2}O_2 \rightleftharpoons O \qquad K_5 = \frac{n_O}{\sqrt{n_{O_2}}}\left(\frac{p_3}{n_t}\right)^{1/2}$

Let

u = moles CO_2 dissociated by reaction 1

w = moles H_2O dissociated by reaction 2

x = moles NO formed by reaction 3

y = moles H_2 dissociated by reaction 4

z = moles of O_2 dissociated by reaction 5

The products at equilibrium are as follows:

$$
\begin{aligned}
CO_2 &= 2 - u \\
CO &= u \\
H_2O &= 3 - w \\
H_2 &= 0.5w - y \\
O_2 &= 0.5w - 0.5x - z \\
N_2 &= 11.29 - 0.5x \\
NO &= x \\
OH &= w \\
H &= 2y \\
O &= 2z \\
\hline
n_t &= 16.29 + 0.5u + 0.5w + y + z
\end{aligned}
$$

The solution of the above five equations requires another equation for establishing the unknown temperature T_3. From temperature T_3, the five equilibrium constants can be obtained and the five unknowns calculated.

For adiabatic combustion taking place at an initial temperature of T_2, it may be conveniently assumed that the combustion first goes completely to CO_2 and H_2O, followed by dissociation according to the five reactions mentioned, to reach equilibrium at temperature T_3. Since the combustion proceeds at constant volume, no work is done. An energy balance gives

$$\Delta U_{c2} = \sum n_i U_{i3} - \sum n_i U_{i2} + \sum n_i \, \Delta U_i + w \tag{10}$$

where ΔU_{c2} = heat of combustion of iso-octane at temperature T_2 and constant volume = −1222.39 k-cal per g-mole

n_i = number of moles of component i resulting from complete combustion of 1 g-mole of iso-octane in air

$n_i = 8$ for CO_2 $\quad n_i = 47.0$ for N_2

$n_i = 9$ for H_2O

ΔU_i = heat of reaction at constant volume and temperature T_3 for reaction i

$\sum$ = the summation for the five reactions

By a trial-and-error procedure the conditions at the end of constant-volume combustion are recorded in column III with a temperature $T_3 = 2813°$ K and pressure $p_3 = 45.88$ atm. The equilibrium products of combustion contain a high content of CO, NO, H, O, and OH.

In the fourth stage of isentropic expansion, work is done, and there is a reversible shift in equilibrium composition due to reversible chemical reaction, as the temperature drops to T_4 and pressure to p_4.

In this stage work is done, and the work term w must be retained in the energy equation. The resultant terminal conditions are recorded in column IV with a final temperature of 2095° K, a pressure of 6.731 atm, and zero change in entropy. The composition changes with a reduction in content of CO, NO, H, O, and OH.

Under actual operation, these high-equilibrium values of temperature and pressure are not attained. Because of heat losses, the actual temperature and pressure approach closer to atmospheric conditions. This loss of heat is irreversible and results in less power and in reduction of both thermal and thermodynamic efficiencies.

In *isentropic compression*, the work done by the engine is equal to the decrease in internal energy:

$$w_2 = -\Delta U = -\frac{189{,}350 - 105{,}900}{1000} = -83.45 \text{ kcal}$$

In the isentropic expansion stroke, the work done by the engine is equal to the decrease in internal energy:

$$w_4 = -\Delta U = -\frac{890{,}030 - 1{,}253{,}840}{1000} = 363.81 \text{ kcal}$$

The net work done by the engine is $\sum w$

$$\sum w = -83.45 + 363.81 = 280.36 \text{ kcal}$$

The heat of combustion of iso-octane gas at constant pressure and 330° K is 1218.8 kcal.

The thermal efficiency of the internal-combustion engine is, hence, 280.36/1218.8 = 0.230.

Much higher efficiencies could be attained by providing for expansion to atmospheric pressure as is done in gas turbines or by operating at higher compression ratios.

Table A. Equilibrium Changes in an Internal-Combustion Engine

Composition, g-moles	I Intake (Isothermal)	II Compression (Isentropic)	III Combustion (Constant Volume)	IV Expansion (Isentropic)
C_8H_{18}	1.0	1.0	0	0
O_2	12.5	12.5	0.559	0.090
N_2	47.0	47.0	46.756	46.974
CO_2	0	0	6.341	7.767
H_2O	0	0	8.509	8.947
CO	0	0	1.659	0.233
H_2	0	0	0.222	0.260
NO	0	0	0.488	0.052
O	0	0	0.054	0.0014
H	0	0	0.048	0.0015
OH	0	0	0.491	0.0530
Total $= n_t$	60.5	60.5	65.126	64.145
Pressure, atm	1.0	8.455	45.88	6.731
Temperature, °K	330.0	558.0	2813.0	2095.0
Volume, liters	1638.2	327.6	327.6	1638.2
U, kcal	105,900	189,350	1,253,840	890,930
H, kcal	145,570	256,440	1,617,890	1,157,970
S, kcal per K°	2979.5	2979.5	3885.0	3885.0

Equilibria in Liquid Solutions

The fundamental thermodynamic principles of equilibrium are universally applicable, and equation 25–6 is rigorous for reactions involving liquid or solid solutions of all types. However, the application of thermodynamic methods to such systems has not been particularly fruitful because of the complex relationships existing between activities and concentrations. When more than one solute is involved, as is commonly so where reactions occur in solution, the complexity of the relationships is compounded to such an extent as to render the methods of little practical value, except where ideal behavior is approximated and where activity coefficients can be assumed to be constant.

Constant activity coefficients can be satisfactorily assumed when dealing with dilute solutions or in systems involving only closely related types of materials such as organic homologs. In such cases the calculation of equilibrium compositions is the same as for gaseous systems except that pressure is not an appreciable factor. Thus, from equation 26–1,

$$K = \left(\frac{N_R{}^r N_S{}^s \cdots}{N_B{}^b N_C{}^c \cdots}\right)\left(\frac{\gamma_R{}^r \gamma_S{}^s \cdots}{\gamma_B{}^b \gamma_C{}^c \cdots}\right) = \left(\frac{N_R{}^r N_S{}^s \cdots}{N_B{}^b N_C{}^c \cdots}\right) K_\gamma \qquad (11)$$

or, in a form similar to equation 26–6,

$$K = K_\gamma \left(\frac{n_R{}^r n_S{}^s \cdots}{n_B{}^b n_C{}^c \cdots}\right)(n_B + n_C + \cdots + n_R + n_S + \cdots + n_I)^{(c+b+\cdots)-(r+s+\cdots)} \qquad (12)$$

In working with liquid solutions it is necessary that the activities and standard free-energy data used in equilibrium calculations be based on the same standard states. The entropy of a substance in solution varies widely with change in concentration. Hence, the standard free energy of formation based on $a_1/N_1 = 1$ when $N_1 = 0$ will be different from that based on $a_1/m_1 = 1$ when $m_1 = 0$, even though both are referred to the state of infinite dilution as explained in Chapter 20. With this precaution, the procedures in calculating equilibrium compositions are the same as those demonstrated for the gaseous systems.

Solubility of Carbonates. The application of thermodynamic principles to the solubility of the metal carbonates has been developed by Kelley and Anderson.[7] These principles are of importance in connection with studies of the recovery and purification of carbonate ores by leaching operations.

When a metal carbonate $MeCO_3$ is dissolved in water in the presence of carbon dioxide, the reactions shown in Table 77 may take place.

TABLE 77. REACTIONS IN THE DISSOLUTION OF A METALLIC CARBONATE

(1)	$MeCO_3(s)$	=	$MeCO_3(aq)$
(2)	$MeCO_3(aq)$	=	$Me^{++} + CO_3{}^{--}$
(3)	$MeCO_3(s) + H_2O$	=	$Me^{++} + OH^- + HCO_3{}^-$
(4)	$MeCO_3(s) + H_2CO_3(aq)$	=	$Me^{++} + 2HCO_3{}^-$
(5)	$MeCO_3(s) + 2H_2O$	=	$Me(OH)_2(s) + H_2CO_3(aq)$
(6)	$Me(OH)_2(s)$	=	$Me(OH)_2(aq)$
(7)	$Me(OH)_2(aq)$	=	$Me^{++} + 2(OH)^-$
(8)	H_2O	=	$H^+ + OH^-$
(9)	$H_2CO_3(aq)$	=	$H^+ + HCO_3{}^-$
(10)	$H_2CO_3(aq)$	=	$H_2O + CO_2(g)$
(11)	$HCO_3{}^- + H_2O$	=	$OH^- + H_2CO_3(aq)$
(12)	$HCO_3{}^-$	=	$H^+ + CO_3{}^{--}$
(13)	$Me(HCO_3)_2(s)$	=	$Me(HCO_3)_2(aq)$
(14)	$Me(HCO_3)_2(aq)$	=	$Me^{++} + 2(HCO_3)^-$
(15)	$MeCO_3(s) + H_2CO_3(aq)$	=	$Me(HCO_3)_2(s)$

All of these possible reactions must be in equilibrium at saturation conditions. Since reaction 10 produces $CO_2(g)$ as a product, it follows

[7] K. K. Kelley and C. T. Anderson, "Contributions to the Data on Theoretical Metallurgy; IV Metal Carbonates—Correlation and Applications of Thermodynamic Properties," *U.S. Bur. Mines Bull.*, **384** (1935).

that a definite equilibrium CO_2 pressure will result from dissolving a carbonate in pure water. Dissolution will at first take place by reactions 1 through 5, and carbon dioxide gas will tend to be formed as a result of reaction 10. If the solution is in contact with an atmosphere containing no partial pressure of carbon dioxide, the gas will be evolved with resultant formation of $Me(OH)_2(aq)$, Me^{++}, and OH^- This reaction will proceed until the $Me(OH)_2(aq)$, Me^{++}, and OH^- concentration reaches saturation and $Me(OH)_2(s)$ is formed. The reaction then proceeds with the over-all formation of $Me(OH)_2(s)$ and $CO_2(g)$ by reactions 5 and 10 until the $MeCO_3(s)$ is completely consumed.

If dissolution were to take place in contact with a CO_2-free atmosphere as previously described, the equilibrium CO_2 pressure of the solution would progressively diminish until the first $Me(OH)_2(s)$ forms, and then remain constant at the value fixed by the equilibrium of reactions 4 and 10. Thus, the partial pressure of CO_2, corresponding to equilibrium in reactions 5 and 10, is the minimum equilibrium CO_2 pressure exerted by a solution in equilibrium with $MeCO_3(s)$, and corresponds to an invariant point at which two solids, one liquid, and one gas phase are in equilibrium.

If a carbonate is dissolved in water in contact with an atmosphere containing a partial pressure of CO_2 greater than the minimum value previously discussed, dissolution will proceed to an equilibrium fixed by reactions 1 through 4 and 7 through 15. If the CO_2 pressure in the atmosphere is less than the equilibrium CO_2 dissolution pressure of the pure carbonate, CO_2 will be evolved as equilibrium is approached, and the final solution will be basic. If the CO_2 pressure in the atmosphere is greater than the dissolution pressure, the solution will absorb CO_2 and be acidic when equilibrium is reached. As the CO_2 pressure in the atmosphere is further increased, the acidity of the solution increases until the concentration of $Me(HCO_3)_2(aq)$ reaches the saturated value, and $Me(HCO_3)_2(s)$ begins to form. The CO_2 pressure at which this occurs is determined by the equilibrium of reactions 15 and 10 and is the maximum pressure that can be in equilibrium with a solution which is in equilibrium with $MeCO_3(s)$. Higher CO_2 pressures will result in the disappearance of the solid carbonate and in the formation of solutions whose compositions are fixed by the CO_2 pressure and the equilibria of reactions 2 and 8 through 14.

In dealing with the sparingly soluble carbonates, the situation is simplified by the fact that the salts are completely dissociated in solution. Thus, reactions 1 and 2, 6 and 7, and 13 and 14 may be combined into single reactions and the dissolved salts, $MeCO_3$, $Me(OH)_2$, and $Me(HCO_3)_2$, disregarded. Furthermore, the solid forms of the acid carbonates of the metals which form sparingly soluble carbonates are not known to exist, and reactions 13 to 15 need not be considered. Additional simplification results from the fact that, in dilute solutions, the activity of water is constant and equal to 1.0.

The equilibrium constant of a reaction in dilute solution is determined from its standard enthalpy and entropy changes. In working with dilute aqueous solutions, it is customary to express compositions in terms of molalities and to select the standard state such that $a/m = 1.0$ where $m = 0$. The standard enthalpy change in dilute solution is obtained from Tables 29, 30, and 31. The standard entropy change is determined from Table 68. It may be noted that values of zero are arbitrarily assigned to the heat of formation and the entropy of the hydrogen ion in dilute solution. The corresponding free-energy change of formation is then 4655 cal per g-mole at 25° C. All other ionic heats of formation and entropies are thus expressed relative to the hydrogen ion since the individual absolute values are unknown.

Illustration 6. Calculate the equilibrium constant of the following reaction at 25° C:

$$CaCO_3(s)\ (\text{ppt}) \rightleftharpoons Ca^{++}(aq) + CO_3^{--}(aq)$$

From Tables 29 and 31, pages 297 and 317,

$$\Delta H_{298} = -160.3 - 129.74 - (-287.8) = -2.24 \text{ kcal per g-mole}$$

From Table 68,

$$\Delta S_{298} = -11.4 - 13.0 - 22.2 = -46.6 \text{ cal/(g-mole)(K°)}$$

$$\left(\frac{\Delta G}{T}\right)_{298} = \frac{-2240}{298} + 46.6 = 39.1$$

$$\log K_{298} = \frac{-39.1}{4.576} = -8.548$$

$$K_{298} = (2.83)10^{-9}$$

In calculating equilibrium constants for ionic reactions, great care must be exercised to use heat of formation and entropy data that are consistent.

The equilibrium composition of a system involving ions is calculated by the procedure demonstrated in the preceding sections dealing with complex reactions. In order to define the equilibrium, it is desirable to consider the smallest number of reactions that include all reactants and products present at equilibrium. An algebraic equation relates the concentrations of the components of each reaction to its equilibrium constant. An additional algebraic equation expresses the electrical neutrality of the solution, that is, the equality of the positive and negative charges of all of the ions. A further equation which may prove useful expresses the material balance of the dissolution. The equilibrium composition is evaluated by simultaneous solution of these equations.

Illustration 7. Calculate the complete composition of the solution obtained when precipitated calcium carbonate is dissolved in water in contact with an atmosphere containing a partial pressure of CO_2 of $3(10^{-4})$ atm.

Solution: It may be assumed that the specified partial pressure of CO_2 is above that of the invariant point and that the system at equilibrium contains $CaCO_3(s)$, Ca^{++}, CO_3^{--}, HCO_3^-, H^+, OH^-, and $H_2CO_3(aq)$. There are, therefore, six unknown compositions to be evaluated, which require five independent equilibrium equations in addition to the equation of electrical neutrality. The material-balance equation cannot be used in this case, because CO_2 may be gained or lost during the dissolution.

Although other selections are possible, it is convenient to consider the following reactions for the establishment of the equilibrium equations. The corresponding equilibrium constants are calculated by the method of Illustration 6.

		K_{298}
(*a*)	$CaCO_3(s) = Ca^{++} + CO_3^{--}$	$(2.9)10^{-9} = (Ca^{++})(CO_3^{--})$
(*b*)	$H_2O(l) = H^+ + OH^-$	$(5.5)10^{-15} = (H^+)(OH^-)$
(*c*)	$H_2CO_3(aq) = H^+ + HCO_3^-$	$(9.3)10^{-8} = \dfrac{(H^+)(HCO_3^-)}{H_2CO_3(aq)}$
(*d*)	$H_2CO_3(aq) = H_2O + CO_2(g)$	$31.7 = \dfrac{CO_2(g)}{H_2CO_3(aq)}$
(*e*)	$HCO_3^- = H^+ + CO_3^{--}$	$(1.4)10^{-11} = \dfrac{(H^+)(CO_3^{--})}{HCO_3^-}$

The equation of electrical neutrality may be written as follows if the chemical formulas are used to denote molalities:

$$2Ca^{++} + H^+ = 2CO_3^{--} + HCO_3^- + OH^- \qquad (f)$$

It should be noted that (*f*) is not a stoichiometric equation.

Since the CO_2 pressure is specified, it follows from (*d*) that the molality of $H_2CO_3(aq)$ is represented by

$$H_2CO_3(aq) = (3)10^{-4}/31.7 = (0.95)10^{-5} \text{ g-moles per 1000 grams } H_2O$$

If y is used to represent the molality of the HCO_3^- ion, the following expressions may be written from equations (*c*), (*b*), (*e*), and (*a*) for the molalities of the other ions:

$$H^+ = (0.95)10^{-5}(9.3)10^{-8}/y = (8.8)10^{-13}/y$$
$$OH^- = (5.5)10^{-15}y/(8.8)10^{-13} = (6.25)10^{-3}y$$
$$CO_3^{--} = (1.4)10^{-11}y^2/(8.8)10^{-13} = 15.9y^2$$
$$Ca^{++} = (2.9)10^{-9}/(15.9)y^2 = (1.82)10^{-10}/y^2$$

These values may be substituted in equation (*f*), resulting in an equation that contains only y as an unknown. Thus,

$$\frac{(3.64)10^{-10}}{y^2} + \frac{(8.8)10^{-13}}{y} = 31.8y^2 + y[1 + (6.25)10^{-3}]$$

This equation is solved graphically or by trial and error.

$$y = HCO_3^- = (7.1)10^{-4} \text{ g-mole per 1000 grams } H_2O$$

The concentrations of the other ions are evaluated from the equations relating them to y. Thus

$$H^+ = (1.2)10^{-9} \text{ g-mole per 1000 grams } H_2O$$
$$OH^- = (4.4)10^{-6} \text{ g-mole per 1000 grams } H_2O$$
$$CO_3^{--} = (7.9)10^{-6} \text{ g-mole per 1000 grams } H_2O$$
$$Ca^{++} = (3.7)10^{-4} \text{ g-mole per 1000 grams } H_2O$$

The procedure of Illustration 7 may be repeated for other CO_2 pressures and curves plotted relating the composition of the solution to the equilibrium partial pressure of CO_2. Such curves for the solubility of $CaCO_3$ in water are shown in Fig. 272. It may be noted that the concentration of the calcium ion passes through a minimum and is increased by either increasing or decreasing the CO_2 pressure from this point. As the CO_2 pressure is increased, CO_2 is absorbed by the solution, but, as the pressure is reduced, CO_2 is evolved and the concentration of the OH^- ion increases until solid $Ca(OH)_2$ is formed. The appearance of the added phase results in an invariant point at which the composition of the solution and the partial pressure of the CO_2 are fixed as long as both solid phases are present. The conditions at the invariant point are established by the following equation in conjunction with those of Illustration 7:

$$(g) \qquad CaCO_3(s) + 2H_2O \rightleftharpoons Ca(OH)_2(s) + H_2CO_3(aq)$$

$$K_{298} = (9.3)10^{-15} = H_2CO_3(aq)$$

If calcium carbonate is dissolved in pure water in a closed liquid system which permits no gain or loss of carbon dioxide, an equilibrium partial pressure of CO_2 is established, which is determined by the equations of Illustration 7, together with the following material-balance equation in which the formulas indicate molalities:

$$(h) \qquad Ca^{++} = CO_3^{--} + HCO_3^- + H_2CO_3$$

This is not a stoichiometric equation, but indicates that each atom of calcium entering the solution is accompanied by an atom of carbon. The solubility of $CaCO_3$ in pure water without gain or loss of CO_2 is determined by simultaneous solution of equations (a through f) and (h). It may be noted from Fig. 272 that minimum solubility is obtained under these conditions in the case of calcium.

These same principles may be applied to calculating the compositions obtained in selective leaching and precipitation operations where several

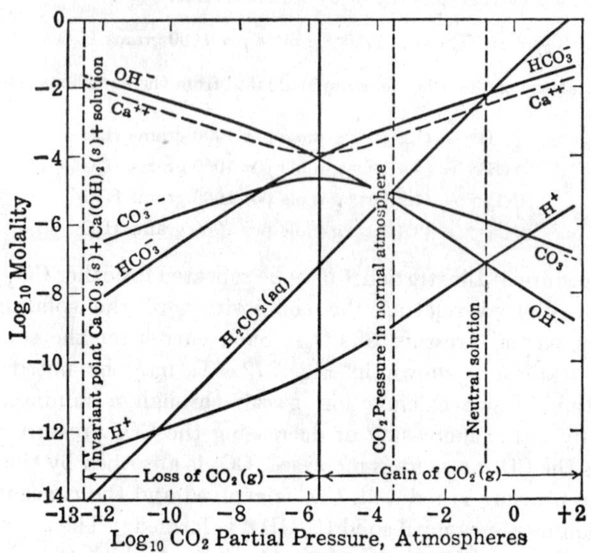

Fig. 272. Solubility of $CaCO_3$ in water

sparingly soluble metals or salts are involved. For each added constituent, an extra equation is added to the group which must be solved simultaneously in order for the equilibrium composition to be evaluated. However, simplification is possible in many instances by neglect of components which are present in the solution in relatively small quantities. For example, from Fig. 272 it is evident that, at CO_2 pressures above 10^{-4} atm, the concentrations of OH^-, CO_3^{--}, and H^+ are negligible in comparison to the Ca^{++}, HCO_3^-, and H_2CO_3. In this range the dissolution may be considered as proceeding by the following stoichiometric equation:

$$(i) \qquad CaCO_3(s) + H_2O(l) + CO_2(g) = Ca^{++} + 2HCO_3^-$$

$$K_{298} = (6.0)10^{-7} = \frac{(Ca^{++})(HCO_3^-)^2}{CO_2(g)} \tag{13}$$

Since it is assumed that $HCO_3^- = 2Ca^{++}$, the foregoing equilibrium equation may be written

$$K_{298} = (6.0)10^{-7} = \frac{4(Ca^{++})^3}{CO_2(g)} \tag{14}$$

or

$$Ca^{++} = (5.3)10^{-3}[CO_2(g)]^{1/3} \tag{15}$$

where the chemical formulas indicate activities in molalities and atmospheres, respectively. In this manner the molalities of the Ca^{++} and HCO_3^- ions are determined directly, and the molalities of the other components may be individually evaluated from the equations of Illustration 7. This procedure is generally applicable to the sparingly soluble carbonates, either singly or in complex mixtures if the CO_2 pressure is equal to or greater than that of the normal atmosphere.

The application of these methods to complex metallurgical problems is discussed in detail by Kelley and Anderson.[7] By assuming that the activity coefficients cancel out of the equilibrium equations, it is possible to solve effectively many such problems.

Problems

1. One pound-mole of producer gas is burned with 7 lb-moles of preheated air. Calculate the equilibrium composition of the products of combustion at 2200° C and 1.0 atm.

	Reactions		Composition of Producer Gas	
(1)	$H_2 + \frac{1}{2}O_2 \rightarrow H_2O$	$K_1 = 167.9$	CO	25.0%
(2)	$CO + \frac{1}{2}O_2 \rightarrow CO_2$	$K_2 = 33.1$	H_2	15.0
(3)	$\frac{1}{2}N_2 + \frac{1}{2}O_2 \rightarrow NO$	$K_3 = 0.0616$	CO_2	5.0
(4)	$\frac{1}{2}H_2 \rightarrow H$	$K_4 = 0.01861$	N_2	55.0
(5)	$\frac{1}{2}O_2 \rightarrow O$	$K_5 = 0.0128$	H_2O in gas = 0.028 mole per mole dry gas	
(6)	$\frac{1}{2}N_2 \rightarrow N$	$K_6 = 0.000115$	H_2O in air = 0.036 mole per mole dry air	

2. Methane gas admixed with the theoretical amount of air required for combustion is initially at 500° K and 10 atm.

(*a*) Calculate the pressure, temperature, and composition of the products following adiabatic combustion at constant volume.

(*b*) Calculate the work of expansion when these gaseous products are expanded isentropically at chemical equilibrium for a five-fold volume increase.

3. Estimate the solubility of $CaCO_3$ in water at 25° C under a partial pressure of 8 atm.

4. Calculate the ionic composition at 25° C in molalities when $CaCO_3$ is dissolved in a carbonic acid solution at a partial pressure of CO_2 of 1 atm.

(1)	$H_2CO_3(aq) \rightarrow H_2O + CO_2(g)$	$K_1 = 31.7$
(2)	$H_2CO_3(aq) \rightarrow H^+ + HCO_3^-$	$K_2 = (9.3)10^{-8}$
(3)	$H_2O \rightarrow H^+ + OH^-$	$K_3 = (5.5)10^{-15}$
(4)	$HCO_3^- \rightarrow H^+ + CO_3^{--}$	$K_4 = (1.4)10^{-11}$
(5)	$CaCO_3(s) \rightarrow Ca^{++} + CO_3^{--}$	$K_5 = (2.9)10^{-9}$

Considering only the following reaction, estimate the concentration of Ca^{++} at 25° C and partial pressure of CO_2 of 1 atm.

$$CaCO_3 + H_2O + CO_2(g) = Ca^{++} + 2HCO_3^- \qquad K = (6.0)10^{-7}$$

5. Calculate the reversible work done when 1 gal of iso-octane (or absolute alcohol) is burned with the theoretical amount of air in an ideal Otto cycle with intake at 68° F, with a compression ratio of 7.0 and with the theoretical air–fuel ratio.

	Alcohol	Iso-octane
Heat of vaporization, g-cal per g-mole at 20° C	11,140	8474
Normal boiling point	351° K	372° K
Specific gravity of liquid at 68° F	0.791	0.700
Heat capacity of gas ($p = 0$), g-cal/(g-mole)(K°)		

$$2.51 + 0.051T - (18.91)10^{-6}T^2 \quad \text{for alcohol}$$
$$16.01 + 0.1046T \quad \text{for iso-octane}$$

6. From the data of Table 69, calculate the complete compositions of the equilibrium mixtures resulting from the polymerization of pure ethylene to the butenes if it is assumed that no other reactions occur at the following conditions:

	Temperature, °C	Pressure, psia
(*a*)	150	300
(*b*)	150	600
(*c*)	200	300
(*d*)	200	600

7. From the data of Table 69, calculate the composition of the equilibrium mixture of the five isomeric hexanes at a temperature of 600° K. Calculate the heat of reaction in converting pure normal hexane to the equilibrium mixture in the ideal-gaseous state at this temperature.

8. When propane is pyrolyzed the following gaseous reactions take place:

$$C_3H_8 \rightarrow C_3H_6 + H_2$$
$$C_3H_8 \rightarrow C_2H_4 + CH_4$$

Assuming that only these two reactions occur, calculate from the data of Table 69 the composition of the equilibrium mixture formed by heating propane at 1400° F and 1 atm.

9. Propane also decomposes to form carbon and hydrogen at high temperatures. Calculate the composition of the equilibrium mixture of problem 6 if this reaction is considered.

10. When propane is pyrolyzed at elevated pressures, a portion of the ethylene formed polymerizes to form butenes according to the following gaseous reaction:

$$2C_2H_4 \rightarrow C_4H_8$$

Assuming that only this reaction and the two of problem 5 occur, calculate the complete composition of the equilibrium mixture formed by heating propane at 1100° F and 1000 psia. The butenes formed will be present in their equilibrium proportions.

11. Magnesium is produced in the carbothermic process by heating briquettes of a mixture of carbon and magnesia. The following reaction results:

$$MgO(s) + C(s) \rightarrow Mg(g) + CO(g)$$

Using the following data for the physical properties[8] of Mg and MgO, calculate the equilibrium temperatures required for the reaction at pressures of 1.0 atm and 0.1 atm. Also calculate the corresponding heats of reaction, at the conditions of operation, per pound of magnesium produced. From these results consider the feasibility of the direct production of liquid magnesium.

Heat capacities:

MgO(s): $c_p = 10.86 + (1.197)10^{-3}T - (2.087)10^5/T^2$ g-cal/(g-mole)(K°)

C(s): $c_p = 2.673 + (2.617)10^{-3}T - (1.169)10^5/T^2$ g-cal/(g-mole)(K°)

Mg(g): $c_p = 4.97$ g-cal/(g-atom)(K°)

Heat of sublimation: Mg(s) = Mg(g), $\Delta H_{298} = 36.12$ kcal per g-atom

Vapor pressure of Mg:

Temperature, °K	789(s)	881(s)	998(l)	1159	1236	1303	1380
Vapor pressure, atm	0.0001	0.001	0.01	0.1	0.25	0.5	1.0

12. Magnesium is produced in the ferrosilicon reduction process by heating a mixture of magnesium oxide and ferrosilicon. The principal reaction may be assumed to be as follows:

$$2MgO(s) + Si(s) = 2Mg(g) + SiO_2(s)$$

The heat capacities of silicon and silica are expressed by the following equation:[1]

Si(s): $c_p = 5.74 + (0.617)10^{-3}T - (1.01)10^5/T^2$ g-cal/(g-atom)(K°)

$SiO_2(s)$: $c_p = 10.87 + (8.712)10^{-3}T - (2.412)10^5/T^2$ g-cal/(g-mole)(K°)

(a) Using the data of problem 11, calculate the equilibrium temperature of the reaction at operating pressures of 1.0, 10^{-3}, and 10^{-6} atm, respectively.

(b) Calculate the heat of reaction, at the operating conditions, per pound of magnesium produced at the three pressures of part (a). Compare these results with those of problem 11.

13. Calculate the compositions of the solutions formed when $MgCO_3(s)$ is dissolved in water in equilibrium with CO_2 at the following partial pressures:

	p_{CO_2}, atm		p_{CO_2}, atm
(a)	10^{-8}	(d)	10^{-2}
(b)	10^{-6}	(e)	1
(c)	10^{-4}	(f)	50

Also calculate the composition of the solution at the invariant point where $MgCO_3(s)$ and $Mg(OH)_2(s)$ are both present. From these data, plot the solubility relationships in the form of Fig. 272, and calculate the composition of the solution formed by dissolving $MgCO_3(s)$ in pure water without gain or loss of CO_2.

14. A low-grade manganese ore consists essentially of a mixture of 35% $MnCO_3$, 25% $CaCO_3$, and 40% $FeCO_3$ by weight. It is desired to concentrate the manganese for recovery. Calculate the composition of the equilibrium solutions resulting if this ore is leached with water in contact with an atmosphere of CO_2 at pressures of (a) 1 atm, (b) 10 atm. The entropy of the Mn^{++} ion may be taken as −15.9.

[8] K. K. Kelley, "High-Temperature Specific Heat Equations for Inorganic Substances," *U.S. Bur. Mines Bull.*, **371** (1934).

15. The ore of problem 14 is leached with a calcium chloride solution having a molality of 1.0. Assuming that the metallic salts are completely dissociated in solution and that the activity coefficients are unity, calculate the calcium, manganese, and iron contents of the solution formed at equilibrium. The relatively small concentrations of carbonate and bicarbonate ions may be neglected as compared with the Cl^- ion concentration, and the over-all reactions considered as

$$Ca^{++} + MnCO_3(s) = CaCO_3(s) + Mn^{++}$$
$$Ca^{++} + FeCO_3(s) = CaCO_3(s) + Fe^{++}$$

Calculate the pounds of leaching solution required to dissolve all the manganese from 1000 lb of ore, and the weight of iron dissolved and calcium precipitated in the operation.

This reaction forms the basis of a proposed process for treating such ores in which the extract solution is treated with $Ca(OH)_2$ and blown with air to precipitate the oxides of Mn and Fe and to regenerate the $CaCl_2$. The complete over-all reaction thus is

$$2MnCO_3 + 2Ca(OH)_2 + \tfrac{1}{2}O_2 = 2CaCO_3 + Mn_2O_3 + 2H_2O$$

Appendix

TABLE A. INTERNATIONAL ATOMIC WEIGHTS—1957

Element	Symbol	Atomic Number	Atomic Weight	Element	Symbol	Atomic Number	Atomic Weight
Actinium	Ac	89	...	Mercury	Hg	80	200.61
Aluminum	Al	13	26.98	Molybdenum	Mo	42	95.95
Americium	Am	95	...	Neodymium	Nd	60	144.27
Antimony	Sb	51	121.76	Neon	Ne	10	20.183
Argon	Ar	18	39.944	Neptunium	Np	93	...
Arsenic	As	33	74.91	Nickel	Ni	28	58.71
Astatine	At	85	...	Niobium	Nb	41	92.91
Barium	Ba	56	137.36	Nitrogen	N	7	14.008
Berkelium	Bk	97	...	Nobelium	No	102	...
Beryllium	Be	4	9.013	Osmium	Os	76	190.2
Bismuth	Bi	83	209.00	Oxygen	O	8	16
Boron	B	5	10.82	Palladium	Pd	46	106.4
Bromine	Br	35	79.916	Phosphorus	P	15	30.975
Cadmium	Cd	48	112.41	Platinum	Pt	78	195.09
Calcium	Ca	20	40.08	Plutonium	Pu	94	...
Californium	Cf	98	...	Polonium	Po	84	...
Carbon	C	6	12.011	Potassium	K	19	39.100
Cerium	Ce	58	140.13	Praseodymium	Pr	59	140.92
Cesium	Cs	55	132.91	Promethium	Pm	61	...
Chlorine	Cl	17	35.457	Protactinium	Pa	91	...
Chromium	Cr	24	52.01	Radium	Ra	88	...
Cobalt	Co	27	58.94	Radon	Rn	86	...
Copper	Cu	29	63.54	Rhenium	Re	75	186.22
Curium	Cm	96	...	Rhodium	Rh	45	102.91
Dysprosium	Dy	66	162.51	Rubidium	Rb	37	85.48
Einsteinium	Es	99	...	Ruthenium	Ru	44	101.1
Erbium	Er	68	167.27	Samarium	Sm	62	150.35
Europium	Eu	63	152.0	Scandium	Sc	21	44.96
Fermium	Fm	100	...	Selenium	Se	34	78.96
Fluorine	F	9	19.00	Silicon	Si	14	28.09
Francium	Fr	87	...	Silver	Ag	47	107.880
Gadolinium	Gd	64	157.26	Sodium	Na	11	22.991
Gallium	Ga	31	69.72	Strontium	Sr	38	87.63
Germanium	Ge	32	72.60	Sulfur	S	16	32.006 [a]
Gold	Au	79	197.0	Tantalum	Ta	73	180.95
Hafnium	Hf	72	178.50	Technetium	Tc	43	...
Helium	He	2	4.003	Tellurium	Te	52	127.61
Holmium	Ho	67	164.94	Terbium	Tb	65	158.93
Hydrogen	H	1	1.0080	Thallium	Tl	81	204.39
Indium	In	49	114.82	Thorium	Th	90	232.05
Iodine	I	53	126.91	Thulium	Tm	69	168.94
Iridium	Ir	77	192.2	Tin	Sn	50	118.70
Iron	Fe	26	55.85	Titanium	Ti	22	47.90
Krypton	Kr	36	83.80	Tungsten	W	74	183.86
Lanthanum	La	57	138.92	Uranium	U	92	238.07
Lead	Pb	82	207.21	Vanadium	V	23	50.95
Lithium	Li	3	6.940	Xenon	Xe	54	131.30
Lutetium	Lu	71	174.99	Ytterbium	Yb	70	173.04
Magnesium	Mg	12	24.32	Yttrium	Y	39	88.92
Manganese	Mn	25	54.94	Zinc	Zn	30	65.38
Mendelevium	Mv	101	...	Zirconium	Zr	40	91.22

[a] Because of natural variations in the relative abundance of the isotopes of sulfur, the atomic weight of this element has a range of ± 0.003.

Table B. Units of Energy*

Carnegie Institute of Technology — American Petroleum Institute Research Project 44 — Pittsburgh, Pa.

To convert the numerical value of a property expressed in one of the units in the left-hand column of the table to the numerical value of the same property expressed in one of the units in the top row of the table, multiply the former value by the factor in the block common to both units.

Units		G-mass (energy equiv)	Joule	Int joule	Cal	I. T. cal	Btu	Kw-hr	Hp-hr	Ft-lb(wt)	Cu ft-lb(wt)/sq in.	Liter-atm
1 g-mass (energy equiv)	=	1	8.987416×10^{13}	8.985933×10^{13}	2.1480444×10^{13}	2.146640×10^{13}	8.518554×10^{10}	2.496505×10^{7}	3.347861×10^{7}	6.628764×10^{13}	4.603308×10^{11}	8.869642×10^{11}
1 joule	=	1.112667×10^{-14}	1	0.999835	0.239006	0.238849	0.947831×10^{-3}	2.777778×10^{-7}	3.72505×10^{-7}	0.737561	5.12195×10^{-3}	9.86896×10^{-3}
1 int joule	=	1.112850×10^{-14}	1.000165	1	0.239045	0.238889	0.947988×10^{-3}	2.778236×10^{-7}	3.72567×10^{-7}	0.737682	5.12279×10^{-3}	9.87058×10^{-3}
1 cal	=	4.655398×10^{-14}	4.1840	4.18331	1	0.999346	3.96573×10^{-3}	1.162222×10^{-6}	1.558562×10^{-6}	3.08595	2.14302×10^{-2}	4.12917×10^{-2}
1 I. T. cal	=	4.658444×10^{-14}	4.18674	4.18605	1.000654	1	3.96832×10^{-3}	1.162983×10^{-6}	1.559582×10^{-6}	3.08797	2.14443×10^{-2}	4.13187×10^{-2}
1 Btu	=	1.173908×10^{-11}	1055.040	1054.866	252.161	251.996	1	2.930667×10^{-4}	3.93008×10^{-4}	778.156	5.40386	10.41215
1 kw-hr	=	4.005601×10^{-8}	3,600,000	3,599,406	860,421	859,858	3412.19	1	1.341019	2,655,218	18439.01	35528.2
1 hp-hr	=	2.986982×10^{-8}	2,684,525	2,684,082	641,617	641,197	2544.48	0.745701	1	1,980,000	13750.	26493.5
1 ft-lb (wt)	=	1.508577×10^{-14}	1.355821	1.355597	0.324049	0.323837	1.285089×10^{-3}	3.766169×10^{-7}	5.05051×10^{-7}	1	6.94444×10^{-3}	1.338054×10^{-2}
1 cu ft-lb (wt)/sq in.	=	2.172351×10^{-12}	195.2382	195.2060	46.6630	46.6325	0.1850529	5.423283×10^{-5}	7.27273×10^{-5}	144.	1	1.926797
1 liter-atm	=	1.127441×10^{-12}	101.3278	101.3111	24.2179	24.2021	0.0960417	2.814662×10^{-5}	3.77452×10^{-5}	74.7354	0.518996	1

* The electrical units are those in terms of which certification of standard cells, standard resistances, etc., is made by the National Bureau of Standards. Unless otherwise indicated, all electrical units are absolute.

Table C. Conversion Factors and Constants

Data on Air

1. Average Dry Analysis by Volume

International Critical Tables, Vol. 1, p. 393 (1926).

N_2	78.03%
A	0.94
O_2	20.99
	99.96%
CO_2	0.03
H_2, Ne, He, Kr, Xe	0.01
	100.00%

2. Values to Use in Combustion Calculations. In combustion calculations, the 0.04% of CO_2, H_2, and rare gases may be ignored. Furthermore, the argon may be lumped with the nitrogen; this is referred to as atmospheric nitrogen.

	% by Volume	% by Weight	Molecular Weight
Atmospheric nitrogen	79.00	76.80	28.16
Oxygen	21.00	23.20	32.00
	100.00	100.00	

Molecular weight = 28.97

Physical Constants

The Gas-Law Constant R

Numerical Value	Units
1.987	g-cal/(g-mole)(K°)
1.987	Btu/(lb-mole)(R°)
82.06	(cm^3)(atm)/(g-mole)(K°)
0.08205	(liter)(atm)/(g-mole)(K°)
10.731	$(ft^3)(lb_f)/(in.)^2$(lb-mole)(R°)
0.7302	$(ft)^3$(atm)/(lb-mole)(R°)

1 faraday = 96,493.1 (abs. coulomb)/(g-equivalent)

Avogadro constant
= 6.02380×10^{23} atoms per gram-atom or molecules per gram-mole

Density

1 g-mole of an ideal gas at 0° C, 760 mm Hg	= 22.4140 liters
	= 22,414.6 cc
1 lb-mole of an ideal gas at 0° C, 760 mm Hg	= 359.05 cu ft
Density of dry air at 0° C and 760 mm Hg	= 1.2929 g per liter
	= 0.080711 lb per cu ft

1 gram per cc = 62.43 lb per cu ft

1 gram per cc = 8.345 lb per U. S. gal

Length

1 in.	2.540 cm
1 micron	10^{-6} meter
1 Ångstrom	10^{-10} meter

Mass

1 lb (avoirdupois)	16 oz
1 lb (avoirdupois)	7000 grains
1 lb (avoirdupois)	453.6 grams
1 ton (short)	2000 lb (Av.)
1 ton (long)	2240 lb (Av.)
1 gram	15.43 grains
1 kilogram	2.2046 lb (Av.)

Mathematical Constants

e	2.7183
π	3.1416
$\ln N$	$2.303 \log N$

Power

1 kw	56.87 Btu per min
1 kw	1.341 hp
1 hp	550 ft-lb per sec
1 watt	44.25 ft-lb per min
1 watt	14.34 g-cal per min

Pressure

1 psi	2.036 in. Hg at 0° C
1 psi	2.311 ft water at 70° F
1 atm	14.696 psi
1 atm	760 mm Hg at 0° C
1 atm	29.921 in. Hg at 0° C

Temperature Scales

Degrees Fahrenheit = 1.8 (degrees centigrade) + 32

Degrees Kelvin = degrees centigrade + 273.16

Degrees Rankine = degrees Fahrenheit + 459.69

Volume

1 cu in.	16.39 cc
1 liter	61.03 cu in.
1 liter	1000.028 cc
1 cu ft	28.32 liters
1 cu meter	1.308 cu yd
1 cu meter	1000 liters
1 U. S. gal	4 qt
1 U. S. gal	3.785 liters
1 U. S. gal	231 cu in.
1 British gal	277.42 cu in.
1 British gal	1.20094 U. S. gal
1 cu ft	7.481 U. S. gal
1 liter	1.057 U. S. qt
1 U. S. fluid oz	29.57 cc

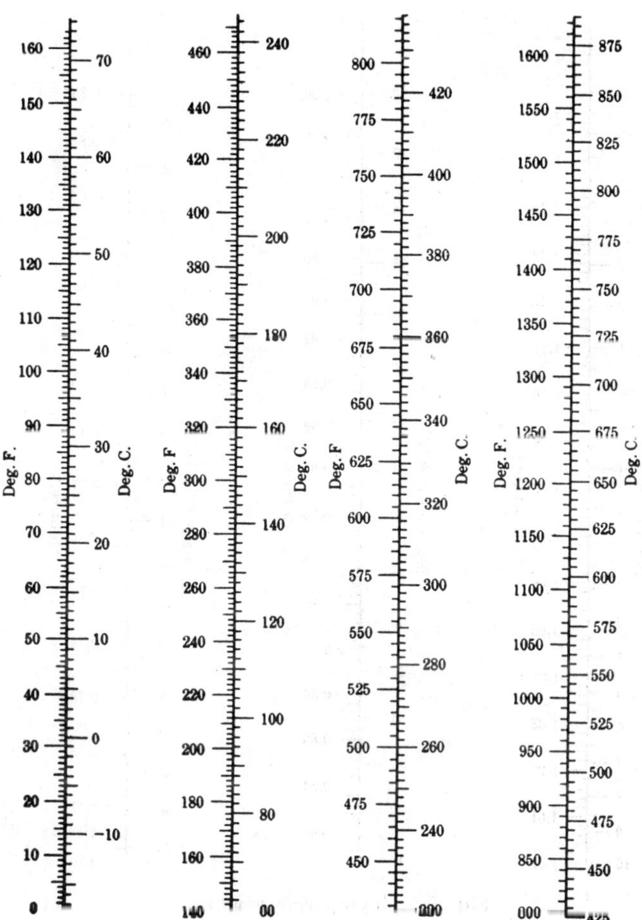

Fig. A. Temperature conversions

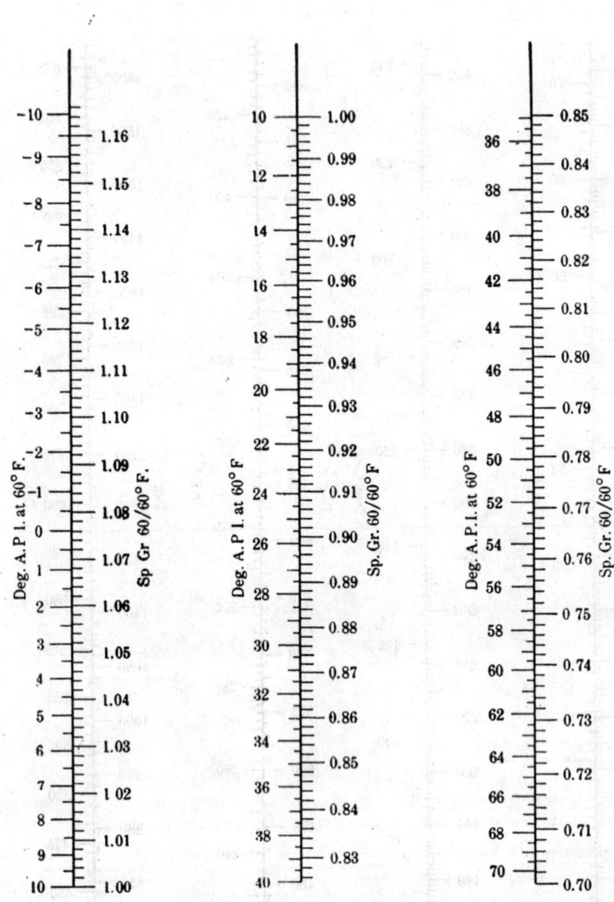

Fig. B. Gravity conversions

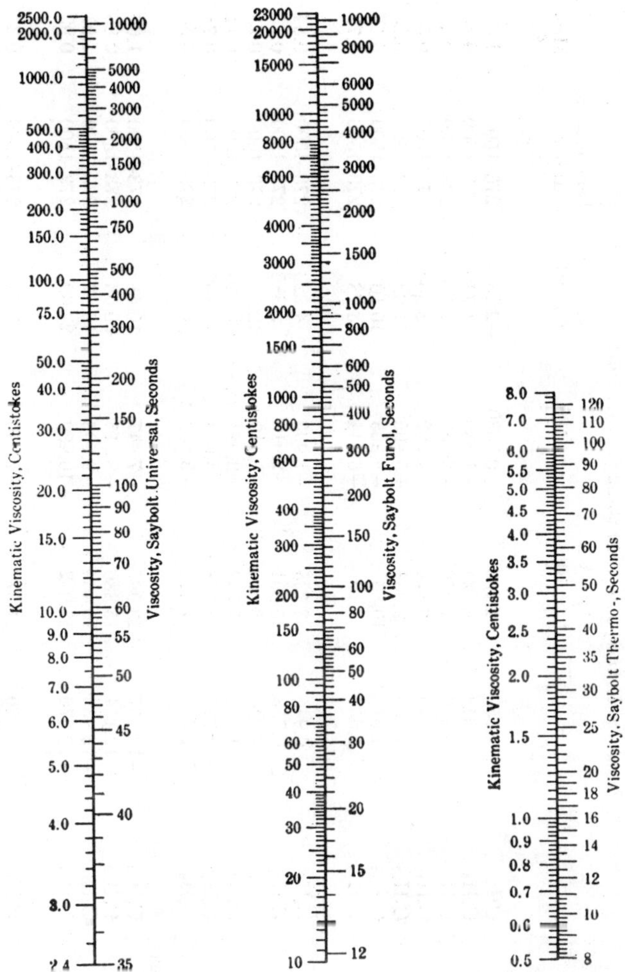

FIG. C. Viscosity conversions

TABLE D. MOLAL HEAT CAPACITIES OF GASES AT ZERO PRESSURE *

$$c°_p = a + bT + cT^2 + dT^3; \ (T = °K)$$

By K. A. Kobe and associates, "Thermochemistry for the Petrochemical Industry," *Petroleum Refiner*, Jan. 1949 through Nov. 1954.

		a	$b \times 10^2$	$c \times 10^5$	$d \times 10^9$	Temperature Range, °K	Error Max. %	Error Avg. %
Paraffinic Hydrocarbons								
Methane	CH_4	4.750	1.200	0.3030	−2.630	273–1500	1.33	0.57
Ethane	C_2H_6	1.648	4.124	−1.530	1.740	273–1500	0.83	0.28
Propane	C_3H_8	−0.966	7.279	−3.755	7.580	273–1500	0.40	0.12
n-Butane	C_4H_{10}	0.945	8.873	−4.380	8.360	273–1500	0.54	0.24
i-Butane	C_4H_{10}	−1.890	9.936	−5.495	11.92	273–1500	0.25	0.13
n-Pentane	C_5H_{12}	1.618	10.85	−5.365	10.10	273–1500	0.56	0.21
n-Hexane	C_6H_{14}	1.657	13.19	−6.844	13.78	273–1500	0.72	0.20
Monoolefinic Hydrocarbons								
Ethylene	C_2H_4	0.944	3.735	−1.993	4.220	273–1500	0.54	0.13
Propylene	C_3H_6	0.753	5.691	−2.910	5.880	273–1500	0.73	0.17
1-Butene	C_4H_8	−0.240	8.650	−5.110	12.07	273–1500	0.25	0.18
i-Butene	C_4H_8	1.650	7.702	−3.981	8.020	273–1500	0.11	0.06
cis-2-Butene	C_4H_8	−1.778	8.078	−4.074	7.890	273–1500	0.78	0.14
trans-2-Butene	C_4H_8	2.340	7.220	−3.403	6.070	273–1500	0.54	0.12
Cycloparaffinic Hydrocarbons								
Cyclopentane	C_5H_{10}	−12.957	13.087	−7.447	16.41	273–1500	1.00	0.25
Methylcyclopentane	C_6H_{12}	−12.114	15.380	−8.915	20.03	273–1500	0.86	0.23
Cyclohexane	C_6H_{12}	−15.935	16.454	−9.203	19.27	273–1500	1.57	0.37
Methylcyclohexane	C_7H_{14}	−15.070	18.972	−10.989	24.09	273–1500	0.92	0.22
Aromatic Hydrocarbons								
Benzene	C_6H_6	−8.650	11.578	−7.540	18.54	273–1500	0.34	0.20
Toluene	C_7H_8	−8.213	13.357	−8.230	19.20	273–1500	0.29	0.18
Ethylbenzene	C_8H_{10}	−8.398	15.935	−10.003	23.95	273–1500	0.34	0.19
Styrene	C_8H_8	−5.968	14.354	−9.150	22.03	273–1500	0.37	0.23
Cumene	C_9H_{12}	−9.452	18.686	−11.869	28.80	273–1500	0.36	0.17

Acetylenes and Diolefins								
Acetylene	C_2H_2	5.21	2.2008	−1.559	4.349	273–1500	1.46	0.59
Methylacetylene	C_3H_4	4.21	4.073	−2.192	4.713	273–1500	0.36	0.13
Dimethylacetylene	C_4H_6	3.54	5.838	−2.760	4.974	273–1500	0.70	0.16
Propadiene	C_3H_4	2.43	4.693	−2.781	6.484	273–1500	0.37	0.19
1 3-Butadiene	C_4H_6	−1.29	8.350	−5.582	14.24	273–1500	0.91	0.47
Isoprene	C_5H_8	−0.44	10.418	−6.762	16.93	273–1500	0.99	0.43
Combustion Gases (Low Range)								
Nitrogen	N_2	6.903	−0.03753	0.1930	−0.6861	273–1800	0.59	0.34
Oxygen	O_2	6.085	0.3631	−0.1709	0.3133	273–1800	1.19	0.28
Air		6.713	0.04697	0.1147	−0.4696	273–1800	0.72	0.33
Hydrogen	H_2	6.952	−0.04576	0.09563	−0.2079	273–1800	1.01	0.26
Carbon monoxide	CO	6.726	0.04001	0.1283	−0.5307	273–1800	0.89	0.37
Carbon dioxide	CO_2	5.316	1.4285	−0.8362	1.784	273–1800	0.67	0.22
Water vapor	H_2O	7.700	0.04594	0.2521	−0.8587	273–1800	0.53	0.24
Combustion Gases (High Range)								
Nitrogen	N_2	6.529	0.1488	−0.02271	—	273–3800	2.05	0.72
Oxygen	O_2	6.732	0.1505	−0.01791	—	273–3800	3.24	1.20
Air		6.557	0.1477	−0.02148	—	273–3800	1.64	0.70
Hydrogen	H_2	6.424	0.1039	−0.007804	—	273–3800	2.14	0.79
Carbon dioxide	CO_2	See footnote † for special equation.				273–3800	2.65	0.54
Carbon monoxide	CO	6.480	0.1566	−0.02387	—	273–3800	1.86	1.01
Water vapor	H_2O	6.970	0.3464	−0.04833	—	273–3800	2.03	0.66
Sulfur Compounds								
Sulfur	S_2	6.499	0.5298	−0.3888	0.9520	273–1800	0.99	0.38
Sulfur dioxide	SO_2	6.157	1.384	−0.9103	2.057	273–1800	0.45	0.24
Sulfur trioxide	SO_3	3.918	3.483	−2.675	7.744	273–1300	0.29	0.13
Hydrogen sulfide	H_2S	7.070	0.3128	0.1364	−0.7867	273–1800	0.74	0.37
Carbon disulfide	CS_2	7.390	1.489	−1.096	2.760	273–1800	0.76	0.47
Carbonyl sulfide	COS	6.222	1.536	−1.058	2.560	273–1800	0.94	0.49

* Reprinted with permission. In the original article, constants are also given for T in degrees centigrade, degrees Fahrenheit, and degrees Rankine.

† Equation for CO_2, 273 to 3800 °K: $c_p^\circ = 18.036 - 0.00004474T - 158.08/\sqrt{T}$.

TABLE D. MOLAL HEAT CAPACITIES OF GASES AT ZERO PRESSURE (*Continued*)

		a	$b \times 10^2$	$c \times 10^5$	$d \times 10^9$	Temperature Range, °K	Error Max. %	Error Avg. %
Halogens and Halogen Acids								
Fluorine	F_2	6.115	0.5864	−0.4186	0.9797	273–2000	0.78	0.45
Chlorine	Cl_2	6.8214	0.57095	−0.5107	1.547	273–1500	0.50	0.23
Bromine	Br_2	8.051	0.2462	−0.2128	0.6406	273–1500	0.43	0.15
Iodine	I_2	8.504	0.13135	−0.10684	0.3125	273–1800	0.11	0.06
Hydrogen fluoride	HF	7.201	−0.1178	0.1576	−0.3760	273–2000	0.37	0.09
Hydrogen chloride	HCl	7.244	−0.1820	0.3170	−1.036	273–1500	0.22	0.08
Hydrogen bromide	HBr	7.169	−0.1604	0.3314	−1.161	273–1500	0.27	0.12
Hydrogen iodide	HI	6.702	0.04546	0.1216	−0.4813	273–1900	0.92	0.39
Chloromethanes								
Methyl chloride	CH_3Cl	3.05	2.596	−1.244	2.300	273–1500	0.75	0.16
Methylene chloride	CH_2Cl_2	4.20	3.419	−2.3500	6.068	273–1500	0.67	0.30
Chloroform	$CHCl_3$	7.61	3.461	−2.668	7.344	273–1500	0.92	0.42
Carbon tetrachloride	CCl_4	12.24	3.400	−2.995	8.828	273–1500	1.21	0.57
Phosgene	$COCl_2$	10.35	1.653	−0.8408	—	273–1000	0.97	0.46
Thiophosgene	$CSCl_2$	10.80	1.859	−1.045	—	273–1000	0.98	0.71
Cyanogens								
Cyanogen	$(CN)_2$	9.82	1.4858	−0.6571	—	273–1000	0.69	0.42
Hydrogen cyanide	HCN	6.34	0.8375	−0.2611	—	273–1500	1.42	0.76
Cyanogen chloride	CNCl	7.97	1.0745	−0.5265	—	273–1000	0.97	0.58
Cyanogen bromide	CNBr	8.82	0.9084	−0.4367	—	273–1000	0.85	0.54
Cyanogen iodide	CNI	9.69	0.7213	−0.3265	—	273–1000	0.75	0.37
Acetonitrile	CH_3CN	5.09	2.7634	−0.9111	—	273–1200	0.45	0.26
Acrylic nitrile	CH_2 CHCN	4.55	4.1039	−1.6939	—	273–1000	0.63	0.41
Oxides of Nitrogen								
Nitric oxide	NO	6.461	0.2358	−0.07705	0.08729	273–3800	2.23	0.54
Nitric oxide	NO	7.008	−0.02244	0.2328	−1.000	273–1500	0.97	0.36
Nitrous oxide	N_2O	5.758	1.4004	−0.8508	2.526	273–1500	0.59	0.26
Nitrogen dioxide	NO_2	5.48	1.365	−0.841	1.88	273–1500	0.46	0.18
Nitrogen tetroxide	N_2O_4	7.9	4.46	−2.71	—	273–600	0.97	0.36

Oxygenated Hydrocarbons								
Formaldehyde	CH_2O	5.447	0.9739	0.1703	−2.078	273–1500	1.41	0.62
Acetaldehyde	C_2H_4O	4.19	3.164	−0.515	−3.800	273–1000	0.40	0.17
Methanol	CH_4O	4.55	2.186	−0.291	−1.92	273–1000	0.18	0.08
Ethanol	C_2H_6O	4.75	5.006	−2.479	4.790	273–1500	0.40	0.22
Ethylene oxide	C_2H_4O	−1.12	4.925	−2.389	3.149	273–1000	0.36	0.14
Ketene	C_2H_2O	4.11	2.966	−1.793	4.22	273–1500	0.48	0.17
Miscellaneous Hydrocarbons								
Cyclopropane	C_3H_6	−6.481	8.206	−5.577	15.61	273–1000	0.94	0.35
Isopentane	C_5H_{12}	−2.273	12.434	−7.097	15.86	273–1500	0.34	0.14
Neopentane	C_5H_{12}	−3.865	13.305	−8.018	18.83	273–1500	0.27	0.12
o-Xylene	C_8H_{10}	−3.789	14.291	−8.354	18.80	273–1500	0.52	0.15
m-Xylene	C_8H_{10}	−6.533	14.905	−8.831	20.05	273–1500	0.67	0.16
p-Xylene	C_8H_{10}	−5.334	14.220	−7.984	17.03	273–1500	0.56	0.18
C_3 Oxygenated Hydrocarbons								
Carbon suboxide	C_3O_2	8.203	3.073	−2.081	5.182	273–1500	1.22	0.31
Acetone	C_3H_6O	1.625	6.661	−3.737	8.307	273–1500	0.56	0.10
i-Propyl alcohol	C_3H_8O	0.7936	8.502	−5.016	11.56	273–1500	0.35	0.18
n-Propyl alcohol	C_3H_8O	−1.307	9.235	−5.800	14.14	273–1500	0.90	0.30
Allyl alcohol	C_3H_6O	0.5203	7.122	−4.259	9.948	273–1500	0.23	0.14
Chloroethenes								
Chloroethene	C_2H_3Cl	2.401	4.270	−2.751	6.797	273–1500	0.46	0.22
1,1-Dichloroethene	$C_2H_2Cl_2$	5.899	4.383	−3.182	8.516	273–1500	0.81	0.40
cis-1,2-Dichloroethene	$C_2H_2Cl_2$	4.336	4.691	−3.397	9.010	273–1500	0.70	0.39
trans-1,2-Dichloroethene	$C_2H_2Cl_2$	5.661	4.295	−3.022	7.891	273–1500	0.50	0.27
Trichloroethene	C_2HCl_3	9.200	4.517	−3.600	10.10	273–1500	1.17	0.50
Tetrachloroethene	C_2Cl_4	15.11	3.799	−3.179	9.089	273–1500	0.94	0.42
Nitrogen Compounds								
Ammonia	NH_3	6.5846	0.61251	0.23663	−1.5981	273–1500	0.91	0.36
Hydrazine	N_2H_4	3.890	3.554	−2.304	5.990	273–1500	1.80	0.50
Methylamine	CH_5N	2.9956	3.6101	−1.6446	2.9505	273–1500	0.59	0.07
Dimethylamine	C_2H_7N	−0.275	6.6152	−3.4826	7.1510	273–1500	0.96	0.15
Trimethylamine	C_3H_9N	−2.098	9.6187	−5.5488	12.432	273–1500	0.91	0.18

Author Index

Part I, pages 1–504 Part II, pages **505–1072**

Subject Index

Part I, pages 1–504; Part II, pages **505–1072**